P9-AQS-326

Translations

of

Mathematical Monographs

Volume 35

Extrinsic Geometry of Convex Surfaces

by

A. V. Pogorelov

American Mathematical Society
Providence, Rhode Island
1973

ВНЕШНЯЯ ГЕОМЕТРИЯ ВЫПУКЛЫХ ПОВЕРХНОСТЕЙ

А. В. Погорелов

Издательство „Наука"
Главная редакция
Физико-Математической Литературы
Москва 1969

Translated from the Russian by
Israel Program for Scientific Translations

AMS (*MOS*) *subject classifications* (1970).
Primary 53C45; Secondary 53A05, 53A35, 35A30.

Library of Congress Cataloging in Publication Data

Pogorelov, Alekseĭ Vasil'evich.
Extrinsic geometry of convex surfaces.

(Translations of mathematical monographs, v. 35)
Translation of Vneshni͡ai͡a geometrii͡a vypuklykh poverkhnosteĭ.
1. Convex surfaces. I. Title. II. Series.
QA649.P6313 516'.36 72-11851
ISBN 0-8218-1585-7

Copyright © 1973 by the American Mathematical Society

Printed in the United States of America

MATH.-SCI.
QA
649
.P6313
1330349
Math-Sci
Sep

TABLE OF CONTENTS

INTRODUCTION

Problems in the theory of convex surfaces, primarily those of geometry "in the large," have engaged the attention of many well-known mathematicians (Cauchy, Minding, Christoffel, Minkowski, Liebmann, Weingarten, Hilbert, Weyl, Bernšteĭn, Blaschke, Cohn-Vossen, Lewy and others). Even today the problems, methods and results due to these authors have not lost their value.

One of the first results in the theory of convex surfaces is due to Cauchy, who proved that closed convex polyhedra with congruent corresponding faces are congruent. This theorem and the celebrated lemma underlying its proof laid the basis for many fundamental results of the modern theory. In particular, it is used in proving the existence of a convex polyhedron realizing a given polyhedral metric, and in other existence and uniqueness theorems for polyhedra.

Minding conjectured that the sphere is rigid. Once this conjecture was proved (Liebmann, Minkowski, Hilbert, Weyl) the general problem of the rigidity and unique definability of an arbitrary convex surface arose in a natural way. For regular surfaces the problem was solved by Cohn-Vossen, who evolved for the purpose methods based on investigation of the index of special vector fields over a surface and the extrema of invariantly constructed auxiliary functions; these methods, and his integral formulas, have found extensive application in the work of later geometers.

Minkowski formulated and solved the problem of whether there exists a closed convex surface with given Gauss curvature. In its analytic interpretation this solution yields a generalized solution of a certain differential equation; it is natural to ask whether the generalized solution is regular.

H. Weyl stated the problem of whether there exists a closed convex surface with given metric, and indicated the solution. This problem has been attacked by many geometers, and an exhaustive solution has been derived in the most general formulation for general metrics and general spaces.

By the early thirties, the capacity of analytical tools for investigating convex surfaces "in the large" had been more or less exhausted. An example is the above-mentioned problem of H. Weyl. Though Weyl indicated the solution, it was completed only twenty years later by H. Lewy, who used subtle results from the theory of analytic Monge-Ampère equations. It became clear that further development of the theory would require new tools: real functions, functional analysis, set-theoretical topology, the theory of differential equations with generalized solutions. To employ these tools implied generalizing the subject matter. Thus, instead of a convex surface as a surface with positive Gauss curvature one considered a convex surface as the boundary of a convex body (or a fragment thereof). These redefined convex surfaces were investigated by Soviet geometers from the thirties onward.

In 1948 A. D. Aleksandrov's monograph *Intrinsic Geometry of Convex Surfaces* was published. This book is a systematic exposition of the author's theory of general convex surfaces. As the name implies, it contains results pertaining mainly to the intrinsic geometry of the surface. The high point of Aleksandrov's theory was a general theorem stating necessary and sufficient conditions for a given metric to be the metric of a convex surface. This theorem largely finalizes the construction of the intrinsic geometry of convex surfaces. Nevertheless, it poses new problems which already belong to "extrinsic" geometry, the study of the form of a surface realizing the metric.

In this context the uniqueness problem for a convex surface with prescribed metric (the problem of monotypy) acquires fundamental importance. Cohn-Vossen's solution of this problem was inadequate, not merely because he confined himself to regular (three times differentiable) surfaces, but mainly because the class of surfaces in which the problem is solved was restricted by this regularity condition. One can effectively apply the monotypy theorem in conjunction with Aleksandrov's realizability theorem only when uniqueness in the class of general convex surfaces is guaranteed without a priori regularity assumptions, even when the metric of the surface is fairly regular.

Just as in other fields of modern mathematics, investigation of the nonregular object (here, the general convex surface) is not an end in itself but rather a means toward a deeper understanding and study of the regular object. The theory of general convex surfaces is a good illustration of this thesis. Deformations of regular surfaces have been studied by numerous geometers of the 19th and 20th centuries. However, all known results in this field prior to Aleksandrov seemed quite modest. Only the formulation of the problem for general convex surfaces led to its final solution and reduced it to an extremely simple problem. Of course, as is the case in other fields of mathematics, consideration of the generalized problem generally means that the solution must be generalized. A natural question in this context is: how regular is the generalized solution of the problem in the regular case? Here one is interested in the regularity of a convex surface with a regular metric. Once this problem is solved, the methods and results of the theory of general convex surfaces prove applicable to many problems of the classical theory.

Closely related to the problem of bending of surfaces is the problem of infinitesimal bending, in particular, the rigidity of convex surfaces, which has attracted many geometers. A solution to this problem is indispensable for the theory of elastic shells. This is because the conditions for structural strength of a body whose elements are elastic shells usually require that the latter be geometrically rigid. An actual shell, projected in the form of a regular surface, never has this property. The convexity conditions, however, are easily satisfied. Hence the problem of rigidity, if it is to be used in applications, must be solved for nonregular convex surfaces. Here, then, is another problem which is justifiably formulated for general convex surfaces.

Aleksandrov's theory of general convex surfaces is easily extended to convex surfaces in spaces of constant curvature—elliptic space and Lobačevskiĭ space. Hence these spaces in turn raise the same problems of monotypy, regularity and the existence of infinitesimal bending. By solving the problems in these more general spaces one achieves a deeper understanding of the results and can extend the potentialities of the methods.

With a deeper understanding of the results of the theory of convex surfaces in euclidean space and spaces of constant curvature, combined with more refined methods, one can extend the formulation of the basic problems to convex surfaces in general Riemann space, and solve them by the methods perfected in treating the simpler spaces. The natural problems here, as in euclidean space, are isometric embedding, monotypy bending and rigidity.

All the problems considered hitherto were connected with the metric of the surface. As far as regular surfaces are concerned, these problems involve the first quadratic form of the surface—its line element (ds^2).

Another group of problems in the regular case involves the third quadratic form – the line element of the spherical image. Here one had a correspondence between the points of two surfaces, under which the exterior normals to the surfaces at corresponding points are parallel and identically oriented. The earliest problems of this type were Christoffel's problem on the existence and uniqueness of a closed convex surface with given sum of principal radii of curvature and the above-mentioned problem of Minkowski on the existence and uniqueness of a closed convex surface with given product of principal radii of curvature. A natural generalization of these problems is the existence and uniqueness of a closed convex surface with an arbitrary given function of the principal radii of curvature.

The theory of convex surfaces makes extensive use of methods and results proper to neighboring fields of mathematics. Conversely, many geometric methods are in turn applicable to these fields, and geometric results are amenable to analytical interpretation. In particular, many questions in the theory of convex surfaces, in their analytical interpretation, lead to existence and uniqueness problems for solutions of partial differential equations. Frequently occurring examples of the latter are the elliptic Monge-Ampère equations. The general existence and uniqueness theorems for convex surfaces enable one to set up a geometric theory of Monge-Ampère equations with extremely general existence and uniqueness theorems.

Metrics of bounded curvature are natural generalizations of the metric of a convex surface. A metric of bounded curvature stands in the same relation to the metric of a convex surface as does the metric of an arbitrary regular surface to that of a regular surface with positive Gauss curvature. The realization of a metric of bounded curvature on surfaces of euclidean space, preserving sufficiently strong relations between the intrinsic metric and the extrinsic form of the surface, is an as yet unsolved problem. What is required here is clearly an investigation of sufficiently general classes of surfaces with metrics of bounded curvature.

We have discussed the formulation of several problems which arise naturally on the road to a theory of convex surfaces. The present book is devoted to a solution of these problems, in most cases a fairly complete solution. The title *Extrinsic Geometry of Convex Surfaces* both reflects its content and stresses its relation to Aleksandrov's monograph *Intrinsic Geometry of Convex Surfaces*, of which it is, in a certain sense, a sequel.

No specialized geometric prerequisites are demanded of the reader. Nevertheless, the diversity of the methods employed necessitates a certain familiarity with neighboring fields of mathematics. The book is aimed primarily at geometers: advanced students, graduate students and scientific workers. Some chapters may hold interest for mathematicians in other fields.

In an Appendix, we formulate a number of unsolved problems which might be attacked by young geometers. These problems differ in difficulty, but in almost all cases we indicate a promising approach to the solution.

CHAPTER I

INTRINSIC GEOMETRY OF CONVEX SURFACES

The intrinsic geometry of a surface studies figures on the surface and related geometric quantities invariant under isometric mappings, i.e. under deformations preserving the length of curves on the surface. For a regular surface these are the mappings that preserve the first fundamental form, the line element of the surface. The objects of the intrinsic geometry of regular surfaces are the angle between curves, the area of the surface, the Gauss curvature of the surface, the geodesic curvature of a curve, geodesic lines.

Many of these concepts become meaningless for nonregular surfaces, at least insofar as the classical definitions are concerned. Hence the first problem in the theory of irregular surfaces is to define the basic concepts of intrinsic geometry as natural generalizations of their classical counterparts.

The fundamental tool in the study of regular surfaces is the analysis of infinitesimals. The success of this method for regular surfaces is due to the fact that they admit a sufficiently regular analytical representation. Nonregular surfaces are far less amenable to treatment by analysis, for the functions appearing in their equations are not regular. The second problem is thus to develop a method for nonregular surfaces.

Like the relevant concepts, many theorems of the intrinsic geometry of regular surfaces become meaningless when carried over to nonregular surfaces. Herein lies the third problem in constructing an intrinsic geometry for nonregular surfaces: to state and prove meaningful theorems generalizing those of the classical theory.

A complete solution of all these problems for general convex surfaces, and even for nonregular surfaces of a more general type (manifolds of bounded curvature) was evolved by Aleksandrov.

Aleksandrov's first achievement was to define the basic concepts: convex surface, shortest join [segment], angle between curves, curvature of a surface, area of a surface, and integral geodesic curvature of a curve. These concepts were the basis for his method, the principle of which was to approximate a general convex surface by a convex polyhedron and the abstractly given convex metric by a polyhedral metric. Using this method Aleksandrov extended the classical integral theorems to general convex surfaces. In particular, he established a general property concerning the mutual positions of segments (inclusion property), and proved the existence of the angle between segments. He studied the intrinsic curvature of the surface and extended Gauss's well-known theorem, which states that the integral curvature and the area of the spherical image are equal, to general convex surfaces. Concerning the integral geodesic curvature of a curve, he generalized the Gauss-Bonnet Theorem. He found necessary and sufficient conditions for the existence of a convex surface realizing an abstractly given metric.

Aleksandrov summarized his investigations of the intrinsic geometry of convex surfaces in the monograph [**2**], which is briefly summarized in this chapter. The exposition in later chapters essentially utilizes the concepts and results of the intrinsic geometry of convex surfaces.

§1. Convex bodies and convex surfaces

A set M in euclidean three-space is said to be *convex* if the straight-line segment joining any two of its points X and Y is entirely contained in M (Figure 1). A closed plane convex set with nonempty interior is called a *convex domain*.

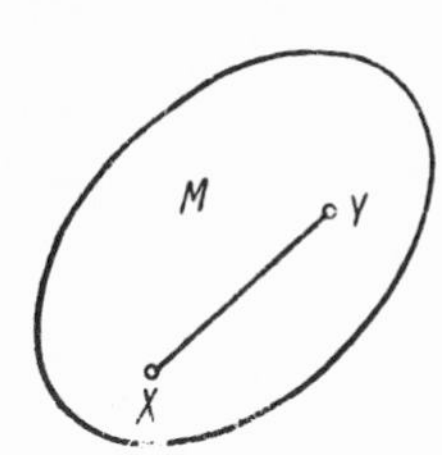

FIGURE 1

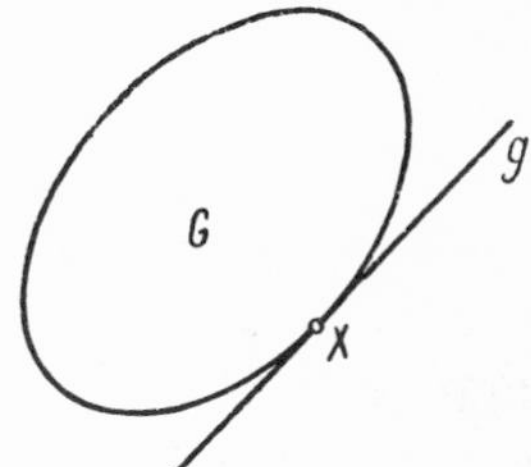

FIGURE 2

Any connected part of the boundary of a convex domain is called a *convex curve*. The boundary of a bounded convex domain is called a *closed convex curve*. Any closed convex curve is homeomorphic to a circle.

A straight line g through a point X on the boundary of a convex domain G is called a *supporting line* if the entire domain lies in one of the halfplanes defined by g (Figure 2). There is at least one supporting line through each boundary point of a convex domain.

If a convex curve γ is the boundary of a convex domain G or part of its boundary, a supporting line to G at a point of γ is also called a supporting line for γ.

The points of a convex curve may be either smooth or corner points. A point X of a convex curve γ is said to be *smooth* if there is a unique supporting line through X. Otherwise X is called a *corner* point. The supporting lines through a corner point describe a vertical angle whose apex is the point and whose sides are supporting lines (Figure 3).

Every convex curve is rectifiable, i.e. it has a definite length. If a convex curve $\bar{\gamma}$ includes a convex curve γ, the length of γ cannot exceed that of $\bar{\gamma}$.

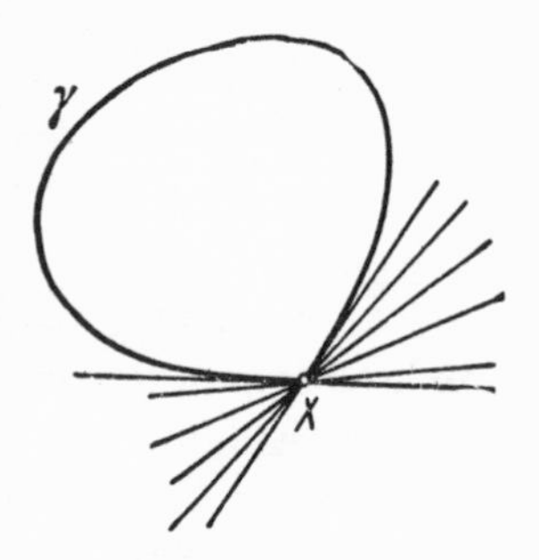

FIGURE 3

A *convex body* is defined to be a convex set in space, with nonempty interior. A closed convex set is a convex body if and only if it is not contained in any plane. The intersection of any family of convex bodies is again a convex body, provided its interior is not empty.

A domain (open connected set) on the boundary of a convex body is called a *convex surface*. A component of the boundary of a convex body is called a *complete convex surface*. Except for the two trivial cases in which the convex body is the entire space or the domain between two parallel planes, a complete convex surface is simply the boundary of a convex body. The boundary of a bounded convex body is homeomorphic to a sphere and is called a *closed convex surface*. Every complete convex surface is homeomorphic to a plane, a sphere or a cylinder. In the latter case the surface itself is a cylinder.

Supporting planes of convex bodies are defined by analogy to the case of convex plane domains: A plane σ through a boundary point X of a body K is a *supporting plane* at that point if all points of the body lie on one side of the plane σ, i.e. in one of the halfspaces that it defines. There is at least one supporting plane through each boundary point of a convex body. The unit vector perpendicular to a supporting plane and directed into the [open] halfspace not containing points of the body is called the *exterior normal* to the plane.

A convex body V generated by rays issuing from a point S is called a *convex cone*, provided V is not the entire space. Particular cases of convex cones by this definition are a dihedral angle and a halfspace. The surface of a convex cone is also usually called a convex cone. In the above particular cases the cone (*qua* surface) is said to degenerate into a dihedral angle or a plane.

Each point S on the boundary of a convex body K can be associated in a natural way with a certain cone $V(S)$, generated by the rays issuing from S and cutting K in at least one point other than S (Figure 4). This cone is called the *tangent cone* at S, and its surface is called the tangent cone of the convex surface bounding the body.

Depending upon the form of the tangent cone, we divide the points of a convex surface into conical, ridge and smooth points. A point X of a convex surface is called a *conical point* if the tangent cone $V(X)$ is not degenerate. If the tangent cone $V(X)$ degenerates into a dihedral angle or a plane, X is called a *ridge point* or a *smooth point*, respectively. Nonsmooth points on a convex surface are in a certain sense exceptional: the set of ridge points has measure 0, while the set of conic points is at most countable.

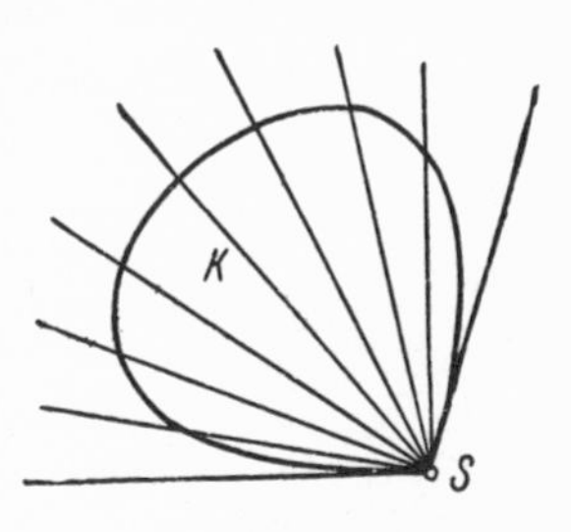

FIGURE 4

The simplest nontrivial convex body is a *convex polyhedron*, defined as the intersection of finitely many halfspaces. The surface of a convex polyhedron consists of convex plane polygons and is also called a convex polyhedron. The polygons making up the surface of a polyhedron are called its *faces*, their sides are called the *edges* of the polyhedron, and their vertices the *vertices* of the polyhedron.

An important concept in the theory of convex bodies is the convex hull. The *convex hull* of a set M is the intersection of all the halfspaces containing M. Consequently the convex hull is a convex set – in fact, the smallest convex set containing M. Every convex polyhedron is the convex hull of its vertices (including points at infinity), and is therefore uniquely determined by them.

We now define convergence of sequences of convex surfaces. A sequence of convex surfaces F_n is said to converge to a convex surface F if any open set G that cuts F must also cut all surfaces F_n for $n > N(G)$. Any convex surface can be represented as the limit of convex polyhedra or regular convex surfaces.

Any infinite family of convex surfaces possesses the important *compactness* property: From any bounded sequence of complete convex surfaces one can always extract a convergent subsequence whose limit is a convex surface, possibly degenerate (the degenerate cases are doubly covered plane domain, straight line, ray or segment).

In conclusion we mention a much used property, the convergence of the supporting planes of a convergent sequence of convex surfaces. Let F_n be a sequence of convex surfaces converging to a convex surface F, X_n a point on F_n and α_n a supporting plane at this point. Then, if the point sequence X_n converges to a point X on F and the supporting planes α_n converge to a plane α, the latter is a supporting plane of F at X. It follows, in particular, that if X_n is a sequence of points on a convex surface F which converges to a point X of F and the supporting planes α_n at X_n converge to a plane α, then the latter is a supporting plane at X.

§2. Manifolds with intrinsic metric

Let R be a metric space. We recall the definition: A metric space is a set of elements (of arbitrary nature) for each pair of which a number $\rho(X, Y)$, called the distance, is defined, satisfying the following conditions:

1. $\rho(X, Y) \geqq 0$.
2. $\rho(X, Y) = 0$ if and only if $X = Y$.
3. $\rho(X, Y) + \rho(Y, Z) \geqq \rho(X, Z)$ (triangle inequality).

For example, euclidean space with the usual distance between points is obviously a metric space. A *curve* in a metric space R is a continuous image of any closed interval. The length of a curve in a metric space is defined in the same way as in ordinary space. If the curve γ is given by a mapping $X = X(t)$, $a \leqq t \leqq b$, the *length* of γ is defined by

$$l_\gamma = \sup \sum \rho(X(t_{k-1}), X(t_k)), \qquad a \leqslant t_1 \leqslant t_2 \ldots \leqslant b,$$

where $\rho(X(t_{k-1}), X(t_k))$ is the distance between the points $X(t_{k-1})$ and $X(t_k)$ in R and the supremum ranges over all finite subdivisions of the interval (a, b) by points t_k. Thus defined, the length of a curve has the usual additivity property: If a curve γ is the union of two curves γ_1 and γ_2 with disjoint interiors, the length of γ is the sum of the lengths of γ_1 and γ_2.

As in euclidean space, any set of curves of bounded length in a compact domain of a metric space is compact, i.e. any infinite sequence of such curves contains a convergent subsequence, and then the length of the limit curve cannot exceed the lower limit of the lengths of the curves in the sequence.

Assume that any two points X, Y of a metric space R can be joined by a rectifiable curve. It is then natural to define an intrinsic distance $\rho^*(X, Y)$ between points X, Y as the infimum of the lengths of curves joining X and Y. It is easy to see that this distance satisfies conditions I–III for a metric space. The metric ρ^* is called the *intrinsic metric* of the space R. If $\rho(X, Y) = \rho^*(X, Y)$ for all X and Y the space R is called a space with intrinsic metric. Throughout the sequel our spaces will be two-dimensional manifolds with intrinsic metric.

Let R be a manifold with intrinsic metric. A curve γ in R is called a *shortest join* or *segment*[1] if its length is equal to the distance between its endpoints, and so does not exceed the length of any other curve joining these points. Every subarc of a segment is a segment. The limit curve of a convergent sequence of segments is a segment.

In the general case, it is not true that every two points in a manifold can be joined by a segment; however, every point of a manifold has a neighborhood in which any two points can be so joined. If R is a metrically complete, i.e. any closed bounded subset is compact, then any two points can be joined by a segment.

A manifold with intrinsic metric is said to satisfy the *inclusion property*[2] if any two distinct segments with common endpoints have no other common points. When the inclusion property holds there are only five possible mutual positions for two segments (see Figure 5):

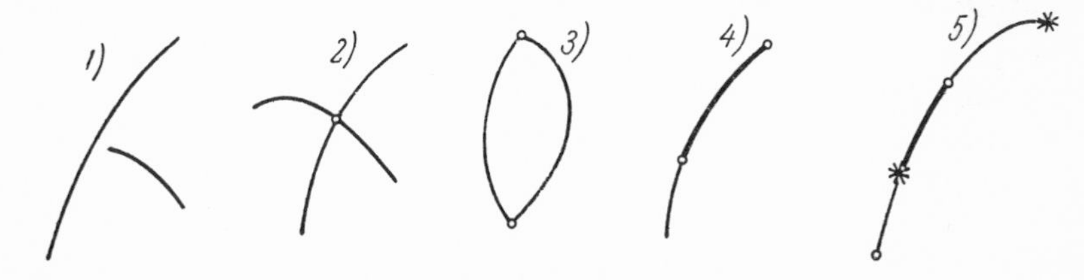

FIGURE 5

1) they have no common points, 2) they have only one common point, 3) they have only two common points and these are their endpoints, 4) one of the segments is included in the other, 5) the segments coincide over a certain subarc; one end of this subarc is an end of one of the segments, and the other end of the subarc is an end of the other segment. It follows from these properties that if AB is a segment and C an interior point of AB, there exists only one segment joining A and C: the subarc of AB between A and C. If A_n and C_n are point sequences converging to A and C, respectively, and each pair A_n, C_n can be joined by a segment, then these segments converge as $n \to \infty$ to the subarc AC of AB.

A set G in a manifold with intrinsic metric is said to be *convex* if any

---

[1] *Translator's note*. We shall follow Busemann (*Convex Surfaces*, Interscience, New York, 1958) in using these two terms more or less interchangeably.

[2] *Translator's note*. So called by Busemann (*Op. cit.*). Literal translation of Russian term: "nonoverlapping of shortest joins."

two of its points can be joined by a segment entirely within G. When the segments in a manifold with intrinsic metric satisfy the inclusion property, convex domains possess several properties analogous to those of convex domains on the plane. We indicate some of these properties.

1) If A and B are points of a convex domain G and A is an interior point, then every segment AB is in the interior of G (except possibly B, if it lies on the boundary of G).

2) If the intersection of two convex domains has a nonempty interior, it is convex.

3) If a segment in the interior of a convex domain divides it into two subsets, each of these is convex.

A *polygon* in a manifold with intrinsic metric is a compact set with nonempty interior whose boundary consists of finitely many segments (Figure 6). If the inclusion property holds, the intersection of any two convex polygons is a convex polygon, provided it has nonempty interior (Figure 7). Each of the subsets into which a segment divides a convex polygon is also a convex polygon. Any polygon admits arbitrarily fine triangulations, i.e. subdivisions into arbitrarily small triangles.

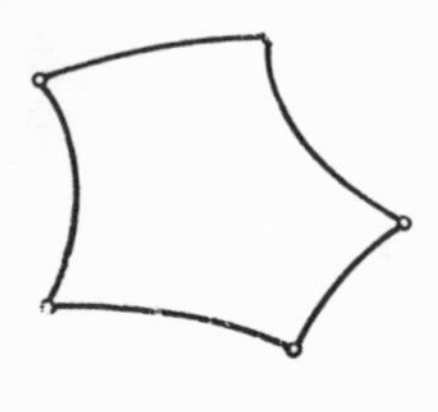

FIGURE 6

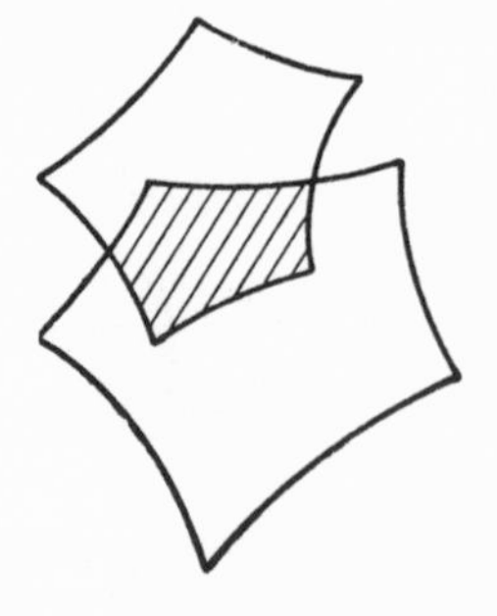

FIGURE 7

Every point of a manifold with intrinsic metric which satisfies the inclusion property has a neighborhood homeomorphic to a disk, in the form of a convex polygon.

A nontrivial example of a manifold with intrinsic metric which satisfies the inclusion property is a convex surface (see §3). Hence all the above properties of these manifolds hold true for convex surfaces.

§3. Convexity property of the intrinsic metric of a convex surface

The basic tool of investigation for the intrinsic metric of a convex surface in Aleksandrov's theory is approximation of a general convex surface by convex polyhedra. The following theorem on the convergence of metrics plays an important part in this connection.

Let F_n be a sequence of closed convex surfaces which converges to a closed convex surface F, X_n and Y_n two point sequences on F_n which converge to points X and Y, respectively, on F. Then the distance between X_n and Y_n in F_n converges to the distance between X and Y in F:

$$\lim_{n\to\infty} \rho_{F_n}(X_n, Y_n) = \rho_F(X, Y).$$

This theorem can be sharpened: the convergence turns out to be uniform. To be precise:

For every closed convex surface F and any $\epsilon > 0$, there exists a $\delta > 0$ such that whenever a convex surface S deviates from F by less than δ[3] *and the distances of any points X and Y on F from points A and B on S are also smaller than δ, then*

$$|\rho_F(X, Y) - \rho_S(A, B)| < \epsilon,$$

where ρ_F and ρ_S are the distances on F and S, respectively.

A corollary of the theorem on convergence of metrics is the following theorem on convergence of segments.

Let F_n be a sequence of closed convex surfaces which converges to a surface F, and (X_n, Y_n) a sequence of pairs of points on F_n which converges to a pair (X, Y) of points on F. Then any sequence of segments $\gamma(X_n, Y_n)$ on F_n contains a subsequence converging to a segment $\gamma(X, Y)$ on F.

The most important property of the intrinsic metric of a convex surface is known as the *convexity property*:

Let γ and γ' be two segments issuing from a point O on a convex surface F, X and X' variable points on these segments, x and x' the distances of X and X' from O, and $z(x, x')$ the distance between X and X' (all distances are on the surface F). Let $\alpha(x, x')$ be the angle opposite the side $z(x, x')$ in a plane triangle with sides $x, x', z(x, x')$. Then $\alpha(x, x')$ is a nonincreasing function of x and x' on any interval $0 < x < x_0$, $0 < x' < x_0'$ on which the segment XX' exists.

The convexity property, as it turns out, characterizes the intrinsic metrics of convex surfaces. Any two-dimensional manifold with intrinsic metric possessing the convexity property is locally isometric to a convex surface.

One first proves that the metric of a closed convex polyhedron possesses the convexity property. The proof depends essentially on the following properties of segments, which are also of independent interest.

---

[3] This means that S is contained in a δ-neighborhood of F and vice versa.

1. A segment issuing from a point O on a convex polyhedron is a euclidean (straight line) segment in some neighborhood of O.

2. A segment on a convex polyhedron cannot pass through a vertex.

3. If A and B are not vertices of the polyhedron, their shortest join $\gamma(A, B)$ on the polyhedron has a neighborhood which can be developed on the plane, and when this is done the segment becomes a straight-line segment.

4. Segments on a convex polyhedron satisfy the inclusion property.

Once the fact that the intrinsic metric of a closed convex surface is convex has been proved using the above properties, the proof is extended to general convex surfaces by approximation with polyhedra and a limit procedure. The theorem runs as follows:

Any closed convex surface satisfies the convexity condition in the large.

The convexity of the intrinsic metric of a nonclosed convex surface is established by imbedding all or part of it in a closed convex surface. The following theorems are obtained:

A (not necessarily closed) convex surface satisfies the convexity condition locally.

The metric of an unbounded complete convex surface satisfies the complexity condition in the large.

The convexity property of the metric of a general convex surface yields a natural definition of the angle between segments, as the limit as $x, x' \to 0$ of the angle $\alpha(x, x')$ figuring in the convexity condition. Thus defined, the angle between segments obviously always exists.

The convexity of the metric implies the inclusion property for segments on a general convex surface. To be precise, if two segments γ and γ' issuing from a point O on a convex surface coincide over a subarc OA, then one of them is included in the other.

Another corollary of the convexity property is that the angles of a triangle formed by segments on a convex surface are not smaller than the corresponding angles of a plane triangle with sides of the same length.

The convexity property of the metric provides the basis for the proof of the following theorem on convergence of the angles between segments:

Let F_n be a sequence of convex surfaces which converges to a convex surface F (the case $F_n = F$ is not excluded), and let γ_n', γ_n'' be segments issuing from a point O_n on F_n, which converge to segments γ', γ'' issuing from O on F. Then the angle $\alpha(\gamma', \gamma'')$ cannot exceed the lower limit of the angles $\alpha(\gamma_n', \gamma_n'')$ as $n \to \infty$:

$$\alpha(\gamma', \gamma'') \leqslant \varliminf_{n\to\infty} \alpha(\gamma_n', \gamma_n'').$$

§4. Basic properties of the angle between segments on a convex surface

Let F be a convex surface, O a point on F, γ_1 and γ_2 segments issuing from O. Let X_1 and X_2 be points on γ_1 and γ_2, respectively, and construct a plane triangle with sides $\rho_F(O,X_1)$, $\rho_F(O,X_2)$, $\rho_F(X_1,X_2)$. Let $\alpha(X_1,X_2)$ be the angle of this triangle opposite the side of length $\rho_F(X_1,X_2)$. By definition, the angle $\alpha(\gamma_1,\gamma_2)$ between the segments γ_1 and γ_2 at O is the limit of the angle $\alpha(X_1,X_2)$ as $X_1,X_2 \to O$. Thus defined, the angle between any two segments issuing from a point exists.

The following theorem concerns the addition of angles between segments.

Let $\gamma_1, \gamma_2, \gamma_3$ be three segments issuing from a point O on a convex surface; let α_{12}, α_{23} and α_{13} be the angles between them. Assume that, for points X_1 and X_3 on α_1 and α_3 arbitrarily close to O, the segment X_1X_3 cuts γ_2. Then $\alpha_{13} = \alpha_{12} + \alpha_{23}$.

An analogous theorem holds for n segments issuing from one point. The proof is based on the inclusion property and the convexity of the metric of the convex surface.

By virtue of the inclusion property, two segments issuing from the same point O on a convex surface divide some neighborhood of this point into two sectors. Let V be one of the sectors bounded by the segments. Let $\gamma_1, \cdots, \gamma_n$ be segments issuing from O in this sector, indexed in order from γ to γ'. Let $\alpha_0, \alpha_1, \cdots, \alpha_n$ be the angles between the neighboring segments γ and γ_1, γ_1 and $\gamma_2, \cdots, \gamma_n$ and γ'. The *angle of the sector* V is defined as the supremum of the sum $\alpha_0 + \alpha_1 + \cdots + \alpha_n$ over all finite sets of segments γ_i within the sector.

The angle of a sector V bounded by segments γ and γ' issuing from a point O cannot be smaller than the angle between the segments γ and γ' themselves. But if the segment XX' joining two points X and X' on γ and γ' is entirely within V for X and X' arbitrarily close to O, the angle of the sector V is equal to the angle between the segments γ and γ'. Consequently, of two complementary sectors, there is always one whose angle equals the angle between the two segments bounding the sector.

Angles of sectors are added in the same way as angles of sectors on the plane: If a sector W splits into sectors U and V, its angle is the sum of the angles of U and V. It follows that the sum of the angles of two complementary sectors is independent of the sectors, depending only on the point which is their common apex. One can thus define the *complete angle about a point* O [or *at* O] on a convex surface as the sum of angles of any two complementary sectors with apex at this point.

The complete angle at a point on a convex surface is always $\leqq 2\pi$. This is obvious for convex polyhedra. For general convex surfaces it is proved by approximating the surface with convex polyhedra and using the theorem on convergence of angles between segments.

If a point O divides a segment γ into segments γ' and γ'', the angle of each of the sectors formed by γ' and γ'' is at least π. Now since the sum of these angles is at most 2π, each of them is exactly π. Hence no segment can pass through points on a convex surface at which the complete angle is less than 2π.

The theorem on convergence of angles between shortest joins (§3) can be used to prove an analogous theorem on the convergence of sector angles:

Let V_n be a sequence of sectors on convex surfaces which converge to a convex surface F, and let V_n converge to a sector V. Then if α and α_n denote the angles of the sectors V and V_n, respectively,

$$\alpha \leqslant \varliminf_{n\to\infty} \alpha_n.$$

This theorem has the following corollary:

Let F_n be a sequence of convex surfaces converging to a convex surface F, and O_n a sequence of points on F_n converging to a point O on F. Then the complete angle ϑ at the point O on F cannot exceed the lower limit of the complete angles ϑ_n at O_n on F_n:

$$\vartheta \leqslant \varliminf_{n\to\infty} \vartheta_n.$$

This in turn implies two important corollaries:

a) *Let V_n, F_n, F and V satisfy the assumptions of the preceding theorem; let the complete angles ϑ_n at the vertices of V_n converge to the complete angle ϑ about the vertex of V. Then, if α_n and α denote the angles of the sectors V_n and V, respectively,*

1) $\lim_{n\to\infty} \alpha_n$ *exists, and*

2) $\lim_{n\to\infty} \alpha_n = \alpha$.

b) *Let V_n, F_n, V and F satisfy the assumptions of the preceding theorem, and let the [complete] angle at the vertex of V be 2π. Then (with α and α_n as before),*

1) $\lim_{n\to\infty} \alpha_n$ *exists, and*

2) $\lim_{n\to\infty} \alpha_n = \alpha$.

These two theorems imply analogous properties of the angles between segments:

Let F_n be a sequence of convex surfaces converging to a convex surface F, and let γ_n' and γ_n'' be segments issuing from points O_n on F_n, which converge to segments γ' and γ'' issuing from O on F, where O_n converge to O. If the complete angles at the points O_n converge to the complete angle at O, or if the complete angle at O is 2π, then the angles between the segments γ_n' and γ_n'' converge to the angle between their limits γ' and γ''.

In particular, if segments issuing from a point O on a convex surface F converge to segments γ' and γ'', then

1) the angles between γ_n' and γ_n'' converge to the angle between γ' and γ'', and

2) the angles of both sectors bounded by γ_n' and γ_n'' converge to the angles of the respective sectors bounded by γ' and γ''.

In §1 of Chapter II we shall prove that there exist semitangents at each point of a segment. This will yield a space interpretation of the angle between segments on a convex surface. Recall that the tangent cone of a convex surface F at a point S is the surface of the solid cone generated by the rays issuing from S and cutting the body bounded by F; an equivalent definition is the following.

Apply to the surface F a similarity mapping with center S and ratio k. As $k \to \infty$ the image F_k of F under this transformation converges to a cone V which is precisely the tangent cone of F at S.

Now let γ' and γ'' be segments issuing from a point S on F; the above similarity mapping of F maps these onto segments γ_k' and γ_k'' on F_k; as $k \to \infty$ the latter converge to the semitangents t' and t'' to the shortest joins γ' and γ'' at S.

Since the complete angle at S on F_k is obviously independent of k, the angle of the sector bounded by γ_k' and γ_k'' on F_k is equal to the angle of the sector bounded by γ' and γ'' on F, and hence to the angle of the sector bounded by the generators t' and t'' in the domain obtained by developing the tangent cone on the plane.

It follows that if S is a smooth point (the tangent cone degenerates into a plane), the angle between shortest joins γ' and γ'' issuing from S is equal to the angle between their semitangents at S.

§5. Curvature of a convex surface

We now define the *intrinsic* and *extrinsic* curvature of a general convex surface.

The intrinsic curvature ω is first defined for the primitive sets – points, open segments and open triangles, as follows.

If M is a point and ϑ the complete angle at M on the surface, the intrinsic curvature of M is $\omega(M) = 2\pi - \vartheta$.

If M is an open segment, i.e. a segment minus its endpoints, then $\omega(M) = 0$.

If M is an open triangle, i.e. a triangle minus its sides and vertices, then $\omega(M) = \alpha + \beta + \gamma - \pi$, where α, β and γ are the angles of the triangle.

Intrinsic curvature is now defined for elementary sets; these are sets expressible as the set-theoretical unions of primitive sets: $M = \sum_1^n B_k$. For such sets, $\omega(M) = \sum_1^n \omega(B_k)$. One proves that this definition of intrinsic curvature of elementary sets is independent of their representation as unions of primitive sets. The proof depends essentially on the following elementary, though important theorem.

Let P be an interior subset of a geodesic polyhedron with angles $\alpha_1, \cdots, \alpha_n$ *and Euler characteristic* χ. *Then the curvature of P is*

$$\omega(P) = 2\pi\chi - \sum_{k=1}^{n} (\pi - \alpha_k).$$

The intrinsic curvature of elementary sets on a convex surface is clearly an additive function.

Let F be a convex surface and M a set of points on F. Drop the exterior unit normals from some point O to all supporting planes of F at the points of M. The endpoints of these normals form a set M^* on the unit sphere about O. This set is called the *spherical image* of the set M. If the area (Lebesgue measure) of M^* exists, it is called the *extrinsic curvature* of the set M.

Extrinsic curvature is a completely additive function on the Borel sets of a convex surface.

The proof of this theorem rests on the following two propositions, which are also of independent interest. The first is fairly easy to prove:

1. *The spherical image of a closed set on a convex surface is a closed set.*

2. *The set of all points of the spherical image of a convex surface which have more than one preimage on the surface has measure zero.*

The following two convergence theorems hold for the extrinsic curvature of convex surfaces:

1. *Let* F_n *be a sequence of convex surfaces converging to a convex surface* F, *and* M_n *a sequence of closed sets on* F_n *converging to a closed set* M *on* F. *Then*

$$\overline{\lim_{n\to\infty}}\, \psi(M_n) \leqslant \psi(M),$$

where ψ *denotes extrinsic curvature.*

2. *Let F_n be a sequence of convex surfaces converging to a convex surface F, G_n and G open sets on the surfaces F_n and F, and $\overline{G}_n$ and $\overline{G}$ their closures. Let the sets $\overline{G}_n$ converge to $\overline{G}$, the sets $F - G_n$ to $F - G$, and the extrinsic curvatures of the sets $\overline{G}_n - G_n$ to the extrinsic curvature of $\overline{G} - G$. Then the extrinsic curvature of G_n converges to that of G.*

In particular, if the boundary of the domain G_n converges to the boundary of G and the extrinsic curvature of the boundary of the domain G is zero, then the extrinsic curvature of G_n converges to that of G.

As yet we have not defined intrinsic curvature for nonelementary sets. We now define the intrinsic curvature of a closed set as the infimum of the intrinsic curvatures of its elementary supersets. The intrinsic curvature of an arbitrary Borel set is defined as the supremum of the intrinsic curvatures of its closed subsets.

The above definition of intrinsic curvature for closed sets and general Borel sets is consistent with the previous definition for elementary sets, as is guaranteed by the following fundamental theorem.

The intrinsic curvature of any Borel set on a convex surface is equal to its extrinsic curvature, i.e. to the measure of its spherical image.

The proof of this theorem is carried out first for primitive sets: points, open segments and open triangles. The principal tool of the proof is approximation of general convex surfaces by convex polyhedra.

Since Borel sets have the same intrinsic and extrinsic curvature, we shall henceforth use the unmodified term "curvature of a surface."

Using the newly defined concept of the curvature of a convex surface, one can augment the above relations between the elements of a triangle on a convex surface and a plane triangle with sides of the same length. In particular:

If α, β, γ are the angles of a triangle on a convex surface and α_0, β_0, γ_0 the corresponding angles of a plane triangle with the same sides, then

$$0 \leqslant \alpha - \alpha_0 \leqslant \omega, \quad 0 \leqslant \beta - \beta_0 \leqslant \omega, \quad 0 \leqslant \gamma - \gamma_0 \leqslant \omega,$$

where ω is the curvature of the triangle.

For the curvature of the triangle is

$$\omega = \alpha + \beta + \gamma - \pi = (\alpha - \alpha_0) + (\beta - \beta_0) + (\gamma - \gamma_0).$$

and the assertion follows from the fact that $\alpha - \alpha_0$, $\beta - \beta_0$ and $\gamma - \gamma_0$ are nonnegative.

Let ABC be a convex triangle on a convex surface and X, Y two points on its sides AB and AC, x, y the lengths of the subarcs AX and AY of the sides AB and AC, and z the distance of X from Y. Construct a plane triangle $A_0B_0C_0$ with sides of the same length ($A_0B_0 = AB$, etc.). Choose points X_0 and Y_0 on its sides A_0B_0 and A_0C_0 such that $A_0X_0 = x$ and $A_0Y = y$, and let $X_0Y_0 = z_0$. Then if ω is the curvature of the triangle ABC and d its greatest side, we have

$$0 \leqslant z - z_0 \leqslant 4\sqrt{xy}\sin\frac{\omega}{4} \leqslant \omega d.$$

A corollary of this is that a convex triangle on a convex surface whose curvature vanishes is isometric to a plane triangle. A domain G on a convex surface is locally isometric to a plane if and only if each of its points has a neighborhood with zero curvature.

§6. Existence of a convex polyhedron with given metric

A metric ρ given on a two-dimensional manifold M is called a *convex polyhedral metric* if each point of the manifold has a neighborhood isometric to a circular cone (which may degenerate into a plane). It is evident that every convex polyhedron has a convex polyhedral metric. In particular, a closed convex polyhedron is a manifold, homeomorphic to the sphere, with convex polyhedral metric.

The following question arises. Can an arbitrary convex polyhedral metric given on a manifold homeomorphic to a sphere be realized as a closed convex polyhedron? The following theorem of Aleksandrov answers this question in the affirmative.

Any convex polyhedral metric given on a sphere or on a manifold homeomorphic to a sphere is realizable as a closed convex polyhedron (possibly degenerating into a doubly covered plane polygon).

Let us describe the idea of Aleksandrov's proof. First note that there are only finitely many points on the manifold at which the total angle is less than 2π (i.e. which have a neighborhood isometric to a nondegenerate cone). We call these points *vertices*.

It is easily seen that for any polyhedral metric given on a manifold homeomorphic to a sphere there are at least three vertices. Furthermore, every convex polyhedral metric with three vertices is realizable as a doubly covered triangle, while a metric with four vertices is realizable as a tetrahedron. The sides of the triangle and the edges of the tetrahedron are equal to the distances between the vertices (in the given metric).

We now assume that every convex polyhedral metric with $l-1$ vertices is realizable, and prove that metrics with l vertices are realizable. To this end it will suffice to establish the following three assertions.

1. *A given metric ρ with l vertices lies in a continuous family of such metrics $\rho(t)$, which contains a metric ρ_0 known to be realizable.*

2. *If a metric $\rho(t_0)$ with l vertices is realizable, then all such metrics $\rho(t)$ sufficiently close to it are also realizable.*

3. *If metrics $\rho(t)$ with l vertices are realizable for all t close to t_0, then $\rho(t_0)$ is also realizable.*

The first assertion is proved in the following manner. Without loss of generality we can assume that the metric ρ has at least five vertices. This implies that there are at least two vertices at which the complete angle ϑ is greater than π; for the manifold is homeomorphic to a sphere and so $\sum (2\pi - \vartheta_k) = 4\pi$. It follows that at least two values of ϑ must exceed π.

Let A and B be vertices at which the complete angle is greater than π. Join these points by a segment γ and cut the manifold along this segment, Now "glue" two congruent triangles to the cut: let each of these triangles have base equal in length to γ, angle at the vertex glued to A equal to $\pi - \vartheta_A/2$, and angle at the vertex glued to B equal to t.

The resulting manifold also has a convex polyhedral metric with the same number (l) of vertices. As t varies from zero to $\pi - \vartheta_B/2$, the metric $\tilde{\rho}(t)$ of the manifold varies continuously; this metric is realizable for $t = \vartheta_B/2$, since the number of vertices is then $l - 1$. (The two vertices A and B in the original metric are replaced by a single new vertex C.)

Suppose that this metric $\tilde{\rho}_1$ is realizable as a polyhedron P_1 with $l - 1$ vertices. Move the point on P_1 corresponding to the vertex A into the exterior of P_1, and construct the convex hull of the moved point and the polyhedron P_1. The convex hull is a polyhedron P_0 with l vertices and metric $\tilde{\rho}_0$ close to $\tilde{\rho}_1$. One can now show that the metrics $\tilde{\rho}(t)$ can be subjected to a small variation under which they become a continuous family of metrics $\rho(t)$ with l vertices "connecting" the given metric ρ with the realizable metric $\tilde{\rho}_0$ (which is either close to or the same as $\tilde{\rho}_0$).

We now consider the second assertion. To simplify matters, assume that all faces of the polyhedron $P(t_0)$ realizing the metric $\rho(t_0)$ are triangles. Place the polyhedron $P(t_0)$ so that the vertices A, B, C of one of its faces Δ_0 fall in the xy plane, with A at the origin, B on the ray $x > 0$ and C in the halfplane $y > 0$. Subject the vertices of $P(t_0)$ to small motions, in such a way that the vertices of the face Δ_0 maintain the positions described above. The edges of the polyhedron also vary in this process. The connection between the coordinates of the vertices and the lengths of the edges of the polyhedron is given by a system of equations

$$(*) \qquad d_{ij} = \{(x_i - x_j)^2 + (y_i - y_j)^2 + (z_i - z_j)^2\}^{1/2} \equiv f_{ij}(x, y, z),$$

where d_{ij} is the length of the edge joining the vertices $A_i(x_i, y_i, z_i)$ and $A_j(x_j, y_j, z_j)$.

To prove the realizability of metrics $\rho(t)$ close to $\rho(t_0)$, it suffices to show that the system (*) is solvable for given d_{ij} close to d_{ij}^0 (where d_{ij}^0 are the edges of the polyhedron $P(t_0)$). The number of equations is the same as the number of edges of $P(t_0)$, while the number of unknowns (x), (y), (z) is $3l - 6$, where l is the number of vertices.

By Euler's formula for the polyhedron $P(t_0)$ we have $l - k + f = 2$, where l is the number of vertices, k the number of edges and f the number of faces of the polyhedron. By assumption, all faces of $P(t_0)$ are triangles; hence $3f = 2k$, and so $3l - 6 = k$. Thus both the number of equations and the number of unknowns in the system (*) are equal to $3l - 6$.

Now a sufficient condition for the system (*) to have a solution in the neighborhood of values (x^0), (y^0), (z^0) corresponding to the polyhedron $P(t_0)$ is that the Jacobian of the system be different from zero. For this, in turn, it suffices that the system of linear equations

$$(**) \qquad df_{ij}|_{P(t_0)} = 0$$

in the differentials dx_i, dy_j, dz_k have only a trivial solution. Geometrically this means that the polyhedron $P(t_0)$, with its face Δ_0 fixed as described, admits no infinitesimal deformations under which its edges are stationary. By the rigidity theorem, a closed convex polyhedron indeed admits no such deformations. This proves the second assertion.

The proof of the third and last assertion is simple. Given a sequence of polyhedra $P(t)$ realizing metrics $\rho(t)$, one can extract a subsequence with convergent vertices. The corresponding polyhedra themselves will then converge, and their limit $P(t_0)$ is a polyhedron realizing the metric $\rho(t_0)$.

§7. Existence of a closed convex surface with given metric

We now consider whether there exists a closed convex surface realizing a given metric. In order to approach the solution of this problem, we first elucidate the conditions to be imposed on the metric.

Since any closed convex surface is homeomorphic to a sphere, the metric to be realized must be given on a manifold homeomorphic to a sphere. It must also be an intrinsic metric, since the metric of a convex surface is intrinsic. Finally, without loss of generality one can require the metric to satisfy the convexity condition (§3), for the metric of any convex surface satisfies this condition.

It turns out that under the above assumptions the metric is indeed realizable as a closed convex surface. However, the last condition (convexity of the metric), though necessary, is too strong to be a satisfactory sufficient condition. It can be replaced by the weaker condition that the curvature be nonnegative; that is, for every sufficiently small triangle whose sides are segments the sum of lower angles is at least π. The *lower angle* between curves always exists and is defined as follows.

Let R be a manifold with intrinsic metric ρ. Let O be an arbitrary point of R and γ_1, γ_2 curves issuing from O. Take points X_1, X_2 on these curves and construct a plane triangle with sides $\rho(O, X_1)$, $\rho(O, X_2)$, $\rho(X_1, X_2)$. Denote the angle of this triangle opposite the side $\rho(X_1, X_2)$ by $\alpha(X_1, X_2)$. We now define the lower angle between the curves γ_1 and γ_2 at O to be

$$\alpha(\gamma_1, \gamma_2) = \varliminf_{X_1, X_2 \to O} \alpha(X_1, X_2).$$

The theorem on the realizability of a given metric as a convex surface runs as follows:

An intrinsic metric of nonnegative curvature, given on a sphere or on a manifold homeomorphic to a sphere, is realizable as a closed convex surface (possibly degenerating into a doubly covered convex plane domain).

One can prove that the convexity condition is satisfied in a manifold with intrinsic metric and nonnegative curvature. Therefore, in order not to complicate the exposition, we shall assume from the start that the metric satisfies the convexity condition and prove the realizability theorem under this assumption.

Let R be a manifold homeomorphic to a sphere, with an intrinsic metric which satisfies the convexity condition. Introduce a sufficiently fine triangulation of the manifold R; that this is possible follows from the fact that the inclusion property holds in R (§1). Let the diameter of each triangle in this triangulation be smaller than δ.

Take any triangle Δ_k of the triangulation and construct a plane triangle Δ_k^0 with sides of the same length. Let f be a homeomorphism of Δ_k onto Δ_k^0 which maps the vertices of Δ_k onto the corresponding vertices of Δ_k^0 and is an isometry on the sides of the triangle. We now construct a manifold R^0 from the triangles Δ_k^0, with polyhedral metric R_0, by identifying corresponding points on the sides of the triangles Δ_k^0; by corresponding points of the sides of triangles Δ_k^0 and Δ_s^0 we mean points which have identical preimages in Δ_k^0 and Δ_s^0 under the homeomorphism f.

Since the angles of the triangle Δ_k^0 do not exceed those of Δ_k and the complete angle at each point of R is at most 2π, the complete angle at each point of the new manifold R^0 is also at most 2π, and so R_0 is a convex polyhedral metric.

By the theorem which states that a convex polyhedral metric on a manifold homeomorphic to a sphere is realizable (§6), there exists a closed convex polyhedron P_0 isometric to R_0. The homeomorphisms f of the triangles Δ_k onto the triangles Δ_k^0 generate a homeomorphism of R onto R_0, and so onto P_0. We continue to denote this homeomorphism by f. It can be shown that for small δ the homeomorphism f of R onto P_0 is near to isometric. In other words, if X and Y are two arbitrary points on R, fX and fY their images on P_0, then

$$|\rho_R(X, Y) - \rho_{P_0}(fX, fY)| < 4\pi\delta.$$

In fact, let X and Y be arbitrary points of R and γ a segment joining them. The homeomorphism f maps this segment onto some curve γ' joining the points fX and fY. Modify γ' by replacing that part γ_k' of it included within the triangle Δ_k^0 by a segment γ_k^0 joining its ends. Call the resulting curve γ^0. Let us compare the lengths l of the curves γ and γ^0.

According to a theorem stated in §5, those subarcs z_k and z_k^0 of the curves included within the corresponding triangles Δ_k and Δ_k^0 satisfy the inequality

$$|z_k - z_k^0| \leqq \omega(\Delta_k)\delta,$$

where $\omega(\Delta_k)$ is the curvature of the triangle Δ_k. Since the total curvature of the manifold R is 4π, the sum of curvatures $\omega(\Delta_k)$ over all triangles Δ_k cutting γ is at most 4π. Hence

$$|l(\gamma) - l(\gamma^0)| \leqq \sum |z_k - z_k^0| \leqq 4\pi\delta.$$

Since γ is a shortest join of X and Y in the manifold R, it follows that

$$\rho_{P_0}(fX, fY) - \rho_R(X, Y) \leqslant 4\pi\delta.$$

Analogous consideration of the segment joining fX and fY on the polyhedron shows that

$$\rho_R(X, Y) - \rho_{P_0}(fX, fY) \leqslant 4\pi\delta.$$

Consequently

$$|\rho_R(X, Y) - \rho_{P_0}(fX, fY)| \leqslant 4\pi\delta.$$

This proves our assertion.

Denote the triangulation introduced in R by T_0. Divide each triangle Δ of this triangulation into smaller triangles in such a way that the sides of the triangles in the new triangulation are smaller than $\delta_1 = \delta/2$. Denote this triangulation by T_1. Similarly, construct a triangulation T_2 in which the sides of the triangles are smaller than $\delta_2 = \delta_1/2$, and so on. Clearly the vertices of the triangulation T are vertices of T_1, those of T_1 are vertices of T_2, and so on. Using the same construction by which the polyhedron P_0 was derived from T_0, use the triangulations $T_1, T_2, \cdots$ to construct polyhedra $P_1, P_2, \cdots$.

Let A_i^k be the vertices of the triangulation T_k of the manifold R. Let the corresponding vertices of the polyhedron P_k be B_i^k; note that some of these may be improper. Now take a subsequence of polyhedra P_k for which the vertices B_i^k converge as $k \to \infty$; construction of such a subsequence is not difficult. Since the net of points B_i^k on the polyhedron P_k becomes arbitrarily fine as k increases, convergence of the vertices B_i^k implies that the polyhedra P_k themselves converge to some convex surface F.

By virtue of the inequality

$$|\rho_R(X, Y) - \rho_{P_k}(fX, fY)| \leqslant 4\pi\delta_k$$

the metrics of the polyhedra P_k converge to the metric of the manifold R. On the other hand, they converge to the metric of the limit surface F. It follows that the metric of the manifold R is realized on the convex surface F. The theorem is proved.

§8. Curves on a convex surface

The angle between arbitrary curves on a convex surface is defined in exactly the same way as the angle between segments. In detail: Let γ_1 and γ_2 be curves through a point O on a convex surface F. Take points X_1 and X_2 on these curves, and construct a plane triangle with sides $\rho(O, X_1)$, $\rho(O, X_2)$, $\rho(X_1, X_2)$. Let $\alpha(X_1, X_2)$ be the angle in this plane triangle opposite the side $\rho(X_1, X_2)$. We shall say that the curves γ_1 and γ_2 form a definite angle at O if the limit of $\alpha(X_1, X_2)$ as X_1, $X_2 \to O$ exists. The limit, if it exists, is called the *angle between the curves.*

Whereas the angle between segments always exists (§4), the angle between arbitrary curves may not exist; examples are easily presented for the plane. To determine conditions for the existence of the angle between curves, we introduce the important concept of the direction of a curve.

A curve γ issuing from a point O on a convex surface is said to *have* *a definite direction at O* if the angle that it forms with itself, in the sense of the above definition, exists (and is then of course zero). It turns out that a curve has a definite direction at a point O if and only if it has a semitangent there. The question of the existence of the angle between curves is settled by the following theorem.

Curves γ_1 and γ_2 issuing from a point O on a convex surface form a definite angle if and only if each of them has a definite direction at O. When it exists, the angle between the curves is equal to the angle between their semitangents when the tangent cone at O is developed on the plane.

We now define the integral geodesic curvature[4] of a curve. The *integral geodesic curvature* (i.g.c.) of a curve is an integral generalization of the geodesic curvature of a regular curve on a regular surface. We first define the i.g.c. of an (open) geodesic polygon. Let γ be some such polygon which is a simple curve, and let $A_1, A_2, \cdots, A_n$ be its interior vertices. Fix some direction on the polygon, so that its left and right sides are defined. Let α_i be the angle of the sector to the right between the segments of the polygon that meet at A_i. Then the right i.g.c. of the curve γ is defined as

$$\varphi_r(\gamma) = \sum_i (\pi - \alpha_i),$$

where the summation extends over all interior vertices. The definition of the left i.g.c. $\varphi_l(\gamma)$ is analogous, with the angles being those of the sectors to the left of γ.

Now let γ be an arbitrary simple curve with endpoints A and B and a definite direction at these points. Fixing a direction on γ, construct a sequence of simple polygons γ_n converging to γ and situated in the right semineighborhood of the curve. Let φ_n be the right i.g.c. of γ_n, and α_n and β_n the angles formed by the first and last segments of the polygon and the curve γ, measured on the right. Then the right i.g.c. of the curve is defined as

$$\varphi_r(\gamma) = \lim_{\gamma_n \to \gamma} (\alpha_n + \beta_n + \varphi_n).$$

The limit in the right-hand side of this equality always exists and is independent of the choice of the sequence of polygons converging to

---

[4] *Translator's note.* In his book *Convex Surfaces* (Interscience, New York, 1958) Busemann coined the term *swerve* for this concept (Russian *povorot*, German *Schwenkung*), but he has since expressed preference (private communication to the translator) for the "more mathematical" term used here.

γ. Thus the existence of the i.g.c. depends solely on the existence of definite directions at the ends of the curve.

The left i.g.c. $\varphi_l(\gamma)$ of the curve γ is defined in analogous fashion, approximating the curve by polygons in its left semineighborhood.

The right and left i.g.c. of a curve are additive functions. That is to say, if a curve γ is divided by a point C into two subarcs γ_1 and γ_2, both of which have definite directions at C, then

$$\varphi(\gamma) = \varphi(\gamma_1) + \varphi(C) + \varphi(\gamma_2),$$

where $\varphi(\gamma)$, $\varphi(\gamma_1)$ and $\varphi(\gamma_2)$ are the right (left) i.g.c.'s of the respective curves and $\varphi(C)$ is the angle of the sector between the curves γ_1 and γ_2 at C, measured on the right (left).

Let γ be a simple closed curve on a convex surface. Fixing a direction on γ, we define its right and left sides. The right (left) i.g.c. of the curve γ is defined as the limit of the right (left) i.g.c.'s of a sequence of closed polygons γ_n converging to γ and situated within the right (left) semineighborhood of γ. If there is a point C on γ at which the curve has a definite direction, the right (left) i.g.c. may be defined as the sum of the right (left) i.g.c. of the curve cut at C and the angle of the sector to the right (left) of γ at C.

We have the following theorem, which generalizes the well-known Gauss-Bonnet Theorem for regular surfaces.

If the curve γ bounds a domain G homeomorphic to a circle, with curvature $\omega(G)$, and the i.g.c. of γ toward the domain G is $\varphi(\gamma)$, then $\omega(G) + \varphi(\gamma) = 2\pi$.

When this curve γ is a polygon this theorem is a simple consequence of the formula for the curvature of a polygonal domain (§5). In the general case it is proved by approximating the domain with polygonal domains inscribed in G.

The Gauss-Bonnet Theorem lays the basis for the proof of an important property of the i.g.c.:

The sum of the right and left i.g.c.'s of a curve is equal to the curvature (area of the spherical image) of the curve minus its endpoints.

We indicate a few frequently used applications of the i.g.c.

A domain G on a convex surface is called a *convex domain* if any pair of its interior points has a shortest join which lies within the domain G. It turns out that every arc of the curve bounding a convex domain has nonnegative i.g.c. toward the domain.

A domain G on a convex surface is said to be convex in itself if any two of its interior points have a shortest join in G, which need not be a shortest join in the entire surface. It turns out that a closed domain G on a convex surface, bounded by a curve γ, is convex in itself if and only if each arc of γ has nonnegative i.g.c. toward the domain.

The i.g.c. concept is the basis for a class of curves which holds interest in many respects: the class of quasigeodesics. A curve γ on a convex surface is called a *quasigeodesic* if its right and left i.g.c.'s over any subarc are nonnegative. Every geodesic is a quasigeodesic, since its right and left i.g.c.'s over any arc vanish. However, there are quasigeodesics which are not geodesics. For example, the circumference of the base of a right circular cone is a quasigeodesic, but it is not a geodesic since no subarc is a shortest join.

We note some properties of quasigeodesics:

1. The limit of quasigeodesics (in particular, geodesics) is a quasigeodesic.

2. Every quasigeodesic is the limit of a sequence of geodesics in the sense of the intrinsic metric.

3. On any convex surface there exist quasigeodesics in any direction, though they need not be unique.

As regards the third property, note that on a general convex surface there may be directions in which no geodesic can be drawn. For example, there are no geodesics through a point on the circumference of the base of a right circular cone, in the direction of the circumference.

The question of quasigeodesics is treated in [**52**].

§9. Area of a convex surface

As we know from differential geometry, the area of a regular surface is an object of the intrinsic geometry of the surface; that is to say, it is determined by the metric of the surface alone. It is therefore natural to define the area of a general convex surface in a purely intrinsic manner. Aleksandrov did this in the following way.

Let P be a polygon on a convex surface. Introduce a triangulation T_n of P into triangles with sides of length less than $1/n$. For each triangle Δ of this triangulation let Δ_0 be a plane triangle with the same sides, and let S_n denote the sum of the areas of these triangles. It can be shown that the sum S_n converges as $n \to \infty$ to a definite limit S, irrespective of the choice of the triangulation T_n in the polygon P. This limit is defined to be the area of the polygon P on the surface.

Now that the area of a polygon has been defined, the area of closed, open, and general Borel sets is defined by the usual methods of measure theory.

The proof that the sequence S_n has a limit is based on the following lemma.

Let Δ be a triangle on a convex surface, d its diameter, $\omega(\Delta)$ the curvature of the interior domain and S_0 the area of a plane triangle Δ^0 with sides of the same length. Then for every $\epsilon > 0$ there exists a partition T of the triangle Δ into smaller triangles Δ_k such that the sum S_T of the areas of plane triangles Δ_k^0 with sides of the same length satisfies the inequality

$$-\varepsilon < S_T - S_0 < \frac{1}{2}\omega(\Delta) d^2 + \varepsilon.$$

This lemma, in turn, is proved on the basis of a lemma on convex polyhedra:

Let Δ be a triangle on a convex polyhedron or, in general, in a manifold with polyhedral metric of positive curvature, Δ_0 a plane triangle with sides of the same length, d the diameter of the triangle Δ, ω the curvature of its interior domain, and S and S_0 the areas of the triangles Δ and Δ_0. Then

$$0 \leqslant S - S_0 \leqslant \frac{1}{2}\omega d^2.$$

The first lemma is proved by approximating the given convex surface by a manifold with polyhedral metric formed by "gluing together" triangles Δ_k^0, and applying the second lemma in an appropriate manner.

To prove the second lemma, note that if the triangle Δ is cut along a segment γ joining two interior vertices and a suitable pair of congruent triangles is glued to the cut, the number of interior vertices may be reduced without thereby changing the total curvature or increasing the area. By means of this device the problem can be reduced to the case of a single interior vertex, and consideration of this case is quite elementary.

Using the first lemma, one shows that the sequence S_n figuring in the definition of area converges in the sense of Cauchy, i.e. $|S_n - S_m| \to 0$ as $m, n \to \infty$, and so the limit S exists. We have thus proved the following theorem:

Every polygon on a convex surface has a definite area. Let S be the area of a polygon P and ω the curvature of its interior domain. Then, if the polygon P is partitioned into triangles of diameter $\leqq d$, the sum S_T of areas of the plane triangles with sides of the same length satisfies the inequality

$$0 \leqslant S - S_T \leqslant \frac{1}{2}\omega d^2.$$

The intrinsic meaning of area follows from the following theorem on convergence of areas.

Let F_n be a sequence of convex surfaces converging to a convex surface F, and P_n a sequence of polygons on F_n converging to a polygon P on F, such that the number of sides of the polygons P_n is bounded by some given number. Then the areas of the polygons P_n converge to the area of P.

This theorem follows easily from the preceding one.

A corollary of the theorem on convergence of areas is that the area of a polygon P on a convex surface F is the limit of the areas of P_n converging to P which lie on convex polyhedra converging to the surface F, provided that the number of sides of the polygons is bounded.

Hence our intrinsic definition of the area of a convex surface is equivalent to the extrinsic definition as the limit of areas of convex polyhedra converging to the surface.

The extrinsic definition of the area of a convex surface yields a formula for the area. If the surface F is represented in rectangular cartesian coordinates x, y, z by the equation $z = z(x, y)$ and has no supporting planes parallel to the z-axis, the area of any polygon P on the surface is defined by the well-known integral

$$S(P) = \int_{P'}\int \sqrt{1 + z_x^2 + z_y^2}\, dx\, dy,$$

where the domain of integration is the projection P' of the polygon on the xy-plane.

§10. Specific curvature of a convex surface

Every domain G on a convex surface has definite area $S(G)$ and curvature $\omega(G)$. The quotient

$$\varkappa(G) = \frac{\omega(G)}{S(G)}$$

is called the *specific curvature* of the domain G. If the specific curvatures of all domains G on a convex surface are bounded by some constant, the surface is known as a *surface of bounded curvature.*

The property of having bounded specific curvature is preserved in passage to a limit. This is implied by the following theorem.

Let F_n be a sequence of convex surfaces with uniformly bounded specific curvatures, converging to a surface F. Then F is a surface of bounded curvature.

The proof is based on the convergence theorems for areas and curvatures of convergent sequences of convex surfaces.

The specific curvature of a convex surface at a point X, that is, the limit of $\kappa(G)$ when the domain G shrinks to a point X, is called the

Gauss curvature of the surface at X. It is easily proved that if the Gauss curvature exists at each point of a surface it is continuous.

Surfaces of bounded curvature possess some of the properties of regular convex surfaces. In particular, there is a segment in any direction through each point of a surface of bounded curvature, over a distance which depends only on the specific curvature of the surface.

The proof is simple: Were the assertion false, this would imply the existence of arbitrarily small lunes with vertex at the given point. The area of such a lune Δ satisfies the estimate $S \leqq \omega d^2/2$ (§9), where ω is the curvature and d the diameter of the lune. Hence

$$\kappa(\Delta) = \omega/S \geqq 2/d^2,$$

and so the specific curvature increases without bound as d decreases.

Since there exists a segment from a point in any direction, for a distance $r_0 > 0$ one can introduce polar coordinates r, ϑ in a neighborhood of the point. If moreover the surface has definite Gauss curvature at each point, the metric of the surface may be given in the parametrized neighborhood by the line element

$$ds^2 = dr^2 + G(r, \vartheta)\, d\vartheta^2,$$

where the coefficient G is a continuous function, and twice differentiable in r. The relation between this coefficient and the Gauss curvature of the surface is given by the well-known formula

$$K = -\frac{1}{\sqrt{G}} (\sqrt{G})_{rr}.$$

If the Gauss curvature of the surface is constant and positive, then, as is easily seen, the coefficient G, which satisfies the equation

$$(\sqrt{G})_{rr} + K\sqrt{G} = 0,$$

must have the form

$$\sqrt{G} = \frac{1}{\sqrt{K}} \sin \sqrt{K} r.$$

Hence a surface with this property is locally isometric to a sphere of radius $1/\sqrt{K}$.

Let the specific curvature of a triangle Δ on a convex surface be $\geqq K$ $(\leqq K)$; then the angles of this triangle are at least (at most) equal to the corresponding angles of a triangle Δ_K with the same sides on a sphere of radius $1/\sqrt{K}$.

Let the specific curvature of a triangle Δ on a convex surface be $\geqq K$ $(\leqq K)$; then the area S of this triangle is at least (at most) equal to the area of a triangle Δ_K with the same sides on a sphere of radius $1/\sqrt{K}$. Moreover, we have the estimates

$$0 \leqslant S(\Delta) - S(\Delta_K) \leqslant \frac{1}{2}(\omega(\Delta) - \omega(\Delta_K))\, d^2,$$

if the specific curvature of the triangle Δ is $\geqq K$, and

$$0 \geqslant S(\Delta) - S(\Delta_K) \geqslant \frac{1}{2}(\omega(\Delta) - \omega(\Delta_K))\, d^2$$

if the specific curvature of Δ is $\leqq K$.

Let γ_1 and γ_2 be two segments issuing from O on a convex surface. Let X_1 and X_2 be variable points on γ_1 and γ_2, with $\rho(O, X_1) = x$, $\rho(O, X_2) = y$, $\rho(X_1, X_2) = z$, and let $\alpha(x, y)$ be the angle opposite z in a triangle with sides x, y, z on a sphere S_K of radius $1/\sqrt{K}$. The metric ρ of the surface is said to satisfy the K-convexity condition, or to be *K-convex*, if for any segments γ_1 and γ_2 the angle $\alpha(x, y)$ is a nonincreasing function in any interval $0 < x < x_0$, $0 < y < y_0$ for which the segment X_1X_2 exists. The metric ρ is said to satisfy the K-concavity condition, or to be *K-concave*, if $\alpha(x, y)$ is a nondecreasing function of x, y in any such interval. We have the following theorem:

If the specific curvature of a convex surface is $\geqq K$ ($\leqq K$), the K-convexity (K-concavity) condition holds on the surface.

Points on a convex surface fall into three categories: conical points, at which the tangent cone is nondegenerate and so the complete angle is less than 2π; ridge points, at which the tangent cone degenerates to a dihedral angle; and plane points, at which the tangent cone degenerates to a plane. It is evident that there can be no conical points on a surface of bounded curvature, since at such points the specific curvature is infinite. There may be ridge points; however, we have the following theorem.

Let A be a point on a convex surface such that the specific curvature of any sufficiently small domain containing A is bounded above by some constant. Then A is either a smooth point or lies on a straight edge of the surface.

It follows as a corollary that any closed convex surface of bounded curvature is smooth. An infinite complete convex surface of bounded curvature, no finite part of which is a cylinder, is smooth.

If a point A on a convex surface lies on a straight-line segment, the surface contains arbitrarily small domains of arbitrarily small specific curvature containing A.

Consequently, if the specific curvature of a convex surface is positive and bounded away from 0 and ∞ in any domain of the surface, the surface is smooth. The proofs of the last two theorems will be presented in the second chapter.

§11. Gluing theorem. Other existence theorems

The "gluing" theorem, which we are about to discuss, is a tool for constructing new manifolds with positive curvature from given manifolds with intrinsic metric and positive curvature, via identification (gluing together) of boundaries. This theorem, which we now state, has numerous important applications.

Let $G_1, \cdots, G_n$ be closed domains in manifolds with intrinsic metric and positive curvature, bounded by a finite number of curves with i.g.c.'s of bounded variation. Let G be the manifold obtained from the domains G_k by so identifying their boundary points as to satisfy the following three conditions:

1. *Identified sections on the boundaries of domains G_i and G_j have equal lengths.*

2. *The sum of i.g.c.'s of any identified sections of the boundaries of domains G_i and G_j toward these domains is nonnegative.*

3. *The sum of the sector angles at identified points of domains $G_\alpha, \cdots, G_\beta$ toward these domains is at most 2π.*

Then the manifold G has an intrinsic metric of positive curvature which coincides with the metrics of the domains G_i in small neighborhoods of the corresponding points.

Let us explain the theorem. First, the theorem deals with curves whose i.g.c.'s have bounded variation. This means that if some finite subarc γ of such a curve is subdivided by points A_k into subarcs γ_k, then

$$\sum_k (\varphi(\gamma_k) + \varphi(A_k)) < c(\gamma),$$

where the constant $c(\gamma)$ is independent of the choice of the points A_k or the number of points. Curves with i.g.c. of bounded variation are recti- fiable. One can therefore speak of the lengths of the identified sections of the boundaries of the domains G_k.

The theorem states that the manifold G has an intrinsic metric of positive curvature which coincides with the metrics of the manifolds G_k in small neighborhoods of corresponding points. The metric in question is defined as follows. Connect two arbitrary points X and Y on G by all possible curves on G. The length of each such curve is defined as the sum of the lengths of its subarcs lying in the domains G_k. The distance between X and Y is defined to be the infimum of the lengths of these curves. The metric thus defined is clearly intrinsic, and moreover it stands in the required relation to the metrics of the domains G_k. Only the fact that this metric has positive curvature is required in the proof.

The gluing theorem is obvious when the domains G_k are polygons on manifolds with convex polyhedral metric. When the domains G_k are polygons on general manifolds with positive curvature, the theorem is proved by dividing the polygons G_k into small triangles Δ. Each of these triangles is replaced by a plane triangle Δ^0. These plane triangles are glued together to give manifolds G_k^0 with convex polyhedral metric. Finally, the manifolds G_k^0 are glued together to give a manifold G^0. It turns out that the manifolds G_k^0 satisfy the assumptions of the gluing theorem. It now suffices to pass to the limit as the triangles Δ into which the G_k are divided become infinitesimal. The proof of the gluing theorem in the general case uses approximation of the domains G_k by polygons and the gluing theorem for polygonal domains.

We consider some applications of the gluing theorem to prove theorems on the realizability of abstractly given convex metrics.

Let R be a manifold with intrinsic metric and positive curvature, G a convex domain in R homeomorphic to a disk. Let G' be a second copy of the domain G; identify corresponding points of the boundaries of these domains. This identification procedure satisfies conditions $1-3$ of the gluing theorem, since the identified sections of the boundaries obviously have the same length and nonnegative i.g.c.'s (since they are convex). By the gluing theorem, the closed manifold $\overline{G}$ obtained by gluing G and G' has positive curvature. By the theorem of §7 this manifold is isometric to a closed convex surface $\overline{F}$. The domain F on $\overline{F}$ which is isometric to G realizes the metric of the manifold R in the domain G. It can be shown that the boundary of F is a plane curve, and that F can be projected in one-to-one fashion onto the plane of its boundary, provided the i.g.c. of the boundary of G over any subarc is positive. A convex surface of this type is called a *convex cap*.

Since every point of a manifold with intrinsic metric and positive curvature has a strictly convex neighborhood, any such manifold has a neighborhood isometric to a convex cap.

Using the gluing theorem, one can prove that *every complete metric with positive curvature, given on a plane, is realizable as an unbounded convex surface*.

We shall sketch the simple proof of this theorem. Let P be a large polygon on the plane on which the metric is given (a polygon with respect to the given metric). Let S be a point outside the polygon. Construct a loop γ of minimal length, with node at S, which encloses the entire polygon P. This loop is a polygonal line bounding a polygon $\overline{P}$ which encloses P. None of the angles of this polygon, except perhaps the angle at the vertex S, can exceed π. Identify those points of the boundary of $\overline{P}$ at equal distances from S (measuring the distances along γ). The assumptions of the gluing theorem are satisfied, and the polygon $\overline{P}$ gives rise to a closed manifold with positive curvature, which is isometric to a closed convex surface $\overline{F}$. It now remains to pass to the limit, letting the polygon P fill out the entire plane.

Considering a convex cone as an example, we easily see that a complete metric of positive curvature given on a plane may be realized by an unbounded convex surface in more than one way. Olovjanišnikov has proved that this is always the case if the total curvature of the manifold to be realized is less than 2π. The theorem follows:

Let F be an unbounded complete convex surface with positive curvature $\omega < 2\pi$ and γ a unilaterally infinite curve on F each subarc of which is a segment [i.e., a geodesic ray] (such a curve can be drawn from any point on F). Let K be a convex cone with curvature (= area of spherical image) equal to the total curvature ω of F, and t a generator of K. Then there exist two unbounded convex surfaces F_1 and F_2 isometric to F, whose limit cone is K,[5] *and such that the similarity mapping contracting them to the cone K maps the rays γ_1 and γ_2 corresponding to γ onto the generator t. The surfaces F_1 and F_2 have different orientations: If the direction on F_1 forms a right-handed screw with the exterior normal, then the corresponding direction on F_2 forms a left-handed screw with the exterior normal.*

Thus, in realization of a complete metric of positive curvature as an unbounded convex surface, the limit cone can be prescribed arbitrarily (except for its curvature, which must be ω), as can the limit direction t of the geodesic ray γ on the surface of the cone.

§12. Convex surfaces in spaces of constant curvature

Theories of convex surfaces may also be developed in elliptic space and Lobačevskiĭ space, defining a convex surface, as in euclidean space, as a domain on the boundary of a convex body. A convex body in

[5] The limit cone of a surface F is the surface of the solid cone consisting of all rays issuing from a given point X on F and contained in the body bounded by F. The limit cone is uniquely defined up to parallel translation depending on the choice of the point X.

Lobačevskiĭ space is defined as in euclidean space: a body which contains together with each pair of points the entire straight-line segment joining them. If Lobačevskiĭ space is mapped geodesically onto the interior of a euclidean sphere (Cayley-Klein interpretation), the convex bodies are mapped onto euclidean convex bodies.

Convex bodies cannot be defined in this way in elliptic space, since the straight-line segment joining two given points is not unique. This difficulty is eliminated by defining a convex body as a body K with the following properties: there exists a plane disjoint from K, and any segment connecting any two points of K and not cutting this plane is contained in K. When elliptic space is interpreted as a three-dimensional sphere with pairwise identified antipodal points, convex bodies are represented by convex sets on a hemisphere.

As in the euclidean case, the first task of the theory of convex surfaces in spaces of constant curvature (elliptic and Lobačevskiĭ space) is to prove that the metrics of convergent convex surfaces converge. One can then study general convex surfaces using approximation by convex polyhedra. The first step in this direction is to prove that the metric of a convex surface satisfies the convexity condition. If the curvature of the space is K, this property is the same as K-convexity, which we considered in §10, and is defined as follows. Let γ_1 and γ_2 be segments issuing from a point O on a convex surface in a space of curvature K, and let X_1 and X_2 be points on these segments. Construct a plane triangle with sides $x = OX_1$, $y = OX_2$ and $z = X_1X_2$ on a K-plane (i.e. a Lobačevskiĭ plane if the space in question in Lobačevskian, or an elliptic plane if the space is elliptic). Let $\alpha(x,y)$ be the angle of this triangle opposite the side z. The convexity property holds if and only if $\alpha(x,y)$ is a nonincreasing function of x, y.

The convexity property of the metric of a convex surface in a space R_K of curvature K implies various corollaries, analogous to those established for the metrics of convex surfaces in euclidean space. In particular, it can be proved that the angles of a triangle on a convex surface in R_K are at least equal to the corresponding angles of a triangle with sides of the same length in a plane of R_K, i.e. on a surface of constant curvature K. It follows that the sum of angles of a triangle on a convex surface in elliptic space is always greater than π. Thus with regard to intrinsic geometry, convex surfaces in elliptic space yield nothing not met with in convex surfaces in euclidean space. Every convex surface in an elliptic space is locally isometric to a convex surface in euclidean space.

The situation is different for convex surfaces in Lobačevskiĭ space. Here there exist convex surfaces which are not isometric to convex surfaces of euclidean space even in the small. One of these is a Lobačevskiĭ plane. It has negative curvature and therefore cannot be isometric to a convex surface in euclidean space, whose Gauss curvature, if it exists, is always nonnegative. Lobačevskiĭ space is remarkable for its greater diversity of topological types of complete convex surfaces. Indeed, every domain on the sphere is homeomorphic to some complete convex surface.

Once the convexity property has been proved for the metric of a convex surface in R_K, the rest of the theory proceeds as in the euclidean case: one studies the properties of the angle between segments, defines the angle of a sector bounded by segments, and determines the extrinsic-geometrical meaning of the angle.

The theory of curvature and area for convex surfaces in a space of constant curvature K is developed as in euclidean space. The only difference is that the curvature theory is constructed from beginning to end, i.e. up to proof of complete additivity, by purely intrinsic means.

The extrinsic curvature of a convex surface in a space of constant curvature K is defined as follows. Divide the surface into small domains G_i. Choose a point P_i in each of these domains. Draw the exterior normals at all points of G_i, and take them to the point P_i by parallel transport (in the sense of Levi-Civita) along rectilinear paths. The transported normals form a solid angle $\psi(G_i)$. The extrinsic curvature of the surface is defined as $\lim \sum \psi(G_i)$ as the subdivisions of the surface into domains G_i becomes progressively finer. This limit always exists and is independent of the specific subdivision into domains G_i.

The connection between extrinsic curvature, intrinsic curvature and the area of the surface is established by a generalized Gauss theorem. If G is a domain on a convex surface in a space of constant curvature K, $\omega(G)$ its intrinsic curvature, $\psi(G)$ its extrinsic curvature and $S(G)$ its area, then $\psi(G) + KS(G) = \omega(G)$.

The theory of the i.g.c. and direction of curves, as developed for curves on convex surfaces in euclidean space, carries over intact to convex surfaces in spaces of constant curvature.

As for the realizability of abstractly given metrics of convex surfaces in spaces of constant curvature, we have theorems analogous to those in the euclidean case. For example:

1. *A metric of curvature* $\geqq K$ *on a sphere is realized in a space of constant curvature* K *as a closed convex surface.*

The condition that the metric has curvature $\geqq K$ means that for any sufficiently small triangle in the manifold in which the metric is given the sum of lower angles is greater than the sum of angles in a K-plane triangle with sides of the same length.

2. *In a manifold with metric of curvature $\geqq K$ any point has a neighborhood isometric to a convex surface in R_K.*

3. *When $K < 0$, a complete metric of curvature $\geqq K$ given in any domain on the sphere is realizable as a complete convex surface in Lobačevskiĭ space of curvature K.*

§13. Manifolds of bounded curvature

Manifolds of bounded curvature are a natural generalization of two-dimensional Riemannian manifolds. They stand in the same relation to manifolds of positive curvature (§3) as do general Riemannian manifolds to Riemannian manifolds with positive Gauss curvature. The theory of manifolds of bounded curvature is due to Aleksandrov, who defined the basic concepts and proved the principal results. A systematic exposition of the theory of manifolds with bounded curvature may be found in the monograph of Aleksandrov and Zalgaller [18]. Though manifolds with bounded curvature are quite general objects, many of the concepts of the theory of regular surfaces remain valid for them. Examples are geodesics, angle between geodesics, area and integral curvature of a compact set.

Let R be a metric manifold with metric ρ (i.e. a metric space which is a manifold). One can define the *length of a curve* in R: Given a curve $x(t)$, $a \leqq t \leqq b$, the length is defined as

$$\sup \sum_s \rho(x(t_{s-1}), x(t_s)), \qquad a = t_0 < t_1 < t_2 \ldots < t_n = b.$$

The metric ρ in the manifold R is said to be *intrinsic* if for any two points X, Y of R the distance $\rho(X, Y)$ is equal to the infimum of the lengths of curves in R connecting X and Y. The first condition that sets manifolds of bounded curvature apart is that R is a *two-dimensional manifold with intrinsic metric*.

A curve γ in a manifold R with intrinsic metric is called a *segment* if it has the smallest length of all curves in R connecting its endpoints. A *geodesic* is a curve each sufficiently small subarc of which is a segment. This should be understood in the sense that each point on a geodesic has a neighborhood which is a segment. Segments in a manifold R with intrinsic metric have the following properties:

1. For any point A and neighborhood U of A there exists a neighborhood $V \subset U$ such that any two points of V can be connected by a segment in R lying in U.

2. Any segment is homeomorphic to a segment on the real line; therefore the neighborhood of any point in the interior of a segment is divided into two sides (given a direction on the segment), left and right.

3. The length of a segment is equal to the distance between its endpoints.

4. Let G be a domain in R. Let $\rho_G(X, Y)$ denote the infimum of the lengths of curves connecting X and Y in G. Then ρ_G satisfies the axioms for a metric space. We say that the metric ρ_G is induced by the metric in G. It can be proved that every point of the domain G has a neighborhood in which ρ_G coincides with ρ.

A *triangle* ABC is defined as a closed set in a domain homeomorphic to a circle, bounded by three segments connecting the points A, B, C. The points are called the vertices of the triangle, the segments, its sides. Two triangles are said to be nonoverlapping if their intersection contains no nondegenerate triangle (i.e. not a segment) and no side of one triangle cuts a side of the other. (Two segments are said to cut each other if they have a common interior point and one of them has points on either side of the other.) A triangle in the sense just defined may lack interior points even when its vertices do not lie on a single segment.

Let γ_1 and γ_2 be two curves issuing from a point O on a manifold R. Choose points A_1, A_2 on them, and construct a plane triangle with sides $\rho(O, A_1)$, $\rho(O, A_2)$, $\rho(A_1, A_2)$. Let $\alpha(A_1, A_2)$ be the angle of this triangle opposite the side $\rho(A_1, A_2)$. The upper angle between the curves γ_1 and γ_2 is defined as the upper limit of the angle $\alpha(A_1, A_2)$ as $A_1, A_2 \to O$, i.e. $\rho(O, A_1) \to 0$, $\rho(O, A_2) \to 0$. The upper angle between curves always exists. The *angle* between curves γ_1, γ_2 at a point O is defined as the ordinary limit of the angle $\alpha(A_1, A_2)$ under the same condition A_1, $A_2 \to O$. Unlike the upper angle, the ordinary angle between curves may not exist. The *upper angle* $\bar{\alpha}$ of a triangle A, B, C at the vertex A is the upper angle between the segments AB and AC. The *excess* of the triangle ABC is the quantity $\bar{\omega} = \bar{\alpha} + \bar{\beta} + \bar{\gamma} - \pi$, where $\bar{\alpha}$, $\bar{\beta}$ and $\bar{\gamma}$ are the upper angles of the triangle.

The second, and last, condition which singles out manifolds of bounded curvature is the following:

For any compact domain G there exists a number $\nu(G)$ such that for *any finite family of pairwise nonoverlapping triangles T_i the sum of absolute values of their excesses satisfies the inequality* $\sum|\bar{\omega}(T_i)| \leqq \nu(G)$.

Zalgaller has shown that it is sufficient to require this inequality to hold only for systems of nondegenerate (i.e. homeomorphic to a circle) nonoverlapping triangles.

The simplest examples of manifolds of bounded curvature are manifolds with polyhedral metric and Riemannian manifolds. A manifold with polyhedral metric is defined by the requirement that each point have a neighborhood isometric to a cone (in particular, isometric to a plane). If we define the curvature at a vertex A of a manifold with polyhedral metric by $\omega(A) = \pi - \vartheta$, where ϑ is the complete angle at the apex of the cone isometric to a neighborhood of A, then $\nu(G) = \sum|\pi - \vartheta_i|$, where the summation extends over all vertices A_i lying in G. For a Riemannian manifold, $\nu(G)$ is the integral of the absolute value of the Gauss curvature over the area of the surface.

The above axiomatic definition of a manifold with bounded curvature can be supplemented by another definition, which is constructive and convenient for application. This definition is a consequence of the following theorem.

A two-dimensional manifold R with intrinsic metric ρ is a manifold with bounded curvature if and only if the metric ρ_G induced by ρ in any compact domain $G \subset R$ may be uniformly approximated by polyhedral or Riemannian metrics for which the absolute curvatures in G are uniformly bounded.

Thanks to this theorem we can use approximation by polyhedra and Riemannian manifolds to study manifolds of bounded curvature.

Let G be an open set in R. We define the positive part of the curvature of G, denoted by $\omega^+(G)$, as the supremum of the sums of positive excesses of pairwise nonoverlapping triangles contained in G. Similarly, the negative part of the curvature of G is the infimum, with reversed sign, of the sums of negative excesses of triangles contained in G; it is denoted by $\omega^-(G)$. For an arbitrary set M,

$$\omega^+(M) = \inf_{G \supset M} \omega^+(G), \qquad \omega^-(M) = \inf_{G \supset M} \omega^-(G).$$

It can be proved that the set functions ω^+ and ω^- are finite on compact sets, nonnegative and completely additive on the ring of Borel sets.

The *curvature of a set* M is defined as the difference $\omega^+(M) - \omega^-(M)$, and the *absolute curvature* as the sum $\omega^+(M) + \omega^-(M)$. If the excess of every triangle is zero, the manifold is locally isometric to a euclidean plane, i.e. each point has a neighborhood isometric to a plane domain.

The proof of all the propositions stated above depends essentially on the following lemma.

Let M be any compact set in a manifold with intrinsic metric. Then, for any $\epsilon > 0$, M has a covering by finitely many pairwise nonoverlapping triangles of diameter less than ϵ.

We now define *area* in manifolds of bounded curvature. For polygons, i.e. domains whose boundary is the union of geodesic polygonal lines, area is defined as follows. Divide the polygon P into triangles Δ_k of sufficiently small diameter; that this is possible follows from the above lemma. For each triangle, construct a plane triangle with the same sides and consider the sum of areas of the plane triangles. It can be proved that when the diameters of the triangles tend to zero the sum approaches a definite limit. This limit is by definition the area of P. The area of an arbitrary set M is defined in the usual manner, on the basis of the definition for polygons. For manifolds with polyhedral metric and Riemannian manifolds the construction yields the usual area concept.

The theory of curves on manifolds of bounded curvature is constructed in the same way as for manifolds with convex metric. One first defines the direction of a curve at a point: a curve γ issuing from a point O has a definite direction at O if it forms a definite angle with itself at O (the angle is of course zero). Any segment issuing from O certainly has a definite direction at O. One then proves the following theorem:

Two curves issuing from a point O form a definite angle at O in the sense of the above definition if and only if each of them has a definite direction at O. Segments issuing from a single point always form a definite angle at the point.

Two curves issuing from a point O in different directions divide the neighborhood of O into two sectors. The *angle of the sector* is defined as the supremum of the angles between disjoint curves subdividing the sector. As usual, sector angles are additive: If a curve γ lies within a sector V bounded by curves γ_1 and γ_2, and V_1 and V_2 are the sectors bounded by γ_1, γ and γ, γ_2, then the angle of the sector V is the sum of the angles of the sectors V_1 and V_2. If curves γ_1 and γ_2 divide the neighborhood of a point A into two sectors, the sum of angles of these sectors is independent of the curves γ_1 and γ_2, and, added to the curvature of the manifold at A, gives 2π.

The concept of i.g.c., generalizing the concept of integral curvature of a curve on a Riemannian manifold, is now introduced for curves on a manifold with bounded curvature, as follows. Let γ be a simple curve connecting points A and B and having definite directions at these points. Designate the two sides of the curve γ, say, right and left. Connect the points A and B by a simple [geodesic] polygon $\bar{\gamma}$ (polygon whose sides are segments) in the right semineighborhood of the curve γ. Let α and β be the angles formed by the first and last segments of the polygon and the curve γ, toward the domain G bounded by γ and $\bar{\gamma}$, and let ϑ_i be the angles at the vertices of the polygon, also toward G. Then the right i.g.c. of γ is defined as the limit of

$$\alpha+\beta+\sum_i (\pi-\vartheta_i)$$

as the polygon $\bar{\gamma}$ approaches the curve γ from the right. The left i.g.c. of γ is defined in an analogous way (taking the polygon in the left semineighborhood of γ). It can be proved that a simple curve which has definite directions at its endpoints always has right and left i.g.c.'s. A segment always has definite right and left i.g.c.'s, but, in contrast to convex surfaces, the i.g.c. need not be positive. The right and left i.g.c.'s of a curve are related to the curvature of the manifold along the curve: the sum of the right and left i.g.c.'s is equal to the curvature of the manifold on the set of interior points of the curve.

The i.g.c. of any closed simple curve is now defined as in the case of manifolds with convex metrics. One proves a theorem generalizing the Gauss-Bonnet Theorem for regular surfaces: *The sum of the curvature of a domain G homeomorphic to a circle and the i.g.c. of the curve γ bounding G toward G is 2π.*

In a series of papers, Rešetnjak [**65**] has developed analytical methods for manifolds of bounded curvature. In particular, he has proved that isometric coordinates can be introduced in manifolds of bounded curvature, and has derived a formula for the line element of the manifold in terms of these coordinates.

Let G be a domain in the complex z-plane and $\lambda(z)$ an arbitrary nonnegative Borel-measurable function defined in G. Using this function one can define a metric s_λ in G by associating with each pair of points z_1 and z_2 the number

$$\rho_\lambda(z_1, z_2)=\inf_\gamma \int_\gamma \sqrt{\lambda}\,|ds|,$$

where the integration is performed along an arc s of the curve connecting z_1 and z_2, and the infimum extends over all rectifiable curves γ. A metric introduced in this way is said to be determined by the line element $ds^2 = \lambda(z)\,|dz|^2$. The metric s_λ is said to be *subharmonic* if the function $\lambda(z)$ has the representation

$$\lambda(z) = \exp\left\{\frac{1}{\pi}\iint_G \ln\left|\frac{1}{z-\zeta}\right| \omega(dE_\zeta) + h(z)\right\},$$

where ω is a completely additive set function in G, $h(z)$ is a harmonic function, and the integration is in the Lebesgue-Stieltjes sense.

It can be proved that *any two-dimensional manifold with subharmonic metric is a manifold with bounded curvature, and then the set function* ω *has a simple geometric interpretation: it is simply the curvature of the manifold. Conversely, any manifold of bounded curvature is locally isometric to a manifold with subharmonic metric.*

CHAPTER II

REGULARITY OF CONVEX SURFACES WITH REGULAR METRIC

One of the most powerful tools for studying deformations of convex surfaces is the gluing method, based on the so-called "Gluing Theorem" of Aleksandrov (Chapter I, §11):

Let F_1 and F_2 be two convex surfaces homeomorphic to a disk, bounded by curves γ_1 and γ_2 of equal length. Assume that the points of γ_1 and γ_2 can be put in an arc-length-preserving correspondence, such that at corresponding points the sum of geodesic curvatures of γ_1 and γ_2 on the surfaces F_1 and F_2 is nonnegative. Then there exists a closed convex surface F consisting of two parts, one isometric to F_1, the other to F_2.

As an illustration, let us use the theorem to prove that a spherical segment is bendable. If the segment is contained in a hemisphere, it is clearly bendable, since a hemisphere can be bent into a spindle-shaped surface of rotation. But if the segment properly contains a hemisphere, its bendability is far from obvious. Liebmann, who first proved that a closed convex surface is rigid, was for a time convinced that a spherical segment containing a hemisphere is rigid, and even published a "proof" of this "fact."

The Gluing Theorem yields an extremely simple solution of this problem. "Close" the segment by adding a disk, thus forming a closed convex surface; deform the disk into an ellipse of nearly the same shape and size, in such a way that the length of the ellipse always remains equal to that of the circumference of the disk. Now the disk and the segment surely satisfy the assumptions of the Gluing Theorem (the sum of geodesic curvatures of the boundary of the segment and the circumference of the disk is strictly positive) and the deviation of the ellipse from the disk is small; hence the ellipse and the segment also satisfy the assumptions of the theorem. The closed convex surface obtained by gluing the ellipse and the segment together contains a domain isometric to the segment but not congruent to it, since the points on the boundary of the segment which are glued to the endpoints of the minor axis of the ellipse are clearly closer together than before.

The above proof of the bendability of a spherical segment is not quite satisfactory from the standpoint of the classical theory of surfaces, which deals only with sufficiently regular surfaces, for it is not known whether the constructed surface, isometric to the spherical segment, is regular. This drawback recurs in all applications of the Gluing Theorem to problems concerning bending of regular convex surfaces.

The following problem arises naturally [**53, 54**]: To what degree does the regularity of the intrinsic metric of a convex surface predetermine the regularity of the surface itself? More precisely, suppose that in some parametrization u, v the coefficients of the line element

$$ds^2 = E du^2 + 2F du dv + G dv^2$$

of a convex surface Φ are sufficiently regular functions of u and v. What can one say about the regularity of the vector-valued function $r(u, v)$ defined by the surface? The principal result of this chapter will be a solution of this problem.

The proof presented here of the regularity of a convex surface with regular metric is extremely complicated. It depends on a penetrating study of the extrinsic geometry of general convex surfaces and convex surfaces with regular metric. Many of the intermediate results are of independent interest. Especially worthy of mention in this respect are Liberman's theorems on geodesics (§1), Aleksandrov's theorems on the smoothness and strict convexity of convex surfaces with bounded specific curvature (§2), theorems on the normal curvatures of convex surfaces, a priori estimates for the normal curvatures and other geometric characteristics of regular surfaces.

Combined with Aleksandrov's theorem on the realizability of convex metrics, our theorem on the regularity of a convex surface with regular metric yields theorems on the realizability of regular metrics with positive Gauss curvature as regular surfaces. Our results thus enter the province of the classical theory of surfaces. Their analytical interpretation leads to general theorems on the solvability of boundary-value problems for the bending equation (the Darboux equation).

§1. Extrinsic properties of geodesic lines on a convex surface

Geodesics, including segments (shortest joins), constitute one of the fundamental elements of our geometrical constructions on convex surfaces. In this section we shall study the fundamental extrinsic properties of geodesics, i.e. their properties as space curves. The intrinsic properties of these important curves were examined in Chapter I, §2. Let us begin our discussion with a theorem of Busemann and Feller [**25**]. Many of our proofs employ this theorem in an essential manner.

THEOREM 1. *Let F be a complete convex surface bounding a body K, and γ a curve situated outside K and connecting points X and Y of F. Then the length l_γ of γ is at least the distance $\rho_F(X, Y)$ between its endpoints on the surface; it is strictly greater than this distance if the curve does not lie entirely on the surface.*

PROOF. We first show that if γ does not lie entirely on the surface F, there exists a curve $\bar{\gamma}$ connecting X and Y, also outside the body K, whose length is less than that of γ.

Indeed, let P be a point on γ not on F. Let α be a plane separating P from F (Figure 8). Let γ_P be the (connected) component of γ containing P, with endpoints on the plane α. Replace γ_P by its projection on the plane α. The resulting curve $\bar{\gamma}$ has the same properties as γ, but its length is smaller.

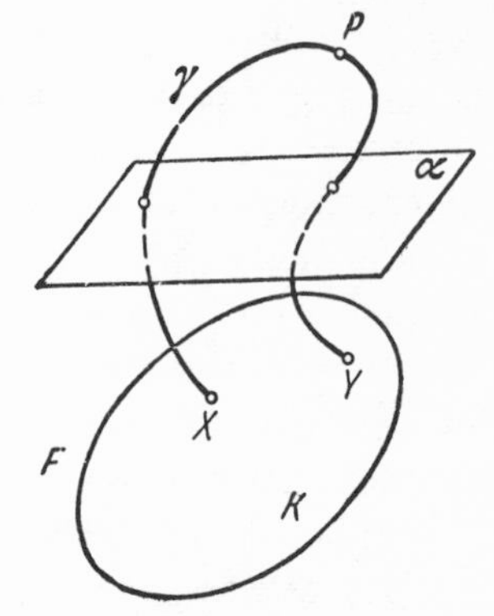

FIGURE 8

Let l_0 be the infimum of the lengths of curves connecting X and Y outside K. Obviously there is a sequence of curves γ_n whose lengths converge to l_0. Without loss of generality we may assume that this sequence converges to some curve γ_0. The curve γ_0 lies on the surface F, since otherwise its length l_0 could be reduced, contradicting the definition of l_0. Since the length of any curve on the surface cannot be less than the distance between its endpoints, the length of γ_0 is at least $\rho(X, Y)$. This is surely the case for the curve γ. Now if γ is not entirely on the surface, then, as we have shown, its length can be reduced. Hence in this case the length of γ is strictly greater than $\rho(X, Y)$. This completes the proof.

Busemann's Theorem (our Theorem 1) yields a simple proof of the following theorem on segments on convex caps. Recall that a convex cap is a convex surface whose boundary is a plane curve and which can be projected in one-to-one fashion onto the plane of its boundary.

THEOREM 2. *Any two points on a convex cap can be joined by a segment.*

PROOF. Complete the convex cap ω to a complete convex surface by adding a semicylinder (Figure 9). Let X and Y be two arbitrary points on ω. They can be joined by a segment γ on the surface consisting of ω and the semicylinder.

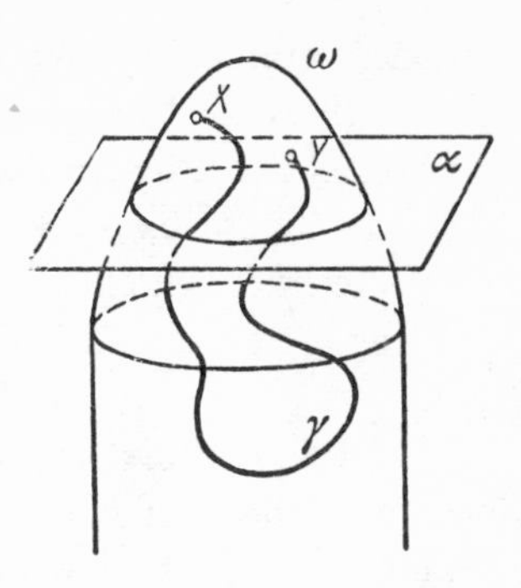

FIGURE 9

Let α be a plane parallel to the base of the cap and separating the points X and Y from its boundary. If the theorem is false, the curve γ must cut the plane α. But then it can be shortened by replacing that section of the curve situated below the plane α (with respect to the base of the cap) by its projection on the plane. This contradicts Busemann's Theorem as applied to the shortened curve.

THEOREM 3. *Let F_0 be a fixed point on a convex surface F, X a point on the surface near X_0, and $\rho(X)$ and $\delta(X)$ the distances between X and X_0 on the surface and in space, respectively. Then $\rho(X)/\delta(X) \to 1$ as X tends to X_0.*

PROOF. Assume the theorem false. Then there exists a sequence of points X_n converging to X_0 such that

$$\rho(X_n)/\delta(X_n) > \lambda > 1.$$

Let F_n denote the convex surface obtained from F by a similarity mapping with center X_0 and ratio $1/\delta(X_n)$. As $n \to \infty$ this sequence of surfaces converges to the tangent cone K_0 of F at X_0. The points X_n on F are mapped onto certain points Y_n on F_n; the distance between Y_n and X_0 in space is unity. As $n \to \infty$, the points Y_n approach the cone K_0 (Figure 10).

Let g be a ray issuing from X_0 whose continuation passes through

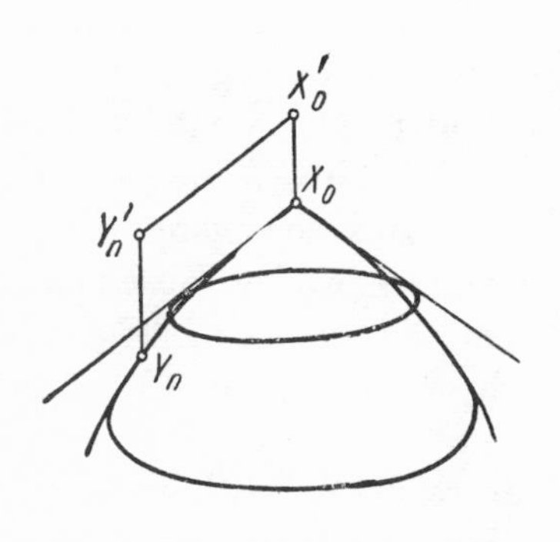

FIGURE 10

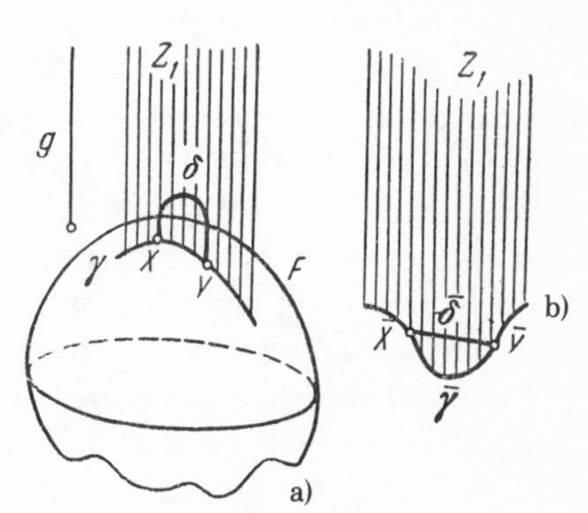

FIGURE 11

the body bounded by the surface F. Displace the straight-line segment $X_0 Y_n$ in the direction of the ray g through a small distance ϵ, to the position $X_0' Y_n'$. For sufficiently large n the polygon $Y_n Y_n' X_0' X_0$ lies within the body bounded by F_n. Therefore the distance between X_0 and Y_n on F_n, which is obviously $\rho(X_n)/\delta(X_n)$, is less than $1+2\epsilon$. By taking $\epsilon < (\lambda - 1)/2$ we get $\rho(X_n)/\delta(X_n) \leq \lambda$, which is a contradiction. Q.E.D.

The fundamental proposition on geodesics on convex surfaces is Liberman's theorem stating that any geodesic is convex on its projecting cylinder [41]. Despite its simplicity, this proposition yields many important corollaries, implying a fairly exhaustive extrinsic characterization in the large of geodesic lines on a convex surface. Here is the theorem.

THEOREM 4. *Let γ be a geodesic on a convex surface F, and g a ray which forms angles less than $\pi/2$ with all exterior normals to the surface along the geodesic. Let Z be the cylinder projecting γ in the direction of g. Then γ is convex on Z and its convexity is in the direction of g (Figures* 11a, b).

PROOF. First observe that it will suffice to prove the theorem for a neighborhood of an arbitrary point P on the geodesic γ, since if a curve is convex in the same direction in a neighborhood of each point, it is convex in the large.

Let γ' be a subarc of the geodesic γ, of length ϵ, containing P in its interior. If ϵ is sufficiently small, γ' is a segment, not only on the surface F but also on any complete convex surface containing F. It will thus suffice to prove the theorem when γ is a segment on a complete convex surface.

Develop the cylinder Z on the plane (Figure 11b). The result is a strip bounded by two parallel straight lines, which is divided by the image $\overline{\gamma}$ of γ into two parts $\overline{Z}_1$ and $\overline{Z}_2$. One of the latter, say $\overline{Z}_1$, corresponds to that part of the cylinder Z situated outside the surface.

Assume the theorem false. Then one can find points $\overline{X}$ and $\overline{Y}$ on $\overline{\gamma}$ which can be joined by a euclidean segment $\overline{\delta}$ within $\overline{Z}_1$. The curve $\overline{\delta}$ is the image of a curve δ on the cylinder Z, joining the points X and Y of F and lying outside the body bounded by F. By Busemann's Theorem the length of this curve is less than the distance between X and Y on G, which is the same as the length of the subarc XY of the geodesic γ. Thus subarc $\overline{X}\overline{Y}$ of $\overline{\gamma}$, which is equal in length to the subarc XY of γ, must be shorter than the euclidean segment $\overline{\delta}$ with endpoints $\overline{X}$ and $\overline{Y}$. This contradiction proves the theorem.

Liberman's Theorem yields the following description of the differential properties of geodesics.

THEOREM 5. *A geodesic γ on a convex surface F has a right and left semitangent at each point; these semitangents are right and left continuous, respectively.*

PROOF. Let g_1, g_2, g_3 be three rays issuing from a point X_0 of the geodesic γ within the body bounded by the surface F. In a small neighborhood of X_0 the interior normals to F along the geodesic form angles smaller than $\pi/2$ with the rays g_i.

Let e_i be unit vectors on the rays g_i and $r(s)$ the radius vector of the point on the geodesic corresponding to arc length s. By Theorem 4 (Liberman's Theorem) each of the vector-functions $e_i r(s)$ is convex in a neighborhood of X_0, and so each of them has right and left derivatives which are right and left continuous, respectively. Now since

$$r(s) = \frac{(re_1)(e_2 \times e_3)}{(e_1 e_2 e_3)} + \frac{(re_2)(e_3 \times e_1)}{(e_2 e_3 e_1)} + \frac{(re_3)(e_1 \times e_2)}{(e_3 e_1 e_2)},$$

this is also true of the function $r(s)$. Hence the right and left semitangents to the geodesic γ exist and have the required properties.

The following theorem amplifies Theorem 5.

THEOREM 5a. *A geodesic on a convex surface has a tangent almost everywhere, except perhaps for at most countably many points. The tangent is continuous on the set of points at which it exists.*

The proof resembles that of Theorem 5. It is based on the differentiability properties of convex functions and the above expression for the radius vector $r(s)$ in terms of the convex functions $(r(s)e_1)$, $(r(s)e_2)$, $(r(s)e_3)$.

A more precise description of the continuity properties of semitangents along a geodesic on a convex surface is given by the following theorem.

THEOREM 6. *If the angles between the exterior normals to a convex surface along a geodesic minus its endpoints are less than $\vartheta < \pi/2$, then the angles between the right (left) semitangents at any two points of the geodesic are at most 2ϑ.*

PROOF. Let X_2 be a point on the geodesic γ to the right of X_1, t_1 the right semitangent to γ at X_1 and n_1 the exterior normal to the supporting plane of the surface at this point. Let g be a ray which forms angles $\pi - \vartheta$ and $\pi/2 - \vartheta$ with the rays t_1 and n_1, respectively. Since the angles that the exterior normals to the surface along γ form with g are less than $\pi/2$, Liberman's Theorem is applicable to the geodesic γ and the ray g. It follows that the right semitangent t_2 to γ at X_2 forms an angle of at least $\pi - \vartheta$ with g.

Since both semitangents t_1 and t_2 form angles of at least $\pi - \vartheta$ with g, it follows that the angle between them is at most 2ϑ. Q.E.D.

THEOREM 6a. *If the angles between the exterior normals to a convex surface along a geodesic minus its endpoints are less than ϑ, the angles between the right and left semitangents at any two points of the geodesic are at least $\pi - 2\vartheta$.*

PROOF. Consider, say, the right semitangent τ_1 at X_1 and the left semitangent τ_2 at X_2. Since a geodesic has a tangent almost everywhere (Theorem 5a), there exists a sequence of points A_n, converging to X_1, at which there exist tangents t_n. Let t_n' and t_n'' be the right and left semitangents to t_n. By Theorem 5 the semitangents t_n' converge to the right semitangent τ_1 at X_1. It now suffices to use Theorem 6, applied to the semitangents t_n'' and τ_2, and to let $A_n \to X_1$.

The following theorem is a corollary of Theorem 6a.

THEOREM 7. *If a point X on a geodesic on a convex surface is a smooth point of the surface, it is also a smooth point of the geodesic.*

In other words, on a smooth convex surface any geodesic is a smooth curve.

THEOREM 8. *Let X and Y be two points on a convex surface and γ a geodesic connecting them. Suppose that the angles between the exterior normals to the surface along γ minus its endpoints do not exceed ϑ. Then the angle between the semitangent to the geodesic at X and the segment XY is at most 2ϑ.*

PROOF. Let $r(s)$ be the radius vector along the geodesic, as a function of arc length s. Since the derivative $r'(s)$ exists almost everywhere, is bounded and $|r'(s)| = 1$ (Theorem 3), we have

$$\overrightarrow{XY} = r(Y) - r(X) = \int_{(X)}^{(Y)} r'(s)\,ds.$$

It is now sufficient to note that the vector $r'(X)$ forms an angle of at most 2ϑ with any vector $r'(s)$, and the statement of the theorem becomes evident.

Theorem 4 gives information on the convergence of the semitangents of a convergent sequence of geodesics; we have the following important theorem.

THEOREM 9. *Let $F_1, F_2, \cdots$ be an infinite sequence of convex surfaces converging to a convex surface F, X_n a point on F_n, γ_n a geodesic on F_n issuing from X_n, and t_n the semitangent to γ_n at X_n.*

Assume that as $n \to \infty$ the sequence X_n converges to a point X on F, the sequence γ_n to a geodesic γ on F, and the sequence t_n to a ray t.

Finally, let t_0 be any ray issuing from X within the body on whose boundary the surface F lies. Then the following assertions are true.

1) *The angle between t and t_0 is at least that between the semitangent to γ at X and the ray t_0.*

2) *The ray t lies in a supporting plane of F at X.*

PROOF. We first make a few observations concerning plane convex curves, for Liberman's Theorem will reduce to the general case to that of the plane. Let C be a convex curve in the xy-plane, defined by an equation $y = f(x)$ in cartesian coordinates x, y. The curve C has right and left semitangents at each point. The angles that they form with the coordinate axes vary monotonically along the curve. It follows that if the coordinates x and y of a point on the curve are regarded as functions of arc length s, each of these functions has monotone right and left derivatives, and so the functions $x(s)$ and $y(s)$ are also convex. Of course, the curve C may be defined by either $x(s)$ or $y(s)$.

Let $C_n : y = f_n(x)$ be an infinite sequence of convex curves issuing from the origin O and convex toward $y < 0$. Assume that as $n \to \infty$ they converge to a convex curve $C: y = f(x)$. For each of the curves C_n, C we have a pair of functions $x_n(s)$ and $y_n(s)$, $x(s)$ and $y(s)$, respectively. Since the curves C_n converge to C, the functions $x_n(s)$ and $y_n(s)$ converge to $x(s)$ and $y(s)$, respectively. Since $x_n(s)$ is a convergent sequence of convex functions,

$$x'(0) \leqslant \varliminf_{n \to \infty} x'_n(0).$$

Hence the angle that the semitangent to the limit curve C at O forms with the positive y-axis cannot exceed the lower limit of the angles formed by the semitangents to C_n with the positive y-axis.

We can now proceed to the proof proper. Since the ray t_0 points into the body bounded by F, there is a neighborhood of X on the geodesic γ along which the interior normals to the surface form angles smaller than $\pi/2 - \epsilon$ (where ϵ is a small positive number) with the direction of t_0. Without loss of generality we may assume that this is the case along the whole of γ. For sufficiently large n, the interior normals to each surface F_n along the geodesic γ_n form angles smaller than $\pi/2$ with the direction of t_0.

Let Z_n denote the cylinder projecting γ_n in the direction t_0, and Z the cylinder projecting γ in this direction. Develop each cylinder Z_n and Z on the xy-plane, in such positions that the projection direction t_0 is that of the positive y-axis, the points x_n and X fall on the origin O, and in such a way that the cylinders are developed on the halfplane $x > 0$. When this is done, the geodesics γ_n become convex curves C_n, with convexity toward $y < 0$, which converge to a convex curve C — the image of γ. As we proved above, the angle that the semitangent to C at O forms with the positive y-axis cannot exceed the lower limit of the angles that the semitangents to C_n at X_n form with the positive y-axis.

Now assume that the theorem is false. Then there exist geodesics γ_n with arbitrarily large n such that the angles α formed by their semitangents at X_n with t_0 satisfy the inequality $\alpha(t_n, t_0) < \alpha(t, t_0) - \epsilon$, where $\epsilon > 0$. Hence

$$\underline{\lim}_{n\to\infty} \alpha(t_n, t_0) < \alpha(t, t_0) - \varepsilon.$$

But this contradicts the convergence of angles for the plane curves C_n. Indeed, $\alpha(t_n, t)$ and $\alpha(t, t_0)$ coincide with the angles that the semitangents to C_n and C, respectively, form with the positive y-axis, and we have proved that necessarily

$$\underline{\lim}_{n\to\infty} \alpha'(t_n, t_0) \geqslant \alpha(t, t_0).$$

The fact that the ray t lies in a supporting plane of the surface F can be seen as follows. Each semitangent t_n lies in a supporting plane σ_n of the surface F_n. The sequence σ_n contains a convergent subsequence, whose limit plane contains t and is a supporting plane of F. This completes the proof of the theorem.

Theorem 9 has two important corollaries:

1. *If the ray t is tangent to the surface F, i.e. coincides with a generator of the tangent cone to F at X, then it is a semitangent to γ at this point.*

In fact, we need only take for t_0 a ray arbitrarily close to t in direction. It follows that the semitangent to γ forms an arbitrarily small angle with t. This can hold only if t is a semitangent.

2. *If X is a smooth point of the surface F, the ray t is a semitangent to the geodesic γ.*

In fact, F has only one supporting plane at X. Therefore any ray issuing from X and lying in this plane touches F at X. This applies, in particular, to the ray t. It now suffices to use the preceding corollary to Theorem 9.

§2. Special resolution of the radius-vector in the neighborhood of an arbitrary initial point

Regular surfaces are easy to investigate in the small, largely thanks to the convenient resolution of the radius-vector in the neighborhood of an arbitrarily chosen initial point. We shall now derive an analogous resolution for general convex surfaces, making no assumptions of regularity.

THEOREM 1. *Let F be a general convex surface, X and Y two points on F and γ a segment connecting them. Lay off a segment $X\overline{Y}$ from X on the semitangent to γ at X, equal in length to the segment γ.*

Then the unit vector in the direction $Y\overline{Y}$ lies in the convex hull of the spherical image of γ minus its endpoints (Figure 12).

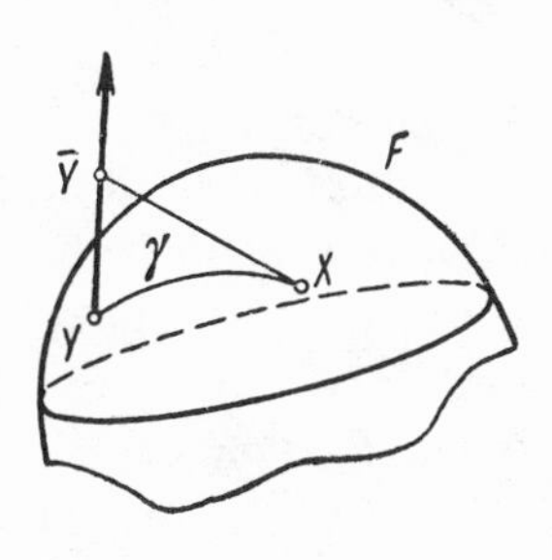

FIGURE 12

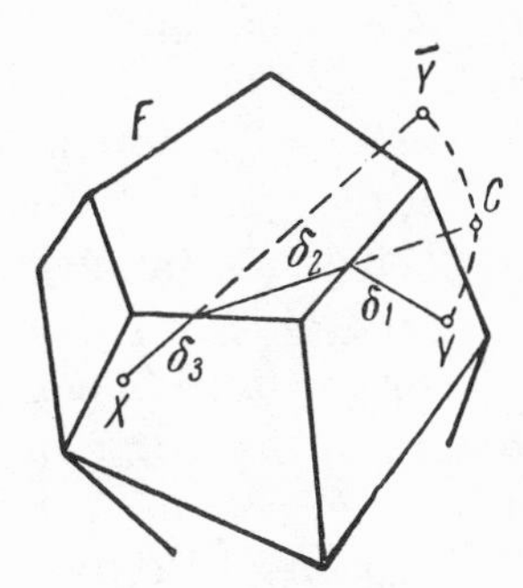

FIGURE 13

PROOF. We first assume that F is a convex polyhedron. In this case the segment γ is a polygon with sides on the faces of the polyhedron and vertices on its edges. Let $\delta_1, \delta_2, \cdots, \delta_n$ denote the sides of the polygon γ, in order from Y to X; let $k_1, k_2, \cdots, k_{n-1}$ denote the edges of the polyhedron which are cut by the polygon, $\vartheta_1, \cdots, \vartheta_{n-1}$ the angles of the polyhedron at these edges, and $\alpha_1, \alpha_2, \cdots, \alpha_n$ the sides of the polyhedron on which lie the sides $\delta_1, \delta_2, \cdots, \delta_n$, respectively.

Deform the polygon γ as follows. First rotate the side δ_1, together with the face α_1 containing it, through an angle $\pi - \vartheta_1$ about the edge k_1. δ_1 will then lie on the continuation of the side δ_2. Then rotate both sides δ_1 and δ_2 through $\pi - \vartheta_2$ about the edge k_2. Both sides δ_1 and δ_2 now lie on the continuation of the side δ_3. Continue the procedure, until rotation about the edge k_{n-1} transforms the entire polygon γ into the straight-line segment $X\overline{Y}$ (Figure 13).

When γ is deformed in this way, the endpoint Y moves along a smooth curve consisting of circular arcs, whose semitangents in the direction $Y\overline{Y}$ are parallel and have the same direction as the exterior normals to the supporting planes of the polyhedron F along γ. This clearly implies that the unit vector in the direction of $Y\overline{Y}$ belongs to the convex hull of the segment γ minus its endpoints.

Now consider the general case. Let X' and Y' be two points on γ close to X and Y, respectively, and let γ' denote the subarc that they define on γ. By the inclusion property (Chapter I, §3) γ' is the unique segment connecting X' and Y'. Lay off a segment equal in length to γ' on the semitangent to γ' at X'; let $\overline{Y}'$ denote the end of this subarc.

We now construct a sequence of convex polyhedra F^n converging to F, all inscribed in F and containing the points X' and Y' as vertices. To do this, we take a sufficiently dense net S_n on the surface F, containing X' and Y', and let Φ be the convex hull of this set. Φ is a convex polyhedron. If the surface F is complete, this polyhedron is precisely F^n. Otherwise F^n is obtained from Φ by suitable cutting.

Connect the points X' and Y' on the polyhedron F^n by a segment γ_n' and let $\overline{Y}_n'$ be the point given by a construction analogous to that giving $\overline{Y}'$ for F and γ'. Since γ' is the unique segment connecting X' and Y' on F, it follows that γ_n' converges to γ' as $n \to \infty$, and the lengths of γ_n' converge to the length of γ'. The semitangents to γ_n' at X' converge to the semitangent to γ' at this point (§1, Corollary 1 to Theorem 9). Hence the points $\overline{Y}_n'$ converge to the point $\overline{Y}'$, and so the direction of the vector $Y'\overline{Y}_n'$ converges to that of the vector $Y'\overline{Y}'$.

Now, for sufficiently large n, the spherical image of γ_n' is contained in an arbitrarily small neighborhood of the spherical image of γ'. Letting $n \to \infty$, we see, therefore, that the unit vector in the direction of $Y'\overline{Y}'$ belongs to the convex hull of the spherical image of γ'.

Now let the points X' and Y' approach arbitrarily close to the points X and Y, respectively, on the segment γ. The length of γ' then tends to that of γ, while the semitangent to γ' at X' tends to the semitangent to γ at X; hence the direction of $Y'\overline{Y}'$ tends to the direction of $Y\overline{Y}$, and consequently the unit vector in the direction of $Y\overline{Y}$ is in the convex hull of the spherical image of γ minus its endpoints. Q.E.D.

THEOREM 2. *Retaining the notation of Theorem* 1, *let* $\tau(Y)$ *be the unit tangent vector to* γ *at* X, $n(Y)$ *the unit vector in the direction of* $\overline{Y}Y$, *and*

$\vartheta(Y)$ the angle between the vectors $\tau(Y)$ and $n(Y)$. If the point Y approaches X in any way, then $\vartheta(Y) \to \pi/2$ and $|Y\overline{Y}|/|XY| \to 0$.

PROOF. Suppose the first assertion false. Then there exists a sequence of points Y_k converging to X such that for every k

$$|\vartheta(Y_k) - \pi/2| > \epsilon, \quad \epsilon > 0.$$

Apply to the surface F a similarity mapping with center X and ratio of similitude $1/s(Y_k)$, where $s(Y_k)$ is the distance between X and Y_k on F. The result is a surface F_k and on it a segment γ_k. By analogy with the surface F and the segment γ, construct vectors τ_k and n_k for the surface F_k and the segment γ_k. These vectors are parallel to $\tau(Y_k)$ and $n(Y_k)$, and therefore the angle between them satisfies the inequality $|\vartheta_k - \pi/2| > \epsilon$.

As $k \to \infty$, the sequence F_k converges to the tangent cone V of F at X. Without loss of generality we may assume that the segments γ_k converge to a segment of unit length on a generator of V; the generator, in turn, is the limit of the semitangents to γ_k at the initial point X. For sufficiently large k the spherical image of γ_k minus its endpoints is contained in an arbitrarily small neighborhood of the spherical image of the above-mentioned generator of V minus its initial point X; consequently for sufficiently large k the angle ϑ_k deviates by an arbitrarily small amount from $\pi/2$. The contradiction proves the first assertion.

Now suppose that the second assertion is false. Then there exists a sequence Y_k converging to X such that for all k

$$|Y_k\overline{Y}_k|/s(Y_k) > \epsilon > 0.$$

Construct the same sequence of convex surfaces F_k as in the proof of the first assertion. Now as $k \to \infty$ the surfaces F_k converge to the tangent cone V of F at X, the segments γ_k converge to a unit segment on a generator of V, and the semitangents to γ_k at X converge to the generator. It follows that the quotient $|Y_k\overline{Y}_k|/s(Y_k)$ approaches zero, a contradiction. This completes the proof of the theorem.

Theorem 1 yields a convenient representation of the radius-vector of an arbitrary point of a convex surface in the neighborhood of a given initial point.

THEOREM 3. *Let X_0 be a smooth point on a convex surface, X a point near X_0, $\tau(X)$ the unit vector on the semitangent at X_0 to the segment connecting X and X_0, and $s(X)$ the length of the segment.*

Then the radius vector $r(X)$ of the point X has the following representation:

$$r(X) = r(X_0) + \tau(X)s(X) + \nu(X),$$

where $\nu(X)$ is a vector whose direction converges as $X \to X_0$ to that of the interior normal to the surface at X_0, and moreover $|\nu(X)|/s(X) \to 0$.

THEOREM 4. *A geodesic line on a convex surface has an osculating plane, normal to the tangent plane, at each smooth point of the surface. If the surface is smooth the osculating plane of a geodesic varies continuously along the curve.*

PROOF. By definition the osculating plane of a curve at a point O is the limiting position of the plane through the tangent to the curve at O and a point X close to O on the curve, as $X \to O$. Repeat the construction described in the proof of Theorem 1. The plane constructed contains a segment OY of the tangent and a segment XY whose direction is near that of the normal to the surface at O. As $X \to O$ this plane tends to the normal plane of the surface, which passes through the tangent to the geodesic at O.

A geodesic on a smooth convex surface is smooth and its osculating plane by its very definition depends continuously on the point at which the tangent touches the geodesic and the normal to the surface; this proves the second assertion. Q.E.D.

We shall use Theorem 3 for the proof of a lemma needed in §6, where we shall prove that convex caps are **monotypic** among convex caps.

LEMMA. *Let F_1 and F_2 be two smooth convex surfaces tangent to the xy-plane at the origin O and convex toward $z < 0$. Assume that there is a one-to-one correspondence between the surfaces, satisfying the following conditions:*

1. *If X_1 and X_2 are corresponding points on the surfaces, then the segments γ_1 and γ_2 connecting them to O on the surfaces F_1 and F_2 have the same length s and a common semitangent at O.*

2. *In a neighborhood of O the z-coordinates of X_1 and X_2 satisfy the inequality*

$$z(X_1) + As^2 \leqq z(X_2) \leqq Bs^2,$$

where A and B are positive constants.

Then in a neighborhood of the point O the surface F_2 is inside the surface F_1. Moreover, the functions $z_1(x,y)$ and $z_2(x,y)$ defining the surfaces in the neighborhood of O satisfy the condition

$$z_1(x,y) + C(x^2+y^2) \leqq z_2(x,y),$$

where C is a positive constant.

PROOF. First note that if points X and Y on a smooth convex surface are sufficiently near O, the straight line connecting them forms an angle close to $\pi/2$ with the normal to the surface at O.

Now repeat the construction in the statement of Theorem 1: On the common semitangent to the segments γ_1 and γ_2, lay off from O a segment s equal in length to γ_1 and γ_2. Connect the resulting point Y to the points X_1, X_2, and consider the triangle X_1X_2Y (Figure 14). By Theorem 3 the sides of this triangle YX_1 and YX_2 form a small angle for sufficiently small s, since their directions are close to that of the normal to the surface at O (the direction of the z-axis). If follows from condition 2 of our lemma that the sides X_1X_2 and YX_2 have the same order of magnitude, and so the segments X_1X_2 and X_2Y also form a small angle.

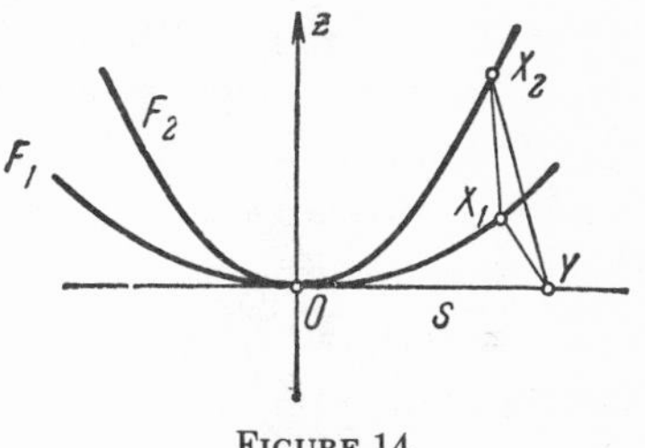

FIGURE 14

Now consider a straight line through X_2 parallel to the z-axis. It cuts F_1 at some point X_1', generally different from X_1. Since the segments X_1X_1' and X_2X_1' meet at the point X_1' at an angle close to $\pi/2$, it follows that $z(X_2) > z(X_1')$. Moreover, the difference $z(X_2) - z(X_1')$ has the same order as X_1X_2. Consequently in a sufficiently small neighborhood of O we have $z(X_2) - z(X_1') > A's^2$, where A' is a constant smaller than A. Now s^2 is clearly equivalent to x^2+y^2, where x and y are the coordinates of the projection of X_2 on the xy-plane. This proves the lemma.

§3. Convex surfaces of bounded specific curvature

Aleksandrov defines a convex surface F to have bounded specific curvature at a point X if X has a neighborhood such that, in any domain G within this neighborhood, the ratio of the curvature of the surface in G to the area of G is bounded by a positive constant independent of G:

$$\omega(G)/S(G) < C < \infty.$$

If the specific curvature is bounded at each point of a surface, the surface is called a *surface of bounded specific curvature*. Since the intrinsic curvature of a surface is equal to its extrinsic curvature (the area of its spherical image), the quantity $\omega(G)$ in the above definition of specific curvature may be replaced by the area of the spherical image of G.

The following important theorem of Aleksandrov [5] is valid for convex surfaces of bounded specific curvature.

THEOREM 1. *Each point X of a convex surface of bounded specific curvature either is smooth or belongs to a straight edge; in the latter case X is not an endpoint of the edge.*

PROOF. If X is not a smooth point or an interior point of a straight edge on the surface, only three possibilities remain:

1) X is a conical point;
2) X is a ridge point but there is no straight edge through X;
3) X is an endpoint of a straight edge.

To prove the theorem it will therefore suffice to show that in each of these three cases the extrinsic specific curvature of the surfaces is not bounded.

If X is a conical point, the spherical image of any arbitrarily small neighborhood of X on the surface covers the spherical image of the tangent cone at X. Consequently the area of the spherical image of any neighborhood of X is at least equal to the area of the spherical image of the cone. Since the point X has neighborhoods of arbitrarily small area, the specific curvature at a conical point is not bounded.

Now let X be a ridge point through which there is no straight edge. The tangent cone of the surface at X must be a dihedral angle. Let α_1 and α_2 be the faces of this angle, g its edge. Let σ be a plane parallel to g which cuts α_1 and α_2 at equal angles; let z be the distance of σ from X. If z is sufficiently small, the cap E which the plane cuts off from F is projected in one-to-one fashion onto the plane σ.

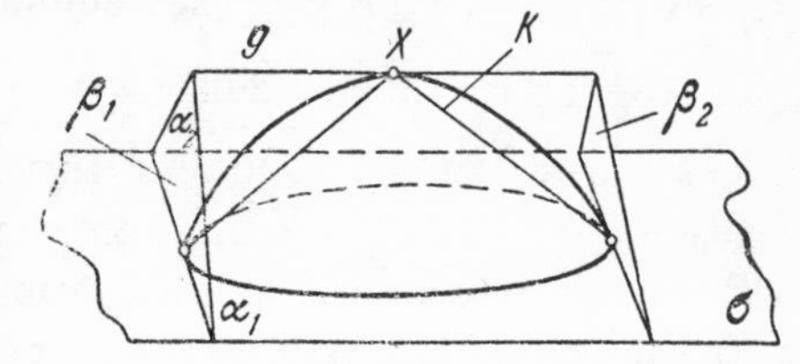

FIGURE 15

Now let β_1 and β_2 be two planes perpendicular to g and supporting the boundary of the cap (Figure 15). The halfplanes α_1 and α_2 and the planes β_1, β_2 and σ form a triangular prism. The surface area of this prism, without the area of the face lying in σ, is not less than the area of the surface of the cap; it is

$$2z \tan \vartheta + \frac{2z}{\cos \vartheta}(x_1 + x_2) = \frac{2z}{\cos \vartheta}(z \sin \vartheta + x_1 + x_2),$$

where x_1 and x_2 are the distances of β_1 and β_2 from X, and ϑ is half the angle between the faces α_1 and α_2. Thus the surface area $S(E)$ of the cap satisfies the inequality

$$S(E) < \frac{2z}{\cos\vartheta}(z + x_1 + x_2).$$

Now let us estimate the area of the spherical image of the cap. Consider the cone K formed by connecting X to the boundary points of the cap. The spherical image of E covers that of K, since for each supporting plane of the cone there is a supporting plane of the cap. For the same reason the spherical image of the tetrahedral angle with apex X and the prism face lying in σ as base is covered by the spherical image of the cone K.

The spherical image of this tetrahedral angle is a spherical quadrangle with diagonals which intersect at right angles and whose lengths are $\pi - 2\vartheta$ and $\pi - 2\varphi$, where 2ϑ is the angle formed by the faces α_1 and α_2, and 2φ the angle formed by the other two opposite faces. If z is sufficiently small, so is $\pi - 2\varphi$, and the area of the spherical quadrangle is approximately $\frac{1}{2}(\pi - 2\vartheta)(\pi - 2\varphi)$. Hence there exists a constant $m \neq 0$ such that the area of the quadrangle is greater than $m(\pi - 2\vartheta) \cdot (\pi - 2\varphi)$.

Since $\pi - 2\varphi = \arctan(z/x_1) + \arctan(z/x_2)$ and the quotients z/x_1 and z/x_2 are small, we may assume that $\pi - 2\varphi > \frac{1}{2}(z/x_1 + z/x_2)$. Thus the area of the spherical image of our tetrahedral angle, and therefore also the area ω of the spherical image of the cap E, satisfies the inequality

$$\omega(E) > \frac{m}{2}(\pi - 2\vartheta)\left(\frac{z}{x_1} + \frac{z}{x_2}\right).$$

Hence the specific curvature of the cap E satisfies the inequality

$$\frac{\omega(E)}{S(E)} > \frac{m\cos\vartheta}{4}(\pi - 2\vartheta)\left(\frac{1}{z + x_1 + 2x_2} + \frac{1}{z + 2x_1 + x_2}\right),$$

so that it increases without bound as the plane σ cutting off the cap E approaches the point X.

Treatment of the case in which X is an endpoint of a straight edge g is similar. One need only consider a plane σ through a point X_1 on the edge g, different from but near X. When this is done the result is that if the point X_1 approaches X and z decreases more rapidly than the distance between X and X_1, then the specific curvature of the cap cut off by σ from the surface F increases without bound.

Thus, if the specific curvature is bounded at a point X on a convex surface F, then either X is a smooth point or there is a straight edge through X, along which the surface has a "break," but X is not an endpoint of this edge. No other singularities can occur at X.

Theorem 1 has important corollaries.

COROLLARY 1. *A convex cap with bounded specific curvature is smooth.*

Indeed, otherwise the cap must contain a straight edge whose endpoints lie on the boundary of the cap; but then the cap would degenerate into a plane convex domain.

COROLLARY 2. *A complete convex surface of bounded curvature that is not a cylinder is smooth. In particular, all closed surfaces of bounded specific curvature are smooth.*

In fact, an edge (the only possible violation of smoothness) must be a straight line. On the other hand, a convex surface containing an entire straight line is cylindrical.

In many cases the following theorem, also due to Aleksandrov [**5**], will prove useful.

THEOREM 2. *Let F be a convex surface, g a straight-line segment on F, and X a point on g. Then there exists a sequence of domains on the surface, converging to the point X, whose specific curvatures become arbitrarily small.*

PROOF. The tangent cone to the surface at X is either a dihedral angle or a plane. In both cases the straight line containing the segment g divides the tangent cone into two halfplanes α_1 and α_2 (Figure 16).

Let σ be a plane through X perpendicular to g. It cuts the surface in a convex curve γ. The semitangents to this curve at X are the rays along which the plane σ cuts the halfplanes α_1 and α_2. Without loss of generality we may assume that each of these semitangents has only one point (X) in common with the curve γ. Otherwise γ would contain a straight-line segment with endpoint X; but then F would be a plane domain and there is no difficulty in constructing the domains whose existence is asserted in our theorem.

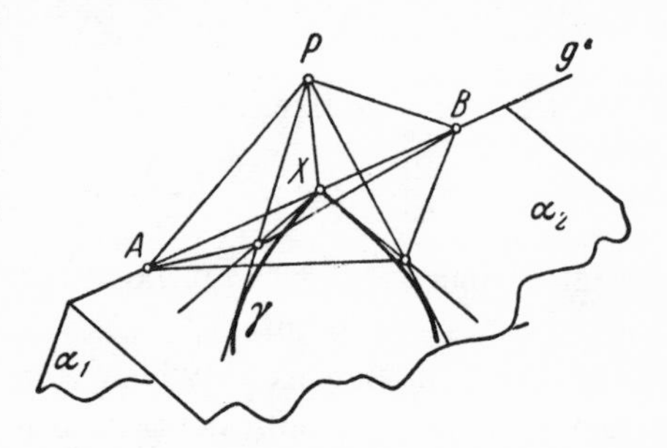

FIGURE 16

Let n be the exterior normal to the surface F at X forming equal angles and the halfplanes α_1 and α_2; let P be a point on n near X, and A and B points on g on different sides of X.

The supporting rays to γ issuing from P form with the rays PA and PB the edge of a convex tetrahedral angle E with apex P.

The angle E cuts out a certain domain G from the surface F; G is a curvilinear quadrangle whose area is obviously not less than that of the corresponding quadrangle $\overline{G}$ on the halfplanes α_1 and α_2, while its curvature is less than that of the angle E. Hence the specific curvature of G cannot exceed the ratio of the curvature of E to the area of the quadrangle $\overline{G}$.

If the distance h between P and X is small in comparison with the distance between X and the points A and B, then the curvature of E satisfies the inequality $\omega(E) \leqq ch/d$, where d is the distance between A and B, and c a constant.

The area of the quadrangle $\overline{G}$ is

$$S(\overline{G}) = \tfrac{1}{2} d(\delta_1(h) + \delta_2(h)),$$

where $\delta_1(h)$ and $\delta_2(h)$ are the distances of X from the vertices of $\overline{G}$ other than A and B.

It is clear that when the point P approaches arbitrarily close to X, i.e. $h \to 0$, we have

$$\frac{h}{\delta_1(h)} \to 0, \quad \frac{h}{\delta_2(h)} \to 0.$$

Therefore the quotient

$$\frac{\omega(E)}{S(\overline{G})} \leqslant \frac{2ch}{d^2(\delta_1(h) + \delta_2(h))}$$

tends to zero, provided the points A, B and P approach X in a suitable manner.

Since the specific curvature of the domain G is smaller than the quotient $\omega(E)/S(\overline{G})$, there must exist a sequence of domains on F which converge to the point P and have specific curvatures which become arbitrarily small. This proves the theorem.

We now define upper and lower curvature at the smooth points of a convex surface.

Let F be a convex surface and X_0 a smooth point on F, X an arbitrary point on F, $h(X)$ the distance between X and the tangent plane α to the surface at X_0, and $d(X)$ the distance between the projection of X on α and the point X_0. We define the *upper* (*lower*) *curvature* of the surface

F at X_0 as the upper (lower) limit of the quotient $2h(X)/d^2(X)$ as $X \to X_0$.

A few obvious properties of upper and lower curvature follow.

1. Let F_1 and F_2 be convex surfaces with a common point X_0 and a common tangent plane at X_0. Then, if a sufficiently small neighborhood of X_0 on F_1 lies in the interior of F_2, the upper and lower curvatures of F_1 at X_0 are not less than the corresponding curvatures of F_2.

2. If X_0 is a point on a regular convex surface, the upper (lower) curvature at X_0 coincides with the maximum (minimum) curvature of the normal cross-sections at X_0. In particular, the upper and lower curvatures at each point on a sphere of radius R are equal to $1/R$.

3. If the lower curvature at a smooth point X_0 of a convex surface F is positive, there exists a sphere, touching the surface F at X_0, whose interior contains a sufficiently small neighborhood of X_0 on F. If the upper curvature at X_0 is bounded, there exists a sphere touching F at X_0 such that a sufficiently small neighborhood of X_0 on the sphere lies in the interior of F.

The next two theorems clarify the relation between the specific curvature of a surface and its upper and lower curvatures.

THEOREM 3. *Let X_0 be a smooth point of a convex surface F such that the lower curvature at X_0 is positive and the specific curvature bounded. Then the upper curvature at X_0 is also bounded.*

THEOREM 4. *If F is a strictly convex surface whose specific curvature at a smooth point X_0 is infinite, then the upper curvature at X_0 is also infinite.*

PROOF OF THEOREM 3. Let α be the tangent plane to the surface at X_0, $L(X)$ the distance from an arbitrary point X on F to α, and $d(X)$ the distance from its projection on α to the point X_0.

Assume that the theorem is false. Then there exists a sequence of points X_k converging to X_0 such that

$$\lambda_k = \frac{h(X_k)}{d^2(X_k)} \to \infty \qquad \text{as} \qquad k \to \infty.$$

We claim that if this is so, there exists a sequence of domains on F which converge to X_0 and whose specific curvatures become arbitrarily large. To see this, we cut off a cap F_k from the surface F by a plane α_k through X_k parallel to α, and estimate the specific curvature of F_k.

Since the lower curvature of F at X_0 is positive, there exists a positive constant c_1 such that for all X sufficiently close to X_0 we have $c_1 < h(X)/d^2(X)$, and so $d(X) < \sqrt{h(X)/c_1}$.

Let K be the cone formed by connecting X_0 to the boundary points of the cap F_k. The extrinsic curvature of K is less than that of the cap.

Describe a trihedral angle about the cone K in the following way. The first face passes through the generator X_0X_k of the cone. The other two faces of the angle touch the cone K in such a way that their intersections with the plane α_k are perpendicular to the intersection of the first face with α_k.

The extrinsic curvature of the trihedral angle is less than the curvature of K, and therefore less than the extrinsic curvature of F_k.

Let us estimate the curvature of the trihedral angle.

Let Y_2 be the point on the base of F_k lying in the second face of the trihedral angle, and h_k the height of the cap. Then $c_1 < h_k/d^2(Y_2)$. Hence

$$\sqrt{c_1h_k} < \frac{h_k}{d(Y_2)} \leqslant \tan\varphi_2,$$

where φ_2 is the angle between the second face of the trihedral angle and the plane α. Similarly, the angle φ_3 between the third face and the plane α is such that $\sqrt{c_1h_k} < \tan\varphi_3$.

Finally, if φ_1 is the angle between the first face of the trihedral angle and the plane α, then $\sqrt{\lambda_kh_k} < \tan\varphi_1$.

Since the angles φ_1, φ_2, and φ_3 are small, we may assume that

$$\sqrt{c_1h_k} < \tan\varphi_3.$$

The spherical image of the trihedral angle is a spherical triangle with base $\varphi_2+\varphi_3$ and height φ_1, and its area is at least

$$\sqrt{\lambda_kh_k} < \tan\varphi_1.$$

If the height h_k of the cap F_k is sufficiently small, the latter's area differs only slightly from that of its base. We may certainly assume that it is at most twice the area of the base. Now since $d(X) < \sqrt{h_k/c_1}$ for all points X on F which belong to the base of the cap F_k, it follows that the area of the base is at most the area of a circle of radius $\sqrt{h_k/c_1}$; hence if h_k is sufficiently small the area of F_k satisfies the inequality $S(F_k) < 2\pi h_k/c_1$.

The curvature $\omega(F_k)$ of F_k is bounded below by the curvature of the trihedral angle, and therefore is bounded by $h_k\sqrt{c_1\lambda_k}/4$. Therefore the specific curvature of the cap satisfies the inequality

$$\varphi_1 > \frac{1}{2}\sqrt{\lambda_kh_k}, \qquad \varphi_2+\varphi_3 > \sqrt{c_1h_k}.$$

and becomes arbitrarily large as $k \to \infty$, since $\lambda_k \to \infty$ by assumption. This contradiction proves the theorem.

PROOF OF THEOREM 4. Suppose that the theorem is false; retain the previous notation. For all X in a sufficiently small neighborhood of X_0 on F we have $h(X)/d^2(X) < c_2$, where c_2 is a constant.

Construct a sphere touching the surface F at X_0 and situated on the same side of the tangent plane α as the surface F, in such a way that some neighborhood of X_0 on the surface is outside the sphere. This is possible, for the upper curvature of F at X_0 is finite by assumption.

Let S be a point on the exterior normal to F at X_0. Construct two cones K_1 and K_2 projecting the sphere and the surface, respectively [from S]. Let $\sigma(S)$ be the area of the part of F which is "visible" from S, $\bar{\sigma}(S)$ the area of the part of the sphere visible from S, $f(S)$ the area of the domain on the plane α contained within K_1, $\omega(S)$ the extrinsic curvature of K_2, and $\bar{\omega}(S)$ the extrinsic curvature of K_1.

If the point S is sufficiently close to X_0, then K_1 is in the interior of K_2. Therefore $\omega(S) < \bar{\omega}(S)$. Moreover,

$$\frac{1}{2}\varphi_1(\varphi_2+\varphi_3) > \frac{1}{4} h_k \sqrt{c_1 \lambda_k}.$$

where c' is an absolute constant and R is the radius of the sphere.

The last four inequalities imply

$$d(X) < \sqrt{\frac{h_k}{c_1}}\,.$$

To derive a contradiction, it is now sufficient to show that that part of F visible from S shrinks to the point X_0 when S approaches X_0. But this is clear, for otherwise F would contain a straight-line segment containing X_0, and this is impossible since F is strictly convex. The proof is complete.

§4. Construction of a convex surface with infinite upper curvature on a given set of points

In this section we shall construct a convex surface F which can be projected in one-to-one fashion onto the xy-plane and has the following properties:

1) F has positive lower curvature at any smooth point.

2) Given a set M of measure zero on the xy-plane, every smooth point of F whose projection lies in M has infinite upper curvature.

We shall need the surface F in our proof of the monotypy of caps (§6). It will also be of use in other uniqueness theorems for surfaces with bounded regularity. The method of construction will be used in proving general existence theorems for surfaces in Chapters VII and VIII.

LEMMA 1. *Let C be a closed polygon in the halfspace $z \geqq 0$ which has a unique projection on the xy-plane as a convex polygon $\overline{C}$ bounding a convex polygonal region G. Let $\overline{A}_i$ $(i = 1, 2, \cdots, n)$ be points in G, each associated with a number $\omega_i > 0$, and let $\omega_1 + \omega_2 + \cdots + \omega_n < 2\pi$.*

Then there exists a convex polyhedron with boundary C which can be projected in one-to-one fashion onto the xy-plane; the projections of the vertices of C are the points $\overline{A}_1, \cdots, \overline{A}_n$, and the extrinsic curvatures at the vertices are $\omega_1, \cdots, \omega_n$, respectively (*Figure* 17).

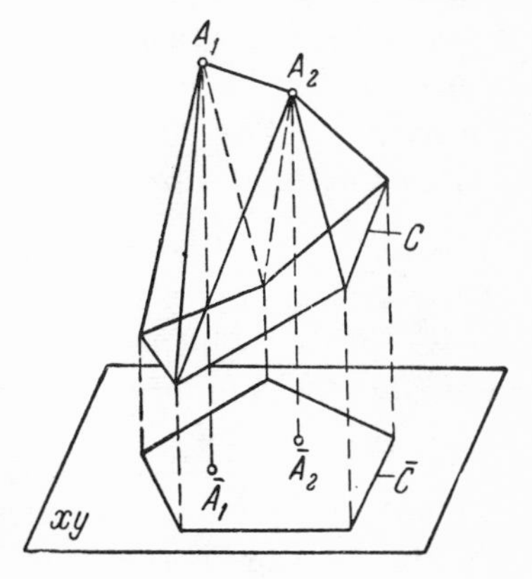

FIGURE 17

PROOF. Let $g_1, \cdots, g_n$ be straight lines parallel to the z-axis, through the points $\overline{A}_1, \cdots, \overline{A}_n$.

Now consider all convex polyhedra with boundary C, convex toward $z > 0$, whose vertices lie on the straight lines $g_1, \cdots, g_n$ and which have extrinsic curvature at most $\omega_1, \cdots, \omega_n$, respectively, at these vertices. The set T of such polyhedra is not empty; it contains the domain bounded by C on the convex hull of C and the negative z-axis (this domain is a polyhedron with boundary C and no interior vertices).

The polyhedra P in T are uniformly bounded. In fact, they are all situated above the xy-plane, within the cylindrical surface through the polygon C whose generators are parallel to the z-axis. Let α be a plane above C, parallel to the xy-plane. Let γ be the convex polygon in which this plane cuts the polyhedron P, and A the vertex of P most distant from α. The curvature at the apex of the polyhedral angle formed by connecting A to the points of γ does not exceed that of the polyhedron P, while the latter in turn does not exceed $\omega_1 + \omega_2 + \cdots + \omega_n < 2\pi$. Now, if A is sufficiently distant from the plane α, the curvature of the above polyhedral angle can be made arbitrarily close to 2π. Thus A cannot be arbitrarily distant from α, and consequently the polyhedra in T are uniformly bounded.

For each polyhedron P in T, let $s(P)$ be the sum of distances from the xy-plane of the points in which the lines $g_1, \cdots, g_n$ cut the polyhedron (even if some of these points are not vertices).

The number $s(P)$ is a maximum for some polyhedron P in T; call this polyhedron P_0. We claim that P_0 is precisely the polyhedron whose existence we wish to prove. Indeed, the boundary of P_0 is the polygon C, its vertices are on the straight lines g_i and so project onto the required points $\overline{A}_i$, and the curvature at the vertex on g_i is at most ω_i. We shall show that it is exactly ω_i.

Assume that the curvature at the vertex A_k of P_0 lying on the line g_k is less than ω_k. Displace this vertex upward along g_k for a small distance δ. In so doing we have not increased the extrinsic curvature at any vertex other than A_k. If δ is chosen sufficiently small, the curvature at A_k, though increased, will remain smaller than ω_k. Since upward displacement of A_k increases $s(P)$, this contradicts the choice of P_0. This completes the proof of Lemma 1.

LEMMA 2. *Let M be any set of measure zero in the xy-plane and $\overline{M}$ any bounded set containing M. Then there exists a strictly convex cap F whose base lies in the xy-plane and covers $\overline{M}$, such that the specific curvature is positive at each point of the cap and infinite at each point whose projection lies in M.*

PROOF. Without loss of generality we may assume that the set $\overline{M}$ is in the interior of the square C_1 bounded by the lines $x=0$, $x=a$, $y=0$, $y=a$.

Construct a sequence of open sets G_k satisfying the following conditions:

1) G_1 is the square C_1 minus its boundary;
2) $G_k \subset G_{k-1}$;
3) the measure (area) of G_k is at most $a^2/4^k$;
4) the set M is contained in each G_k.

Now divide the square C_1 into small squares of side $\delta_n = a/2^n$ by straight lines parallel to its sides. Call the set of all squares of this partition T_n.

Assign the number 2^{k-2n} to each point in the xy-plane which is the center of a square in T_n which is contained in G_k but not in G_{k+1}. It is easy to see that the sum of all numbers assigned to centers of squares in T_n is at most unity, and so less than 2π.

Let C_2 be any convex closed polygon in the xy-plane whose interior contains the square C_1. By Lemma 1 there exists a convex polyhedron P_n with boundary C_2 whose vertices project onto the xy-plane at the centers of the squares of T_n and whose curvatures equal the numbers assigned to these centers.

Now consider the sequence of polyhedra $P_1, P_2, \cdots$. Since the polyhedra P_n are uniformly bounded, we can extract a convergent subsequence; without loss of generality let us assume that the sequence P_n itself converges, to a convex surface, say F.

We claim that the specific curvature of F is infinite at any point X whose projection on the xy-plane is a point of the set M.

Let X be a point on the surface F whose projection on the xy-plane lies in M, and G an arbitrary domain on F situated in a small neighborhood H of X. Let $\overline{G}$ denote the projection of G on the xy-plane. Let $\overline{R}$ be a closed set in $\overline{G}$, of area at least half that of $\overline{G}$, and R the closed set in F whose projection is $\overline{R}$. For sufficiently large n the closed set $\overline{U}_n$, defined as the union of all squares in T_n that cut $\overline{R}$, is in the interior of $\overline{G}$. Let U_n be the closed set on the polyhedron P_n whose projection is $\overline{U}_n$.

Since $U_n \to R$ as $n \to \infty$ and R is a closed set, it follows that for sufficiently large n the spherical image of U_n is in any sufficiently small neighborhood of the spherical image of R. Hence

$$\overline{\lim_{n\to\infty}}\, \omega(U_n) \leqslant \omega(R) \leqslant \omega(G). \tag{1}$$

Let us estimate $\omega(U_n)$. If the neighborhood H of the point X is sufficiently small, the domain $\overline{G}$ is a subset of an open set G_k for sufficiently large k. The number of squares of T_n contained in $\overline{U}_n$ is at least $S(\overline{G})\,2^{2n}/2a^2$, since they cover more than half the area of $\overline{G}$. Since one of the vertices of P_n, with curvature at least 2^{k-2n}, lies above the center of each such square, we have

$$\omega(U_n) \geqslant \frac{1}{2a^2} S(\overline{G})\, 2^k.$$

Furthermore, $S(G) \leqq S(\overline{G})/\cos\vartheta_0$, where ϑ_0 is the largest angle between the xy-plane and the supporting planes of F at points of H. Therefore

$$\omega(U_n) \geqslant \frac{\cos\vartheta_0}{2a^2} S(G)\, 2^k,$$

so that

$$\overline{\lim_{n\to\infty}}\, \omega(U_n) \geqslant \frac{\cos\vartheta_0}{2a^2} S(G)\, 2^k. \tag{2}$$

Inequalities (1) and (2) imply

$$\frac{\omega(G)}{S(G)} \geqslant \frac{\cos\vartheta_0}{2a^2}\, 2^k. \tag{3}$$

If the domain G shrinks to a point X, its projection $\overline{G}$ will ultimately fall within the domain G_k for any k, whence it follows that $\omega(G)/S(G) \to 0$ as $G \to X$.

Another result of inquality (3) is that the specific curvature at any point X of the surface F is positive in the sense that

$$\varliminf_{G \to X} \frac{\omega(G)}{S(G)} > 0.$$

According to Theorem 2 of §3, this implies that the cap F is strictly convex, and the proof of Lemma 2 is thus complete.

THEOREM 1. *Let M be an arbitrary set of measure zero in the xy-plane, and $\overline{M}$ any bounded set containing M. Then there exists a convex cap Ω, whose base is in the xy-plane and covers $\overline{M}$, such that the lower curvature of the cap is positive at all smooth points and the upper curvature is infinite at every point whose projection is in the set M.*

PROOF. By Lemma 2 there exists a cap whose base is in the xy-plane and covers an ϵ-neighborhood of $\overline{M}$, such that the specific curvature of the cap is infinite at every smooth point whose projection lies in M. Let $z = f(x, y)$ be the equation of this cap.

Consider the convex surface defined by the equation

$$z = f(x, y) - \epsilon'(x^2 + y^2), \quad \epsilon' > 0.$$

For sufficiently small ϵ' the part of this surface lying above the xy-plane is a convex cap Ω whose base lies in the xy-plane and covers the set $\overline{M}$. We shall show that it has the required properties.

Let $\Phi : z = \varphi(x, y)$ be a convex surface and $z = \psi(x, y)$ its tangent plane at a point $X_0(x_0, y_0)$. Let $X_1(x_1, y_1)$ be a point on Φ close to X_0. Let d denote the distance between X_0 and the projection $\overline{X}_1$ of X_1 on the tangent plane at X_0, and δ the distance between X_1 and $\overline{X}_1$. Then, if X_1 and X_0 are sufficiently close together, the quantity d^2 is of the same order as $(x_1 - x_0)^2 + (y_1 - y_0)^2$, while δ is of the same order as $|\varphi(x_1, y_1) - \psi(x_1, y_1)|$. It follows that the surface Φ has positive lower curvature at X_0 if

$$\varliminf_{X_1 \to X_0} \frac{|\varphi(x_1, y_1) - \psi(x_1, y_1)|}{(x_1 - x_0)^2 + (y_1 - y_0)^2} > 0.$$

Φ has infinite upper curvature at X_0 if

$$\varlimsup_{X_1 \to X_0} \frac{|\varphi(x_1, y_1) - \psi(x_1, y_1)|}{(x_1 - x_0)^2 + (y_1 - y_0)^2} = \infty.$$

It follows from these properties of the upper and lower curvatures of a convex surface that if at least one of two convex surfaces $z = f_1(x,y)$ and $z = f_2(x,y)$, convex in the same direction, has positive lower curvature at a point (x_0, y_0), the same holds for the surface $z = f_1(x,y) + f_2(x,y)$. If at least one of these surfaces has infinite upper curvature at (x_0, y_0), the same holds for the surface $z = f_1(x,y) + f_2(x,y)$.

Now the surfaces $z = -\epsilon'(x^2 + y^2)$ and $F: z = f(x,y)$ are both convex toward $z > 0$. The first has positive curvature at each point. Consequently the surface $\Omega: z = f(x,y) - \epsilon'(x^2 + y^2)$ has positive lower curvature at each smooth point.

The surface $z = f(x,y)$ has infinite upper curvature at all points whose projections lie in M, since at any such point the surface has infinite specific curvature, therefore also infinite upper curvature (Theorem 4 of §3). Hence at any smooth point whose projection is in M the surface Ω has infinite upper curvature. Q.E.D.

THEOREM 2. *Let $\Omega: z = \omega(x,y)$ be the convex surface whose existence was established in Theorem* 1. *Then every surface $\Omega_\lambda: z = \lambda\omega(x,y)$ $(\lambda \neq 0)$ has the same property as Ω, i.e. its lower curvature is positive at every smooth point, and its upper curvature is infinite at any point with projection in M.*

The proof is obvious.

§5. Auxiliary surface Φ and some of its properties

In this section we shall construct a certain auxiliary surface Φ, starting from two given isometric surfaces F_1 and F_2. This surface will also be needed in our proof of the monotypy of convex caps in §6. We have allotted a special section to Φ so as not to obscure the above-mentioned proof with details.

LEMMA 1. *Let F be a convex surface with regular metric and positive Gauss curvature. Let (u,v) be a regular coordinate net on the surface, regular in the sense that the coefficients of the line element of the surface*

$$ds^2 = E\,du^2 + 2F\,du\,dv + G\,dv^2$$

are regular (twice differentiable) functions of u and v.

Then the vector-valued function $r(u,v)$ defining the surface is smooth, i.e. has continuous first derivatives r_u, r_v such that $r_u \times r_v \neq 0$.

PROOF. By a theorem of Aleksandrov (Theorem 1 of §3) the surface F is smooth. Choose an arbitrary point (u,v) on F and construct a geodesic γ tangent to the u-curve ($v = \text{const}$) at this point. Lay off segments of small length s along the u-curve and the geodesic γ from the point (u,v); call the endpoints of these segments A and B. Since the parametrization u, v is regular, the distance $\rho(A,B)$ between A and B on the surface is of smaller order than s when $s \to 0$. Hence there exists a derivative r'_s along the u-curve, since the derivative r'_s along the geodesic exists. Since the parametrization u, v of the surface is regular, the existence of r'_s for $v = \text{const}$ implies the existence of the derivative r_u. The existence of r_v is proved in analogous fashion.

We now prove that the derivative r_u is continuous. To this end it is clearly sufficient to show that the unit vector τ tangent to the u-coordinate curve is a continuous function of u and v. Let $A(u,v)$ be an arbitrary point on the surface F, and $B(u+\Delta u, v+\Delta v)$ a point close to $A(u,v)$. Consider a segment connecting A and B; let $\tau'(A)$ and $\tau'(B)$ denote the unit vectors tangent to this segment at A and B (Figure 18). Since the parametrization u, v is regular, the angles α and β are approximately equal when A and B are sufficiently close together. The tangent vectors $\tau'(A)$ and $\tau'(B)$ are close together by Theorem 8 of §1. Now note that all four vectors $\tau(A)$, $\tau(B)$, $\tau'(A)$ and $\tau'(B)$ form angles close to $\pi/2$ with the normal to the surface at A; this implies that the vectors $\tau(A)$ and $\tau(B)$ are also close together, and hence the vector-valued function τ is continuous. As indicated above, since the intrinsic metric of the surface is regular, the continuity of τ implies that of the derivative $r_u(u,v)$. The proof that the derivative $r_v(u,v)$ is continuous is similar. The condition $r_u \times r_v \neq 0$ follows from the fact that the quadratic form ds^2 is positive definite, since $(r_u \times r_v)^2 = EG - F^2$. Q.E.D.

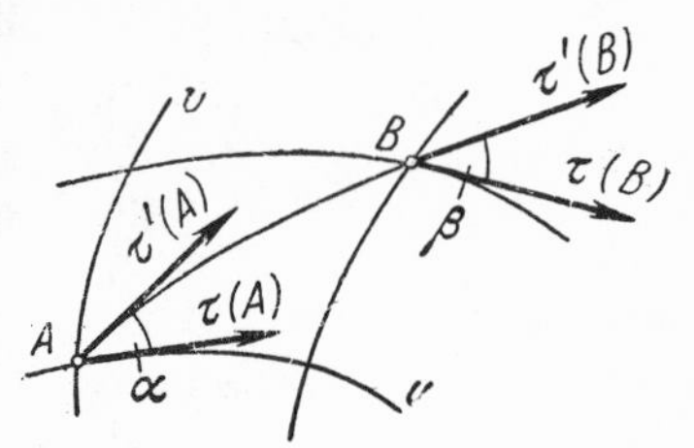

FIGURE 18

Let F be a convex surface defined by the equation $z = f(x,y)$. In general the function $f(x,y)$ need not be regular. At points where the surface is not geometrically smooth, the function does not even have first derivatives. However, the following theorem of Busemann and Feller is valid.[1)]

A function $f(x,y)$ defining a convex surface has a second differential almost everywhere, i.e.

---

1) Aleksandrov has extended this theorem to the case of convex hypersurfaces [7].

$$f(x+\Delta x,\ y+\Delta y)=f(x,\ y)+a_1\Delta x+a_2\Delta y$$
$$+\frac{1}{2}(a_{11}\Delta x^2+2a_{12}\Delta x\,\Delta y+a_{22}\Delta y^2)+\varepsilon\cdot(\Delta x^2+\Delta y^2),$$

where $\epsilon(x, y, \Delta x, \Delta y)\to 0$ *as* $\Delta x^2+\Delta y^2\to 0$.

We shall now define upper and lower curvatures for an arbitrary smooth surface, by analogy with the definition for smooth convex surfaces in §3. Call a ray issuing from a point O on a smooth surface Φ an interior normal if it is perpendicular to the tangent plane and there are points of the surface arbitrarily close to O on the side of the tangent plane into which the ray points. If O has a plane neighborhood on the surface, any ray perpendicular to the tangent plane will be considered an interior normal.

Let X be a point on Φ close to O; let $d(X)$ denote its distance from O and $\delta(X)$ its directed distance from the tangent plane: $\delta(X)$ is positive if X is on the side of the tangent plane at O into which the interior normal points, negative otherwise. We define the lower curvature of Φ at O as

$$\underline{\lim}_{X\to 0} 2\delta(X)\, d^2(X),$$

and the upper curvature as

$$\overline{\lim}_{X\to 0} 2\delta(X)/d^2(X).$$

It is clear that the lower curvature cannot exceed the upper curvature.

LEMMA 2. *Let F_1 and F_2 be two isometric convex surfaces with regular metric and positive Gauss curvature. Suppose that the origin O lies on both surfaces F_1 and F_2 and corresponds to itself under the isometry of F_1 and F_2; furthermore, the directions of both surfaces at O that correspond under the isometry coincide. Both surfaces lie on one side of their common tangent plant at O, and the latter is not perpendicular to the xy-plane.*

Introduce a common regular parametrization in the surfaces F_1 and F_2 (points corresponding under the isometry having the same coordinates u, v). Let $r_1(u,v)$ and $r_2(u,v)$ be vector-valued functions defining the surfaces F_1 and F_2. Let $r_2^*(u,v)$ denote the reflection of the vector $r_2(u,v)$ in the xy-plane.

Then in a sufficiently small neighborhood of the point O the vector-valued function $r=r_1(u,v)+r_2^*(u,v)$ defines a smooth surface Φ which can be projected in one-to-one fashion onto the xy-plane. There are no points on

Φ which have both positive lower curvature and infinite upper curvature. If F_1 and F_2 are twice differentiable at points $X_1(u,v)$ and $X_2(u,v)$, then Φ cannot have positive lower curvature at the corresponding point $X(u,v)$.

PROOF. First note that, considered in a neighborhood of O, the vector-valued function $r = r_1(u,v) + r_2(u,v)$ clearly defines a smooth surface Φ'. Indeed, at O we have $r_u \times r_v = 4r_{1u} \times r_{1v} \neq 0$. By continuity, $r_u \times r_v \neq 0$ in a neighborhood of O. The surface Φ' can be projected in one-to-one fashion onto the xy-plane, since the z-component of the vector $r_u \times r_v$ is not zero at O or, consequently, near O. Since the vectors $r_1 + r_2$ and $r_1 + r_2^*$ differ only in their z-components, the vector-function $r_1(u,v) \times r_2^*(u,v)$ also defines a certain smooth surface Φ which can be projected in one-to-one fashion onto the xy-plane. The xy-plane is obviously the tangent plane of Φ at O.

Let $\overline{X}$ be an arbitrary fixed point of F_1 close to O. As was shown in §2 (Theorem 3), the vector-valued function $r_1(X)$ defining F_1 has the representation

$$r_1(X) = r_1(\overline{X}) + \tau_1(X)s(X) + \nu_1(X),$$

where $s(X)$ is the distance between x and $\overline{X}$ on the surface, and $\tau_1(X)$ the unit vector tangent at $\overline{X}$ to the segment connecting x and $\overline{X}$. When $X \to \overline{X}$, the direction of the vector $\tau_1(X)$ tends to the interior normal of the surface at $\overline{X}$, and $|\nu_1(X)|$ approaches zero more rapidly than $s(X)$, i.e. $|\nu_1(X)|/s(X) \to 0$. The vector $r_2(X)$ has a similar representation. Thus the vector-valued function $r(X)$ defining the surface Φ has the representation

$$r(X) = r(\overline{X}) + (\tau_1(X) + \tau_2^*(X))\, s(X) + \nu_1(X) + \nu_2^*(X).$$

Since $\tau_1(X) + \tau_2^*(X) \neq 0$ at the point O, and so also in a neighborhood of O, there exists a constant c_1 such that

$$|r(X) - r(\overline{X})| > c_1 s(X). \tag{1}$$

Let n_1, n_2, n_2^* and n be the unit vectors of the exterior normals to the surfaces F_1, F_2, F_2^* and Φ (F_2^* denotes the reflection of F_2 in the xy-plane). We shall show that $n_1 n = -n_2^* n \neq 0$.

In fact, let τ_1 and τ_1' be two mutually perpendicular unit vectors tangent to the surface F_1 at $\overline{X}$, and τ_2^* and $\tau_2'^*$ the tangent vectors of F_2^* corresponding to τ_1 and τ_1' under isometry. Set

$$A = n_1 n = (\tau_1, \tau_1', n),$$
$$B = -n_2^* n = (\tau_2^*, \tau_2'^*, n).$$

Then

$$A^2 = 1 - (\tau_1 n)^2 - (\tau_1' n)^2,$$
$$B^2 = 1 - (\tau_2^* n)^2 - (\tau_2'^* n)^2.$$

Since the vectors $\tau_1 + \tau_2^*$ and $\tau_1' + \tau_2'^*$ lie in a plane tangent to Φ, it follows that

$$\tau_1 n + \tau_2^* n = (\tau_1 + \tau_2^*) n = 0,$$
$$\tau_1' n + \tau_2'^* n = (\tau_1' + \tau_2'^*) n = 0.$$

Hence $A^2 = B^2$.

We claim that A and B are not zero. Suppose that $A = 0$. Then the vector n touches the surface F_1 at $\overline{X}$. Set n equal to τ_1. Since $\tau_1 n + \tau_2^* n = 0$, necessarily $\tau_2^* = -n$. But then $\tau_1 + \tau_2^* = 0$; but this is impossible at O and therefore near O; in particular, at $\overline{X}$. Since $A(\overline{X})$ and $B(\overline{X})$ are continuous functions of $\overline{X}$ and clearly $A = B$ for $\overline{X} = 0$, it follows that $A^2 = B^2 \neq 0$ implies $A = B$ for all $\overline{X}$ close to O.

Now let the surface Φ have positive lower curvature at $\overline{X}$ (that is, at the point with vector $r(X)$). Then

$$(r(X) - r(\overline{X}))\, n = (\nu_1(X) + \nu_2^*(X))\, n > c_1 (r(X) - r(\overline{X}))^2,$$

where c_1 is a positive constant. Hence, in view of inequality (1),

$$\nu_1(X) n + \nu_2^*(X) n > \beta_1 s^2(X), \tag{2}$$

where β_1 is another positive constant.

Now as $X \to \overline{X}$ the directions of the vectors $\nu_1(X)$ and $\nu_2^*(X)$ approach the directions of the normals n_1 and n_2^*, and $nn_1 = -n_2^* n \neq 0$; hence for X sufficiently close to $\overline{X}$ the quantities $\nu_1(X) n$ and $\nu_2^*(X) n$ do not change sign. Without loss of generality we assume that $\nu_1(X) n > 0$ and $\nu_2^*(X) n \leqq 0$. That the first of these inequalities is strict follows from the fact that $\nu_1(X) n$ and $\nu_2^*(X) n$ have unlike signs and satisfy inequality (2). Since $\nu_1(X) n > 0$ and $\nu_2^*(X) n \leqq 0$, inequality (2) implies

$$\nu_1(X) n > \beta_1 s^2(X). \tag{3}$$

Thus for sufficiently small nonzero $s(X)$ we have $|\nu_1(X)| \neq 0$.

We now express the vector $\nu_1(X)$ as the sum of two vectors:

$$\nu_1(X) = n_1 \delta_1(X) + t_1(X),$$

where $\delta_1(X)$ is the distance of X from the tangent plane to F_1 at $\overline{X}$. Clearly $n\nu_1(X) = nn_1\delta_1(X) + nt_1(X)$. Substitute this expression for $n\nu_1(X)$ in inequality (3):

$$\delta_1(X) = \frac{\beta_1 s^2(X)}{nn_1 + \dfrac{nt_1(X)}{\delta_1(X)}}.$$

Since $nn_1 \neq 0$ and $t_1(X)/\delta_1(X) \to 0$ as $s(X) \to 0$, there exists a positive constant β_2 such that

$$\delta_1(X) > \beta_2 s^2(X). \tag{4}$$

Inequality (4) means that the surface F_1 has positive lower curvature at $\overline{X}$. Since F_1 has bounded specific curvature at $\overline{X}$ (the metric of the surface is regular), the fact that the lower curvature at $\overline{X}$ is positive implies that the upper curvature is bounded there (Theorem 3, §2). Hence $\delta_1(X) < \beta_3 s^2(X)$, and so

$$n\nu_1(X) < \beta_3' s^2(X). \tag{5}$$

Since $n\nu_1(X) > 0$, $n\nu_2^*(X) \leqq 0$ and $(r(X) - r(\overline{X}))n = \nu_1(X)n + \nu_2^*(X)n$, we have

$$(r(X) - r(\overline{X}))n \leqq n\nu_1(X) < \beta_3' s^2(X).$$

In view of inequality (1), this implies

$$(r(X) - r(\overline{X}))n \leqq c'(r(X) - r(\overline{X}))^2.$$

Now this inequality means that the upper curvature of the surface Φ is bounded at $\overline{X}$. Thus at no point of Φ can the lower curvature be positive and the upper curvature infinite.

Now let the surface F_1 be twice differentiable at $\overline{X}$, and the surface F_2 twice differentiable at the point corresponding to $\overline{X}$ under the isometry (twice differentiable in the sense of the definition given earlier). We claim that then the lower curvature of Φ at $r(\overline{X})$ cannot be positive.

Take the point $\overline{X}$ of F_1 as the origin O, the tangent plane at this point as the xy-plane, and the interior normal as the positive z-axis. Since the surface is twice differentiable at $\overline{X}$, there is a neighborhood of this point in which F_1 is defined by an equation

$$z = \varphi_2(x, y) + \varepsilon \cdot (x^2 + y^2), \quad \varphi_2 = \frac{1}{2}(a_{11}x^2 + 2a_{12}xy + a_{22}y^2),$$

where $\epsilon(x,y) \to 0$ as $x^2 + y^2 \to 0$. By assumption the surface F_1 has positive Gauss curvature. We shall show that at $\overline{X} = 0$ the Gauss curvature is $K = a_{11}a_{22} - a_{12}^2$.

First note that the quadratic form $\varphi_2(x,y)$ is nonnegative. Indeed, suppose that $\varphi_2(\xi, \eta) < 0$ for some ξ and η. Since F_1 is in the halfspace $z \geqq 0$, it follows that for any $\lambda \neq 0$

$$\frac{1}{\lambda^2} z(\lambda\xi, \lambda\eta) = \varphi(\xi, \eta) + \varepsilon(\xi^2 + \eta^2) \geqslant 0.$$

If $\lambda \to 0$, then $\epsilon \to 0$. Consequently for small λ the sign of $z(\lambda\xi, \lambda\eta)$ is determined by that of the form $\varphi_2(\xi, \eta)$. Now since $\varphi_2(\xi, \eta) < 0$ it follows that $z(\lambda\xi, \lambda\eta) < 0$ for sufficiently small λ, and this is a contradiction. Hence the form $\varphi_2(x, y)$ never assumes negative values.

We shall now prove that since F_1 has positive Gauss curvature at O, the form $\varphi_2(x, y)$ is positive definite. Consider the paraboloid P defined by

$$z = \frac{1}{2}(a_{11}x^2 + 2a_{12}xy + a_{22}y^2) + \frac{\varepsilon_0}{2}(x^2 + y^2),$$

where ϵ_0 is a small positive number. At $\overline{X}$ (the origin), the paraboloid P touches the surface F_1, and near $\overline{X}$ it lies in the interior of the surface (Figure 19). Indeed, for sufficiently small $x^2 + y^2$ we have $\epsilon(x, y) < \epsilon_0$, i.e. $z(x, y)|_{F_1} \leqq z(x, y)|_P$, with equality (in a sufficiently small neighborhood of $\overline{X}$) only at $\overline{X}$. Without loss of generality we may assume that the entire paraboloid P lies in the interior of F_1. This can always be achieved by considering not the entire surface but a sufficiently small neighborhood of $\overline{X}$, in which all our construction will be carried out.

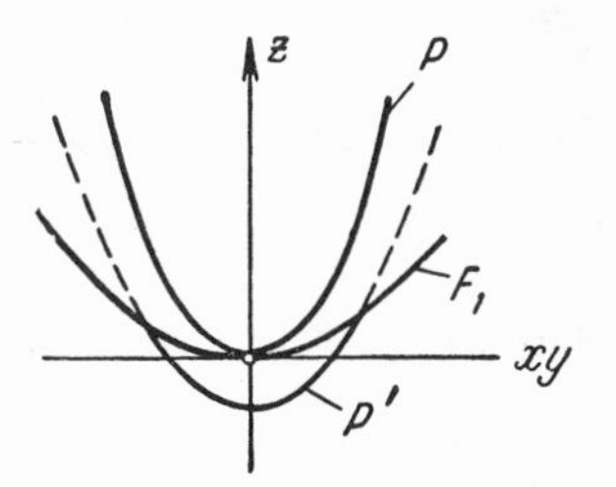

FIGURE 20

Displace the paraboloid P toward $z < 0$ for a small distance h; a certain part P' of the paraboloid will then lie in the exterior of the surface (Figure 19). Let F_1' be that part of the surface which is projected onto P' by straight lines parallel to the z-axis. The curvature of the paraboloid in the region P' is at least that of the surface F_1 in the region F_1', since for any tangent plane to the surface in F_1' there is a parallel tangent plane to the paraboloid in P'. As $h \to 0$ the region P' on the paraboloid and the region F_1' on the surface shrink to the point $\overline{X}$. Moreover, since both surfaces are tangent to the xy-plane when $h = 0$, the ratio of areas $\sigma(P')/\sigma(F_1')$ tends to unity as $h \to 0$. Since the curvatures of P' and F_1' satisfy the inequality $\omega(P') \geqq \omega(F_1')$, we have

$$\lim_{h \to 0} \frac{\omega(P')}{\sigma(P')} \geqslant \lim_{h \to 0} \frac{\omega(F_1')}{\sigma(F_1')}.$$

Thus the Gauss curvature of the surface F_1 at $\overline{X}$ is at most the Gauss curvature of the paraboloid P at $\overline{X}$. Were the form $a_{11}x^2 + 2a_{12}xy + a_{22}y^2$ not positive definite, we would have $a_{11}a_{22} - a_{12}^2 = 0$, and so the curvature of the paraboloid

$$K = (a_{11} + \epsilon_0)(a_{22} + \epsilon_0) - a_{12}^2$$

would decrease with ϵ_0. By making ϵ_0 sufficiently small we derive a contradiction, since the curvature of the surface F_1 is strictly positive Hence the form $a_{11}x^2 + 2a_{12}xy + a_{12}y^2$ is positive definite.

Now construct two paraboloids:

$$P^+ : z = \frac{1}{2}(a_{11}x^2 + 2a_{12}xy + a_{22}y^2) + \frac{\varepsilon_0}{2}(x^2 + y^2),$$

$$P^- : z = \frac{1}{2}(a_{11}x^2 + 2a_{12}xy + a_{22}y^2) - \frac{\varepsilon_0}{2}(x^2 + y^2),$$

where ϵ_0 is a small positive number. In a neighborhood of $\overline{X}$ the paraboloid P^+ is in the interior of the surface F_1, while P^- is in its exterior. Repeating the preceding arguments word for word, we conclude that the Gauss curvature of F_1 at $\overline{X}$ is at most that of the paraboloid P^+ and at least that of the paraboloid P^- at the same point $\overline{X}$:

$$(a_{11} - \varepsilon_0)(a_{22} - \varepsilon_0) - a_{12}^2 \leqslant K_{F_1}(\overline{X}) \leqslant (a_{11} + \varepsilon_0)(a_{22} + \varepsilon_0) - a_{12}^2.$$

Since ϵ_0 is arbitrary, the Gauss curvature of F_1 at $\overline{X}$ is

$$K_{F_1}(\overline{X}) = a_{11}a_{22} - a_{12}^2.$$

We now return to the two surfaces F_1 and F_2. In view of inequalities (2) and (5),

$$0 \leqq -\nu_2^*(X)n < \beta_4' s^2(X).$$

Express the vector $\nu_2^*(X)$ as the sum of two vectors, in the same way as for $\nu_1(X)$:

$$\nu_2^*(X) = n_2^* \delta_2^*(X) + t_2^*(X).$$

Then

$$0 \leqq -nn_2^*\delta_2^*(X) - nt_2^*(X) < \beta_4' s^2(X).$$

Hence $\delta_2^*(X) < \beta_4 s^2(X)$, where β_4 is a constant. Substitute the expression

$$\nu_1(X) = n_1\delta_1(X) + t_1(X), \quad \nu_2^*(X) = n_2^*\delta_2^*(X) + t_2^*(X)$$

in inequality (2); recalling that $nn_1 = -n_2^* n \neq 0$ and that $t_1(X)/\delta_1(X) \to 0$ and $t_2(X)/\delta_2(X) \to 0$ as $s(X) \to 0$, while $\delta_1(X)/s^2(X)$ and $\delta_2^*(X)/s^2(X)$ are bounded, we derive the inequality

$$\delta_1(X) - \delta_2^*(X) > cs^2(X), \tag{6}$$

where c is a positive constant.

Identify the point $\overline{X}$ on F_1 with the point of F_2 corresponding to it under the isometry, in such a way that the directions corresponding under the isometry at these points also coincide. Take the common tangent plane to F_1 and F_2 at $\overline{X}$ as the xy-plane; we can then assume that the equations of the surfaces are

$$F_1:\ z = \frac{1}{2}(a_{11}x^2 + 2a_{12}xy + a_{22}y^2) + \varepsilon_1 (x^2 + y^2),$$

$$F_2:\ z = \frac{1}{2}(b_{11}x^2 + 2b_{12}xy + b_{22}y^2) + \varepsilon_2 (x^2 + y^2),$$

where $\epsilon_1, \epsilon_2 \to 0$ as $x^2 + y^2 \to 0$. Since both surfaces have the same Gauss curvature at the origin (for they are isometric!),

$$a_{11}a_{22} - a_{12}^2 = b_{11}b_{22} - b_{12}^2. \tag{7}$$

Let X_1 be a point on F_1 close to $\overline{X}$, and X_2 the point on F_2 corresponding to X_1 under the isometry. Connect both points to $\overline{X}$ by segments γ_1 and γ_2 on F_1 and F_2. Since the directions on F_1 and F_2 that correspond under the isometry coincide at $\overline{X}$, the segments γ_1 and γ_2 have a common semitangent t at $\overline{X}$. Lay off a subarc s on t, equal in length to the segments γ_1 and γ_2. Let X be the endpoint of this subarc. By a theorem of §2 the directions of the segments XX_1 and XX_2 are close to the direction of the normal to F_1 and F_2 at $\overline{X}$, i.e. close to the direction of the z-axis. It follows that the coordinates x_1, y_1 of X_1 and the coordinates x_2, y_2 of X_2 differ from the coordinates x, y of X by a quantity of order $\bar{\epsilon}\sqrt{x^2+y^2}$, where $\bar{\epsilon} \to 0$ as $x^2 + y^2 \to 0$. This being so, the distances δ_1 and δ_2 of X_1 and X_2 from the xy-plane are

$$\delta_1 = \frac{1}{2}(a_{11}x^2 + 2a_{12}xy + a_{22}y^2) + \bar{\varepsilon}_1 (x^2 + y^2),$$

$$\delta_2 = \frac{1}{2}(b_{11}x^2 + 2b_{12}xy + b_{22}y^2) + \bar{\varepsilon}_2 (x^2 + y^2).$$

By inequality (6), $\delta_1 - \delta_2 > cs^2$, where c is a positive constant. Since the quantity s is of order $\sqrt{x^2+y^2}$, this inequality implies that

$$\frac{1}{2}(a_{11}x^2 + 2a_{12}xy + a_{22}y^2) - \frac{1}{2}(b_{11}x^2 + 2b_{12}xy + b_{22}y^2) > \frac{\varepsilon_0}{2}(x^2 + y^2),$$

where ϵ_0 is a positive constant. Hence

$$a_{11}a_{22} - a_{12}^2 > (b_{11} + \epsilon_0)(b_{22} + \epsilon_0) - b_{12}^2.$$

Now this contradicts (7), since b_{11}, b_{22} and ϵ_0 are positive. Thus if the surfaces F_1 and F_2 are twice differentiable at $\overline{X}$ and its counterpart on F_2, the surface Φ cannot have positive lower curvature at the point with vector $r(\overline{X}) = r_1(\overline{X}) + r_2(\overline{X})$. This completes the proof of the lemma.

§6. Monotypy of convex caps with regular metric

In this section we shall prove that convex caps with regular metric are monotypic among convex caps. This theorem plays a central role in the proof of the regularity of convex surfaces with regular metric.

THEOREM. *Isometric convex caps with regular metric and positive Gauss curvature are congruent.*

To prove this theorem we need two lemmas.

LEMMA 1. *Let F_1 and F_2 be two isometric convex surfaces which can be projected in one-to-one fashion onto the xy-plane. Let $z_1(X)$ be the z-coordinate of a point X on F_1 and $z_2(X)$ the z-coordinate of the corresponding point (under the isometry) on F_2. Then, if $z_1(X) = z_2(X)$ for all points X, the surfaces F_1 and F_2 are congruent.*

PROOF. Consider a plane parallel to the xy-plane through two arbitrary points X_1 and Y_1 of the surface F_1. This plane cuts F_1 in a convex curve. Let γ_1 be the subarc of this curve between X_1 and Y_1. Let γ_2 be the curve on F_2, connecting points X_2 and Y_2, which corresponds to γ_1 under the isometry. Let Z be a cylinder whose generators are parallel to the z-axis and project the curve γ_2 onto the xy-plane. If the cylinder Z is developed on the plane, the curve γ_2 becomes a plane curve congruent to γ_1, since $z_1(X) = z_2(X)$.

When the cylinder Z is developed the distance in space between X_2 and Y_2 is not reduced. Hence the distance in space between X_2 and Y_2 on the surface F_2 cannot exceed the distance in space between X_1 and Y_1 on F_1. Interchanging the roles of F_1 and F_2, we get the converse inequality: the distance in space between X_2 and Y_2 is not less than the distance in space between X_1 and Y_1. Hence the distances are equal, i.e. the isometry preserves distances in space, and so the surfaces F_1 and F_2 are congruent, Q.E.D.

LEMMA 2. *If there exist two noncongruent isometric convex caps F_1 and F_2 with regular metric and positive Gauss curvature, they may be brought by a rigid motion or a rigid motion plus a reflection to positions in which they satisfy the following conditions:*

1) *A point of F_1 and a point of F_2 which correspond under the isometry coincide at the origin O.*

2. *At the common point O the directions on F_1 and F_2 that correspond under the isometry also coincide.*

3. *The surfaces lie on the same side of their common tangent plane at O, and the latter is not perpendicular to the xy-plane.*

4. *If $z_1(X)$ is the z-coordinate of an arbitrary point X on F_1 and $z_2(X)$ the z-coordinate of the corresponding (under isometry) point on F_2, then $z_1(X) = z_2(X) \leqq 0$; moreover, if the end of the vector $r_1(X) + r_2(X)$ is projected onto a point of the halfplane $x \geqq 0$ other than O, then $z_1(X) - z_2(X) < 0$.*

PROOF. Let the boundaries of the caps F_1 and F_2 lie in the xy-plane, and the caps themselves in the halfspace $z < 0$. Consider the difference $z(X) = z_1(X) - z_2(X)$. Since the caps F_1 and F_2 are not congruent and $z_1 - z_2 = 0$ along the boundary, there exists a point X such that $z_1(X) - z_2(X) \neq 0$ (Lemma 1). Without loss of generality we may assume that $\max z(X) = m > 0$. In fact, if $m = 0$ we need only interchange the roles of F_1 and F_2. Let M_1 denote the set of all points of F_1 such that $z(X) = m$. M_1 is clearly a closed set containing no points of the cap boundary (on the boundary $z(X) = 0$). Let M_2 denote the set of all points of F_2 corresponding to points of M_1 under the isometry. Consider the planes $z = h_1$ and $z = h_2$ supporting the caps F_1 and F_2 from below. There are two possible cases:

1. The set M_1 lies in the plane $z = h_1$.
2. There are points of M_1 above the plane $z = h_1$.

Consider the first case.

F_1, as a surface with positive Gauss curvature, is strictly convex (Theorem 2, §3). Hence, the set M_1 consists of a single point X_1—the point at which the plane $z = h_1$ touches F_1. At this point $dz_1 = 0$, $dz = 0$, and therefore $dz_2 = 0$. Hence the tangent plane to F_2 at X_2 is also parallel to the xy-plane. We may assume that the surfaces F_1 and F_2 are similarly oriented; this may always be achieved by reflecting one of them in a plane perpendicular to the xy-plane. Then a suitable translation and rotation about the z-axis make F_1 and F_2 coincide at X_1 and X_2, together with the directions at these points which correspond under the isometry. Now a simultaneous translation of F_1 and F_2 brings their common point $X_1 = X_2$ to the origin O. The surfaces F_1 and F_2 in their new position satisfy all the conditions of the lemma.

Now consider the second case: There are points of M_1 lying above the plane $z = h_1$. Denote

$$c_1 = \max z_1(X), \quad c_2 = \max z_2(X) \text{ for } X \subset M_1.$$

Consider the planes $z = c_1$ and $z = c_2$. By definition of the number c_1

there are no points of M_1 above the plane $z = c_1$. But since M_1 is a closed set it must have a closed subset $\overline{M}_1$ in the plane $z = c_1$. Similarly, there are no points in M_2 above the plane $z = c_2$, but there are points *in* this plane; denote the set of all such points by $\overline{M}_2$. Let $X \in \overline{M}_1$. Then $z_1(X) = c_1$ and $z_2(X) \leqq c_2$. Therefore $m = \max(z_1(X) - z_2(X)) \geqq c_1 - c_2$. On the other hand, let X be a point in M_1 which the isometry onto F_2 maps onto a point of $\overline{M}_2$. Then $z_1(X) \leqq c_1$ and $z_2(X) = c_2$, and so $m \leqq c_1 - c_2$. Hence $m = c_1 - c_2$. Since $Z_1(X) = c_1$ for all points X of $\overline{M}_1$, and $m = c_1 - c_2$, it follows that the set $\overline{M}_1$ on the surface F_1 is mapped by isometry onto the set $\overline{M}_2$ on the surface F_2.

The set $\overline{M}_1$ lies on the curve γ_1 which is the intersection of the plane $z = c_1$ and the surface F_1. γ_1 is strictly convex since F_1 is strictly convex. Let g_1 be the tangent to γ_1 at a point X_1 of $\overline{M}_1$. The plane α parallel to the z-axis and passing through the straight line g_1 is clearly a supporting plane of M_1, and X_1 is the only point of M_1 in this plane (Figure 20).

Since X_1 is in M_1, we have $d(z_1 - z_2)/ds = 0$ at X_1, where d/ds denotes differentiation with respect to arc length on any geodesic on F_1 issuing from X_1. But $dz_1/ds = \cos\vartheta_1$ and $dz_2/ds = \cos\vartheta_2$, where ϑ_1 and ϑ_2 are the angles formed with the positive z-axis by the directions at X_1 and X_2 that correspond under the isometry of F_1 and F_2.

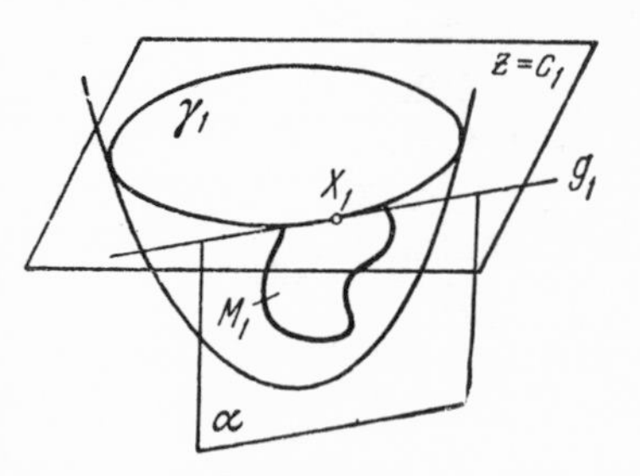

FIGURE 19

The plane $z = c_2$ cuts the surface F_2 in a curve γ_2 which contains the point X_2 corresponding to X_1 under the isometry. The plane parallel to the z-axis through the tangent to γ_2 at X_2 is a supporting plane for the set M_2, and again X_2 is the only point of M_2 in this plane.

Since the directions on F_1 and F_2 at X_1 and X_2, respectively, that correspond under the isometry for the same angles with the positive z-axis, the points X_1 and X_2, together with the corresponding directions at X_1 and X_2, can be made to coincide by subjecting F_1 and F_2 to a translation plus a rotation about the z-axis. Both surfaces will then be on the same side of their common tangent plane.

The next stage of the transformation is to translate both surfaces F_1 and F_2 (together) until their common point $X_1 = X_2$ coincides with the origin O, and then to rotate them about the z-axis until the straight line g_1 coincides with the y-axis and the set M_1 can be projected into the halfplane $x \leqq 0$. In this situation the set M_2 is also projected into the halfplane $x \leqq 0$. Now since the halfplane $x \geqq 0$ contains no projections of points of M_1 and M_2 other than $X_1 \equiv X_2 \equiv O$, the surfaces F_1 and F_2 satisfy the requirements of the lemma. This proves Lemma 2.

PROOF OF THE THEOREM. Let F_1 and F_2 be isometric convex caps with regular metric and positive Gauss curvature. If the theorem is false, F_1 and F_2 are not congruent. Then by Lemma 2 they can be brought into a position such that the following conditions hold:

1. Two points of F_1 and F_2 that correspond under the isometry coincide at the origin O.

2. The directions of F_1 and F_2 that correspond under the isometry also coincide at O.

3. The surfaces are on the same side of their common tangent plane at O; this tangent plane is not perpendicular to the xy-plane.

4. If $z_1(X)$ is the z-coordinate of an arbitrary point of F_1 and $z_2(X)$ the z-coordinate of the corresponding point on F_2, then $z_1(X) - z_2(X) \leqq 0$, and the inequality is strict if the projection of the end of the vector $r_1(X) + r_2(X)$ onto the xy-plane is in the halfplane $x \geqq 0$ and is not O.

Let $r_1(X)$ be the vector of an arbitrary point X on F_1, and $r_2(X)$ the vector of the corresponding point on the surface F_2^* obtained by reflecting F_2 in the xy-plane. By Lemma 2 of §5, in the neighborhood of the point O the vector-valued function $r(X) = r_1(X) + r_2^*(X)$ defines a smooth surface Φ which has the following properties:

1. There are no points on Φ at which the lower curvature is positive and the upper curvature infinite.

2. If F_1 is twice differentiable at X and F_2 twice differentiable at the point corresponding to X under the isometry, then Φ cannot have positive lower curvature at $r(X)$.

In the neighborhood of the origin O, the surface Φ has a unique projection on the xy-plane, and it can therefore be defined by a smooth function $z = \varphi(x, y)$. By Property 4 of the mutual position of the caps F_1 and F_2, the function $\varphi(x, y)$ has the property that $\varphi(x, y) \leqq 0$ for $x \geqq 0$, and equality holds only at the origin, $x = y = 0$.

Call a point X on the surface F_1 *singular* if F_1 is not twice differentiable there or if F_2 is not twice differentiable at the point corresponding to X under the isometry. The set M of singular points of F_1 has measure zero on the surface. Denote the corresponding set on Φ by $r(M)$. We claim that this set also has measure zero on Φ. Indeed, introduce a common parametrization u, v of the surfaces F_1 and F_2^*. Then the vector-valued functions $r_1(u,v)$ and $r_2^*(u,v)$ defining F_1 and F_2^* are smooth functions (Lemma 1 of §5), and moreover $r_{1u} \times r_{1v} \neq 0$ near the point O, which is the region of interest for us. It follows that the preimage M' of the set M on the u,v-plane also has measure zero. Since the vector-function $r(u,v)$ is smooth, the set $r(M)$ has measure zero on Φ.

Let $\overline{M}$ denote the projection of $r(M)$ on the xy-plane. The surface Φ has the following properties in relation to the set $\overline{M}$. If the projection of a point P of Φ on the xy-plane lies in $\overline{M}$, then Φ cannot have both positive lower curvature and infinite upper curvature at P. If the projection of P on the xy-plane is not in $\overline{M}$, the lower curvature cannot be positive at P.

Consider a region on Φ whose projection on the xy-plane is a small semidisk $\kappa: x^2+y^2 \leqq \epsilon$, $x \geqq 0$. If x, y are coordinates of a point on Φ whose projection is on the semicircle $x^2+y^2=\epsilon$, $x \geqq 0$, then $z(x,y) < -\epsilon' < 0$, where ϵ' is a small positive number. Consider the half-plane $z = -\epsilon' x/2\epsilon$, $x \geqq 0$. Since the surface Φ touches the xy-plane at O, this halfplane cuts off a small "hump" $\overline{\Phi}$ from Φ; the projection of this "hump" is the semidisk κ.

By Theorem 1 of §4 there exists a convex cap Ω with boundary in the xy-plane, the boundary enclosing the semidisk κ, the cap convex toward $z>0$ and possessing the following properties:

If the projection of a smooth point X of Ω onto the xy-plane is not in the set $\overline{M}$, then the lower curvature of the cap is positive at X; if the projection of X *is* in $\overline{M}$, then the lower curvature is positive there, but the upper curvature is infinite. Let the equation of the cap be $z=\omega(x,y)$. Consider the surface defined by the equation

$$z = \lambda\omega(x,y) - \epsilon' x/2\epsilon.$$

This surface (call it $\overline{\Omega}$) has the same properties as Ω (Theorem 2 of §4). Its boundary lies in the plane $z = -\epsilon' x/2\epsilon$ and encloses the boundary of the surface $\overline{\Phi}$. For sufficiently large λ, the surface $\overline{\Omega}$ lies above the surface $\overline{\Phi}$.

By reducing λ, we subject $\overline{\Omega}$ to an affine transformation compressing it toward $\overline{\Phi}$; for some value of λ, $\overline{\Omega}$ will be tangent to $\overline{\Phi}$ at a point A. Since $\overline{\Phi}$ is tangent to $\overline{\Omega}$ and is in the interior of the latter, the lower curvature of $\overline{\Phi}$ at A is at least that of $\overline{\Omega}$ there, and is therefore positive. If we now project A onto the xy-plane, onto a point of the set $\overline{M}$, the upper curvature of $\overline{\Phi}$ (which is at least that of $\overline{\Omega}$) is infinite. This is a contradiction, since Lemma 2 of §5 states that the lower and upper curvatures cannot stand in this relation. This proves the theorem.

§7. Intrinsic estimates for some geometric quantities along the boundary of an analytic cap

The main theorem of this chapter states that a convex surface with regular metric is regular. The idea of the proof is roughly as follows. Let F be a convex surface with regular metric and positive Gauss curvature. F is smooth and strictly convex (Theorems 1 and 2 of §3). Let P be an arbitrary point on F. Cut off a small cap ω from the surface F by a plane parallel to the tangent plane at P. It can be shown that there exists a regular cap $\bar{\omega}$ isometric to ω. By the theorem of §6, ω and $\bar{\omega}$ are congruent. It follows that F is regular in the neighborhood of an arbitrary point P, which proves the theorem.

To prove that a regular cap exists, one uses a theorem of Bernšteĭn [21] on the solvability of the Darboux equation; the problem of constructing a surface realizing the metric in question can be reduced to consideration of the Darboux equation. This theorem states that the equation is solvable provided one has certain a priori estimates for the postulated solution and its derivatives up to order 2. It is easy to establish such estimates if intrinsic estimates (i.e. estimates depending only on the intrinsic metric) are available for the normal curvatures of the cap and the inclination angle of its tangent planes to the base plane. This section and the next are devoted to derivation of these estimates.

THEOREM. *Let ω be an analytic convex cap with positive Gauss curvature, whose boundary has positive geodesic curvature. Then there exists an estimate, depending only on the intrinsic metric of the cap (i.e. its line element), for the normal curvatures along the boundary of the cap and also for the tangent of the inclination angle of the tangent planes to the base plane.*

PROOF. Let ω be a convex cap with analytic metric and positive Gauss curvature. We may assume that its base lies in the xy-plane and the z-axis is so oriented that the cap is convex toward $z < 0$. Introduce an analytic parametrization u, v on the surface. Let $r(u,v)$ be the vector-valued function defining the surface ω, and $z(u,v)$ its z-component. By the Gauss equations of surface theory,

$$\begin{aligned} z_{uu} &= \Gamma_{11}^1 z_u + \Gamma_{11}^2 z_v + L\zeta, \\ z_{uv} &= \Gamma_{12}^1 z_u + \Gamma_{12}^2 z_v + M\zeta, \\ z_{vv} &= \Gamma_{22}^1 z_u + \Gamma_{22}^2 z_v + N\zeta, \end{aligned}$$

where L, M, N are the coefficients of the second fundamental form of the surface, ζ the z-component of the unit vector normal to the surface, and Γ_{ij}^k the Christoffel symbols of the second kind.

The component ζ is easily expressed in terms of the coefficients of the line element $ds^2 = E\,du^2 + 2F\,du\,dv + G\,dv^2$ and the first derivatives z_u, z_v. Indeed, if $d\sigma$ is the area element of the surface and $d\bar{\sigma}$ its projection on the xy-plane, then $\zeta = d\bar{\sigma}/d\sigma$. Since the line element of the xy-plane may be expressed as $d\bar{s}^2 = ds^2 - dz^2$, it follows that ζ^2 is the quotient of the discriminants of the forms $d\bar{s}^2$ and ds^2:

$$\zeta^2 = \frac{(E - z_u^2)(G - z_v^2) - (F - z_u z_v)^2}{EG - F^2}.$$

Compute the expression $LN - M^2$ from the Gauss equations, and recall that $(LN - M^2)/(EG - F^2)$ is the Gauss curvature K; we get

$$(z_{uu} - \Gamma_{11}^1 z_u - \Gamma_{11}^2 z_v)(z_{vv} - \Gamma_{22}^1 z_u - \Gamma_{22}^2 z_v) \\ - (z_{uv} - \Gamma_{12}^1 z_u - \Gamma_{12}^2 z_v)^2 - K\frac{(E - z_u^2)(G - z_v^2) - (F - z_u z_v)^2}{EG - F^2} = 0.$$

This is precisely the Darboux equation. Denoting its left-hand side by Φ, we get

$$\frac{\partial \Phi}{\partial z_{uu}} \frac{\partial \Phi}{\partial z_{vv}} - \frac{1}{4}\left(\frac{\partial \Phi}{\partial z_{uv}}\right)^2 = K\zeta^2.$$

Thus the Darboux equation is elliptic at each point of a surface with positive Gauss curvature where the tangent plane is not perpendicular to the xy-plane.

If the parametrization of the surface is semigeodesic (line element $ds^2 = du^2 + c^2 dv^2$), the Darboux equation becomes

$$r(t + \alpha) - (s - \beta)^2 + \gamma = 0, \tag{1}$$

where

$$\alpha = cc_u p - \frac{c_v}{c} q, \qquad \beta = \frac{c_u}{c} q, \qquad \gamma = \frac{c_{uu}}{c}(c^2 - c^2 p^2 - q^2),$$

and p, q, r, s, t denote the first and second derivatives of the function $z(u, v)$.

Let P be an arbitrary point on the surface and τ an arbitrary direction at P. Let γ be a geodesic issuing from P in the direction τ and introduce semigeodesic coordinates such that the geodesic γ is a u-curve. Then $\Gamma_{11}^1 = 0$, $\Gamma_{11}^2 = 0$, and the first Gauss equation gives $z''_{ss} = k_\gamma \cos\vartheta$, where z is differentiated with respect to arc length along the geodesic γ, k_γ is the normal curvature of the surface in the direction of the geodesic, and ϑ is the angle between the tangent plane to the surface and the xy-

plane. It is clear from this formula that in order to derive intrinsic estimates for the normal curvatures of the surface we need only estimate the first and second derivatives of $z(u, v)$. To this end it is convenient to introduce a semigeodesic parametrization on the cap ω, taking the boundary γ of the cap as the curve $u = 0$, the geodesics perpendicular to γ as the u-curves, and the orthogonal trajectories of these geodesics as the v-curves. The parameters u and v will be arc length along γ and arc length along the u-curves. With this parametrization of the surface, the Darboux equation has the form (1). Let us estimate the first and second derivatives of $z(u, v)$ along the boundary of the cap.

First note that there is a number u_0, which depends only on the metric of the cap and on the minimum geodesic curvature κ_0 of the boundary curve γ, such that for all points of the cap ω satisfying $u \leqq u_0$ we have $|p| < 1$ and $|q| < 2$.

Indeed, $|p| = \sin\vartheta$, where ϑ is the angle between the tangent to the u-curve and the xy-plane. This angle is always less than $\pi/2$, since all tangent planes to the cap ω form angles less than $\pi/2$ with the xy-plane. Next, $|q| = c^{-1}\sin\varphi$, where φ is the angle between the tangent to the v-curve and the xy-plane, and c defines the line element $ds^2 = du^2 + c^2 dv^2$. Since $c = 1$ on the curve γ, there surely exists a number u_0 such that $c > \frac{1}{2}$ whenever $u \leqq u_0$. And then, obviously, $|q| < 2$.

Introduce rectangular cartesian coordinates u, v in the plane, and let G denote the infinite strip between the axis $u = 0$ and the line $u = u_0$. We define a function $\zeta(u, v)$ in the strip G as follows:

1) $\zeta(u, v)$ is periodic in v with period l, where l is the length of the curve γ.

2) $\zeta(u, v) = q$ for $0 \leqq u \leqq u_0$, $0 \leqq v < l$; $\zeta = 0$ on the u-axis, and $|\zeta| < 2$ on the line $u = u_0$.

Consider the surface Φ defined by the equation $\zeta = \zeta(u, v)$ in the strip $0 \leqq u \leqq u_0$. Let σ be the plane through the line $u = 0$, forming the smallest possible angle with the u, v-plane such that the surface Φ lies below σ. There are two possibilities: the plane σ may either support the boundary of Φ, whose projection is the straight line $u = u_0$, or touch the surface Φ at some point. In the first case $\zeta_u < 2/u_0$ for $u = 0$, and so $s < 2/u_0$ along the boundary of ω.

Consider the second case. Differentiating equation (1) with respect to v, we get

$$q_{uu}(t + \alpha) - 2q_{uv}(s - \beta) + q_{vv}r + r\alpha_v + 2(s - \beta)\beta_v + \gamma_v = 0. \tag{2}$$

At the point of contact of σ with the surface Φ, we have $q_v = t = 0$. Hence $\alpha > 0$ at this point, since the discriminant

$$\Delta = (t + \alpha)r - (s - \beta)^2$$

of equation (1) is positive. We can therefore solve equation (1) for r at the above point:

$$r = \frac{(s-\beta)^2 - \gamma}{\alpha}.$$

Turning now to equation (2), we see that at the point where σ touches Φ the sum of the first three terms is nonpositive. For the form

$$\xi^2(t + \alpha) + 2\xi\eta(s - \beta) + \eta^2 r$$

is positive definite, since $r > 0$ and $\Delta > 0$. Furthermore, the form

$$\xi^2 q_{uu} + 2\xi\eta q_{uv} + \eta^2 q_{vv}$$

is nonpositive, since the surface Φ lies below the plane σ. Hence

$$q_{uu}(t + \alpha) - 2q_{uv}(s - \beta) + q_{vv} r \leqq 0.$$

The sum of the remaining terms in (2) may be written as

$$\frac{s^3 c c_u + O_2(s)}{\alpha},$$

where $O_2(s)$ is a quadratic polynomial in s with coefficients whose absolute values can easily be bounded above in terms of the supremum of the absolute values of the function $c(u, v)$ and its derivatives of order $\leqq 3$.

Since $\alpha > 0$, $c > \frac{1}{2}$ and $c_u < -\kappa_0$, it follows that, when s is greater than a suitable number s_0 which depends only on the intrinsic metric of the surface and the geodesic curvature κ of γ, the sum of the remaining terms of (2) is negative. The number s_0 is precisely the limit which the derivative s at the point of contact cannot exceed. But in any case the derivative $s = q_u$ along $u = 0$ is not greater than its value at the point of contact. Therefore s_0 is the upper limit of the derivative s along the curve γ. Obviously a lower bound for s is $-s_0$.

It is easy to find bounds for the derivative r on the boundary of the cap, provided the lower limit of $|p|$ is known along the boundary. This is easily derived by geometric arguments.

The curve γ bounds a certain domain $\bar{\omega}$ in the xy-plane. Since the geodesic curvature of γ on the surface of the cap is $\geqq \kappa_0$ everywhere, the ordinary curvature is also $\geqq \kappa_0$.[2)]

[2)] The ordinary curvature of a curve is related to its geodesic curvature by the equality $k \cos\vartheta = k_g$, where ϑ is the angle between the osculating plane of the curve (here the base plane of the cap) and the tangent plane to the surface.

We shall now show that the domain $\bar{\omega}$ contains a disk $\tilde{\omega}$ whose radius depends only on κ_0 and the length of the curve γ. Let A and B be two points on a diameter of $\bar{\omega}$ near its endpoints (Figure 21). Now construct a lune consisting of two circular arcs of radius $1/\kappa_0$, with endpoints A and B. This lune is in the interior of $\bar{\omega}$. Indeed, if the radius of the circular arcs is sufficiently large, the lune clearly lies within $\bar{\omega}$. Now reduce the radius continuously. The lune then becomes larger, but remains within $\bar{\omega}$ as long as the radius of the circular arcs is greater than $1/\kappa_0$; this is because no arc of radius greater than $1/\kappa_0$ can touch the curve γ, whose curvature is at least κ_0. Obviously within the lune one can construct a disk $\tilde{\omega}$ of radius at least ρ_0, where ρ_0 depends only on κ_0 and the diameter of $\bar{\omega}$. Now since the diameter of $\bar{\omega}$ is at least l/π, where l is the length of γ, it is indeed true that ρ_0 depends only on κ_0 and the length l of γ.

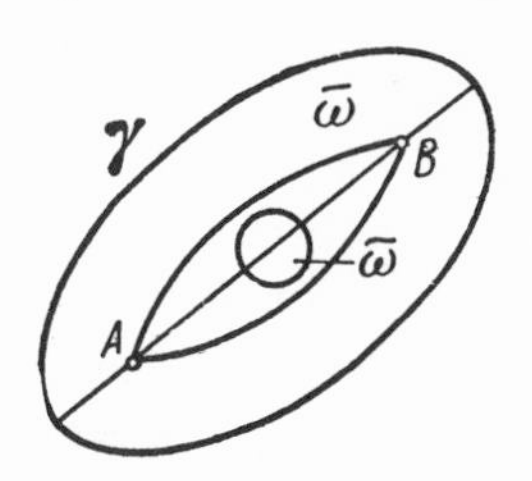

FIGURE 21

Now let g be a ray issuing from the center of the disk $\tilde{\omega}$, perpendicular to the base plane of ω, in the direction opposite to that of the convexity. Construct a sphere Ω at a sufficiently large distance from the cap, with center on g and curvature less than the minimum curvature of the cap ω. Delete the disk $\tilde{\omega}$ from the domain $\bar{\omega}$. The remainder of $\bar{\omega}$ plus the cap ω form a surface Φ whose boundary is the circumference of the disk $\tilde{\omega}$. Now displace the sphere Ω along g toward the surface Φ. At some stage of the procedure Ω will touch either the boundary of Φ, i.e. the circumference of $\tilde{\omega}$, or the cap ω. The second case is impossible, since the curvature of Ω is smaller than the minimum curvature of the cap. Let h be the distance at which the sphere reaches the position in which it touches the circumference of $\tilde{\omega}$. Construct a cone whose base is the base of ω and whose apex is a point lying within the cap, at distance h from the base plane, whose projection is the center of $\tilde{\omega}$. This cone is in the interior of the cap. Therefore its tangent planes along γ form with the base plane angles greater than those formed by the tangent planes of the cap. We thus arrive at a suitable lower bound for $|p|$: the number $\delta = h/\sqrt{h^2 + (l/2)^2}$, where l is the length of γ.

We now estimate the derivative r. We have $t = 0$, $q = 0$ and $\delta < |p| < 1$ along the curve γ. It follows from (1) that along γ we have

$$r = \frac{(s-\beta)^2 - \gamma}{cc_u p}.$$

Since $c_u < -\kappa_0$, $|p| > \delta$ and we have an intrinsic estimate for $|s|$, we can now derive an estimate for $|r|$.

We shall now derive an estimate for the angle between the tangent planes to the cap and the plane of its boundary for sufficiently small analytic caps; this will be done with an eye to future applications. Let F be a convex surface with analytic metric and positive Gauss curvature, P an arbitrary fixed point on F and G a small convex domain on the surface, bounded by an intrinsically analytic curve with everywhere positive geodesic curvature. Let ω be an analytic convex cap isometric to G on F. We assert that if G is sufficiently small, the angle of inclination of the tangent planes to ω along its boundary to the cap's base plane is bounded above by a number smaller than $\pi/2$ which depends solely on the line element of F and the geodesic curvature of the cap boundary.

Let Q be any point of the boundary of the cap ω, Q' the point on F corresponding to Q under the isometry, and γ a geodesic on F tangent to the domain G at Q'. We introduce semigeodesic coordinates u, v on F such that γ is a u-curve (corresponding, say, to $v = 0$). Consider the Darboux equation for the line element $ds^2 = du^2 + c^2 dv^2$ of F (equation (1)). By the Cauchy-Kowalewski Theorem there exists a solution $\bar{z}(u, v)$ of this equation such that $\bar{z}(u, 0) = u^2$ and $q(u, 0) = 0$. If the domain is sufficiently small, the existence domain of this solution obviously covers G (as the domain of variation of the parameters u, v) irrespective of the choice of the point Q. It is well known that this solution may be supplemented by two other analytic functions $\bar{x}(u, v)$ and $\bar{y}(u, v)$ such that the equations

$$x = \bar{x}(u, v), \quad y = \bar{y}(u, v), \quad z = \bar{z}(u, v)$$

define an analytic surface $\bar{F}$ isometric to F. If the domain G is sufficiently small, the derivatives of $\bar{x}$, $\bar{y}$ and $\bar{z}$ with respect to u and v (up to any fixed order) are uniformly bounded in the image of G on $\bar{F}$ (under the isometry). Since the directions of the coordinate curves u and v along $v = 0$ are conjugate (the coefficient M of the second fundamental form of $\bar{F}$ is proportional to $s - c_v q / c = 0$) and the coordinate net is orthogonal, it follows that the geodesic $v = 0$ is a line of curvature and hence a plane curve.

Let $\bar{Q}$ and $\bar{\gamma}$ be the point and geodesic, respectively, on $\bar{F}$ corresponding under the isometry to the point Q' and geodesic γ on F. The plane $\bar{\alpha}$ containing $\bar{\gamma}$ is a normal plane of the surface $\bar{F}$ at each point of the geodesic. Since the geodesic curvature of the boundary of the domain $\bar{\omega}$ corresponding under the isometry to the domain G on F is strictly positive, exceeding some $\epsilon > 0$, a rotation of the plane $\bar{\alpha}$ through a small angle $\vartheta(\epsilon)$ about the tangent to $\bar{\gamma}$ at $\bar{Q}$ will bring $\bar{\omega}$ to a position in which it again lies to one side of this plane and has a unique projection on it. We may assume that if G is sufficiently small, then $\vartheta(\epsilon)$ is independent of the form of G and the choice of Q' on its boundary, depending only on the lower limit of the geodesic curvature of the boundary of G.

Now place the surface $\bar{\omega}$ in such a position that it is convex toward $z < 0$, the plane $\bar{\alpha}$ rotated through the angle $\vartheta(\epsilon)$ as above coincides with the xy-plane, and the point $\bar{Q}$ coincides with the origin O and the tangent to $\bar{\gamma}$ at $\bar{Q}$ is the y-axis. Place the cap ω in a similar position: convex toward $z < 0$, base plane in the xy-plane, and tangent to the base plane along the y-axis. Let $\bar{z}$ and z be the coordinates along the axis Oz of points of $\bar{\omega}$ and ω (in these positions) that correspond under the isometry. The function $z - \bar{z}$ has a minimum on the boundary of the surface (see the proof of the theorem in §6). Consequently this minimum is attained at the origin, where $z - \bar{z} = 0$, since $z - \bar{z} \geqq 0$ at other boundary points. Differentiating in the direction perpendicular to the boundary of the surface at O, we get $z' - \bar{z}' \geqq 0$, or $-\sin\beta + \sin(\pi/2 - \vartheta(\epsilon)) \geqq 0$. Hence we have the estimate $\beta \leqq \pi/2 - \vartheta(\epsilon)$ for the angle β between the tangent plane of the cap ω at O and its base plane (the xy-plane). This proves the theorem.

§8. Estimate for normal curvatures at interior points of a regular convex cap

THEOREM 1. *Let the normal curvature of a convex surface with regular metric and positive Gauss curvature have a maximum at an interior point of the surface. Then there is an estimate for this maximum which depends only on the intrinsic metric of the surface. To be precise, if K_0 is the maximum and $\bar{K}_0$ the minimum Gauss curvature of the surface and λ the maximum of the second derivative of the Gauss curvature along any geodesic on the surface, then the normal curvature cannot exceed*

$$\sqrt{K_0 + \frac{2\lambda}{\bar{K}_0}}.$$

PROOF. Let X_0 be a point of maximum upper curvature; there are two possibilities:

1) X_0 is an umbilical point, i.e. the curvatures of the normal sections in all directions at X_0 are $\sqrt{K(X_0)}$, where $K(X_0)$ is the Gauss curvature at X_0;

2) X_0 is not an umbilical point.

In the first case it is obvious that the maximum upper curvature is bounded by $\sqrt{K_0}$, where K_0 is the maximum Gauss curvature of the surface.

In the second case, since X_0 is not an umbilical point, in a sufficiently small neighborhood of X_0 one can introduce a coordinate net consisting of lines of curvature. As the parameters u and v we take arc length along the coordinate curves through the point X_0, $v = 0$ being the direction in which maximum normal curvature is attained.

By our choice of the parameters u and v, we have $ds^2 = du^2$ along the curve $v = 0$ and $ds^2 = dv^2$ along $u = 0$. Hence the coefficients of the fundamental form

$$ds^2 = E\,du^2 + 2F\,du\,dv + G\,dv^2$$

at the point X_0 satisfy the relations

$$E = G = 1, \quad F = 0, \quad E_u = G_v = 0.$$

The mean and Gauss curvatures of the surface are given by

$$H = \frac{EN + GL}{2EG}, \quad K = \frac{LN}{EG} = -\frac{1}{\sqrt{EG}}\left\{\frac{\partial}{\partial v}\frac{(\sqrt{E})_v}{\sqrt{G}} + \frac{\partial}{\partial u}\frac{(\sqrt{G})_u}{\sqrt{E}}\right\},$$

where L and N are the first and third coefficients of the second fundamental form of the surface.

The Codazzi equations are also simple in form:

$$L_v = \frac{1}{2}E_v\left(\frac{L}{E} + \frac{N}{G}\right), \qquad N_u = \frac{1}{2}G_u\left(\frac{L}{E} + \frac{N}{G}\right).$$

Let κ and $\bar{\kappa}$ be the maximum and minimum normal curvatures, respectively, of the surface at an arbitrary point near X_0. Then we easily see from the formulas for the mean and Gauss curvatures that

$$\varkappa = \frac{L}{E}, \qquad \bar{\varkappa} = \frac{K}{\varkappa} = \frac{N}{G}.$$

The Codazzi equations now give

$$\varkappa_v = \frac{1}{2}\frac{E_v}{E}(-\varkappa + \bar{\varkappa}), \qquad \bar{\varkappa}_u = \frac{1}{2}\frac{G_u}{G}(\varkappa - \bar{\varkappa}).$$

Solving these equations for E_v and G_u, we get

$$E_v = \frac{2E\varkappa_v}{-\varkappa + \dfrac{K}{\varkappa}}, \qquad G_u = \frac{2G\left(\dfrac{K}{\varkappa}\right)_u}{\varkappa - \dfrac{K}{\varkappa}}. \tag{1}$$

We now derive an expression for the curvature K at X_0. First,

$$\frac{\partial}{\partial v}\left(\frac{(\sqrt{E})_v}{\sqrt{G}}\right)_{(X_0)} = \frac{1}{2}E_{vv} - \frac{1}{4}E_v^2,$$

$$\frac{\partial}{\partial u}\left(\frac{(\sqrt{G})_u}{\sqrt{E}}\right)_{(X_0)} = \frac{1}{2}G_{uu} - \frac{1}{4}G_u^2.$$

It follows from (1) that

$$E_{vv} = E_v^2 + \frac{\partial}{\partial v}\left(\frac{2\varkappa_v}{-\varkappa + \frac{K}{\varkappa}}\right),$$

$$G_{uu} = G_u^2 + \frac{\partial}{\partial u}\left(\frac{2\left(\frac{K}{\varkappa}\right)_u}{\varkappa - \frac{K}{\varkappa}}\right). \tag{2}$$

Since κ is maximum at X_0, we have $\kappa_u = \kappa_v = 0$ there, and equations (2) yield

$$E_{vv} = E_v^2 + \frac{2\varkappa_{vv}}{-\varkappa + \frac{K}{\varkappa}}, \qquad G_{uu} = G_u^2 + \frac{2K\frac{\varkappa_{uu}}{\varkappa^2}}{-\varkappa + \frac{K}{\varkappa}} + 2\,\frac{\frac{K_{uu}}{\varkappa}}{\varkappa - \frac{K}{\varkappa}}.$$

Substituting these expressions for E_{vv} and G_{uu} in the formula for the Gauss curvature K at X_0, we get

$$K = -\left(\frac{1}{4}E_v^2 + \frac{1}{4}G_u^2 + \frac{2\varkappa_{vv}}{-\varkappa + \frac{K}{\varkappa}} + \frac{2K\varkappa_{uu}}{\varkappa^2\left(-\varkappa + \frac{K}{\varkappa}\right)}\right) - \frac{2K_{uu}}{\varkappa^2 - K}.$$

Since the maximum of κ is attained at X_0, which is not an umbilical point, we have $\kappa > K/\kappa$, $\kappa_{uu} \leqq 0$ and $\kappa_{vv} \leqq 0$ at X_0. Hence the expression in parentheses is nonnegative, and so

$$K \leqslant -\frac{2K_{uu}}{\varkappa^2 - K}. \tag{3}$$

The curve $v = 0$ has geodesic curvature zero at X_0. Indeed, the geodesic curvature is

$$k_g = \frac{1}{W}\frac{\Gamma}{(Eu'^2 + 2Fu'v' + Gv'^2)^{3/2}},$$

where

$$W = \sqrt{EG - F^2},$$

$$\begin{aligned}\Gamma = {} & W^2(u'v'' - v'u'') \\ & + (Eu' + Fv')\left[\left(F_u - \frac{1}{2}E_v\right)u'^2 + G_u u'v' + \frac{1}{2}G_v v'^2\right] \\ & - (Fu' + Gv')\left[\frac{1}{2}E_u u'^2 + E_v u'v' + \left(F_v - \frac{1}{2}G_u\right)v'^2\right]\end{aligned}$$

This expression for the geodesic curvature clearly shows that the geodesic curvature of the line $v = 0$ vanishes at X_0. Therefore $K_{uu}(X_0) = K_{ss}(X_0)$.

Now let λ denote the maximum of the second derivative of the Gauss curvature with respect to arc length along the geodesic, taken over all geodesics on the surface. Inequality (3) then implies that for X_0 we have $K \leqq 2\lambda/(\kappa^2 - K)$.

Hence

$$\varkappa \leqslant \sqrt{K + \frac{2\lambda}{K}}.$$

Let K_0 and $\overline{K}_0$ denote the maximum and minimum Gauss curvature, respectively, of the surface; then the upper curvature of the surface satisfies the estimate

$$\varkappa \leqslant \sqrt{K_0 + \frac{2\lambda}{\overline{K}_0}}.$$

This completes the proof.

If the surface is closed, its maximum Gauss curvature is evidently attained at some interior point, and therefore cannot be greater than

$$\sqrt{K_0 + \frac{2\lambda}{\overline{K}_0}},$$

where K_0, $\overline{K}_0$ and λ have the same meaning as before.

One corollary of Theorem 1 and the theorems of §7 is the following theorem on the existence of intrinsic estimates over the entire surface of an analytic convex cap with positive Gauss curvature.

THEOREM 2. *Let F be an analytic cap with positive Gauss curvature whose boundary has positive geodesic curvature. Then the normal curvatures of the cap and the tangent of the inclination angle of its tangent planes to the base plane satisfy estimates that depend only on the intrinsic metric of the cap.*

The estimate given by Theorem 2 for the normal curvatures of the cap utilizes the positivity of the geodesic curvature of the boundary in an essential way. Without assuming that the geodesic curvature of the boundary is strictly positive one cannot guarantee the existence of estimates for the normal curvatures over the whole cap. Nevertheless, the following theorem is valid.

THEOREM 3. *Let ω be a regular convex cap with positive Gauss curvature. Let P be an arbitrary point on the cap whose distance from the plane of the boundary is at least $\delta > 0$. Then there are estimates for the normal* *curvatures of the cap at P which depend only on the intrinsic metric of the cap and the number δ.*

PROOF. Let X be an arbitrary point of the cap and γ an arbitrary geodesic issuing from X. Let $\kappa_\gamma(X)$ denote the normal curvature of the cap at X in the direction of γ, and consider the function $w_\gamma(X) = h(X)\kappa_\gamma(X)$, where $h(X)$ is the distance of X from the base plane of the cap. The function $w_\gamma(X)$ is nonnegative and vanishes on the boundary of the cap. It has a maximum at some interior point X_0 of the cap, for a geodesic γ_0 through X_0. Let w_0 denote this maximum.

Let $\bar{\gamma}_0$ be a geodesic through X_0 perpendicular to γ_0. Introduce semi-geodesic coordinates u, v in a neighborhood of X_0, taking the geodesics perpendicular to $\bar{\gamma}_0$ as the u-curves. The parameters u, v will be arc length along the geodesics γ_0 and $\bar{\gamma}_0$, with X_0 as origin. Let $\gamma(X)$ denote the geodesic u-curve through a point X close to X_0, and consider the function $w(X) = w_{\gamma(X)}(X)$. This function clearly also attains a maximum at X_0, and the maximum is precisely w_0.

Set up the Darboux equation for the cap, with the tangent plane to the cap at X_0 as the plane $z = 0$. To fix ideas, let us assume that the cap lies in the halfspace $z > 0$. Since the line element of the cap in the coordinates u, v is $ds^2 = du^2 + c^2dv^2$, the Darboux equation for the function $z(u,v)$ is (§7)

$$r(t+\alpha) - (s-\beta)^2 + \gamma = 0.$$

where

$$\alpha = cc_u p - \frac{c_v}{c} q, \qquad \beta = \frac{c_u}{c} q,$$

$$\gamma = \frac{c_{uu}}{c}(c^2 - c^2p^2 - q^2).$$

As was shown in §7, the normal curvature along a geodesic γ on the surface is given by the formula

$$\varkappa_\gamma = \frac{z''_\gamma(X)}{\cos\varphi},$$

where z is differentiated with respect to arc length along the geodesic γ and φ is the angle between the tangent plane to the cap and its base plane. Since

$$\cos^2\varphi = 1 - p^2 - q^2/c^2,$$

we see that the function $w(X)$ can be expressed as

$$w = \frac{hr}{\sqrt{1 - p^2 - \frac{q^2}{c^2}}}.$$

Solving this equation for r, we get

$$r=\frac{w}{h}\sqrt{1-p^2-\frac{q^2}{c^2}}.$$

Using this expression, let us compute the first and second derivatives of r with respect to u and v at X_0, recalling throughout that by our choice of the coordinate system u, v we have $c=1$, $c_u=c_v=c_{uv}=c_{vv}=0$, $c_{uu}=-K$ at X_0, where K is the Gauss curvature of the surface, the derivatives p and q vanish at X_0 (since the plane $z=0$ touches the surface there), the derivative s vanishes at X_0 (since $M=0$: the directions of the coordinate curves at X_0 are principal directions), and finally $w_u=w_v=0$ at X_0 (since w has a maximum there).

The final results are

$$r_u=\left(\frac{1}{h}\right)_u w, \qquad r_v=\left(\frac{1}{h}\right)_v w,$$

$$r_{uu}=\frac{w_{uu}}{h}+w\left(\frac{1}{h}\right)_{uu}-\frac{w}{h}r^2, \qquad r_{vv}=\frac{w_{vv}}{h}+w\left(\frac{1}{h}\right)_{vv}-\frac{w}{h}t^2.$$

The following observations concerning the values of these derivatives at X_0 are essential. Since u and v are arc length along geodesics at X_0 and h is the distance to the base of the cap, it follows that $|h_u|\leqq 1$ and $|h_v|\leqq 1$. Hence $|r_u|<r/\delta$ and $|r_v|<r/\delta$. Using the formula $\kappa_\gamma=h''_\gamma/\cos\varphi$ for the normal curvature, we get

$$r=-\frac{h_{uu}}{\sqrt{1-h_u^2-h_v^2}}, \qquad t=-\frac{h_{vv}}{\sqrt{1-h_u^2-h_v^2}}.$$

Hence $|h_{uu}|\leqq r$ and $|h_{vv}|\leqq t$. Consequently

$$r_{uu}=\frac{w_{uu}}{h}-\frac{r^3}{h}+Ar^2+Br, \qquad r_{vv}=\frac{w_{vv}}{h}-\frac{t^3}{h}+Cr+Drt,$$

where A, B, C, D are quantities having simple estimates in terms of δ.

Differentiating the Darboux equation with respect to u at X_0, we find that $rt_u+tr_u-K_u=0$. Hence $t_u=(K_u-tr_u)/r$. Differentiating again with respect to u, we get

$$r_{uu}t+t_{uu}r+2r_ut_u-2s_u^2-K_{uu}=0.$$

Now substitute the known values of r_u, r_v, $s_u=r_v$, t_{uu} and $t_{uu}=r_{vv}$ into this equation:

$$(*)\qquad \frac{1}{h}(w_{uu}t+w_{vv}r)-\frac{Kr^2}{h}+Pr+Q=0,$$

where P and Q are numbers which satisfy estimates in terms of δ, the

Gauss curvature K and its first and second derivatives. Since the function w has a maximum at X_0, it follows that $w_{uu}t + w_{vv}r \leqq 0$ at the point X_0. Hence, dropping all terms known to be positive on the left-hand side of (*), we get

$$-Kr^2/h + Pr + Q \geqq 0.$$

Now, setting $w = r/h$, we get

$$-Kw^2 + P'w + Q' \geqq 0.$$

This yields a simple estimate for w and hence for the normal curvature κ. The proof is complete.

§9. Existence of an analytic convex cap realizing a given analytic metric

LEMMA 1. *An analytic metric with positive Gauss curvature*

$$ds_\alpha^2 = E_\alpha du^2 + 2F_\alpha du\,dv + G_\alpha dv^2,$$

where the coefficients are analytic functions of the parameter α, is given in an ϵ-neighborhood of the unit disk $\bar{\omega}: u^2 + v^2 \leqq 1$.

Assume that for some value of the parameter $\alpha = \alpha_0$ the metric ds_α^2 is realized by a convex analytic cap ω_{α_0} which has an analytic continuation across its boundary. Then ds_α^2 is realized for all α sufficiently close to α_0 by an analytic convex cap ω_α which has an analytic continuation across its boundary.

PROOF. Assume that there is an analytic cap ω_α realizing the metric ds_α^2. Let ω_α be placed so that its base lies in the xy-plane and the cap itself in the halfspace $z > 0$. The z-coordinate of a point (u, v) of ω_α satisfies a certain elliptic partial differential equation in the disk $\bar{\omega}$, namely the Darboux equation

$$\Phi_\alpha(t, s, t, p, q, u, v) = 0$$

and vanishes on the circumference of $\bar{\omega}$.

When $\alpha = \alpha_0$ the equation $\Phi_\alpha = 0$ clearly has an analytic solution z_{α_0} which is analytically continuable across the boundary of the disk $\bar{\omega}$, since there exists an analytic cap ω_{α_0} realizing the metric $ds_{\alpha_0}^2$. By the well-known lemma of Bernšteĭn ([**21**], Chapter VII), for α close to α_0 there exists a solution $z_\alpha(u, v)$ of the equation $\Phi_\alpha = 0$, which is analytic in u, v, α, is analytically continuable across the circumference of $\bar{\omega}$ and vanishes on the circumference. We claim that if $|\alpha - \alpha_0|$ is sufficiently small every such solution z_α may be associated with a cap

ω_α, analytically continuable across its boundary, which realizes the metric ds_α^2.

Consider the quadratic form $d\sigma_\alpha^2$ defined in the disk $\bar{\omega}$ and in a sufficiently small neighborhood thereof by $d\sigma_\alpha^2 = ds_\alpha^2 - dz_\alpha^2$.

When $\alpha = \alpha_0$ this form is positive definite and is the line element of the base plane of the cap ω_{α_0}, if the coordinates u, v of a point P of the base are defined to be the coordinates of the point of the cap directly above it. Thus the metric $d\sigma_{\alpha_0}^2$ in the disk $\bar{\omega}$ is realized by a plane convex domain: the base of ω_{α_0}. The curvature of the boundary of the domain, which is the same as the geodesic curvature of the circumference of $\bar{\omega}$ with respect to the metric $d\sigma_{\alpha_0}^2$, is positive.

Since the metric $d\sigma_\alpha^2$ is an analytic function of α, the geodesic curvature of the circle $u^2 + v^2 = 1$ with respect to the metric $d\sigma_\alpha^2$ is also positive for small $|\alpha - \alpha_0|$, and the form $d\sigma_\alpha^2$ is positive definite in $\bar{\omega}$. Computing the Gauss curvature in terms of the coefficients of the form $d\sigma^2$, one sees that it is the same as Φ_α (up to a nonzero factor), and therefore vanishes. It follows that for small $|\alpha - \alpha_0|$ the metric $d\sigma_\alpha^2$, viewed in a sufficiently small neighborhood of the disk $\bar{\omega}$, is realized by a domain on the xy-plane, the disk $\bar{\omega}$ corresponding to a convex domain bounded by an analytic contour with positive curvature.

Let $x_\alpha(u, v)$ and $y_\alpha(u, v)$ be the coordinates of the point on the xy-plane corresponding to the point (u, v) of $\bar{\omega}$ under the isometry of $\bar{\omega}$ with metric $d\sigma_\alpha^2$ into the xy-plane. ds_α^2 is the line element of the analytic surface defined by the equations

$$x = x_\alpha(u, v), \quad y = y_\alpha(u, v), \quad z = z_\alpha(u, v),$$

since

$$dx_\alpha^2 + dy_\alpha^2 + dz_\alpha^2 = d\sigma_\alpha^2 + dz_\alpha^2 = ds_\alpha^2,$$

and this surface is the analytic convex cap whose existence was to be proved.

LEMMA 2. *An analytic metric with positive Gauss curvature*

$$ds_\alpha^2 = E_\alpha du^2 + 2F_\alpha du dv + G_\alpha dv^2,$$

where the coefficients are analytic functions of α $(\alpha_0 \leqq \alpha \leqq \alpha_1)$, *is given in an* ϵ*-neighborhood of the disk* $\bar{\omega} : u^2 + v^2 \leqq 1$. *Assume that the circumference of* $\bar{\omega}$ *has positive Gauss curvature with respect to the metric* ds_α^2 *for all* α *in the above interval.*

Assume that for $\alpha = \alpha_0$ *the metric* ds_α^2 *is realized by an analytic convex cap* ω_{α_0} *which is analytically continuable across its boundary. Then the same holds for* $\alpha = \alpha_1$.

PROOF. Let α' denote the supremum of the numbers α satisfying the following condition: For all α ($\alpha_0 \leqq \alpha < \alpha'$) the metric ds_α^2 is realized by an analytic cap which is analytically continuable across its boundary. By Lemma 1, $\alpha_0 < \alpha'$. We wish to show that $\alpha' = \alpha_1$.

Let $\beta_1, \beta_2, \cdots$ be a sequence of numbers in the interval (α_0, α') converging to α'. For each β_k there is an analytic cap ω_k realizing the metric $ds^2_{\beta_k}$. Let $r_k(u,v)$ be the radius-vector of the point on ω_k corresponding to the point (u,v) of the disk $\bar{\omega}$. Without loss of generality we may assume that the vector-valued functions $r_k(u,v)$ are uniformly bounded, the base of each cap ω_k lies in the xy-plane and the cap itself in the halfspace $z > 0$. By Theorem 2 of §8, both the normal curvature of each cap ω_k and the angle between its tangent planes and the xy-plane are bounded by intrinsic quantities. It follows from the assumptions of the lemma that these estimates may be assumed independent of k, and moreover that the estimate for the angles is less than $\pi/2$. Using the estimates for the curvature, we can derive estimates for the second derivatives of the function $r_k(u,v)$.

Now let g_1 and g_2 be two axes through the origin, neither of which is coplanar with the z-axis, forming with the latter angles not exceeding $\frac{1}{2}(\pi/2 - \vartheta_0)$, where ϑ_0 is the maximum of the angles between the tangent planes of all caps ω_k to the xy-plane. Let $\xi_k(u,v)$ and $\eta_k(u,v)$ be the projections of the vector $r_k(u,v)$ on these axes. Each of the functions $z_k(u,v)$, $\xi_k(u,v)$ and $\eta_k(u,v)$ satisfies a second-order elliptic partial differential equation, namely the Darboux equation. In §11 of this chapter we shall prove a theorem on estimates for the derivatives of a solution of a general elliptic equation. According to the theorem, the fourth derivatives of the functions z_k, ξ_k and η_k in the disk $u^2 + v^2 \leqq 1 - \epsilon_1$ ($\epsilon_1 > 0$) satisfy estimates which depend only on estimates for the first and second derivatives. Because of the way in which the equations for z_k, ξ_k and η_k depend on β_k, we may assume that the estimates for the fourth derivatives of these functions are independent of k. The estimates for the derivatives of ξ_k, η_k and z_k may be used to obtain estimates for the derivatives of the vector-valued function $r_k(u,v)$. Thus we may assume that the fourth derivatives of the function $r_k(u,v)$ in the disk $u^2 + v^2 \leqq 1 - \epsilon_1$ are uniformly bounded (with respect to k).

Since the sequence of functions $r_k(u,v)$ is uniformly bounded and equicontinuous, it contains a convergent subsequence, so that without loss of generality we may assume that the sequence $r_k(u,v)$ itself converges to a function $r(u,v)$. Correspondingly, the caps ω_k converge to a convex cap ω with radius-vector $r(u,v)$, which realizes the metric $ds^2_{\alpha'}$.

Since the fourth derivatives of the functions $r_k(u,v)$ are uniformly bounded in the disk $u^2+v^2 \leqq 1-\epsilon_1$, it follows that the limit function $r(u,v)$ is thrice differentiable in this disk and hence in the entire disk $\bar{\omega}$ except (perhaps) on its boundary.

Since $r(u,v)$ is thrice differentiable inside the disk, so are its projections $z(u,v)$, $\xi(u,v)$ and $\eta(u,v)$ on the axes z, g_1 and g_2. Now each of these functions satisfies an elliptic analytic equation, and so, by the well-known theorem of Bernšteĭn which states that the solutions of an elliptic equation are analytic, the functions ξ, η and z are analytic in the interior of $\bar{\omega}$.

The discriminant of the Darboux equation for z is strictly positive throughout the disk $\bar{\omega}$, since the tangent planes to the cap ω form angles of at most ϑ_0 with the xy-plane and the Gauss curvature of the cap is strictly positive.

Hence by Bernšteĭn's theorem ([**21**], Chapter VII) the function z is analytically continuable across the boundary of the disk $\bar{\omega}$. Therefore the analytic metric $d\sigma^2_{\alpha'}$ used to define the other two coordinates (x and y) of the vector $r(u,v)$ is defined not only in $\bar{\omega}$, but also in a neighborhood of $\bar{\omega}$. It follows that the functions $x(u,v)$, $y(u,v)$ and $z(u,v)$ are analytic and are analytically continuable across the boundary. By Lemma 1, α' cannot be less than α_1. Q.E.D.

THEOREM. *An analytic metric with positive Gauss curvature is defined in an ϵ-neighborhood of the disk $\bar{\omega}: u^2+v^2 \leqq 1$ by the line element ds^2. Assume that the circumference γ of $\bar{\omega}$ has positive geodesic curvature with respect to the metric ds^2, and each geodesic (with respect to ds^2) through the center of $\bar{\omega}$ meets the circumference of the disk in a point whose distance from the center along the geodesic is less than $1/\sqrt{K_2}$, where K_2 is the maximum Gauss curvature in $\bar{\omega}$.*

Then there exists an analytic cap ω realizing the metric ds^2 defined in $\bar{\omega}$.

PROOF. In the disk $\bar{\omega}$ we introduce polar geodesic coordinates ρ, ϑ relative to the metric ds^2, taking the center O of $\bar{\omega}$ as pole. In terms of these coordinates the line element ds^2 is

$$ds^2 = d\rho^2 + c^2(\rho,\vartheta)\,d\vartheta^2,$$

where $c(0,\vartheta)=0$ and $c_\rho(0,\vartheta)=1$.

Since the Gauss curvature K is $-c_{\rho\rho}/c$, the coefficient c^2 appearing in the expression for ds^2 satisfies the ordinary differential equation $y''+Ky=0$ along every geodesic issuing from O, with initial conditions $y=0$, $y'=1$ for $\rho=0$. If $K_1 \leqq K \leqq K_2$, then the integral curve of this equation in the strip $0 \leqq \rho \leqq \pi/\sqrt{K_2}$ lies between the integral curves of the equations

$$y'' + K_1 y = 0, \qquad y'' + K_2 y = 0$$

with the same initial conditions. It follows that $c(\rho, \vartheta)$, which is a convex function of ρ for fixed ϑ, is monotone increasing at least in the interval $(0, 1/\sqrt{K_2})$, and hence the coordinate net ρ, ϑ can be introduced throughout the disk $\bar{\omega}$ and in a certain neighborhood of $\bar{\omega}$.

Let $\rho = \rho(\vartheta)$ be the equation of the circumference γ of $\bar{\omega}$ in the coordinates ρ, ϑ. We shall show that the geodesic curvature of the curve γ_α defined by the equation $\rho = \alpha\rho(\vartheta)$, $\alpha < 1$, is greater than that of γ (at points corresponding to the same ϑ). Indeed, consider the quadrangle $ABA_\alpha B_\alpha$ (Figure 22) bounded by the coordinate geodesics $\vartheta = \vartheta_0$, $\vartheta = \vartheta_0 + \Delta\vartheta$ and arcs of γ and γ_α. The length of the arc $A_\alpha B_\alpha$ is less than that of the arc AB, for these lengths are

$$\int_{\vartheta_0}^{\vartheta_0+\Delta\vartheta} \sqrt{(\alpha\rho')^2 + c^2(\alpha\rho, \vartheta)}\, d\vartheta,$$

$$\int_{\vartheta_0}^{\vartheta_0+\Delta\vartheta} \sqrt{\rho'^2 + c^2(\rho, \vartheta)}\, d\vartheta$$

respectively, and $c(\rho_1, \vartheta) < c(\rho_2, \vartheta)$ if $\rho_1 < \rho_2 < 1/\sqrt{K_2}$. Applying the Gauss-Bonnet Theorem to the quadrangle $A_\alpha B_\alpha AB$ and noting that the geodesics $\vartheta = \text{const}$ cut the curves γ_α and γ at right angles, we get

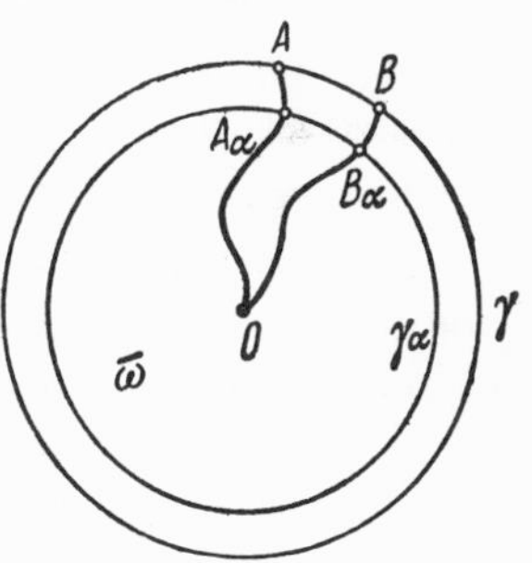

FIGURE 22

$$\int_{(A_\alpha)}^{(B_\alpha)} \varkappa(\vartheta, \alpha)\, ds_\alpha - \int_{(A)}^{(B)} \varkappa(\vartheta)\, ds = \iint_{(A_\alpha B_\alpha AB)} K(\rho, \vartheta)\, d\sigma,$$

where $\kappa(\vartheta, \alpha)$ and $\kappa(\vartheta)$ are the geodesic curvatures of γ_α and γ, and K is the Gauss curvature. Hence

$$\kappa_\alpha^* \Delta s_\alpha - \kappa^* \Delta s = K^* \Delta\sigma,$$

where κ_α^* and κ^* are the mean geodesic curvatures of γ_α and γ over the arcs $A_\alpha B_\alpha$ and AB, respectively, and K^* is the mean Gauss curvature of the quadrangle. Dividing this equality by Δs_α and noting that $\Delta s/\Delta s_\alpha > 1$, we see that $\kappa_\alpha(\vartheta) - \kappa(\vartheta) \geqq 0$ when $\Delta s \to 0$.

Let $\bar{\omega}_\alpha$ be the domain bounded by γ_α in the disk $\bar{\omega}$. Map this domain onto the disk $\bar{\omega}$ by mapping the point (ρ, ϑ) of $\bar{\omega}_\alpha$ onto the point $(\rho/\alpha, \vartheta)$

of $\bar{\omega}$. Under this mapping the curve γ_α goes into the circumference γ. The mapping can evidently be extended to a mapping of a neighborhood of ω_α into a neighborhood of $\bar{\omega}$. The metric ds^2, considered in the neighborhood of ω_α, goes into a metric ds_α^2 defined in a neighborhood of $\bar{\omega}$. Since ds_α^2 is analytic in the parameter α and the corresponding geodesic curvature of the circumference is positive, it follows from Lemma 2 that the metric ds^2 in $\bar{\omega}$ is realizable by an analytic convex cap which is analytically continuable across its boundary, provided ds_α^2 is so realizable for at least one nonzero value of the parameter α, or, equivalently, provided ds^2 is realizable when considered in the domain $\bar{\omega}_\alpha$.

Let $d\tilde{s}^2 = d\rho^2 + \tilde{c}^2 d\vartheta^2$ be the line element, in polar geodesic coordinates, of a sphere of radius $1/\sqrt{K_0}$, where K_0 is the Gauss curvature at the center of $\bar{\omega}$ relative to the metric ds^2. Consider the metric $d\tilde{s}_\beta^2$ defined in the disk $\bar{\omega}$ by the line element

$$d\tilde{s}_\beta^2 = d\rho^2 + (\beta c + (1-\beta)\tilde{c})^2 d\vartheta^2, \qquad 0 \leqslant \beta \leqslant 1.$$

At the point O the functions $c(\rho, \vartheta)$ and $\tilde{c}(\rho, \vartheta)$ coincide, as do their first and second derivatives. Thus for sufficiently small $\alpha = \alpha_0$ the metric $d\tilde{s}_\beta^2$ has positive Gauss curvature in a neighborhood of the domain $\bar{\omega}_{\alpha_0}$, while the curve γ_{α_0} has positive geodesic curvature not only relative to the metric ds^2 but also relative to each metric $d\tilde{s}_\beta^2$.

It follows that, considered in a sufficiently small neighborhood of the domain $\bar{\omega}_{\alpha_0}$, the metric ds^2 is realizable by an analytic cap which is analytically continuable across its boundary, provided the metric $d\tilde{s}^2$ is so realizable. Let $\tilde{\omega}_{\alpha_0}$ be the domain on the sphere S bounded by the curve $\tilde{\gamma}_{\alpha_0}$ with the same equation in polar geodesic coordinates ρ, ϑ as the curve γ_{α_0}; then the metric $d\tilde{s}^2$, considered in ω_α, is realizable by an analytic cap analytically continuable across its boundary if and only if there exists such a cap isometric to $\tilde{\omega}_{\alpha_0}$.

Let σ be the tangent plane to the sphere S at the point $\rho = 0$; project the curve $\tilde{\gamma}_{\alpha_0}$ from the center of S onto σ. The result is a convex curve $\tilde{\gamma}'_{\alpha_0}$ with positive curvature. Introduce polar coordinates r, φ in the plane σ, taking the direction $\vartheta = 0$ on S as the direction $\varphi = 0$ on σ. Let $r = f(\varphi)$ be the equation of $\tilde{\gamma}'_{\alpha_0}$ in polar coordinates r, φ. Consider the curve λ_t $(0 \leqq t \leqq 1)$ defined in σ by the equation

$$r = \frac{f(\varphi)}{(1-t) + tf(\varphi)}.$$

This curve has positive curvature everywhere for every t in the interval considered; when $t=0$ it coincides with the curve $\tilde{\gamma}'_{\alpha_0}$, and when $t=1$, with the circumference $r=1$.[3)]

Project the curve λ_t onto the sphere S from its center, and let $\tilde{\omega}_{\alpha_0,t}$ denote the convex domain on the sphere bounded by the projection.

An analytic cap, analytically continuable across its boundary and isometric to the domain $\tilde{\omega}_{\alpha_0}$, will exist if such a cap exists for at least one domain $\tilde{\omega}_{\alpha_0,t}$. But for $\tilde{\omega}_{\alpha_0,1}$ a cap of this description exists, namely the projection of the disk $r \leqq 1$ in the plane σ onto the sphere S. Hence there exists an analytic cap, analytically continuable across its boundary, isometric to the domain $\tilde{\omega}_{\alpha_0}$ on S, and so there exists a cap of this type realizing the metric $ds^2_{\alpha_0}$ defined in $\bar{\omega}$. This in turn implies the existence of an analytic cap realizing the metric ds^2 in $\bar{\omega}$. This completes the proof of the theorem.

§10. Regularity of convex surfaces with regular metric

The main result of this chapter is the following theorem.

THEOREM 1. *A convex surface with a regular metric of class C^n $(n \geqq 2)$ and positive Gauss curvature is of class $C^{n+1+\alpha}$ for any α, $0<\alpha<1$. If the metric of the surface is analytic, the surface is analytic.*

PROOF. Let Φ be a convex surface satisfying the assumptions of the theorem. By Theorems 1 and 2 of §3, it is smooth and strictly convex, in the sense that every tangent plane of Φ has exactly one point, the point of contact, in common with Φ. Let O be an arbitrary point of Φ. There is a plane, parallel to the tangent plane at O, which cuts off from Φ a small convex cap ω with interior diameter less than $1/\sqrt{K_2}$, where K_2 is the maximum Gauss curvature of the cap. This follows from the fact that the surface Φ is strictly convex and so, if the secant plane is sufficiently close to O, the cap ω that it cuts out from Φ will have arbitrarily small interior diameter. Let γ denote the boundary of ω.

Inscribe a geodesic polygon with small sides in the curve γ; "smooth"

---

3) The statement made here that the curve λ_t has positive curvature may not be immediately obvious. It follows easily from the fact that the curvature of a curve given in polar coordinates r, φ by the equation $r=p(\varphi)$ is

$$\frac{\frac{1}{p}+\left(\frac{1}{p}\right)''}{\left(1+(\ln p)'^2\right)^{3/2}}.$$

the vertices of the polygon by means of geodesic disks with small radii. We get a smooth convex curve γ_1, close to and inside γ. Now displace each point X of this curve by a distance $\epsilon_1 d$ along the segment connecting it to O, where d is the original distance between X and O on Φ. This transformation increases the geodesic curvature of γ_1, and we get a curve with the property that the geodesic curvature exists and is positive to the right and left of each point. Obviously this curve, call it γ_2, may be taken arbitrarily close to γ: we need only make the sides of the polygon inscribed in γ sufficiently small, and the number ϵ_1 sufficiently small.

Now construct an analytic curve $\tilde{\gamma}$ close to γ_2 with positive geodesic curvature. Since γ_2 is smooth and has strictly positive geodesic curvature at all but a finite number of its points, this curve is easily constructed. Let $\tilde{\omega}$ be the domain on Φ bounded by $\tilde{\gamma}$.

Let u, v be a coordinate net on Φ, and

$$ds^2 = E\,du^2 + 2F\,du\,dv + G\,dv^2$$

the line element of the surface, where E, F and G are thrice differential functions of u and v. Construct a sequence of analytic functions E_n, F_n and G_n converging to E, F and G uniformly together with their derivatives of order $\leqq 3$ in the domain ω on Φ. Consider the metric defined in ω by the quadratic form

$$ds_n^2 = E_n du^2 + 2F_n du\,dv + G_n dv^2.$$

If n is sufficiently large, the Gauss curvature based on the coefficients of ds_n^2 can be made arbitrarily close to the Gauss curvature of ω, and the geodesic curvature of $\tilde{\gamma}$ relative to the metric ds_n^2 is positive at each point. By the theorem of §9 there exists an analytic cap $\tilde{\omega}_n$ realizing the metric ds_n^2 defined in $\tilde{\omega}$.

Letting the curve $\tilde{\gamma}$ tend to γ, construct a sequence of analytic caps $\tilde{\omega}_n'$ converging to a cap $\tilde{\omega}'$ isometric to ω. By the theorem of §6 the caps $\tilde{\omega}'$ and ω are congruent. Hence it remains only to prove that the limit cap $\tilde{\omega}'$ has the required regularity property.

In [35], Heinz proved the following theorem. *Let $r(u,v)$ be a twice differentiable vector-valued function defined in a domain Ω of the uv-plane, and let $ds^2 = Edu^2 + 2Fdudv + Gdv^2$ be the line element of the surface $\Phi : r = r(u,v)$. Let the coefficients of the form ds^2, their derivatives of order $\leqq 3$, and the reciprocal of the discriminant of the form be bounded by a constant a; the Gauss curvature of the surface Φ is everywhere greater than $b > 0$, and the integral mean curvature is bounded by a constant $c < \infty$.*

Then the absolute values of the second derivatives of $r(u,v)$ *on the set of interior points of* Φ *whose distances from the boundary are at least* $\rho > 0$ *satisfy an estimate which depends solely on* a, b, c, ρ. *Moreover, for any positive number* $\nu < 1$ *the Hölder constants of the second derivatives of* r *relative to the exponent* ν *satisfy an estimate depending on the same quantities and on* ν.

Applying this theorem to the sequence of caps $\tilde{\omega}'_n$ and letting $n \to \infty$, we conclude that their limit $\tilde{\omega}'$ must be of class $C^{2+\nu}$, $0 < \nu < 1$, i.e. the function $r(u,v)$ that defines $\tilde{\omega}'$ has second derivatives which satisfy a Hölder condition with any exponent ν, $0 < \nu < 1$.

That the limit cap $\tilde{\omega}'$ has the required regularity properties follows from Nirenberg's theorem on the regularity of twice differentiable solutions of an elliptic equation with regular coefficients. Applying this theorem to the Darboux equation, we see that, since the surface $\tilde{\omega}'$ is twice differentiable and its metric n times differentiable, it follows that $\tilde{\omega}'$ is at least $n-1$ times differentiable and the $(n-1)$th derivatives of the function $r(u,v)$ defining the surface satisfy a Hölder condition with any exponent λ, $0 < \lambda < 1$ [**48**].

Thus for $n \geqq 3$ a convex surface with metric of class C^n and positive Gauss curvature belongs (at least) to class $C^{n-1+\alpha}$ for any α, $0 < \alpha < 1$.

An analogous result for $n = 2$ follows from another theorem of Heinz [**36**]. Heinz proved that *if a twice differential function* $z(x,y)$ *in a domain* Ω *of the* xy*-plane satisfies the condition*

$$0 < \alpha < z_{xx} z_{yy} - z_{xy}^2 < \beta < \infty,$$

then the first derivatives of $z(x,y)$ *on the set of interior points of* Ω *whose distance from the boundary is at least* d *satisfy a Hölder condition with exponent* $\nu = \sqrt{\beta/\alpha}$ *and constant depending only on* α, β, d *and*

$$\delta = d \max_{\Omega} \left(z_x^2 + z_y^2\right)^{1/2}.$$

Applying this theorem to the caps $\tilde{\omega}'_n$ converging to $\tilde{\omega}'$, we conclude that in a neighborhood of O the first derivatives of the functions $z_n(x,y)$ defining the caps satisfy a Hölder condition uniformly, with any exponent $\nu < 1$ (for a small neighborhood the quotient β/α can be made arbitrarily close to unity, since the Gauss curvature is continuous and the limit surface is smooth). Letting $n \to \infty$, we see that the cap $\tilde{\omega}'$ also has this property. Q.E.D.

Theorem 1, together with Aleksandrov's theorem on the realizability of a convex metric, defined on a manifold homeomorphic to a sphere, as a convex surface (§7 of Chapter I), yields a complete solution to Weyl's problem:

THEOREM 2. *A regular metric with positive Gauss curvature, defined on a manifold homeomorphic to a sphere, is realized by a closed regular surface. Indeed, if the metric is of class* C^n, $n \geqq 2$, *the surface is (at least) of class* $C^{n-1+\alpha}$, $0 < \alpha < 1$. *If the metric is analytic, so is the surface.*

Using Aleksandrov's theorem on the realizability of complete convex metrics (Chapter I, §11) in conjunction with Theorem 1, we obtain the following theorem.

THEOREM 3. *A complete regular metric with positive Gauss curvature, defined in a domain homeomorphic to a disk, is realizable as an unbounded regular convex surface. If the metric is of class* C^n, $n \geqq 2$, *the surface is of class* $C^{n-1+\alpha}$, $0 < \alpha < 1$. *If the metric is analytic, so is the surface.*

The following theorem is a corollary of Theorem 1 and Aleksandrov's theorem on the realizability of noncomplete convex metrics by a convex cap.

THEOREM 4. *A regular metric with positive Gauss curvature is defined in a domain G which is homeomorphic to a disk and bounded by a curve γ. Let the geodesic curvature of γ relative to the metric be nonnegative. Then there exists a regular convex cap realizing the metric. If the metric is in class* C^n, $n \geqq 2$, *the cap is in class* $C^{n-1+\alpha}$. *If the metric is analytic, so is the cap.*

THEOREM 5. *A regular metric with positive Gauss curvature is defined in a domain G which is homeomorphic to a disk and bounded by a curve γ. Let the geodesic curvature of γ relative to the metric be positive. Let Ω be an arbitrary convex domain on the unit sphere, of area equal to the integral curvature of the metric. Then there exists a regular convex surface realizing the metric defined in G, and the spherical image of the surface is Ω. If the metric is in class* C^n, $n \geqq 2$, *then the surface is in class* $C^{n-1+\alpha}$, $0 < \alpha < 1$. *If the metric is analytic, so is the surface.*

PROOF. By Aleksandrov's theorem the metric defined in G is realized by a convex cap F. Let A_k be a sufficiently fine net of points on the surface F. The convex hull of the set of points A_k and the base of the cap F is a closed convex surface, comprising a certain cap F' and the base of F. As the net A_k is made finer, the cap F' converges to F. The integral geodesic curvature of the curve γ' bounding F' converges to the i.g.c. of the boundary γ of F. The entire curvature of the cap F' is concentrated at its vertices A_k; away from these points F' is locally isometric to a plane.

Take two points $A = A_i$ and $B = A_j$. Let α and β be the curvatures at these points. Connect A and B by a segment on the surface F', of length l, say. Cut the cap F' along the segment AB; along the cut, glue two triangles with base l and base angles $\alpha/2$ and $\beta/2$. The result is a manifold with convex metric, in which the vertices A and B have been replaced by a new vertex C with curvature $\alpha + \beta$. Continuing this gluing procedure, we finally obtain a manifold F'' with a single vertex, at which the curvature is equal to the total curvature of F'. The i.g.c. of the boundary of F'' is clearly equal to that of the boundary of F'. Now make the net A_k infinitesimally fine; the limit is a manifold F''' isometric to a domain G on a cone V whose curvature at the apex is equal to the curvature of F. Corresponding parts of the boundaries of G and F have equal lengths and i.g.c.'s. By Aleksandrov's Gluing Theorem (Chapter I, §11) we can construct a complete manifold H from the cap F and the manifold $V - G$. The entire curvature of this manifold is concentrated in the part isometric to F; the curvature in the remainder of H is zero.

By a theorem of Olovjanišnikov (Chapter I, §11) the intrinsic metric of the manifold H is realizable on an unbounded convex surface Φ whose limit cone has spherical image Ω. The surface Φ has bounded specific curvature and is therefore smooth. Let Φ' be a domain on Φ isometric to F, and Φ'' the remainder of Φ. The curvature of Φ'' is zero. Choose an arbitrary point Q in the interior of Φ'' and construct the tangent plane α to Φ at G. Let M be the intersection of α and Φ. M is clearly a convex set. We shall show that it is unbounded.

First M must contain points other than Q, since otherwise a suitable plane parallel to α would cut out a small cap from Φ''; this is impossible, because Φ'' has zero curvature. We now show that M is not a straight-line segment. Indeed, suppose M were a straight-line segment. We claim that it cannot have more than one point in common with the boundary γ' of Φ'. Otherwise, let δ be that part of the segment whose endpoints are on γ'. δ together with γ' bounds a domain on Φ'' which is homeomorphic to a disk and has zero curvature. Applying the Gauss-Bonnet Theorem (Chapter I, §8) to this domain, we immediately conclude that γ' has zero i.g.c. toward this domain. But the left and right i.g.c.'s of γ' are equal in absolute value, and the i.g.c. toward the domain

Φ' is surely positive, since the geodesic curvature is positive. This contradiction proves that the segment M and the boundary of Φ' cannot have more than one common point. It follows that one of the endpoints of M is an interior point of Φ''. Let Q' be the boundary point of Φ' in the set M, if such a point exists. Let g be a straight line in the plane α separating Q and Q'. Then a sufficiently small rotation of the plane α about g will cut out a cap from the surface Φ''; and this is impossible. If M contains no boundary points of Φ', the cap can be cut out by a plane parallel to and near α.

A similar argument shows that M cannot be a bounded domain. Hence M is an unbounded convex set. Since the normals to the surface Φ at all points of M are parallel, the spherical image of every point Q in Φ'' belongs to the boundary of the domain Ω on the unit sphere. The spherical image of the surface Φ' is an open set, since the Gauss curvature of this surface is positive. By what we have proved, it contains all interior points of the domain and only these points. The theorem is proved.

Using the Gluing Theorem and Theorem 1, we can establish a very general theorem on the solvability of the first boundary-value problem (Dirichlet problem) for the Darboux equation.

THEOREM 6. *Let G be a domain homeomorphic to a disk, bounded by a curve γ. Let a regular metric with positive Gauss curvature be defined in G by the line element*

$$ds^2 = Edu^2 + 2Fdudv + Gdv^2.$$

Let

$$\Phi(z_{uu}, z_{uv}, z_{vv}, z_u, z_v, u, v) = 0$$

be the Darboux equation based on the line element ds^2. Let h be an arbitrary function defined on the boundary of G.

Then, if the geodesic curvature κ of the curve γ relative to the metric ds^2 is positive and satisfies the condition

$$(*) \qquad \varkappa > \frac{|h''|}{(1-h'^2)^{1/2}}, \qquad |h'| < 1,$$

the first boundary-value problem for the equation $\Phi = 0$ in the domain G, with boundary values h along γ, is always solvable. If the metric ds^2 is of class C^n, $n \geqq 2$, the solution is of class $C^{n-1+\alpha}$, $0 < \alpha < 1$. If the metric is analytic, the solution is analytic.

PROOF. Let $h(s)$ be the function defined on γ (where s is arc length along γ with respect to the metric ds^2). Construct the following curve

γ in the xy-plane:

$$x = x(s), \quad y = h(s), \qquad 0 \leqq s \leqq l,$$

where s is arc length along this curve and l is the length of γ in the metric ds^2. The curve $\bar{\gamma}$ is completely determined by the function $h(s)$. It is easy to see that its curvature is $k = h''/(1 + h'^2)^{1/2}$. Thus condition (*) of the theorem is none other than $\kappa > |k|$. Construct straight lines parallel to the y-axis through the endpoints of $\bar{\gamma}$, and fold the resulting rectangular strip into a cylinder. This cylinder is divided by $\bar{\gamma}$ into two semicylinders; denote one of them by Z.

Let F be a convex cap realizing the metric defined by the line element ds^2 in G. Since $\kappa > |k|$ by assumption, the cap F and the semicylinder Z satisfy the assumptions of the Gluing Theorem. Hence there exists an unbounded complete convex surface Φ comprising two parts $\widetilde{F}$ and $\widetilde{Z}$ isometric to F and Z, respectively. The limit cone of this surface obviously degenerates into a ray. Let x, y, z be cartesian coordinates in space, and place the surface Φ so that its limit cone coincides with the positive z-axis. In this position of Φ the domain $\widetilde{F}$ can be projected in one-to-one fashion onto the xy-plane, since it has strictly positive Gauss curvature. We claim that the domain $\widetilde{Z}$ on Φ is a semicylinder (of course, not necessarily circular).

Let $\tilde{\gamma}$ be the common boundary of the domains $\widetilde{F}$ and $\widetilde{Z}$ on Φ, and Q a point on $\tilde{\gamma}$. Choose a point Q' in $\widetilde{Z}$ close to Q. Let α be a supporting plane of Φ at Q'. Let M be the intersection of α and Φ. M is a convex set. We shall show that it cannot be bounded. First, M must contain more than one point (the point Q'), since otherwise a plane parallel to and sufficiently near α would cut out a "hump" from $\widetilde{Z}$, and $\widetilde{Z}$ would have positive curvature; this is impossible.

Now suppose that M is a bounded set containing points other than Q'. The set of points at which the boundary of M is strictly convex contains at least two points, and all of them belong to the curve $\tilde{\gamma}$. Indeed, let P be a point at which the boundary of M is strictly convex, and suppose it is not on γ. Then P is an interior point of Z. Let g be a supporting line of M at P; its only point in common with M is P. Displace this line parallel to itself in the direction of M, over a small distance, so that all points of M belonging to $\tilde{\gamma}$, if such exist, are separated from P. Now a small rotation of the plane α about g will cut out a hump from the surface Z, which is impossible. Hence all points at which the boundary of M is strictly convex lie on the curve $\tilde{\gamma}$.

This being so, there are only two possibilities:

1) Some part of the boundary of M coincides with $\tilde{\gamma}$.

2) Some straight-line segment δ_1 of the boundary of M connects two points A_1 and B_1 of $\tilde{\gamma}$ and lies on the surface $\widetilde{Z}$.

In the first case one can find arbitrarily close points of the curve $\tilde{\gamma}$ which can be joined by a straight-line segment on $\widetilde{Z}$: these are the points lying on the intersection of the boundary of M and the curve $\tilde{\gamma}$. In the second case the segment δ_1 together with the arc A_1B_1 of $\tilde{\gamma}$ bounds a domain G_1 homeomorphic to a disk. Let C_1 be the point half way between A_1 and B_1 on the curve $\tilde{\gamma}$, and choose C_1' a point in G_1 close to C_1. Let α_1 be a supporting plane of the surface Φ at C_1', and let M_1 denote the intersection of this plane with Φ. M_1 is a convex subset of G_1. The points at which this set is strictly convex are also on $\tilde{\gamma}$. We now conclude that there exists a straight-line segment δ_2 on $\widetilde{Z}$, joining points A_2 and B_2 on one of the arcs A_1C_1 or B_1C_1. Repeating the construction, we see that in any case the curve $\tilde{\gamma}$ contains arbitrarily close points A_n and B_n which can be connected on $\widetilde{Z}$ by a straight-line segment δ_n. Let l_n be the length of the shortest join of A_n and B_n in the domain F_n, and $\bar{l}_n$ the length of δ_n. We intend to show that when A_n is sufficiently near B_n we have $l_n < \bar{l}_n$. This contradiction will prove that M cannot be a bounded set.

In proving Theorem 5 for a convex cap F, we constructed a conical surface V on which corresponding sections of the boundary had the same length and i.g.c. as the boundary of F. It is clear from the construction of this surface by gluing triangles to the cap that the distance between two given points on the boundary of the cap is at most the distance between the corresponding points on the boundary of the conical surface (the distances considered here are those on the surface).

Let C be a point on the boundary of V, and A and B points of the boundary near C. Since V is locally isometric to a plane, the distance between A and B on V is

$$\rho = s - \frac{1}{24}\varkappa^2 s^3 + (*),$$

where s is the distance between A and B along the boundary, κ is the geodesic curvature at C, and (*) stands for lower-order terms.

By what has been proved above, there exists a sequence of straight-line segments δ_n whose lengths become arbitrarily small and which join points A_n and B_n of $\tilde{\gamma}$ on the surface $\widetilde{Z}$. Without loss of generality, we may assume that the sequences of points A_n and B_n both converge to the point C. If A_n and B_n are sufficiently close to C, the distance between them on F is less than

$$s-\left(\frac{1}{24}\varkappa^2(C)-\varepsilon\right)s^3,$$

where ϵ is a small positive number. On the other hand, the length of the straight-line segment A_nB_n is at least

$$s-\left(\frac{1}{24}k^2(C)+\varepsilon\right)s^3.$$

Since the length of the shortest join of A_n and B_n on the surface F is at least that of the straight-line segment A_nB_n, it follows that

$$s-\left(\frac{1}{24}\varkappa^2(C)-\varepsilon\right)s^3\geqslant s-\left(\frac{1}{24}k^2(C)+\varepsilon\right)s^3.$$

But if ϵ is sufficiently small this is impossible, since by assumption $\kappa>|k|$. Thus the assumption that M is bounded has led to a contradiction, and so M must be unbounded.

Since M is an unbounded convex set, it contains a ray issuing from the point Q'. Letting $Q'\to Q$, we see that the surface $\widetilde{Z}$ contains a ray g_Q issuing from Q. Since the limit cone of Φ degenerates into a ray, namely, the positive z-axis, the ray g_Q is parallel to the z-axis. If Q moves along the curve $\tilde{\gamma}$, the ray g_Q describes the surface $\widetilde{Z}$. Consequently $\widetilde{Z}$ is a semicylinder. Obviously by displacing the surface Φ along the z-axis we can guarantee the condition $z=h$ along the curve $\tilde{\gamma}$. The z-coordinate of a point of the surface F, viewed as a function of the parameters u and v, furnishes the required solution $z(u,v)$. This completes the proof.

Let F be a convex surface which is in a starlike position relative to the point O, i.e. no ray issuing from O is tangent to the surface. Let $\rho(u,v)$ be the distance between O and the point of the surface with coordinates u,v. It can be shown that the function $\rho(u,v)$ satisfies a Monge-Ampère equation with coefficients depending only upon the coefficients of the line element ds^2 of the surface F and their derivatives. This equation is also known as the Darboux equation. It is an elliptic equation, provided the Gauss curvature of F is positive.

THEOREM 7. *Let G be a domain homeomorphic to a disk, bounded by a curve γ and lying on the unit sphere. Let a regular metric*

$$ds^2=Edu^2+2Fdudv+Gdv^2$$

with positive Gauss curvature be defined in G. Let

$$\psi(\rho_{uu},\rho_{uv},\rho_{vv},\rho_u,\rho_v,\rho,u,v)=0$$

be the Darboux equation based on the line element ds^2. Let h be an arbitrary *function defined on the boundary of G. If the geodesic curvature* κ *of* γ *relative to the metric* ds^2 *is positive and satisfies the condition*

$$\varkappa > \sqrt{1-h'^2}\left|\frac{1}{h}-\frac{h''}{1-h'^2}\right|, \qquad |h'|<1,$$

then the first boundary-value problem for the equation $\psi = 0$ *in the domain* G, *with boundary values* h *along* γ, *is always solvable. If the metric* ds^2 *is in class* C^n, $n \geqq 2$, *then the solution is in class* $C^{n-1+\alpha}$. *If the metric is analytic, so is the solution.*

The proof of the theorem is based on the same arguments as those used to prove Theorem 6. One first constructs a plane curve $\bar{\gamma}$, defined in polar coordinates ρ, ϑ by the equations $\vartheta = \vartheta(s)$ and $\rho = h(s)$, where s is arc length along $\bar{\gamma}$. The curvature of this curve is

$$k = \sqrt{1-h'^2}\left(\frac{1}{h}-\frac{h''}{1-h'^2}\right).$$

Draw rays from the origin O to the endpoints of the curve $\bar{\gamma}$; rotate the angle formed by these rays to form a cone. The curve $\bar{\gamma}$ divides this cone into two sections. Let V be the part that contains the apex O. The metric ds^2 defined in the domain G is realized by a convex cap F. This cap and the cone V satisfy the assumptions of the Gluing Theorem. Hence follows the existence of a closed convex surface Φ consisting of two parts: $\bar{F}$, isometric to F, and $\bar{V}$, isometric to V. One then shows that the surface $\bar{V}$ is also a cone. The distance ρ between the apex of this cone and the points of $\bar{F}$, viewed as a function of the coordinates u, v on the surface, is the required solution of the boundary-value problem.

§11. Estimates for the derivatives of the solution of a second-order elliptic partial differential equation

Our proof that a convex surface with a regular metric is regular (Theorem 1 of §10) involved estimates for the derivatives of the solution of a second-order elliptic partial differential equation. These estimates will be established in this section.

THEOREM 1. *Let*

$$F(x,y,z,p,q,r,s,t) = 0 \tag{1}$$

be an elliptic partial differential equation and $z = z(x,y)$ *a regular solution of* (1) *in a domain* G *of the* xy*-plane. Then the absolute values of the third derivatives of* z *at a point* (x,y) *of* G *satisfy estimates which depend only* *on the suprema of the absolute values of z and its derivatives of order* $\leqq 2$, *the suprema of the absolute values of the partial derivatives of F of order* $\leqq 3$, *the suprema of the absolute values of the quantities*

$$\frac{1}{F_r}, \quad \frac{1}{F_t}, \quad \frac{1}{F_rF_t - \frac{1}{4}F_s^2}$$

and the distance of the point (x, y) *from the boundary of G.*[4)]

PROOF. Differentiating equation (1) with respect to x, we get

$$F_r r_x + F_s s_x + F_t t_x = M_1. \tag{2}$$

Differentiating (2) again with respect to x, we get

$$F_r r_{xx} + F_s s_{xx} + F_t t_{xx} = M_2, \tag{3}$$

where M_2 is a quadratic polynomial in the third derivatives r_x, s_x and t_x, with coefficients involving derivatives of z of order at most two. Using (2), we can express t_x in terms of r_x and s_x, and substitute the resulting expression in M_2. M_2 then becomes a polynomial of degree at most two in the third derivatives r_x and s_x, with coefficients involving derivatives of z of order at most two.

Let $n_1, \cdots, n_5$ be the derivatives of z with respect to x and y of order $\leqq 2$, and $N_1, N_2, \cdots$ all derivatives of F with respect to its arguments of order $\leqq 3$. To simplify the discussion we shall call any expression of the type

$$\sum \frac{n_1^{\alpha_1} \cdots n_5^{\alpha_5} N_1^{\beta_1} \cdots N_k^{\beta_k}}{F_r^\lambda F_t^\mu \left(F_rF_t - \frac{1}{4}F_s^2\right)^\nu}, \tag{ω}$$

where $\alpha_1, \alpha_2, \cdots, \beta_1, \beta_2, \cdots, \lambda, \mu$ and ν are arbitrary nonnegative integers, an expression of type (ω).

Set $r = -M + \alpha \ln \ln u$, where M is the supremum of the absolute value of the second derivative r. We then find from (3) that

$$\begin{aligned} &Ar_1 + 2Bs_1 + Ct_1 \\ &= \frac{1+\ln u}{u \ln u}(Ap_1^2 + 2Bp_1q_1 + Cq_1^2) + \frac{\alpha}{u \ln u} P_2 + P_1 + \frac{u \ln u}{\alpha} P_0 = Q_1, \end{aligned} \tag{4}$$

---

4) It is assumed that the derivatives of F are first computed formally, z, p, q, r, s and t are then replaced by the solution $z(x, y)$ and its derivatives, and the suprema of the respective quantities are computed.

where p_1, q_1, r_1, s_1 and t_1 denote the first and second derivatives of the function $u(x,y)$; $A = F_r$, $2B = F_s$ and $C = F_t$; P_2, P_1 and P_0 are polynomials in p_1 and q_1 of degrees two, one and zero, respectively, with coefficients of type (ω).

Following Bernšteĭn [21], we now consider the function $w(x,y)$ defined by

$$w = Ap_1^2 + 2Bp_1q_1 + Cq_1^2.$$

Suppose that this function has a maximum at some point (x_0, y_0). Then, at this point,

$$w_x = 0, \qquad w_y = 0. \tag{5}$$

Solving these equations together with equation (4) for r_1, s_1 and t_1, we get

$$r_1 = \frac{1}{w\Delta}\left\{w\,(Bp_1 + Cq_1)^2\frac{1+\ln u}{u\ln u} + \frac{\alpha g_1}{u\ln u} + h_1 + \frac{l_1 u\ln u}{\alpha}\right\},$$

$$s_1 = -\frac{1}{w\Delta}\left\{w\,(Bp_1 + Cq_1)(Ap_1 + Bq_1)\frac{1+\ln u}{u\ln u} + \frac{\alpha g_2}{u\ln u} + h_2 + \frac{l_2 u\ln u}{\alpha}\right\},$$

$$t_1 = \frac{1}{w\Delta}\left\{w\,(Ap_1 + Bq_1)^2\frac{1+\ln u}{u\ln u} + \frac{\alpha g_3}{u\ln u} + h_3 + \frac{l_3 u\ln u}{\alpha}\right\},$$

where $\Delta = AC - B^2$; g_1, g_2 and g_3 are polynomials of degree at most four in p_1 and q_1; h_1, h_2 and h_3 are polynomials of degree at most three; l_1, l_2 and l_3 are polynomials of degree at most two; all polynomials have coefficients of type (ω).

Consider the expression

$$K(w) = \tfrac{1}{2}\,(Aw_{xx} + 2Bw_{xy} + Cw_{yy}).$$

Without loss of generality we may assume that $A > 0$; for, if this is not so, we need only multiply equation (1) by -1. Now $K(w) \leq 0$ at the point at which w is a maximum, since the form $A\xi^2 + 2B\xi\eta + C\eta^2$ is positive definite, while the form $w_{xx}\xi^2 + 2w_{xy}\xi\eta + w_{yy}\eta^2$ is never positive. Substituting $w = Ap_1^2 + 2Bp_1q_1 + Cq_1^2$ in $K(w)$, we find that

$$\begin{aligned} K(w) = {} & (Ap_1 + Bq_1)(Ap_{1xx} + 2Bp_{1xy} + Cp_{1yy}) \\ & + (Bp_1 + Cq_1)(Aq_{1xx} + 2Bq_{1xy} + Cq_{1yy}) + A\,(Ar_1^2 + 2Br_1s_1 + Cs_1^2) \\ & + 2B\,(Ar_1s_1 + B\,(r_1t_1 + s_1^2) + Cs_1t_1) + C\,(As_1^2 + 2Bs_1t_1 + Ct_1^2) \\ & + H_1 + H_2 + \frac{\alpha}{u\ln u}H_3 + \alpha\frac{1+\ln u}{(u\ln u)^2}H_4 + \frac{\alpha^2}{(u\ln u)^2}H_5, \end{aligned}$$

where H_1 is linear in p_1r_1, p_1s_1, p_1t_1, q_1s_1, q_1r_1 and q_1t_1; H_2 is a quadratic polynomial in p_1 and q_1; H_3 is linear in $p_1^2r_1$, $p_1^2s_1$, $p_1^2t_1$, $p_1q_1r_1$, $p_1q_1s_1$, $p_1q_1t_1$, $q_1^2r_1$, $q_1^2s_1$ and $q_1^2t_1$; and H_4 and H_5 are fourth-degree polynomials in p_1 and q_1. The coefficients of all these polynomials are of type (ω).

Differentiating (4) with respect to x, we get

$$Ap_{1xx}+2Bp_{1xy}+Cp_{1yy}$$
$$=-\frac{1+\ln u+(\ln u)^2}{(u\ln u)^2}p_1w+\frac{\alpha(1+\ln u)}{(u\ln u)^2}G_1+\frac{\alpha}{u\ln u}G_2$$
$$+\frac{\alpha^2}{u\ln u}G_3+G_4+\frac{u\ln u}{\alpha}G_5+\frac{1+\ln u}{u\ln u}G_6,$$

where G_1 and G_3 are third-degree polynomials; G_5 is a linear polynomial in p_1 and q_1; G_2 is a linear polynomial in p_1r_1, p_1s_1, p_1t_1, q_1r_1, q_1s_1, q_1t_1, p_1^2, p_1q_1 and q_1^2; G_4 is a linear polynomial in r_1, s_1, t_1, p_1 and q_1; and G_6 is a quadratic polynomial in p_1 and q_1; all coefficients are of type (ω).

Similarly, differentiation of equation (4) with respect to y yields

$$Aq_{1xx}+2Bq_{1xy}+Cq_{1yy}$$
$$=-\frac{1+\ln u+(\ln u)^2}{(u\ln u)^2}q_1w+\frac{\alpha(1+\ln u)}{(u\ln u)^2}G_{(1)}+\frac{\alpha}{u\ln u}G_{(2)}$$
$$+\frac{\alpha^2}{(u\ln u)^2}G_{(3)}+G_{(4)}+\frac{u\ln u}{\alpha}G_{(5)}+\frac{1+\ln u}{u\ln u}G_{(6)},$$

where $G_{(1)},\cdots,G_{(6)}$ are polynomials with properties analogous to those of the polynomials $G_1,\cdots,G_6$.

Now consider $K(w)$; substituting in $K(w)$ the expressions just found for $Ap_{1xx}+2Bp_{1xy}+Cp_{1yy}$ and $Aq_{1xx}+2Bq_{1xy}+Cq_{1yy}$, we get

$$w^2K(w)=-\frac{1+\ln u+(\ln u)^2}{(u\ln u)^2}w^4+\left(\frac{1+\ln u}{u\ln u}\right)^2w^4$$
$$+\alpha\frac{1+\ln u}{(u\ln u)^2}+wP_6+\frac{\alpha^2}{(u\ln u)^2}P_8+\frac{1+\ln u}{u\ln u}P_7$$
$$+\frac{\alpha}{u\ln u}P_7'+\frac{u\ln u}{\alpha}P_6'+P_6''+\left(\frac{u\ln u}{\alpha}\right)P_4,$$

where P_k, P_k' and P_k'' denote polynomials of degree k in p_1 and q_1 with coefficients of type (ω).

All terms of degree eight in p_1 and q_1 that appear in $w^2K(w)$ are given by the expression

$$T_8=\frac{w^4}{u^2\ln u}+\alpha\frac{1+\ln u}{(u\ln u)^2}wP_6+\frac{\alpha^2}{(u\ln u)^2}P_8.$$

Since $\ln u>1$, one can find a number α so small that when $|p|$

$+|q| > 1$,

$$T_8 > \frac{1}{2}\frac{w^4}{u^2 \ln u}.$$

Choose α with this property.

Then, if $|p| + |q| < 1$, the function

$$w = Ap_1^2 + 2Bp_1q_1 + Cq_1^2$$

cannot exceed the maximum $|A| + 2|B| + |C|$. But if $|p| + |q| > 1$, we have

$$w^2 K(w) > \frac{1}{2}\frac{w^4}{u^2 \ln u} + T_7.$$

Since $K(w) \leqq 0$, one can easily find an upper bound w_0 for w at the point (x_0, y_0).

Now that we have an estimate for w, there is no difficulty in deriving estimates for the third derivatives r_x and s_x, since

$$w = \alpha^2 e^{2\left(\frac{r+M}{\alpha}\right)} e^{2e^{\left(\frac{r+M}{\alpha}\right)}} (Ar_x^2 + 2Br_x s_x + Cs_x^2).$$

In analogous fashion we find estimates for the third derivatives s_y and t_y.

In deriving estimates for the maximum of w, we have assumed that the maximum is attained in the interior of G for sufficiently small values of α. We shall now estimate w at an interior point of G whose distance from the boundary is ϵ, without assuming that the maximum of w is attained in the interior. Without loss of generality, we may assume that the point of interest is the origin.

Consider the function $\tilde{w} = \lambda w$, where $\lambda = (\epsilon^2 - x^2 - y^2)^2$ and w is Bernšteĭn's auxiliary function, defined above. The function $\tilde{w}$ is nonnegative in the disk ω_ϵ $(x^2 + y^2 \leqq \epsilon^2)$; it vanishes on the circumference of the disk, and so has a maximum at some interior point P of ω_ϵ. At P we have

$$\tilde{w}_x = w_x\lambda + w\lambda_x = 0, \quad \tilde{w}_y = w_y\lambda + w\lambda_y = 0.$$

Hence

$$w_x = -\frac{\lambda_x}{\lambda} w, \quad w_y = -\frac{\lambda_y}{\lambda} w.$$

Now, again at the point P,

$$K(\tilde{w}) = \frac{1}{2}(A\tilde{w}_{xx} + 2B\tilde{w}_{xy} + C\tilde{w}_{yy}) \leqslant 0.$$

Express $K(\tilde{w})$ in the following form:

$$K(\tilde{w}) = \frac{1}{2}\lambda(Aw_{xx} + 2Bw_{xy} + Cw_{yy}) + \frac{1}{2}w(A\lambda_{xx} + 2B\lambda_{xy} + C\lambda_{yy})$$
$$+ (A\lambda_x w_x + B(\lambda_x w_y + \lambda_y w_x) + C\lambda_y w_y).$$

Replacing w_x and w_y by the expressions just found for their values at P and noting that $|\lambda^{-1}\lambda_x^2|$, $|\lambda^{-1}\lambda_x\lambda_y|$ and $|\lambda^{-1}\lambda_y^2|$ are bounded, we get

$$K(\tilde{w}) = \lambda K(w) + wL,$$

where L is a function bounded in absolute value by a number L_0 which depends on the suprema of the absolute values of A, B and C.

Since $K(\tilde{w}) \leqq 0$ at P, we get

$$K(\tilde{w}) - wL_0/\lambda \leqq 0$$

at P.

We find the second derivatives r_1, s_1 and t_1 of the function u at P, which is the maximum point for w, from the system

$$Ar_1 + 2Bs_1 + Ct_1 = Q_1;$$
$$w_x = -\frac{\lambda_x}{\lambda}w; \quad w_y = -\frac{\lambda_y}{\lambda}w.$$

If the right-hand sides of the last two equations are replaced by zero, the result is the system (4), (5) considered above. Hence the solution of our new system is

$$r_1 = \beta_1 + n_1, \quad s_1 = \beta_2 + n_2, \quad t_1 = \beta_3 + n_3,$$

where β_1, β_2, β_3 is a solution of the system (4), (5) and n_1, n_2, n_3 a solution of the system

$$Ar_1 + 2Bs_1 + Ct_1 = 0,$$
$$r_1(Ap_1 + Bq_1) + s_1(Bp_1 + Cq_1) = -\frac{\lambda_x w}{2\lambda},$$
$$s_1(Ap_1 + Bq_1) + t_1(Bp_1 + Cq_1) = -\frac{\lambda_y w}{2\lambda}.$$

The solutions of this system have the form

$$n_1 = \frac{\lambda_x k_1 + \lambda_y l_1}{\lambda}, \quad n_2 = \frac{\lambda_x k_2 + \lambda_y l_2}{\lambda}, \quad n_3 = \frac{\lambda_x k_3 + \lambda_y l_3}{\lambda},$$

where k_1, l_1, k_2, l_2, k_3 and l_3 are linear expressions in p_1 and q_1 with known bounded coefficients.

Set $K(w) = K_\beta(w) + K_n(w)$, where $K_\beta(w)$ is the value of $K(w)$ when r_1, s_1 and t_1 are replaced by β_1, β_2 and β_3 and $K_n(w)$ is the increment to $K_\beta(w)$ when r_1, s_1 and t_1 are made equal to $\beta_1 + n_1$, $\beta_2 + n_2$ and $\beta_3 + n_3$, respectively.

All the terms of $K(w)$ contained in $K_n(w)$ appear in the expression

$$(Ap_1+Bq_1)\left(\frac{\alpha G_2}{u\ln u}+G_4\right)+(Bp_1+Cq_1)\left(\frac{\alpha G_{(2)}}{u\ln u}+G_{(4)}\right)$$
$$+A\left(Ar_1^2+2Br_1s_1+Cs_1^2\right)+2B\left(Ar_1s_1+B\left(r_1t_1+s_1^2\right)+Cs_1t_1\right)$$
$$+C\left(As_1^2+2Bs_1t_1+Ct_1^2\right)+H_1+\frac{\alpha H_3}{u\ln u}.$$

Now observe that for large w the functions β_1, β_2 and β_3 are of the same order as w, while n_1, n_2 and n_3 have order at most that of $\lambda^{-1}\sqrt{w}(|\lambda_x|+|\lambda_y|)$. Hence we easily conclude that for large w

$$|K_n(w)|<N\left\{\frac{w^{3/2}}{\lambda}(|\lambda_x|+|\lambda_y|)+\frac{w}{\lambda^2}(\lambda_x^2+\lambda_y^2)\right\},$$

where N is a constant which depends only on the suprema of the absolute values of the first and second derivatives of z, the suprema of the absolute values of the derivatives of F of order $\leqq 2$, and the suprema of the functions $1/F_r$, $1/F_t$ and $1/(F_rF_t-F_s^2/4)$ in G.

For sufficiently small α and sufficiently large w,

$$K_\beta(w)>\frac{1}{4}\frac{w^2}{u^2\ln u}.$$

But $K(w)-L_0w/\lambda<0$, and hence for large w

$$\frac{1}{4}\frac{w^2}{u^2\ln u}-N\left(\frac{w^{3/2}}{\lambda}(|\lambda_x|+|\lambda_y|)+\frac{w}{\lambda^2}(\lambda_x^2+\lambda_y^2)\right)-\frac{wL_0}{\lambda}<0.$$

It follows that for a sufficiently large number R_0, which depends only on the suprema of the absolute values of the derivatives of F and z in G, the function w is bounded above by $w_0=R_0/\lambda$ at the point P.

Consequently $\tilde{w}\leqq R_0$ in the disk ω_ϵ, and in particular at its center. We thus have the estimate $w\leqq R_0/\epsilon^4$ for w at a point of G whose distance from the boundary is ϵ, and R_0 depends only on the suprema of the absolute values of the derivatives of z and F of orders up to two and three, respectively, and on the suprema of the absolute values of

$$\frac{1}{F_r},\ \frac{1}{F_t},\ \frac{1}{F_rF_t-\frac{1}{4}F_s^2}$$

in the domain G.

Once estimates have been derived for w, the third derivatives of z are easily estimated. This completes the proof.

Estimates for the fourth and higher derivatives of the solution $z(x,y)$ of equation (1) may be derived on the basis of the following theorem of Schauder [**67**] for linear elliptic equations:

Consider the following linear elliptic partial differential equation in a bounded domain G of the x_1, x_2-plane:

$$a_{11}(x_1, x_2)u_{11}+2a_{12}(x_1, x_2)u_{12}+a_{22}(x_1, x_2)u_{22}=f(x_1, x_2),$$

where $a_{11}a_{22} - a_{12}^2 = 1$. Assume that the coefficients a_{ij} and the function f satisfy a Hölder condition with exponent $\alpha + \epsilon$ $(0 < \alpha < 1,\ \epsilon > 0)$ and constant M in the closed domain $\overline{G}$.

Let the second derivatives of the solution $u(x_1, x_2)$ satisfy a Hölder condition with exponent α. Then the suprema of the absolute values of the first and second derivatives of $u(x_1, x_2)$ in a domain B whose closure is contained in G, and the smallest Hölder constants for the second derivatives of $u(x_1, x_2)$ in the domain B for exponent α, satisfy estimates depending only on M, the maximum modulus of $u(x_1, x_2)$ in G, and the distance of B from the boundary of G.

Let

$$F(x, y, z, p, q, r, s, t) = 0 \tag{6}$$

be an elliptic partial differential equation in the disk ω_ϵ $(x^2 + y^2 \leqq \epsilon^2)$ and $z(x, y)$ a solution. Differentiating equation (6), say with respect to x, we obtain an equation for the function $u = z_x$:

$$\frac{A}{\Delta} u_{xx} + \frac{2B}{\Delta} u_{xy} + \frac{C}{\Delta} u_{yy} = \frac{D}{\delta}, \tag{7}$$

where $A = F_r$, $2B = F_s$, $C = F_t$, $\Delta^2 = AC - B^2$ and D is an expression in the first derivatives of F and the first and second derivatives of $z(x, y)$.

If we regard the coefficients of equation (7) as unknown functions of x and y, the assumptions of Schauder's Theorem hold. In fact, we can find an estimate for the Hölder constant for exponent α $(0 < \alpha < 1)$ depending only on the suprema of the absolute values of the second derivatives of F, the third derivatives of $z(x, y)$, and the infimum of the discriminant $AC - B^2$ in the disk ω_ϵ.

By Schauder's Theorem we can find an estimate of the Hölder constant for the second derivatives of u (i.e. the third derivatives of z), for exponent $\alpha' < \alpha$, in a disk $\omega_{\bar{\epsilon}}$ contained in ω_ϵ. Thus, having estimates for the third derivatives of the solution, we can estimate the smallest Hölder constants for these derivatives for any exponent α $(0 < \alpha < 1)$ in a disk $\omega_{\bar{\epsilon}} \subset \omega_\epsilon$, using only second derivatives of F.

Differentiating equation (6) twice, say, with respect to x, we obtain an equation of the same type as (7) for the function $v = z_{xx}$, except that now the right-hand side involves the second derivatives of F and the third derivatives of $z(x,y)$. Again regarding the coefficients as unknown functions of x and y, we apply Schauder's Theorem. This yields estimates for the second derivatives of v and their smallest Hölder constants for exponent $\alpha'' < \alpha'$ in a disk $\omega_{\bar{\bar{\epsilon}}}$ contained in $\omega_{\bar{\epsilon}}$.

Thus step by step we can derive estimates for derivatives of z of arbitrary order and their smallest Hölder constants for any positive exponent smaller than unity. The final result may be stated in the following form.

THEOREM 2. *Let*

$$F(x,y,z,p,q,r,s,t) = 0$$

be an elliptic partial differential equation, and $z(x,y)$ *a solution regular in the disk* $\omega_\epsilon : x^2 + y^2 \leqq \epsilon^2$. *Then one can find estimates for the kth derivatives of the solution* $z(x,y)$ *in the interior of a disk* $\omega_{\bar{\epsilon}}$ $(\bar{\epsilon} < \epsilon)$ *which depend only on the suprema of the absolute values of* z *and its first and second derivatives in* ω_ϵ, *the suprema of the quantities*

$$\frac{1}{|F_r|}, \quad \frac{1}{|F_t|}, \quad \frac{1}{F_r F_t - \frac{1}{4} F_s^2}$$

and the suprema of the absolute values of the derivatives of F *of order* $\leqq s$, *where* $s = 3$ *for* $k = 3$ *and* $s = k - 1$ *for* $k > 3$.

Moreover, one can find estimates, depending on the same quantities, for the smallest Hölder constants of the kth derivatives of z *for any exponent* α $(0 < \alpha < 1)$.

CHAPTER III

MONOTYPY OF CONVEX SURFACES

A convex surface F is said to be monotypic in a class K of convex surfaces if any surface in K isometric to F is congruent to F. If K is the class of all convex surfaces F is said to be monotypic.[1] A typical monotypy theorem is the following:

Any two isometric closed convex polyhedra are congruent.

The significance of the monotypy problem for convex surfaces was considerably enhanced by Aleksandrov's theorems on the realizability of abstractly defined convex metrics by convex surfaces. As tools for investigation of convex surfaces, these theorems become more powerful when monotypy can be established. Realizability and monotypy theorems stand in the same relation in the theory of convex surfaces as do existence and uniqueness theorems in the theory of differential equations.

The monotypy problem has attracted the efforts of numerous geometers. The earliest result is Cauchy's theorem (1813), which states that two closed convex polyhedra whose faces are congruent and arranged in the same way are congruent.

Somewhat later, in 1838, Minding conjectured that the sphere is rigid. However, only in 1899 were Liebmann and Minkowski able to prove this theorem. They in fact proved a stronger result: the sphere admits no nontrivial isometric images, i.e. it is monotypic.

Subsequently questions of rigidity and monotypy were studied by Hilbert, Weyl, Blaschke and Cohn-Vossen, whose joint efforts yielded proofs of the rigidity of closed regular convex surfaces (Liebmann, Blaschke, Weyl) and the monotypy of closed regular (thrice continuously differentiable) [convex] surfaces with positive Gauss curvature in the class of surfaces with this degree of regularity (Cohn-Vossen [**27**]). Cohn-Vossen's theorem is the following:

---

[1] *Translator's note.* This paragraph has been slightly rephrased, to conform with Busemann's terminology.

If F_1 is a thrice continuously differentiable closed convex surface with positive Gauss curvature and F_2 a thrice continuously differentiable surface isometric to F_1, then F_2 is congruent either to F_1 or to its mirror image.

In 1941, S. P. Olovjanišnikov proved an essentially new result concerning the monotypy of convex surfaces: Any convex surface F isometric to a closed convex polyhedron P (without the assumption that F is a polyhedron) is a polyhedron congruent to P [**50**].

The definitive solution of the monotypy problem for (not necessarily closed) convex surfaces was first obtained by the present author for convex surfaces of bounded specific curvature [**55**], and subsequently for general convex surfaces [**56**]. This chapter is devoted to an exposition of these results.

§1. Curves with i.g.c. of bounded variation

Curves with integral geodesic curvature of bounded variation constitute a fairly large class of curves which are, in a certain sense, "well-behaved" on a general convex surface. In the sequel they will occupy a central position in our proof of the fundamental monotypy theorem for closed convex surfaces (§6).

A *curve* is a continuous and single-valued image in space of a closed interval [of the real axis]. We now define the variation of the i.g.c. of a curve.

Let γ be a space curve. Let c be a polygon inscribed in γ and denote by $\omega(c)$ the sum of complements of the angles of the polygon. The upper limit $\omega(\gamma)$ of $\omega(c)$ over all polygons c inscribed in γ will be called the *variation of the i.g.c.* of γ. If $\omega(\gamma)$ is finite, we call γ a *curve with i.g.c. of bounded variation.*

We consider some properties of curves with i.g.c. of bounded variation. First note the following lemma.

LEMMA 1. *Let c be a polygon in space. If a side a of c is replaced by two sides, then the i.g.c. ω of the polygon does not decrease, and indeed it will increase if the sides of c adjacent to a are not coplanar.*

PROOF. Let e_1, e_2, e_3, e_2' and e_2'' be vectors representing sides of the polygon (Figure 23). Let $\vartheta(x,y)$ denote the angle between vectors x and y. Then the i.g.c. of the original polygon over AB is

$$\omega(c) = \vartheta(e_1, e_2) + \vartheta(e_2, e_3).$$

When a has been replaced by two sides, the i.g.c. of the polygon over AB becomes

$$\omega(c') = \vartheta(e_1, e_2') + \vartheta(e_2', e_2'') + \vartheta(e_2'', e_3).$$

Since $e_2 = e_2' + e_2''$, we have

$$\vartheta(e_2', e_2) + \vartheta(e_2, e_2'') = \vartheta(e_2', e_2'').$$

Now the sum of two plane angles of a trihedral angle is not less than the third, and is equal to it only when the trihedral angle is degenerate. Hence

$$\vartheta(e_1, e_2) \leqslant \vartheta(e_1, e_2') + \vartheta(e_2', e_2),$$
$$\vartheta(e_2, e_3) \leqslant \vartheta(e_2, e_2'') + \vartheta(e_2'', e_3).$$

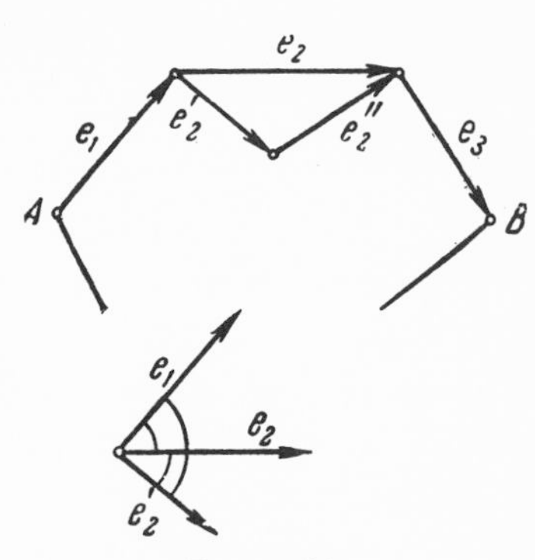

FIGURE 23

Consequently $\omega(c) \leqq \omega(c')$, and strict inequality holds here if the three sides in question are not coplanar. This proves the lemma.

LEMMA 2. *Let V_1 and V_2 be two circular cones with common apex O and no other common points. Let A and B be points of V_1, C a point of V_2. Then there exists a positive number ϵ such that for any position of A, B, C the angle $\vartheta(C)$ of the triangle ABC is at most $\pi - \epsilon$.*

PROOF. When the apex A is displaced along the ray OA in a suitable direction, the angle $\vartheta(C)$ increases monotonically. Similarly, when B is displaced along the ray OB in a suitable direction the angle $\vartheta(C)$ again increases monotonically. It follows that $\vartheta(C)$ is bounded above either by the maximum angle α between rays within the cone V_1 or by $\pi - \beta$, where β is the minimum angle between a ray within V_1 and a ray within V_2. The number ϵ is the smaller of $\pi - \alpha$ and β. Q.E.D.

THEOREM 1. *A curve with i.g.c. of bounded variation has a left and right semitangent at each point.*

PROOF. Suppose the assertion false, and let γ be a curve with i.g.c. of bounded variation which has no right semitangent at a point O. Then there exist two sequences of points on the curve which converge from the right at O in two different directions t_1 and t_2. Describe a cone with apex O about each of the rays t_1 and t_2, so that the two cones have no common points other than O. Now inscribe in the curve γ a polygon c with so many sides that all its vertices are in the cones, and such that any two neighboring vertices lie in different cones. By Lemma 2 the i.g.c. of the polygon c may be made arbitrarily large by increasing the number of its sides. But this is impossible, since by assumption the curve γ has i.g.c. of bounded variation. Q.E.D.

THEOREM 2. *Let γ be a curve with i.g.c. of bounded variation, divided by a point P into two curves γ' and γ''. Let $\vartheta(P)$ be the angle between the right and left semitangents at P. Then $\omega(\gamma) = \omega(\gamma') + \omega(\gamma'') + (\pi - \vartheta(P))$.*

PROOF. Let c_n', c_n'' and c_n be sequences of polygons inscribed in the curves γ', γ'' and γ such that $\omega(c_n') \to \omega(\gamma')$, $\omega(c_n'') \to \omega(\gamma'')$ and $\omega(c_n) \to \omega(\gamma)$. By Lemma 1 we mav assume without loss of generality that the sides of the polygons approach zero. To the vertices of each of the polygons c_n', c_n'' and c_n, add those vertices of the two others that belong to the curve in question. Call the new polygons $\bar{c}_n'$, $\bar{c}_n''$ and $\bar{c}_n$. They possess the same properties as c_n', c_n'' and c_n: as $n \to \infty$,

$$\omega(\bar{c}_n') \to \omega(\gamma'), \quad \omega(\bar{c}_n'') \to \omega(\gamma''), \quad \omega(\bar{c}_n) \to \omega(\gamma).$$

Since $\omega(\bar{c}_n) = \omega(\bar{c}_n') + \omega(\bar{c}_n'') + (\pi - \vartheta_n(P))$ (where $\vartheta_n(P)$ is the angle between the sides of the polygon $\bar{c}_n$ at P) and $\vartheta_n(P) \to \vartheta(P)$, we get

$$\omega(\gamma) = \omega(\gamma') + \omega(\gamma'') + (\pi - \vartheta(P)).$$

This proves the theorem.

THEOREM 3. *A curve with i.g.c. of bounded variation cannot contain more than countably many corner points, i.e. points at which the right and left semitangents form an angle different from π.*

PROOF. By Theorem 2, the set M_n of corner points of the curve at which the semitangents form an angle less than $\pi - 1/n$ is finite. Hence the set $M = \sum_n M_n$ of all corner points is at most countable. Q.E.D.

THEOREM 4. *Let P be a point on a curve γ with i.g.c. of bounded variation. Then there is a point $Q \neq P$ on γ, to the right (left) of P, such that the variation of the i.g.c. of the arc PQ of γ is arbitrarily small.*

PROOF. Let γ' be the part of γ to the right of the point P. Inscribe a polygon c in γ' such that $\omega(\gamma') - \omega(c) < \epsilon$. Let A be the point of c closest to P. Let Q be another point of γ', between P and A, so close to P that when it is added to c as a new vertex the angle between the sides at A varies by at most ϵ. Obviously the variation of the arc PQ of γ is at most 2ϵ. This proves the theorem.

THEOREM 5. *A curve with i.g.c. of bounded variation can be divided into a finite number of subarcs with i.g.c. of arbitrarily small variation.*

PROOF. By Theorem 4 each point P of the curve has a closed neighborhood $U(P)$ such that the right and left semineighborhoods $U'(P)$ and $U''(P)$ have i.g.c. less than ϵ. There exists a finite covering of the curve consisting of neighborhoods $U(P)$. Using the corresponding semineighborhoods, one easily constructs a partition of the curve with the required property. Q.E.D.

THEOREM 6. *Any curve with i.g.c. of bounded variation is rectifiable.*

PROOF. By Theorem 5 it will suffice to prove that a curve with i.g.c. of variation less than ϵ is rectifiable. Let c be a polygon inscribed in such a curve. The sides of this polygon form angles less than ϵ with the first side. Therefore the lengths of all possible polygons c are bounded by a number l which depends only on ϵ and the distance between the endpoints of the curve. It follows that the curve is rectifiable.

THEOREM 7. *Let P be a point on a curve γ with i.g.c. of bounded variation, Q a point on γ close to P, s_{PQ} the length of the arc PQ of γ, and d_{PQ} the distance between P and Q. Then s_{PQ}/d_{PQ} tends to unity as $Q \to P$.*

PROOF. The variation $\omega(P, Q)$ of the i.g.c. of the arc PQ of γ tends to zero as $P \to Q$. Now each side of a polygon c inscribed in the arc PQ forms an angle at most $\omega(P, Q)$ with the first side; hence the length of the polygon c, and therefore of the arc PQ, is at most $d_{PQ}/\cos\omega(P, Q)$. Since $s_{PQ}/d_{PQ} > 1$, it follows that $s_{PQ}/d_{PQ} \to 1$ as $Q \to P$.

THEOREM 8. *Let γ be a curve with i.g.c. of bounded variation, and $r(s)$ the point on the curve corresponding to arc length s. Then the vector-valued function $r(s)$ has right and left derivatives for all s, and these are equal to unity in absolute value.*

The assertion is obvious, by Theorems 1 and 7.

For curves which have a right (or left) semitangent at each point, the variation of the i.g.c. may be defined in a different manner. Let γ be a curve having a right semitangent at each point. Take finitely many points P_k on γ and let t_k be the right semitangent at P_k. The supremum of the sum of angles between consecutive semitangents, taken over all finite sequences P_k, is called the *variation of the i.g.c.* of γ. It is not yet evident that this definition is equivalent to the preceding one. In the following theorems, all references to i.g.c. of bounded variation follow this new definition.

THEOREM 9. *Let γ be a curve with i.g.c. of bounded variation, P a point on γ and Q a point on γ close to P. Then if $Q \to P$ from the right, the right semitangent at Q tends to the right semitangent at P; if $Q \to P$ from the left, the semitangent at Q tends to a ray which need not coincide with the right semitangent at P.*

PROOF. The fact that the semitangent at Q tends to a well-defined limit when Q tends to P from the right (left) follows directly from the boundedness of the i.g.c. of γ. We shall show that the semitangent at Q tends to the semitangent at P when $Q \to P$ from the right. Suppose this assertion false. Let t_1 be the right semitangent at P and t_2 the limit of the semitangent at Q as $Q \to P$ from the right. Let P' be a point close to P. Construct the plane α through P and P' perpendicular to to the plane of the rays t_1 and t_2, and then rotate α through a small angle about the midpoint of the segment PP', into a position α' such that P' is on one side of α', P and t_2 on the other (Figure 24). Let P'' be the first point of the arc PP' of the curve γ, counting from P' to P, which is in the plane α'. The right semitangent to γ at P'' either points into the halfspace defined by α' which contains P'', or lies in α'. Hence there exists a point P'' arbitrarily close to P at which the right semitangent to γ forms an angle less than $\vartheta - \epsilon$ with the ray t_2, where ϑ is the angle between t_1 and t_2 and ϵ an arbitrary positive number. This contradicts the fact that t_2 is a limit of right semitangents, proving the theorem.

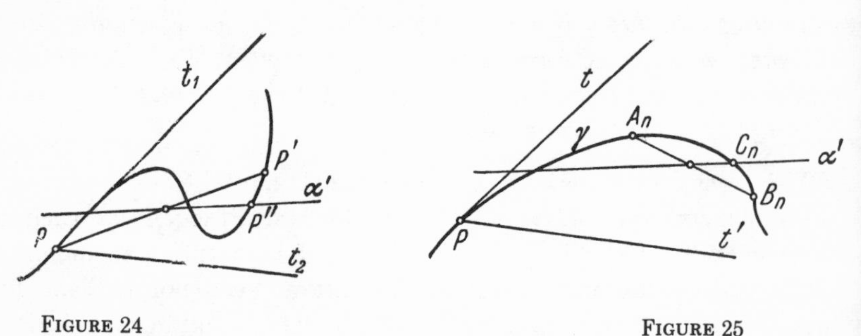

FIGURE 24 FIGURE 25

THEOREM 10. *Let γ be a curve with i.g.c. of bounded variation, P a point on the curve, A and B points on γ to the right of P such that A lies between P and B. Then when A and B tend to P the direction of the secant AB tends to the direction of the right semitangent t to γ at P.*

PROOF. Suppose the assertion false. Without loss of generality we may assume that there exists a sequence of pairs of points A_n and B_n such that the direction of the secant A_nB_n converges to a direction t' different from t. Construct the plane α parallel to t' and perpendicular to the plane of t and t'. Displace α parallel to itself until it passes through the midpoint of A_nB_n; then rotate it through a small angle to a position α' in which the points A_n and B_n lie on different sides of α', A_n being on the side into which the ray t points (Figure 25). Now let C_n be the first point on the arc B_nP of γ, counting from B_n to P, which lies in α'. If the angle between the planes α and α' is small in comparison with the angle between t and t', then the right semitangent at C_n does not converge to t, the right semitangent at P. This contradiction proves the theorem.

REMARK TO THEOREM 10. Analogous reasoning will obviously prove that if A and B lie to the left of P and B lies between A and P, then when A and B approach P the direction of the secant AB tends to the limiting direction of the right semitangent at Q when Q tends to P from the left.

THEOREM 11. *Any curve with i.g.c. of bounded variation is rectifiable.*

PROOF. Let P be an arbitrary point of the curve and Q a point on the curve to the right of and close to P. By Theorem 10, if Q is sufficiently close to P, then all sides of a polygon c inscribed in the arc PQ of the curve form angles less than $\epsilon < \pi/2$ with the right semitangent at P. It follows that the length of the polygon c is at most $PQ/\cos\epsilon$. Similarly, using the Remark to Theorem 10 one shows that this is also true for an arc PQ of the curve when Q lies to the left of and sufficiently close to P. Hence each point of the curve has a rectifiable neighborhood, and so the entire curve is rectifiable.

THEOREM 12. *Let γ be a curve with i.g.c. of bounded variation, and $r(s)$ the radius-vector of the point defined on the curve by arc length s. Then the right and left derivatives of $r(s)$ with respect to s exist at each point of the curve, and they are equal to unity in absolute value. For almost all s, except perhaps a countable set, the right and left derivatives coincide.*

PROOF. Let Δs be a small positive number. Inscribe a polygon c in the arc $(s, s+\Delta s)$ of the curve, with sides so small that the length of c is greater than $\Delta s(1-\epsilon)$. By Theorem 10, when Δs is sufficiently small the sides of the polygon form angles less than ϵ with the right semitangent at s. Consequently the direction of the vector $(r(s+\Delta s) - r(s))/\Delta s$

tends to the right semitangent at s, and its absolute value tends to unity.

Analogous reasoning proves the existence of the left derivative of $r(s)$. That the points at which these derivatives are different form an at most countable set is proved by arguments similar to those of Theorem 3.

THEOREM 13. *The two definitions of variation of i.g.c. are equivalent.*

PROOF. Let γ be a curve with i.g.c. of bounded variation in the sense of the first definition. By Theorem 1 it has right and left semitangents at each point. Let $P_1, \cdots, P_n$ be a finite sequence of points on γ and t_k the right semitangent at P_k. Let Q_k be a point to the right of and close to P_k, and c a polygon inscribed in γ with vertices $P_1, Q_1, \cdots, P_n, Q_n$. The i.g.c. $\omega(c)$ of this polygon is at most $\omega(\gamma)$. If we let Q_k approach P_k, the directions of the segments P_kQ_k tend to the directions of the semitangents t_k. Hence the sum of the angles $\omega_k'(\gamma)$ between consecutive semitangents t_k cannot exceed $\omega(\gamma)$. Thus the curve γ has i.g.c. of bounded variation in the sense of the second definition, and moreover $\omega'(\gamma) \leqq \omega(\gamma)$.

Now let γ have i.g.c. of bounded variation in the sense of the second definition. We shall show that it has this property in the sense of the first definition and moreover $\omega(\gamma) \leqq \omega'(\gamma)$. Let c be an arbitrary polygon inscribed in γ. Enlarge the set of vertices of c by adding points at which the right and left semitangents form angles greater than ϵ (where ϵ is a small positive number). Now extend the polygon c by adding vertices in such a way that the right semitangents to γ at points lying between neighboring vertices of the polygon form angles less than ϵ_1 ($\epsilon_1 \ll \epsilon$). Let us estimate the i.g.c. of the resulting polygon c'.

Let $\tau_i(s)$ be a unit vector on the right semitangent to γ, over the arc corresponding to the ith side of the polygon, δ_i the length of this side. Then the i.g.c. of c' is

$$\omega(c') = \sum_i \vartheta\left(\frac{1}{\delta_i}\int \tau_i\, ds,\ \frac{1}{\delta_{i+1}}\int \tau_{i+1}\, ds\right),$$

where ϑ denotes the angle between its arguments. It is convenient to split the sum in the right-hand side into sums $\sum'$ and $\sum''$; the first will include terms corresponding to the vertices of the polygon at which the right and left semitangents form angles greater than ϵ, the second all other terms. Now the terms of the original sum satisfy the inequality

$$\vartheta\left(\frac{1}{\delta_i}\int \tau_i\, ds,\ \frac{1}{\delta_{i+1}}\int \tau_{i+1}\, ds\right) \leqslant (1+\eta_1)\left|\frac{1}{\delta_i}\int \tau_i\, ds - \frac{1}{\delta_{i+1}}\int \tau_{i+1}\, ds\right|$$

$$\leqslant (1+\eta_2)\left|\frac{1}{s_i}\int \tau_i\, ds - \frac{1}{s_{i+1}}\int \tau_{i+1}\, ds\right|,$$

where η_1 and η_2 are arbitrarily small positive numbers, provided ϵ_1 is sufficiently small. Now parametrize each arc of the curve γ by assigning to each point X the value λ equal to the distance of X from the initial point of the arc divided by the length of the entire arc. The terms of $\sum''$ then satisfy the inequality

$$\left|\frac{1}{s_i}\int \tau_i\, ds - \frac{1}{s_{i+1}}\int \tau_{i+1}\, ds\right| = \left|\int_0^1 \tau_i\, d\lambda - \int_0^1 \tau_{i+1}\, d\lambda\right|$$

$$= \left|\int_0^1 (\tau_i - \tau_{i+1})\, d\lambda\right| \leqslant (1+\eta_3)\int_0^1 \vartheta(\tau_i, \tau_{i+1})\, d\lambda,$$

where η_3 tends to zero with ϵ_1.

As for the terms of $\sum'$,

$$\vartheta\left(\frac{1}{\delta_i}\int \tau_i\, ds, \frac{1}{\delta_{i+1}}\int \tau_{i+1}\, ds\right) \leqslant (1+\eta_3')\int_0^1 \vartheta(\tau_i, \tau_{i+1})\, d\lambda.$$

Using these estimates for the sums $\sum'$ and $\sum''$, we get

$$\omega(c') \leqslant (1+\eta)\int_0^1 \sum_i \vartheta(\tau_i, \tau_{i+1})\, d\lambda,$$

where η tends to zero with ϵ_1. It follows that $\omega(c') \leqq \omega'(\gamma)$, and hence $\omega(\gamma) \leqq \omega'(\gamma)$. Since we have already proved the reverse inequality, it follows that $\omega(\gamma) = \omega'(\gamma)$. Q.E.D.

THEOREM 14. *Let $\gamma_1, \gamma_2, \cdots$ be an infinite sequence of curves with i.g.c. of bounded variation such that the variation of the i.g.c. of each curve is at most V. If the sequence γ_n converges to a curve γ, then γ has i.g.c. of bounded variation which is also bounded above by V.*

PROOF. Inscribe an arbitrary polygon c in γ. If n is sufficiently large, we can inscribe a polygon c_n close to c in γ_n in such a way that $|\omega(c) - \omega(c_n)| < \epsilon$. Since the variation of the i.g.c. of the curves is bounded by V, we have $\omega(c_n) \leqq V$. Since c is arbitrary and ϵ can be made arbitrarily small, it follows that the curve γ has i.g.c. of bounded variation at most V.

THEOREM 15. *If the diameter of the curve γ is at most d and the vari-*

ation at most V, then its length is at most $2([3V/\pi]+1)d$, where (as usual) $[3V/\pi]$ denotes the greatest integer $\leqq 3V/\pi$.

PROOF. Divide the curve γ into $[3V/\pi]+1$ arcs γ_k, each with i.g.c. of variation at most $\pi/3$. Inscribe an arbitrary polygon c in γ_k. The angles between the sides of c and the initial side are at most $\pi/3$. Therefore the length of the polygon c, and hence of the curve γ_k, is at most $d/\cos(\pi/3)$. Since the number of curves γ_k is $[3V/\pi]+1$, the length of the entire polygon is at most $2([3V/\pi]+1)d$. This proves the theorem.

In §8 of Chapter I we defined the i.g.c. of a curve on a convex surface and indicated its fundamental properties. This concept will now be used to define the class of curves with i.g.c. of bounded variation *on a surface*, which will play an essential role in the sequel.

Let γ be a curve on a convex surface which has a well-defined i.g.c. on each subarc. Divide it by points $P_1, \cdots, P_k$ into subarcs $\gamma_1, \cdots, \gamma_{k+1}$ and consider the sum

$$\sum_i |\varphi(\gamma_i)| + \sum_j |\pi - \vartheta(P_j)|,$$

where $\varphi(\gamma_i)$ is the right i.g.c. of γ_i and $\vartheta(P_i)$ the angle of the right sector at P_i. The supremum of all such sums over all possible subdivisions of the curve γ will be called the *variation of the right i.g.c. of* γ on the surface. It turns out that if the variation of the right i.g.c. is bounded, so is the variation of the left i.g.c.

THEOREM 16. *Let F be a convex surface lying above the xy-plane such that the supporting planes to F form angles $\leqq \alpha < \pi/2$ with the xy-plane. Let γ be a curve on F whose projection on the xy-plane is a curve $\bar{\gamma}$ with i.g.c. of bounded variation (for plane curves both definitions yield the same variation of i.g.c.). Then γ has i.g.c. of bounded variation in space.*

PROOF. Inscribe a polygon with small sides in the curve $\bar{\gamma}$. The vertices of this curve can be subjected to small displacements so that they become projections of smooth points of F and the i.g.c. of the polygon varies by at most ϵ. Let c denote the resulting polygon. Let γ_c be the curve on the surface whose projection is c, and Z_c the cylinder projecting the curve γ_c onto the xy-plane. If the cylinder Z_c is developed on a plane, the curve γ_c becomes a plane curve $\tilde{\gamma}_c$ consisting of arcs of convex curves $\tilde{\gamma}_1, \cdots, \tilde{\gamma}_n$ whose endpoints $P_1, \cdots, P_{n-1}$ correspond to the vertices of c. When the curve $\tilde{\gamma}_c$ is traversed, the right semitangent turns in the same sense over each subarc $\tilde{\gamma}_i$. Let $\tilde{\vartheta}_i$ be the change of direction of the right semitangent at P_i, and $\bar{\vartheta}_i$ the change of direction of the side of c at the vertex corresponding to P_i. Since the tangent planes to the surface F form angles $\leqq \alpha < \pi/2$ with the xy-plane, we can find a constant m such that $|\tilde{\vartheta}_i| \leqq m|\bar{\vartheta}_i|$. Hence $\sum|\tilde{\vartheta}_i| \leqq m\sum|\bar{\vartheta}_i|$, or, in terms of the variation of the i.g.c. of $\bar{\gamma}$, $\sum|\tilde{\vartheta}_i| < m\omega(\bar{\gamma}) + \epsilon$.

To summarize: When $\tilde{\gamma}_c$ is traversed, the right semitangent turns in the same sense over each subarc $\tilde{\gamma}_i$, and the sum of absolute values of the angles through which the right semitangents at P_i turn is less than $m\omega(\bar{\gamma}) + \epsilon$; hence the variation of the i.g.c. of $\tilde{\gamma}_c$ is at most $\pi + m\omega(\bar{\gamma}) + \epsilon$.

It follows that if the cylinder projecting the curve γ is developed on a plane, the curve γ becomes a plane curve $\tilde{\gamma}$ with i.g.c. of variation at most $\pi + m\omega(\bar{\gamma})$.

Let Q_1 and Q_2 be two points on the curve γ. Let t_1 and t_2 be the right semitangents at these points, $\bar{t}_1$ and $\bar{t}_2$ the right semitangents to $\bar{\gamma}$, and $\tilde{t}_1$ and $\tilde{t}_2$ the right semitangents to $\tilde{\gamma}$ at the points corresponding to Q_1 and Q_2. Obviously the sum of the angles between $\bar{t}_1$ and $\bar{t}_2$, $\tilde{t}_1$ and $\tilde{t}_2$ is not less than the angle between t_1 and t_2. Therefore the variation of the i.g.c. of the curve γ in space is at most the sum of variations of i.g.c. of $\bar{\gamma}$ and $\tilde{\gamma}$, i.e. at most $\pi + (m+1)\omega(\bar{\gamma})$. This completes the proof of Theorem 16.

THEOREM 17. *If a curve γ on a convex surface F has right i.g.c. of bounded variation, then it has i.g.c. of bounded variation in space.*

PROOF. We first observe that γ can be divided into a finite number of subarcs each of which has an ϵ-neighborhood whose spherical image lies on a segment of the unit sphere smaller than a hemisphere. It will therefore suffice to prove the theorem for each of these subarcs. Without loss of generality we may assume that the curve γ itself has an ϵ-neighborhood whose spherical image lies in a segment smaller than a hemisphere.

Let g' be a ray which forms angles $\leqq \pi/2 - \delta$ with the exterior normals to F at points of the ϵ-neighborhood of γ, where $\delta > 0$.

By a theorem of Zalgaller [72] there exists a sequence of geodesic polygons c_n on the surface which converges uniformly to γ, whose right i.g.c. have variations which converge to that of γ, and whose lengths converge to that of γ. We may assume that the vertices of these polygons are smooth points of F.

Project the geodesic polygon c_n in the direction of g', and develop the projecting cylinder Z' on a plane. The polygon c_n then becomes a curve $\tilde{c}_n$. By Liberman's theorem on the convexity of geodesics each side of $\tilde{c}_n$ on the developed cylinder is a convex curve, and all these curves are convex toward the direction of g'. Let us estimate the variation of the i.g.c. of $\tilde{c}_n$. Let $\tilde{c}_n^1, \cdots, \tilde{c}_n^m$ be the subarcs of this curve corresponding to the sides of the geodesic polygon, and $\widetilde{P}_1, \cdots, \widetilde{P}_{m+1}$ the points corresponding to its vertices. The curve c_n has i.g.c. of variation $\sum|\vartheta_i|$ on the surface F, where ϑ_i is the angle through which the right semitangent to c_n turns at P_i. Let $\tilde{\vartheta}_i$ be the angle through which the right semitangent to $\tilde{c}_n$ turns at $\widetilde{P}_i$. Since the exterior normals to the tangent planes of F at the vertices of c_n form angles less than $\pi/2 - \delta$ with g', there exists a constant m which depends only on δ such that $|\tilde{\vartheta}_i| \leqq m|\vartheta_i|$. Hence for sufficiently large n we have $\sum|\tilde{\vartheta}_i| < m\omega_F(c_n) + \epsilon'$, where $\omega_F(c_n)$ is the variation of the right i.g.c. of c_n on F and ϵ' tends to zero as $n \to \infty$.

Now when we move along the curve c_n, the right semitangent turns in the same sense over each subarc c_n^i, and the sum of absolute values of the angles through which the right semitangents turn at $\widetilde{P}_i$ is less than $m\omega_F(c_n) + \epsilon'$; hence the variation of the i.g.c. of $\tilde{c}_n$ is less than $\pi + m\omega_F(c_n) + \epsilon'$.

It follows that when the cylinder projecting γ in the direction of g' is developed on a plane, the curve γ becomes a curve $\tilde{\gamma}'$ with i.g.c. of variation at most $\pi + m\omega_F(\gamma)$.

We now construct two more rays g'' and g''' with the same property as g', but not coplanar with g'. Construct curves $\tilde{\gamma}''$ and $\tilde{\gamma}'''$ in the same way as $\tilde{\gamma}'$. Let Q_1 and Q_2 be two arbitrary points on γ, t_1 and t_2 the right semitangents at these points, and t_1' and t_2', t_1'' and t_2'', t_1''' and t_2''' the right semitangents at the corresponding points of $\tilde{\gamma}'$, $\tilde{\gamma}''$, $\tilde{\gamma}'''$. Elementary arguments show that there exists a constant κ, depending only on δ and the angles between the rays g', g'' and g''', such that

$$\vartheta(t_1, t_2) \leqq \kappa(\vartheta(t_1', t_2') + \vartheta(t_1'', t_2'') + \vartheta(t_1''', t_2''')).$$

It follows that the variation of the i.g.c. of γ in space satisfies the inequality $\omega(\gamma) \leqq 3\kappa(\pi + m\omega_F(\gamma))$. This completes the proof.

THEOREM 18. *Let F be a convex surface lying above the xy-plane, whose supporting planes form angles $\leqq \pi/2 - \delta$, $\delta > 0$, with the xy-plane. Let γ be a curve on F, $\bar{\gamma}$ its projection on the xy-plane. If $\bar{\gamma}$ has i.g.c. of bounded variation, then γ has right (left) i.g.c. of bounded variation on F.*

PROOF. Let F_n be a sequence of convex polyhedra converging to F, and $\bar{c}_n$ a sequence of polygons on the xy-plane converging from the right to $\bar{\gamma}$ such that the variation of the i.g.c. of $\bar{c}_n$ converges to that of $\bar{\gamma}$ and no vertex of any polyhedron F_n is projected onto a vertex of a polygon $\bar{c}_n$. Let c_n be the curve on the polyhedron F_n whose projection is $\bar{c}_n$. The variation of the right i.g.c. of c_n on F_n is clearly not greater than that of c_n in space, and the latter can be bounded in terms of the variation of the i.g.c. of $\bar{c}_n$ and the greatest angle between the supporting planes of F_n and the xy-plane.

Thus the variations of the right i.g.c. of the curves c_n on the polyhedra F_n are uniformly bounded. It follows that the curve γ has right i.g.c. of bounded variation on F.

§2. On convergence of isometric convex surfaces

Throughout the sequel, we shall often have to consider convergent sequences of isometric convex surfaces. Most frequently we shall have to study how directions that correspond under the isometry converge, and ascertain their relation to the direction on the limit surface.[2)] To avoid repetitious and laborious proofs of auxiliary theorems, in this section we shall prove certain general propositions.

THEOREM 1. *Let F be a convex surface, F_n a sequence of convex surfaces isometric to F and converging to F, X_0 a point on F, X_n points on F_n corresponding under the isometry to X_0, and $s_1, s_2, \cdots$ a decreasing sequence of positive numbers converging to zero. Let $\bar{F}_n$ be the surface obtained from F by a similarity mapping with center X_n and ratio of similitude $1/s_n$.*

Then, if the sequence of surfaces $\bar{F}_n$ converges as $n \to \infty$, the limit surface is:

1) a cone congruent to the tangent cone of F at X_0, if the latter is a conical point;

2) a cylinder with generators parallel to the edge of F at X_0, if X_0 is a ridge point; this cylinder then contains the dihedral angle at X_0;

3) a plane, if X_0 is a smooth point of F.

PROOF. The curvature of the surface $\bar{F}_n$ at X_n is independent of n and equal to the curvature of F at X_0. The curvature of any bounded region of $\bar{F}_n$ not including X_n decreases without bound when $n \to \infty$. It follows that the limit $\bar{F}$ of the sequence $\bar{F}_n$ is either a cone (if X_0 is a conical point of F), or a cylinder (if X_0 is a smooth or ridge point of F).

The supporting planes of $\bar{F}$ at X_0 must be supporting planes of F, since they are limits of supporting planes of F_n. The tangent cone of

[2)] Throughout the sequel, the phrase "corresponding under the isometry" will often be abbreviated to "corresponding" when the meaning is clear from the context.

$\overline{F}$ at X_0 contains the tangent cone of F. Since the curvatures of F and $\overline{F}$ at X_0 are equal, it follows that $\overline{F}$ is precisely the tangent cone of F at X_0, if X_0 is a conical point. If X_0 is a ridge point, then $\overline{F}$ is a cylinder, one of whose generators is the edge of the dihedral angle at X_0 in F, and which contains the entire dihedral angle. If X_0 is a smooth point of F, then $\overline{F}$ is a plane—the tangent plane to F at X_0. This completes the proof.

THEOREM 2. *Let F be a convex surface, X_0 a point on F, F_n a sequence of convex surfaces isometric to F and converging to F, and X_n the point on F_n corresponding to X_0 under the isometry.*

Then, if X_0 is a smooth or conical point, the directions at X_n on F_n converge uniformly as $n \to \infty$ to the directions on F at X_0 that correspond to them under the isometry.

If X_0 is a ridge point, this statement is valid for at least the two directions corresponding to the directions of the edge at X_0 on F.

PROOF. Let Y_0 be a point on F close to X_0; join these two points by a segment γ_0. Let γ_n be the corresponding segment on F_n. As $n \to \infty$ we have $F_n \to F$ and $\gamma_n \to \gamma_0$.

If X_0 is smooth or conical, it follows from Theorem 1 that the semitangent t_n to γ_n at X_n converges to a ray lying in the tangent plane to F at X_0 or to one of the generators of the tangent cone. By the corollaries to Liberman's Theorem (Chapter I, §11), this implies that the limit of the semitangents t_n is the semitangent to the limit segment γ_0 at X_0. Since the curvature of F at X_0 is equal to that of F_n at X_n and the surfaces $\overline{F}_n$ converge to $\overline{F}$ as described in Theorem 1, we see that all directions at X_n on F_n converge uniformly to the directions at X_0 on F that correspond to them under the isometry.

Now let X_0 be a ridge point of F. Let Y_0 be a point on F such that the angle between the segment γ_0 joining X_0 and Y_0 and the direction of the edge through X_0 is less than ϵ. Without loss of generality we may assume that the semitangents t_n to γ_n at X_n converge to a ray t_0. It follows from Theorem 1 that t_0 cannot lie in the interior of the dihedral angle of F at X_0; consequently, by Theorem 9 of Chapter II, §1, the angle between t_0 and the direction of the semitangent to γ_0 at X_0 must be of the order of ϵ. Hence the directions on F_n at X_n that correspond under the isometry to the directions of the edge through X_0 on F converge to these directions. This completes the proof.

The next theorem is a slight generalization of Theorem 2: instead of a single point X_0 we consider an arbitrary closed set.

THEOREM 3. *Let F be a convex surface, M a closed set on F, and F_n a sequence of convex surfaces isometric to and converging to F. Then for sufficiently large n every tangent cone to F at points of M has two directions which bisect the cone and form angles less than ϵ with the directions on F_n that correspond to them under the isometry.*

PROOF. Suppose the assertion false. That is to say, for every n there exists a point $X_n \in M$ on F at which the assertion is false. Without loss of generality we may assume that the sequence X_n converges to a point X_0 of F. There are three possibilities:

1) X_0 is a smooth point;
2) X_0 is a conical point;
3) X_0 is a ridge point.

Let s_n be the distance between X_0 and X_n on F. Let $F^n = F/s_n$ be the surface obtained from F by a similarity mapping with center X_0 and ratio of similitude $1/s_n$, and F_n^n the analogous surface for F_n, the center of similitude being the point corresponding to X_0 under the isometry. Let γ_n be the segment on F joining X_0 and X_n, and γ^n and γ_n^n the corresponding segments on F^n and F_n^n. Without loss of generality we may assume that the sequences F^n and F_n^n converge to surfaces F^0 and F_0^0, and the sequences of segments γ^n and γ_n^n converge to segments γ^0 and γ_0^0, respectively.

Now assume that X_0 is a smooth or conical point of F. Then the surfaces F^0 and F_0^0 both coincide with the tangent plane to F at X_0 in the smooth case, or with the tangent cone at X_0 in the conical case. The segments γ^0 and γ_0^0 are euclidean segments with endpoint X_0 which lie in the tangent plane (cone) of F at X_0. By Theorem 2 the euclidean segments γ^0 and γ_0^0 coincide. An application of Theorem 9 of Chapter II, §1, shows that the semitangents to γ^n and γ_n^n at the points corresponding to X_n converge to the same ray t_0, which contains the euclidean segment $\gamma^0 \equiv \gamma_0^0$. Hence for sufficiently large n there are a direction on F at X_n and a direction on F_n at the corresponding point which correspond under the isometry and form an arbitrarily small angle. It is also clear that those directions on F and F_n which, together with the above directions, bisect the corresponding tangent cones, correspond under the isometry and also form a small angle. This is a contradiction.

Now let X_0 be a ridge point of F. Then, if the sequence of points X_n

converges to X_0 in such a way that the semitangents to γ^n at X_0 converge to the direction of the edge through X_0 (without loss of generality we may assume that the semitangents do indeed converge), we need only repeat the arguments used for smooth or conical points. Now if the limit semitangent does not lie along the edge of the dihedral angle (i.e. the tangent cone) at X_0, the limit segments γ^0 and γ_0^0 need not coincide, but each of the semitangents to them at X_0 can be obtained from the other by a rotation through a certain angle about the edge of F through X_0; in other words, they form the same angle ϑ_0 with the direction of the edge. Hence, since F^0 is a dihedral angle and F_0^0 is a cylinder containing the edge of this angle, it follows that one of the directions on F at X_n which forms the angle ϑ_0 with γ_n will form an arbitrarily small angle with the corresponding direction on F_n for sufficiently large n. Together with the other direction with this property, this direction divides the tangent cone into two parts. This contradiction completes the proof of the theorem.

Our proof of the fundamental theorem (monotypy of closed convex surfaces) will start from the assumption that there exist two closed convex surfaces, close to each other and isometric but not congruent, and a contradiction will be derived. In this connection we shall need the following lemma.

LEMMA. *Let F be a closed convex surface. Then, if there exists a closed convex surface F' isometric but not congruent to F, there exist such surfaces arbitrarily close to F (in this context, doubly covered convex plane domains are also regarded as closed convex surfaces).*

PROOF. We first show that a closed convex surface F can be partitioned into convex triangles with the following properties:

1. The sides of each triangle are less than ϵ.
2. Each side of a triangle is the unique segment joining its endpoints.
3. The angles of each triangle are less than π.
4. The curvature of each triangle is less than an arbitrary [but fixed] positive number ω.

There exists a triangulation T of F into triangles Δ with properties 1 and 3 (Chapter I, §2). This triangulation will now be refined to a triangulation satisfying property 2. To this end, consider the geodesic medians of each triangle Δ in T. If these intersect at one point, the six resulting triangles satisfy conditions 1, 2 and 3. If the medians do not intersect at a point, they divide Δ into four triangles and three quadrangles. By drawing the diagonal of each quadrangle, we divide Δ into sixteen triangles which satisfy conditions 1, 2 and 3 (Figure 26). The fact that condition 2 indeed holds in each of these cases follows from the inclusion property for segments (Chapter I, §2).

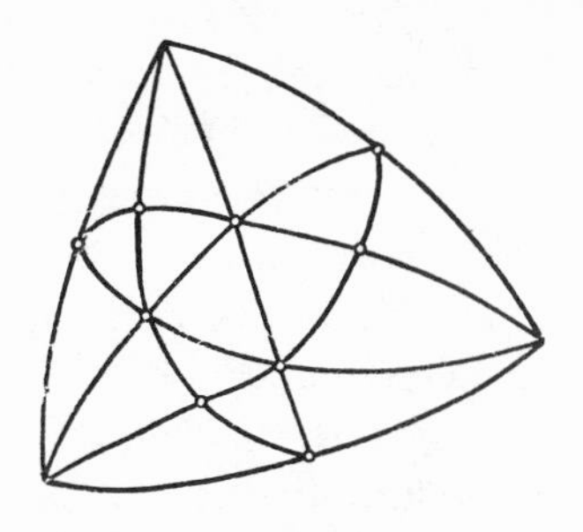

FIGURE 26

We must now prove the existence of a triangulation which satisfies conditions 1, 2, 3 *and* 4. First note that there exists a number δ, depending on ω, such that the curvature of any triangle on the surface with sides less than δ, minus at most one interior point, is less than ω. Indeed, any point X on F has a neighborhood $U(X)$ of points whose distance from X is less than $\delta(X)$, such that the curvature of the set $U(X)-X$ is less than ω. The sets $U(X)$ form an open covering of the surface; by a standard theorem we can extract a finite covering $U_1, \cdots, U_s$. Using this covering we can find a number $\delta > 0$ such that any triangle of diameter less than δ belongs to at least one set $U_1, \cdots, U_s$. Now take a triangulation T so fine that the diameter of each triangle Δ is less than δ. The curvature of each triangle Δ minus one of its interior points is less than ω. Join this point to the vertices of the triangle. The new triangulation of the surface satisfies all four conditions, as required.

We now construct the surface whose existence is asserted in the lemma. Let T be a triangulation of the surface into geodesic triangles satisfying conditions 1 to 4. Let S be a sufficiently dense net of points on F, containing all vertices of T. Denote the points of F' corresponding to S under the isometry by S'.

The convex hulls of the sets S and S' are closed convex polyhedra P and P', respectively. Join the vertices of P and P' corresponding to the vertices of the triangulation T by segments corresponding to the sides of the triangles Δ. If the net S is sufficiently dense, these segments will divide P and P' into triangles with the same properties as the triangulation T of F, and the curvatures of these triangles decrease with decreasing ω.

Let $\sigma_{\Delta,P}$ and $\sigma_{\Delta,P'}$ denote the geodesic triangles on P and P' corresponding to Δ on F. Each of these triangles is a convex polyhedron with boundary. Let X and Y be any two vertices of P in the interior of $\sigma_{\Delta,P}$. Let α and β be the curvatures at these vertices, γ the segment joining X and Y in $\sigma_{\Delta,P}$ and l its length. Construct two plane triangles with base l and base angles x and y. If $x \leqq \alpha/2$ and $y \leqq \beta/2$, these triangles and the polyhedron P cut along the segment γ will satisfy the assumptions of the Gluing Theorem. Let $P_{x,y}$ be the closed convex polyhedron obtained by gluing the two triangles to P cut along γ. By the monotypy theorem for closed convex polyhedra we may assume that when x and y increase continuously the polyhedra $P_{x,y}$ form a continuous family of polyhedra "connecting" P (for $x=y=0$) to a polyhedron with fewer vertices than P: the two vertices X and Y of P are replaced by one, at which the curvature is $\alpha+\beta$.

By finitely many repetitions of this procedure the polyhedron P is deformed continuously into a polyhedron such that each of the domains corresponding to the triangles contains at most one vertex. Let $\overline{P}$ be this polyhedron and $\bar{\sigma}_{\Delta,\overline{P}}$ the triangle on it corresponding to Δ. The latter is isometric either to a plane triangle or to the lateral surface of a trihedral pyramid. Consider the second case. Cut out the triangle $\bar{\sigma}_{\Delta,\overline{P}}$, develop it on the lateral surface of a trihedral pyramid, and then deform it continuously into a triangle lying in the base of the pyramid by displacing its apex along a straight line to some interior point of the base.

Now the surface obtained by removing the triangle $\bar{\sigma}_{\Delta,\overline{P}}$ from the polyhedron $\overline{P}$ (call it $\overline{P}-\bar{\sigma}_{\Delta,\overline{P}}$) and the deformed lateral surface of the pyramid satisfy the assumptions of the Gluing Theorem at each stage of the deformation. Applying the Gluing Theorem to these surfaces, and at the same time continuously deforming the lateral surface of the pyramid, we get a continuous family of surfaces "connecting" $\overline{P}$ to the polyhedron obtained by gluing $\overline{P}-\bar{\sigma}_{\Delta,\overline{P}}$ to a plane triangle with the same sides as $\bar{\sigma}_{\Delta,\overline{P}}$.

Repeating this procedure finitely many times, we finally get a closed convex polyhedron $\widetilde{P}$ glued together from plane triangles with sides equal to the corresponding sides of the triangles $\sigma_{\Delta,P}$.

We now define a concept of proximity for the closed convex polyhedra obtained by the above deformation of P. Let us say that two polyhedra P_1 and P_2 are *ϵ-close* if they can be so placed that the distances between their vertices that correspond to the vertices of T on F are at most ϵ, but they cannot be so placed that the distances between these vertices are less than ϵ.

Now apply the same deformation to P' as to P. The result is a closed convex polyhedron $\widetilde{P}'$ glued together from plane triangles with sides equal to the sides of the triangles $\sigma_{\Delta,P'}$ corresponding to the triangles Δ on F.

If the net S is sufficiently dense, the sides of the triangles $\sigma_{\Delta,P}$ and $\sigma_{\Delta,P'}$ will differ in length by an arbitrarily small amount, and so the polyhedron $\widetilde{P}$ is close either to $\widetilde{P}'$ or to its mirror image.

Hence an important conclusion: For a sufficiently dense net S, the family of polyhedra obtained by the deformations described above from P and P' contains polyhedra ϵ-close to P and ϵ-close to P', for arbitrarily small $\epsilon > 0$. Let P_ϵ and P'_ϵ be any two such polyhedra.

If we now make the triangulations progressively finer, decreasing ω and making the net S denser, we may assume that the sequences of polyhedra P_ϵ and P'_ϵ converge to convex surfaces isometric to F, close to F and F', respectively, but not congruent to them. This proves the lemma

§3. Mixture of isometric surfaces

Let F_0 and F_1 be two figures whose points are in one-to-one correspondence. The mixture F_λ of F_0 and F_1 is defined as the figure consisting of all points in space dividing the segments joining corresponding points of F_0 and F_1 in the ratio $\lambda : (1-\lambda)$. This operation proves to be useful in studying isometric surfaces.

For the moment, we consider mixtures of the simplest isometric surfaces: planes, dihedral angles and cones. The results will then be utilized to study mixtures of general isometric convex surfaces.

We shall say that two isometric surfaces F and F' have the same orientation if for any corresponding points P and P' of these surfaces the senses of traversal about these points that correspond under the isometry both form right-handed (left-handed) screws with the directions of the rays t and t' pointing from P and P' into the interior of the bodies bounded by F and F', respectively. A surface F and its mirror image F^* are oriented in opposite directions. Therefore, given any two isometric surfaces, one is always similarly oriented either to the other or to its mirror image. In this section all pairs of isometric surfaces will be assumed to have the same orientation.

Let α and β be two isometric convex cones (the degenerate case of dihedral angles or planes is not excluded), P and Q their apices, and ω a ball in the interior of both cones. Let t' and t'' be two directions at the apex P of α which bisect the angle at P; assume that the angle between these directions in space is greater than $\pi - \epsilon$, and the angles between them and the directions corresponding to them on β at Q under the isometry are less than ϵ.

LEMMA 1. *If the curvatures of the cones α and β are sufficiently small, their vertices are close, and ϵ and $|\lambda - 1/2|$ are small, then any mixture γ of the cones α and β is a convex cone containing the ball ω in its interior.*

PROOF. If α and β are planes, their mixture is also a plane. Indeed, if e_1 and e_2 are two independent unit vectors in α, then the latter is defined by a vector-valued function $r_\alpha = e_1 u + e_2 v$, $-\infty < u$, $v < \infty$. If e_1' and e_2' are the corresponding unit vectors in the plane β, the latter is defined by the vector-valued function $r_\beta = e_1' u + e_2' v$. The mixture γ of the planes α and β is defined by the function $r_\gamma = \lambda(e_1 u + e_2 v) + (1-\lambda)(e_1' u + e_2' v)$, and so is a plane. For planes the assertion of the lemma is obvious for $\epsilon = 0$ and $\lambda = 1/2$, since γ is the bisector plane for α and β. Hence the assertion remains valid for planes α and β for sufficiently small ϵ and $|\lambda - 1/2|$.

If α is a plane and β a dihedral angle, the mixture γ is clearly a dihedral angle. Its faces are obtained by mixing the faces of β and the halfplanes of α_2 corresponding to the faces of β under the isometry. The assertion is obvious for $\epsilon = 0$ and $\lambda = 1/2$, and so remains valid for small ϵ and $|\lambda - 1/2|$.

Now let α and β be nondegenerate cones. Suppose the assertion false. Then there exist infinite sequences of cones α_n and β_n with apices P_n and Q_n, and numbers $\lambda_n \to 1/2$ and $\epsilon_n \to 0$, for which the assertion is false. Without loss of generality we may assume that the cones α_n and β_n converge to dihedral angles (or planes) α_0 and β_0, and their vertices to a point S. The isometric mapping of α_n onto β_n becomes an isometric mapping of α_0 onto β_0 under which the edges correspond and the isometry coincides with the identity mapping on the edges.

For sufficiently large n the cone γ_n is convex. To verify this it suffices to show that it is locally convex in the sense that there is a plane through each of its generators t_n such that all generators close to t_n lie in one of the halfspaces defined by the plane, namely in the halfspace containing the ball ω. Suppose that each cone γ_n contains a generator t_n not satisfying this local convexity condition. Without loss of generality we may assume that the corresponding generators t_n' on α_n converge to t_0' on α_0. Let A_n' be the point on t_n' at unit distance from the apex of α_n, k_n' the tangent dihedral angle at this point and k_n'' the tangent dihedral angle at the point A_n'' corresponding to A_n' on β_n. For sufficiently large n the mixture of the angles k_n' and k_n'' is a dihedral angle k_n whose interior contains ω. There exists a supporting plane through the edge of k_n whose interior normal $\bar{n}$, laid off from the points A_n' and A_n'', points into the cones α_n and β_n. Let B_n' be a point near A_n'; join B_n' to A_n' by a segment γ' on α_n. Let $r_1(s)$ be the radius-vector of the point on γ' corresponding to arc length s ($s = 0$ corresponds to A_n'), and $r_2(s)$ the radius-vector of the point on β_n corresponding to the former under the isometry. When $s = 0$ we have $(d/ds)(\lambda_n r_1 + (1-\lambda_n) r_2)\bar{n} \geqq 0$. By Liberman's Theorem on the convexity of geodesics, applied to the segment γ' on α_n and the corresponding segment γ'' on β_n, it follows that $(d/ds)(\lambda_n r_1 + (1-\lambda_n) r_2)\bar{n} \geqq 0$ for all s along γ'.

Integrating this inequality, we conclude that all points on the cone γ_n close to the image of A_n' are situated on one side of the supporting plane with interior normal $\bar{n}$. This contradiction proves that the cone γ_n is indeed locally convex. Note that the cone γ can degenerate into a dihedral angle only if α and β are both dihedral angles and their edges correspond under the isometry. This completes the proof of Lemma 1.

LEMMA 2. *For sufficiently small ϵ and $|\lambda - 1/2|$ the mapping of α onto γ taking each point $X \in \alpha$ onto the point $X_\lambda \in \gamma$ which divides the distance between X and the corresponding point on β in the ratio $\lambda : (1-\lambda)$ is a homeomorphism.*

PROOF. If the assertion is false, there exist infinite sequences of cones α_n and β_n, pairs of points A_n, A_n' and B_n, B_n' on these cones, and numbers $\lambda_n \to 1/2$ and $\epsilon \to 0$, such that the points C_n and C_n' corresponding to A_n and A_n' on the cones γ_n coincide. Without loss of generality we may assume that the sequences α_n, β_n of cones and the point sequences A_n, A_n'; B_n, B_n'; C_n, C_n' all converge, and moreover $\lim A_n = A_0 \neq S$, where S is the apex of the limit cones α_0 and β_0. This may always be achieved by extracting suitable subsequences of cones and applying a similarity mapping. If we now assume that $\lim A_n' = A_0' \neq A_0$, then C_0 and C_0' must be different, which is impossible since $C_n' \equiv C_n$ for all n.

Assume that $A_0' \equiv A_0$. Apply to the cones α_n and β_n a similarity mapping with ratio of similitude $1/s_n$, where s_n is the distance between A_n and A_n'. Join the points A_n and B_n to a fixed point of space O and let $n \to \infty$. Without loss of generality we may assume that the surfaces α_n and β_n converge to cylinders α_0 and β_0 (Theorem 1 of §2). The isometry between α_n and β_n induces an isometry of the cylinders α_0 and β_0. The directions of the generators of these cylinders are parallel and correspond under the isometry. Since the cylinders α_0 and β_0 are convex in the same direction and similarly oriented, it follows that $C_0 \not\equiv C_0'$, contradicting the assumption $C_n \equiv C_n'$ and proving the lemma.

LEMMA 3. *Assume that the angle between any direction at the apex P of the cone α and the corresponding direction on the cone β is less than ϵ, and that the curvatures at the apices of the cones are at least $\vartheta_0 > 0$.*

Then for sufficiently small ϵ and $|\lambda - 1/2|$ the mixture of the cones α and β is a convex cone whose interior contains the ball ω, and the natural mapping of α onto γ defined in Lemma 2 is a homeomorphism.

Lemma 3 is proved by the same arguments as Lemmas 1 and 2. That the mapping is a homeomorphism is proved as in Lemma 2, while local convexity is proved as in Lemma 1. The details will be omitted.

LEMMA 4. *Let F be a closed nondegenerate convex surface and F_n a sequence of convex surfaces isometric to F and converging to F.*

Then for sufficiently large n and small $|\lambda - 1/2|$ the mixture $F_{n\lambda}$ of F and F_n is a closed convex surface. The mapping of F onto $F_{n\lambda}$ defined by mapping each point X of F onto the point $X_{n\lambda}$ on $F_{n\lambda}$ which divides the segment joining X to its counterpart on F_n in the ratio $\lambda : (1-\lambda)$ is a homeomorphism.

PROOF. Let $V(X)$ denote the tangent cone of F at X and $V_n(X)$ the tangent cone of F_n at the corresponding point. The isometry of F and F_n induces a natural isometry of the cones V and V_n.

Let Q be a point within the surface F. We claim that if n is sufficiently large and $|\lambda - 1/2|$ sufficiently small, the mixture of the cones $V(X)$ and $V_n(X)$ is a convex cone $V_{n\lambda}(X)$ (possibly a dihedral angle or a plane) for all X, and the point Q is within this cone.

Assume the assertion false. Then there exist an infinite sequence of points X^n on F and a sequence of numbers λ_n converging to $1/2$ as $n \to \infty$ such that each cone $V_{n\lambda_n}(X^n)$ does not satisfy one of the above conditions.

Without loss of generality we may assume that the sequence X^n converges to a point X^0 on F, and the cones $V(X^n)$ and $V_n(X^n)$ converge to cones V and V_0 with apex X^0 which contain the tangent cone of F at this point. The isometry of the cones $V(X^n)$ and $V_n(X^n)$ becomes an isometry of the limit cones V and V_0.

One can always extract from the sequence X^n a subsequence satisfying one of the following conditions:

1. The subsequence of X^n contains only conical points with curvature at least $\vartheta_0 > 0$.

2. The curvature at each point of the subsequence of X^n is less than ϑ_0.

Applying Theorem 3 of §2 and then Lemma 3 or Lemmas 1 and 2 of this section, we derive a contradiction in both cases.

We now claim that for sufficiently large n and small $|\lambda - 1/2|$ the set $F_{n\lambda}$ is locally convex in the sense that there is a plane through each point $X_{n\lambda}$ of $F_{n\lambda}$ such that all points $Y_{n\lambda}$ corresponding to points Y on F close to X lie on the same side of the plane, namely the side containing the point Q. Suppose this assertion false. Without loss of generality we may assume that the point at which local convexity fails corresponds to X^n.

Now local convexity cannot fail at the point $X^n_{n\lambda}$ on $F_{n\lambda}$ unless the cone $V_{n\lambda}(X^n)$ degenerates into a plane or a dihedral angle. It follows that only the second of the above-mentioned possibilities can hold for the subsequence X^n. We may assume that if X^n and the corresponding point on F_n are both ridge points, the directions of their edges correspond under the isometry. Then V and V_0 are dihedral angles or planes. It follows from Theorem 2 (§2) that they have a straight line in common, on which corresponding points of the angles V and V_0 coincide. Each of these angles contains the tangent cone to F at X^0.

Let V_λ be the mixture of V and V_0. In the general case V_λ is a dihedral angle. Let α_λ be a plane through its edge such that the interior normal to α_λ points into the interior of V and V_0. It is easily seen that such a plane exists provided $|\lambda - 1/2| < \chi$, where $\chi > 0$ depends only on the smallest dihedral angle which can contain the tangent cone to F at X^0. It follows that for sufficiently large n and small $|\lambda - 1/2|$ there is a plane $\alpha_{n\lambda}$ through the edge of the angle $V^n_{n\lambda}$ such that rays in the direction of the interior normal to this plane, issuing from points on the edges of the angles $V(X^n)$ and $V_n(X^n)$, point into the interior of these angles.

Let Y be a point close to X^n on F, and γ a segment joining it to X^n on F.

Let $r(s)$ be the radius-vector of a point on γ, and $r_n(s)$ the radius-vector of the corresponding point on F_n. If $\bar{n}$ is the unit vector of the interior normal to the plane $\alpha_{n\lambda}$, then $(d/ds)(\lambda r + (1-\lambda) r_n)\bar{n} \geqq 0$ at the initial point of γ (i.e. X^n). By Liberman's Theorem on the convexity of geodesics, applied to the segment γ and the corresponding segment on F_n, we see that $(d/ds)(\lambda r + (1-\lambda) r_n)\bar{n} \geqq 0$ for all points of γ. It follows that for all points Y close to X^n

$$\{\lambda(r(Y) - r(X^n)) + (1-\lambda)(r_n(Y) - r_n(X^n))\}\bar{n} \geqq 0.$$

Now this means that the surface is locally convex at $X^n_{n\lambda}$, which is a contradiction.

Thus for sufficiently large n and small $|\lambda - 1/2|$ the surface $F_{n\lambda}$ is locally convex and the cone $V_{n\lambda}(X)$ is convex and contains the point Q.

We now claim that the mapping of F onto $F_{n\lambda}$ taking each point X of F to the point $X_{n\lambda}$ of $F_{n\lambda}$ is a homeomorphism.

If this is false, then for every n there exist λ_n which converge to $1/2$ as $n \to \infty$ and pairs of points X^n, Y^n on F such that $X^n_\lambda \equiv Y^n_\lambda$. Without loss of generality we may assume that the sequences X^n and Y^n converge. It is clear that these sequences have the same limit, X^0 say. Apply to the surfaces F and F_n similarity mappings about X^0 and X_n^0 with ratio of similitude $1/s_n$, where s_n is the distance between X^0 and the farther of the points X^n and Y^n. Denote the resulting surfaces by F^n and F_n^n. Without loss of generality we may assume that F^n and F_n^n converge as $n \to \infty$ to F^0 and F_0^0. The isometry of F^n and F_n^n becomes an isometry of F^0 and F_0^0. Moreover, if X^0 is a conical point, then F_0^0 is a cone coinciding with F^0, so that the corresponding directions of F^0 and F_0^0 at X_0^0 coincide. If X^0 is a ridge point, then F^0 is a dihedral angle and F_0^0 is generally a cylinder, one of whose generators is the edge of F^0; this cylinder contains the angle F^0 and the edge of the angle coincides with the generator to which it corresponds under the isometry. If X^0 is a smooth point, then F^0 and F_0^0 are planes which coincide. Proceeding to the surfaces F^n and F_n^n as $n \to \infty$, we readily see that for sufficiently large n the points $X_{\lambda_n}^n$ and $Y_{\lambda_n}^n$ cannot coincide, and this is a contradiction.

Thus the mapping of F onto $F_{n\lambda}$ taking each $X \in F$ to $X_{n\lambda} \in F_{n\lambda}$ is a homeomorphism.

Finally, we shall prove that $F_{n\lambda}$ is a closed convex surface. To this end, cut the surface $F_{n\lambda}$ by a plane through the point Q. We claim that the curve γ cut out by this plane is convex. In fact, were this false, there would be a point P on γ and a straight line through P separating Q from points on γ close to P. This contradicts the relative position of points on $F_{n\lambda}$ close to P vis-à-vis the local supporting plane at P.

Since any such curve is convex, the surface $F_{n\lambda}$ is a closed convex surface, and this completes the proof of the lemma.

Let F_0 and F_1 be two closed isometric convex surfaces, X_0 any point on F_0, and X_1 the corresponding point on F_1. Let X_λ be the point dividing the segment X_0X_1 in the ratio $\lambda : (1-\lambda)$. We have proved that if F_1 is sufficiently close to F_0 and $|\lambda - 1/2|$ sufficiently small, the locus F_λ of the points X_λ is a closed convex surface and the mapping of F onto F_λ which takes a point X_0 of F_0 to X_λ is a homeomorphism. We shall say that the surface F_λ is a mixture of the surfaces F_0 and F_1.

If F_1 and F_0 are sufficiently close to each other, $|\lambda - 1/2|$ is small and $\mu = 1-\lambda$, then F_λ and F_μ are isometric convex surfaces; the mapping of F_λ onto F_μ taking each $X_\lambda \in F_\lambda$ to $X_\mu \in F_\mu$ is an isometry.

Indeed, let X_λ and Y_λ be two arbitrary points of F_λ, and γ_λ a segment joining them. Let X_μ, Y_μ and γ_μ be the corresponding points and curve on

F_μ; finally, let X_0, Y_0, γ_0 and X_1, Y_1, γ_1 be the corresponding points and curves on F_0 and F_1.

If the surface F_1 is sufficiently near F_0, the directions of the segments connecting corresponding points of F_0 and F_1 form angles less than $\pi - \delta$, $\delta > 0$. Hence, obviously, if the curve γ_λ is rectifiable, then so are γ_0 and γ_1, and so is γ_μ. Let $r_0(s)$, $r_1(s)$, $r_\lambda(s)$ and $r_\mu(s)$ denote the radius-vectors of corresponding points on the curves γ_0, γ_1, γ_λ and γ_μ (where s is arc length on γ_0). For almost all s the vector-valued functions $r_0(s)$ and $r_1(s)$ have bounded derivatives, and therefore the same is true of $r_\lambda(s) = \lambda r_0(s) + \mu r_1(s)$ and $r_\mu(s) = \mu r_0(s) + \lambda r_1(s)$. Hence the lengths of the curves γ_λ and γ_μ are respectively

$$\int \sqrt{(r'_\lambda(s))^2}\, ds, \qquad \int \sqrt{(r'_\mu(s))^2}\, ds.$$

But a direct check shows that the integrands in these integrals are equal for almost all s, and therefore the integrals are equal. Since γ_μ is a segment, this means that the distance between X_λ and Y_λ on F_λ is not less than the distance between X_μ and Y_μ on F_μ. Conversely, analogous reasoning shows that the distance between X_μ and Y_μ on F_μ is at least the distance between X_λ and Y_λ on F_λ. Hence these distances are equal, and this implies that the surfaces F_λ and F_μ are isometric, proving our assertion.

Rotate the surface F_μ through a small angle ϑ about an arbitrary axis, and call the resulting surface F_μ^ϑ. Construct the mixed surfaces $F_{\lambda\lambda_1}^\vartheta = \lambda_1 F_\lambda + \mu_1 F_\mu^\vartheta$ and $F_{\lambda\mu_1}^\vartheta = \mu_1 F_\lambda + \lambda_1 F_\mu^\vartheta$ ($\mu_1 = 1 - \lambda_1$). When F_1 is sufficiently close to F_0, λ and λ_1 sufficiently near $1/2$ and ϑ sufficiently small, the surfaces $F_{\lambda\lambda_1}^\vartheta$ and $F_{\lambda\mu_1}^\vartheta$ are isometric convex surfaces, and the mapping of $F_{\lambda\lambda_1}^\vartheta$ onto $F_{\lambda\mu_1}^\vartheta$ taking $X_{\lambda\lambda_1}^\vartheta \in F_{\lambda\lambda_1}^\vartheta$ to $X_{\lambda\mu_1}^\vartheta \in F_{\lambda\mu_1}^\vartheta$ is an isometry.

The proof of this proposition is analogous to that of Lemma 4. We shall not go into detail but only indicate the main steps of the proof.

1. Let V_λ be the tangent cone of the convex surface F_λ at an arbitrary point X_λ, V_μ^ϑ the tangent cone at the corresponding point X_μ^ϑ of F_μ^ϑ, and Q a point within the surface $F_{(1/2)}$. If F_1 and F_0 are sufficiently close to each other, λ and λ_1 are sufficiently near $1/2$ and ϑ is small, then the mixed cones $V_{\lambda\lambda_1}^\vartheta$ and $V_{\lambda\mu_1}^\vartheta$ are convex for all X_λ, and they contain Q in their interior.

2. Each of the sets $F_{\lambda\lambda_1}^\vartheta$ and $F_{\lambda\mu_1}^\vartheta$ is locally convex if F_1 is sufficiently close to F_0, if λ and λ_1 are sufficiently near $1/2$ and if ϑ is small.

3. For F_1 sufficiently close to F_0, λ and λ_1 near $1/2$ and small ϑ, the mapping taking each $X_0 \in F_0$ to $X_{\lambda\lambda_1}^\vartheta \in F_{\lambda\lambda_1}^\vartheta$ is a homeomorphism.

4. Under the same assumptions on λ, λ_1, ϑ and the closeness of F_1 to F_0, the sets $F^{\vartheta}_{\lambda\lambda_1}$ and $F^{\vartheta}_{\lambda\mu_1}$ are closed isometric convex surfaces; the mapping of the first onto the second taking each point $X^{\vartheta}_{\lambda\lambda_1} \in F^{\vartheta}_{\lambda\lambda_1}$ to the point $X^{\vartheta}_{\lambda\mu_1} \in F^{\vartheta}_{\lambda\mu_1}$ is an isometry.

§4. On isometric convex surfaces in canonical position

Let F be a convex surface. We shall say that F is in *canonical position* relative to the xy-plane if it can be projected one-to-one onto the plane, is convex toward it and the angles between its supporting planes and the xy-plane are less than $\vartheta_0 < \pi/2$.

Two isometric surfaces F_0 and F_1 are said to be in *canonical position* if they satisfy the following conditions:

1) The surfaces are similarly oriented and each of them is in canonical position relative to the xy-plane.

2) Any segment connecting two arbitrary points of one surface forms an angle less than $\varphi_0 < \pi$ with the segment connecting the corresponding points on the second surface; in particular, directions on F_0 and F_1 that correspond under the isometry form angles less than φ_0.

3) If $|\lambda - 1/2|$ is sufficiently small, and if the mixed surface $F_\lambda = \lambda F_0 + (1-\lambda) F_1$ is convex and is in canonical position relative to the xy-plane, then the mapping of F_0 onto F_λ which takes each $X_0 \in F$ to the point $X_\lambda = \lambda X_0 + (1-\lambda) X_1$ (where X_1 is the point on F_1 corresponding under the isometry to X_0) is a homeomorphism.

Curves γ_0 and γ_1 on isometric surfaces F_0 and F_1 in canonical position are said to be *equidistant* (from the xy-plane) if they correspond under the isometry of F_0 and F_1 and corresponding points on these curves are at equal distances from the xy-plane.

Equidistant curves γ_0 and γ_1 on convex surfaces F_0 and F_1 in canonical position are said to be *normal* if they satisfy the following conditions:

1) γ_0 and γ_1 have i.g.c. of bounded variation and do not contain conical points of F_0 and F_1, respectively.

2) If a point X on γ_i $(i = 0, 1)$ is a ridge point of F_i, neither of the semitangents to γ_i at X can coincide in direction with the direction of the edge of the dihedral angle at X.

3) If X_0 is an arbitrary point on γ_0, t_0 a semitangent to γ_0 at X_0, and X_1 and t_1 the corresponding point and semitangent of γ_1, then the angle between the xy-plane and the supporting plane to F_0 at X_0 containing the ray t_0 is greater than the angle between the xy-plane and the supporting plane to F_1 at X_1 containing the ray t_1.

We consider some properties of isometric convex surfaces in canonical position.

Let F_0 and F_1 be isometric convex surfaces in canonical position, γ_0 and g_1 equidistant normal curves on these surfaces, and $\bar{\gamma}_0$ and $\bar{\gamma}_1$ their projections on the xy-plane. Then the variation of the i.g.c. of $\bar{\gamma}_\lambda$, the projection of the curve $\gamma_\lambda = \lambda\gamma_0 + (1-\lambda)\gamma_1$ on the xy-plane, is bounded uniformly (in λ) for sufficiently small $|\lambda - 1/2|$.

Indeed, $\bar{\gamma}_\lambda = \lambda\bar{\gamma}_0 + (1-\lambda)\bar{\gamma}_1$. Let $\bar{\tau}_0(s)$ be the unit vector of the right semitangent to $\bar{\gamma}_0$ at the point corresponding to arc length s, and $\bar{\tau}_1(s)$ the unit vector of the right semitangent at the corresponding point of $\bar{\gamma}_1$. Then the variation of the i.g.c. of $\bar{\gamma}_0$ ($\bar{\gamma}_1$) is precisely the variation of the vector-valued function $\bar{\tau}_0(s)$ ($\bar{\tau}_1(s)$), and the variation of the i.g.c. of the curve $\bar{\gamma}_\lambda$ is equal to the variation of the vector-valued function

$$\bar{\tau}_\lambda(s) = \frac{\lambda\bar{\tau}_0(s) + (1-\lambda)\bar{\tau}_1(s)}{|\lambda\bar{\tau}_0(s) + (1-\lambda)\bar{\tau}_1(s)|}$$

and so uniformly bounded (in λ), since $\bar{\tau}_0(s)$ and $\bar{\tau}_1(s)$ have bounded variation and $|\lambda\bar{\tau}_0(s) + (1-\lambda)\bar{\tau}_1(s)|$ cannot vanish for any λ and s.

Let Z_λ be a cylinder through the curve γ_λ on F_λ, with generators parallel to the z-axis. Then, if $|\lambda - 1/2|$ is sufficiently small, the variations of the i.g.c. of the curve γ_λ in space, on the surface F_λ and on the cylinder Z_λ are bounded uniformly (in λ).

This follows from the uniform boundedness of the variation of the i.g.c. of $\bar{\gamma}_\lambda$ and from the results of §1.

Let F_0 and F_1 be isometric convex surfaces in canonical position. Then, if $|\lambda - 1/2|$ is small, the isometric convex surfaces F_λ and $F_{(1-\lambda)}$ are also in canonical position.

The proof of this assertion is a direct verification of the conditions in the definition of surfaces in canonical position. We only remark that any surface $\lambda_1 F_\lambda + (1-\lambda_1)F_{(1-\lambda)}$ is a surface F_ν for suitable ν.

Throughout the sequel we shall assume that $|\lambda - 1/2|$ is small and denote $\mu = 1 - \lambda$.

Let F_0 and F_1 be isometric convex surfaces in canonical position, and γ_0 and γ_1 equidistant normal curves on them. Then the curves γ_λ and γ_μ on F_λ and F_μ are also equidistant and normal.

This again reduces to an easy verification of the relevant conditions 1)–3) above.

Let p_0 be the i.g.c. of the curve $\bar{\gamma}_0$ over a subarc Δ_0, and p_1 and p_λ the i.g.c. of the curves $\bar{\gamma}_1$ and $\bar{\gamma}_\lambda$ over the corresponding subarcs Δ_1

and Δ_λ. If $p_0 \neq p_1$, then $\partial p_\lambda/\partial\lambda = 0$, and if $\lambda = 1/2$, then $|\partial p_\lambda/\partial\lambda| = 2|\tan(\vartheta'/2) - \tan(\vartheta''/2)|$, where ϑ' and ϑ'' are the angles through which the semitangents at the endpoints of the curve Δ_0 must be rotated to make them coincide with the corresponding semitangents at the endpoints of Δ_1.

The proof of this assertion, which is fairly simple, will be omitted.

Let F be a convex surface in canonical position relative to the xy-plane, and γ a curve of bounded variation on F. A partition of γ into finitely many subarcs Δ_i and δ_j will be called an *ϵ-subdivision* if the following conditions hold:

1) The endpoints of each subarc Δ are smooth points of F.

2) The angle between supporting planes of F at any two points of each subarc Δ is less than ϵ.

3) The angle between the right (left) semitangents to γ at any two points of each subarc Δ is less than ϵ.

4) The sum of variations of the i.g.c. of γ over all subarcs δ on F and on the projecting cylinder Z of γ is at most ϵ; the same holds for the sum of variations of i.g.c. of the subarcs $\bar{\delta}$ of the curve $\bar{\gamma}$ (the projections of the subarcs δ of γ).

For any $\epsilon > 0$, there exists an ϵ-subdivision of γ into subarcs Δ and δ.

Indeed, each point X on γ has a neighborhood $\omega(X)$ on the curve which is divided by X into subarcs satisfying conditions 2) and 3). By a standard theorem we can extract a finite covering $\omega(X_1), \cdots, \omega(X_n)$ from the family of neighborhoods $\omega(X)$. The points X_k and the endpoints of the subarcs $\omega(X_k)$ divide γ into finitely many open subarcs satisfying conditions 2) and 3). Now lay off small subarcs δ to the right and left of each point of this subdivision of γ, such that the sum of variations of i.g.c. of all these subarcs (in the sense of condition 4)) is less than ϵ, and such that the other endpoint of each subarc is a smooth point of F. The result is a subdivision of γ into subarcs δ satisfying condition 4) and subarcs Δ (i.e. all the remaining subarcs) satisfying 1), 2) and 3).

Let F_1 and F_2 be isometric convex surfaces in canonical position, and γ_1 and γ_2 equidistant normal curves on these surfaces. Then for any $\epsilon > 0$ there exist ϵ-subdivisions of γ_1 and γ_2 which correspond under the isometry.

The proof is similar to that of the preceding statement.

Let F be a convex surface in canonical position relative to the xy-plane, and γ a curve of bounded variation on F not containing conical points of the surface, with the property that the direction of the edge of the tangent dihedral angle at a ridge point X on the curve (if such exist) does not coincide with the direction of either of the semitangents to γ at X. Assume that an ϵ-subdivision of the curve γ into subarcs Δ and δ is given. Prescribe a direction on the curve, so that the expression "to the right of the curve" has a rigorous meaning.

Using the theorem of Zalgaller [72] cited in §1, one easily proves that there exists a simple geodesic polygon $\bar{\gamma}$ on F, situated to the right of γ, satisfying the following conditions:

1. The subarcs $\bar{\Delta}$ and $\bar{\delta}$ cover $\bar{\gamma}$ in the same way as Δ and δ cover γ.

2. The angles between the supporting planes to F at points of corresponding subarcs Δ and $\bar{\Delta}$ are less than 2ϵ.

3. The angles between the (right or left) semitangents at points of corresponding subarcs Δ and $\bar{\Delta}$ are less than 2ϵ.

4. The right i.g.c. of any two corresponding subarcs Δ and $\bar{\Delta}$ or δ and $\bar{\delta}$ on γ and $\bar{\gamma}$ differ by at most ϵ' (where ϵ' is an arbitrarily small positive number).

5. The variation of the right i.g.c. of the polygon $\bar{\gamma}$ exceeds the variation of the right i.g.c. of γ by at most ϵ'.

6. The vertices of the polygon $\bar{\gamma}$ are smooth points of F, and each side of $\bar{\gamma}$ is the unique segment joining its endpoints on the surface.

Now construct a regular convex surface $\widetilde{F}$, sufficiently close to F, with positive Gauss curvature. Project the vertices of the polygon $\bar{\gamma}$ onto this surface and connect the resulting points on F in the same way as the vertices of $\bar{\gamma}$ are connected by its sides. Call the resulting polygon $\tilde{\gamma}'$ and denote the sides corresponding to $\bar{\Delta}$ and $\bar{\delta}$ by $\tilde{\Delta}'$ and $\tilde{\delta}'$.

If $\widetilde{F}$ is sufficiently close to F, we may assume that the polygon $\tilde{\gamma}'$ satisfies the following conditions:

1. The angles between supporting planes to $\widetilde{F}$ at the points of each subarc $\tilde{\Delta}'$ and the supporting planes to F at points of the corresponding subarc $\bar{\Delta}$ are less than 3ϵ.

2. The angles between the (right and left) semitangents at points of corresponding subarcs $\bar{\Delta}$ and $\tilde{\Delta}'$ of $\bar{\gamma}$ and $\tilde{\gamma}'$ are less than 3ϵ.

3. The right i.g.c. of any two corresponding subarcs $\bar{\Delta}$ and $\tilde{\Delta}'$ or $\bar{\delta}$ and $\tilde{\delta}'$ differ by at most $2\epsilon'$.

4. The variation of the right i.g.c. of $\tilde{\gamma}'$ exceeds the variation of the right i.g.c. of $\bar{\gamma}$ by at most $2\epsilon'$.

These properties are easily proved by *reductio ad absurdum*, making use of the following facts.

Let F be a convex surface and X a point on F, F_n a sequence of convex surfaces converging to F, X_n a point on F_n and α_n a supporting plane to F_n at X_n. If $X_n \to X$ as $n \to \infty$ and the sequence of planes α_n converges to a plane α, then α is a supporting plane of F at X.

Let X_n and Y_n be two points on each surface F_n, and γ_n a segment connecting them. If X_n and Y_n converge to points X and Y of F as $n \to \infty$ which can be joined by a unique segment, then γ_n converges to γ.

If X is a smooth point of the surface F, then the semitangents to the segments γ_n at X_n converge to the semitangent to γ at X.

If the diameter of the spherical image of the segment is less than η, then the angles between the right semitangents at any two points of the segment are less than 3ϵ.

Smooth out the polygon $\tilde{\gamma}'$ at the vertices of each of its sides $\tilde{\Delta}'$, ensuring the while that the right i.g.c. of these subarcs and the variation of the entire curve $\tilde{\gamma}'$ vary by at most ϵ'. Call the resulting curve and the corresponding subarcs $\tilde{\gamma}$, $\tilde{\Delta}$ and $\tilde{\delta}$.

Let γ_{xy} and $\underset{\sim}{\tilde{\gamma}_{xy}}$ be the projections of the curves γ and $\tilde{\gamma}$ on the xy-plane, Z and $\tilde{Z}$ the projecting cylinders, and Δ_{xy} and δ_{xy}, $\tilde{\Delta}_{xy}$ and $\tilde{\delta}_{xy}$ the subarcs of γ_{xy} and $\tilde{\gamma}_{xy}$ corresponding to the subarcs Δ and δ, $\tilde{\Delta}$ and $\tilde{\delta}$ of γ and $\tilde{\gamma}$.

We now introduce certain formulas involving the i.g.c. of the regular curve $\tilde{\Delta}$ on the regular surface $\widetilde{F}$.

We have prescribed a certain sense of traversal on the curve γ on F. This defines a sense of traversal on $\tilde{\gamma}$, and accordingly we may speak of the "right" and "left" sides of this curve. Let σ and σ_z be tangent planes to $\widetilde{F}$ and $\widetilde{Z}$, respectively, at an arbitrary point X of the curve $\tilde{\Delta}$. The tangent to $\tilde{\Delta}$ at X divides each of these planes into two halfplanes, σ', σ'' and σ_z' and σ_z''. To fix ideas, assume that σ' is the halfplane of σ close to the region of $\widetilde{F}$ lying to the right of $\tilde{\Delta}$ at X, and σ_z' the halfplane of σ_z close to the region of the cylinder $\widetilde{Z}$ lying above $\tilde{\Delta}$.

Let $\tilde{p}$ be the right i.g.c. of $\tilde{\Delta}$ on $\widetilde{F}$, $\tilde{p}_1$ the i.g.c. of $\tilde{\Delta}$ on the semicylinder $\widetilde{Z}_1$, and $\tilde{\kappa}$ and $\tilde{\kappa}_1$ the geodesic curvatures of $\tilde{\Delta}$ on $\widetilde{F}$ and $\widetilde{Z}$, respectively. Let $\tilde{\beta}$ be the angle between the halfplanes σ' and σ_z'', and $\tilde{\alpha}$ the angle between the principal normal $\bar{n}$ to $\tilde{\Delta}$ and the halfplane σ_z' (this angle is assumed to be positive if the normal $\bar{n}$ points into the halfspace defined by σ_z which contains the halfplane σ'). Finally, let $\bar{\tilde{p}}$ denote the right i.g.c. of the curve $\bar{\tilde{\Delta}}$ (the projection of $\tilde{\Delta}$ on the xy-plane), $\tilde{\gamma}_{xy}$ the projection of the curve $\tilde{\gamma}$ on the xy-plane, $\bar{\tilde{\kappa}}$ the curvature of this projection, and $\tilde{\varphi}$ the angle between the tangent to $\tilde{\gamma}$ and the xy-plane.

The ordinary curvature $\tilde{k}$ of the curve $\tilde{\Delta}$ is related to its geodesic curvatures on $\widetilde{F}$ and $\widetilde{Z}$, and also to the curvature $\bar{\tilde{\kappa}}$ of its projection $\bar{\tilde{\Delta}}$ on the xy-plane, by standard formulas:

$$\tilde{k}\cos(\tilde{\alpha}+\tilde{\beta}) = -\tilde{\varkappa},$$
$$\tilde{k}\cos\tilde{\alpha} = \tilde{\varkappa}_1,$$
$$\tilde{k}\sin\tilde{\alpha} = \bar{\tilde{\varkappa}}\sin^2\tilde{\varphi}.$$

An immediate consequence of these formulas is

$$\bar{\tilde{\varkappa}} = \frac{\tilde{\varkappa}_1\cos\tilde{\beta}+\tilde{\varkappa}}{\sin\tilde{\beta}\sin^2\tilde{\varphi}}.$$

Consequently

$$\bar{\tilde{p}} = \int\limits_{\bar{\tilde{\Delta}}} \frac{\tilde{\varkappa}_1\cos\tilde{\beta}+\tilde{\varkappa}}{\sin\tilde{\beta}\sin^2\tilde{\varphi}}\, d\bar{\tilde{s}}.$$

We claim that

$$\frac{d}{d\beta}\bar{\tilde{\varkappa}} = -\frac{\tilde{\varkappa}_1+\tilde{\varkappa}\cos\tilde{\beta}}{\sin^2\tilde{\varphi}\sin^2\tilde{\beta}} < 0.$$

Indeed,

$$\tilde{\varkappa}_1+\tilde{\varkappa}\cos\tilde{\beta} = \tilde{k}(\cos\tilde{\alpha}-\cos\tilde{\beta}\cos(\tilde{\alpha}+\tilde{\beta})) = \frac{\tilde{k}}{2}\sin\tilde{\beta}\sin(\tilde{\alpha}+\tilde{\beta}).$$

Since $\tilde{k}>0$, $0<\tilde{\beta}<\pi$, $0<\tilde{\alpha}+\tilde{\beta}<\pi$, we have $\tilde{\kappa}_1+\tilde{\kappa}\cos\tilde{\beta}>0$. Therefore

$$-\frac{\varkappa_1+\tilde{\varkappa}\cos\tilde{\beta}}{\sin^2\tilde{\varphi}\sin^2\tilde{\beta}} < 0.$$

We now introduce analogous notation for the curve Δ on the surface F:

p is the i.g.c. of Δ on F;

p_1 is the i.g.c. of Δ on the cylinder Z projecting γ onto the xy-plane;

$\bar{p}$ is the i.g.c. of the projection $\bar{\Delta}$ of Δ on the xy-plane;

β and φ are the angles that the tangent plane to F and the tangent to γ, respectively, at some point of Δ which is a smooth point of F, form with the xy-plane.

We claim that the sum

$$\sum_{(\Delta)}\left(-\frac{p_1+p\cos\beta}{\sin^2\varphi\sin^2\beta}\right)\cos\varphi,$$

where the summation extends over all subarcs Δ in an ϵ-subdivision of γ, can be made smaller than any positive number, provided ϵ is sufficiently small.

Indeed, if ϵ is sufficiently small, the sum

$$\left|\sum_{(\Delta)}\left(-\frac{p_1+p\cos\beta}{\sin^2\varphi\sin^2\beta}\right)\cos\varphi - \sum_{(\tilde{\Delta})}\left(-\frac{\tilde{p}_1+\tilde{p}\cos\beta}{\sin^2\varphi\sin^2\beta}\right)\cos\varphi\right|$$

may be made as small as desired, since $|p_1 - \tilde{p}_1|$ and $|p - \tilde{p}|$ are arbitrarily small. Further, the quantity

$$\left|\sum_{(\tilde{\Delta})}\left(-\frac{\tilde{p}_1+\tilde{p}\cos\beta}{\sin^2\varphi\sin^2\beta}\right)\cos\varphi - \sum_{(\tilde{\Delta})}\left(\int_{\tilde{\Delta}}\frac{\tilde{\varkappa}_1+\tilde{\varkappa}\cos\tilde{\beta}}{\sin^2\tilde{\varphi}\sin^2\tilde{\beta}}\cos\tilde{\varphi}\,d\tilde{s}\right)\right|$$
$$= \left|\sum_{(\tilde{\Delta})}\int_{\tilde{\Delta}}\frac{\tilde{\varkappa}_1+\tilde{\varkappa}\cos\beta}{\sin^2\varphi\sin^2\beta}\cos\tilde{\varphi}\,d\tilde{s} - \sum_{(\tilde{\Delta})}\int_{\tilde{\Delta}}\frac{\tilde{\varkappa}_1+\tilde{\varkappa}\cos\tilde{\beta}}{\sin^2\tilde{\varphi}\sin^2\tilde{\beta}}\cos\tilde{\varphi}\,d\tilde{s}\right|$$

is also arbitrarily small, since $|\varphi - \tilde{\varphi}|$ and $|\beta - \tilde{\beta}|$ tend to zero with ϵ, and $\sum_{(\tilde{\Delta})}\int_{\tilde{\Delta}}|\tilde{\kappa}_1|\,d\tilde{s}$ and $\sum_{(\tilde{\Delta})}\int_{\tilde{\Delta}}|\tilde{\kappa}|\,d\tilde{s}$ are bounded above by constants independent of ϵ.

Recalling that

$$-\frac{\tilde{\varkappa}_1+\tilde{\varkappa}\cos\tilde{\beta}}{\sin^2\tilde{\varphi}\sin^2\tilde{\beta}} < 0,$$

we see that the sum

$$\sum_{\Delta}\left(-\frac{p_1+p\cos\beta}{\sin^2\varphi\sin^2\beta}\right)\cos\varphi$$

is smaller than any positive number if ϵ is sufficiently small.

Let w and w' be the i.g.c. of the curve $\bar{\gamma}$ over subarcs $\bar{\Delta}$ and $\bar{\delta}$, respectively, and w'' the geodesic curvature of this curve at the common point P of two subarcs of the subdivision of $\bar{\gamma}$. Then the total i.g.c. of $\bar{\gamma}$ is

$$\bar{p} = \sum_{(\bar{\Delta})} w + \sum_{(\bar{\delta})} w' + \sum_{(P)} w''.$$

We claim that *if ϵ is sufficiently small, then $\bar{p}$ is arbitrarily close to the sum*

$$\sum_{(\bar{\Delta})}\frac{p_1\cos\beta+p}{\sin\beta\sin^2\varphi}\cos\varphi + \sum_{(P)} w''.$$

Indeed, for sufficiently small ϵ', $\bar{p}$ is arbitrarily close to the sum

$$\sum_{(\bar{\Delta})}\int_{\tilde{\Delta}}\frac{\tilde{\varkappa}_1\cos\tilde{\beta}+\tilde{\varkappa}}{\sin\tilde{\beta}\sin^2\tilde{\varphi}}\cos\tilde{\varphi}\,d\tilde{s} + \sum_{(\bar{\delta})} w' + \sum_{(P)} w''.$$

Now, using the previous arguments, we can show that

$$\left|\sum_{(\tilde{\Delta})}\int_{\tilde{\Delta}}\frac{\tilde{\varkappa}_1\cos\tilde{\beta}+\tilde{\varkappa}}{\sin\tilde{\beta}\sin^2\tilde{\varphi}}\cos\tilde{\varphi}\,d\tilde{s} - \sum_{(\bar{\Delta})}\frac{p_1\cos\beta+p}{\sin\beta\sin^2\varphi}\cos\varphi\right.$$

is small when ϵ is sufficiently small. The sum $\sum_{(\bar{\delta})} w'$ is also small when ϵ is small.

Thus $\bar{p}$ differs by an arbitrarily small amount from the sum

$$\sum_{(\bar{\Delta})} \frac{p_1 \cos\beta + p}{\sin\beta \sin^2\varphi} \cos\varphi + \sum_{(P)} w''.$$

Let F_0 and F_1 be isometric convex surfaces in canonical position relative to the xy-plane, and γ_0 and γ_1 equidistant normal curves on these surfaces. Construct the isometric surfaces F_λ and F_μ ($|\lambda - 1/2|$ is small; $\mu = 1 - \lambda$) and the curves γ_λ and γ_μ on them. As we proved previously, the surfaces F_λ and F_μ are in canonical position and the curves γ_λ and γ_μ are equidistant and normal.

Consider ϵ-subdivisions Δ_λ, δ_λ and Δ_μ, δ_μ of the curves γ_λ and γ_μ such that the subarcs Δ_λ and δ_λ, Δ_μ and δ_μ correspond under the isometry. We now introduce notation for F_λ analogous to that used above for F:

p_λ is the i.g.c. of Δ_λ on F_λ;

$p_{1\lambda}$ is the i.g.c. of Δ_λ on the cylinder Z_λ projecting γ_λ onto the xy-plane;

β_λ and φ_λ are the angles that the tangent plane to F_λ and the tangent to Δ_λ, respectively, at a point of Δ_λ which is a smooth point of F_λ, form with the xy-plane;

w_λ and w'_λ are the i.g.c.'s of the subarcs $\bar{\Delta}_\lambda$ and $\bar{\delta}_\lambda$ of the projection $\bar{\gamma}_\lambda$ of γ_λ on the xy-plane which correspond to the subarcs Δ_λ and δ_λ;

w''_λ is the geodesic curvature of the curve $\bar{\gamma}_\lambda$ at the common endpoint P_λ of two subarcs of the subdivision of $\bar{\gamma}_\lambda$.

Analogous notation is introduced for F_μ and γ_μ.

Let γ_0 and γ_1 be closed Jordan curves. We claim that *all points of these curves are smooth points of the surfaces F_0 and F_1, respectively.*

Let P_0 be a point of γ_0 which is a ridge point of F_0. Then $P_{0\lambda}$ and $P_{0\mu}$ are ridge points of F_λ and F_μ, respectively. The corresponding points $\bar{P}_{0\lambda}$ and $\bar{P}_{0\mu}$ of $\bar{\gamma}_\lambda$ and $\bar{\gamma}_\mu$ are necessarily corner points. Let $w''_{0\lambda}$ and $w''_{0\mu}$ be the geodesic curvatures at the points $\bar{P}_{0\lambda}$ and $\bar{P}_{0\mu}$ of the curves $\bar{\gamma}_\lambda$ and $\bar{\gamma}_\mu$, respectively. Then $w''_{0\lambda} > w''_{0\mu}$ ($\lambda < \mu$); moreover, when $\lambda \to 1/2$ the quotient $(w''_{0\mu} - w''_{0\lambda})/(\mu - \lambda)$ approaches a negative limit w_0.

Since the curves $\bar{\gamma}_\lambda$ and $\bar{\gamma}_\mu$ are closed, the i.g.c. of each is 2π.

On the other hand, for sufficiently small ϵ these i.g.c. differ by an arbitrarily small amount from

$$\sum_{(\Delta_\lambda)} \frac{p_{1\lambda} \cos\beta_\lambda + p_\lambda}{\sin\beta_\lambda \sin^2\varphi_\lambda} \cos\varphi_\lambda + \sum_{(P_\lambda)} w''_\lambda$$

and

$$\sum_{(\Delta_\mu)} \frac{p_{1\mu} \cos\beta_\mu + p_\mu}{\sin\beta_\mu \sin^2\varphi_\mu} \cos\varphi_\mu + \sum_{(P_\mu)} w''_\mu$$

respectively.

Thus for fixed λ and μ we can find ϵ so small that

$$R = \sum_{(\Delta_\lambda, \Delta_\mu)} \frac{1}{\mu - \lambda}\left(\frac{p_{1\mu}\cos\beta_\mu + p_\mu}{\sin\beta_\mu \sin^2\varphi_\mu}\cos\varphi_\mu - \frac{p_{1\lambda}\cos\beta_\lambda + p_\lambda}{\sin\beta_\lambda \sin^2\varphi_\lambda}\cos\varphi_\lambda\right) + \sum_{(\overline{P}_\lambda, \overline{P}_\mu)} \frac{w''_\mu - w''_\lambda}{\mu - \lambda}$$

is arbitrarily small in absolute value.

Now

$$\frac{\cos\beta_\mu}{\sin\beta_\mu} = \frac{\cos\beta_\lambda}{\sin\beta_\lambda} - (\mu - \lambda)\left(\frac{1}{\sin^2\beta_\lambda} + \xi_1\right),$$

$$\frac{1}{\sin\beta_\mu} = \frac{1}{\sin\beta_\lambda} - (\mu - \lambda)\left(\frac{\cos\beta_\lambda}{\sin^2\beta_\lambda} + \xi_2\right),$$

where ξ_1 and ξ_2 are arbitrarily small, provided $|\lambda - 1/2|$ is small. Since $p_{1\lambda} = p_{1\mu}$, $p_\lambda = p_\mu$, $\varphi_\lambda = \varphi_\mu$ and the sums $\sum_{\Delta_\lambda} |p_{1\lambda}|$ and $\sum_{\Delta_\lambda} |p_\lambda|$ are uniformly bounded (in λ), the expression

$$\sum_{(\Delta_\lambda, \Delta_\mu)} \frac{1}{\mu - \lambda}\left(\frac{p_{1\mu}\cos\beta_\mu + p_\mu}{\sin\beta_\mu \sin^2\varphi_\mu}\cos\varphi_\mu - \frac{p_{1\lambda}\cos\beta_\lambda + p_\lambda}{\sin\beta_\lambda \sin^2\varphi_\lambda}\cos\varphi_\lambda\right)$$

differs by an arbitrarily small amount from the sum

$$\sum_{(\Delta_\lambda)} \left(-\frac{p_{1\lambda} + p_\lambda \cos\beta_\lambda}{\sin^2\beta_\lambda \sin^2\varphi_\lambda}\right)\cos\varphi_\lambda,$$

if $|\lambda - 1/2|$ is small, and this sum in turn is arbitrarily small for sufficiently small ϵ.

Now all the terms in the sum

$$\sum_{(P_\lambda, P_\mu)} \frac{w''_\mu - w''_\lambda}{\mu - \lambda}$$

are nonpositive, and moreover if ϵ is so small that $P_{0\lambda}$ and $P_{0\mu}$ are points of the subdivisions of γ_λ and γ_μ and $|\lambda - 1/2|$ is small, then at least one term (that corresponding to the points $P_{0\lambda}$ and $P_{0\mu}$) is certainly smaller than $w_0/2$.

It follows that when ϵ and $|\lambda - 1/2|$ are sufficiently small, R cannot be arbitrarily small in absolute value. This contradiction proves that all points of the curves γ_0 and γ_1 on the surfaces F_0 and F_1 are smooth, and the curves γ_0 and γ_1 are thus smooth curves.

We now claim that *the curves γ_0 and γ_1 on F_0 and F_1 cannot be closed.*

Without loss of generality we may assume that there are no subarcs

δ_λ and δ_μ in the ϵ-subdivisions of γ_λ and γ_μ. When $\epsilon \to 0$, the above expression R approaches zero, and in the limit it may be expressed as a Stieltjes integral:

$$\frac{1}{\mu-\lambda}\left\{\int \frac{dp_{1\mu}\cos\beta_\mu + dp_\mu}{\sin\beta_\mu \sin^2\varphi_\mu}\cos\varphi_\mu - \int \frac{dp_{1\lambda}\cos\beta_\lambda + dp_\lambda}{\sin\beta_\lambda \sin^2\varphi_\lambda}\cos\varphi_\lambda\right\} = 0.$$

Hence, as in the preceding argument, we conclude that

$$-\int_{(\gamma_\lambda)} \frac{dp_{1\lambda} + dp_\lambda \cos\beta_\lambda}{\sin^2\beta_\lambda \sin^2\varphi_\lambda}\cos\varphi_\lambda$$

is arbitrarily small in absolute value. Now since this integral has a fixed sign over any subarc of γ_λ, it must be small for any subarc, and may be made arbitrarily small by making λ approach $1/2$. Thus for any two corresponding subarcs m_λ and m_μ of the curves γ_λ and γ_μ, if λ is sufficiently close to $1/2$, then the expression

$$\frac{1}{\mu-\lambda}\left\{\int_{(m_\mu)} \frac{dp_{1\mu}\cos\beta_\mu + dp_\mu}{\sin\beta_\mu \sin^2\varphi_\mu}\cos\varphi_\mu - \int_{(m_\lambda)} \frac{dp_{1\lambda}\cos\beta_\lambda + dp_\lambda}{\sin\beta_\lambda \sin^2\varphi_\lambda}\cos\varphi_\lambda\right.$$

is arbitrarily small. And since the integrals

$$\int_{(m_\mu)} \frac{dp_{1\mu}\cos\beta_\mu + dp_\mu}{\sin\beta_\mu \sin^2\varphi_\mu}\cos\varphi_\mu \quad\text{and}\quad \int_{(m_\lambda)} \frac{dp_{1\lambda}\cos\beta_\lambda + dp_\lambda}{\sin\beta_\lambda \sin^2\varphi_\lambda}\cos\varphi_\lambda$$

are precisely the i.g.c. of the projections of m_μ and m_λ, respectively, it follows, as shown above, that this is possible only if the curves $\bar{\gamma}_\lambda$ and $\bar{\gamma}_\mu$ have equal i.g.c. over the corresponding subarcs, i.e. they are congruent.

Since $\bar{\gamma}_0$ and $\bar{\gamma}_1$ are congruent, so are γ_0 and γ_1.

Hence the curves γ_1 and γ_0 can be made to coincide by rotating the surface F_1 about an axis parallel to the z-axis. When this is done, since the angles between the tangent planes to F_0 along γ_0 and the xy-plane are greater than the angles between the tangent planes to F_1 and the xy-plane, it follows that the curvature of that part of F_0 which is bounded by γ_0 is strictly greater than the curvature of the corresponding region on F_1. But this is impossible, since these regions correspond under isometry. Hence the curves γ_0 and γ_1 cannot be closed.

§5. The auxiliary surface Ω and its plane sections

Let F_0 and F_1 be isometric convex surfaces in canonical position relative to the xy-plane. Let $r_0(X)$ be the radius-vector of an arbitrary point X on F_0, and $r_1(X)$ the radius-vector of the point corresponding to X on F_1 under the isometry. Let $\overline{F}$ denote the mixed surface with equation

$$r = \tfrac{1}{2}(r_0(X) + r_1(X)) = \bar{r}(X).$$

Let F_1^* be the surface obtained by reflecting F_1 in the xy-plane, and $r_1^*(X)$ the radius-vector of the point on F_1^* corresponding to the point X of F_0. The surface with equation

$$r = \tfrac{1}{2}(r_0(X) + r_1^*(X)) = r(X)$$

will be denoted by Ω. We shall consider plane sections of this surface.

The surface $\overline{F}$ can be projected in one-to-one fashion onto the surface Ω, by straight lines parallel to the z-axis. Under this correspondence a point $\bar{r}(X)$ on $\overline{F}$ will correspond to the point $r(X)$ on Ω. If $\bar{r}(X)$ is a smooth or ridge point of $\overline{F}$, then $r(X)$ is a smooth or ridge point of Ω. Indeed, corresponding points $r_0(X)$ and $r_1(X)$ on F_0 and F_1 are either both smooth or both ridge points, and in the latter case the directions of the edges must correspond under the isometry.

The set U of all points on Ω which correspond to conical points of $\overline{F}$ is at most countable. Therefore almost all planes cutting Ω do not contain such points.

Let α be a plane cutting Ω and containing no points of the set U. We shall say that the plane α *essentially cuts* Ω, or that it never touches Ω, if: a) α is not the tangent plane to Ω at any smooth point of Ω contained in α; b) at ridge points, the plane α either separates the faces of the tangent dihedral angle or cuts its edge. We shall prove that almost all planes α cutting the surface Ω cut it essentially.

We first define the thickness of a region of Ω. Let ω be a region of Ω, and P and Q two points in ω such that neither P nor the point of $\overline{F}$ lying above it are conical points. This means that either the points on F_0 and F_1 corresponding to P are both smooth, or one is smooth and the other a ridge point, or both are ridge points and the directions of their edges correspond under the isometry. If P is a smooth point, let α be the tangent plane at P. If P is a ridge point, let α be a plane through the edge of the tangent dihedral angle at P which does not separate the faces of the angle. Let $d_\alpha(P, Q)$ be the distance from Q to the plane α. We define the thickness d of the region ω as the supremum of the quantities $d_\alpha(P, Q)$ for all P, Q and α satisfying the above conditions.

We now construct a partition of the surface Ω into regions ω as follows. First divide the xy-plane into small squares with side δ; then project the resulting net onto Ω by straight lines parallel to the z-axis. Proceed in the same way for $\bar{F}$. Let ω_P denote the region on Ω which is the projection of the square with center P. Let $\bar{\omega}_P^k$ denote the region on $\bar{F}$ which is the projection of the square with center P and side $k\delta$.

We claim that *for sufficiently large but fixed k there exists a number $\epsilon > 0$ so small that when $\delta < \epsilon$ we have $d_{\omega_P} < \lambda d_{\bar{\omega}_P^k}$ for all P (centers of squares on the xy-plane) with λ independent of δ.*

Let K_φ be a circular cone with opening angle $\pi - \varphi$, where φ is sufficiently small. This cone is "almost" a plane.

Let $\eta > 1$. Then for every region ω_P there are points X and Y on the region and a plane α such that $\eta d_\alpha(X, Y) \geqq d_{\omega_P}$.

Divide the set of regions ω_P into two classes, as follows. Let $\bar{X}_P$ and X_P be the points of $\bar{F}$ and Ω whose projection is the center P of a square. Consider any supporting plane to $\bar{F}$ at $\bar{X}_P$, and place the cone K_φ in such a position that its axis coincides with the normal to the supporting plane, its apex is at $\bar{X}_P$ and the cone itself lies on the same side of the supporting plane as the surface $\bar{F}$. If no point of the region $\bar{\omega}_P^{2k}$ is within the cone K_φ, the region ω_P is in class I; otherwise ω_P is in class II.

It is clear that if a region ω_P is in class II, then $d_{\bar{\omega}_P^k} > c\delta$, where $c > 0$ may be assumed independent of δ. And since $d_{\omega_P} < c'\delta$ (c' may be assumed independent of δ), it follows that there is a constant λ, independent of δ, for regions ω_P of class II, such that $d_{\omega_P} < \lambda d_{\bar{\omega}_P^k}$.

Now consider the regions ω_P in class I. Let $\omega_P'^{2k}$ and $\omega_P''^{2k}$ be the regions on F_0 and F_1 corresponding to $\bar{\omega}_P^{2k}$. Since ω_P is in class I, the spherical image of the region $\bar{\omega}_P^k$ is small when φ is small. Hence the spherical images of $\omega_P'^k$ and $\omega_P''^k$ are also small.

Let X_P' be an arbitrary point of the region $\omega_P'^k$, corresponding to a nonconical point $\bar{X}_P$ of $\bar{F}$, and Y_P' another point of the same region. Join the points X_P' and Y_P' by a segment γ_0 on F_0. Let X_P'' and Y_P'' be the corresponding points under the isometry, and γ_1 the corresponding segment on F_1. The vectors $\overline{X_P'Y_P'}$ and $\overline{X_P''Y_P''}$ have the following representations:

$$\overline{X_P'Y_P'} = \tau_0 s + \nu_0, \qquad \overline{X_P''Y_P''} = \tau_1 s + \nu_1,$$

where τ_0 and τ_1 are the unit vectors of the segments γ_0 and γ_1 at the points X_P' and X_P'', and s is arc length along γ_0 and γ_1 (see Chapter II, §2). Consider the supporting plane to $\bar{F}$ at $\bar{X}_P$ which contains the vector $\tau_0 + \tau_1$. Then the distance from the point $\bar{Y}_P$ to this plane is of the order of $|\nu_0| + |\nu_1|$. Choice of any other supporting plane at $\bar{X}_P$ will increase this distance by a quantity of the order of $\bar{\psi}\chi s$, where $\bar{\psi}$ is the supplement of the angle between the extreme supporting planes to $\bar{F}$ at $\bar{X}_P$ and χ the angle between the direction of τ_0 and the edge of F_0, or, what is the same, the angle between the direction of τ_1 and the edge of F_1 at X''_P (that these angles are the same follows from the fact that the directions of the edges at X'_P and X''_P correspond under the isometry). Thus

$$d(\bar{X}_P, \bar{Y}_P) \geqq \bar{\mu}(|\nu_0| + |\nu_1| + \bar{\psi}\chi s).$$

Now consider the region ω_P^k on Ω. The distance of the point Y_P (corresponding to $\bar{Y}_P$ on $\bar{F}$) from the face of the tangent angle at the point X_P (corresponding to $\bar{X}_P$) is at most $|\nu_0| + |\nu_1|$. If the faces of the angle are replaced by any plane through its edge which does not separate the faces, the distance of Y_P from this plane may be greater than $|\nu_0| + |\nu_1|$ by a quantity of the order of $\psi\chi s$, where ψ is the angle through which one of the faces turns when the angle is developed on a plane, and χ and s are as before. Hence

$$d(X_P, Y_P) \leqq \mu(|\nu_0| + |\nu_1| + \psi\chi s).$$

Since $\psi \leqq m\bar{\psi}$ (m is a constant independent of δ), it follows that $\lambda d_{\bar{\omega}_P^k} \geqq d_{\omega_P^k} \geqq d\omega_P$ for regions ω_P of class I. This completes the proof of our assertion.

We shall now estimate $d_{\bar{\omega}_P^k}$. Let $\bar{X}$ and $\bar{Y}$ be any two points on $\bar{\omega}_P^k$. Consider a section through these points by a plane parallel to the z-axis. Let $\bar{d}(\bar{X}, \bar{Y})$ be the distance of $\bar{Y}$ from the supporting line of $\bar{\gamma}$ at $\bar{X}$ in the direction of the z-axis, and $\bar{d}_{\bar{\omega}_P^k}$ the supremum of these distances. Obviously $\bar{d}_{\bar{\omega}_P^k} \geqq d_{\bar{\omega}_P^k}$. It will therefore suffice to estimate $\bar{d}_{\bar{\omega}_P^k}$.

Define the discrete thickness $d^*_{\bar{\omega}_P^s}$ of the region $\bar{\omega}_P^s$ by $d^*_{\bar{\omega}_P^s} = \sup \bar{d}(\bar{X}, \bar{Y})$ where $\bar{X}$ and $\bar{Y}$ belong to $\bar{\omega}_P^s$ and their projections on the xy-plane are vertices of the squares of the net.

Let l be a large but fixed number. We claim that $d^*_{\omega_P^{kl}} \geqq \lambda' \bar{d}_{\bar{\omega}_P^k}$ where λ' is a constant independent of both δ and P.

Draw $2(2t+1)$ rays from the point P in the xy-plane, at equal angles to each other, and number the $2(2t+1)$ angles consecutively.

Define a set M in the region $\bar{\omega}_P^{kl}$ as follows:

1) The set M contains one point $\bar{X}$ of $\bar{\omega}_P^k$, for which there exists a point $\bar{Y}$ in the same region such that $\bar{d}(\bar{X}, \bar{Y}) \geqq \frac{1}{2}\bar{d}_{\bar{\omega}_P^k}$.

2) The set M contains all points on the boundary of the region $\bar{\omega}_P^{kl}$ whose projections on the xy-plane are vertices of squares lying in even-numbered angles (in the above enumeration).

The convex hull of the set M is a convex polyhedron. Let Q be the part of the polyhedron facing the xy-plane; $\bar{X}$ may be the only vertex of Q. Let $\overline{Y}'$ be the point of Q lying above $\overline{Y}$. Obviously $\bar{d}_{\bar{F}}(\overline{X},\overline{Y}) \leqq \bar{d}_Q(\overline{X},\overline{Y}')$. Moreover, if A and B are two points of M and the plane curve connecting them passes within the region $\bar{\omega}_P^k$, then $d_{\bar{F}}^*(A,B) \geqq d_Q^*(A,B)$.

The polyhedron Q can be developed on a polyhedral angle V at $\overline{X}$. When this is done, if the numbers k, l and t satisfy the inequality $k \ll t \ll l$, then the projection of the region $\bar{\omega}_P^k$ on the polyhedron Q remains in its previous position. When Q is replaced by the polyhedral angle V, we have

$$\bar{d}_Q(\bar{X},\ \bar{Y}') = \bar{d}_V(\bar{X},\ \bar{Y}'),\quad \overset{*}{d}_Q(A',\ B') \geqslant \overset{*}{d}_V(A'',\ B''),$$

where A'' and B'' are the points of V lying below A' and B'. Now the angle V clearly contains two admissible points A'' and B'' for which $\bar{d}_V(\overline{X},\overline{Y}') \leqq \lambda'' d_V^*(A'',B'')$. It follows that when $k \ll l$ there exists a constant λ' independent of δ and P such that $d^*_{\omega_P^{kl}} \geqq \lambda' d_{\bar{\omega}_P^k}$.

We now estimate $\sum_P d^*_{\omega_P^{kl}}$. First,

$$\sum_P \overset{*}{d}_{\bar{\omega}_P^{kl}} \leqslant \sum_{(\bar{X},\bar{Y})} \overset{*}{d}(\bar{X},\ \bar{Y}),$$

where the summation ranges over all pairs of points $\overline{X}$ and $\overline{Y}$ whose projections on the xy-plane are vertices of squares the distance between which is less than $2k/\delta$. To estimate the sum on the right-hand side, we note that if the points $\overline{X}_1,\cdots,\overline{X}_n$ lie in one vertical plane (a plane parallel to the z-axis), then $\sum d^*(\overline{X}_k,\overline{X}_j) < c\delta$, where c may be assumed independent of δ. It follows that for fixed k and l the sum $\sum_{(\overline{X},\overline{Y})} d^*(\overline{X},\overline{Y})$ is bounded by a constant independent of δ. But this means that $\sum_P \omega_P$ is also bounded by a constant independent of δ.

We can now estimate the measure of the set of planes which do not cut the surface Ω essentially. We shall say that these planes "touch" the surface. The measure $\psi(\Omega)$ of this set of planes cannot exceed the sum of the measures $\sum \psi(\omega_P)$ for individual regions. If the regions are sufficiently small, the area of the spherical image of each of them may be assumed less than ϵ (not counting conical points). But if the area of the spherical image of ω_P is less than ϵ, the measure of the set of planes "touching" it is less than ϵd_{ω_P}. And since the sum of thicknesses d_{ω_P} is bounded by a constant independent of δ, it follows that the measure of the set of planes "touching" Ω is zero.

Thus *almost all planes which have points in common with the surface* Ω *essentially cut it.*

Let α be a plane cutting Ω, X an arbitrary point of their intersection, and suppose that the following conditions hold:

1. If one of the points X_0 on F_0 or X_1 on F_1 corresponding to X, say X_0, is a ridge point, and the other (X_1) is smooth, then the direction on Ω at X corresponding to the direction of the edge at X_0 is not in the plane α.

2. If both X_0 and X_1 are ridge points and the directions of their edges correspond under the isometry, then the direction on Ω at X corresponding to the direction of the edges on F_0 and F_1 at X_0 and X_1 is not in the plane α.

We shall then say that the plane α *cuts* the edges of the surface Ω.

We claim that *almost all planes α cutting the surface cut the edges of the surface.*

Define a set G_0 on F_0 as follows. A point X_0 is a member of G_0 if one of the following conditions holds:

1) X_0 is a conical point.

2) X_0 is a ridge point, the point on F_1 corresponding to it under the isometry is also a ridge point and the directions of the edges do not correspond under the isometry.

The set G_0 is clearly at most countable, since this is the case for the corresponding set on the mixed surface $\overline{F}$.

Let G_1 be the set on F_1 corresponding to G_0 under the isometry.

Cover the set G_0 with open geodesic disks, with total sum of diameters less than ϵ. Call the covering G_0^ϵ; the corresponding covering of G_1 on F_1 will be called G_1^ϵ. Let G^ϵ be the set on the surface Ω corresponding to the sets G_0^ϵ and G_1^ϵ.

Obviously the measure of the set of planes containing at least one point of G^ϵ can be made arbitrarily small if ϵ is sufficiently small.

We now define a set E_0^ϑ on F_0. A point X_0 on F_0 belongs to E_0^ϑ if it is not in G_0 and either it is a ridge point at which the plane angle of the tangent dihedral is at least ϑ or the corresponding point on F_1 has this property. The corresponding set on F_1 will be called E_1^ϑ. E_0^ϑ and E_1^ϑ are clearly closed sets.

Partition the xy-plane into small squares with side δ and project them onto the surface F_0 by straight lines parallel to the z-axis. For convenience the sets on F_0 corresponding to these squares will also be called squares; they will be denoted by ω_0, and the sets on F_1 corresponding to them under the isometry by ω_1.

If the diameters of the squares ω_0 are small, i.e. δ is small, then the angles between the directions of the edges at points of E_0^ϑ belonging to one square may be made arbitrarily small. Indeed, one easily proves that otherwise there exist conical points on F_0 not belonging to G_0, and this is impossible.

If the diameters of the squares ω_0 are small, then the angles between the directions of the edges at points of E_0^{ϑ} belonging to ω_0 and the directions corresponding to the edges of ω_1 are also arbitrarily small. Otherwise the surface $\overline{F}$ would contain conical points corresponding to points of F_0 not in G_0.

Analogous considerations apply to the directions of edges on F_1 at points of E_1^{ϑ} in a single square ω_1.

Let Σ_0 denote the set of squares ω_0 containing points of the set E_0^{ϑ}, Σ_1 the set of squares ω_1 on F_1 corresponding to the set Σ_0 under the isometry.

Let n be the number of squares in Σ_0. We claim that *the number* $n\delta$ *is bounded by a constant* $c(\vartheta)$ *independent of* δ. To prove this we recall the definition of the integral of mean curvature for general convex surfaces.

If F is a regular convex surface, the integral of mean curvature of a set M is defined as

$$H(M) = \int_M \frac{1}{2}(k_1 + k_2)\, dM,$$

where k_1 and k_2 are the principal curvatures of the surface. There is another definition: Lay off segments of length h along the exterior normals to the surface at the points of M. The endpoints of these segments form a set M_h on a surface F_h parallel to F. Let S_0 be the area of the set M on F, S_h the area of M_h on F_h. Then

$$H(M) = \lim_{h \to 0} \frac{1}{2} \frac{S_h - S_0}{h}.$$

This definition of $H(M)$ is also meaningful for general convex surfaces.

To abbreviate the exposition, the integral of mean curvature will be called simply the mean curvature. We list some of its properties.

1. Mean curvature is an additive set function on a convex surface; it is defined for all closed, open and, in general, Borel sets.

2. The mean curvature of a closed convex surface with diameter d is at most $2\pi d$.

3. If a convex surface is deformed, preserving its boundary, in such a way that the points of the surface are displaced into the interior of the body bounded by the surface and the new surface is convex, then the mean curvature of the surface is not increased.

4. Let F be a conical convex surface, d the distance of its apex from the nearest boundary point of F, and φ the greatest angle between the exterior normals to the supporting planes at the apex. Then the mean curvature of F is at least $c\varphi d$, where c is a constant independent of the surface.

Let ω_0 be a square in Σ_0 containing a ridge point X_0 of F_0 at which the plane angle is at least ϑ. Let ω_0' denote the "tripled" square, i.e. the square consisting of ω_0 and the eight adjacent squares. Remove the square ω_0' from the surface F_0 and construct the convex hull of the remaining part of F_0 and the point X_0. Let ω_0'' be the region on the new surface which corresponds to the square ω_0'. The mean curvature of ω_0'' is at most that of ω_0'. If the surface ω_0'' is developed on the tangent cone at X_0, this does not increase its mean curvature, and the mean curvature of this cone (which may also be a dihedral angle) is at least $c'\delta\vartheta$, where c' may be assumed independent of δ.

Let n_0 be the number of squares in Σ_0 which contain ridge points at which the plane angle is at least ϑ, and $H(F_0)$ the mean curvature of F_0. Then it is clear that $c'n_0\delta\vartheta \leqq 9H(F_0)$. Similarly, if n_1 is the number of squares in Σ_1 which contain ridge points of F_1 at which the plane angle is at least ϑ and $H(F_1)$ the mean curvature of F_1, then $c'n_1\delta\vartheta \leqq 9H(F_1)$. Since $n \leqq n_0 + n_1$, we have

$$c'n\delta\vartheta \leqslant 9\,(H\,(F_0) + H\,(F_1))$$

and so $n\delta$ is bounded by a constant independent of δ. Q.E.D.

Let β be a plane cutting the surface Ω which contains at least one point X of Ω corresponding to a point X_0 on F_0 in the set E_0^ϑ and to a point X_1 on F_1 in the set E_1^ϑ. At least one of these points, say X_0, is a ridge point at which the plane angle is at least ϑ. Suppose that the direction on Ω at X which corresponds to the direction of the edge at X_0 on F_0 lies in the plane β. We claim that *the measure of the set of planes β is zero.*

Let ω denote the regions on Ω corresponding to the squares on F_0 in the set Σ_0. Let us estimate the measure of the set of planes β for a single region ω. It follows from the properties of the directions of the edges in squares ω_0 and ω_1 on F_0 and F_1, as described above, that the directions on the region ω which are contained in the planes β form arbitrarily small angles, provided the squares ω_0 are sufficiently small, or, what is the same, δ is sufficiently small. Since moreover the diameters of the regions ω are less than $\lambda\delta$ (where λ is a constant independent of δ), the measure of the set of planes β for one region is less than $\epsilon'(\delta)$, where

$\epsilon'(\delta)$ tends to zero with δ. Hence the measure of the set of planes β for the entire surface Ω is less than $\epsilon'(\delta)n\delta$. Since $n\delta$ is bounded, the set of planes β has measure zero.

Since the measure of the set of planes β is zero for every $\vartheta > 0$ and the measure of the set of planes intersecting the set G^{ϵ} is arbitrarily small if ϵ is small, it follows that almost all planes cutting the surface Ω cut the edges of the surface.

This completes the proof of the main assertion.

Assume that the surface Ω and the xy-plane satisfy the following conditions:

1) The xy-plane essentially cuts Ω and cuts the edges of Ω.

2) The boundary of the surface Ω lies above the xy-plane.

Let M be the set of all points of Ω in the xy-plane, M_0, M_1 and M_1^* the corresponding sets on the surfaces F_0, F_1 and F_1^*. We claim that *the sets M_0 and M_1 are unions of equidistant normal curves* (§4).

Let $r_0(Y)$ denote the radius-vector of a point Y on F_0, and $r_1(Y)$ and $r_1^*(Y)$ the radius-vectors of the corresponding points on F_1 and F_1^* (recall that the latter is the reflection of F_1 in the xy-plane).

Let X_0 be any point of M_0. In a neighborhood of this point the vector-valued functions $r_0(Y)$, $r_1(Y)$ and $r_1^*(Y)$ have the following convenient representations: (Chapter II, §2):

$$r_0(Y) = r_0 + \tau_0(Y)\, s(Y) + \nu_0(Y),$$
$$r_1(Y) = r_1 + \tau_1(Y)\, s(Y) + \nu_1(Y),$$
$$r_1^*(Y) = r_1^* + \tau_1^*(Y)\, s(Y) + \nu_1^*(Y).$$

The vectors $\tau_0(Y)$, $\tau_1(Y)$ and $\tau_1^*(Y)$ lie on the faces of the tangent cones to F_0 at X_0 and to F_1 and F_1^* at the corresponding points $\overline{X}_1$ and X_1^*.

If Y is a point of M_0, then the vector $r_0(Y) + r_1^*(Y)$ lies in the xy-plane. Hence for sufficiently small $s(Y)$ the angle between the vector $\tau_0(Y) + \tau_1^*(Y)$ and the xy-plane is arbitrarily small. Since the xy-plane essentially cuts the surface Ω and cuts the edge of this surface at the point X corresponding to X_0 (if at least one of the points X_0 or X_1 is a ridge point), we conclude that there are two directions τ_0' and τ_0'' on F_0 at X_0 satisfying the following conditions.

1) The directions on Ω corresponding to τ_0' and τ_0'' lie in the xy-plane.

2) The directions τ_0' and τ_0'', and also the corresponding directions on F_1, lie in different faces of the tangent dihedral angles to F_0 and F_1 at X_0 and X_1, respectively, if the latter are ridge points.

3) If a point Y in M_0 is sufficiently close to X_0, i.e. if $s(Y)$ is sufficiently small, then the angle between the vector $\tau_0(Y)$ and one of the vectors τ_0' or τ_0'' is arbitrarily small.

Let a, b, c and d be segments on F_0 issuing from the point X_0 and satisfying the following conditions:

1. The semitangents to the segments a and b at X_0 form small angles with the direction τ_0', the latter lies between the semitangents, and the semitangents themselves lie on the same face of the tangent dihedral angle to F_0 at X_0 (if the latter is a ridge point) as the direction τ_0'.

2. The semitangents to the segments c and d at X_0 form small angles with the direction τ_0'', the latter lies between the semitangents, and the semitangents lie on the same face of the tangent dihedral angle at τ_0''.

3. The segments on F_1 corresponding to a, b, c and d satisfy the above conditions 1 and 2 relative to the directions corresponding to τ_0' and τ_0''.

Obviously, if Y is a point in M_0 sufficiently close to X_0, it must lie either between a and b or between c and d.

Let Y be a point on F_0 between the segments a and b, and let it approach arbitrarily close to X_0. Then the supporting plane to F_0 at Y tends to the face of the dihedral angle at X_0 which contains the direction τ_0'. But if the point Y approaching X_0 lies between c and d, then the supporting plane tends to the face of the angle which contains τ_0''. The analogous conclusion holds true for the surface F_1 and the corresponding segments and directions.

Draw an arc of a geodesic circle k on F_0 between the segments a and b, with endpoints A and B on a and b, respectively. If the radius of k is sufficiently small, the arc has a semitangent $t_0(Z)$ at each point Z. The points on Ω corresponding to A and B lie on different sides of the xy-plane if the radius of the circle k is sufficiently small. Hence there is at least one point Z_0 on k such that the corresponding point on Ω lies on the xy-plane. But it is not difficult to see that there is only one such point Z_0, since if the radius of k is small the angle between the semitangent $t_0(Z)$ and the face of the tangent dihedral angle to F_0 at X_0 which contains the direction t_0' may be made arbitrarily small, and the angle between this semitangent and the direction τ_0' may be made arbitrarily close to $\pi/2$. Similar conclusions hold true for the semitangent $t_1(Z)$ to the circle k_1 corresponding to k on F_1. It follows that the subset of M_0 in a sufficiently small neighborhood of X_0 between the segments a and b is a simple arc. The same applies to the subset of M_0 situated between the segments c and d.

It is now obvious that the entire set M_0 is the union of a finite number of simple closed curves. Without loss of generality, we may assume that M_0 itself is a closed curve.

The curve M_0 has right and left semitangents at each point; the right semitangent is continuous from the right, and the left semitangent is continuous from the left.

Essentially we have already proved the existence of a right and left semitangent at each point of M_0: the semitangents at a point X_0 of M_0 are τ_0' and τ_0''. We now prove the second part of the assertion. To fix ideas, let τ_0' be the right semitangent to M_0 at X_0. Let Y be a point on M_0 to the right of X_0. Let $\tilde{\tau}_0(Y)$ be the unit vector of the right semitangent at this point, and $\tilde{\tau}_1^*(Y)$ the unit vector of the semitangent to M_1^* on F_1^* at the corresponding point. We wish to prove that $\tilde{\tau}_0(Y) \to \tau_0'$ as Y approaches X_0 from the right. Suppose that this is false. Then there exists a sequence of points Y_k converging to X_0 such that $\tilde{\tau}_0(Y_k) \to \tilde{\tau}_0 \neq \tau_0'$ and $\tilde{\tau}_1^*(Y_k) \to \tilde{\tau}_1^*$. Join the points Y_k and X_0 by a segment γ_0. Let $\bar{\tau}_0(Y_k)$ be the unit vector of the semitangent to this segment at Y_k, and $\bar{\tau}_1^*(Y_k)$ the unit vector of the semitangent to the corresponding segment on F_1^* at the corresponding point. As $k \to \infty$ the angles between the vectors $\tilde{\tau}_0(Y_k)$ and $\bar{\tau}_0(Y_k)$, $\tilde{\tau}_1^*(Y_k)$ and $\bar{\tau}_1^*(Y_k)$ approach the same limit, which is greater than $\pi/2$. The angles between the vectors $\bar{\tau}_0(Y_k)$ and $\tau_0'^*$, $\bar{\tau}_1^*(Y_k)$ and $\tau_1'^*$ approach π. It follows that the limit vectors $\tilde{\tau}_0$ and $\tilde{\tau}_1^*$ form equal angles with the vectors τ_0' and $\tau_1'^*$. Now, since the two limit vectors lie in the same faces of the tangent dihedral angles as the vectors τ_0' and $\tau_1'^*$, the vector $\tilde{\tau}_0 + \tilde{\tau}_1^*$ cannot be parallel to the xy-plane. But this contradicts the fact that the vector $\tilde{\tau}_0(Y_k) + \tilde{\tau}_1^*(Y_k)$ is parallel to the xy-plane for every point Y_k. This proves that the right semitangent to M_0 is continuous from the right. The left continuity of the left semitangent is proved in a similar manner. The right continuity of the right semitangent implies that the curve is rectifiable.

Thus the curves M_0 and M_1 on F_0 and F_1 are rectifiable, and have right and left semitangents which are continuous from the right and left, respectively.

We shall now show that M_0 and M_1 are curves with i.g.c. of bounded variation. It will obviously suffice to prove that each point of the curve has right and left semineighborhoods with i.g.c. of bounded variation. Let m_0 denote a sufficiently small right semineighborhood of a point X_0. Let m_1 and m_1^* be the corresponding curves on F_1 and F_1^*.

If the arc m_0 on the curve M_0 is sufficiently small, the angles between the right (left) semitangents at any two points of m_0 are small, as are the angles between the supporting planes to F_0 at any two points of m_0. The same holds for the curves m_1 and m_1^* on F_1 and F_1^*.

Let $\bar{m}_0$, $\bar{m}_1$ and $\bar{m}_1^*$ denote the projections of m_0, m_1 and m_1^* on the xy-plane. The isometry of the surfaces F_0, F_1 and F_1^* induces a natural arc-length-preserving correspondence between the points of the curves $\bar{m}_0$, $\bar{m}_1$ and $\bar{m}_1^*$.

Let F_2 be the reflection of the surface F_1 in some plane σ perpendicular to the xy-plane; let m_2 and $\bar{m}_2$ be the curves in F_2 corresponding to m_1^* and its projection on the xy-plane. The surface F_2 is obtained from F_1 by rotation through the angle π about a suitable axis in the xy-plane.

Construct the mixed curve $\bar{m} = \frac{1}{2}(\bar{m}_0 + \bar{m}_2)$. We claim that the curve $\bar{m}$ is convex if the arc m_0 is sufficiently small and the plane σ is suitably chosen. To be precise, σ must be so chosen that the semitangents to m_0 form small angles with it.

We first show that the curve $\bar{m}$ is locally convex. Let $\bar{Y}$ be an arbitrary point on $\bar{m}$, Y_0 the corresponding point on m_0. Let $V_0(Y_0)$ be the tangent dihedral angle to F_0 at Y_0, $V_2(Y_0)$ the tangent dihedral angle at the corresponding point on F_2. Construct the mixed dihedral angle $V(Y_0) = \frac{1}{2}(V_0(Y_0) + V_2(Y_0))$. The edge of the angle $V(Y_0)$ essentially cuts the xy-plane. Let β be a plane through the edge of $V(Y_0)$ which does not separate the faces of the angle (if Y_0 and the corresponding point of F_2 are smooth points, β simply coincides with $V(Y_0)$). β is a local supporting plane of the curve $\bar{m}$ at $\bar{Y}$, and the points of $\bar{m}$ close to $\bar{Y}$ lie in the halfspace defined by β which contains the angle $V(Y_0)$. In fact, the radius-vector of the point on $\bar{m}$ corresponding to Y on m_0 may be expressed as

$$r_0(Y_0) + r_2(Y_0) + s(Y)(\tau_0(Y) + \tau_2(Y)) + \nu_0(Y) + \nu_2(Y).$$

The point with radius-vector

$$r(Y) = r_0(Y_0) + r_2(Y_0) + s(Y)(\tau_0(Y) + \tau_2(Y))$$

lies on the dihedral angle $V(Y_0)$. Lay off a vector $\nu_0(Y) + \nu_2(Y)$ from this point. We claim that each of the vectors $\nu_0(Y)$ and $\nu_2(Y)$, laid off from the point $P(Y)$, points into the angle $V(Y_0)$. Let e_1 and e_2 be two mutually perpendicular vectors on one face of $V_0(Y_0)$. Let $e_3 = e_1 \times e_2$ be the unit vector of the interior normal to this face. The vectors e_1 and e_2 are mapped by the isometry onto vectors e_1' and e_2' on one face of the angle $V_2(Y_0)$, and the vector $e_3' = e_1' \times e_2'$ is the unit vector of the interior normal to this face of $V_2(Y_0)$. The plane of the vectors $e_1 + e_1'$ and $e_2 + e_2'$ contains one of the faces of $V(Y_0)$. To prove that the vectors $\nu_0(Y)$ and $\nu_2(Y)$ point into the angle $V(Y_0)$, we must prove that the triple scalar products

$$\left(e_1+e_1',\ e_2+e_2',\ \frac{\nu_0(Y)}{|\nu_0(Y)|}\right),\ \left(e_1+e_1',\ e_2+e_2,\ \frac{\nu_2(Y)}{|\nu_2(Y)|}\right)$$

are nonnegative, irrespective of the face of $V_0(Y_0)$ on which the vectors e_1 and e_2 are chosen. We may assume that the vectors $\nu_0(Y)$ and $\nu_2(Y)$ are not zero, for if one of them is zero we may regard it as pointing into $V(Y_0)$ by definition.

We first show that $(e_1+e_1',\ e_2+e_2',\ e_3)>0$. Set

$$f=(e_1+e_1',e_2+e_2',e_3+e_3'),\ g=(e_1+e_1',e_2+e_2',e_3-e_3').$$

Let α_{ij} be the ith component of the vector e_j' relative to the basis e_1, e_2, e_3. Then

$$f=\begin{vmatrix}1+\alpha_{11} & \alpha_{12} & \alpha_{13}\\ \alpha_{21} & 1+\alpha_{22} & \alpha_{23}\\ \alpha_{31} & \alpha_{32} & 1+\alpha_{33}\end{vmatrix}=|E+\alpha|,$$

where E is the unit matrix and α an orthogonal matrix. Let C be a nonsingular matrix such that

$$C^{-1}\alpha C=\begin{pmatrix}1 & 0 & 0\\ 0 & \cos\vartheta & \sin\vartheta\\ 0 & -\sin\vartheta & \cos\vartheta\end{pmatrix}.$$

Then

$$f=|C^{-1}(E+\alpha)C|=\begin{vmatrix}2 & 0 & 0\\ 0 & 1+\cos\vartheta & \sin\vartheta\\ 0 & -\sin\vartheta & 1+\cos\vartheta\end{vmatrix}=4(1+\cos\vartheta)\geqslant 0.$$

Now f cannot vanish, since otherwise the face of the angle $V_0(Y_0)$ would coincide with a face of $V_2(Y_0)$, suitably oriented, after rotation of the latter face through the angle π, and this is impossible. Thus f is strictly positive.

Similarly one can prove that $g>0$. Hence

$$(e_1+e_1',e_2+e_2',e_3)=f+g>0.$$

When the point Y_0 approaches X_0, the direction of the vector $\nu_0(Y_0)$ tends to the direction of the interior normal to the face of the angle $V_0(X_0)$ which contains the vector τ_0' (the unit vector of the semitangent to m_0 at X_0). And since we may assume that

$$(e_1+e_1',e_2+e_2',e_3)>c_0>0,$$

it follows that for Y_0 close to X_0 and Y close to Y_0

$$\left(e_1 + e_1', \ e_2 + e_2', \ \frac{\nu_0(Y)}{|\nu_0(Y)|}\right) > 0.$$

Similarly one proves that

$$\left(e_1 + e_1', \ e_2 + e_2', \ \frac{\nu_2(Y)}{|\nu_2(Y)|}\right) > 0.$$

Thus β is a local supporting plane for $\bar{m}$ at $\bar{Y}$. The intersection of β and the xy-plane is a local supporting line for $\bar{m}$ at $\bar{Y}$. It now follows easily from the local convexity of $\bar{m}$ and the position of the curve relative to its local supporting lines that $\bar{m}$ is convex.

Let s be arc length along the curve $\bar{m}_0$, $\bar{r}_0(s)$ the radius-vector of the representative point of $\bar{m}_0$, and $\bar{r}_2(s)$ the radius-vector of the corresponding point of $\bar{m}_2$. Since the curve $\bar{m}$ is convex, the vector-valued function

$$(*) \qquad \varphi_\sigma(s) = \frac{\bar{r}_0^+(s) + \bar{r}_2^+(s)}{|\bar{r}_0^+(s) + \bar{r}_2^+(s)|}$$

has bounded variation (the index σ signifies that the surface F_2 is obtained by reflection of the surface F_1^* in the plane σ).

If we consider two different planes σ, we get two vector-valued functions φ_{σ_1} and φ_{σ_2}. The two vector equalities (*) for σ_1 and σ_2 yield regular expressions for the components of the vectors $\bar{r}_0^+$ and $\bar{r}_2^+$ in terms of the components of φ_{σ_1} and φ_{σ_2}. It follows that the components of the vectors $\bar{r}_0^+$ and $\bar{r}_2^+$, hence also the vectors themselves, have bounded variation. But this implies that the curves $\bar{m}_0$ and $\bar{m}_2$ have i.g.c. of bounded variation.

This in turn implies that the curves m_0, m_2 and m_1 have i.g.c. of bounded variation (§1), and so, as indicated previously, the curves M_0 and M_1 have i.g.c. of bounded variation.

We have already remarked that the semitangents to M_0 and M_1 do not coincide with the directions of the tangent dihedral angles to F_0 and F_1 at the corresponding points. The fact that the angles between the faces of the tangent dihedral angles to F_0 at points of M_0 and the xy-plane are either always greater or always smaller than the angles between the corresponding faces of the dihedral angles tangent to F_1 and the xy-plane easily follows from the fact that the xy-plane essentially cuts the surface Ω.

Thus the curves M_0 and M_1 on the surfaces F_0 and F_1 are normal and equidistant.

§6. Monotypy of closed convex surfaces

We have now prepared all the material required to prove the fundamental theorem of this chapter: the monotypy of closed convex surfaces.

THEOREM 1. *Isometric closed convex surfaces are congruent.*

PROOF. Let F_0 be a closed convex surface and F_1 a convex surface isometric to F_0 (in this context a convex surface may degenerate into a doubly covered convex plane domain). We wish to prove that F_0 and F_1 are congruent. First note that if each of the surfaces is a doubly covered convex plane domain, the assertion of the theorem is fairly obvious. Indeed, it is readily seen that the boundaries of the plane convex domains F_0 and F_1 correspond under the isometry, and so corresponding subarcs of the boundary have equal length and i.g.c. Hence the boundaries of F_0 and F_1 are congruent, and therefore so are the domains themselves.

Thus if F_0 and F_1 are not congruent, then at least one of them, say F_0, is not degenerate.

As was shown in §2, there exists a closed convex surface F_1' isometric to F_0, arbitrarily close to F_0 but not congruent to it. Let $r_0(X)$ be the radius-vector of an arbitrary point X on F_0, and $r_1(X)$ the radius-vector of the corresponding point on F_1'.

If $|\lambda - 1/2|$ is small, then the equation $r = \lambda r_0(X) + (1-\lambda) r_1(X)$ defines a closed convex surface F_λ, and the mapping of F_0 onto F_λ taking the point $r_0(X)$ of F_0 to the point $r_\lambda(X) = \lambda r_0(X) + (1-\lambda) r_1(X)$ of F_λ is a homeomorphism (§3).

It will be convenient to set $r_\lambda(X) = \bar{r}(X)$ and $F_\lambda = \bar{F}$ for $\lambda = 1/2$.

Describe a sphere of minimal radius about the surface $\bar{F}$. Let $\bar{r}(X_0)$ be a point of the surface on the sphere, and α_0 the tangent plane to the sphere at this point (which is therefore a supporting plane for $\bar{F}$). Introduce rectangular cartesian coordinates in space, with $\bar{r}(X_0)$ as origin, α_0 as xy-plane and the positive z-axis directed into the halfspace defined by a_0 which contains the surface $\bar{F}$.

Without loss of generality we may assume that

$$r_0(X_0) = r_1(X_0) = \bar{r}(X_0) = 0.$$

We claim that the equality $|r_0(X)| = |r_1(X)|$ cannot hold identically. Indeed, consider the plane defined by two arbitrary points X and Y on F_0 and the origin. The plane cuts the surface in a plane convex curve;

let γ denote the arc of this curve connecting X and Y which does not contain the origin. Let γ' be the curve on F_1' corresponding to γ, and X' and Y' its endpoints, which correspond to X and Y under the isometry.

Consider the cone formed by connecting the origin to all points of γ' by straight lines, and develop this cone on a plane. The curve γ' then becomes a plane curve $\bar{\gamma}'$ congruent to γ. Since the distance between the endpoints of $\bar{\gamma}'$ is at least the extrinsic distance between the endpoints of γ', it follows that the extrinsic distance between X and Y on F_0 is at least the extrinsic distance between X' and Y' on F_1'.

Interchanging the roles of F_0 and F_1', we get the converse. It follows that the isometry of F_0 and F_1' preserves extrinsic distances, and so the surfaces are congruent—a contradiction.

Thus the equality $|r_0(X)| = |r_1(X)|$ is not an identity.

Let G_0 be a (connected) component of the set of points on F_0 for which $|r_0(X)| \neq |r_1(X)|$. Without loss of generality we may assume that $|r_0(X)| > |r_1(X)|$ in G_0. We now define the vector-valued function

$$r(X) = \frac{r_0(X) + r_1(X)}{r_0^2(X) - r_1^2(X)}$$

on G_0. Let $\bar{G}$ be the domain on $\bar{F}$ corresponding to the domain G_0 on F_0, and Φ the surface defined by the equation $r = r(X)$ in G_0. The surface Φ is uniquely projected onto the domain $\bar{G}$ on $\bar{F}$ by rays issuing from the origin.

We claim that if a point X of G_0 approaches the boundary of G_0, then the distance between the point $r(X)$ of Φ and the xy-plane becomes arbitrarily large.

Indeed, let Y_0 be the boundary point of G_0 to which X tends. If Y_0 is different from X_0, our assertion is obvious, since then $r_0^2(X) - r_1^2(X) \to 0$, and $X \to Y_0$, but $\bar{z}(X) = z_0(X) + z_1(X)$ tends to a positive limit.

Now assume that $Y_0 = X_0$. As was shown in §2 of Chapter II, in a neighborhood of X_0 the vector-valued functions $r_0(X)$ and $r_1(X)$ have representations

$$r_0(X) = s(X)\tau_0(X) + \varepsilon_0(X)s(X),$$
$$r_1(X) = s(X)\tau_1(X) + \varepsilon_1(X)s(X),$$

where $s(x)$ is the intrinsic distance between X_0 and X on F_0, $\tau_0(X)$ is the unit vector of the semitangent at X_0 to the segment connecting X_0 and X on F_0, $\tau_1(X)$ is the unit vector of the semitangent to the corresponding segment on F_1' at the point corresponding to X_0, and $\epsilon_0(X)$ and $\epsilon_1(X)$ are vectors tending to zero as $X \to X_0$.

If the surfaces F_1' and F_0 are sufficiently close together, the angle between the vectors $r_0(X)$ and $r_1(X)$ will be less than $\vartheta_0 < \pi$. Therefore if X is sufficiently close to X_0 we have $|r_0(X) + r_1(X)| > c_1 s(X)$, where c_1 is a positive constant. Hence in view of the way in which X_0 was chosen we see that $z_0(X) + z_1(X) > c_2 s^2(X)$. Now the difference $r_0^2(X) - r_1^2(X)$ is equal to $s^2(X)\epsilon(X)$, where $\epsilon(X) \to 0$ as $X \to X_0$. Thus the distance of the point $r(X)$ from the xy-plane is greater than $c_2/\epsilon(X)$, so that it approaches infinity as $X \to X_0$.

Cut the surface Φ by the plane $z = h > 0$. The part $\bar{\Phi}$ of Φ lying below $z = h$, i.e., in the halfspace $z \leqq h$, is bounded.

Let P be a "flat" paraboloid, $z = k(x^2 + y^2)$, such that the entire surface $\bar{\Phi}$ is in the interior of P. Subject this paraboloid to an affine transformation, "compressing" it toward the plane $z = h$ by displacing its points along straight lines parallel to the z-axis toward the plane $z = h$, in proportion to their original distance from this plane. At some instant the paraboloid will touch the surface $\bar{\Phi}$, at a point whose radius-vector we denote by $r(Y_0)$.

Let $\bar{n}$ be the unit vector of the interior normal to the tangent plane of the paraboloid at $r(Y_0)$. Since the surface F_1' is close to F_0, we see that when X is close to Y_0 we have

$$|r(X) - r(Y_0)| > c_3 s(X),$$

where $s(X)$ is the distance between X and Y on F_0 and c_3 is a positive constant. Therefore

$$(r(X) - r(Y_0))\bar{n} > c_4 s^2(X), \tag{*}$$

where c_4 is a positive constant.

We introduce the abbreviations $r_0(Y_0) = a_0$, $r_1(Y_0) = a_1$ and $2r(Y_0) = a$. Then $r_0(X) = a_0 + \bar{r}_0$, $r_1(X) = a_1 + \bar{r}_1$ and

$$r(X) - r(Y_0) = \frac{a_0 + a_1 + \bar{r}_0 + \bar{r}_1}{(a_0 + \bar{r}_0)^2 - (a_1 + \bar{r}_1)^2} - \frac{a_0 + a_1}{a_0^2 - a_1^2}$$

$$= \frac{(\bar{r}_0 + \bar{r}_1)(a_0^2 - a_1^2) - 2(a_0 + a_1)(a_0\bar{r}_0 - a_1\bar{r}_1)}{(a_0^2 - a_1^2)(a_0^2 - a_1^2 + 2(a_0\bar{r}_0 - a_1\bar{r}_1))} + \varepsilon s^2(X).$$

Hence, using the inequality (*), we get

$$(\bar{r}_0 + \bar{r}_1 - a(a_0\bar{r}_0 - a_1\bar{r}_1))\bar{n} > cs^2(X),$$

or

$$-\bar{r}_0(-\bar{n} + a_0(a\bar{n})) + \bar{r}_1(\bar{n} + a_1(a\bar{n})) > cs^2(X).$$

Set

$$-\bar{n}+a_0(a\bar{n})=A_0,\quad \bar{n}+a_1(a\bar{n})=A_1.$$

Then

$$-A_0\bar{r}_0+A_1\bar{r}_1>cs^2(X)$$

(where c is a positive constant).

We claim that the vector a is not on the surface $\bar{F}$ (i.e. it points into the surface) and that the scalar product $(a\bar{n})$ is positive. Indeed, the paraboloid P is by assumption "flat". Hence the normal $\bar{n}$ forms a small angle with the positive z-axis, and the vector a ends at a point of the surface $\bar{\Phi}$, which is bounded and lies at a positive distance from the xy-plane. Hence the angle between this vector and the positive z-axis cannot be arbitrarily close to $\pi/2$. It follows that the scalar product $(a\bar{n})$ may be assumed positive. It is also clear that the vector a cannot lie on the surface $\bar{F}$, since otherwise the euclidean segment a on $\bar{F}$ would correspond to euclidean segments on F_0 and F_1', and so we would have $r_0^2(Y_0)-r_1^2(Y_0)=0$, which is impossible.

Now replace the surfaces F_0 and F_1' by mixed surfaces F_λ and F_μ ($\mu=1-\lambda$) and construct surfaces $\bar{F}_\lambda$ and Φ_λ for them in the same way as $\bar{F}$ and Φ were constructed for F_0 and F_1'. It is easy to see that $\bar{F}_\lambda$ coincides with $\bar{F}$ and the surface Φ_λ is obtained from Φ by a similarity mapping about the origin with ratio of similitude $1/(\lambda-\mu)^2$. Therefore in a neighborhood of Y_0 we have

$$(**)\qquad -A_\lambda\bar{r}_\lambda+A_\mu\bar{r}_\mu>c_\lambda s_\lambda^2(X),$$

where

$$A_\lambda=-\bar{n}+(\lambda a_0+\mu a_1)\frac{(a\bar{n})}{(\lambda-\mu)^2},$$

$$A_\mu=\bar{n}+(\mu a_0+\lambda a_1)\frac{(a\bar{n})}{(\lambda-\mu)^2}.$$

When $\lambda\to 1/2$, the directions of the vectors A_λ and A_μ approach the direction of the vector a. Therefore, if $|\lambda-1/2|$ is sufficiently small, the vectors $-A_\lambda$ and $-A_\mu$, laid off from the points $r_\lambda(Y_0)$ and $r_\mu(Y_0)$, point into the tangent cones of F_λ and F_μ at these points, since the vector a points into the tangent cone to $\bar{F}$ at $\bar{r}(X_0)$, as noted above.

Displace the surface F_μ parallel to itself so that its point $r_\mu(Y_0)$ coincides with the point $r_\lambda(Y_0)$ of F_λ, and then rotate it about an axis through this point until the vectors A_λ and A_μ coincide. Denote the surface F_μ in its new position by F_μ'. We now introduce new rectangular coordinates in space, with the point $r_\lambda(Y_0)$ as origin and the direction of the vector A_λ as the direction of the negative z-axis.

Let G be a neighborhood of Y_0 on F_0. Denote the corresponding domains on F_λ and F'_λ by $\overline{F}_\lambda$ and $\overline{F}'_\lambda$.

If $|\lambda - 1/2|$ and ϵ are sufficiently small, then the mixed surface $\frac{1}{2}(\overline{F}_\lambda + \overline{F}'_\mu)$ is convex, has a unique projection on the xy-plane and is convex toward $z < 0$.

Consider the surface Φ' defined for $X \in G$ by the equation

$$r = \tilde{r}(X) = r_\lambda(X) + r'^*_\mu(X),$$

where $r'^*_\mu(X)$ is the reflection of the vector $r'_\mu(X)$ in the xy-plane. The point $\tilde{r}(Y_0)$ of this surface lies in the xy-plane, and all its other points lie in the halfspace $z > 0$. The latter statement follows from inequality (**), since $|A_\lambda| = |A_\mu|$ (indirect proof).

It was shown in §5 that almost all planes α not disjoint from Φ' cut it essentially and cut its edges. Let α be a plane of this type, forming a sufficiently small angle with the xy-plane and not cutting the boundary of Φ'.

Reflect the surface F'^* defined by the equation $r = r'^*_\mu(X)$ in the plane α. Call the resulting surface F'^{**}_μ. If the angle between the plane α and the xy-plane is sufficiently small, the surfaces F_λ and F'^{**}_μ are in canonical position relative to the xy-plane.

The plane α cuts the surface $\frac{1}{2}(F_\lambda + F'^*_\mu)$ in a closed curve c (or several closed curves). The corresponding curves c_λ and c'^{**}_μ on F_λ and F'^{**}_μ are closed normal curves equidistant from the plane α. But it was proved in §4 that isometric convex surfaces in canonical position cannot contain closed normal equidistant curves.

This contradiction completes the proof of Theorem 1.

Using Theorem 1, we easily prove the following theorem.

THEOREM 2. *Let F_1 be a convex surface with total curvature 4π bounded by contours $\gamma_1, \cdots, \gamma_n$ with i.g.c. of bounded variation. Then any convex surface F_2 isometric to F_1 is congruent to F_1.*

PROOF. Let F'_1 be the convex hull of the surface F_1. F_1 is a domain on the closed surface F'_1. Since the curvature of this domain is 4π, the remainder of F'_1 must have zero curvature and is therefore made up of locally isometric plane domains G_k bounded by the curves γ_k. Similar conclusions hold for the convex hull F'_2 of the surface F_2.

Since the curvature of F'_1 along a curve γ_k is zero, the i.g.c. of this curve toward the domain G_k differs only in sign from its i.g.c. toward the domain F_1 (the sum of i.g.c. is equal to the curvature of the surface along the curve; see Chapter I, §8). A similar conclusion holds for the corresponding domain G'_k on the surface F'_2. Since the surfaces F_1 and F_2 are isometric, the i.g.c. of the curves γ_k and γ'_k bounding the domains G_k and G'_k are equal over subarcs which correspond under the isometry. It follows that the isometry of the domains F_1 and F_2 can be extended to the interiors of G_k and G'_k, i.e. to the entire closed surfaces F'_1 and F'_2. Now the isometric surfaces F'_1 and F'_2 must be congruent. Since the domains F_1 and F_2 on the surfaces F'_1 and F'_2 correspond under the isometry, they are congruent. Q.E.D.

The convex surfaces whose rigidity is established by Theorem 2 are obtained from closed convex surfaces by removing domains with zero curvature. It is natural to ask what can be said of the rigidity of such surfaces if the missing domains have positive curvature. The following theorem of Leĭbin [**39**] answers this question.

THEOREM 3. *A convex surface F obtained from a closed convex surface by removing a domain of positive curvature is not rigid, i.e. there exist surfaces isometric to F but not congruent to it.*

PROOF. Assume that F is obtained from a closed convex surface F' by removing a domain G of positive curvature. Let Φ be the convex hull of the surface F. It consists of the surface F and a developable surface $\bar{F}$. Suppose that $\bar{F}$ is not a plane domain. Obviously there exists a point S inside F' but outside Φ from which not all of the domain $\bar{F}$ on Φ is visible from without. Construct a conical surface V' projecting the surface Φ from the point S. The surface V' touches $\bar{F}$ along some generator g [of $\bar{F}$] whose endpoints are on the boundary of F, and the plane triangle Δ with base g and apex S lies in V' but not in $\bar{F}$. Let A be an interior point of the segment g and O a point in Δ close to A. Let Φ' be the convex hull of O and F. This surface consists of F and a certain surface V, isometric to the cone, with conical point O. The segment g lies in this surface.

Let γ_1 be a geodesic issuing from O on the surface V, perpendicular to g. Let B be the point at which this geodesic cuts g, and let B' be a point on the geodesic close to B but beyond g. Let γ_2 be a geodesic issuing from O which together with γ_1 bisects the complete angle at O. Let C' be a point on γ_2 close to O. The points B' and C' can be joined on the surface V by two segments γ' and γ'' (Figure 27). Remove the domain bounded by these segments and identify the points of γ' and

γ'' corresponding to equal arc length (measured from B'). By the Gluing Theorem there exists a closed surface, cut along a certain curve, which is isometric to Γ' cut along the contour $\gamma' + \gamma''$. The domain $\tilde{F}$ on this surface isometric to F is not congruent to F, since the distance between the endpoints of the euclidean segment g is smaller in $\tilde{F}$ than in F.

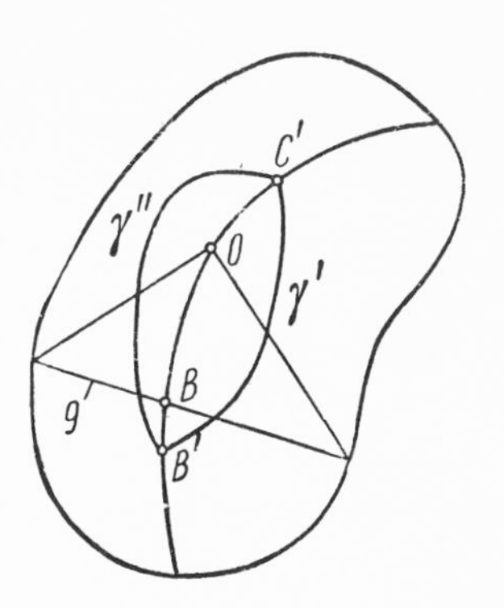

FIGURE 27

Now let $\bar{F}$ be a plane domain. Drop a perpendicular to $\bar{F}$ from an interior point of $\bar{F}$, which cuts the surface F' at a point O not on F. Let Φ' be the convex hull of F and O. It consists of F and a conical surface V with apex O. Let B be a smooth point on the boundary of F (i.e. a smooth point of the boundary). Since the boundary of the surface V has nonnegative i.g.c., it contains a point C which can be joined to B by two segments on V. Remove the domain on V bounded by these segments and identify corresponding points on the segments. The Gluing Theorem yields a closed convex surface containing a domain isometric to F but not congruent to it. Indeed, the point B on the surface we have constructed is surely conical. No two generators of the tangent cone at this point can form an angle π, whereas on the original surface F the semitangents to the boundary at B form an angle π. This completes the proof

We shall now use Theorem 1 to prove two theorems on simultaneous approximation of isometric general convex surfaces by isometric convex polyhedra and isometric analytic convex surfaces with positive Gauss curvature. In many cases these theorems will enable us to reduce monotypy theorems for general convex surfaces to analogous theorems for convex polyhedra and analytic convex surfaces.

THEOREM 4. *Let F_1 and F_2 be bounded isometric convex surfaces, G_1 and G_2 domains on them that correspond to each other under the isometry and whose closures do not contain boundary points of the surfaces. Then for any positive ϵ there exist isometric convex polyhedra P_1 and P_2 and homeomorphisms f_1 and f_2 of F_1 and F_2 onto these polyhedra which satisfy the following conditions*:

1. *If X_1 and X_2 are corresponding points in G_1 and G_2, then the points $f_1(X_1)$ and $f_2(X_2)$ on the polyhedra correspond under the isometry.*

2. *The distances between the points X_1 and $f_1(X_1)$, X_2 and $f_2(X_2)$ are at most ϵ.*

Proof. Since the surfaces F_1 and F_2 are bounded, they may be regarded as domains on two closed convex surfaces F_1' and F_2'.

Construct sufficiently fine triangulations of F_1' and F_2'; this may be done in such a way that the triangulations in G_1 and G_2 correspond under the isometry of F_1 and F_2. Now construct polyhedral metrics, replacing each triangle T of the triangulation by a plane triangle with the same sides. By Aleksandrov's theorem, these metrics are realizable by closed convex polyhedra P_1 and P_2. It is easy to construct homeomorphisms f_1 and f_2 of the surfaces F_1' and F_2' onto the polyhedra P_1 and P_2 which satisfy the following conditions:

1. If A is a point in a triangle T on F_i', then its image $f_i(A)$ under the homeomorphism f_i is in the corresponding triangle T^0 on the polyhedron P_i.

2. If points A_1 and A_2 in G_1 and G_2 correspond under the isometry, then the points $f_1(A_1)$ and $f_2(A_2)$ on the polyhedra P_1 and P_2 also correspond under the isometry.

By condition 1 the metric of the polyhedron P_i tends to the metric of the surface F_i if the triangulation is made progressively finer. Hence by Theorem 1 the polyhedra P_i tend to surfaces congruent to F_i'. Without loss of generality we may assume that the polyhedra P_i converge to the surfaces F_i' themselves. Condition 2 of the theorem obviously holds if the polyhedra P_i are sufficiently close to the surfaces F_i'. This completes the proof.

Theorem 5. *Let F_1 and F_2 be bounded isometric convex surfaces, and G_1 and G_2 domains on them that correspond under the isometry and whose closures contain no boundary points of the surfaces. Then for any positive ϵ there exist isometric analytic convex surfaces Φ_1 and Φ_2 with positive Gauss curvature and homeomorphisms f_1 and f_2 of F_1 and F_2 onto Φ_1 and Φ_2 satisfying the following conditions:*

1. *If X_1 and X_2 are corresponding points in G_1 and G_2, then the points $f_1(X_1)$ and $f_2(X_2)$ on Φ_1 and Φ_2 also correspond under the isometry.*

2. *The distances between the points X_1 and $f_1(X_1)$, X_2 and $f_2(X_2)$ are smaller than ϵ.*

Proof. The isometric convex polyhedra P_1 and P_2 whose existence was proved in Theorem 4 are domains on closed convex polyhedra P_1' and P_2' (see the proof of Theorem 4). Without loss of generality we may assume that P_1' and P_2' have positive curvature at the vertices of the triangulation T. This may always be achieved by a small deformation of the metrics of P_1' and P_2'.

Replace each plane triangle in the triangulation of P_1' and P_2' by a spherical triangle with the same sides and curvature $1/R$. If R is sufficiently large, the resulting metrics will be convex, and their realizations, which are convex surfaces Φ_1' and Φ_2', will satisfy conditions 1 and 2 of the theorem. Now the metrics of the surfaces Φ_1' and Φ_2' are locally isometric to a sphere of radius R everywhere except at the vertices of the triangulations. Cut out an ϵ'-neighborhood of each vertex of the triangulation on Φ_1' and Φ_2', and "glue" to the resulting cut a spherical segment of total curvature equal to that of the removed neighborhood. The resulting metrics are realized by convex surfaces Φ_1'' and Φ_2'' which also satisfy conditions 1 and 2.

Finally, "smooth" the metrics of Φ_1'' and Φ_2'' in such a way that they are analytic in the domains corresponding to G_1 and G_2, have positive Gauss curvature, and "merge smoothly" with the metrics of the remaining regions of the surfaces. If the resulting metrics are sufficiently close to those of Φ_1'' and Φ_2'' in the domains corresponding to G_1 and G_2, the surfaces Φ_1 and Φ_2 that realize them satisfy conditions 1 and 2, and will be analytic in G_1 and G_2 (see the theorem in §10 of Chapter II). This completes the proof.

§7. Monotypy of convex surfaces with boundary. Maximum principle

Using Theorem 1 of §6, we shall now prove a maximum principle for general isometric convex surfaces (Theorems 1 and 3). The fundamental monotypy theorems for convex surfaces with boundary (Theorems 2 and 4) are simple corollaries of the maximum principle.

THEOREM 1. *Let F_1 and F_2 be two isometric convex surfaces which can be projected uniquely onto the xy-plane and are convex in the same direction, say toward $z < 0$. Let $z_1(X)$ be the z-coordinate of a point X on F_1, $z_2(X)$ the z-coordinate of the corresponding point on F_2. Then the function $\Delta(X) = z_1(X) - z_2(X)$ assumes its maximum and minimum on the boundary of the surface.*

PROOF. We first prove Theorem 1 for isometric convex polyhedra. Let M be the set of points on the polyhedron F_1 at which the function $\Delta(X)$ assumes its minimum. If the assertion is false, M does not contain boundary points of F_1. Also, $\Delta' \geqq 0$ at each point of M, differentiation being performed with respect to arc length along any segment issuing from the point. We claim that the set M contains a vertex of the polyhedron F_1, and there are directions from this vertex along which $\Delta' > 0$. Indeed, the boundary of the set M can contain only vertices or edges of F_1 or segments corresponding under the isometry to edges of F_2. Because of this structure of the boundary of M, it must contain at least one vertex X_0 of F_1. The inequality $\Delta' > 0$ must hold for some direction (pointing away from M) at this vertex. Let X_0' be the corresponding vertex of F_2. We claim that if $\Delta' > 0$ at X_0 the polyhedral angles V_1 and V_2 at the vertices X_0 and X_0' cannot be isometric. To prove this assertion we need some properties of convex polygons on a sphere.

Let a and b be two convex polygons on a sphere, in star-like position relative to a point O and convex away from O. Assume that corresponding sides $A_{k-1}A_k$ and $B_{k-1}B_k$ of the polygons are equal, and the distances of the vertices from O satisfy the conditions

$$(*) \qquad OA_k \geqq OB_k,$$

where strict inequality ($OA_k > OB_k$) holds for at least one pair of vertices. Then, with the angles α and β as in Figure 28, we have $\alpha < \beta$.

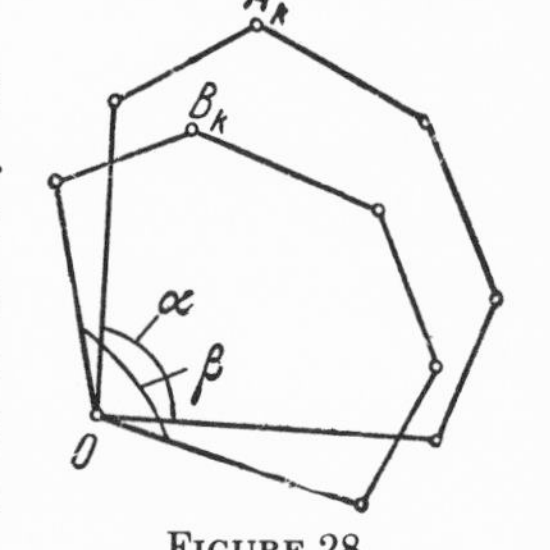

FIGURE 28

Indeed, the set of all polygons a satisfying (*) contains at least one for which α is a maximum. We claim that for this polygon $OA_k = OB_k$ for all k, and so $\alpha = \beta$. In fact, if $OA_k > OB_k$ and the angle at the vertex A_k of a is not π, then the angle α can be increased by suitably varying the distance OA_k. Thus in the extremal polygon a the angle at all vertices A_k for which $OA_k > OB_k$ must be π.

Let A' and A'' be two essential vertices of the polygon, between which there is a vertex A_k for which $OA_k > OB_k$. It is obvious that $OA' = OB'$ and $OA'' = OB''$. Transform the polygon b as follows. Preserving the lengths of the sides, make one of its essential vertices approach the point O, without changing the distance of the other vertices from O, until the angle at this vertex becomes equal to π. Then do the same for another vertex, and so on. At each step of this deformation the distance OB_k either remains the same or decreases. When the polygon b finally becomes an arc of a great circle, we have $OA_k = OB_k$. But this is impossible, since $OA_k > OB_k$ for the original polygon and throughout the deformation the distance OB_k cannot increase. Thus the extremal polygon must satisfy the equality $OA_k = OB_k$ at all its vertices. Consequently for any polygons a and b we have $\alpha \leqq \beta$, and equality can hold only when $OA_k = OB_k$ at all vertices. When this condition holds, the polygons are clearly congruent.

Now consider two closed polygons a and b. Let O be a point in the interior of the domains bounded by the polygons, such that $OA_k \geqq OB_k$ at all vertices A_k. Then, if the corresponding sides $A_{k-1}A_k$ and $B_{k-1}B_k$ are equal, we have $OA_k = OB_k$ and so the polygons are congruent. Indeed, let A' and A'' be two vertices of a. They divide the polygon into two parts a' and a''. Let α' and α'' be the corresponding angles. The points B' and B'' corresponding to A' and A'' divide the polygon b into two parts b' and b'', with corresponding angles β' and β''. Since $\alpha' + \alpha'' = 2\pi$ and $\beta' + \beta'' = 2\pi$, it follows that either $\alpha' \geqq \beta'$ or $\alpha'' \geqq \beta''$. Let $\alpha' \geqq \beta'$. Then, by what was proved above, we have $OA_k = OB_k$ at all vertices of the polygon a', and so $\alpha' = \beta'$. Hence $\alpha'' = \beta''$ and the equality $OA_k = OB_k$ also holds for the polygon a''. This proves our assertion.

We now return to the polyhedral angles V_1 and V_2 on our polyhedra. Identifying the apices of these angles, describe a unit sphere about them and let a and b denote the closed polygons thus defined on the sphere. To fix ideas, assume that V_1 and V_2 are convex toward $z < 0$. Let O denote the point at which the sphere is cut by the positive z-axis. Because of the minimum condition imposed on the function $\Delta(X)$, the directions on V_1 issuing from its vertex form with the positive z-axis angles not less than those between the positive z-axis and the corresponding directions on V_2. By what was proved above for closed polygons on a sphere, the polygons a and b are congruent, and hence so are the polyhedral angles V_1 and V_2. Thus the angles between corresponding directions on V_1 and V_2 and the positive z-axis are equal. It follows that $\Delta'(X_0) = 0$ along all directions from the point X_0. This is a contradiction, since there are directions from X_0 such that $\Delta' > 0$. This proves Theorem 1 for polyhedra.

Now let F_1 and F_2 be general convex surfaces. If the set M on which the function $\Delta(X)$ assumes its minimum contains no boundary points of F_1, then the latter surface contains a domain G_1 which contains M and its closure contains no boundary points of F_1. Let G_2 be the corresponding domain on F_2.

By Theorem 4 of §6 there exist isometric polyhedra P_1 and P_2 arbitrarily close to the surfaces F_1 and F_2. Since the set M is in the interior of G_1, it follows that if the polyhedra P_1 and P_2 are sufficiently close to F_1 and F_2 then the function $\Delta(X)$, defined on the polyhedron P_1, assumes its minimum on some set M on P_1 which contains no boundary points. But this contradicts what we have already proved for convex polyhedra, and the proof is complete.

Theorem 1 implies the following monotypy theorem for convex surfaces with boundary.

THEOREM 2. *Let F_1 and F_2 be isometric convex surfaces which can be projected uniquely onto the xy-plane and are convex in the same direction, say toward $z<0$. Let $z_1(X)$ be the z-coordinate of a point X on F_1, $z_2(X)$ the z-coordinate of the corresponding point on F_2. If $z_1(X)=z_2(X)$ along the boundary of the surface, the surfaces are congruent. In particular, isometric convex caps are congruent.*

PROOF. Since the maximum and minimum of the difference $z_1(X) - z_2(X)$ are assumed on the boundary, and there this difference vanishes, it follows that $z_1(X) \equiv z_2(X)$ (for all X). Let X_1 and X_2 be any two points on F_1. Cut the surface F_1 by a plane [through X_1 and X_2] perpendicular to the xy-plane. Let γ denote the curve on the resulting section joining X_1 and X_2. Let X_1' and X_2' be the corresponding points on F_2, and γ' the curve joining them which corresponds to γ. If the cylinder Z' projecting γ' onto the xy-plane is developed on a plane, it follows from the equality $z_1 \equiv z_2$ that the curve γ' becomes a curve congruent to γ. Since the distance between the endpoints of γ' is not changed when Z' is developed, the extrinsic distance between X_1 and X_2 is not less than the extrinsic distance between X_1' and X_2'. Interchanging the roles of F_1 and F_2, we get the converse. Hence the extrinsic distances between corresponding points on F_1 and F_2 are equal, and so the surfaces are congruent. Q.E.D.

THEOREM 3. *Let F_1 and F_2 be isometric convex surfaces, situated in the halfspace $z>0$ in star-like position relative to the origin O and convex in the same direction, i.e. they are visible from O either from without or from within. Let X be an arbitrary point on F_1, $z_1(X)$ its z-coordinate, and $\rho_1(X)$ its distance from O. Let $z_2(X)$ and $\rho_2(X)$ be the z-coordinate and the distance from O of the corresponding point on F_2.*

Then the function

$$\delta(X) = \frac{\rho_1^2(X) - \rho_2^2(X)}{z_1(X) + z_2(X)}$$

assumes a positive maximum and a negative minimum on the boundary of the surface.

PROOF. First let F_1 and F_2 be analytic surfaces with positive Gauss curvature. Suppose that the assertion is false, and let M be the set of points X at which the function $\delta(X)$ assumes a positive maximum. The set M is closed and disjoint from the boundary of F_1. Let $X_1 \in M_1$ and let X_2 be the corresponding point on F_2. Without loss of generality we may assume that the surfaces F_1 and F_2 are similarly oriented (otherwise we need only reflect one of them in the xz-plane). Let $r_1(X)$ be the radius-vector of an arbitrary point X on F_1, and $r_2(X)$ the radius-vector of the corresponding point on F_2. Without loss of generality we may assume that the vector-valued function

$$r = \frac{r_1(X) + r_2(X)}{r_1^2(X) - r_2^2(X)}$$

defines an analytic surface Φ for X close to X_1. The necessary condition $dr \neq 0$ for this to be true may always be satisfied by rotating one of the surfaces through a small angle about the z-axis.

We claim that the surface Φ has nonpositive Gauss curvature in a neighborhood of $r(X_1)$. Let X_0 be a point on F_1 close to X_1. Join X and X_0 by a segment γ. Let $s(X)$ be the length of the segment, $\tau(X)$ the unit vector of the tangent to the segment at X_0, n_1 the unit vector of the interior normal to the surface at X_0, and $k_1(X)$ the normal curvature of the surface at X_0 in the direction of γ. Then the vector-valued function $r_1(X)$ has the representation

$$r_1(X) = r_1(X_0) + \tau_1(X)\, s(X) + \frac{k_1(X)}{2}\, s^2(X)\, n_1 + s^2(X)\, \varepsilon(X),$$

where $\epsilon(X)$ tends to zero with $s(X)$. Reproducing the construction on F_2 corresponding to the above under the isometry, we get a representation of the function $r_2(X)$:

$$r_2(X) = r_2(X_0) + \tau_2(X)\, s(X) + \frac{k_2(X)}{2}\, s^2(X)\, n_2 + s^2(X)\, \varepsilon_2(X).$$

Using these representations of $r_1(X)$ and $r_2(X)$, we get

$$r(X) - r(X_0) = \alpha(X)s + \beta(X)s^2 + \gamma(X) + s^2\epsilon,$$

where α, β and γ are defined as follows:

$$\alpha(X) = \lambda(\tau_1 + \tau_2 - a(a_1\tau_1 - a_2\tau_2)),$$
$$\beta(X) = -2\lambda(a_1\tau_1 - a_2\tau_2)\,\alpha(X),$$
$$\gamma(X) = \frac{\lambda}{2}(k_1n_1 + k_2n_2 - a(a_1n_1k_1 - a_2n_2k_2)),$$
$$a_1 = r_1(X_0), \quad a_2 = r_2(X_0), \quad a = 2r(X_0), \quad \lambda = \frac{1}{a_1^2 - a_2^2}.$$

Let n be the normal to the surface Φ at $r(X_0)$. Then

$$(r(X) - r(X_0))\, n = \frac{\lambda s^2}{2} (-A_1k_1 + A_2k_2) + \varepsilon' s^2,$$

where $A_1 = -n + (an)a_1$, $A_2 = n + (an)a_2$ and ϵ' tends to zero with s. As shown in §5, $A_1 = A_2 \neq 0$. Therefore

$$(r(X) - r(X_0))n = \lambda' s^2(k_1 - k_2) + \epsilon' s^2.$$

Since the surfaces F_1 and F_2 are isometric and have positive Gauss curvature, when X encircles the point X_0 the difference $k_1 - k_2$ either is identically zero or changes sign. In the former case the Gauss curvature of Φ at $r(X_0)$ is zero; in the latter it is negative. Now if $k_1(X) \equiv k_2(X)$ at X_0, then X_0 and the corresponding point on F_2 are points of congruence. That is to say, when the points and the corresponding directions on F_1 and F_2 at these points are superimposed, the surfaces F_1 and F_2 are in second-order contact.

Now consider the structure of the set M of maximum points of $\delta(X)$. First, this set cannot contain interior points. Indeed, if X_1 is an interior point of M it has a neighborhood in which $\delta(X) = c = \text{const}$. But since the function δ is analytic we must have $\delta(X) \equiv c$ on the entire surface F_1, and δ assumes its maximum on the boundary of F_1, contrary to assumption.

We now claim that M cannot contain isolated points. Let X_1 be an isolated point of M. In a neighborhood of the point $r(X_1)$, the surface Φ must lie on one side of the tangent plane $z = 1/c$, where $c = \max \delta(C)$, and moreover the point $r(X_1)$ is the only point of this neighborhood in the plane $z = 1/c$. Hence the surface Φ contains elliptic points arbitrarily close to $r(X_1)$; but this is impossible, since Φ has nonpositive curvature. Thus M contains no isolated points.

Since F_1 and F_2 are analytic surfaces and M contains neither interior nor isolated points, it must consist of arcs of analytic curves. Let γ_1 be one of them and γ_2 the corresponding curve on F_2. By what was proved above, the curves γ_1 and γ_2 consist of points of congruence of the surfaces F_1 and F_2, and they are therefore congruent curves. If F_1 and F_2 are superimposed along the curves γ_1 and γ_2, the tangent planes along these curves will also coincide. By the Cauchy-Kowalewski Theorem applied to the Darboux equation, this implies that certain neighborhoods of the curves γ_1 and γ_2 on F_1 and F_2 will also coincide. This means that a neighborhood of the curve $r(\gamma_1)$ on Φ consists entirely of flat points, and so contains a plane domain. But then M must contain interior points, which is impossible. The negative minimum of the function $\delta(X)$ is treated in analogous fashion. Thus Theorem 3 is proved for analytic surfaces with positive Gauss curvature.

Using Theorem 5 of §6, we now extend this result to general convex surfaces. Let F_1 and F_2 be general convex surfaces satisfying the assumptions of our theorem. Suppose that the theorem is false and that $\delta(X)$ assumes a positive maximum on a set M which contains no boundary points. Since M is a closed set, there exists a domain G_1 on F_1, containing M, whose closure also contains no boundary points. Let G_2 be the corresponding domain on the surface F_2. By Theorem 5 of §6 there exist isometric analytic surfaces Φ_1 and Φ_2 arbitrarily close to F_1 and F_2 in the domains G_1 and G_2. If Φ_1 and Φ_2 are sufficiently close to F_1 and F_2, then $\delta(X)$, defined on Φ_1, will also assume its maximum on a set M containing no boundary points. But this contradicts what was proved above for analytic surfaces, and completes the proof of Theorem 3.

The following theorem of Aleksandrov and Sen′kin [17] is a corollary of Theorem 3.

THEOREM 4. *Let F_1 and F_2 be isometric convex surfaces in the half-space $z > 0$, situated in star-like position relative to the origin O and convex in the same direction, i.e. visible from O either from within or from without. If corresponding points of the boundaries of F_1 and F_2 are situated at equal distances from O, the surfaces are congruent.*

PROOF. Let X be any point on F_1, $z_1(X)$ its z-coordinate and $\rho_1(X)$ its distance from O. Let $z_2(X)$ and $\rho_2(X)$ be the z-coordinate and distance from O of the corresponding point of F_2. By assumption the function

$$\delta(X) = \frac{\rho_1^2(X) - \rho_2^2(X)}{z_1(X) + z_2(X)}$$

vanishes on the boundary of F_1. By Theorem 3 this function must vanish identically on the boundary, i.e. $\rho_1(X) \equiv \rho_2(X)$.

Let X_1 and Y_1 be arbitrary points on F_1, and cut F_1 by the plane through X_1, Y_1 and O. Denote the arc of the resulting curve joining X_1 and Y_1 by γ_1. Let γ_2 be the corresponding curve on F_2, X_2 and Y_2 its endpoints. Construct the cone V projecting the curve γ_2 from the point O. Develop the cone V on a plane. Then, since $\rho_1(X) \equiv \rho_2(X)$, the curve γ_2 becomes a curve congruent to γ_1. In the process the distance between the points X_2 and Y_2 on γ_2 can only increase, and so the extrinsic distance between X_1 and Y_1 is not less than the extrinsic distance between X_2 and Y_2. Interchanging the roles of F_1 and F_2, we get the converse assertion. It follows that the extrinsic distances between corresponding pairs of points on F_1 and F_2 are equal. Consequently the surfaces are congruent. Q.E.D.

§8. Monotypy of unbounded convex surfaces with total curvature 2π

THEOREM. *Unbounded isometric convex surfaces with total curvature 2π are congruent.*

Before proving this theorem we need a number of lemmas. Let F be an unbounded convex surface with total curvature 2π, and O a point on F. Let t be a ray issuing from O within the body bounded by F. Make O the origin of a cartesian coordinate system xyz, the ray z being the z-axis. Throughout the sequel we shall assume that the surface occupies this position relative to the xy-plane.

LEMMA 1. *Let $s(X)$ denote the distance of a point X on the surface F from O on the surface, and $\vartheta(X)$ the angle between the supporting plane at X and the xy-plane. Then $\vartheta(X) \to \pi/2$ as $s(X) \to \infty$.*

PROOF. Suppose the assertion false. Then there exists a sequence of points X_k on F such that $s(X_k) \to \infty$, but $\vartheta(X_k) < \pi/2 - \epsilon$, where ϵ is some positive number. The angle between the straight-line segment OX_k and the xy-plane is less than $\pi/2 - \epsilon$ for every k. Extract a convergent subsequence of segments OX_k. The limit of this subsequence is a ray t' issuing from O and contained in the body bounded by the surface. The angle that it forms with the z-axis is at least ϵ. It follows that the limit cone of the surface contains two different rays: the positive z-axis and the ray t'. Since the angle between these rays is at least ϵ, the curvature of the limit cone is at most $2\pi - 2\epsilon$. Hence the curvature of the surface, which is equal to that of the limit cone, is less than 2π. This contradiction proves the lemma.

LEMMA 2. *Let $\alpha(z)$ be a plane parallel to the xy-plane at a distance z from it, $\gamma(z)$ the curve in which $\alpha(z)$ cuts the surface F, and $l(z)$ the length of this curve. Then $l(z)/z \to 0$ as $z \to \infty$.*

PROOF. Suppose the assertion false. Then there exists a sequence of planes $\alpha(z_k)$ such that $z_k \to \infty$ but $l(z_k)/z_k > \epsilon$, where ϵ is some positive number. The curve $\gamma(z_k)$ is convex, and therefore contains a point X_k whose distance from the z-axis is at least $l(z_k)/2\pi$. The angle ϑ_k between the straight-line segment OX_k and the positive z-axis satisfies the inequality $\tan \vartheta_k > \epsilon/2\pi$. Hence, reasoning as in the proof of Lemma 1, we conclude that the surface F has curvature less than 2π, and this is a contradiction.

LEMMA 3. *Let $\gamma(X)$ be a segment joining a point X on F to the origin O, and $\vartheta(X)$ the angle between the semitangent to $\gamma(X)$ at X and the xy-plane. Then $\vartheta(X) \to \pi/2$ as $s(X) \to \infty$.*

PROOF. Suppose the assertion false. Then there exists a sequence of points X_k on the surface such that $s(X_k) \to \infty$ but $\vartheta(X_k) < \pi/2 - \epsilon$, where ϵ is some positive number. Let $\bar{\alpha}$ be the supporting plane to F parallel to the xy-plane. Let $z(X)$ be the distance of X from the plane $\bar{\alpha}$, and $\delta(X)$ the distance between the projections of X and O on the plane $\bar{\alpha}$. Let c be the polygon connecting X and O, whose sides are the perpendiculars dropped from X and O to the plane $\bar{\alpha}$ and the straight-line segment connecting the feet of these perpendiculars. This polygon lies within the body bounded by F. By Busemann's Theorem (§1 of Chapter II) the length of c is at least $s(X)$. It follows that $z(X_k) + \delta(X_k) \to \infty$ as $k \to \infty$. Since $\delta(X_k)/z(X_k) \to 0$ (Lemma 2), we have $z(X_k) \to \infty$.

Let $\bar{\gamma}(X_k)$ be the curve in which the plane $z = \text{const}$ through the point X_k cuts the surface F. Let $l(X_k)$ be its length. Develop the cylinder projecting the segment $\gamma(X_k)$ on a plane. By Liberman's Theorem (Chapter II, §1), the curve $\gamma(X_k)$ then becomes convex (Figure 29). Let us estimate the length d_k of the hypotenuse of the triangle OPX_k. First, it is obvious that $d_k < s(X_k)$. Further, $s(X_k) < z(O) + \delta(X_k) + z(X_k) + 2a$, $\delta(X_k) < l(X_k)$, since the curve $\bar{\gamma}(X_k)$ encloses the z-axis. Therefore

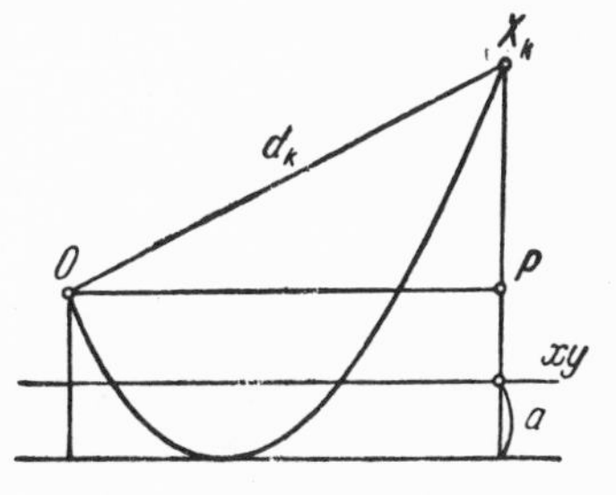

FIGURE 29

$$\sin \vartheta_k > \frac{z(X_k) - z(O)}{s(X_k)} > \frac{z(X_k) - z(O)}{z(X_k) + \delta(X_k) + z(O) + 2a}.$$

As $k \to \infty$ the right-hand side of this inequality tends to unity. Consequently $\vartheta_k \to \pi/2$, and this contradiction proves the lemma.

LEMMA 4. *Let F_1 and F_2 be two isometric unbounded convex surfaces situated above the xy-plane, which are convex toward the xy-plane in the sense that the exterior normals to their supporting planes form angles $\leq \pi/2$ with the positive z-axis. Let $\alpha(z)$ be a plane parallel to the xy-plane at a distance z from it, $\bar{\gamma}_1(z)$ the curve in which this plane cuts the surface*

F_1 and $\overline{\gamma}_2(z)$ the corresponding curve on F_2. Then for sufficiently large z the projection of the curve $\overline{\gamma}_2(z)$ on the xy-plane is a convex curve.

PROOF. By Lemma 1, for sufficiently large z the angles between the xy-plane and the supporting planes to F_1 and F_2 along $\overline{\gamma}_1(z)$ and $\overline{\gamma}_2(z)$ are greater than $\pi/2 - \epsilon$. Let O_1 and O_2 be two fixed points of F_1 and F_2, corresponding to each other under the isometry, X_1 and X_2 two arbitrary corresponding points of $\overline{\gamma}_1(z)$ and $\overline{\gamma}_2(z)$, and $\gamma_1(X_1)$ and $\gamma_2(X_2)$ corresponding segments connecting the points X_1 and X_2 to O_1 and O_2, respectively. By Lemma 3 the angles between the semitangents to $\gamma_1(X_1)$ and $\gamma_2(X_2)$ at X_1 and X_2 and the xy-plane are greater than $\pi/2 - \epsilon$, provided z is large. It follows that for sufficiently large z the angles between the semitangents to the curve $\overline{\gamma}_2(z)$ and the xy-plane are less than 2ϵ.

Now take z so large that the angles between the xy-plane and the supporting planes to F_2 along the curve $\gamma_2(z)$ are greater than $\pi/2 - \epsilon$, while the angles between the xy-plane and the semitangents to $\gamma_2(z)$ are less than ϵ, where ϵ is a small positive number. We claim that for sufficiently small ϵ the projection $\tilde{\gamma}_2(z)$ of the curve $\overline{\gamma}_2(z)$ on the xy-plane bounds a convex domain $\tilde{\sigma}_2(z)$. Suppose that $\tilde{\sigma}_2(z)$ is not convex. Let $\tilde{b}$ be its convex hull, and $\tilde{c}$ the boundary of $\tilde{b}$. Since $\tilde{\gamma}_2(z)$ is a simple closed curve and $\tilde{b} \not\equiv \tilde{\sigma}_2(z)$, the curve $\tilde{c}$ must contain a point not on $\tilde{\gamma}_2(z)$. This point lies on a straight-line segment t of the curve $\tilde{c}$ whose endpoints belong to $\tilde{\gamma}_2(z)$ (Figure 30). Let m denote the arc of the curve $\tilde{\gamma}_2(z)$ corresponding to t. The closed curve whose components are m and t bounds a domain G lying within $\tilde{\sigma}_2(z)$.

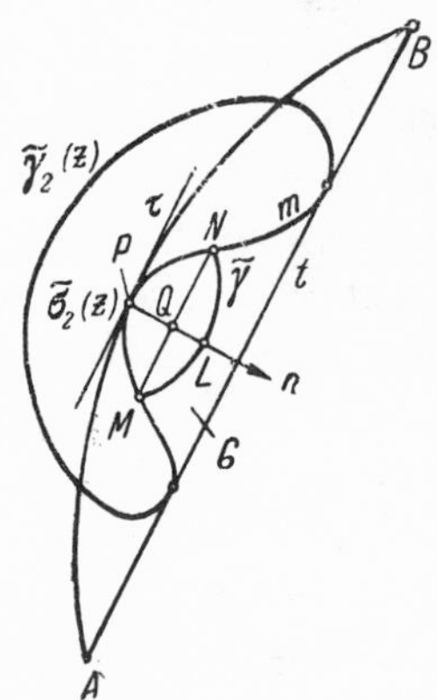

FIGURE 30

Let A and B be two sufficiently distant points on the continuation of t, on different sides of t. Describe a circle of maximum radius containing the domain such that its center and the domain G lie on different sides of the line containing t. This circle touches the curve m at some point P. Let τ be a supporting line to m at P, and n the normal to the circle at P pointing into G. Let Q be a point on n close to P, and τ_Q a straight line through Q parallel to τ; let M and N be the points nearest to Q at which this line cuts m. Connect the points of the curve $\overline{\gamma}_2(z)$ whose projections are M and N by a segment $\overline{\gamma}$ in the domain bounded by $\overline{\gamma}_2(z)$ on the surface F_2. This is possible, since the domain in question is intrinsically convex. Apply Liberman's Theorem to the segment $\bar{\gamma}$, with the direction of projection that of the normal n. Since the projections of the endpoints of $\bar{\gamma}$ on the xy-plane are M and N, the projection $\tilde{\gamma}$ of the geodesic $\bar{\gamma}$ on the xy-plane lies on one side of the line τ_Q, viz. the side into which the normal n points. It lies a fortiori on one side of the line τ. If Q is sufficiently close to P, the length of $\tilde{\gamma}$ can be made arbitrarily small. Therefore the point of intersection L of $\tilde{\gamma}$ and the normal n will also be arbitrarily close to P. But all points of the ray n close to P lie in the domain G, i.e. outside $\tilde{\sigma}_2(z)$. This is a contradiction, since the curve $\tilde{\gamma}$ must lie in $\tilde{\sigma}_2(z)$. This completes the proof of the lemma.

PROOF OF THE THEOREM. We first consider regular twice differentiable surfaces. Let F be a twice differentiable unbounded convex surface with total curvature 2π and F' an isometric convex surface. The surfaces F and F' may be assumed to occupy positions such that two of their points that correspond under the isometry coincide with the origin O and the positive z-axis lies within each of the surfaces.

Let h be sufficiently large, and consider the plane $z=h$. It cuts the surface F in a curve γ which bounds a cap G on F. Let the corresponding curve and domain on F' be γ' and G'. As we have shown above (Lemmas 1, 2 and 4), for sufficiently large h the curves γ and γ' satisfy the following conditions:

1. The tangent planes to F along γ and the tangent planes to F' along γ' form arbitrarily small angles with the z-axis.

2. The tangents to γ' form arbitrarily small angles with the xy-plane.

3. The projection of G' on the xy-plane is a convex domain.

Introduce a coordinate net u, v on the surface F and consider its image under the isometry on the surface F'. Since F and F' are isometric, the coefficients e, f, g of the first fundamental forms of F and F' coincide. Let l, m, n and l', m', n' be the coefficients of their second fundamental forms, and let λ, μ, ν and λ', μ', ν' be the coefficients of the second fundamental forms divided by the discriminant $\sqrt{eg-f^2}$ of the first fundamental form. Then

$$(*)\qquad \iint_G \begin{vmatrix} \lambda-\lambda', & \mu-\mu', \\ \mu-\mu', & \nu-\nu', \end{vmatrix} n\, d\sigma = \int_\gamma \{[(\mu'-\mu)r_u-(\lambda'-\lambda)r_v]\,du+[(\nu'-\nu)r_u-(\mu'-\mu)r_v]\,dv\},$$

where n is the normal to F and r the radius-vector of a point on F. The derivation of this formula is analogous to that of the Herglotz formula for isometric surfaces (see Efimov [**31**]).

Consider the scalar product of (*) with a unit vector e in the direction of the negative z-axis; let us estimate the right-hand side of the resulting equality. It is clearly invariant with respect to the choice of the coordinate net u, v. If we assume that the latter is a semigeodesic coordinate net based on γ, the right-hand side becomes

$$\int_{\gamma} (K - K')(e\xi)\, ds, \tag{**}$$

where K' and K are the normal curvatures of F' and F along the curves γ and γ', ξ is a unit vector tangent to the surface F perpendicular to the tangent of γ and directed toward G, and ds is the element of arc length on γ.

Of course, the value of the integral (**) depends on the plane $z = h$ cutting out the cap G. We claim that if h is sufficiently large then both the integrals

$$\int_{\gamma} K(e\xi)\, ds \text{ and } \int_{\gamma'} K'(e\xi)\, ds$$

approach arbitrarily close to 2π, and so their difference (*) can be made arbitrarily small.

Let α denote the angle between the normal plane of F' through the tangent to γ' and the osculating plane of F', β the angle between the tangent plane of F' and the tangent plane of the cylinder Z' projecting γ' onto the xy-plane, and ϑ the angle between the tangent to γ' and the xy-plane.

The geodesic curvature of γ' is $k' \tan\alpha$. By the Gauss-Bonnet Theorem, applied to the domain G' on F',

$$\omega_{G'} + \int_{\gamma'} k' \tan\alpha\, ds = 2\pi,$$

where $\omega_{G'}$ is the integral curvature of the domain G'.

The normal curvature of the cylinder Z' in the direction perpendicular to its generators is $k'\cos(\alpha - \beta)/\cos\alpha\cos\vartheta$. Therefore

$$\int_{\gamma'} \frac{k'\cos(\alpha-\beta)}{\cos\alpha\cos\vartheta}\cos\vartheta\, ds = 2\pi.$$

Hence

$$\int_{\gamma'} k'\cos\beta\, ds + \int_{\gamma'} k'\tan\alpha\sin\beta\, ds = 2\pi.$$

For sufficiently large h the angles β are arbitrarily small, and since $k' \tan \alpha \geqq 0$ and

$$\int_{\gamma'} k' \tan \alpha ds \leqq 2\pi,$$

it follows that $\int_{\gamma} k' \cos \beta ds$ is arbitrarily close to 2π for sufficiently large h. Finally, since $\cos \beta$ and $(e\xi)$ are arbitrarily close to unity for sufficiently large h, we see that $\int_{\gamma'} (e\xi) k' ds$ is also arbitrarily close to 2π. Similarly, for sufficiently large h the integral $\int_{\gamma} (e\xi) |k| ds$ is arbitrarily close to 2π.

Thus it follows that the integral (**) is arbitrarily small for sufficiently large h. Hence, letting $h \to \infty$ in formula (*), we get

$$(***) \qquad \iint_F \begin{vmatrix} \lambda - \lambda' & \mu - \mu' \\ \mu - \mu' & \nu - \nu' \end{vmatrix} (ne) d\sigma = 0.$$

Since the surfaces F and F' are isometric and their Gauss curvatures nonnegative, we have $\lambda\nu - \mu^2 = \lambda'\nu' - \mu'^2 \geqq 0$. A standard argument then shows that

$$\begin{vmatrix} \lambda - \lambda' & \mu - \mu' \\ \mu - \mu' & \nu - \nu' \end{vmatrix} \leqslant 0.$$

Since $(ne) \geqq 0$ everywhere on F, the integrand in (***) does not change sign. Hence

$$\begin{vmatrix} \lambda - \lambda' & \mu - \mu' \\ \mu - \mu' & \nu - \nu' \end{vmatrix} (ne) = 0.$$

Let the Gauss curvature of the surface F be positive at a point X. Then X is an elliptic point and therefore its spherical image is an interior point. Hence $(ne) > 0$. Thus at the point X

$$\begin{vmatrix} \lambda - \lambda' & \mu - \mu' \\ \mu - \mu' & \nu - \nu' \end{vmatrix} = 0.$$

Since $\lambda\nu - \mu^2 > 0$ at this point, it follows that $\lambda = \lambda'$, $\mu = \mu'$ and $\nu = \nu'$. Thus, for any two points with positive Gauss curvature which correspond under the isometry, both the first *and* second fundamental forms of F and F' coincide. It follows that domains on F and F' that correspond under the isometry must be congruent.

Let H be a component of the set of points of positive Gauss curvature on F, and H' the corresponding domain on F'. By what we have just proved, the surfaces F and F' can be superimposed on each other in the domains H and H'. The tangent planes along the boundaries of H and H' form developable surfaces. These surfaces contain the domains of zero curvature on F and F' adjacent to H and H', respectively. It follows that these domains of zero curvature, L and L' say, will also coincide. If K and K' are the domains of positive curvature adjacent to L and L', then the parts of their boundaries adjacent to L and L' coincide, and so the entire domains K and K' coincide. Thus by alternate consideration of domains of positive and zero Gauss curvature we finally conclude that the surface F and F' are congruent. This completes the proof for regular surfaces.

Extension of this result to general convex surfaces is quite laborious, and we shall therefore confine ourselves to the main idea of the proof. Let F and F' be general convex surfaces.

Construct a sequence of analytic surfaces F_n converging to F, and find a homeomorphism of F_n onto F such that the intrinsic metrics of F_n converge to that of F (for example, if F is nondegenerate, a suitable homeomorphism is obtained by projecting F onto F_n from some point within F).

Let h and H be two numbers, the first sufficiently large and the second larger by several orders of magnitude ($h \ll H$). Cut the surfaces F and F' by the plane $z = H$, forming caps Φ and Φ'. By virtue of the isometry of F and F' and the homeomorphism of F onto F_n, these correspond to certain domains on F_n. Let $\bar{\Phi}_n$ and $\bar{\Phi}'_n$ denote the convex hulls of these domains on F_n. (The convex hull of a set M on a convex surface is the minimal convex domain containing M.) We know (Chapter II, §10) that there exist analytic caps Φ_n and Φ'_n isometric to $\bar{\Phi}_n$ and $\bar{\Phi}'_n$. The monotypy theorem for convex caps (Theorem 2 of §7) implies that as $n \to \infty$ the surfaces Φ_n and Φ'_n converge to surfaces congruent to Φ and Φ', respectively. Without loss of generality we may assume that $\Phi_n \to \Phi$ and $\Phi'_n \to \Phi'$.

Now cut a cap ω_n from the surface Φ_n by the plane $z = h$, and let ω'_n denote the corresponding domain on Φ'_n. Consider the integral

$$\Omega = \iint_{\omega_n} \begin{vmatrix} \lambda - \lambda' & \mu - \mu' \\ \mu - \mu' & \nu - \nu' \end{vmatrix} (\bar{n}e)\, d\sigma.$$

As in the proof for regular surfaces (see above), one proves that for sufficiently large h, H and n the quantity Ω is arbitrarily small.

To simplify the exposition, we now assume that the surfaces F and F' are smooth and strictly convex. Let X_0 be any point on F. Call the relative position of the surfaces F and F' normal if the point X_0 coincides with the corresponding point X'_0 on F', the corresponding directions at these points also coincide (the surfaces are assumed to be similarly oriented), and the common tangent plane of the surface at $X_0 \equiv X_0'$ is the xy-plane.

Let the surfaces F and F' be in almost normal position. Let $r(X)$ denote the radius-vector of an arbitrary point X on F, $r'(X)$ the radius-vector of the corresponding point on F', and $r_n(X)$ and $r_n'(X)$ the radius-vectors of the corresponding points on F_n and F_n'. Consider the surfaces S and S_n defined in the neighborhood of X_0 by the vector equations $\bar{r} = r(X) + r'^*(\bar{X})$ and $r = r_n(X) + r_n'^*(X)$, where $r'^*(X)$ and $r_n'^*(X)$ are the reflections of the vectors $r'(X)$ and $r_n'(X)$ in the xy-plane. The first of these surfaces is smooth, and the second is analytic and has everywhere nonpositive Gauss curvature (see the proof of the theorem in §6 of Chapter II). The absolute curvature of the surface S_n is bounded by Ω, and since the latter is arbitrarily small for sufficiently large h and n, the absolute curvature of S_n is also arbitrarily small.

Now let γ be a closed contour on S. Let $\bar{\gamma}$ be its spherical image, γ_n the corresponding contour on S_n and $\bar{\gamma}_n$ the spherical image of the latter. Suppose that the contour $\bar{\gamma}$ encloses some point P of the unit sphere. Then it is clear that for sufficiently large n the contour $\bar{\gamma}_n$ will also enclose P. If d is the distance of P from $\bar{\gamma}$, then for sufficiently large n the distance of P from $\bar{\gamma}_n$ is greater than $d/2$. It follows that the absolute curvature of S_n is greater than the area of a geodesic disk of radius $d/2$. But this is impossible, since we have already seen that the absolute curvature of S can be made arbitrarily small. Thus $\bar{\gamma}$ cannot enclose any point P.

It follows that the surface S is developable and has the usual property of developables: straight-line generators and stationary tangent plane along each generator (Theorem 1 in §4 of Chapter IX). Now a generator of S corresponds to congruent curves on F and F'. It is easily seen that if isometric surfaces F and F' contain families of congruent curves, then the surfaces themselves are congruent. The proof of our theorem in the most general case, without the assumption that the surfaces are smooth and strictly convex, proceeds along similar lines, but the details are far more complicated.

§9. Monotypy of unbounded convex surfaces with total curvature less than 2π

One consequence of Olovjanišnikov's theorem (see Chapter I, §11) on the realization of a complete metric by an unbounded convex surface is that unbounded convex surfaces with total curvature less than 2π admit nontrivial isometric mappings (bendings). To be precise, the following theorem holds.

Let F be an unbounded convex surface with total curvature $\omega < 2\pi$, and let γ be a ray on F. Then, if K is an arbitrary convex conical surface whose curvature at the apex is the curvature of F, and t is one of its generators, there exists a convex surface F^* isometric to F, with the same orientation, such that K is the limit cone of F^* and t the limit generator of the ray γ^*, where γ^* is the image of γ on F^*.

A natural question in this context is to what degree an unbounded convex surface is determined by its metric, limit cone and limit generator of a given ray. The answer to this question is given by the following theorem.

THEOREM 1. *Let F_1 and F_2 be unbounded, complete, convex, isometric, similarly oriented surfaces, with total curvature $\omega < 2\pi$, which have a common limit cone K; assume that a generator t of K is the limit generator of two rays γ_1 and γ_2 on F_1 and F_2 which correspond under the isometry. Then the surfaces F_1 and F_2 are congruent and in parallel position, i.e. they may be made to coincide by a translation. More concisely, an unbounded convex surface with total curvature less than 2π is uniquely determined by its metric, orientation, limit cone and the limit generator of any ray on the surface.*

We shall prove this theorem under the assumption that the surfaces and the cone are smooth.

Let O be the apex of the cone K. Construct the minimal circular cone $\bar{K}$ with apex O containing K. The axis of $\bar{K}$ is strictly within the cone K. Bring the surfaces F_1 and F_2 together so that the initial points of the rays γ_1 and γ_2 coincide with the apex O of the cones K and $\bar{K}$, and refer them to a rectangular cartesian coordinate system with O as origin, the axis of $\bar{K}$ as the z-axis, and the plane through O perpendicular to the latter as the xy-plane. The halfspace $z > 0$ is the halfspace defined by the xy-plane which contains the cone $\bar{K}$.

Assume that the limit cone of the unbounded convex surface F is K. Let $s(X)$ denote the intrinsic distance of a point X on F from the origin O, $\rho(X)$ the extrinsic distance of X from O, and $h(X)$ the distance of X from the xy-plane.

LEMMA 1. *When $s(X) \to \infty$, we have $h(X)$, $\rho(X) \to \infty$ and*

$$\rho(X)/s(X) \to 1, \quad 0 < c_1 < h(x)/\rho(X) < c_2 < 1,$$

where c_1 and c_2 are independent of X.

PROOF. We first show that $h(X) \to \infty$ as $s(X) \to \infty$. In fact, were this not true, there would be a sequence of points X_k such that $s(X_k) \to \infty$ but $h(X_k) < h_0$. But then the sequence of points X_k would lie on the cap cut out from F by the plane $z = h_0$. If d is the diameter of the base of this cap and h_1 its height, then the distance between any two points of the cap, in particular the distance $s(X_k)$ between O and X_k, is at most $d + 2h_0$, and therefore bounded. This contradiction proves that $h(X) \to \infty$ as $s(X) \to \infty$. The fact that $\rho(X) \to \infty$ as $s(X) \to \infty$ follows from $h(X) \to \infty$ and the obvious inequality $h(X) < \rho(X)$.

We now show that for sufficiently large $s(X)$ there are constants $c_1 > 0$ and $c_2 < 1$ such that $c_1 < h(X)/\rho(X) < c_2$. To prove this we construct two cones of revolution κ_1 and κ_2 with apex O and axis of rotation z such that κ_1 is within the cone K and κ_2 encloses the cone $\overline{K}$. All points X on F sufficiently distant from the origin O lie within the region between the cones κ_1 and κ_2. These points of the surface obviously satisfy the inequality $c_1 < h(X)/\rho(X) < c_2$, where $c_1 > 0$ and $c_2 < 1$ are constants depending only on the opening angles of the cones κ_1 and κ_2.

To prove that $\rho(X)/s(X) \to 1$ as $s(X) \to \infty$, displace the segment OX in the direction of the negative z-axis over a distance $\lambda h(X)$ $(\lambda > 0)$. The straight-line segments described in the process by the points X and O and the segment XO in its new position form a polygon connecting the points X and O of the surface F (Figure 31). Now the surfaces $F_n = F/n$ converge to the cone K as $n \to \infty$; hence for sufficiently small but fixed λ the polygon we have constructed lies outside the body bounded by F, provided $s(X)$ is sufficiently large.

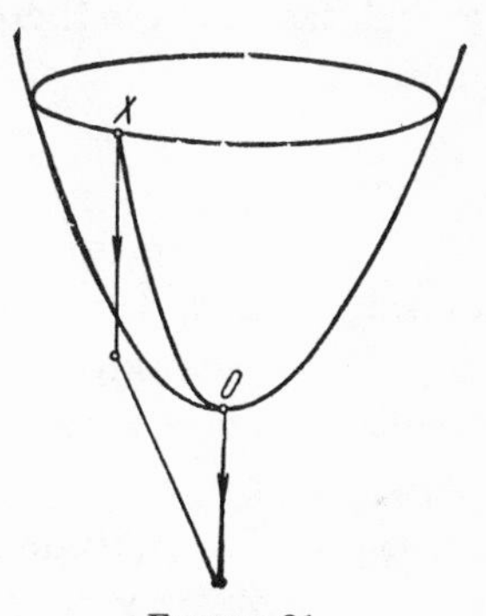

FIGURE 31

Applying Busemann's Theorem to this polygon, we get $s(X) < \rho(X) + 2\lambda h(X)$. In combination with the obvious inequality $\rho(X) \leq s(X)$, simple manipulations yield the inequality $1 > \rho(X)/s(X) > 1/(1 + 2\lambda c_2)$ for sufficiently large $s(X)$. But this is equivalent to the statement that $\rho(X)/s(X) \to 1$ as $s(X) \to \infty$. This completes the proof of Lemma 1.

LEMMA 2. *Let $r(s)$ be the radius-vector of the point on the ray γ corresponding to arc length s, and τ_0 the unit vector of the limit generator of γ. Then $(r(s) - s\tau_0)/s \to 0$ as $s \to \infty$.*

Proof. Let $\tau(s)=r'(s)$ denote the unit vector tangent to the ray γ at the point s. Then by Olovjanišnikov's theorem $\tau(s)\to\tau_0$ as $s\to\infty$. Further,

$$r(s)=\int_0^s \tau(s)\,ds=\int_0^s \tau_0\,ds+\int_0^s (\tau-\tau_0)\,ds=s\tau_0+s\varepsilon(s),$$

where $\epsilon(s)\to 0$ as $s\to\infty$, since $\tau(s)\to\tau_0$ as $s\to\infty$. Hence $(r(s)-s\tau_0)/s\to 0$ as $s\to\infty$, and the proof is complete.

Subject the surfaces F_1 and F_2 to a translation so that the initial points of the rays γ_1 and γ_2 coincide, and introduce rectangular cartesian coordinates as described above. Let $r_1(X)$ denote the radius-vector of a point X on F_1, $r_2(X)$ the radius-vector of the corresponding point on F_2.

Lemma 3. *Let F_1 and F_2 be convex surfaces satisfying the assumptions of Theorem* 1. *Then* $(r_1(X)-r_2(X))/s\to 0$ *as* $s(X)\to\infty$.

Proof. If the point X recedes to infinity along the ray γ_1, the assertion follows directly from Lemma 2. Suppose that the lemma is false in the general case. Then there exist a positive number ϵ and a sequence of points X_k such that $s(X_k)\to\infty$ but

$$|r_1(X_k)-r_2(X_k)|>\epsilon s(X_k).$$

The plane through X_k parallel to the xy-plane cuts the ray γ_1 at a point Y_k. Let Δ_k denote the triangle formed by the segments OX_k, X_kY_k and OY_k on the surface F_1. The corresponding triangle on the surface F_2 will be denoted by Δ_k'; its vertices are O, X_k' and Y_k'.

Apply to the surfaces F_1 and F_2 a similarity mapping with center O and ratio of similitude $1/s(Y_k)$. Denote the resulting surfaces by F_1^k and F_2^k, the triangles on them corresponding to Δ_k and Δ_k' by $\overline{\Delta}_k$ and $\overline{\Delta}_k'$, and the vertices of these triangles by O, $\overline{X}_k$, $\overline{Y}_k$ and O, $\overline{X}_k'$, $\overline{Y}_k'$, respectively. As $k\to\infty$ both surfaces F_1^k and F_2^k converge to the limit cone K of F_1 and F_2. Without loss of generality we may assume that the sequences of triangles $\overline{\Delta}_k$ and $\overline{\Delta}_k'$ converge, for otherwise we can extract subsequences of surfaces F_1^k and F_2^k (choosing the same indices in both subsequences) with the required property.

By Lemma 2 the sequences of vertices $\overline{Y}_k$ and $\overline{Y}_k'$ converge as $k\to\infty$ to a point $\overline{Y}$ on the limit generator of γ_1 and γ_2, which is situated at unit distance from the origin O; the sequences of vertices $\overline{X}_k$ and $\overline{X}_k'$ converge to points $\overline{X}$ and $\overline{X}'$ on the limit cone K; and the sequences of triangles $\overline{\Delta}_k$ and $\overline{\Delta}_k'$ converge to triangles $\overline{\Delta}$ and $\overline{\Delta}'$ with vertices O, $\overline{X}$, $\overline{Y}$ and O, $\overline{X}'$, $\overline{Y}'$, respectively.

Corresponding sides of the triangles $\bar{\Delta}$ and $\bar{\Delta}'$ are equal, and the triangles have zero curvature; they are therefore congruent. Since $|r_1(X_k) - r_2(X_k)| > \epsilon s(X_k)$, it follows that the vertices $\bar{X}$ and $\bar{X}'$ cannot coincide. But the triangles $\bar{\Delta}$ and $\bar{\Delta}'$ have a common side OY, and hence they are intrinsically symmetric on the cone K with respect to the limit generator of the rays γ_1 and γ_2. This contradicts the assumption that F_1 and F_2 are similarly oriented. If the triangle $\bar{\Delta}$ degenerates into a segment, the point $\bar{X}$ lies on the generator $O\bar{Y}$, the point $\bar{X}'$ on K at the same distances from $\bar{Y}$ and O as $\bar{X}$ must coincide with $\bar{X}$, and this contradicts the assumption. The proof is complete.

Lemma 4. *Let X and X' be corresponding points on F_1 and F_2. Then the angles between the directions at X and X' that correspond under the isometry are arbitrarily small if $s(X)$ is sufficiently large.*

Proof. If this is not so, there exist sequences of corresponding points X_n and X_n' on F_1 and F_2, and corresponding directions at these points, such that $s(X_n) \to \infty$ but the angle ϑ between the directions is greater than $\epsilon > 0$.

Connect the point X_n to the origin O by a segment γ_n on F_1, and let t_n denote the semitangent to γ_n at X_n. Suppose that the ray through OX_n converges as $n \to \infty$ to a generator t of the limit cone K. We claim that then $t_n \to t$.

Let F_n denote the surface obtained from F by a similarity mapping with center O and ratio of similitude $1/s(X_n)$. Let $\bar{X}_n$ and $\bar{\gamma}_n$ be the point and segment corresponding to X_n and γ_n on F_1. As $n \to \infty$, the points $\bar{X}_n$ converge to some point A on the generator t of the limit cone K; the segments $\bar{\gamma}_n$ converge to a segment OA of the ray t. It follows from Theorem 9 in §1 of Chapter II that the semitangents to $\bar{\gamma}_n$ at $\bar{X}_n$ converge as $n \to \infty$ to the direction of the segment OA. By similarity the semitangents $\bar{t}_n$ converge to the same limit.

Since the rays through the segments OX_k' converge to the same generator t of the limit cone, it follows from what we have proved that for sufficiently large n there exist directions at X_n and X_n' which form an arbitrarily small angle; these are the directions of the semitangents to γ_n at X_n and the corresponding segment γ_n' at X_n' on F_2.

Since the limit cone K is smooth, the fact that the rays through OX_n and OX_n' converge to the generator t implies that the exterior normals to F_1 and F_2 at the points X_n and X_n' converge to the exterior normal of the cone K along the generator t. Since the normals and the two corresponding directions at X_n and X_n' are close together, the same applies to the other corresponding directions. This contradiction proves Lemma 4.

Let F_1 and F_2 be unbounded convex surfaces satisfying the assumptions of Theorem 1. We define a vector-valued function $r(X)$ on the surface F_1 by $r(X) = r_1(X) + r_2^*(X)$, where $r_1(X)$ is the radius-vector of a point X on F, $r_2^*(X)$ the radius-vector of the corresponding point on the reflection F_2^* of F_2 in the xy-plane. Let $z(X)$ denote the distance of the point $r(X)$ from the xy-plane, $d(X)$ the distance of its projection from the origin O.

Lemma 5. *$z(X)/d(X) \to 0$ as $s(X) \to \infty$.*

Proof. Let $h_1(X)$, $\rho_1(X)$ and c_2 have the same meaning as in Lemma 1, and let $d_1(X)$ be the distance of the projection of X on the xy-plane from the origin. We first observe that

$$d(X) > 2d_1(X) - |r_1(X) - r_2(X)|,$$
$$z(X) < |r_1(X) - r_2(X)|, \quad d_1(X) > \rho_1(X) - h_1(X).$$

Further, by Lemma 1, if $s(X)$ is sufficiently large, then

$$h_1(X)/\rho_1(X) < c_2 < 1,$$

and by Lemma 2

$$|r_1(X) - r_2(X)| < \epsilon(X)s(X),$$

where $\epsilon(X) \to 0$ as $s(X) \to \infty$. Hence

$$\frac{z(X)}{d(X)} < \frac{\varepsilon(X)}{2(1-c_2)\dfrac{\rho_1(X)}{s(X)} - \varepsilon(X)}.$$

Since $\epsilon(X) \to 0$, while $\rho_1(X)/s(X) \to 1$ as $s(X) \to \infty$, it follows that $z(X)/d(X) \to 0$ as $s(X) \to \infty$. Q.E.D.

Lemma 6. *Let G_0 be the domain on F_1 defined by $s(X) > s_0$. Then for sufficiently large s_0 the vector-valued function $\bar{r}(X)$ defined in G_0 by $\bar{r}(X) = r_1(X) + r_2(X)$ determines a two-dimensional manifold $\bar{r}(G_0)$ which can be uniquely projected onto the xy-plane.*

Proof. We first observe that if for arbitrarily large s_0 the set $\bar{r}(G_0)$ is not a manifold, or is a manifold which cannot be uniquely projected onto the xy-plane, then there exists a sequence of pairs of points X_k and Y_k, $X_k \neq Y_k$, which recede to infinity, while the points $\bar{r}(X_k)$ and $\bar{r}(Y_k)$ have identical projections on the xy-plane. We wish to prove that this is impossible. We can extract an infinite subsequence such that either $s(X_k) \leqq s(Y_k)$ for all pairs of points in the subsequence, or $s(X_k) \geqq s(Y_k)$ for all pairs of points. Without loss of generality we may assume that all pairs of points of the original sequence satisfy the first inequality.

As in Lemma 4, construct sequences of surfaces $F_1^k = F_1/s(Y_k)$ and $F_2^k = F_2/s(Y_k)$. Both these sequences converge to the limit cone K of the surfaces F_1 and F_2. Without loss of generality we may assume that the sequences of points $\overline{X}_k$, $\overline{Y}_k$ and $\overline{X}'_k$, $\overline{Y}'_k$ corresponding to X_k and Y_k under the isometric and similarity mappings on the surfaces F_1^k and F_2^k converge. Let $\overline{X}$, $\overline{Y}$, $\overline{X}'$ and $\overline{Y}'$ be the limits of these sequences. By Lemma 3, $\overline{X} = \overline{X}'$ and $\overline{Y} = \overline{Y}'$. Now, since the points $\bar{r}(X_k)$ and $\bar{r}(Y_k)$ lie on the same vertical straight line (parallel to the z-axis) for each k, it follows that $\overline{X} = \overline{Y}$. Moreover, $\overline{Y}$ lies at unit distance from the apex of the cone K and is therefore a smooth point thereof.

Because of the special choice of coordinate axes, there exists a number $\delta > 0$ such that the projection of any unit segment tangent to the surface at any point is greater than δ. Join the points $\overline{X}_k$ and $\overline{Y}_k$ on the surface F_1^k by a segment $\bar{\gamma}_k$, and the points $\overline{X}'_k$ and $\overline{Y}'_k$ on the surface F_2^k by a segment $\bar{\gamma}'_k$. Let $\bar{\tau}_k$ and $\bar{\tau}'_k$ be the unit tangent vectors to the segments $\bar{\gamma}_k$ and $\bar{\gamma}'_k$ at corresponding points. Letting $\tilde{r}_{1k}$ and $\tilde{r}_{2k}$ denote the radius-vectors of corresponding points on F_1^k and F_2^k, we have

$$\tilde{r}_{1k}(\overline{Y}_k) - \tilde{r}_{1k}(\overline{X}_k) = \int_{\bar{\gamma}_k} \bar{\tau}_k\, ds, \quad \tilde{r}_{2k}(\overline{Y}_k) - \tilde{r}_{2k}(\overline{X}_k) = \int_{\bar{\gamma}'_k} \bar{\tau}'_k\, ds.$$

Since the points $\bar{r}(X_k)$ and $\bar{r}(Y_k)$ lie on the same vertical straight line, the projection of the vector

$$\overline{m}_k = (\tilde{r}_{1k}(Y_k) + \tilde{r}_{2k}(Y_k)) - (\tilde{r}_{1k}(X_k) - \tilde{r}_{2k}(Y_k)) = \int_{\bar{\gamma}_k} \bar{\tau}_k\, ds + \int_{\bar{\gamma}'_k} \bar{\tau}'\, ds$$

on the xy-plane must vanish for every k. But this is impossible for sufficiently large k. Indeed,

$$\overline{m}_k = (\bar{\tau}_k(\overline{X}_k) + \bar{\tau}'_k(\overline{X}_k))\bar{s}_k + \int_{\bar{\gamma}_k} (\bar{\tau}_k - \bar{\tau}_k(X_k))\, ds + \int_{\bar{\gamma}'_k} (\bar{\tau}'_k - \bar{\tau}'_k(\overline{X}_k))\, ds,$$

where $\bar{\tau}_k(X_k)$ and $\bar{\tau}'_k(X_k)$ are the unit vectors tangent to the segments $\bar{\gamma}_k$ and $\bar{\gamma}'_k$ at the points $\overline{X}_k$ and $\overline{X}'_k$, respectively, and $\bar{s}_k$ is the length of the segment $\bar{\gamma}_k$. By Lemma 4, for sufficiently large k,

$$|\bar{\tau}_k(\overline{X}_k) - \bar{\tau}'_k(\overline{X}_k)| < \delta/3;$$

by the theorem in §1 of Chapter II,

$$|\bar{\tau}_k - \bar{\tau}_k(\overline{X}_k)| < \delta/3 \quad \text{and} \quad |\bar{\tau}'_k - \bar{\tau}'_k(\overline{X}_k)| < \delta/3.$$

If we now represent the vector $\overline{m}_k$ in the form

$$\overline{m}_k = 2\overline{\tau}_k(\overline{X}_k)\overline{s}_k + (\overline{\tau}'_k(\overline{X}_k) - \overline{\tau}_k(\overline{X}_k))s_k + \int_{\overline{\gamma}_k} (\overline{\tau}_k - \overline{\tau}_k(\overline{X}_k))\,ds + \int_{\overline{\gamma}'_k} (\overline{\tau}'_k - \overline{\tau}'_k(\overline{X}_k))\,ds$$

and recall that the projection of the vector $2\tau_k(\overline{X}_k)$ on the xy-plane is a segment of length at least 2δ, we see that the projection of the vector $\overline{m}_k$ on this plane is a segment of length greater than $\delta\overline{s}_k$. This is a contradiction, since by assumption the vector $\overline{m}_k$ is perpendicular to the xy-plane. This completes the proof of Lemma 6.

Lemma 7. *Let G_0 be the domain on F_1 defined in Lemma 6. Then the vector-valued function $r(X)$ defined in G_0 by the equation $r(X) = r_1(X) + r_2(X)$ defines a smooth two-dimensional manifold F_0 which can be projected in one-to-one fashion on the xy-plane. The angles between the tangent planes to the manifold and the xy-plane are uniformly small, provided s_0 is sufficiently large. The manifold F_0 is not part of a plane, for arbitrarily large s_0, if the surface F_1 and F_2 are not congruent.*

Proof. The fact that F_0 is a manifold which can be projected onto the xy-plane follows from Lemma 6. Since the surfaces F_1 and F_2 are smooth, so is F_0. The angles between the tangent planes to F_0 and the xy-plane are small, since the angles between the directions in G_0 on F_1 and the corresponding directions in the corresponding domain G'_0 on F_2 are small, and they form angles less than $\pi/2$ with the xy-plane. It remains to prove that F_0 is not part of a plane.

Suppose that F_0 is a plane, and therefore a plane parallel to the xy-plane. Then by a suitable displacement of one of the surfaces in the direction of the z-axis we can ensure that for sufficiently large $s(X)$ the z-coordinates of corresponding points on F_1 and F_2 satisfy the condition $z_1(X) = z_2(X)$. But then Theorem 2 of §7 implies that F_1 and F_2 are congruent. This proves Lemma 7.

Proof of Theorem 1. Let π be a plane sufficiently distant from the origin O and perpendicular to the xy-plane. It cuts the manifold F_0 in a smooth curve γ which recedes to infinity in both directions. This curve has a unique projection on the xy-plane, and the angles between its tangents and the xy-plane are small. By Lemma 7, we may assume that γ is not a straight line. For if every sufficiently distant vertical plane were to cut F_0 in a straight line, and therefore in a straight line parallel to the xy-plane (Lemma 5), it would follow that F_0 is part of a horizontal plane for sufficiently large s_0, contrary to Lemma 7. Let P be a point on γ whose projection on the xy-plane is close to the origin.

It follows from Lemma 5 that there is a straight line g through P which forms an arbitrary small but nonzero angle with the xy-plane, cuts the curve γ in at least two points and is never tangent to γ. Let σ be a plane through g, not perpendicular to the xy-plane. If the surface F_1 contains a bounded connected domain G such that every point $r(X)$, $X \in G$, lies on one side of the plane σ, and lies in the plane σ when X is on the boundary of G, we shall say that the plane σ cuts off a hummock.[3]

We claim that under our assumptions on F_0 no plane can cut off a hummock. Indeed, in any given bounded domain the isometric convex surfaces F_1 and F_2 can be simultaneously approximated by isometric analytic surfaces F_1' and F_2' with positive Gauss curvature (Theorem 4 in §6). Now a hummock can be cut off from the manifold F_0' constructed for F_1' and F_2' in the same way as F_0 was constructed for F_1 and F_2, provided F_1' and F_2' are sufficiently close to F_1 and F_2. However, as we shall now prove, the manifold F_0' has nonpositive curvature, and hence no hummock can be cut from it.

The vector-valued function $r_1'(X)$ defining the surface F_1' has the following representation in the neighborhood of X_0:

$$r_1'(X) = r_1'(X_0) + \tau_1' s' + \frac{k_1 n_1}{2} s'^2 + \varepsilon_1 s'^2,$$

where s' is the length of the segment γ_1' joining X and X_0, τ_1' the unit vector tangent to γ_1' at X_0, n_1 the unit vector normal to the surface at X_0, k_1 the normal curvature of F_0' at X_0 in the direction τ_1', and ϵ_1 tends to zero with s'. The vector-valued function $r_2'(X)$ defining the surface F_2' has an analogous representation. Hence we get the following formula for the vector-valued function $r'(X)$ defining the manifold F_0':

$$r'(X) - r'(X_0) = (\tau_1' + \tau_2'^*) s' + \frac{1}{2} s'^2 (k_1 n_1 + k_2 n_2^*) + \varepsilon s'^2.$$

Let n denote the unit vector normal to the manifold F_0'. Then

$$(r'(X) - r'(X_0)) n = \frac{s'^2}{2} (k_1 n_1 n + k_2 n_2^* n) + \bar{\varepsilon} s'^2.$$

As was proved in §6 of Chapter II, $n_1 n = -n_2^* n \neq 0$. Hence

$$(r'(X) - r'(X_0)) n = c s'^2 (k_1 - k_2) + \bar{\epsilon} s'^2.$$

Since F_1' and F_2' are isometric surfaces, the difference $k_1 - k_2$ either

---

[3] *Translator's note*. The Russian word used here, *горбушка*, means the crust at the end of a loaf of bread.

changes sign when the point X describes a contour around X_0, or vanishes identically. Consequently the manifold F_0' has nonpositive Gauss curvature, and so no hummock can be cut from it.

Let $\bar{A}_1$ and $\bar{B}_1$ be two consecutive points of intersection of the straight line g with the curve γ (the intersection of F_0 and the plane π). Let $\bar{C}$ be some point on γ between $\bar{A}_1$ and $\bar{B}_1$. Suppose that this point lies, say, below the straight line g, and therefore below the plane σ, irrespective of how this plane passes through g. Let $\bar{A}$ and $\bar{B}$ be two more points on γ, close to $\bar{A}_1$ and $\bar{B}_1$, respectively, and lying above the straight line g, hence above the plane σ. Let A, B and C be points on F_1 such that $r(A) = \bar{A}$, $r(B) = \bar{B}$ and $r(C) = \bar{C}$. Let G_A, G_B and G_C be the maximal connected domains on F_1 containing A, B and C, respectively, such that each of the sets $r(G_A)$, $r(G_B)$ and $r(G_C)$ lies on one side of the plane σ. Each of the domains G_A, G_B and G_C is unbounded, since otherwise the plane σ would cut off a hummock, which is impossible.

If G_A and G_B are not disjoint, they must coincide. We claim that there exists a plane σ through g such that G_A and G_B are disjoint. Suppose that for every choice of a plane σ through g the sets G_A and G_B coincide, and so the points A and B may be joined by a curve κ on F_1 whose image $r(\kappa)$ lies above the plane σ. If κ_1 is another curve with these properties, it can be continuously deformed to the curve κ without leaving the domain $G_A \equiv G_B$. In fact, otherwise the closed curve $\kappa + \kappa_1$ (or part of it) would bound a certain domain G. The image $r(G)$ divides the plane σ in such a way that the image of the boundary lies on one side of the plane. This would mean that σ cuts off a hummock, which is impossible.

We may assume that the curve κ lies in a bounded region of the surface F_1 for any position of the plane σ. Indeed, this is certainly true for planes σ forming small angles with the plane π. Suppose that there is a sequence of planes σ_k for which the points A and B can be connected only by curves κ_k which cannot be confined to a bounded region of the surface F_1 (bounded simultaneously for all k). Extract a convergent subsequence from σ_k, with limit plane σ_0. It is clear that $\sigma_0 \neq \pi$. Relative to σ_0, the points A and B can be connected by a curve κ_0 which lies in a bounded region of the surface F_1 by assumption, since $G_A \equiv G_B$ for any plane σ. If k is large, the plane σ_k is close to σ_0; the curve $r(\kappa_0)$ lies above the plane σ_0 and therefore also above σ_k for sufficiently large k. This is a contradiction. Thus, if $G_A \equiv G_B$ for any position of the plane σ, we may assume that the curve κ connecting A and B on F, whose image $r(\kappa)$ lies above the plane σ, is situated in a bounded region of the surface F_1 (bounded simultaneously for all σ).

Remove the point C from F_1, and call the resulting surface $\widetilde{F}_1$. The domains G_A and G_B obviously belong to $\widetilde{F}_1$. The curve κ in the domain $G_A \equiv G_B$ lies within the domain $\widetilde{F}_1$. We proved above that if A and B are connected by two paths in the domain $G_A \equiv G_B$, these paths are homotopic in this domain. They are therefore homotopic in $\widetilde{F}_1$. Let us associate with every plane σ through g a path κ_σ connecting A and B in $G_A \equiv G_B$, lying in a bounded region of the surface F_1; the existence of such paths has just been proved. Not all paths κ_σ are homotopic in the domain $\widetilde{F}_1$. To verify this, consider two planes σ_1 and σ_2 obtained from the plane σ by rotation about g in different directions, through small angles. The corresponding paths κ_{σ_1} and κ_{σ_2} are not homotopic. It follows that there exists a plane σ for which there are planes σ' arbitrarily close to σ (i.e. the angle between σ' and σ is arbitrarily small) such that the paths $\kappa_{\sigma'}$ are not homotopic to κ_σ in $\widetilde{F}_1$. But then the contour $\kappa_{\sigma'} + \kappa_\sigma$ encloses the point C, and since the contours κ_σ and $\kappa_{\sigma'}$ are uniformly bounded and the angle between σ and σ' is small, the plane σ may be displaced parallel to itself in such a way that the contour $\kappa_{\sigma'} + \kappa_\sigma$ lies entirely above the plane σ, and the point $\overline{C} = r(C)$ below σ. But this means that the plane we have constructed cuts off a hummock. This contradiction shows that there exists a plane σ through g for which the sets G_A and G_B are disjoint.

By assumption the straight line g is not parallel to the xy-plane; therefore a point moving along g in a suitable direction will recede from the xy-plane in the direction $z < 0$. By Lemma 5 a certain ray g_1 on g lies below the manifold F_0. Rotate this ray about its initial point in the plane σ through a small angle, in both directions, in such a way that the angle V_1 described by g_1 in σ lies entirely below the manifold F_0. Let $\overline{V}_1$ denote the preimage on F_1 of the set of points of the manifold F_0 lying above the angle V_1. By Lemma 5 there is a ray g_2 on g which lies above the manifold F_0. Rotate this ray through small angles in both directions to form an angle V_2 in the plane σ lying above the manifold F. Since the set $\overline{V}_1$ is connected, its intersection with one of the sets G_A or G_B is empty. Assume that $\overline{V}_1 \cap G_A = 0$. Now the image $r(G_A)$ of G_A lies above the domain $\bar{\sigma}$ of the plane σ which is obtained from the latter by removing the angles V_1 and V_2. Since the domain G_A is unbounded, $r(G_A)$ is also unbounded. Let π' be a plane parallel to π, intersecting $r(G_A)$, and sufficiently distant from the origin. The intersection of $r(G_A)$ and the plane π' is either a single curve lying above σ, with endpoints in σ, or several or even infinitely many such curves. Let $\tilde{\gamma}$ denote one of these curves. There is a tangent to the curve $\tilde{\gamma}$ which is parallel to g, say the tangent at a point P. The angle between the tangent plane to F_0 at P and the xy-plane is not smaller than the angle between g and the xy-plane. But by Lemma 6 this is impossible if the plane π' is sufficiently distant from the origin. Thus our assumption that the surfaces F_1 and F_2 are not congruent has led to a contradiction, and this completes the proof of Theorem 1.

THEOREM 2. *Let F_1 and F_2 be surfaces satisfying all the assumptions of Theorem* 1, *except that the limit cones are now assumed to be smooth but different. Then the surface F_1 can be continuously bent into the surface F_2.*

PROOF. It is clearly sufficient to prove that each of the surfaces F_1 and F_2 can be bent into a surface with circular limit cone. We shall construct a deformation of this type, say for the surface F_1. Introduce rectangular cartesian coordinates such that the origin O is the apex of the limit cone K_1 of F_1, the positive z-axis lies within the cone K_1, and the entire cone lies in the halfspace $z > 0$. Consider the plane $z = 1$. It cuts the cone K_1 in a convex curve κ_1. Construct a curve *parallel* to κ_1 at a distance λ from it, and then dilate it by a similarity mapping in such a way that the cone with apex O through this curve has the same curvature as K_1. Denote this cone by $K_{1\lambda}$. Let $t_{1\lambda}$ be the generator on $K_{1\lambda}$ obtained by cutting $K_{1\lambda}$ by a halfplane through the z-axis that contains the limit generator of the ray γ_1.

By Olovjanišnikov's theorem there exists a surface $F_{1\lambda}$ isometric to F_1, with the same orientation, whose limit cone is $K_{1\lambda}$, such that the generator $t_{1\lambda}$ is the limit generator of the ray $\gamma_{1\lambda}$ corresponding to γ_1 under the isometry, and O is the initial point of the ray $\gamma_{1\lambda}$.

By Theorem 1, when λ is varied continuously the surfaces $F_{1\lambda}$ form a continuous family of surfaces "connecting" F_1 to a surface with circular limit cone. This completes the proof of Theorem 2.

CHAPTER IV

INFINITESIMAL BENDINGS OF CONVEX SURFACES

An *infinitesimal bending* is an infinitesimal deformation of a surface under which the lengths of curves on the surface are stationary. The velocity field of the deformation is called the bending field. The bending field is said to be trivial if it is the velocity field of a rigid motion of the surface. If the surface admits no nontrivial infinitesimal bendings, it is said to be *rigid*. The problem of infinitesimal bendings of convex surfaces comprises several components:

1. To prove the existence of infinitesimal bendings, given a deformation of the boundary.
2. To prove the uniqueness of bending fields; in particular, to prove that the surface is rigid.
3. To prove that bending fields of regular surfaces are regular.

Many well-known geometers have studied the problem of infinitesimal bending. In particular, Blaschke [22] proved that a closed regular convex surface with positive Gauss curvature admits no nontrivial infinitesimal bendings.

The problem of infinitesimal bendings arose naturally in the context of the general theory of nonregular convex surfaces. The first results are due to Aleksandrov [12], who found necessary and sufficient conditions for a vector field given on a general convex surface to be a bending field.

Starting from this result, the present author has elaborated a fairly complete solution of the problem of infinitesimal bendings for general convex surfaces [57]. In particular, he has proved general existence theorems for infinitesimal bendings, given a deformation of the boundary of the surface, and corresponding uniqueness theorems for the bending fields (including rigidity theorems). He has determined the degree of regularity of the bending field as a function of the regularity of the surface. These results will be presented in the present chapter.

The complete solution of the rigidity problem for general complex surfaces furnishes a new solution of the monotypy problem, which will be indicated at the end of the chapter for the case of closed convex surfaces.

§1. Bending fields of general convex surfaces

In this section, we shall follow Aleksandrov and extend the well-known concept of infinitesimal bendings of regular surfaces to general convex surfaces, proving Aleksandrov's characterization of bending fields of these surfaces.

Let F be a regular surface, and let $r = r(u,v)$ $(r_u \times r_v \neq 0)$ be any smooth parametrization of F. Let $\tau(u,v)$ be a smooth vector field on F ($\tau(u,v)$ is continuously differentiable with respect to u,v). Consider the deformation of F whose image at time t is the surface F_t defined by the equation

$$r = r(u,v) + t\tau(u,v).$$

This equation indeed defines a surface for sufficiently small t, since $r_u \times r_v \neq 0$.

Let γ be any smooth curve on F. With respect to arc length s as parameter, it is defined on the surface by equations

$$u = u(s), \quad v = v(s) \quad (u'^2 + v'^2 \neq 0),$$

and in space by an equation

$$r = r(u(s), v(s)) \qquad (r_s'^2 = 1).$$

Under the above deformation of F, the curve goes into a smooth curve γ_t on the surface F_t, given by the equation

$$r = r(u(s), v(s)) + t\tau(u(s), v(s)).$$

Let $l_\gamma(t)$ denote the length of this curve. By the standard formula,

$$l_\gamma(t) = \int_0^{l_\gamma} \left(1 + 2tr'\tau' + t^2\tau'^2\right)^{1/2} ds.$$

It is clear that for any smooth curve γ the function $l_\gamma(t)$ is differentiable at $t = 0$ and

$$\frac{dl_\gamma(t)}{dt} = \int_0^{l_\gamma} r'\tau'\, ds.$$

We now define an *infinitesimal bending* as a deformation of the above type under which the lengths of all curves γ on F are stationary at $t = 0$, i.e. for any curve γ

$$dl_\gamma(t)/dt = 0$$

when $t=0$. The vector field τ which defines the infinitesimal bending of the surface is called a *bending field*.

Under what conditions is a vector field given on a surface a bending field? By definition, a bending field satisfies $dl_\gamma(t)/dt=0$ at $t=0$ for any smooth curve γ. Since any subarc of γ is also a smooth curve, it follows that if we integrate along the curve we have $\int_0^s r'\tau' ds=0$ for any s. And this means that necessarily $r'\tau'=0$ along γ, or, equivalently,

$$(r_u\tau_u)u'^2+(r_u\tau_v+r_v\tau_u)u'v'+(r_v\tau_v)v'^2=0.$$

Since there exist smooth curves γ through any point (u,v) on the surface and in any direction $u':v'$, it follows that a bending field $\tau(u,v)$ must satisfy the following conditions over the entire surface:

$$(*) \qquad r_u\tau_u=0, \quad r_u\tau_v+r_v\tau_u=0, \quad r_v\tau_v=0.$$

The converse is also true: if a smooth vector field τ on a smooth surface F satisfies conditions (*), it is a bending field. To prove this it suffices to note that the integrand $r'\tau'$ in the expression for $dl_\gamma(t)/dt$ vanishes by (*), and so $dl_\gamma(t)/dt=0$ at $t=0$.

We now want to extend the concept of infinitesimal bending to general convex surfaces. But before doing this we must extend the definition of curve and surface. By a curve we shall mean the image of a closed interval of the real axis under a continuous mapping into space; a surface will be the continuous image of a two-dimensional manifold.

Of course, the curves and surfaces thus defined may be very different from our usual conceptions of curves and surfaces. For example, a curve may fill out an entire domain in space, while a surface may degenerate into a curve.

Let $x(t)$, $y(t)$ and $z(t)$ be continuous functions defined on a closed interval $g: a \leqq t \leqq b$. Then the mapping of g into space under which the point t goes into the point of space with cartesian coordinates $x(t)$, $y(t)$, $z(t)$ defines a curve in the above sense. Conversely, any curve may be defined analytically by equations of this sort. The system of equations

$$x=x(t), \quad y=y(t), \quad z=z(t)$$

will be called the equations of the curve.

We shall say that a curve γ_1, which is the image of a closed interval g_1 under a continuous mapping f_1, coincides with a curve γ_2, which is the image of an interval g_2 under a continuous mapping f_2, if there exists a homeomorphism of g_1 onto g_2 under which the images of corresponding points on the intervals under f_1 and f_2 coincide. It follows that all parametrizations of a curve can be derived from one parametrization by a monotone parameter transformation.

We now define rectifiable curves and the length of a curve. Let γ be a curve which is a continuous image of an interval g. A polygon Γ is said to be *regularly inscribed* in γ if the order of its vertices is the same as the order of their preimages on g. Obviously, this property of the polygon is independent of the parametrization of the curve. The curve is said to be *rectifiable* if the lengths of all regularly inscribed polygons are uniformly bounded; the supremum of the lengths of these polygons is called the length of the curve.

We now define an infinitesimal bending of a general convex surface. Let F be a convex surface, τ a continuous vector field on F. Let x be an arbitrary point on the surface, $r(x)$ its radius vector and $\tau(x)$ the field vector at x. Consider the deformation of the surface F whose image at time t is the surface F_t defined by the equation $r = r(x) + t\tau(x)$,

Let γ be an arbitrary rectifiable curve on F. The above deformation takes it into a curve γ_t on F_t. Suppose that for sufficiently small $|t|$ the curve γ_t is also rectifiable. Let l_γ denote the length of γ, and $l_\gamma(t)$ the length of γ_t for sufficiently small $|t|$. Set $l_\gamma(t) = l_\gamma + t\epsilon_\gamma(t)$.

The above deformation of F into F_t is called an *infinitesimal bending* if $\epsilon_\gamma(t) \to 0$ as $t \to 0$ for any rectifiable curve γ. The vector field τ is then called a *bending field*.

At first sight, one might think that the investigation of infinitesimal bendings of general convex surfaces as defined here is a hopeless affair, for no a priori smoothness conditions have been imposed on the surface or on the bending field. This is not so. Aleksandrov [12] has proved that any bending field of a general convex surface satisfies a Lipschitz condition in any compact domain on the surface, and moreover equation (*) holds almost everywhere. Because of the importance of this theorem for the sequel, we present a proof.

Before we proceed to study the differential properties of bending fields some remarks are in order regarding convex surfaces and rectifiable curves on convex surfaces.

Let F be a convex surface and K a body whose boundary contains F. Let P be an arbitrary point on the surface and Q a point within the body K. Then a sufficiently small neighborhood ω of P on F has a unique projection in the direction PQ. If we introduce rectangular cartesian coordinates x, y, z in space with the straight line PQ as z-axis, the neighborhood ω of P on F will be defined by an equation $z = \varphi(x,y)$, where $\varphi(x,y)$ is a convex function.

If ω is sufficiently small, there clearly exists a number $\vartheta_0 < \pi/2$ such that all supporting planes of F at points of ω form angles smaller than ϑ_0 with the xy-plane. Then, if $\bar{\omega}$ is the projection of ω on the xy-plane, at each point of $\bar{\omega}$ the derivative of φ, which exists in any direction, is bounded by the constant $k = \tan\vartheta_0$.

Now let γ be a curve defined by equations

$$x = x(t), \quad y = y(t), \quad z = z(t) \qquad (a \leqq t \leqq b).$$

A standard theorem states that the curve γ is rectifiable if and only if each of the functions $x(t)$, $y(t)$, $z(t)$ has bounded variation.

Let the curve γ lie on the convex surface ω. If it is rectifiable, its projection $\bar{\gamma}$ on the xy-plane is clearly also rectifiable. We claim that the converse holds: if $\bar{\gamma}$ is rectifiable, so is γ.

Inscribe an arbitrary polygon Γ in γ. Its projection $\bar{\Gamma}$ is a polygon inscribed in $\bar{\gamma}$. Let us compare the lengths of the sides δ and $\bar{\delta}$ of these polygons. Let $\tilde{\delta}$ be the length of the convex curve on ω whose projection is $\bar{\delta}$. Obviously $\bar{\delta} \leqq \delta \leqq \tilde{\delta}$. And since

$$\tilde{\delta} = \int_{(\bar{\delta})} \sqrt{1 + z_{\bar{\delta}}'^2}\, d\bar{\delta},$$

and $|z_{\bar{\delta}}'| \leqq k$, it follows that

$$\delta \leqslant \tilde{\delta} \leqslant \sqrt{1 + k^2}\,\bar{\delta}.$$

Consequently the lengths of the polygons Γ and $\bar{\Gamma}$ satisfy the inequality

$$l_\Gamma \leqslant \sqrt{1 + k^2}\, l_{\bar{\Gamma}}$$

and hence, if $\bar{\gamma}$ is rectifiable, so is γ.

Let the projection $\bar{\omega}$ of the surface ω on the xy-plane be a convex domain. Let $\psi(X)$ be a function defined on the surface ω. Since the position of a point on the surface is uniquely determined by its coordinates x and y, we may also assume that this function is defined in the domain $\bar{\omega}$ of the xy-plane. We claim that if the function $\psi(X)$ satisfies a Lipschitz condition on the surface, then the function $\psi(x,y)$ satisfies a Lipschitz condition in the domain $\bar{\omega}$, and vice versa.

To prove this, it will suffice to show that the distance $\rho(X, Y)$ between any two points X and Y on the surface and the distance $\rho(\bar{X}, \bar{Y})$ between their projections on the xy-plane satisfy the inequalities

$$\rho(\overline{X}, \overline{Y}) \leqq c_1 \rho(X, Y), \quad \rho(X, Y) \leqq c_2 \rho(\overline{X}, \overline{Y}),$$

where c_1 and c_2 are constants independent of the points X and Y.

It is not difficult to see that the constants c_1 and c_2 exist. Indeed, the distance between X and Y on the surface is at least equal to the extrinsic distance between them, and the latter is at least the distance between $\overline{X}$ and $\overline{Y}$. Hence $\rho(\overline{X}, \overline{Y}) \leqq \rho(X, Y)$. Furthermore, the intrinsic distance between X and Y cannot exceed the length of the convex curve whose projection on the xy-plane is the segment $\overline{X}\overline{Y}$. The latter, as follows from previous arguments, is at most $\sqrt{1+k^2}\rho(\overline{X}, \overline{Y})$. Hence

$$\rho(X, Y) \leqq \sqrt{1+k^2}\rho(\overline{X}, \overline{Y}).$$

This proves our assertion.

Now consider a bending field $\tau(X)$ on the surface F. We first examine it in the small neighborhood ω of P considered above.

Let $\overline{\gamma}$ be an arbitrary rectifiable curve in the domain $\overline{\omega}$ on the xy-plane; let γ be the projection of $\overline{\gamma}$ on the surface ω by straight lines parallel to the z-axis. As we have seen, γ is rectifiable. The infinitesimal bending of F defined by the bending field τ transforms the curve γ into a curve γ_t. By assumption, this curve must be rectifiable for sufficiently small $|t|$. We shall see that this alone is a sufficient condition for the vector field τ to satisfy a Lipschitz condition in a neighborhood of P, and therefore in any compact domain on F. Aleksandrov [12] proved this statement rigorously, using a theorem of Lebesgue concerning rectifiable surfaces. We present his proof in detail.

We first find the equation of the curve γ_t. Assume that the surface ω is defined by the equation

$$z = z(x, y), \qquad (x, y) \in \overline{\omega},$$

and the curve $\overline{\gamma}$ by the equations

$$x = x(\alpha), \quad y = y(\alpha) \qquad (0 \leqq \alpha \leqq b);$$

let $\xi(x,y)$, $\eta(x,y)$ and $\zeta(x,y)$ be the components of the vector τ along the x-, y- and z-axes, respectively. It is easily seen that the curve γ_t is defined by the equations

$$\begin{aligned} x &= x(\alpha) + t\xi(x(\alpha), y(\alpha)), \\ y &= y(\alpha) + t\eta(x(\alpha), y(\alpha)), \\ z &= z(x(\alpha), y(\alpha)) + t\zeta(x(\alpha), y(\alpha)). \end{aligned}$$

Suppose that one of the three functions $\xi(x,y)$, $\eta(x,y)$ or $\zeta(x,y)$, say ξ, does not satisfy a Lipschitz condition in any neighborhood of $\overline{P}$ (the projection of P on the xy-plane). We construct a sequence of pairs of interior points A_n, B_n of $\bar{\omega}$ as follows. Describe a disk of radius $1/n^3$ about $\overline{P}$, and let A_n, B_n be two points within the disk such that

$$|\xi(A_n) - \xi(B_n)| > n|A_nB_n|.$$

The existence of these points follows from our assumption that ξ does not satisfy a Lipschitz condition in any neighborhood of $\overline{P}$. Without loss of generality, we may assume that the unit disk about $\overline{P}$ is contained in $\bar{\omega}$, and so all points A_n and B_n lie in $\bar{\omega}$.

Now consider the following polygon $\bar{\gamma}$ in $\bar{\omega}$:

$$A_1B_1A_1B_1 \cdots A_2B_2A_2B_2 \cdots A_3B_3A_3B_3 \cdots,$$

where for sufficiently large n the side A_nB_n is repeated m_n times, with m_n such that

$$\frac{1}{n^2} \leqslant m_n |A_nB_n| \leqslant \frac{2}{n^2}.$$

It follows from the convergence of the infinite series $\Sigma 1/n^2$ that the polygon $\bar{\gamma}$ is rectifiable.

We have already proved that a curve γ on ω whose projection on the xy-plane is $\bar{\gamma}$ is rectifiable. Since τ is a bending field, the function $\lambda = x + t\xi$ must have bounded variation along the curve $\bar{\gamma}$ for sufficiently small $|t|$.

Since the polygon $\bar{\gamma}$ is rectifiable, the function x has bounded variation along $\bar{\gamma}$. Since x and $x + t\xi$ have bounded variation along $\bar{\gamma}$, it follows by a standard theorem that their difference $t\xi$ has bounded variation, and hence so has ξ. We shall now show directly that the function ξ has unbounded variation along the polygon $\bar{\gamma}$.

Indeed, on each side A_nB_n of the polygon $\bar{\gamma}$ the variation of ξ is at least

$$|\xi(A_n) - \xi(B_n)| > n|A_nB_n|.$$

Since the number m_n of sides A_nB_n satisfies the inequality

$$1/n^2 \leqq m_n|A_nB_n|,$$

the variation of ξ over all sides A_nB_n is greater than $1/n$. But variation is additive and the series $\Sigma 1/n$ is divergent, and so the variation of ξ along $\bar{\gamma}$ is infinite. This contradiction proves that the bending field τ satisfies a Lipschitz condition in a sufficiently small neighborhood of any point P on F.

Now let G be any compact domain on the surface F, and $\bar{G}$ its closure. Each point $P \in \bar{G}$ has a neighborhood ω_P in which the bending field satisfies a Lipschitz condition, and the neighborhoods ω_P form a covering of $\bar{G}$. By the Heine-Borel Theorem, we can extract a finite covering by neighborhoods ω_P. It follows that the bending field τ satisfies a Lipschitz condition in $\bar{G}$, and therefore in G.

Thus, *a bending field of a general convex surface satisfies a Lipschitz condition in any compact domain on the surface.*

Let ω be a sufficiently small neighborhood of an arbitrary point P on a convex surface F, and introduce a system of rectangular cartesian coordinates x, y, z as before. The convex surface ω has a unique projection on the xy-plane and its supporting planes form angles less than $\vartheta_0 < \pi/2$ with the xy-plane. We may assume that the bending field τ on F satisfies a Lipschitz condition in the domain ω.

Let γ be a curve in ω whose projection $\bar{\gamma}$ on the xy-plane is a segment parallel to the x-axis. The image of γ under the infinitesimal bending defined by the field τ at time t is a curve γ_t: $r = r(x) + t\tau(x)$.

Each of the functions $r(x)$ and $\tau(x)$ satisfies a Lipschitz condition on the segment $\bar{\gamma}$. Hence the curve γ_t is rectifiable and its length $l_\gamma(t)$ is given by the standard formula

$$l_\gamma(t) = \int_{(\bar{\gamma})} |r'_x + t\tau'_x|\, dx.$$

Since r and τ satisfy a Lipschitz condition along the segment $\bar{\gamma}$, it follows that the derivatives r'_x and τ'_x, wherever they exist, are bounded by a constant c, and moreover $|r'_x| \geqq 1$. Hence,

$$|r'_x + t\tau'_x| = \left(r'^2_x + 2tr'_x\tau'_x + t^2\tau'^2_x\right)^{1/2} = |r'_x| + \frac{r'_x\tau'_x}{|r'_x|} + t^2R,$$

where R is bounded by some constant depending only on c.

Substituting this expression for $|r'_x + t\tau'_x|$ in the formula for $l_\gamma(t)$, we get

$$l_\gamma(t) = l_\gamma + t\int_{(\bar{\gamma})} \frac{r'_x\tau'_x}{|r'_x|}\, dx + O(t^2).$$

Hence, by our definition of an infinitesimal bending,

$$\int_{(\bar{\gamma})} \frac{r'_x\tau'_x}{|r'_x|}\, dx = 0.$$

Since this equality holds not only for the curve γ but for any subarc, it follows that $r'_x\tau'_x = 0$ almost everywhere on $\bar{\gamma}$. Finally, since γ was arbitrary we may assume that its projection on the xy-plane is any line $y = \text{const}$; it follows that $r'_x \tau'_x = 0$ almost everywhere in $\bar{\omega}$.

Similarly, one proves that $r'_y \tau'_y = 0$ almost everywhere in $\bar{\omega}$.

Now let γ be a plane curve on the surface ω, whose projection is a segment $\bar{\gamma}$ of an arbitrary straight line $x - y = \text{const}$. Let s denote length along $\bar{\gamma}$. Then, reasoning as before, we see that $r'_s \tau'_s = 0$ at almost all points of $\bar{\gamma}$. Again, since γ was arbitrary it follows that $r'_s \tau'_s = 0$ at almost all points of $\bar{\gamma}$ in the direction $x - y = \text{const}$.

A theorem of Rademacher states that a function which satisfies a Lipschitz condition has a total differential almost everywhere. At points where the total differentials of r and τ exist, we have

$$r'_s = r_x \frac{dx}{ds} + r_y \frac{dy}{ds} = \frac{1}{\sqrt{2}} (r_x + r_y),$$

$$\tau'_s = \tau_x \frac{dx}{ds} + \tau_y \frac{dy}{ds} = \frac{1}{\sqrt{2}} (\tau_x + \tau_y).$$

At these points, therefore,

$$r'_s \tau'_s = \frac{1}{2} r_x \tau_x + \frac{1}{2} (r_x \tau_y + r_y \tau_x) + \frac{1}{2} (r_y \tau_y).$$

Hence, almost everywhere in $\bar{\omega}$,

$$r_x \tau_x + (r_x \tau_y + r_y \tau_x) + r_y \tau_y = 0.$$

Now since, as we have proved, $r_x \tau_x = 0$ and $r_y \tau_y = 0$, it follows that almost everywhere

$$r_x \tau_y + r_y \tau_x = 0,$$

At points where the total differentials of r and τ exist,

$$dr d\tau = (r_x \tau_x) dx^2 + (r_x \tau_y + r_y \tau_x) dx dy + (r_y \tau_y) dy^2.$$

It follows that almost everywhere in $\bar{\omega}$, or, equivalently, almost everywhere on the surface ω, $dr d\tau = 0$.

To summarize: *A bending field on a general convex surface satisfies a Lipschitz condition in any compact domain, and $dr d\tau = 0$ almost everywhere.*

Conversely, let τ be a vector field defined on a convex surface F, satisfying a Lipschitz condition in any compact domain on the surface and such that $dr d\tau = 0$ almost everywhere. We claim that τ is a bending field.

To prove this statement, it is sufficient to show that τ is a bending field in a sufficiently small neighborhood of any point on the surface. Indeed, suppose that τ is a bending field in a sufficiently small neighborhood ω_P of any point P. Since any curve γ is compact, it can be covered by finitely many neighborhoods ω in each of which τ is a bending field. The curve γ can be divided into finitely many subarcs γ_k each of which is contained in one of the neighborhoods ω.

Since τ is a bending field in ω, the curve γ_k contained in ω satisfies the equality

$$l_{\gamma_k}(t) = l_{\gamma_k} + t\varepsilon_k,$$

where ϵ_k tends to zero as $t \to 0$. Adding these equalities, we get

$$l_\gamma(t) = l_\gamma + t\epsilon,$$

where $\epsilon = \sum \epsilon_k$ tends to zero as $t \to 0$. Since γ was an arbitrary curve, it follows that τ is a bending field on the entire surface.

Now choose a neighborhood ω of an arbitrary point as in the previous arguments. We wish to prove that τ is a bending field in ω.

Let γ be an arbitrary curve in ω. As before, let $\bar{\gamma}$ be its projection on the xy-plane. We already know that the lengths of the curves γ_t and γ satisfy the relation

$$l_\gamma(t) = l_\gamma + t \int_{\bar{\gamma}} \frac{(r'\tau')}{|r'|} ds + t^2 R,$$

where r and τ are differentiated with respect to arc length along the curve $\bar{\gamma}$, and R is bounded by some constant which depends essentially only on the Lipschitz constants of the vector-valued functions r and τ and on the length of $\bar{\gamma}$.

Inscribe a polygon Γ with small sides in the curve γ. Let $\bar{\Gamma}$ and $\tilde{\Gamma}$ denote the projections of this polygon on the xy-plane and the surface ω, respectively, by straight lines parallel to the z-axis. By subjecting the vertices of the polygon Γ to small displacements, we may assume that $r'\tau' = 0$ almost everywhere on $\bar{\Gamma}$. Then, as before, we have

$$l_{\tilde{\Gamma}}(t) = l_{\tilde{\Gamma}} + t^2 \widetilde{R}.$$

Note that the integral in the right-hand side is dropped, since $r'\tau' = 0$ almost everywhere on $\bar{\Gamma}$.

Now we know that $l_{\tilde{\Gamma}} \to l_\gamma$ as $\Gamma \to \gamma$. Hence, taking the lower limit of the above relation and noting that the lower limit of the length of convergent curves is not less than the length of the limit curve, we get

$$l_\gamma(t) \leqq l_\gamma + t^2 \widetilde{R}^*,$$

where $\widetilde{R}^*$ is bounded by a constant.

Comparing the two relations between $l_\gamma(t)$ and l_γ, we see that

$$t \int\limits_{(\bar{\gamma})} \frac{(r'\tau')}{|r'|} ds + t^2 (R - \tilde{R}^*) \leqslant 0$$

for sufficiently small $|t|$. Since R and $\tilde{R}^*$ are bounded as $t \to 0$, it follows that

$$\int\limits_{(\bar{\gamma})} \frac{(r'\tau')}{|r'|} ds = 0$$

and consequently $l_\gamma(t) = l_\gamma + t^2 R$. Now this means that τ is a bending field in ω. As we have proved, it now follows that τ is a bending field on the entire surface F.

The contents of this section can be summarized in the following theorem of Aleksandrov:

A vector field on a general convex surface F is a bending field if and only if it satisfies a Lipschitz condition in any compact domain on F and $dr d\tau = 0$ almost everywhere.

§2. Fundamental Lemma on bending fields of convex surfaces

In this section we shall state one of the fundamental results of the chapter, which we call the Fundamental Lemma. It will play an essential role in our proofs of theorems on the rigidity of general convex surfaces and the regularity of bending fields of regular surfaces. The proof of the Fundamental Lemma is extremely intricate if no a priori assumptions are made concerning the smoothness of the surface and the bending field. In this section, therefore,we shall only outline the proof; the essential details of the proof will be presented in §§3, 4 and 5, and the proof itself in §6.

Let F be a regular (thrice continuously differentiable) convex surface containing no flat pieces, which can be projected in one-to-one fashion onto the xy-plane, and let τ be a regular (thrice continuously differentiable) bending field on F. As was shown in §1, the field τ satisfies a system of differential equations:

$$r_x \tau_x = 0, \quad r_x \tau_y + r_y \tau_x = 0, \quad r_y \tau_y = 0.$$

If p and q denote the first derivatives of the function $z(x, y)$ defining the surface F, and ξ, η and ζ the x-, y- and z-components of the field τ, this system of equations is

$$\xi_x + p \zeta_x = 0, \quad \xi_y + p \zeta_y + \eta_x + q \zeta_x = 0, \quad \eta_y + q \zeta_y = 0.$$

ξ and η are easily eliminated from this system: Differentiate the first equation twice with respect to y, add the third equation differentiated twice with respect to x, and then subtract the second equation differentiated with respect to x and y. The result is the following equation for the z-component ζ of the field τ:

$$r\zeta_{yy} - 2s\zeta_{xy} + t\zeta_{xx} = 0,$$

where r, s, t denote the second derivatives of $z(x,y)$.

An important consequence of this equation is the following assertion.

The surface $\Phi: z = \zeta(x,y)$ has nonpositive curvature.

In fact, suppose that Φ has positive curvature at some point (x,y). Then the curvature is also positive in a neighborhood ω of the point. The second differential d^2z cannot vanish identically in ω, since then the corresponding piece of the surface F would be flat. Hence there exists a point (x_0, y_0) at which the curvature of Φ is positive and d^2z does not vanish identically.

Choose the directions of the x- and y-axes in such a way that $s = z_{xy} = 0$ at the point (x_0, y_0). Then at this point the equation for ζ has the form $r\zeta_{yy} + t\zeta_{xx} = 0$. Since the surface F is convex, r and t cannot have different signs, and since d^2z does not vanish identically at (x_0, y_0) it follows that r and t cannot both be zero. Without loss of generality we may assume that $r > 0$ and $t \geqq 0$.

Since the curvature of Φ is positive at (x_0, y_0), it follows that at this point

$$\zeta_{xx}\zeta_{yy} - \zeta_{xy}^2 > 0,$$

and so ζ_{xx} and ζ_{yy} are not zero and have the same sign.

Combining this with our assumptions on r and t at the point (x_0, y_0), we see that at this point the equation $r\zeta_{yy} + t\zeta_{xx} = 0$ cannot hold, and this is a contradiction.

Thus, if the convex surface F contains no flat pieces the surface Φ has nonpositive curvature.

The Fundamental Lemma extends this well-known result for regular convex surfaces and regular bending fields to general convex surfaces and bending fields, relaxing the regularity condition. Before stating the lemma, we must define surfaces of nonpositive curvature with no regularity assumptions.

Let Φ be any surface, and $z = z(x,y)$ its equation. The only condition imposed on $z(x,y)$ is continuity. We shall say that the surface Φ is *strictly convex at a point P* if there is a plane α_P through P such that all points sufficiently close to P are not in the plane, and therefore lie on one side of it. A surface containing no points of strict convexity will be called a surface of nonpositive curvature. It is obvious that in the regular case (thrice differentiable surface) this definition is equivalent to nonpositive Gauss curvature.

FUNDAMENTAL LEMMA 1. *If $F: z = z(x, y)$ is a convex surface containing no flat pieces, and $\zeta(x, y)$ the z-component of a bending field on F, then the surface $\Phi: z = \zeta(x, y)$ is a surface of nonpositive curvature.*

The condition that F contain no flat pieces is essential, as is easily seen from the following example. Let F be a domain in the xy-plane. Obviously, F is a convex surface. Consider the following functions on $F: \xi \equiv 0$, $\eta \equiv 0$, and $\zeta(x, y)$ an arbitrary function satisfying a Lipschitz condition. It is easily seen that the vector field on F with components ξ, η, ζ satisfies the assumptions of Aleksandrov's theorem (§1) and is therefore a bending field on F. But the surface $z = \zeta(x, y)$ need not have nonpositive curvature, since it is arbitrary.

In order to extend the Fundamental Lemma to convex surfaces containing flat pieces, we generalize the concept of a surface of non-positive curvature. Let H be any set of points which can be projected one-to-one onto the xy-plane. A point P of H will be called a point of strict convexity if there is a plane α_P through P, not perpendicular to the xy-plane, such that all points of H sufficiently close to P lie outside and on one side of α_P. A set H containing no points of strict convexity is said to have nonpositive curvature. We can now state the Fundamental Lemma for convex surfaces which contain flat pieces.

FUNDAMENTAL LEMMA 2. *Let $F: z = z(x, y)$ be a convex surface containing flat pieces G_α. Let $\zeta(x, y)$ be the z-component of a bending field τ on F. Then the surface $\Phi: z = \zeta(x, y)$ has nonpositive curvature on the set H of all points whose projections on F by straight lines parallel to the z-axis lie outside the domains G_α.*

As already mentioned, here we shall only outline the main idea of the proof, which will be given in full in §6.

Let $F: z = z(x, y)$ be a regular surface which can be projected in one-to-one fashion onto the xy-plane, and $\tau(\xi, \eta, \zeta)$ a regular bending field on F. Let G be an arbitrary domain on F, homeomorphic to a disk and bounded by a regular curve γ. Let $\bar{G}$ denote the projection of G on the xy-plane and $\bar{\gamma}$ the boundary of $\bar{G}$.

Since τ is a bending field, it satisfies the system of equations

$$\xi_x + p\zeta_x = 0, \quad \xi_y + p\zeta_y + \eta_x + q\zeta_x = 0, \quad \eta_y + q\zeta_y = 0.$$

Set $\lambda = \xi + p\zeta$ and $\mu = \eta + q\zeta$, and consider the line integral

$$\frac{1}{2}\oint_{\bar{\gamma}} (\lambda\, d\mu - \mu\, d\lambda) = \frac{1}{2}\oint_{\bar{\gamma}} (\lambda\mu_x - \mu\lambda_x)\, dx + (\lambda\mu_y - \mu\lambda_y)\, dy.$$

Using the Green-Ostrogradskiĭ formula, we replace this by an integral over the domain $\bar{G}$:

$$\frac{1}{2}\oint_{\bar{\gamma}} (\lambda\, d\mu - \mu\, d\lambda) = \iint_{\bar{G}} (\lambda_x\mu_y - \lambda_y\mu_x)\, dx\, dy.$$

The integrand in the right-hand side of this equality may be expressed as follows:

$$\lambda_x\mu_y - \lambda_y\mu_x = \frac{1}{4}(\lambda_y - \mu_x)^2 + \left\{\lambda_x\mu_y - \frac{1}{4}(\lambda_y + \mu_x)^2\right\}.$$

Using the equations of the bending field, we find

$$\lambda_x = \xi_x + p\zeta_x + r\zeta = r\zeta,$$
$$\mu_y = \eta_y + q\zeta_y + t\zeta = t\zeta,$$
$$\lambda_y + \mu_x = \xi_y + p\zeta_y + \eta_x + q\zeta_x + 2s\zeta = 2s\zeta.$$

Substituting these relations into the expression in curly brackets in the equation for $\lambda_x\mu_y - \lambda_y\mu_x$, we get

$$\lambda_x\mu_y - \lambda_y\mu_x = \frac{1}{4}(\lambda_y - \mu_x)^2 + (rt - s^2)\zeta^2.$$

Hence

$$\frac{1}{2}\oint_{\bar{\gamma}} (\lambda\, d\mu - \mu\, d\lambda) = \iint_{\bar{G}} (rt - s^2)\zeta^2\, dx\, dy + \frac{1}{4}\iint_{\bar{G}} (\lambda_y - \mu_x)^2\, dx\, dy.$$

Both this formula and its proof have been borrowed from a paper of Minagawa [**43**], in which the rigidity of closed convex surfaces is proved under the assumption that the surface is twice piecewise differentiable and the bending field smooth. We have included the formula here as a heuristic argument toward derivation of the integral relation to be used in our proof of the Fundamental Lemma.

Retaining the above notation, we drop the requirement that τ be a bending field. The quadratic form

$$\sigma = dr\, d\tau = \sigma_{11}dx^2 + 2\sigma_{12}dx\,dy + \sigma_{22}dy^2$$

is no longer identically zero. Its coefficients are obviously

$$\sigma_{11}=\xi_x+p\zeta_x,\quad 2\sigma_{12}=\xi_y+p\zeta_y+\eta_x+q\zeta_x,\quad \sigma_{22}=\eta_y+q\zeta_y.$$

As before, set $\lambda=\xi+p\zeta$ and $\mu=\eta+q\zeta$, and repeat the preceding arguments word for word, to get

$$\frac{1}{2}\oint_{\bar\gamma}(\lambda\,d\mu-\mu\,d\lambda)$$
$$=\frac{1}{4}\iint_{\bar G}(\lambda_y-\mu_x)^2\,dx\,dy+\iint_{\bar G}\left\{\lambda_x\mu_y-\frac{1}{4}(\lambda_y+\mu_x)^2\right\}dx\,dy.$$

Transforming the second integral in the right-hand side by means of the equalities

$$\lambda_x=\xi_x+p\zeta_x+r\zeta=\sigma_{11}+r\zeta,$$
$$\mu_y=\eta_y+q\zeta_y+t\zeta=\sigma_{22}+t\zeta,$$
$$\lambda_y+\mu_x=\xi_y+p\zeta_y+\eta_x+q\zeta_x+2s\zeta=2\sigma_{12}+2s\zeta,$$

we get the formula

$$\frac{1}{2}\oint_{\bar\gamma}(\lambda\,d\mu-\mu\,d\lambda)$$
$$=\frac{1}{4}\iint_{\bar G}(\lambda_y-\mu_x)^2\,dx\,dy+\iint_{\bar G}\zeta^2(rt-s^2)\,dx\,dy$$
$$+\iint_{\bar G}\zeta\Delta(\sigma,\ d^2z)\,dx\,dy+\iint_{\bar G}\Delta(\sigma,\ \sigma)\,dx\,dy,$$

where

$$\Delta(\sigma,d^2z)=\sigma_{11}t-2\sigma_{12}s+\sigma_{22}r$$

is the mixed discriminant of the two forms σ and d^2z, and

$$\Delta(\sigma,\sigma)=\sigma_{11}\sigma_{22}-\sigma_{12}^2$$

the discriminant of the form σ.

Before proceeding with our outline, some remarks are in order concerning the integrals in the right-hand side of the above formula.

Consider the integral

$$\iint_{\bar G}(\lambda_y-\mu_x)^2\,dx\,dy\geqslant 0;$$

its integrand

$$(\lambda_y-\mu_x)^2=(\xi_y+p\zeta_y-\eta_x-q\zeta_x)^2$$

remains bounded if our regular convex surface F approaches arbitrarily close to a general convex surface, while the varying vector field τ remains bounded in the C^1-norm.

The second integral can be expressed as a Stieltjes integral:

$$\int_{\bar{G}}\int \zeta^2 (rt - s^2)\, dx\, dy = \int_{\Omega}\int \zeta^2\, d\Omega,$$

where Ω is the conditional curvature of the surface: Given a set M on the surface, $\Omega(M)$ is the area (Lebesgue measure) of the image of M in the pq-plane under the mapping

$$p = \partial z/\partial x, \qquad q = \partial z/\partial y.$$

The concept of conditional curvature extends in an obvious manner to general convex surfaces, in exactly the same way as the concept of integral curvature (area of the spherical image). The conditional curvature possesses many properties of ordinary curvature; in particular, it is totally additive on the ring of Borel sets.

The above transformation of the integral is important for the generalization from regular to general convex surfaces.

As for the remaining two integrals,

$$\int_{\bar{G}}\int \zeta\Delta(\sigma,\ d^2z)\, dx\, dy, \qquad \int_{\bar{G}}\int \Delta(\sigma,\ \sigma)\, dx\, dy,$$

for the moment the only essential point is that their integrands, being discriminants, are invariant. This facilitates the transition from one coordinate system to another. We shall have to do this repeatedly in deriving estimates of the integrals for general convex surfaces and bending fields.

Essentially, the exposition of the next three sections prepares the ground for the proof of the Fundamental Lemma in §6. To clarify the goals and problems of this preparatory work not only as a whole but also in regard to details, we now present the basic idea of the proof.

Let F be a general convex surface which can be projected one-to-one onto the xy-plane, and $\zeta(x, y)$ the z-coordinate of a bending field on F. If the surface contains no flat pieces, the Fundamental Lemma states that $z = \zeta(x, y)$ defines a surface of nonpositive curvature; that is, at no point P is this surface strictly convex.

Without loss of generality, we may assume that the projection of P on the xy-plane is the origin. Describe a disk $\bar{\omega}: x^2 + y^2 < \epsilon^2$ of sufficiently small radius such that the projection of F on the xy-plane covers this disk together with its circumference $\bar{\gamma}$. Let ω denote the domain on the surface whose projection is the disk $\bar{\omega}$.

Approximate the surface F by a strictly convex analytic surface $\tilde{F}$, and let $\tilde{\omega}$ denote the domain on $\tilde{F}$ whose projection is $\bar{\omega}$.

Now approximate the bending field τ of F by a regular field $\bar{\tau}$, by averaging τ with respect to a sufficiently regular kernel (§4).

Construct a regular bending field $\tilde{\tau}$ of the surface $\tilde{\omega}$, whose vertical component coincides with that of the field $\bar{\tau}$ on the boundary of $\tilde{\omega}$, i.e. $\tilde{\zeta} = \bar{\zeta}$ (§3).

This is accomplished by solving the first boundary-value problem for the linear equation

$$\tilde{r}\tilde{\zeta}_{yy} - 2\tilde{s}\tilde{\zeta}_{xy} + \tilde{t}\tilde{\zeta}_{xx} = 0,$$

where r, s and $\tilde{t}$ are the second derivatives of the function $\tilde{z}(x, y)$ defining the surface $\tilde{F}$, in the disk $x^2 + y^2 < \epsilon^2$ with boundary condition $\tilde{\zeta} = \bar{\zeta}$. Once the vertical component of $\tilde{\tau}$ has been determined, the other components $\tilde{\xi}$ and $\tilde{\eta}$ are determined by integration.

One now proves that, for a suitable choice of a sequence of surfaces $\tilde{F} \to F$ and a sequence of averaged fields $\bar{\tau} \to \tau$, the corresponding fields $\tilde{\tau}$ converge to a certain bending field of the surface F (§3).

We now consider the vector field $\bar{\tau} - \tilde{\tau}$ on the surface $\tilde{\omega}$. By construction, its vertical component $\bar{\zeta} - \tilde{\zeta}$ vanishes on the boundary of $\tilde{\omega}$. Its horizontal component is modified in a special way, by addition of a regular field τ^* such that $\bar{\tau} - \tilde{\tau} + \tau^*$ is a bending field on the boundary of $\tilde{\omega}$ (§4).

We now apply the integral formula derived above to the surface $\tilde{\omega}$ and the vector field $\bar{\tau} - \tilde{\tau} + \tau^*$ (§6). We prove that by suitable choice of the bending field (which is defined up to a trivial term with vanishing vertical component) one can ensure that the contour integral in the left-hand side of the integral formula, namely

$$\oint_{\gamma} (\lambda\, d\mu - \mu\, d\lambda),$$

vanishes. That this is possible follows from the fact that $\bar{\tau} - \tilde{\tau} + \tau^*$ is a bending field on the boundary of the surface and has vanishing vertical component. This is the reason for adding a "correction" τ^* to the field $\bar{\tau} - \tilde{\tau}$.

After this operation, the integral formula becomes

$$\frac{1}{4}\iint\limits_{\bar{\omega}} (\lambda_y - \mu_x)^2\, dx\, dy + \iint\limits_{\bar{\omega}} \zeta^2 (\tilde{r}\tilde{t} - \tilde{s}^2)\, dx\, dy$$
$$+ \iint\limits_{\bar{\omega}} \zeta \Delta(\sigma,\ d^2\tilde{z})\, dx\, dy + \iint\limits_{\bar{\omega}} \Delta(\sigma,\ \sigma)\, dx\, dy = 0,$$

where λ and μ, ζ and σ refer to the surface $\tilde{\omega}$ and the vector field $\bar{\tau} - \tilde{\tau} + \tau^*$.

The next step of the proof is to let $\tilde{F} \to F$ and $\bar{\tau} \to \tau$ in the above integral formula. We prove that each of the two integrals

$$\iint\limits_{\bar{\omega}} \zeta \Delta(\sigma,\ d^2\tilde{z})\, dx\, dy, \qquad \iint\limits_{\bar{\omega}} \Delta(\sigma,\ \sigma)\, dx\, dy$$

tends to zero as $\tilde{F} \to F$ and $\bar{\tau} \to \tau$ (§5). This assertion is comparatively easy to prove for the second integral, since, because of the special approximation of the field τ by $\bar{\tau}$, the integrand $\Delta(\sigma, \sigma)$ is bounded and tends to zero in measure. On the other hand, the proof that the integral

$$\iint\limits_{\bar{\omega}} \zeta \Delta(\sigma,\ d^2\tilde{z})\, dx\, dy$$

tends to zero involves quite intricate constructions, a description of which would not clarify this account.

After passing to the limit, we get the following relation:

$$\frac{1}{4}\iint\limits_{\bar{\omega}} (\xi_y^0 + p\zeta_y^0 - \eta_x^0 - q\zeta_x^0)^2\, dx\, dy + \iint\limits_{\Omega} (\zeta^0)^2\, d\Omega = 0,$$

where ξ^0, η^0 and ζ^0 are the components of the vector field $\tau^0 = \tau - \lim \tilde{\tau}$. An immediate consequence of this relation is that $\zeta^0 = 0$ at each point of strict convexity of the surface ω. If ω is a strictly convex surface, i.e. it contains neither flat pieces nor straight-line segments, the proof of the lemma is easily completed. In fact, in this case $\zeta = \lim \tilde{\zeta}$. And since the surface $z = \tilde{\zeta}(x, y)$ is a surface of nonpositive curvature, the same is true of the limit surface.

If the surface F contains no flat pieces but does contain straight-line segments, we must use the fact that

$$\xi_y^0 + p\zeta_y^0 - \eta_x^0 - q\zeta_x^0 = 0$$

almost everywhere in $\bar{\omega}$. Combining this equation with the other equations satisfied by the bending field τ^0 of F, and recalling that ζ^0 vanishes on the boundary of ω and at all points of strict convexity of the surface, we finally prove, using a certain integral relation, that ζ^0 vanishes on the straight-line segments of ω as well.

In order to prove the Fundamental Lemma for a surface F containing flat pieces, we utilize the fact that the bending field is arbitrary on the flat pieces. This enables us to modify τ in these domains, leaving it unchanged at all other points, in such a way that ζ^0 is strictly negative on the flat pieces. It then follows that the set of points of the surface $z=\zeta(x,y)$ whose projections are not in flat pieces of ω cannot contain points at which the surface is strictly convex toward $z>0$; that is to say, there is no plane through any point of this set such that neighboring points lie below the plane.

Similarly, we prove that the set can contain no points at which the surface is strictly convex toward $z<0$.

§3. Bending field of a convex surface with prescribed vertical component along the boundary

As we indicated in our outline of the proof of the Fundamental Lemma, the proof is based on comparison of the given bending field with a specially constructed bending field possessing the properties stated in the lemma. This construction is the subject of the present section. We shall prove that a convex surface which can be projected in one-to-one fashion onto a convex domain of the xy-plane has a bending field with prescribed vertical component along the boundary of the surface.

Let ω be an analytical surface with positive Gauss curvature which can be projected in one-to-one fashion into the disk $\bar{\omega}: x^2+y^2<R^2$. Let $z=z(x,y)$ be the equation of this surface. The function $z(x,y)$ is analytic in the disk $\bar{\omega}$ and on its boundary.

Let h be a sufficiently regular function defined on the boundary of $\bar{\omega}$. We wish to prove that there exists a bending field whose vertical component $\zeta(x,y)$ (i.e. the component along the z-axis) coincides with h on the boundary of $\bar{\omega}$.

As was shown in §2, the vertical component ζ of a bending field satisfies the differential equation

$$(*) \qquad r\zeta_{yy}-2s\zeta_{xy}+t\zeta_{xx}=0,$$

where r, s and t are the second derivatives of the function $z(x,y)$ defining the surface ω. It follows that, if the required bending field exists, the vertical component ζ is a solution of the first boundary-value problem for equation (*) in the disk $\bar{\omega}$, with boundary condition $\zeta=h$.

We shall show below that if $\zeta(x,y)$ is a function satisfying equation (*), there exist two other functions $\xi(x,y)$ and $\eta(x,y)$ such that the vector field $\tau(\xi,\eta,\zeta)$ is a bending field for the surface ω. Thus, to prove our geometrical problem it will suffice to prove that the above boundary-value problem for equation (*) is solvable.

A well-known theorem of Bernšteĭn [21] states that the boundary-value problem for equation (*) has a solution if a priori estimates can be established for the solution and its first derivatives. Such estimates are easily found.

The maximum modulus of $\zeta(x,y)$ does not exceed the maximum modulus of h. For, suppose this were false. Then either $\max \zeta > \max h$, or $\min \zeta < \min h$. To fix ideas, let $\max \zeta > \max h$. Cut the surface $z = \zeta(x,y)$ by the plane $\alpha : z = \frac{1}{2}(\max \zeta + \max h)$. Let Φ_α denote that part of the surface lying above the plane α. Construct a sphere σ containing the surface Φ_α together with its boundary. It cuts the plane α in a certain circle κ. Now vary the radius of the sphere in such a way that the sphere always passes through the circle κ but the spherical segment defined by the plane α and containing the surface Φ_α becomes smaller. At some stage of this process, the sphere will touch the surface Φ_α at some point. Now Φ_α is clearly strictly convex at this point, which is impossible. This contradiction proves the existence of an a priori estimate for $\max|\zeta|$.

We now prove the existence of a priori estimates for the derivatives of ζ. First we show that the derivatives ζ_x and ζ_y assume their maximum and minimum on the boundary of the disk $\bar{\omega}$. Otherwise, let $\max \zeta_x = m$ at an interior point P of $\bar{\omega}$, while $\zeta_x \leqq m' < m$ on the boundary.

Now ζ_x is a regular function; hence, for almost all c in the interval $m' < c < m$, the set M_c of points of $\bar{\omega}$ such that $\zeta_x(x,y) = c$ consists of a finite number of regular curves [37]. Since M_c separates the point P at which ζ_x assumes its maximum from the circumference of $\bar{\omega}$, it follows that M_c contains a closed curve γ bounding a domain G which contains the point P.

Consider the line integral

$$I = \frac{1}{2} \oint_\gamma (\zeta_x \, d\zeta_y - \zeta_y \, d\zeta_x).$$

This integral vanishes, since $\zeta_x = \text{const}$ along the contour. Applying the Ostrogradskiĭ-Green formula to I, we get

$$I = \int_G\int (\zeta_{xx}\zeta_{yy} - \zeta_{xy}^2)\, dx\, dy.$$

Now, since the integrand does not change sign (it is nonpositive) while $I = 0$, it follows that $\zeta_{xx}\zeta_{yy} - \zeta_{xy}^2 = 0$ everywhere in G. Hence that part of the surface $z = \zeta(x,y)$ above the domain G is developable.

Through each point of a developable surface there is a straight-line generator issuing from a point on the boundary of the surface. The tangent plane along this generator is stationary. Hence $\zeta_x = m$ on the boundary of G, and we have derived a contradiction. Thus the minimum and maximum of ζ_x and ζ_y are assumed on the circumference of the disk $\bar{\omega}$.

In order to estimate ζ_x and ζ_y on the boundary of $\bar{\omega}$, it will suffice to estimate the maximum angle of inclination of the tangent planes along the boundary of the surface $z = \zeta(x,y)$. Let γ be the curve defined by the given function h on the circumference of $\bar{\omega}$.

Let us choose an arbitrary point P on γ and estimate the angle of inclination of the tangent plane to the surface $z = \zeta(x,y)$ at this point. Without loss of generality, we may assume that the projection of P on the xy-plane is the point $(R,0)$. Let α be a plane through the tangent to γ at P, such that the curve γ lies below α. Since the surface $z = \zeta(x,y)$ has nonpositive curvature, no part of it can protrude above the plane α, and so it lies entirely below the plane. The same conclusion holds for a plane β through the tangent to γ at P such that γ is above β. It follows that the angle of inclination of the tangent plane to $z = \zeta(x,y)$ at P cannot exceed the maximum angle of inclination of the planes passing through the tangent to γ at P and cutting γ in at least one more point. It is not difficult to estimate this maximum.

Parametrize the curve γ with respect to the polar angle ϑ. The equation of the plane through the tangent to γ at P and a point $Q(\vartheta)$ on γ (distinct from P) is

$$\begin{vmatrix} x-R & y & z-h(0) \\ 0 & R & h'(0) \\ R\cos\vartheta - R & R\sin\vartheta & h(\vartheta)-h(0) \end{vmatrix} = 0.$$

Its gradients are

$$\frac{\partial z}{\partial x} = \frac{h(\vartheta)-h(0)-h'(0)\sin\vartheta}{R(1-\cos\vartheta)}, \quad \frac{\partial z}{\partial y} = \frac{h'(0)}{R}.$$

It is readily seen that they are bounded by a constant which depends on the maximum modulus of the first and second derivatives of $h(\vartheta)$. This is obvious for $\partial z/\partial y$. For $\partial z/\partial x$, we have

$$\frac{\partial z}{\partial x} = \frac{h(\vartheta)-h(0)-\vartheta h'(0)}{R(1-\cos\vartheta)} + \frac{(\vartheta-\sin\vartheta)h'(0)}{R(1-\cos\vartheta)}$$
$$= \frac{\vartheta^2 u(\vartheta)}{2R(1-\cos\vartheta)} + \frac{\vartheta-\sin\vartheta}{R(1-\cos\vartheta)} h'(0),$$

where $u(\vartheta)$ is bounded by a constant which depends on the second derivatives: $|u(\vartheta)| \leqq \max|h''(\vartheta)|$.

Since the expressions

$$\frac{\vartheta^2}{1-\cos\vartheta}, \quad \frac{\vartheta-\sin\vartheta}{1-\cos\vartheta},$$

are obviously bounded, we thus have an estimate for $\partial z/\partial x$ which depends only on the maximum modulus of the derivatives h' and h''. This completes the derivation of an estimate for the angle of inclination of the tangent planes to the surface $z=\zeta(x,y)$, and hence an estimate for the first derivatives ζ_x and ζ_y on the boundary of the disk $\bar{\omega}$.

Bernšteĭn's theorem [**21**] now implies that the required boundary-value problem for ζ is solvable. A useful remark concerning the solution is the following. Using Bernšteĭn's method, we can establish a priori estimates for the posited solution for the derivatives of second and higher orders in the closed disk $\bar{\omega}$, once estimates for the solution itself and its first derivatives are known. Now this implies that the solution whose existence we have just proved is sufficiently regular in the closed disk, provided the boundary values h are sufficiently regular.

We have defined the vertical component ζ of the bending field of the surface ω by solving the boundary-value problem for the equation

$$r\zeta_{yy}-2s\zeta_{xy}+t\zeta_{xx}=0,$$

whose solution is ζ. However, it is not yet clear that we can find two other functions ξ and η such that the vector field $\tau(\xi,\eta,\zeta)$ is a bending field of ω. We now consider this question.

Consider the system of differential equations for the bending field:

$$(**) \qquad \xi_x+p\zeta_x=0, \quad \xi_y+p\zeta_y+\eta_x+q\zeta_x=0, \quad \eta_y+q\zeta_y=0.$$

If a bending field with vertical component ζ indeed exists, the other components ξ and η must be found by solving this system. This we now proceed to do.

Differentiate the first equation of (**) with respect to y:

$$\xi_{xy}+(p\zeta_x)_y=0.$$

Differentiating the second equation with respect to y and subtracting the third equation differentiated with respect to x, we get

$$\xi_{yy}+(p\zeta_y)_y+(q\zeta_x)_y-(q\zeta_y)_x=0.$$

Set

$$A=(p\zeta_x)_y, \quad B=(p\zeta_y)_y-(q\zeta_x)_y-(q\zeta_y)_x.$$

A direct check shows that

$$B_x - A_y = r\zeta_{yy} - 2s\zeta_{xy} + t\zeta_{xx} = 0,$$

and so ξ_y may be expressed as a line integral

$$\xi_y + \int (A\,dx + B\,dy) + C_1 = 0.$$

Now set

$$A_1 = p\zeta_x, \quad B_1 = \int (A\,dx + B\,dy) + C_1.$$

It is easy to see that $(B_1)_x - (A_1)_y = 0$. Now, by equations (**), $\xi_x + A_1 = 0$, and so we get the following representation for ξ:

$$\xi + \int (A_1\,dx + B_1\,dy) + C_2 = 0$$

Analogous reasoning yields an expression for the function $\eta(x, y)$. We omit the intermediate arguments and present the final result. First, the function η_x is expressed as a line integral:

$$\eta_x + \int (\overline{A}\,dy + \overline{B}\,dx) + \overline{C}_1 = 0,$$

where

$$\overline{A} = (q\zeta_y)_x, \quad \overline{B} = (q\zeta_x)_x + (p\zeta_y)_x - (p\zeta_x)_y.$$

Finally,

$$\eta + \int (\overline{A}_1\,dy + \overline{B}_1\,dx) + \overline{C}_2 = 0,$$

where

$$\overline{A}_1 = q\zeta_y, \quad \overline{B}_1 = \int (\overline{A}\,dy + \overline{B}\,dx) + \overline{C}_1.$$

We have thus found the two functions $\xi(x, y)$ and $\eta(x, y)$, expressed as line integrals of known functions. It remains to verify that the functions ξ, η and ζ indeed satisfy the system (**).

The expression for ξ implies $\xi_x + A_1 = 0$. And since $A_1 = p\zeta_x$, the first equation of (**) holds.

Differentiating the expression for η with respect to y, we obtain $\eta_y + \overline{A}_1 = 0$. But $\overline{A}_1 = q\zeta_x$. Hence the third equation is satisfied.

Differentiate the expression for ξ with respect to y and the expression for η with respect to x; add the results:

$$\xi_y + \eta_x + B_1 + \overline{B}_1 = 0.$$

But

$$B_1 + \overline{B}_1 = \int \{(A + \overline{B})\,dx + (B + \overline{A})\,dy\} + C_1 + \overline{C}_1,$$

$$A + \overline{B} = (q\zeta_x)_x + (p\zeta_y)_x,$$
$$B + \overline{A} = (p\zeta_y)_y + (q\zeta_x)_y.$$

Consequently

$$B_1 + \bar{B}_1 = p\zeta_y + q\zeta_x + C.$$

Setting $C = 0$, we see that

$$\xi_y + \eta_x + p\zeta_y + q\zeta_x = 0,$$

and thus the second equation of (**) is also satisfied.

We have thus proved the existence of a bending field τ of the surface ω, with prescribed vertical component on the boundary.

We now claim that the bending field in question is uniquely defined, up to a trivial term—the velocity field of a rigid translation of the surface ω parallel to the xy-plane and the velocity field of a rotation of the surface about the z-axis.

Let $\bar{\tau}$ be another (regular) bending field with the same vertical component on the boundary of ω as the field τ just constructed. The vector field $\tau - \bar{\tau}$ is obviously a bending field for the surface ω.

The boundary of the surface $z = \zeta(x,y) - \bar{\zeta}(x,y)$ lies in the xy-plane. Since this surface has nonpositive curvature, it must lie entirely in the xy-plane, i.e. necessarily $\zeta(x,y) - \bar{\zeta}(x,y) \equiv 0$.

To simplify the notation, set

$$\tilde{\xi} = \xi - \bar{\xi}, \quad \tilde{\eta} = \eta - \bar{\eta}, \quad \tilde{\zeta} = \zeta - \bar{\zeta}.$$

Since $\tilde{\zeta} \equiv 0$, the equations for the bending field $\tau - \bar{\tau}$ give

$$\tilde{\xi}_x = 0, \quad \tilde{\xi}_y + \tilde{\eta}_x = 0, \quad \tilde{\eta}_y = 0.$$

Hence $\tilde{\xi} = \varphi(y)$ and $\tilde{\eta} = \psi(x)$, and moreover $\varphi'(y) + \psi'(x) = 0$. Now this means that $\varphi'(y)$ and $\psi'(x)$ are constants which differ only in sign. Thus

$$\bar{\xi} = \omega y + C_1, \quad \tilde{\eta} = -\omega x + C_2.$$

Here ωy and $-\omega x$ are the components of the angular velocity of the surface as a rigid body about the z-axis, while C_1 and C_2 are the components of the translational velocity of the surface parallel to the xy-plane. Our assertion is proved.

We are now going to consider the existence of a bending field for a general convex surface F with prescribed vertical component on the boundary of F. The proof will be based on our result for regular surfaces, employing approximation of F by regular surfaces.

This procedure raises the following question. If a sequence of convex surfaces F_n converges to a convex surface F and a corresponding sequence of bending fields τ_n converges to a field τ, is τ a bending field of the limit surface F? In the general case, the answer is negative. Nevertheless, we have the following result.

THEOREM 1. *If a sequence of convex surfaces F_n converges to a convex surface F which can be projected in one-to-one fashion onto the xy-plane, a corresponding sequence of bending fields τ_n converges to a field τ, and moreover the fields τ_n satisfy a Lipschitz condition* uniformly *in any compact domain, then τ is a bending field of the surface F.*

PROOF. Since the fields τ_n satisfy a Lipschitz condition uniformly in every compact domain, the same is true of their limit τ. Hence, by Aleksandrov's theorem (§1), a sufficient condition for τ to be a bending field is that $dr d\tau = 0$ almost everywhere on F.

Let $\bar{F}$ denote the projection of F on the xy-plane. The functions r and τ have a total differential almost everywhere in this domain. The theorem will be proved if we can show that $dr d\tau = 0$ at every point $\bar{P}$ of $\bar{F}$ at which r and τ are differentiable.

Describe a small disk $\bar{\omega}$ about the point $\bar{P}$. Now the equality $dr_n d\tau_n = 0$ holds almost everywhere in $\bar{\omega}$ for any n, and the sequence F_n is countable; hence there is a subset $\tilde{\omega}$ of $\bar{\omega}$, with measure equal to that of $\bar{\omega}$, in which $dr_n d\tau_n = 0$ for all n.

Since the measure of $\tilde{\omega}$ is equal to that of $\bar{\omega}$, it follows that almost all points of almost all diameters of the disk $\bar{\omega}$ belong to $\tilde{\omega}$. Let δ be one of these diameters: $dr_n d\tau_n = 0$ for all n, almost everywhere on δ.

Parametrize δ with respect to signed distance s from the point $\bar{P}$. We claim that $r'\tau' = 0$ at $\bar{P}$.

Let P denote the point on F whose projection is $\bar{P}$, and $Q_n(s)$ the point of F_n whose projection is the point $\bar{Q}(s)$ on the straight line δ. Let α be a vertical plane through δ. This plane cuts the surfaces F and F_n in convex curves γ and γ_n. As $n \to \infty$, the curves γ_n converge to γ. The point P lies on the curve γ, the points $Q_n(s)$ on the curves γ_n. Let g be the tangent to γ at P, and g_n a semitangent to γ_n at $Q_n(s)$. We claim that if $|s|$ is sufficiently small and n sufficiently large, the straight lines g and g_n are arbitrarily close together, i.e. they form an arbitrarily small angle.

Indeed, since P is a smooth point of the curve γ, the supporting lines to γ at points close to P are close to g. For small $|s|$ and $n \to \infty$ we can extract from any sequence of straight lines g_n a convergent subsequence whose limit is a supporting line to γ at some point close to P. Hence for small $|s|$ and large n the straight lines g and g_n are arbitrarily close together.

This being so, we can state that

$$r'(0) = r'_n(s) + \epsilon_n(s),$$

where ϵ_n is arbitrarily small for small s and sufficiently large n. (At points where r_n is not differentiable, the symbol $r'_n(s)$ denotes the right derivative.)

Since τ has a differential at $\overline{P}$, we have

$$\tau'(0) = \frac{1}{s}(\tau(s) - \tau(0)) + \varepsilon(s),$$

where $\epsilon(s)$ tends to zero with $|s|$.

Since the fields τ_n converge to τ,

$$\frac{1}{s}(\tau(s) - \tau(0)) = \frac{1}{s}(\tau_n(s) - \tau_n(0)) + \varepsilon'_n,$$

where ϵ'_n is arbitrarily small for large n. Hence

$$\tau'(0) = \frac{1}{s}(\tau_n(s) - \tau_n(0)) + \varepsilon''_n(s),$$

where $\epsilon''_n(s)$ is arbitrarily small for small s and large n.

Since the function $\tau_n(s)$ satisfies a Lipschitz condition, it is absolutely continuous and so has the representation

$$\tau_n(s) - \tau_n(0) = \int_0^s \tau'_n(s)\,ds.$$

We thus get the following final expression for $\tau'(0)$:

$$\tau'(0) = \frac{1}{s}\int_0^s \tau'_n(s)\,ds + \varepsilon''_n(s).$$

Substituting the expressions for $r'(0)$ and $\tau'(0)$ in $r'\tau$, we get

$$r'(0)\tau'(0) = \frac{1}{s}\int_0^s r'_n(s)\tau'_n(s)\,ds + \frac{1}{s}\int_0^s \varepsilon_n(s)\tau'_n(s)\,ds + r'(0)\varepsilon''_n(s).$$

Since $r'_n(s)\tau'_n(s) = 0$ for almost all s, the first integral on the right-hand side is zero. For small s and large n, the integrand in the second integral is arbitrarily small together with $\epsilon_n(s)$, since $\tau'_n(s)$ is bounded uniformly in n and s. Therefore the second integral is arbitrarily small for small $|s|$ and large n. The last term on the right-hand side, $r'(0)\epsilon''_n(s)$, is arbitrarily small for small $|s|$ and large n.

Since the right-hand side of the equality can be made arbitrarily small by suitable choice of s and n, while the left-hand side is independent of both s and n, it must vanish. Thus $dr d\tau = 0$ at the point $\overline{P}$, in the direction of the line δ.

Since almost all straight lines δ through $\overline{P}$ have this property, i.e. $dr_n d\tau_n = 0$ almost everywhere on δ, it follows that $dr d\tau = 0$ at $\overline{P}$ in almost all directions. And since r and τ have total differentials at $\overline{P}$, it follows that $dr d\tau = 0$ at $\overline{P}$ in *all* directions. This completes the proof of the theorem.

THEOREM 2. *Let F be a general convex surface whose projection on the xy-plane is a strictly convex domain $\overline{F}$. Let f be a function defined on the boundary of $\overline{F}$ and satisfying a Lipschitz condition there. Then there exists a bending field τ of F whose vertical component ζ is equal to f on the boundary of F.*

As far as the proof of the Fundamental Lemma is concerned, we shall need this theorem only for the case in which $\overline{F}$ is a disk. We shall therefore prove the theorem for this case only.

Approximate the surface F by an analytic surface ω with everywhere positive Gauss curvature. Approximate the function f by an analytic function h. Without loss of generality, we may assume that the derivative of h with respect to the polar angle is at most twice the Lipschitz constant of the function f.

Construct a bending field τ of the surface ω with vertical component h on the boundary, which satisfies the conditions $\xi = \eta = 0$ and $d\xi = d\eta = 0$ at the center of the disk $\overline{F}: x^2 + y^2 < R^2$. This is easily done by suitable choice of the trivial term of the solution. We claim that as $\omega \to F$ and $h \to f$ the bending field τ of ω converges to a bending field of F whose vertical component ζ on the boundary is f.

To prove this, we first show that as $\omega \to F$ and $h \to f$ the bending field τ satisfies a Lipschitz condition uniformly in any disk $\bar{\omega}_\epsilon: x^2 + y^2 \leqq R^2 - \epsilon$ $(\epsilon > 0)$. We begin with the vertical component $\zeta(x,y)$ of τ.

Since the Gauss curvature of the surface ω is strictly positive, so that $rt - s^2 > 0$, the quadratic form

$$r\alpha^2 - 2s\alpha\beta + t\beta^2$$

is positive definite, unless $\alpha = \beta = 0$ (when it vanishes identically). It follows that there can be no parabolic points on the surface $\Phi: z = \zeta(x,y)$. Indeed, at a parabolic point we would have $\zeta_{xx}\zeta_{yy} - 2\zeta_{xy} = 0$, and so

$$r\zeta_{yy}^2 - 2s\zeta_{yy}\zeta_{xy} + t\zeta_{xy}^2 = \zeta_{yy}\,(r\zeta_{yy} - 2s\zeta_{xy} + t\zeta_{xx}) = 0,$$

$$r\zeta_{xy}^2 - 2s\zeta_{xy}\zeta_{xx} + t\zeta_{xx}^2 = \zeta_{xx}\,(r\zeta_{yy} - 2s\zeta_{xy} + t\zeta_{xx}) = 0.$$

But this implies $\zeta_{xx} = \zeta_{xy} = \zeta_{yy} = 0$, i.e. the point is not parabolic but flat.

Let γ denote the boundary of the surface Φ. The curve γ lies on the circular cylinder $x^2 + y^2 = R^2$. Since the derivative h_ϑ is bounded by a constant, the inclination of the curve γ to the xy-plane cannot be too large.

Since the surfaces Φ have nonpositive curvature, we may assume that they all lie between the parallel planes $z = \pm c$, where c depends on $\max|f|$. Let V_ϵ denote the cylinder defined by

$$x^2 + y^2 \leqq R^2 - \epsilon, \quad |z| \leqq c.$$

Let P be an arbitrary point on V_ϵ and α a plane through P. Since the inclination of the curve γ to the xy-plane is uniformly bounded, there exists ϵ' such that whenever the angle of inclination of the plane α to the xy-plane exceeds $\pi/2 - \epsilon'$, it will cut the curve γ in only two points.

Let Φ_ϵ denote the part of the surface Φ whose projection lies in the disk $\bar{\omega}$. We claim that the inclinations of the tangent planes to Φ_ϵ are uniformly bounded as $\omega \to F$ and $h \to f$.

Let P be a hyperbolic point of Φ_ϵ. It lies on the cylinder V_ϵ. If the inclination of the tangent plane α at P is sufficiently large, it will cut the curve γ (the boundary of Φ) at only two points. But we can prove that there must be at least four points of intersection.

Indeed, since P is hyperbolic, the plane α divides the neighborhood of P into four "sectors." Two of them lie on one side of the plane α, the other two, on the other side. Sectors on the same side of α have no common points, other than points of their boundaries, in the plane α. For were there a path from one sector to the other, lying entirely outside the plane α, one of the other two sectors would have its boundary entirely within the plane α; this is impossible, since Φ is a surface of nonpositive curvature. Thus the four sectors belong to different components of the partition of Φ by the plane α.

Since each of the sectors reaches the boundary of the surface Φ (the curve γ), it follows that γ divides α into at least four parts. Thus we have a contradiction, and our assertion is proved for the case of a hyperbolic point P.

Now let P be a flat point of the surface Φ_ϵ. The tangent plane α at P may also touch the surface at other points. Let M_P be the set of all these points. If the inclination of the plane α is sufficiently large, M_P cannot be the entire surface. Moreover, there are hyperbolic points arbitrarily close to M_P. Indeed, let Q be a boundary point of M_P. If all points near Q are flat, it follows that each (connected) component of the set of these points lies in a plane; this plane must obviously be α, which is impossible.

Now let P' be a hyperbolic point close to M_P. Then the tangent plane at P' is close to α, and we can derive a contradiction by following the above arguments for a hyperbolic point P. Thus, at all points of the surface Φ_ϵ, the inclinations of the tangent planes to the xy-plane are uniformly bounded as $\omega \to F$ and $h \to f$.

In analytical terms, this means that as $\omega \to F$ and $h \to f$, the derivatives ζ_x and ζ_y are uniformly bounded in each disk $\bar{\omega}_\epsilon$. Hence the sequence of functions $\zeta(x,y)$ contains a subsequence which converges uniformly in each disk $\bar{\omega}_\epsilon$. The limit function $\zeta^0(x,y)$ will be continuous in the open disk $\bar{\omega}$ and satisfy a Lipschitz condition in each disk $\bar{\omega}_\epsilon$. We shall now show that the function $\zeta^0(x,y)$ is extended continuously to the boundary of the disk $\bar{F}$ by the function f.

Let γ^0 denote the curve defined by the function f on the cylinder $x^2+y^2=R^2$. Let Ω be the convex hull of this curve. Ω consists of two developable surfaces Ω_1 and Ω_2, separated by the curve γ^0. We claim that each of the surfaces Ω_1 and Ω_2 can be projected in one-to-one fashion onto the xy-plane.

If this is not so, the only possible reason is that the surface Ω_i contains a straight-line segment parallel to the z-axis. One of the ends of this segment clearly lies on γ^0, while the other (Q, say) is an interior point of Ω_i. Since the surface Ω_i is developable, there is a generator through Q whose endpoints lie on the boundary of the surface—the curve γ^0. But this is impossible, for the projection of γ^0 on the xy-plane is strictly convex. This proves the assertion.

Now suppose that f does not extend $\zeta^0(x,y)$ continuously to the boundary of the disk $\bar{F}$. This means that there exist points P on the surfaces Φ which as $\omega \to F$ and $h \to f$ draw arbitrarily close to a point P^0 of the cylinder $x^2+y^2=R^2$ which is not on the curve γ^0. We shall show that this is impossible.

Indeed, the point P^0 is not on the surface Ω, and lies outside it. Hence there exists a plane s separating P^0 from Ω and therefore also from the curve γ^0. When ω is sufficiently close to F and h sufficiently close to f, the curve γ (the boundary of Φ) is close to γ^0, and the point P of Φ is close to P^0. Therefore the point P and the curve γ are also separated by the plane σ. But this is impossible, since Φ has nonpositive curvature.

Thus the function $\zeta^0(x,y)$ is extended continuously by the function f to the circumference of the disk $\bar{F}$.

We now prove that the derivatives of the horizontal components ξ and η of the bending field τ of ω are uniformly bounded as $\omega \to F$ and $h \to f$ in any disk $\bar{\omega}_\epsilon$. For the derivatives ξ_x and η_y this follows from the equations of the bending field:

$$\xi_x + p\zeta_x = 0, \quad \eta_y + q\zeta_y = 0.$$

Now consider ξ_y and η_x. The expression derived above for ξ_y as a line integral can be brought to the following form:

$$-\xi_y(Q) = \int^{Q} \{d(p\zeta_y) + (s\zeta_x - r\zeta_y)\,dx + (t\zeta_x - s\zeta_y)\,dy\},$$

where P is the point on ω whose projection is the center of $\bar{F}$. Alternatively, we can write

$$-\xi_y(Q) = p\zeta_y\Big|_P^Q + \int_P^Q (\zeta_x\,dq - \zeta_y\,dp).$$

Take some positive $\epsilon' < \epsilon$. It can be shown that, for ω sufficiently close to F, the points P and Q may be joined by a curve on the surface $\omega_{\epsilon'}$, the length of whose spherical image is bounded by a constant $l(\epsilon, \epsilon')$ depending only on ϵ and ϵ'. The proof of this assertion is quite complicated, and will be given below. For the moment, assume that it is true, and let the path of integration in the above integral be a curve connecting P and Q on the surface $\omega_{\epsilon'}$, with spherical image at most $l(\epsilon, \epsilon')$. Then

$$|\xi_y(Q)| \leqslant \left| p\zeta_y\Big|_P^Q \right| + \int_P^Q \sqrt{\zeta_x^2 + \zeta_y^2}\,\sqrt{dp^2 + dq^2}.$$

The integral

$$\int_P^Q \sqrt{dp^2 + dq^2}$$

can be estimated in terms of $l(\epsilon, \epsilon')$ and the maximum moduli of p and q in $\omega_{\epsilon'}$. We thus obtain an estimate for $|\xi_y|$ in ω_ϵ which depends on the maximum moduli of ζ_x, ζ_y, p and q in the disk $\omega_{\epsilon'}$, and on the constant $l(\epsilon, \epsilon')$.

The estimate for η_x is established by analogous arguments.

To complete the proof of Theorem 2, it is now sufficient to use Theorem 1, after extracting a convergent subsequence of bending fields τ of the surfaces ω. The proof is complete.

We now prove the above property of analytic convex surfaces approximating the general convex surface F.

Let F be a general convex surface whose projection lies in the disk $\bar{\omega}: x^2+y^2<R^2$ on the xy-plane, and ω a strictly convex analytic surface close to F. Let ω_ϵ denote the part of ω whose projection lies in the disk $\bar{\omega}_\epsilon: x^2+y^2<R^2-\epsilon$.

We wish to prove that if the surfaces ω and F are sufficiently close together then any two points of ω_ϵ can be connected by a curve on a surface $\omega_{\epsilon'}$, $\epsilon'<\epsilon$, such that the length of the spherical image of this curve is at most $l(\epsilon,\epsilon')$.

Let $\gamma_{\epsilon'}$ denote the boundary of the surface $\omega_{\epsilon'}$: it is an analytic curve whose projection is the circumference of the disk $\bar{\omega}_{\epsilon'}$. Let Ω be that part of the convex hull of $\gamma_{\epsilon'}$ which is convex in the same sense as the surface ω. Ω is a developable surface with boundary $\gamma_{\epsilon'}$, and it consists of a finite number of analytic surfaces.

Let A and B be two arbitrary points on ω_ϵ. Project these points onto the xy-plane and the surface Ω by straight lines parallel to the z-axis. Denote the projections by $\bar{A}$, $\bar{B}$ and A_1, B_1, respectively. Let α and β be tangent planes to the surface Ω at the points A_1 and B_1. These planes cut out convex caps ω_α and ω_β, respectively, from the surface $\omega_{\epsilon'}$. Let A_2 and B_2 denote the points of these caps most distant from the planes α and β.

The points A and A_2 on the cap ω_α can be joined by the line of the shadow contour, i.e. the line along which the tangent planes are parallel to a fixed direction. Obviously, the spherical image of this line, as an arc of a great circle smaller than a semicircle, is smaller than π. The points B and B_2 can be joined in similar fashion on the cap ω_β.

To prove our assertion, it remains to join the points A_2 and B_2 in a suitable manner.

Connect the points A_1 and B_1 on the surface Ω by the curve κ whose projection on the xy-plane is the straight-line segment $\overline{AB}$. For each tangent plane of Ω, there is a parallel tangent plane of the surface $\omega_{\epsilon'}$. It follows that there exists a curve on $\omega_{\epsilon'}$ which has the same spherical image as κ. This curve joints A_2 and B_2, since by construction the tangents to the surface $\omega_{\epsilon'}$ at these points are parallel to the tangents of Ω at A_1 and B_1.

Let us estimate the length of the spherical image of κ. Let $\delta(P)$ denote the length of the generator of Ω passing through a point P, and consider the integral along the curve κ:

$$I = \frac{1}{2} \int_{\varkappa} \delta(P) \, | dn(P) |,$$

where $n(P)$ is the normal to the surface Ω at P. We claim that this integral can be estimated in terms of the integral mean curvature of the surface Ω.

Consider a small "rectangle" on the surface Ω, with side $\Delta\delta$ along a generator and side Δs perpendicular to the generator. The integral mean curvature of Ω in the rectangle is $\sim \frac{1}{2} k \Delta\delta \Delta s$, where k is the normal curvature of the surface in the direction perpendicular to the generator.

By the formula of Rodrigues, $k\Delta s \cong |\Delta n|$. Hence the integral mean curvature of the rectangle is $\frac{1}{2}\Delta\delta|\Delta n|$. It follows that I is the integral mean curvature of the part of the surface Ω described by straight-line generators cutting the curve κ. Thus I cannot exceed the integral mean curvature H of the surface Ω. Let us estimate H. Now,

$$H = \frac{1}{2} \iint_{\bar{\omega}_{\varepsilon'}} \left| \frac{(1+p^2)t - 2pqs + (1+q^2)r}{(1+p^2+q^2)} \right| dx \, dy.$$

Obviously,

$$H \leqslant \frac{1}{2} \iint_{\bar{\omega}_{\varepsilon'}} (|r| + 2|s| + |t|) \, dx \, dy.$$

Now $rt - s^2 = 0$, so that $2|s| \leqq |r| + |t|$, and therefore

$$H \leqslant \iint_{\bar{\omega}_{\varepsilon'}} (|r| + |t|) \, dx \, dy.$$

Since the function $p(x,y)$ is monotone on the straight lines $y = \text{const}$, while the function $q(x,y)$ is monotone on the lines $x = \text{const}$ (because Ω is a convex surface), it follows that

$$\int_{x_1}^{x_2} | r(x, y) | dx = | p(x_2, y) - p(x_1, y) |,$$

$$\int_{y_1}^{y_2} | t(x, y) | dy = | q(x, y_2) - q(x, y_1) |.$$

Hence, if $\max|p|$ and $\max|q|$ do not exceed c on Ω, we have

$$\iint_{\bar{\omega}_{\varepsilon'}} | r | dx \, dy \leqslant 4cR, \quad \iint_{\bar{\omega}_{\varepsilon'}} | t | dx \, dy \leqslant 4cR,$$

and thus $H \leqq 8cR$.

The maxima of $|p|$ and $|q|$ on $\omega_{\epsilon'}$ cannot exceed the maxima of $|p|$ and $|q|$ on $\omega_{\epsilon'}$, since for each tangent plane of Ω there is a parallel tangent plane of $\omega_{\epsilon'}$. Now, if the surface ω is sufficiently close to F, the maxima of $|p|$ and $|q|$ on $\omega_{\epsilon'}$ can be estimated in terms of the angle of inclination of the supporting planes to F on the set $F_{\epsilon'}$ of all points whose projections lie in the closed disk $\bar{\omega}_{\epsilon'}$. Thus we may assume that the constant c in the estimate for H depends only on ϵ', and $H \leqq 8c(\epsilon')R$.

It is now easy to estimate the length l^* of the spherical image of the curve κ. We have $l^*(\kappa) = \int_\kappa |dn|$.

Since the projection of each point P of κ lies within the disk $\bar{\omega}_\epsilon$, and the generator through P has endpoints on the boundary of $\omega_{\epsilon'}$ provided $dn(P) \neq 0$, it follows that the length of each such generator is at least that of the chord of the disk $\bar{\omega}_{\epsilon'}$ touching the disk $\bar{\omega}_\epsilon$. Let δ_0 be the length of this chord.

Consider the integral

$$I = \frac{1}{2} \int_{\varkappa} \delta(P)\, |dn(P)|.$$

Since $\delta(P) \geqq \delta_0$ at each point P where $dn \neq 0$, we have

$$I \geqslant \frac{\delta_0}{2} \int_{\varkappa} |dn| = \frac{\delta_0}{2} l^*(\varkappa).$$

But $I \leqq H$ and $H \leqq 8c(\epsilon)R'$. We thus obtain the estimate

$$l^*(\varkappa) \leqslant \frac{16}{\delta_0} c(\varepsilon') R$$

Thus, if the surfaces ω and F are sufficiently close together, any two points of ω_ϵ can be joined by a curve on the surface $\omega_{\epsilon'}$, the length of whose spherical image is at most $16c(\epsilon')R/\delta_0 + 2\pi$.

This proves our assertion.

§4. Special approximation of a bending field of a general convex surface

This section is devoted to certain properties of the "Steklov average" of a bending field of a general convex surface; these properties will be used mainly in estimation of certain integrals (§5) figuring in the proof of the Fundamental Lemma (§6).

Let F be a general convex surface which can be projected in one-to-one fashion onto the xy-plane, F_ω the compact domain on F whose projection is the disk $\bar{\omega}: x^2 + y^2 \leqq R^2$. Thus the projection of the entire surface F includes the disk $\bar{\omega}$ together with a certain neighborhood of $\bar{\omega}$.

Let τ be a bending field of F. Consider the vector field $\bar{\tau}$ defined by

$$\bar{\tau}(x, y) = \iint_{\bar{F}} \varphi(x-u, y-v)\, \tau(u, v)\, du\, dv,$$

where the domain of integration is the projection $\bar{F}$ of F and the function φ is defined by

$$\varphi(x-u, y-v) = \frac{1}{H_\delta} \exp \frac{(x-u)^2 + (y-v)^2}{(x-u)^2 + (y-v)^2 - \delta^2} \quad \text{for} \quad (x-u)^2 + (y-v)^2 < \delta^2$$

$$\varphi(x-u, y-v) = 0 \quad \text{for} \quad (x-u)^2 + (y-v)^2 \geqslant \delta^2,$$

$$H_\delta = \iint_{u^2+v^2 \leqslant 1} \exp \frac{u^2+v^2}{u^2+v^2-\delta^2}\, du\, dv.$$

We shall call $\bar{\tau}$ the *average* of the field τ, or simply the *average field*. Investigation of this field is based on the following easily verified properties of the function $\varphi(x-u, y-v)$.

1. The function φ is differentiable to any order with respect to both arguments.

2. The function φ is nonnegative, and for fixed x and y it vanishes outside a disk of radius δ about the point (x, y).

3

$$\iint \varphi(x-u, y-v)\, du\, dv = 1,$$

where the domain of integration is the disk $(x-u)^2 + (y-v)^2 \leqq \delta^2$.

We mention some properties of the average field in the disk $\bar{\omega}$:

a) The field $\bar{\tau}$ converges uniformly to τ as $\delta \to 0$.

b) The field $\bar{\tau}$ is infinitely differentiable, i.e. has derivatives of arbitrary order.

c) As $\delta \to 0$, the first derivatives of $\bar{\tau}$ remain bounded and have representations

$$\bar{\tau}_x(x, y) = \iint_{\bar{F}} \varphi(x-u, y-v)\, \tau_u(u, v)\, du\, dv,$$

$$\bar{\tau}_y(x, y) = \iint_{\bar{F}} \varphi(x-u, y-v)\, \tau_v(u, v)\, du\, dv.$$

Indeed, let δ be so small that the δ-neighborhood of the disk $\bar{\omega}$ is contained in the domain $\bar{F}$. By the mean-value theorem,

$$\bar{\tau}(x, y) = \tau(u^*, v^*) \iint_{\bar{F}} \varphi(x-u, y-v)\, du\, dv = \tau(u^*, v^*),$$

where u^*, v^* are the coordinates of a point whose distance from (x,y) is less than δ. Now the fact that $\bar{\tau}$ converges uniformly to τ as $\delta \to 0$ follows from the continuity of τ in $\bar{F}$.

Property b) follows from the fact that φ is infinitely differentiable.

To prove property c), note that in a sufficiently small neighborhood $\bar{\omega}_\epsilon$ of the disk $\bar{\omega}$ the function $\varphi(x-u, y-v)\tau(u,v)$ is absolutely continuous in u for $v=\text{const}$ and in v for $u=\text{const}$. It is therefore an indefinite integral of its derivative. When $(x,y) \in \bar{\omega}$, on the boundary of $\bar{\omega}_\epsilon$ we have $\varphi(x-u, y-v)\tau(u,v)=0$; hence

$$\int \frac{d}{du}(\varphi(x-u, y-v)\tau(u, v))\, du = 0.$$

Integrating with respect to v, we get

$$\int\int_{\bar{\omega}_\varepsilon} \frac{d}{du}(\varphi(x-u, y-v)\tau(u, v))\, du\, dv = 0.$$

Now the domain of integration $\bar{\omega}_\epsilon$ may be replaced by $\bar{F}$, since the function $\varphi(x-u, y-v)\tau(u,v)$ vanishes outside $\bar{\omega}_\epsilon$. Differentiating the integrand with respect to u and observing that

$$\bar{\tau}_x(x, y) = \int\int_{\bar{F}} \varphi_x(x-u, y-v)\tau(u, v)\, du\, dv,$$

$$\varphi_x(x-u, y-v) = -\varphi_u(x-u, y-v),$$

we get the required representation:

$$\bar{\tau}_x(x, y) = \int\int_{\bar{F}} \varphi(x-u, y-v)\tau_u(u, v)\, du\, dv.$$

The formula for $\bar{\tau}_y(x,y)$ is proved in analogous fashion.

The fact that $\bar{\tau}_y(x,y)$ is bounded follows from the fact that the derivative $\tau_u(u,v)$ is bounded, which in turn is a consequence of the Lipschitz condition for the vector field τ. Similarly, the derivative $\bar{\tau}_y(x,y)$ is bounded as $\delta \to 0$.

We now define a set M_ϑ in the closed disk $\bar{\omega}$. A point $\bar{P}$ is in M_ϑ if the point P on F whose projection is $\bar{P}$ has at least two supporting planes which form an angle at least ϑ. M_ϑ is obviously a closed set. Let G_ϑ be an open set containing M_ϑ.

Approximate F by an analytic surface ω; let $\bar{r}(x,y)$ and $r(x,y)$ denote the radius vectors of the surfaces ω and F.

We claim that *if the surfaces ω and F are sufficiently close together and the field $\bar{\tau}$ is sufficiently close to τ (i.e. for small δ), then $|\bar{r}'\bar{\tau}'| < \epsilon_\vartheta$ at each* *point of the set* $\bar{\omega} - G_\vartheta$ *(differentiating along any direction), where* $\epsilon_\vartheta \to 0$ *as* $\vartheta \to 0$. Let us prove this.

Let P be an arbitrary point of the surface F whose projection $\bar{P}$ is in the set $\bar{\omega} - G_\vartheta$, and Q an arbitrary point of F_ω with projection $\bar{Q}$. Then there exists $\epsilon' > 0$, independent of P and Q, such that whenever the distance between $\bar{P}$ and $\bar{Q}$ is less than ϵ', the angle between the supporting planes at P and Q is at most 2ϑ.

Indeed, otherwise one could easily conclude from the compactness of $\bar{\omega} - G_\vartheta$ that this set contains a point with two supporting planes forming an angle at least 2ϑ, and this is impossible.

For the same reason, when ω is sufficiently close to F the angle between their supporting planes at any two points having the same projection in $\bar{\omega} - G_\vartheta$ is also at most 2ϑ.

We may therefore assume that for any point $\bar{P}$ in the set $\bar{\omega} - G_\vartheta$ and any point $\bar{Q}$ whose distance from $\bar{P}$ is at least ϵ' we have

$$\bar{r}'(\bar{P}) = r'(Q) + \epsilon'',$$

differentiating along any direction, where ϵ'' is arbitrarily small provided d is small.

We now use the integral representation of $\bar{\tau}'$. We have

$$\bar{\tau}'(\bar{P}) = \int\int \varphi(\bar{P} - \bar{Q})\,\tau'(\bar{Q})\,d\bar{Q},$$

$$\bar{r}'(\bar{P})\,\bar{\tau}'(\bar{P}) = \int\int \varphi(\bar{P} - \bar{Q})\,\bar{r}'(\bar{P})\,\tau'(\bar{Q})\,d\bar{Q}.$$

Substitute the above formula for $\bar{r}'(\bar{P})$ in the right-hand side of this formula, and note that $r'(Q)\,\tau'(Q) = 0$ almost everywhere; the result is

$$\bar{r}'(\bar{P})\,\bar{\tau}'(\bar{P}) = \int\int \varphi(\bar{P} - \bar{Q})\,\varepsilon''\,d\bar{Q}.$$

This implies that $\bar{r}'(\bar{P})\,\bar{\tau}'(\bar{P})$ is arbitrarily small for sufficiently small ϑ, proving our assertion.

Let G^* be an open set containing all points of the disk $\bar{\omega}$ which are projections of conic points of the surface F. The set $M_\vartheta - G^*$ consists of the projections of the ridge points of F whose plane angle is at least ϑ. Let us associate with each point of the set $M_\vartheta - G^*$ the direction defined by the projection of its edge. (By the edge of a surface at a ridge point we mean the straight line contained in all the supporting planes to the surface at the point.) The field of directions thus defined on the set $M_\vartheta - G^*$ is continuous, even uniformly continuous.

Indeed, otherwise it would follow that the set of points on the surface whose projections lie in $M_\vartheta - G^*$ contains at least one conic point, which is impossible.

Let ϵ and ϵ_1 be two positive numbers. Let $\overline{P}$ be an arbitrary point of the disk $\bar{\omega}$, whose ϵ-neighborhood contains points of $M_\vartheta - G^*$. Let g be any direction at $\overline{P}$ which forms an angle smaller than ϵ_1 with the "direction of the edge" at a point $\overline{Q} \in M_\vartheta - G^*$ in the ϵ-neighborhood of $\overline{P}$.

We claim that, if ω is sufficiently close to F, the field $\bar{\tau}$ sufficiently close to τ, and ϵ and ϵ_1 are sufficiently small, then the product $\bar{r}'(\overline{P})\bar{\tau}_g'(\overline{P})$ is arbitrarily small, and this is true uniformly in P and g.

To prove this, let S be any point of F, and S_ω a point on ω whose projection $\overline{S}$ lies in the ϵ-neighborhood of $\overline{P}$. If ϵ is sufficiently small, the supporting plane at S forms a small angle with the direction of the edge at Q. Hence it follows that for sufficiently small ϵ_1 the derivatives $r_g'(\overline{S})$ and $r_g'(\overline{Q})$ are close together, and this holds uniformly in $\overline{P}$, $\overline{Q}$ and g.

If ω is sufficiently close to F, the tangent plane to ω at the point S_ω also forms a small angle with the edge at Q, and this angle is uniformly small with respect to the choice of $\overline{P}$. Thus the difference between $\bar{r}_g'(\overline{P})$ and $r_g'(\overline{S})$ is small, and we may therefore write

$$\bar{r}_g'(\overline{P}) = r_g'(\overline{S}) + \epsilon_2,$$

where ϵ_2 is small independently of the choice of $\overline{P}$ and $\overline{Q}$, provided ϵ and ϵ_1 are small.

Now, using the integral representation of $\bar{\tau}_g'$ and reasoning as before, we conclude that $\bar{r}_g'(\overline{P})\bar{\tau}_g'(\overline{P})$ is small, as required.

The vector field $\bar{\tau}$ that we have constructed is not a bending field of the surface ω. In the proof of the Fundamental Lemma, we shall have to modify this field near the boundary of ω to obtain a bending field for the curve bounding the surface. In analytical terms, this means that the field $\bar{\tau}$ must be replaced by a field $\tilde{\tau}$ such that $d(\bar{\tau}+\tilde{\tau})d\bar{r}=0$ at points of the curve bounding ω, where differentiation is performed along the curve.

Let γ be the boundary of ω, $\bar{\gamma}$ its projection on the xy-plane, i.e. the circumference of the disk $\bar{\omega}$. Refer the xy-plane to polar coordinates ρ, ϑ and set $\bar{r}_\vartheta\bar{\tau}_\vartheta = \epsilon(\vartheta)$ on the circle $\bar{\gamma}$.

We shall try to find a vector field $\tilde{\tau}$ along the curve γ in the form

$$\tilde{\tau} = \lambda(\vartheta)t + \mu(\vartheta)n,$$

where t and n are the unit vectors tangent and normal to the circle

$\bar{\gamma}$. A necessary condition for the field $\bar{\tau}+\tilde{\tau}$ to be a bending field on γ is that $\tilde{\tau}$ satisfy the condition $\bar{r}_\vartheta \tilde{\tau}_\vartheta = -\epsilon(\vartheta)$.

Now

$$\tilde{\tau}_\vartheta = (\lambda_\vartheta - \mu)t + (\lambda + \mu_\vartheta)n, \quad \bar{r}_\vartheta = Rt + \cdots,$$

where R denotes the radius of the disk $\bar{\omega}$ (we have omitted the component of the vector $\bar{r}_\vartheta$ along the z-axis). Hence

$$\bar{r}_\vartheta \tilde{\tau}_\vartheta = R(\lambda_\vartheta - \mu).$$

Thus the field $\bar{\tau}+\tilde{\tau}$ can be a bending field on γ only if λ and μ satisfy the equation $\lambda_\vartheta - \mu = -\epsilon(\vartheta)/R$.

Let $\mu(\vartheta)$ be a constant, so chosen that the function λ defined by

$$\lambda = \int_0^\vartheta \left(\mu - \frac{1}{R}\varepsilon(\vartheta)\right) d\vartheta$$

is periodic. Obviously, this constant μ must satisfy the condition

$$\int_0^{2\pi} \left(\mu - \frac{1}{R}\varepsilon(\vartheta)\right) d\vartheta = 0,$$

whence we obtain

$$\mu = \frac{1}{2\pi R}\int_0^{2\pi} \varepsilon(\vartheta)\, d\vartheta.$$

This completes the construction of the field $\tilde{\tau}$ on the circumference $\bar{\gamma}$ of the disk $\bar{\omega}$. We now extend it to the interior of $\bar{\omega}$ by defining

$$\tilde{\tau} = \psi(\rho)(\lambda(\vartheta)t + \mu n),$$

where $\psi(\rho)$ is a twice differentiable function such that $\psi(R) = 1$ and $\psi(\rho) = 0$ for $\rho \leqq R' < R$.

We now consider some properties of the vector field $\tilde{\tau}$. Note that its vertical component is zero.

Recall that F_ω is the domain on the surface F whose projection is the disk $\bar{\omega}$. Suppose that the boundary of F_ω contains no conic points and that almost all points of the boundary are smooth. Since the set of conic points of a convex surface is at most countable while the set of nonsmooth points has measure zero, it follows that this condition holds for almost all disks $\bar{\omega}$, i.e. for almost all R.

Let $\bar{\omega}'$ be a disk whose radius is somewhat larger than that of $\bar{\omega}$, and consider the set M_ϑ of points in $\bar{\omega}'$ which are projections of points of F with supporting planes forming an angle not less than ϑ. (This set was considered at the beginning of this section.) Let M_ϑ^ϵ denote the ϵ-neighborhood of the set M_ϑ. For sufficiently small ϵ, the measure of the intersection of the circle $\bar{\gamma}$ and the set M_ϑ^ϵ is arbitrarily small, since the measure of the intersection of $\bar{\gamma}$ and M_ϑ is zero and the set M_ϑ is closed.

It now follows from the properties of the average field $\bar{\tau}$ that when ω is sufficiently close to F and $\bar{\tau}$ sufficiently close to τ the measure of the set of points on $\bar{\gamma}$ at which $|\bar{r}_\vartheta \bar{\tau}_\vartheta| > \bar{\epsilon} > 0$ can be made arbitrarily small. In other words, the function $\epsilon(\vartheta)$ defining the field $\tilde{\tau}$ converges to zero in measure as $\omega \to F$ and $\bar{\tau} \to \tau$.

This property of the function $\epsilon(\vartheta)$ implies that the constant μ and function $\lambda(\vartheta)$ defining the field τ, viz.,

$$\mu = \frac{1}{2\pi R} \int_0^{2\pi} \varepsilon(\vartheta)\, d\vartheta, \qquad \lambda = \int_0^{\vartheta} \left(\mu - \frac{1}{R}\varepsilon(\vartheta)\right) d\vartheta,$$

tend to zero as $\omega \to F$ and $\bar{\tau} \to \tau$. Hence, in view of our expression for $\tilde{\tau}$,

$$\tilde{\tau} = \psi(\rho)\,(\lambda(\vartheta) t + \mu n),$$

it follows that as $\omega \to F$ and $\bar{\tau} \to \tau$ we have $\tilde{\tau}_\rho \to 0$ uniformly in the disk $\bar{\omega}$, and $\tilde{\tau}_\rho \to 0$ in measure.

One further remark: as $\omega \to F$ and $\bar{\tau} \to \tau$, the derivative $\tilde{\tau}_\vartheta$ is uniformly bounded.

§5. Estimation of certain integrals

The proof of the Fundamental Lemma, as outlined in §2, involves estimation of certain integrals figuring in the fundamental integral relation. The relevant estimates will be derived in this section.

We retain the notation of the preceding sections: F is a convex surface which can be projected in one-to-one fashion onto the xy-plane; F_ω is the compact domain on F whose projection is the disk $\bar{\omega}: x^2 + y^2 < R^2$; ω is a strictly convex analytic surface approximating the surface F. In order to avoid new notation, the domain on ω whose projection is $\bar{\omega}$ will also be denoted by ω.

Since the projection of the surface F on the xy-plane contains the disk $\bar{\omega}$ together with its boundary $\bar{\gamma}$, the angle of inclination of its supporting planes to the xy-plane in the domain F_ω and on its boundary is bounded. It follows that when ω is sufficiently close to F the inclination of the tangent planes of ω to the xy-plane is bounded uniformly with respect to the distance between ω and F. Thus, for ω sufficiently close to F, the derivatives p and q of the function $z(x, y)$ defining the surface

ω are uniformly bounded. Henceforth we shall assume that $|p|$ and $|q|$ are bounded by a constant c_1.

Consider the three integrals

$$\int_{\bar{\omega}}\int |r|\,dx\,dy, \qquad \int_{\bar{\omega}}\int |s|\,dx\,dy, \qquad \int_{\bar{\omega}}\int |t|\,dx\,dy,$$

where r, s and t are the second derivatives of the function $z(x,y)$. Since the function p is strictly monotone in x along any chord AB ($y = \text{const}$) of the disk $\bar{\omega}$, we have

$$\int_A^B |r|\,dx = |p(B) - p(A)| \leqslant 2c_1,$$

and so

$$\int_{\bar{\omega}}\int |r|\,dx\,dy \leqslant 4c_1 R.$$

Similarly,

$$\int_{\bar{\omega}}\int |t|\,dx\,dy \leqslant 4c_1 R.$$

Since the surface ω is convex and so $s^2 \leqq rt$, it follows that $2|s| \leqq |r| + |t|$. Hence

$$\int_{\bar{\omega}}\int |s| \leqslant 4c_1 R.$$

Now consider the integral mean curvature H of the surface ω:

$$H = \frac{1}{2}\int_{\bar{\omega}}\int \left| \frac{(1+q^2)r - 2pqs + (1+p^2)t}{1+p^2+q^2} \right| dx\,dy.$$

This integral is independent of the coordinate system x, y, since it is geometrically invariant. It is easy to estimate the integral H:

$$H \leqslant \frac{1}{2}\int_{\bar{\omega}}\int (|r| + |s| + |t|)\,dx\,dy \leqslant 6c_1 R.$$

Now let $\bar{\omega}^*$ be some domain in the disk $\bar{\omega}$. Let us estimate the integral

$$\int_{\bar{\omega}^*}\int |r|\,dx\,dy$$

in terms of the integral mean curvature $H_{\bar{\omega}^*}$ over the domain ω^* of the surface ω.

We have

$$\frac{(1+q^2)r - 2pqs + (1+p^2)t}{1+p^2+q^2} = \frac{r}{1+p^2+q^2} + \frac{rq^2 - 2spq + tp^2}{1+p^2+q^2} + \frac{t}{1+p^2+q^2}.$$

Since the quadratic form $r\alpha^2 - 2s\alpha\beta + t\beta^2$ is positive definite $(rt - s^2 > 0)$, it follows that r, t and $rq^2 - 2spq + tq^2$ have the same sign. Hence

$$\frac{|r|}{1+p^2+q^2} \leqslant \left| \frac{(1+q^2)r - 2pqs + (1+p^2)t}{1+p^2+q^2} \right|.$$

This implies the desired estimate:

$$\iint_{\bar{\omega}^*} |r|\, dx\, dy \leqslant H_{\bar{\omega}^*},$$

where $H_{\bar{\omega}^*}$ is the mean curvature normalized by the factor $1 + 2c_1^2 > 1 + p^2 + q^2$.

An essential feature of this estimate is that it is valid for any coordinate system x, y.

Let M_α denote the set of all ridge points on the surface at which the plane angle is at least α. Let $\overline{M}_\alpha$ denote the projection of this set on the xy-plane. Cover the disk $\bar{\omega}$ with a net of small squares Δ, and let G_α denote the set of all squares Δ containing points of $\overline{M}_\alpha$.

Let Δ be any square in G_α. It contains at least one point of $\overline{M}_\alpha$, which is the projection of some ridge point of the surface F. Define a new coordinate system by taking the edge at this point as the y-axis, and any perpendicular to it as the x-axis.

In this coordinate system, we shall estimate the three integrals

$$\iint_\Delta |r|\, dx\, dy, \quad \iint_\Delta |s|\, dx\, dy, \quad \iint_\Delta |t|\, dx\, dy.$$

As indicated previously, the first of these integrals does not exceed the integral mean curvature of that part of the surface ω whose projection lies in the square Δ. Hence the sum of all these integrals

$$\sum_\Delta \iint_\Delta |r|\, dx\, dy$$

cannot exceed the integral mean curvature of the entire surface ω, and is therefore bounded as $\omega \to F$. It is essential to note here that the special coordinate system x, y is chosen anew within each square Δ.

We shall use the term *cell* for a square Δ' consisting of a square Δ and the eight squares having a side or a vertex in common with Δ. As before, one shows that

$$\sum_\Delta \iint_{\Delta'} |r|\, dx\, dy$$

does not exceed the integral mean curvature of the surface multiplied by 9, and is therefore bounded as $\omega \to F$.

We now prove the following assertion. *There exists a constant* $\bar{\epsilon}(\alpha) > 0$ *such that, if the sides of the squares* Δ *are sufficiently small and the surface* ω *is sufficiently close to* F, *then in each cell* Δ' *the variation of* p *along any line* $y =$ const *which cuts the square* Δ *is at least* $\bar{\epsilon}(\alpha)$, *and the variation of* q *along any line* $x =$ const *within the square* Δ *may be made arbitrarily small* (*i.e. smaller than any preassigned* $\epsilon > 0$).

We shall first prove this assertion not for the approximating surface ω but for the surface F itself.

Suppose the assertion false. Then there exists a sequence of cells Δ_n' with sides δ_n tending to zero, each of which contains either a line g_n^1 ($y =$ const) cutting Δ_n along which the variation of p is less than $1/n$, or a line g_n^2 ($x =$ const) along which the variation of q in Δ_n is greater than ϵ. Let $F_{\Delta_n'}$ be the domain on F whose projection is Δ_n'.

Subject the surface $F_{\Delta_n'}$ to a similarity mapping with factor of similitude $1/\delta_n$, and call the new surface $F^\delta_{\Delta_n'}$. The sequence of convex surfaces $F^\delta_{\Delta_n'}$ contains a convergent subsequence; without loss of generality, we may assume that the entire sequence $F^\delta_{\Delta_n'}$ converges. We claim that the limit surface Z of this sequence is a cylinder with an edge, and the plane angle at this edge is at least α.

Let A_n denote a ridge point on the surface $F_{\Delta_n'}$, and h_n the corresponding edge. Refer the cell Δ_n' to a coordinate system in which the direction of the x-axis is perpendicular to the projection of the edge h_n on the xy-plane. Let the point and edge on $F^\delta_{\Delta_n'}$ corresponding to A_n and h_n be A_n^δ and h_n^δ, respectively. We may assume that A_n^δ and h_n^δ converge as $n \to \infty$; the limit point $A^0 = \lim A_n^\delta$ is a ridge point of the surface Z, and the angle at the edge $h^0 = \lim h_n^\delta$ is at least α.

Suppose that the surface Z is not a cylinder. Then it must contain a point B^0 which has a supporting plane β^0 forming an angle greater than some $\epsilon' > 0$ with the direction of the edge h^0. It follows that, for sufficiently large n, the surface $F^\delta_{\Delta_n'}$ contains a point B_n at which there is a supporting plane β_n^δ forming an angle greater than ϵ' with the edge h_n^δ. Hence the surface $F_{\Delta_n'}$ contains a point B_n with a supporting plane β_n forming an angle greater than ϵ' with the edge h_n.

Without loss of generality, we may assume that the sequence of points A_n on F converges to some point A. Obviously, A is also the limit of the sequence B_n. Since each point A_n is a ridge point and the plane angle at the edge h_n is at least α, while the supporting plane β_n at B_n cuts the edge h_n at an angle greater than ϵ', it follows that A is a conic point. But this is impossible, since A is in the set M_α. Thus Z

must be a cylinder, and it has an edge h^0 through the point A_0.

Now, if we suppose that each cell Δ_n' contains a line g_n^1 ($y = \text{const}$) cutting the square Δ_n, along which the variation of p is less than $1/n$, then it is readily shown that the cylinder Z contains a straight-line segment cutting the edge h^0, and this is impossible.

On the other hand, if we suppose that each Δ_n' contains a line g_n^2 ($x = \text{const}$) along which the variation of q within Δ_n is greater than ϵ, then there exists a vertical plane through a generator of the cylinder containing two (not necessarily distinct) points at which Z has distinct supporting lines lying in the plane. But this is impossible, since such lines must coincide with the generator along which the plane cuts Z.

We have thus proved that if the sides of the squares Δ are sufficiently small, then in any cell Δ', the variation of p is at least $\bar{\epsilon}(\alpha)$ along any line $y = \text{const}$ cutting Δ, while the variation of q is less than $\epsilon > 0$ along any line $x = \text{const}$ within Δ.

Since the number of cells is finite, we can choose a surface ω so close to F that the variations of p and q in the cells on ω have the same property.

Thus there exists a constant $\bar{\epsilon}(\alpha) > 0$, which depends only on α, such that, if the sides of the squares Δ are sufficiently small and the surface ω is sufficiently near F, then in any cell Δ' the variation of p is at least $\bar{\epsilon}(\alpha)$ along any line $y = \text{const}$ cutting the square Δ, while the variation of q is arbitrarily small (less than some preassigned $\epsilon > 0$) along any line $x = \text{const}$ within each square Δ.

Take the number ϵ so small that $\bar{\epsilon}(\alpha) > N^2\epsilon$, where N is a large integer. Then, since

$$\int\limits_{\Delta'}\int |r|\,dx\,dy > \delta\bar{\varepsilon}(\alpha), \quad \int\limits_{\Delta}\int |t|\,dx\,dy < 2\delta\varepsilon$$

(where δ is the side of Δ), it follows that

$$\frac{2}{N^2}\int\limits_{\Delta'}\int |r|\,dx\,dy > \int\limits_{\Delta}\int |t|\,dx\,dy$$

and so

$$\sum_{\Delta}\int\limits_{\Delta}\int |t|\,dx\,dy \leqslant \frac{18}{N^2} H_\omega,$$

where H_ω is the integral mean curvature of the surface ω.

Since $|s|^2 \leqq |r| + |t|$, we have

$$|s| \leqslant \frac{1}{2}\left(\frac{|r|}{N} + N|t|\right).$$

Hence

$$\iint_{\Delta} |s|\,dx\,dy \leqslant \frac{1}{2N}\iint_{\Delta} |r|\,dx\,dy + \frac{N}{2}\iint_{\Delta} |t|\,dx\,dy,$$

and so

$$\sum_{\Delta}\iint_{\Delta} |s|\,dx\,dy \leqslant \frac{10}{N} H_{\omega}.$$

Thus, if the sides of the squares are sufficiently small and the surface ω sufficiently close to F, then

$$\sum_{\Delta\subset G_{\alpha}}\iint_{\Delta} |r|\,dx\,dy \leqslant H_{\omega},$$

$$\sum_{\Delta\subset G_{\alpha}}\iint_{\Delta} |s|\,dx\,dy \leqslant \frac{10}{N} H_{\omega},$$

$$\sum_{\Delta\subset G_{\alpha}}\iint_{\Delta} |t|\,dx\,dy \leqslant \frac{18}{N^2} H_{\omega}.$$

Now assume that the surface ω approximating F is given by an equation in polar coordinates: $z = z(\rho, \vartheta)$.

Suppose that almost all points of F whose projections lie on the circumference of the disk $\bar{\omega}$ are smooth. Let $\bar{\omega}_{\epsilon}$ denote the annulus defined by $R - \epsilon < \rho < R$, where R is the radius of $\bar{\omega}$. We claim that if ω is sufficiently close to F then the integral

$$\iint_{\bar{\omega}_{\varepsilon}} |z_{\rho\rho}|\rho\,d\rho\,d\vartheta$$

tends to zero with ϵ.

Let $\bar{m}_{\alpha}$ be the set of all points on the circumference $\bar{\gamma}$ of $\bar{\omega}$ which are projections of points of F having at least two supporting planes forming an angle at least α. The set $\bar{m}_{\alpha}$ is closed and its linear measure on γ is zero.

Let $\bar{m}_{\alpha}^{\epsilon'}$ be an ϵ'-neighborhood of the set $\bar{m}_{\alpha}$ on $\bar{\gamma}$. For sufficiently small ϵ', the linear measure of this set on $\bar{\gamma}$ can be made arbitrarily small. For sufficiently small ϵ'', a plane ϵ''-neighborhood $n_{\alpha}^{\epsilon''}$ of the set $\gamma - \bar{m}_{\alpha}^{\epsilon'}$ will contain no points which are projections of points on F having two supporting planes forming an angle at least α.

Hence it follows that, if ω is sufficiently close to F, the variation of z_{ρ} along any line $\vartheta = \text{const}$ within $n_{\alpha}^{\epsilon''}$ will be less than some $\epsilon(\alpha)$ which is arbitrarily small for small α. And since the variation of z_{ρ} along any line $\vartheta = \text{const}$ is clearly bounded, we see that, for sufficiently small ϵ and ω sufficiently close to F,

$$\iint_{\bar{\omega}_\varepsilon} |z_{\rho\rho}|\rho\, d\rho\, d\vartheta$$

is arbitrarily small.

Now consider the integral

$$\int_{\rho=\text{const}} |z_{\vartheta\vartheta}|\, d\vartheta$$

for ρ close to R.

Since ω is a convex surface, its normal curvature along a line $\rho = \text{const}$ preserves sign, and the same is therefore true of the expression

$$\begin{vmatrix} x_{\vartheta\vartheta} & y_{\vartheta\vartheta} & z_{\vartheta\vartheta} \\ x_\rho & y_\rho & z_\rho \\ x_\vartheta & y_\vartheta & z_\vartheta \end{vmatrix} = \rho z_{\vartheta\vartheta} + \rho^2 z_\rho$$

To fix ideas, let $z_{\vartheta\vartheta} + \rho z_\rho > 0$. As $\omega \to F$, the quantity $|\rho z_\rho|$ remains bounded, less than some constant c. Therefore $|z_{\vartheta\vartheta}| < z_{\vartheta\vartheta} + C$. Hence

$$\int |z_{\vartheta\vartheta}|\, d\vartheta \leqslant \int (z_{\vartheta\vartheta} + C)\, d\vartheta = 2\pi C.$$

Thus as $\omega \to F$ the integral $\int |z^{\vartheta\vartheta}|\, d\vartheta$ is bounded by some constant.

It follows that the integral

$$\iint_{\bar{\omega}_\varepsilon} |z_{\vartheta\vartheta}|\rho\, d\rho\, d\vartheta$$

has the same order of magnitude as ϵ as $\omega \to F$.

Finally, consider the integral

$$\iint_{\bar{\omega}_\varepsilon} |z_{\rho\vartheta}|\rho\, d\rho\, d\vartheta.$$

The sign of the Gauss curvature of the surface ω is determined by the expression

$$z_{\rho\rho}(z_{\vartheta\vartheta} + \rho z_\rho) - (z_{\rho\vartheta} - z_\rho)^2.$$

Now, since the Gauss curvature of the surface is positive,

$$(z_{\rho\vartheta} - z_\rho)^2 \leqq z_{\rho\rho}(z_{\vartheta\vartheta} + \rho z_\rho).$$

Hence

$$|z_{\rho\vartheta}| \leqq |z_\rho| + |z_{\rho\rho}| + |z_{\vartheta\vartheta}| + |\rho z_\rho|,$$

and so the integral

$$\iint_{\bar{\omega}_\varepsilon} |z_{\rho\vartheta}| \rho \, d\rho \, d\vartheta$$

can be made arbitrarily small for small ϵ and ω sufficiently close to F.

Retaining the previous notation, let σ denote the quadratic differential form

$$d\bar{r}\,(d\bar{\tau} + d\tilde{\tau}) = \sigma_{11}\, dx^2 + 2\sigma_{12}\, dx\, dy + \sigma_{22}\, dy^2.$$

This form is expressible as the sum of two forms, $\sigma = \sigma' + \sigma''$, where

$$\sigma' = d\bar{r}\, d\bar{\tau} = \sigma'_{11}\, dx^2 + 2\sigma'_{12}\, dx\, dy + \sigma'_{22}\, dy^2,$$

$$\sigma'' = d\bar{r}\, d\tilde{\tau} = \sigma''_{11}\, dx^2 + 2\sigma''_{12}\, dx\, dy + \sigma''_{22}\, dy^2.$$

In the sequel, we shall need three integrals related to the surface ω and the vector fields $\bar{\tau}$, $\tilde{\tau}$:

$$I_1 = \iint_{\bar{\omega}} (\sigma_{11}\sigma_{22} - \sigma_{12}^2)\, dx\, dy,$$

$$I_2 = \iint_{\bar{\omega}} \zeta\,(r\sigma'_{22} - 2s\sigma'_{12} + t\sigma'_{11})\, dx\, dy,$$

$$I_3 = \iint_{\bar{\omega}} \zeta\,(r\sigma''_{22} - 2s\sigma''_{12} + t\sigma''_{22})\, dx\, dy,$$

where $\zeta(x, y)$ is a bounded function.

To be precise, we wish to study the behavior of these integrals as $\omega \to F$ and $\bar{\tau} \to \tau$. We begin with the integral I_1.

Let M_α be the set of all points on F at which there are at least two supporting planes forming an angle at least α. Let $\overline{M}_\alpha$ be the projection of M_α in the disk $\bar{\omega}$. Almost all points of a convex surface are smooth; the set M_α certainly contains no smooth points and therefore has measure zero. Let G_α be a closed set of measure less than ϵ which contains the set $\overline{M}_\alpha$.

As was shown in §4, when ω is sufficiently close to F and $\bar{\tau}$ sufficiently close to τ we have the following inequality in the set $\bar{\omega} - G_\alpha$, differentiating along any direction:

$$|\bar{r}'\bar{\tau}'| < \epsilon(\alpha),$$

where $\epsilon(\alpha)$ is small together with α. Moreover, the product $|\bar{r}'\bar{\tau}'|$ remains bounded throughout the disk $\bar{\omega}$.

It was also shown in §4 that if ω is sufficiently close to F and $\bar{\tau}$ to τ, the measure of the set of all ϑ such that $|\tilde{\tau}_\vartheta| > \epsilon'$ is arbitrarily small.

As for $|\tilde{\tau}_\rho|$, it is arbitrarily small for all ϑ, provided ω is close to F. To this we can add that $\tilde{\tau} \equiv 0$ for $\rho < R'$, and therefore $\tilde{\tau}_\vartheta = \tilde{\tau}_\rho = 0$.

Observe that

$$\tilde{\tau}_x = \tilde{\tau}_\rho \cos\vartheta - \tilde{\tau}_\vartheta \frac{\sin\vartheta}{\rho}, \quad \tilde{\tau}_y = \tilde{\tau}_\rho \sin\vartheta + \tilde{\tau}_\vartheta \frac{\cos\vartheta}{\rho}.$$

This implies that for ω sufficiently close to F and $\bar{\tau}$ sufficiently close to τ the set M'' of all points at which either $|\tilde{\tau}_x|$ or $|\tilde{\tau}_y|$ is greater than ϵ has arbitrarily small measure.

When ω is sufficiently close to F, the maxima of $|p|$ and $|q|$ do not exceed a certain constant C_1. Therefore

$$|\bar{r}_x| \leqq 1 + C_1, \qquad |\bar{r}_y| \leqq 1 + C_1,$$

and so, in the set $\bar{\omega} - M''$,

$$|\sigma''_{11}| = |\bar{r}_x \tilde{\tau}_x| \leqslant (1 + C_1)\varepsilon,$$

$$|\sigma''_{12}| = \frac{1}{2}|\bar{r}_x \tilde{\tau}_y + \bar{r}_y \tilde{\tau}_x| \leqslant (1 + C_1)\varepsilon,$$

$$|\sigma''_{22}| = |\bar{r}_y \tilde{\tau}_y| \leqslant (1 + C_1)\varepsilon.$$

Moreover, in the set M'' the quantity $|\sigma''_{ij}|$ is (at least) uniformly bounded (as $\omega \to F$ and $\bar{\tau} \to \tau$).

Now consider the form σ'. Since $|\bar{r}'\bar{\tau}'| < \epsilon(\alpha)$ in the set $\bar{\omega} - G_\alpha$, with differentiation in anv direction, we have

$$|\sigma'_{11}| = |\bar{r}_x \bar{\tau}_x| < \varepsilon(\alpha), \quad |\sigma'_{22}| = |\bar{r}_y \bar{\tau}_y| < \varepsilon(\alpha)$$

Differentiating in tne direction $dy = dx$, we see that

$$\bar{r}' = \frac{1}{\sqrt{2}}(\bar{r}_x + \bar{r}_y), \quad \bar{\tau}' = \frac{1}{\sqrt{2}}(\bar{\tau}_x + \bar{\tau}_y).$$

Hence

$$\frac{1}{2}|\bar{r}_x\bar{\tau}_x + \bar{r}_x\bar{\tau}_y + \bar{r}_y\bar{\tau}_x + \bar{r}_y\bar{\tau}_y| < \varepsilon\alpha,$$

and consequently

$$|\sigma'_{12}| = \frac{1}{2}|\bar{r}_x\bar{\tau}_y + \bar{r}_y\bar{\tau}_x| \leqslant 2\varepsilon(\alpha).$$

The above arguments can be summarized as follows: For any positive numbers ϵ_1 and ϵ_2, if ω is sufficiently close to F and $\bar{\tau}$ sufficiently close to τ, there exists a set G of measure less than ϵ_2 such that $|\sigma'_{ij}| < \epsilon_1$ in the set $\bar{\omega} - G$, and in the set G itself $|\sigma'_{ij}|$ and $|\sigma''_{ij}|$ are bounded by a constant C.

It follows that when ω is sufficiently close to F and $\bar{\tau}$ sufficiently close to τ, the integral

$$I_1 = \int\limits_{\bar{\omega}}\int (\sigma_{11}\sigma_{22} - \sigma_{12}^2)\, dx\, dy$$

is arbitrarily small.

We now show that the integral

$$I_2 = \int\limits_{\bar{\omega}}\int \zeta(r\sigma'_{22} - 2s\sigma'_{12} + t\sigma'_{11})\, dx\, dy$$

is also arbitrarily small, under the same assumptions.

Let $\bar{M}^*$ denote the set of all points on the xy-plane which are projections of conic points of the surface F. The set $\bar{M}^*$ is at most countable. Consequently it can be covered by a set G^* which is the union of disks, the total sum of whose diameters is at most ϵ. We claim that the integral

$$I^* = \int\limits_{G^*}\int \zeta(r\sigma'_{22} - 2s\sigma'_{12} + t\sigma'_{11})\, dx\, dy$$

is arbitrarily small for small ϵ. Now

$$|I^*| \leqslant \int\limits_{G^*}\int |\zeta|(|r||\sigma'_{22}| + 2|s||\sigma'_{12}| + |t||\sigma_{11}|)\, dx\, dy.$$

Since $2|s| \leqq |r| + |t|$, we get

$$|I^*| \leqslant \int\limits_{G^*}\int |\zeta|\{|r|(|\sigma'_{22}| + |\sigma'_{12}|) + |t|(|\sigma'_{12}| + |\sigma'_{11}|)\}\, dx\, dy.$$

Consider the integrals

$$\int\limits_{G^*}\int |r|\, dx\, dy, \quad \int\limits_{G^*}\int |t|\, dx\, dy.$$

As was shown previously, $\int |r|\, dx$ and $\int |t|\, dy$ can be estimated in terms of the maxima of $|p|$ and $|q|$. Moreover, for ω sufficiently close to F we may assume that these integrals are bounded by some constant c. Since the measure of the projections of the set G^* on the x- and y-axes is at most ϵ, it follows that

$$\int\limits_{G^*}\int |r|\, dx\, dy \leqslant c\varepsilon, \quad \int\limits_{G^*}\int |t|\, dx\, dy \leqslant c\varepsilon.$$

As $\omega \to F$ and $\bar{\tau} \to \tau$, the quantities $|\zeta|\ |\sigma'_{ij}|$ are bounded by a constant c'; therefore

$$|I^*| \leqslant 2c' \int\limits_{G^*}\int (|r| + |t|)\, dx\, dy \leqslant 4cc'\varepsilon.$$

Consequently the integral I^* is small together with ϵ.

Let $\overline{M}_\alpha$ denote the set of all points of $\bar{\omega} - G^*$ which are projections of points on F at which there are at least two supporting planes forming angles less than α. $\overline{M}_\alpha$ is a closed set, consisting of projections of ridge points of the surface F. Cover the disk $\bar{\omega}$ by small squares Δ, and let G_α denote the union of all squares Δ containing points of $\overline{M}_\alpha$.

We saw in §4 that, if α is small, ω close to F and $\bar{\tau}$ close to τ, then the product $|\bar{r}' \bar{\tau}'|$ is arbitrarily small in $\bar{\omega} - G_\alpha - G^*$, irrespective of the direction of differentiation. This means that $|\sigma'_{ij}|$ is arbitrarily small in $\bar{\omega} - G_\alpha - G^*$.

Now, since the integrals

$$\iint\limits_{\bar{\omega}} |r|\, dx\, dy, \qquad \iint\limits_{\bar{\omega}} |s|\, dx\, dy, \qquad \iint\limits_{\bar{\omega}} |t|\, dx\, dy$$

remain bounded by a constant as $\omega \to F$, it follows that for sufficiently small α, ω close to F and $\bar{\tau}$ close to τ, the integral

$$\iint\limits_{\omega - G_\alpha - G^*} \zeta (r\sigma'_{22} - 2s\sigma'_{12} + t\sigma'_{11})\, dx\, dy$$

is arbitrarily small.

It remains to study the behavior of the integral

$$\iint\limits_{G_\alpha} \zeta (r\sigma'_{22} - 2s\sigma'_{12} + t\sigma'_{11})\, dx\, dy$$

as $\omega \to F$ and $\bar{\tau} \to \tau$.

This integral can be expressed as a sum of integrals

$$\sum_{\Delta \subset G_\alpha} \iint\limits_{\Delta} \zeta (r\sigma'_{22} - 2s\sigma'_{12} + t\sigma'_{11})\, dx\, dy,$$

and we shall calculate each integral in a special coordinate system. The coordinate system in a square Δ is defined as follows.

The domain F_Δ on F whose projection is the square Δ contains a ridge point A with plane angle at least α. We take the direction of the x-axis perpendicular to the projection of the edge at A on the xy-plane.

If the sides of Δ are sufficiently small and the surface ω is sufficiently close to F, then in the interior of each square Δ the quantity $|\sigma'_{22}| = |\bar{r}_y \bar{\tau}_y|$ will be smaller than ϵ (§4) and

$$\sum_{\Delta \subset G_\alpha} \iint\limits_{\Delta} |r|\, dx\, dy \leqslant H_\omega, \qquad \sum_{\Delta \subset G_\alpha} \iint\limits_{\Delta} |s|\, dx\, dy \leqslant \frac{10}{N} H_\omega,$$

$$\sum_{\Delta \subset G_\alpha} \iint\limits_{\Delta} |t|\, dx\, dy \leqslant \frac{18}{N^2} H_\omega,$$

where H_ω is the mean curvature of the surface ω and N may be taken arbitrarily large.

Since $|\sigma_{12}|$ and $|\sigma_{11}|$ are bounded, it follows that if the sides of the squares Δ are small and ω is close to F, then the integral

$$\int\int_{G_\alpha} \zeta(r\sigma'_{22} - 2s\sigma'_{12} + t\sigma'_{11})\, dx\, dy$$

is arbitrarily small.

Summarizing the foregoing arguments, we conclude that if ω is sufficiently close to F and $\bar{\tau}$ sufficiently close to τ, the integral

$$I_2 = \int\int_{\bar{\omega}} \zeta(r\sigma'_{22} - 2s\sigma_{12} + t\sigma'_{11})\, dx\, dy$$

is arbitrarily small.

Finally, we estimate the integral

$$I_3 = \int\int_{\bar{\omega}} \zeta(r\sigma''_{22} - 2s\sigma''_{12} + t\sigma''_{11})\, dx\, dy.$$

This will be done in polar coordinates ρ, ϑ instead of cartesian coordinates x, y.

Since the integrand here is the mixed discriminant of the quadratic differential forms d^2z and $\sigma'' = d\bar{r}d\tilde{\tau}$, the form of the integral I_3 in polar coordinates is

$$I_3 = \int\int_{\bar{\omega}} \zeta(z_{\rho\rho}\bar{\sigma}_{22} - 2z_{\rho\vartheta}\bar{\sigma}_{12} + z_{\vartheta\vartheta}\bar{\sigma}_{11}) \frac{d\rho\, d\vartheta}{\rho},$$

where

$$\bar{\sigma}_{22} = \bar{r}_\vartheta \tilde{\tau}_\vartheta, \quad \bar{\sigma}_{12} = \bar{r}_\vartheta \tilde{\tau}_\rho + \bar{r}_\rho \tilde{\tau}_\vartheta, \quad \bar{\sigma}_{11} = \bar{r}_\rho \tilde{\tau}_\rho.$$

We wish to prove that if the surface ω is sufficiently close to F and the field $\bar{\tau}$ sufficiently close to τ, then I_3 is arbitrarily small.

Recall the expression derived in §4 for $\tilde{\tau}$:

$$\tilde{\tau} = \psi(\rho)(\lambda(\vartheta)t + \mu n),$$

where $\psi(\rho)$ is a twice differentiable function defined by $\psi(R) = 1$ and $\psi(\rho) = 0$ for $\rho \leqq R' < R$. We now set $R' = R - \epsilon$, and impose the conditions $\psi(\rho) \leqq 1$ and $|\psi'(\rho)| \leqq 2/\epsilon$. This is clearly legitimate.

Recall that the functions μ and λ occurring in the formula for $\tilde{\tau}$ tend uniformly to zero as $\omega \to F$, while $\lambda'(\vartheta)$ tends to zero in measure.

Consider the integral

$$I_3' = \iint\limits_{\bar{\omega}_\varepsilon} |\zeta||z_{\rho\rho}||\bar{r}_\vartheta||\tilde{\tau}_\vartheta| \frac{d\rho\, d\vartheta}{\rho}.$$

Now the integral

$$\iint\limits_{\bar{\omega}_\varepsilon} |z_{\rho\rho}|\rho\, d\rho\, d\vartheta$$

is small for sufficiently small ϵ and ω close to F; moreover, $|\zeta|\,|\bar{r}_\vartheta|$ remains bounded as $\omega \to F$ and $\bar{\tau} \to \tau$. Hence the integral I_3' is also small under these conditions.

Consider the integral

$$I_3''' = \iint\limits_{\bar{\omega}_\varepsilon} |\zeta||z_{\vartheta\vartheta}||\bar{r}_\rho||\tilde{\tau}_\rho| \frac{d\rho\, d\vartheta}{\rho}.$$

We have

$$I_3''' \leqslant \frac{2}{\varepsilon} \iint\limits_{\bar{\omega}_\varepsilon} |\zeta||z_{\rho\rho}|\,|\bar{r}_\rho||\lambda t + \mu n| \frac{d\rho\, d\vartheta}{\rho}.$$

Since $|\zeta|\,|\bar{r}_\rho|\,|\lambda t + \mu n| \to 0$ as $\omega \to F$, while the order of magnitude of the integral

$$\iint\limits_{\bar{\omega}} |z_{\vartheta\vartheta}|\rho\, d\rho\, d\theta$$

is ϵ, it follows that I_3''' is arbitrarily small when ω is close to F and $\bar{\tau}$ close to τ.

Finally, consider the integral

$$I_3'' = \iint\limits_{\bar{\omega}_\varepsilon} |\zeta||z_{\rho\vartheta}|(|\bar{r}_\vartheta\tilde{\tau}_\rho| + |\bar{r}_\rho\tilde{\tau}_\vartheta|) \frac{d\rho\, d\vartheta}{\rho}.$$

Split this integral into two integrals:

$$\iint\limits_{\bar{\omega}_\varepsilon} |\zeta||z_{\rho\vartheta}||\bar{r}_\vartheta\tilde{\tau}_\rho| \frac{d\rho\, d\vartheta}{\rho}, \quad \iint\limits_{\bar{\omega}_\varepsilon} |\zeta||z_{\rho\vartheta}||\bar{r}_\rho\tilde{\tau}_\vartheta| \frac{d\rho\, d\vartheta}{\rho}.$$

When ω is sufficiently close to F and $\bar{\tau}$ sufficiently close to τ, the first of these integrals is arbitrarily small, since as $\omega \to F$ and $\bar{\tau} \to \tau$ the quantity $|\tilde{\tau}_\rho|$ converges uniformly to zero in $\bar{\omega}_\epsilon$, while $|\bar{r}_\vartheta|$ and the integral

$$\iint\limits_{\bar{\omega}_\varepsilon} |z_{\rho\vartheta}|\rho\, d\rho\, d\vartheta$$

remain bounded.

Now, for the second integral we have

$$(z_{\rho\vartheta} - z_\rho)^2 \leqq z_{\rho\rho}(z_{\vartheta\vartheta} + \rho z_\rho).$$

Hence

$$|z_{\rho\vartheta}| \leqslant |z_\rho| + N|z_{\rho\rho}| + \frac{1}{N}(|z_{\vartheta\vartheta}| + |\rho z_\rho|).$$

Substitute the right-hand side of this inequality for $|z_{\rho\vartheta}|$ in the second integral—this does not decrease the integral. Split the resulting integral into three terms:

$$\iint_{\bar{\omega}_\varepsilon} |\zeta||z_\rho|\left(1 + \frac{\rho}{N}\right)|\bar{r}_\rho||\tilde{\tau}_\vartheta|\frac{d\rho\, d\vartheta}{\rho},$$

$$N\iint_{\bar{\omega}_\varepsilon} |\zeta||z_{\rho\rho}||\bar{r}_\rho||\tilde{\tau}_\vartheta|\frac{d\rho\, d\vartheta}{\rho}, \quad \frac{1}{N}\iint_{\bar{\omega}_\varepsilon} |\zeta||z_{\vartheta\vartheta}||\bar{r}_\rho||\tilde{\tau}_\vartheta|\frac{d\rho\, d\vartheta}{\rho}.$$

Let N be a large fixed number. When ω is sufficiently close to F and $\bar{\tau}$ sufficiently close to τ, the first of these integrals is small, since $|\tilde{\tau}_\vartheta|$ converges to zero in measure. The second is small for small ϵ, since

$$\iint_{\bar{\omega}_\varepsilon} |z_{\rho\rho}|\rho\, d\rho\, d\vartheta$$

is small together with ϵ. Finally, the third integral is small because the integral

$$\iint_{\bar{\omega}_\varepsilon} |z_{\vartheta\vartheta}|\rho\, d\rho\, d\vartheta$$

is bounded, and N can be chosen sufficiently large.

Thus we see that for ω sufficiently close to F and $\bar{\tau}$ sufficiently close to τ the integral I_3'' is also arbitrarily small.

Since $|I_3| \leqq I_3' + I_3'' + I_3'''$, and each of the integrals I_3', I_3'' and I_3''' is small for ω sufficiently close to F and $\bar{\tau}$ sufficiently close to τ, the integral I_3 is also small under these conditions.

§6. Proof of the Fundamental Lemma

We have now prepared all material necessary for the proof of the Fundamental Lemma, which is presented in this section. We restate the lemma:

FUNDAMENTAL LEMMA 1. *Let $F: z = z(x, y)$ be a convex surface which contains no flat pieces, and let $\zeta(x, y)$ be the z-component of a bending field of F. Then the surface $\Phi: z = \zeta(x, y)$ is a surface of nonpositive curvature.*

This means that the surface Φ contains no points of strict convexity, i.e. points P through which there is a plane such that all points in the neighborhood of P lie on one side of the plane.

FUNDAMENTAL LEMMA 2. *Let $F: z = z(x, y)$ be a convex surface containing flat pieces G_α, and let $\zeta(x, y)$ be the z-component of a bending field of F. Then the surface $\Phi: z = \zeta(x, y)$ has nonpositive curvature on the set H of points whose projections onto F by straight lines parallel to the z-axis lie outside the domains G_α.*

That is to say, H contains no points P through which there is a plane such that all points of H near P lie on one side of the plane.

Let F be a general convex surface which can be projected in one-to-one fashion onto the xy-plane, P an arbitrary point on F and $\bar{P}$ its projection on the xy-plane. Without loss of generality, we may assume that $\bar{P}$ is the origin of the coordinate system. Consider the disk $\bar{\omega}: x^2 + y^2 < R^2$, where R is so small that the projection of the surface F on the xy-plane contains $\bar{\omega}$ together with its circumference. In addition, choose R so that almost all points on F whose projections on the xy-plane lie on the circumference $\bar{\gamma}$ of $\bar{\omega}$ are smooth.

Approximate the surface F by a strictly convex analytic surface; let ω be the domain on this surface above $\bar{\omega}$. Approximate the bending field τ of F by a regular vector field (the Steklov average of τ) (§4). Denote the approximating field by $\bar{\tau}$.

Construct a vector field $\tilde{\tau}$ with zero vertical component, such that the vector field $\bar{\tau} + \tilde{\tau}$ is a bending field for the boundary γ of the surface ω. This construction is not unique; we assume that the field has been constructed as in §4.

Construct a regular bending field τ^* of the surface ω, whose vertical component on the boundary γ is the same as that of $\bar{\tau}$. The existence of such a field was proved in §3.

Finally, let τ^0 be a trivial bending field (as yet arbitrary) of the surface ω, with zero vertical component.

Recall the integral relation of §2 for the surface ω and the vector field $v = \bar{\tau} + \tilde{\tau} + \tau^0 - \tau^*$:

$$\frac{1}{2} \oint_{\bar{\gamma}} (\lambda\, d\mu - \mu\, d\lambda) = \iint_{\bar{\omega}} \zeta^2 (rt - s^2)\, dx\, dy + \frac{1}{4} \iint_{\bar{\omega}} (\lambda_y - \mu_x)^2\, dx\, dy$$

$$+ \iint_{\bar{\omega}} \zeta \Delta(\sigma, d^2 z)\, dx\, dy + \iint_{\bar{\omega}} \Delta(\sigma, \sigma)\, dx\, dy.$$

Here $\lambda = \xi + p\zeta$ and $\mu = \eta + q\zeta$, where ξ, η and ζ are the components of the field v along the x-, y- and z-axes; p and q are the first derivatives of the function $z(x, y)$ defining the surface ω; r, s and t are the second derivatives of z; $\Delta(\sigma, \sigma)$ is the discriminant of the quadratic differential form

$$\sigma = d\bar{r}dv = d\bar{r}(d\bar{\tau} + d\tilde{\tau}) \quad (d\bar{r}d\tau^0 = 0, d\bar{r}d\tau^* = 0),$$

and $\Delta(\sigma, d^2z)$ is the mixed discriminant of the forms σ and d^2z.

We claim that the trivial bending field τ^0 can be so chosen that the contour integral

$$\frac{1}{2}\oint_{\bar{\gamma}} (\lambda\, d\mu - \mu\, d\lambda)$$

in the left-hand side of this formula is nonpositive.

To prove this, consider a vector field u along the circumference $\bar{\gamma}$ of the disk $\bar{\omega}$, with x- and y-components ξ and η, respectively. u is a bending field on the circumference. Indeed, along the boundary γ of ω

$$d\bar{r}du = dxd\xi + dyd\eta + dzd\zeta.$$

But $d\zeta = 0$ along the curve γ. Hence $dxd\xi + dyd\eta = 0$, and this means that $u(\xi, \eta)$ is a bending field for the circle $\bar{\gamma}$.

Let us denote the z-component of the vector product $a \times b$ by $[a, b]$. The contour integral under investigation is

$$\frac{1}{2}\oint_{\bar{\gamma}} (\lambda\, d\mu - \mu\, d\lambda) = \frac{1}{2}\oint_{\bar{\gamma}} (\xi\, d\eta - \eta\, d\xi) = \frac{1}{2}\oint_{\bar{\gamma}} [u,\, du].$$

The vector field u is defined up to the velocity field τ^0. Our problem is to show that suitable choice of τ^0 will ensure that the integral

$$\frac{1}{2}\oint_{\bar{\gamma}} [u,\, du]$$

is never positive.

We retain the notation u for some fixed field, and express the general field u as $u + \Omega$, where Ω is the velocity field of a rotation about the z-axis. We claim that for suitable Ω

$$I = \frac{1}{2}\oint_{\bar{\gamma}} [u + \Omega,\, du + d\Omega] \leqslant 0.$$

Let r_1 denote the radius vector of an arbitrary point on $\bar{\gamma}$, and r_2 the radius vector obtained from r_1 by rotation through an angle $\pi/2$

toward the right. Since $du dr_1 = 0$, the vectors du and r_2 are parallel. Assume that the parameter along the circle $\bar{\gamma}$ is arc length.

Let $|u'| < M$. Set $\bar{u} = u + Mr_2$, $\Omega = (M + c) r_2$. Then

$$I = \frac{1}{2} \oint_{\bar{\gamma}} [\bar{u} + cr_2, \; d\bar{u} + c \, dr_2].$$

The vector equation $r = \bar{u}(s)$ defines a smooth closed curve C without singularities $(u' \neq 0)$. Its tangent vector $\bar{u}'(s)$, being parallel to the vector $r_2(s)$, revolves in the same sense while the representative point is moving along the curve. Hence C is a closed convex curve.

If the length of C is denoted by L and the area of the domain that it bounds by S, then

$$\frac{1}{2} \oint_{\bar{\gamma}} [\bar{u}, \; d\bar{u}] = S, \qquad \frac{1}{2} \oint_{\bar{\gamma}} [r_2, \; dr_2] = \pi R^2;$$

$$\oint_{\bar{\gamma}} [\bar{u}, \; dr_2] = \oint_{\bar{\gamma}} [d\bar{u}, \; r_2] = RL.$$

Hence

$$I = S + cRL + \pi c^2 R^2.$$

Since L and S satisfy the isoperimetric inequality $4\pi S \leqq L^2$, there exists at least one value of c for which the quadratic trinomial $I(c)$ is not positive.

We have thus proved that the trivial vector field τ^0 may always be chosen so that the contour integral

$$\frac{1}{2} \oint_{\bar{\gamma}} (\lambda \, d\mu - \mu \, d\lambda)$$

is not positive. Henceforth we shall assume that this is indeed the case.

Now let the surface ω approach arbitrarily close to F, and the field $\bar{\tau}$ to τ. Then, as we proved in §5,

$$\iint_{\bar{\omega}} \zeta \Delta (\sigma, \; d^2 z) \, dx \, dy \to 0, \qquad \iint_{\bar{\omega}} \Delta (\sigma, \; \sigma) \, dx \, dy \to 0.$$

Consider the first two integrals on the right-hand side of the formula. We have

$$\iint_{\bar{\omega}} \zeta^2 (rt - s^2) \, dx \, dy = \iint_{\Omega} \zeta^2 \, d\Omega,$$

where the integration in the right-hand side is performed with respect to the area of the conditional spherical image of ω, which is obtained by mapping each point of the surface onto the point (p,q) of the plane with coordinates $p=\partial z/\partial x$, $q=\partial z/\partial y$.

The second integral is

$$\frac{1}{4}\iint_{\bar{\omega}}(\lambda_y-\mu_x)^2\,dx\,dy=\frac{1}{4}\iint_{\bar{\omega}}(\xi_y+p\zeta_y-\eta_x-q\zeta_x)^2\,dx\,dy.$$

Without loss of generality, we may assume that the field $\tau^*-\tau^0$ converges as $\omega\to F$ and $\bar{\tau}\to\tau$. The field $\tilde{\tau}$, for its part, converges to zero by construction. The limit $\tau'=\lim(\tau^*-\tau^0)$ is a bending field of the surface F in the domain $F_{\bar{\omega}}$ whose projection is the disk $\bar{\omega}$.

Passing to the limit, we get the following integral inequality:

$$\iint_{\Omega}\zeta^2\,d\Omega+\frac{1}{4}\iint_{\bar{\omega}}(\xi_y+p\zeta_y-\eta_x-q\zeta_x)^2\,dx\,dy\leqslant 0,$$

where ξ, η and ζ are the components of the bending field $\tau\dot{-}\tau'$, p and q are the first derivatives of the function $z(x,y)$ defining the surface F, and the domain of integration in the first term is the conditional spherical image of the surface $F_{\bar{\omega}}$.

Proof of Fundamental Lemma 1. Suppose that the surface F contains no flat pieces. We claim that in this case the component $\zeta(x,y)$ of the bending field $\tau-\tau'$ of F is identically zero.

Indeed, it follows from the above fundamental integral inequality that

$$\iint_{\Omega}\zeta^2\,d\Omega=0,\qquad \iint_{\bar{\omega}}(\xi_y+p\zeta_y-\eta_x-q\zeta_x)^2\,dx\,dy=0.$$

Let P be a point of strict convexity of F, i.e. a point at which there is a supporting plane whose only point in common with F is P. The point P has an arbitrarily small neighborhood $G(P)$ whose conditional spherical image has nonzero area. Since

$$\iint_{G(P)}\zeta^2\,d\Omega=0,$$

it follows from the continuity of ζ that $\zeta(P)=0$. Thus the function ζ vanishes at all points of strict convexity of F.

Again by the continuity of ζ, it follows that ζ vanishes at all limits of points of strict convexity of F. Thus, if ζ is at all different from zero, this can only be so on an open set G containing no points of strict convexity. This set G consists of developable convex surfaces containing no flat pieces. We wish to prove that ζ vanishes in G as well.

We have

$$\iint_{\bar{\omega}} (\xi_y + p\zeta_y - \eta_x - q\zeta_x)^2\, dx\, dy = 0.$$

Hence, almost everywhere in $\bar{\omega}$,

$$\xi_y + p\zeta_y - \eta_x - q\zeta_x = 0.$$

Moreover, the equations of the bending field hold almost everywhere:

$$\xi_x + p\zeta_x = 0, \quad \xi_y + p\zeta_y + \eta_x + q\zeta_x = 0, \quad \eta_y + q\zeta_y = 0.$$

Hence it follows that, almost everywhere,

$$(*) \qquad \begin{array}{ll} \xi_x + p\zeta_x = 0, & \eta_x + d\zeta_x = 0, \\ \xi_y + p\zeta_y = 0, & \eta_y + q\zeta_y = 0. \end{array}$$

Consider the integral $\oint_C \zeta d\xi$ along a closed contour C in the interior of the disk $\bar{\omega}$.[1] Applying the Ostrogradskiĭ-Green formula and noting that $\xi_x\zeta_y - \xi_y\zeta_x = 0$ almost everywhere by virtue of (*), we see that $\oint_C \zeta d\xi = 0$.

By formulas (*), $d\xi = p d\zeta$. Therefore

$$\oint_C \zeta\, d\xi = \oint_C p\zeta\, d\zeta = \oint_C p\, d(\zeta^2).$$

It is of course assumed that $d\xi$, $d\zeta$ and dz exist almost everywhere on C.

We now choose the contour C in a special way. Let g be an arbitrary generator of the developable surface G. Without loss of generality, we may assume that it is parallel to the y-axis. Let g_1 and g_2 be two generators close to g, on different sides of g, such that the surface G is smooth along both generators. There is no difficulty in finding such generators, since almost all generators have this property. The contour C will consist of segments of the generators g_1 and g_2 and intersections of the surface with planes $y = \text{const}$. Let c_1 and c_2 denote the sides of the quadrilateral C on g_1 and g_2, c_3 and c_4 the other two sides.

Integrating by parts, we conclude from $\oint_C pd(\zeta^2) = 0$ that $\oint_C \zeta^2 dp = 0$.

Since $dp = 0$ along the sides c_1 and c_2 of the quadrangle C (because the tangent plane along the generators g_1 and g_2 is stationary), it follows that

$$\int_{c_3} \zeta^2\, dp = \int_{c_4} \zeta^2\, dp.$$

---

[1] This integral was introduced by Minagawa in [47].

Again using the fact that the tangent planes to G along g_1 and g_2 are stationary, we see that $\int_{c_3} dp = \int_{c_4} dp$. Hence by the continuity of ζ we have $\zeta^2(A_3) = \zeta^2(A_4)$, where A_3 and A_4 are certain points on the curves c_3 and c_4.

Letting $g_1 \to g$ and $g_2 \to g$, we get $\zeta^2(A) = \zeta^2(B)$, where A and B are points on the generator g

Thus $\zeta = \text{const}$ on the generator g. Now the endpoint of the generator g is either a boundary point of $F_{\bar{\omega}}$, a point of strict convexity, or the limit of points of strict convexity. In all these cases we have $\zeta = 0$ at the endpoint of g. Thus $\zeta = 0$ along the entire generator g.

Now the generator g was chosen arbitrarily, and hence $\zeta = 0$ everywhere in $\bar{\omega}$.

Let $\zeta_1(x,y)$ be the vertical component of the field τ, and $\zeta_2(x,y)$ the vertical component of the field τ'. Since the surface $z = \zeta_2(x,y)$ is a limit of surfaces of nonpositive curvature, it has nonpositive curvature. But $\zeta_1(x,y) \equiv \zeta_2(x,y)$, and so the surface $z = \zeta_1(x,y)$, where $\zeta_1(x,y)$ is the vertical component of the given field τ, is a surface of nonpositive curvature.

This completes the proof of Fundamental Lemma 1.

Proof of Fundamental Lemma 2. Let F be a general surface containing flat pieces, τ a bending field of F and $\zeta(x,y)$ its vertical component. Let Φ denote the surface defined by the equation $z = \zeta(x,y)$, and H the set of all points on Φ whose projections by straight lines parallel to the z-axis lie outside the flat pieces of F. Fundamental Lemma 2 states that the set H has nonpositive curvature, i.e. if P is a point of H, there is no plane through P such that all points of H near P lie strictly on one side of the plane.

Suppose the assertion false, and let α_P be a plane through some point P such that all points of H near P lie on one side of α_P, say below α_P. Let $\bar{P}$ denote the projection of P on the xy-plane and $\bar{\omega}$ a disk of sufficiently small radius about $\bar{P}$, such that all points of H whose projections lie in $\bar{\omega}$ are below α_P, except P itself which lies in α_P.

Let Φ_ω be the part of the surface Φ whose projection is the disk $\bar{\omega}$. We shall now define a special transformation of Φ and the bending field τ.

Let β be a plane not parallel to the z-axis, which lies above the surface Φ_ω. Move this plane downward parallel to itself until it touches the set H. It may happen that in this position the plane β cuts the surface Φ_ω. Let $\bar{\Phi}_\omega$ denote the set of all points of Φ_ω lying above the plane β.

Straight lines parallel to the z-axis project the set $\bar{\Phi}_\omega$ into flat pieces of the surface F, since there are no points of H above the plane β.

Each component $\overline{\Phi}'_\omega$ of the set $\overline{\Phi}_\omega$ is projected into a single flat piece of F, since the regions of Φ corresponding to the flat regions of F are separated by the set H, and the latter lies below the plane β.

Let F'_ω be the flat region of F_ω corresponding under projection to the component $\overline{\Phi}'_\omega$; modify the field τ in this region as follows. Let n be the unit normal to F'_ω, chosen so that $ne_z > 0$. Then, in the set F'_ω, we replace the field τ by

$$\tau - n(\zeta - \overline{\zeta})/\cos\alpha,$$

where α is the angle between the plane of F'_ω and the xy-plane, and $\overline{\zeta}$ is the z-coordinate of the point of β which is the projection of the point of the surface at which the field is being defined. The modified field τ will obviously satisfy a Lipschitz condition; it is a bending field since the "correction" $n(\zeta - \overline{\zeta})/\cos\alpha$ is perpendicular to the plane of F'_ω. Modify the field τ in this way on all flat pieces of the surface F_ω corresponding to $\overline{\Phi}_\omega$. On the surface Φ_ω this transformation amounts to replacing that part of the surface lying above the plane β by the corresponding parts of the plane β.

After finitely many operations of this kind, the boundary of the surface Φ_ω will lie strictly below the plane α_P, and all parts of Φ_ω protruding above this plane are truncated by planes β. Let τ_1 denote the bending field of the surface F_ω obtained by these transformations.

Now let $\bar{\omega}'$ be a disk contained in $\bar{\omega}$, such that the projection of F_ω on the xy-plane covers $\bar{\omega}'$ together with its circumference. Now repeat all the reasoning in the proof of Fundamental Lemma 1, with $F_{\omega'}$ playing the role of the surface F_ω.

We conclude that there exists a bending field τ' of the surface $F_{\omega'}$ satisfying the following conditions:

1) The vertical component of τ' coincides with that of the field τ_1 everywhere outside the flat pieces of the surface F_ω and on the boundary of this surface.

2) If $\zeta'(x,y)$ is the vertical component of the field τ', the surface $\Phi_{\omega'}: z = \zeta'(x,y)$ is a surface of nonpositive curvature.

Since the fields τ and τ_1 on F_ω have the same vertical components outside the flat pieces of the surface, the set H lies on the surface $\Phi'_{\omega'}$. The boundary of $\Phi'_{\omega'}$ lies below the plane α_P. Thus the boundary of the surface $\Phi'_{\omega'}$ may be separated from the point P by a plane parallel to α_P. This is impossible, since $\Phi'_{\omega'}$ is a surface of nonpositive curvature. This contradiction completes the proof of Fundamental Lemma 2.

§7. Rigidity of convex surfaces

A bending field on a convex surface is said to be *trivial* if it is the velocity field of a rigid motion of the surface. The most general form of a trivial bending field is $\tau = a \times r + b$, where a and b are constant vectors and r is the radius vector of the surface.

A convex surface is said to be *rigid* if it admits no nontrivial bending fields.

Let F be a convex surface with boundary γ, which can be projected in one-to-one fashion onto a plane α. We shall say that the surface F is anchored to its boundary relative to the plane α if the distances of the points of γ from the plane α are stationary under all admissible deformations of F.

THEOREM 1. *Let F be a convex surface with boundary, which contains no flat pieces and can be projected in one-to-one fashion onto a plane α. If F is anchored to its boundary relative to this plane, then F is rigid. If F contains flat pieces, it is rigid outside these pieces.*

PROOF. Without loss of generality we may assume that α is the xy-plane. Let $\zeta(x,y)$ be the vertical component of a bending field τ of F. If F contains no flat pieces, the surface $\Phi: z = \zeta(x,y)$ is a surface of negative curvature. And since $\zeta(x,y)$ vanishes on the boundary of F, it follows that $\zeta(x,y) = 0$ everywhere on F.

Let F contain flat pieces. Then, by Fundamental Lemma 2, $\zeta(x,y) = 0$ outside the flat pieces of F. Using the transformation described in the proof of Fundamental Lemma 2 (§6), we can modify the field τ so that ζ vanishes everywhere on F. Denote the new bending fleld by τ'.

Thus we have $\zeta \equiv 0$ for the field τ itself if F contains no flat pieces, and for the field τ' if it does.

Now consider the differential equations of the bending field. Since $\zeta \equiv 0$, we have

$$\xi_x = 0, \quad \xi_y + \eta_x = 0, \quad \eta_y = 0.$$

The first and third equations imply that ξ depends only on y, and η only on x. Hence the second equation gives $\xi_y = -\eta_x = \text{const}$. Consequently,

$$d\xi = c\,dy, \quad d\eta = -c\,dx,$$

whence

$$\xi = cy + c_1, \quad \eta = -cx + c_2.$$

The bending field $(cy+c_1, -cx+c_2, 0)$ is trivial; it consists of a rotation about an axis and a translation parallel to the xy-plane.

Thus if the surface contains no flat pieces, then the field τ is trivial and hence the surface is rigid. If the surface contains flat pieces, we have proved that the field τ' is trivial. Now since τ' coincides with τ outside the flat pieces, this means that the surface is rigid there. Q.E.D.

Note that *if each of the flat pieces of F is rigid separately, then the surface F is rigid*. This follows directly from Theorem 1.

A convex surface with boundary is said to be a convex cap if its boundary lies in a plane and the surface can be projected one-to-one onto this plane.

A corollary of Theorem 1 is that *a convex cap is rigid with respect to the class of deformations under which its boundary remains plane.*

A surface is said to be in starlike position relative to a center O if any ray through O cuts the surface in at most one point. A surface F with boundary γ is said to be anchored to its boundary relative to a center O if the distances of the points of γ from O are stationary under all admissible deformations of F.

THEOREM 2. *Let F be a convex surface with boundary, which contains no flat pieces, is in starlike position relative to a point O separated from the surface by a plane α, and is anchored to its boundary relative to O. Then F is rigid. If F contains flat pieces, it is rigid outside these pieces.*

PROOF. For regular surfaces and regular bending fields, we have the following theorem of Darboux and Sauer.

Let F be a surface and $\tau(\xi, \eta, \zeta)$ a bending field of F. Then, if F' is the image of F under the projective mapping

$$x' = x/z, \quad y' = y/z, \quad z' = 1/z, \tag{*}$$

then the vector field $\tau'(\xi', \eta', \zeta')$ defined by

$$\xi' = \xi/z, \quad \eta' = \eta/z, \quad \zeta' = -(x\xi + y\eta + z\zeta)/z, \tag{**}$$

is a bending field of F'.

Indeed, a direct check shows that

$$dx'\,d\xi' + dy'\,d\eta' + dz'\,d\zeta' = \frac{1}{z^2}(dx\,d\xi + dy\,d\eta + dz\,d\zeta).$$

and thus $dr\,d\tau = 0$ implies that $dr'\,d\tau' = 0$.

It is also directly verifiable that the field τ' of F' is trivial if and only if the field τ is trivial on F.

This theorem is clearly valid for general convex surfaces and general bending fields. For τ' satisfies a Lipschitz condition, this being so for the field τ. Moreover, $dr' d\tau' = 0$ almost everywhere on F' since $dr d\tau = 0$ almost everywhere on F.

It is now easy to prove Theorem 2. Introduce a coordinate system whose xy-plane is a plane through O parallel to α, and O is the origin. Without loss of generality, we may assume that the surface F lies in the halfspace $z > 0$. Let F' be the image of F under the projective mapping (*), and define the bending field τ' by (**).

Now the distances from the center O are stationary under all admissible deformations of F, and so $\zeta' = 0$ on the boundary of F', which can be projected in one-to-one fashion onto the xy-plane. By Theorem 1, the field τ' is trivial outside the flat pieces of F'. Therefore τ is trivial outside the flat pieces of F. Q.E.D.

Another corollary of the Darboux-Sauer theorem and the Fundamental Lemma is that the surface $\Phi: r = \bar{r}/\bar{r}\tau$, where $\bar{r}$ is the radius vector of a convex surface F and τ a bending field of F, has negative curvature if F contains no flat pieces. If F does contain flat pieces, this surface has nonpositive curvature on the set H of points whose projections from the origin do not lie in flat pieces of F.

We now consider unbounded complete convex surfaces which can be projected in one-to-one fashion onto some plane α. Without loss of generality, we may assume that this plane is the xy-plane. Let us say that the surface is anchored to infinity relative to the xy-plane if all its admissible deformations must satisfy the condition

$$\frac{\zeta(P)}{(x^2(P) + y^2(P))^{1/2}} \to 0 \quad \text{as} \quad P \to \infty,$$

where $\zeta(P)$ is the vertical component of the velocity of deformation.

THEOREM 3. *Let F be an unbounded convex surface which can be projected in one-to-one fashion onto the xy-plane, is not a cylinder, contains no flat pieces and is anchored to infinity relative to the xy-plane. Then F is rigid.*

PROOF. The projection G of F on the xy-plane is a convex domain, which may be bounded or unbounded, and in the latter case may or may not coincide with the entire xy-plane.

Let $\zeta(x, y)$ be the vertical component of a bending field of F. Since F contains no flat pieces, the surface $\Phi: z = \zeta(x, y)$ is a surface of nonpositive curvature. If the domain G is not the entire xy-plane, the boundary of the surface Φ lies in the xy-plane and coincides with that of G. This is an immediate consequence of the condition

$$\frac{\zeta(P)}{(x^2(P)+y^2(P))^{1/2}} \to 0 \quad \text{as} \quad P\to\infty.$$

In this case the surface Φ can be extended so that its projection on the xy-plane will coincide with the entire xy-plane and the resulting surface will also have negative curvature. We need only extend the function $\zeta(x,y)$ beyond G by defining $\zeta(x,y)=0$ there. The resulting extension of Φ will be denoted by $\bar{\Phi}$.

By Bernštein's theorem on surfaces of negative curvature, extended to nonregular surfaces by Adel'son-Vel'skiĭ [1], the surface $\bar{\Phi}$ is a cylinder. Therefore so is Φ.

Without loss of generality, we may assume that the generators of the surface Φ are parallel to the x-axis, and so $\zeta=\gamma_1(y)$.

Consider the equations of the bending field:

$$\xi_x+p\zeta_x=0,\quad \xi_y+\eta_x+p\zeta_y+q\zeta_x=0,\quad \eta_y+q\zeta_y=0.$$

The first equation implies that $\xi_x=0$, and so $\xi=\gamma_2(y)$.

Integrating the second equation $\gamma_2'+\eta_x+p\gamma_1'=0$ along the straight lines $y=\text{const}$, we get

$$(***)\qquad x\gamma_2'+\eta+z\gamma_1'+\gamma_3=0,$$

where γ_3 is a function of y.

Consider the second difference of equation (***) along a straight line $y=\text{const}$:

$$\Delta\Delta\eta+\gamma_1'\Delta\Delta z=0.$$

Since the surface F contains no complete straight lines, we may assume that $\Delta\Delta z\neq 0$. Hence the function $\gamma_1'=\Delta\Delta\eta/\Delta\Delta z$ is differentiable almost everywhere, since $\Delta\Delta\eta$ and $\Delta\Delta z$ satisfy Lipschitz conditions.

Computing the first difference of equation (***), we get

$$\gamma_2'+\Delta\eta+\Delta z\gamma_1'=0.$$

Hence, as before, we see that γ_2' is differentiable almost everywhere. Finally, we conclude that γ_3 is also differentiable almost everywhere.

Differentiating equation (***) with respect to y and substracting $\eta_y+q\zeta_y=0$, we obtain

$$x\gamma_2''+z\gamma_1''+\gamma_3'=0.$$

If $\gamma_1''\neq 0$ for at least one y_0, then z is a linear function of x along the line $y = y_0$ on the surface. This means that the surface contains a straight line and is therefore a cylinder, which is absurd. But if γ_1'' vanishes wherever it exists, then the Lipschitz condition for γ_1' implies that γ_1 is a linear function; hence ζ is a linear function. Since the surface is anchored to infinity, we have $\zeta = \text{const}$.

Hence, reasoning as in the proof of Theorem 1, we see that the bending field τ is trivial.

THEOREM 4. *Any closed convex surface containing no flat pieces is rigid. A convex closed surface containing flat pieces is rigid outside these pieces.*

PROOF. Let F be a closed convex surface and τ a bending field on F. Let Ω be a ball enclosing F whose surface has at least one point, O say, in common with F. Introduce rectangular cartesian coordinates x, y, z, with the point O as origin and the tangent plane to the ball at O as xy-plane.

Let V denote the tangent cone of F at O. (Recall that the tangent cone of the surface at O is the surface of the smallest solid cone with apex O enclosing the surface F.) V is a proper cone only if O is a conic point of F. The tangent cone at a ridge point degenerates into a dihedral angle; at a smooth point, into the tangent plane to F.

Let V_F be the intersection of the surface F and the tangent cone V. In the general case, V_F will consist of segments with one endpoint at O, and none of these segments can lie in the xy-plane.

The part of the surface F not in V, i.e. the domain $F - V_F$, is cut in at most one point by any ray through the origin O.

Consider the following projective mapping of F:

$$x' = x/z, \quad y' = y/z, \quad z' = 1/z,$$

and the field defined by

$$\xi' = \xi/z, \quad \eta' = \eta/z, \quad \zeta' = -(x\xi + y\eta + z\zeta)/z.$$

By the Darboux-Sauer theorem, the field $\tau'(\xi', \eta', \zeta')$ is a bending field of the image F'.

Let F'' denote the image on F' of the domain $F - V_F$ under the above projective mapping. Let us see how the vertical component ζ' of the field τ' behaves at points approaching the boundary of F'', or going to infinity if F'' is unbounded.

If a point Q of the surface F'' approaches the boundary or goes to infinity, the corresponding point P of F approaches the boundary of the domain $F - V_F$.

Without loss of generality, we may assume that the bending field τ of F vanishes at O. This can always be achieved by superimposing on τ the velocity field of a suitable rigid translation of F.

Assuming, therefore, that $\tau(O)=0$, we claim that when the point Q of F'' approaches the boundary of F'' or goes to infinity, then $\zeta'(Q)\to 0$.

Suppose the assertion false, i.e. there exists a sequence of points Q approaching the boundary of F'' or going to infinity such that $\zeta'(Q) > \epsilon > 0$. At the same time, the sequence of corresponding points P on F approaches the boundary of $F - V_F$. Without loss of generality, we may assume that the points P converge to some point P_0. Then P_0 is either O or the endpoint of a segment g on a generator of the cone V. This segment lies on the surface F but not in the xy-plane, since the surface is inside the ball Ω.

Assume that P_0 is the endpoint of a segment g. We have $r'\tau'=0$ almost everywhere along g. Since $r'=\text{const}$, we have $r'(\tau(P_0)-\tau(O))=0$. But $\tau(O)=0$, and so $r'\tau(P_0)=0$. Now since r' and $r(P_0)$ are parallel, it follows that $r(P_0)\tau(P_0)=0$. Thus the limit of ζ' for the points of the sequence Q, which is $-r(P_0)\tau(P_0)/z(P_0)$, is zero, and this is impossible.

Now let $P_0\equiv 0$. We claim that then $\zeta'(Q)\to 0$.

Without loss of generality, we may assume that the straight lines P_0P converge to a generator of the cone V. Connect the points P_0 and P by a segment (shortest join) γ_P. If P is sufficiently close to P_0, the semitangents to γ_P form arbitrarily small angles with the direction of P_0P. Therefore

$$r'(s)=r(P)+\epsilon(s,P)\,|r(P)|,$$

where the differentiation is with respect to arc length s along the segment γ_P and $\epsilon(s,P)\to 0$ as $P\to P_0$.

Now $r'\tau'=0$ almost everywhere on γ_P. Substituting $r'=r(P)+\epsilon|r(P)|$ in this equality, we get

$$r(P)\tau'(s)+\tau'(s)\epsilon(s,P)\,|r(P)|=0.$$

Integrating this equality along γ_P, we get

$$r(P)\tau(P)+\bar{\epsilon}(P)\,|r(P)|^2=0,$$

where $\bar{\epsilon}(P)\to 0$ as $P\to P_0$.

Now consider the function $\zeta'(Q)$:

$$\zeta'(Q)=-\frac{r(P)\tau(P)}{z(P)}=\bar{\varepsilon}(P)\frac{|r(P)|^2}{z(P)}.$$

Within the ball Ω, the quotient $|r(P)|^2/z(P)$ is bounded by a constant which depends on the radius of the ball. And since $\bar{\epsilon}(P) \to 0$, we have $\zeta'(Q) \to 0$, giving a contradiction for the case $P_0 \equiv 0$ as well.

Thus, if the point Q approaches the boundary of F'' or goes to infinity, then $\zeta'(Q) \to 0$.

Now assume that F contains no flat pieces; then the same is true of F'. Consequently the surface $\Phi' : z = \zeta'(x,y)$ has negative curvature. It follows from the behavior of the function $\zeta'(Q)$ established above that Φ' is either the xy-plane or a domain on the xy-plane. Hence the bending field is trivial in the domain $F - V_F$.

Now, since the ball Ω was arbitrary, it follows that the field τ is trivial on the entire surface.

When the surface F contains flat pieces, a suitable modification of the field τ' on the surface F'' will ensure that $\zeta' \equiv 0$. Hence the field is trivial outside the flat pieces of the surface $F - V_F$, and this conclusion can then be extended to the entire surface F. This completes the proof.

As a corollary of Theorem 4, we see that *if the flat pieces of a closed convex surface are rigid, then the surface is rigid*. For example, a closed convex polyhedron with rigid faces if rigid.

§8. Applications of the rigidity theorems

In this section we present a few important applications of rigidity theorems for general convex surfaces. We shall consider the regularity of infinitesimal bendings of regular surfaces. New approaches will be indicated to the proof of monotypy theorems for general convex surfaces.

THEOREM 1. *A bending field of a regular convex surface with positive Gauss curvature is regular. More precisely, if the surface is of class $C^{n+\alpha}$ ($n \geqq 2$, $\alpha > 0$), i.e. it is n times differentiable and its nth derivatives satisfy a Lipschitz condition with exponent α, then every bending field is in class $C^{n+\alpha'}$ for any $\alpha' < \alpha$. If the surface is analytic, all its bending fields are analytic.*

PROOF. Let F be a convex surface of class $C^{n+\alpha}$ with positive Gauss curvature, and τ a bending field of F. We shall study the regularity of τ in the neighborhood of an arbitrary point P on F.

Introduce a coordinate system x, y, z, with P as the origin and the xy-plane so chosen that some neighborhood of P on F can be projected in one-to-one fashion onto the xy-plane. For sufficiently small R, the neighborhood of P on F is defined by an equation $z = z(x,y)$, where $z(x,y)$ is $C^{n+\alpha}$ in the disk $\bar{\omega} : x^2 + y^2 \leqq R^2$.

Construct a sequence of analytic functions $\tilde{z}(x,y)$ converging to $z(x,y)$ together with their derivatives of order up to n, whose nth derivatives satisfy a Lipschitz condition with exponent α uniformly in the domain $\bar{\omega}' : x^2+y^2 < R^2+\epsilon'$ $(\epsilon' > 0)$. Construct a sequence of analytic functions $\tilde{\zeta}(x,y)$ with uniformly bounded first derivatives which converges to the vertical component $\zeta(x,y)$ of the bending field τ in the domain $\bar{\omega}'$. The existence of these two sequences is ensured by the fact that $z(x,y)$ is $C^{n+\alpha}$ and $\tilde{\zeta}(x,y)$ satisfies a Lipschitz condition.

Construct an analytic bending field $\tilde{\tau}$ of the surface $\tilde{F}_{\bar{\omega}} : z = \tilde{z}(x,y)$ $(x^2+y^2 < R^2)$ with vertical component $\tilde{\zeta}(x,y)$ on the boundary, which satisfies the conditions $\tilde{\xi} = \tilde{\eta} = 0$ and $\tilde{\xi}_y = 0$ at P. The existence of this field was proved in §3.

The vertical component $\tilde{\zeta}(x,y)$ satisfies the equation

$$(*) \qquad \tilde{z}_{xx}\tilde{\zeta}_{yy} - 2\tilde{z}_{xy}\tilde{\zeta}_{xy} + \tilde{z}_{yy}\tilde{\zeta}_{xx} = 0.$$

A theorem of Schauder [**67**] states that the derivatives of $\tilde{\zeta}(x,y)$ of order up to n and the Lipschitz coefficients for these derivatives for exponent $\alpha' < \alpha$, in any compact domain G in the disk $\bar{\omega}$, satisfy estimates which depend only on the maximum moduli of the coefficients of equation (*), their derivatives of order up to n, and the Lipschitz coefficients of their nth derivatives for exponent α. Recalling the formulas for the horizontal components of $\tilde{\tau}$ in terms of $\tilde{\zeta}(x,y)$ (see §3), we see that similar estimates are valid for $\tilde{\xi}$, $\tilde{\eta}$, their derivatives of order up to n and the Lipschitz coefficients of these derivatives for exponent α'.

The sequence of bending fields $\tilde{\tau}$ contains a convergent subsequence. In view of the above-mentioned estimates for the fields $\tilde{\tau}$ the limit field $\bar{\tau}$ of this subsequence is in class $C^{n+\alpha'}$. By the convergence theorem for bending fields, $\bar{\tau}$ is a bending field on the surface $F_{\bar{\omega}}$.

Since the vertical components of the bending fields τ and $\bar{\tau}$ coincide on the boundary of $F_{\bar{\omega}}$, it follows from Theorem 1 of §7 that the field $\tau - \bar{\tau}$ is trivial. Hence the field τ is also in class $C^{n+\alpha}$.

If the surface F is analytic, it follows from what we have just proved that τ is in C^3. Now since the vertical component $\zeta(x,y)$ of τ satisfies an elliptic equation with analytic coefficients,

$$r\zeta_{yy} - 2s\zeta_{xy} + \zeta_{xx} = 0,$$

Bernšteĭn's theorem tells us that $\zeta(x,y)$ is an analytic function. Hence $\xi(x,y)$ and $\eta(x,y)$, which are expressible in terms of $\zeta(x,y)$, must also be analytic. This completes the proof.

THEOREM 2. *Isometric closed convex surfaces are congruent.*

PROOF. Let F_1 and F_2 be two isometric closed convex surfaces. Suppose that they are not congruent. Then there exists a surface F_2' arbitrarily close to F_1 which is isometric but not congruent to it (Chapter III, §2).

Let $r_1(X)$ be the radius vector of an arbitrary point X on F_1, and $r_2(X)$ the radius vector of the corresponding point on F_2'. If F_2' is sufficiently near F_1, the surface F defined by the equation $r = r_1(X) + r_2(X)$ is convex and $dr_1 + dr_2 \neq 0$.

Consider the vector field $\tau = r_1(X) - r_2(X)$ on the surface F. It is readily shown that it satisfies a Lipschitz condition, and $dr d\tau = 0$ almost everywhere, and so τ is a bending field on F.

Since the convex surface F is rigid outside its flat pieces (Theorem 4 of §7), it follows that τ is trivial there.

As for the flat pieces of F, it can be proved that the corresponding regions on F_1 and F_2 are also flat. Hence one easily concludes that the field τ is trivial on the entire surface F, and therefore

$$\tau = r_1 - r_2 = a \times (r_1 + r_2) + b,$$

where a and b are constant vectors.

Let X and X' be two arbitrary points on the surface F_1, and set $r_1' - r_1 = \Delta r_1$. Let Δr_2 denote the corresponding vector for the surface F_2'. Then

$$\Delta r_1 - \Delta r_2 = a \times (\Delta r_1 + \Delta r_2).$$

Hence $\Delta r_1^2 - \Delta r_2^2 = 0$.

Thus the extrinsic distance $|\Delta r_1|$ between the points X and X' of the surface F_1 is equal to the extrinsic distance $|\Delta r_2|$ between the corresponding points of F_2', i.e. the surfaces F_1 and F_2' are congruent. This contradiction completes the proof.

THEOREM 3. *Let F_1 and F_2 be two isometric convex surfaces which can be projected in one-to-one fashion onto the xy-plane, and are convex in the same sense. Assume that two points of these surfaces which correspond under the isometry coincide at the origin. Let $z_1(P)$ be the z-coordinate of an arbitrary point P of F_1 and $z_2(P)$ the z-coordinate of the corresponding point on F_3. Then the function $z_1(P) - z_2(P)$ cannot assume a strict maximum or minimum at the origin O.*

A proof based on the Fundamental Lemma on infinitesimal bendings of general convex surfaces runs as follows.

Let $r_1(P)$ and $r_2(P)$ be the radius vectors of corresponding points on F_1 and F_2. By a rotation about the z-axis we can bring the surface F_2 to a position such that the surface F defined by the equation $r=r_1(P)+r_2(P)$ is convex near the origin O, and $dr_1+dr_2\neq 0$ there.

The vector field $\tau=r_1(P)+r_2(P)$ is a bending field on F. By the Fundamental Lemma, its vertical component $z_1(P)-z_2(P)$ can assume neither a strict maximum nor a strict minimum at the origin O, unless O is an interior point of a flat region of F.

Now if O is an interior point of a flat region on F, it is an interior point of flat regions on F_1 and F_2. And in this case the function $z_1(P)-z_2(P)$ clearly cannot have a strict maximum or minimum at O. This completes the proof.

The Fundamental Lemma on infinitesimal bendings of general convex surfaces may also be used to prove the following theorem.

Let F_1 and F_2 be two isometric convex surfaces with a common point P_0, each cut in at most one point by any ray from the origin O, which are convex in the same sense relative to the direction OP_0. Let $\rho_1(P)$ and $\rho_2(P)$ be the distances of corresponding points of the surfaces from the origin. Then the function $\rho_1(P)-\rho_2(P)$ cannot have a strict maximum or minimum at the point P_0.

CHAPTER V

CONVEX SURFACES IN SPACES OF CONSTANT CURVATURE

A. D. Aleksandrov carried over the basic results of his theory of convex surfaces to convex surfaces in spaces of constant curvature (see Chapter I, §12). Among other things, he proved his theorems on the realizability of abstract convex metrics by surfaces in these spaces.

It is therefore natural to try to carry over our results on monotypy of convex surfaces, regularity of convex surfaces with regular metrics, and infinitesimal bendings of convex surfaces (Chapters II, III and IV) to convex surfaces in spaces of constant curvature. This is the object of the present chapter.

Our solution of this problem will be based on a geodesic mapping of a space R of constant curvature into a euclidean space E, under which pairs of isometric objects in R go into pairs of isometric objects in E. Let us describe the basic idea of this mapping for elliptic space.

Let R be an elliptic space with curvature $K = 1$. Define Weierstrass coordinates x_i $(i = 0,1,2,3)$ in R and associate with each point of R the pair of points in euclidean four-space with cartesian coordinates x_i and $-x_i$. These points fill out the unit sphere, since the Weierstrass coordinates satisfy the condition

$$x^2 = x_0^2 + x_1^2 + x_2^2 + x_3^2 = 1.$$

Let E_0 denote the euclidean space $x_0 = 0$. Let F' and F'' be two isometric figures in the elliptic space R. Let X' be an arbitrary point of F', and X'' the corresponding point of F''. Then the equations

$$y = \frac{x' - e_0(x'e_0)}{e_0(x' + x'')}, \qquad y = \frac{x'' - e_0(x''e_0)}{e_0(x' + x'')},$$

where e_0 is the unit vector along the x_0-axis, define two isometric figures Φ' and Φ'' in the euclidean space E_0. If the figures F' and F'' are congruent, so are Φ' and Φ''. Conversely, if Φ' and Φ'' are congruent, so are F' and F''. Thus, the transformation reduces the problem of monotypy in elliptic space to the same problem for euclidean space.

Since the study of infinitesimal bendings of a surface is related to the study of infinitesimally close isometric surfaces, the former problem is also reduced to the analogous problem in euclidean space.

To keep the exposition as simple and clear as possible, we shall restrict ourselves to convex surfaces in elliptic space. Convex surfaces in hyperbolic (Lobačevskiĭ) space will be considered only in regard to the problem of the regularity of a convex surface with regular metric.

§1. Elliptic space

In order to make the following exposition more accessible, let us recall some of the elements of the geometry of elliptic space. Elliptic space can be interpreted in terms of objects of euclidean space. In particular, it can be regarded as a euclidean four-sphere in which antipodal points are identified. We shall repeatedly use this interpretation. We therefore begin with a few well-known concepts of euclidean four-space.

A four-vector is any ordered quadruple of real numbers (x_0, x_1, x_2, x_3). The numbers x_i are called the components of the vector. Addition and subtraction of vectors and multiplication by a scalar are defined as usual: the sum (difference) of two vectors (x_i) and (y_i) is the vector $(x_i \pm y_i)$, the product of a vector (x_i) and a number λ is the vector (λx_i).

Four-vectors may be interpreted as directed segments in euclidean four-space. The formally defined operations of addition, subtraction and multiplication by a scalar then correspond to the usual geometric operations for three-vectors.

Apart from the above three vector operations, we define the scalar product of two vectors, vector triple product of three vectors, and mixed product of four vectors.

The *scalar product* of two vectors x and y is the number

$$(x, y) = x_0 y_0 + x_1 y_1 + x_2 y_2 + x_3 y_3,$$

where x_i are the components of x and y_i those of y.

The vector triple product of three vectors x, y, z, in this order, is the vector (xyz) whose ith component is $(-1)^i$ times the 3×3 minor of the matrix

$$\begin{pmatrix} x_0 & x_1 & x_2 & x_3 \\ y_0 & y_1 & y_2 & y_3 \\ z_0 & z_1 & z_2 & z_3 \end{pmatrix}$$

obtained by deleting the ith column $(i = 0, 1, 2, 3)$.

Finally, the mixed product $(abcd)$ of four vectors a, b, c, d is a number, defined as the determinant of the components of the vectors:

$$(abcd) = \begin{vmatrix} a_0 & a_1 & a_2 & a_3 \\ b_0 & b_1 & b_2 & b_3 \\ c_0 & c_1 & c_2 & c_3 \\ d_0 & d_1 & d_2 & d_3 \end{vmatrix}$$

The vector triple and mixed products are obviously invariant under cyclic permutations of the factors, and are multiplied by -1 when two factors are interchanged. It follows that the vector triple and mixed products vanish when the factors are linearly dependent.

All three products are distributive in each factor.

Note the following identities:

$$(abc)\, d = (abcd);$$

$$(abc)\,(a'b'c') = \begin{vmatrix} (aa') & (ab') & (ac') \\ (ba') & (bb') & (bc') \\ (ca') & (cb') & (cc') \end{vmatrix};$$

$$(abcd)\,(a'b'c'd') = \begin{vmatrix} (aa') & (ab') & (ac') & (ad') \\ (ba') & (bb') & (bc') & (bd') \\ (ca') & (cb') & (cc') & (cd') \\ (da') & (db') & (dc') & (dd') \end{vmatrix}.$$

The first follows directly from the definitions, the third from the multiplication theorem for determinants. To prove the second identity, note that both sides are linear in each of the vectors $a, b, \cdots, c'$. It is therefore sufficient to verify the identity for the basis vectors $(1,0,0,0)$, $(0,1,0,0)$, $(0,0,1,0)$, $(0,0,0,1)$. For these vectors, if the triples a, b, c and a', b', c' are identical up to a cyclic permutation, both sides of the equality are either $+1$ or -1; if the triples are different, both sides vanish. (The vector triple product of three different basis vectors is equal to the fourth basis vector up to sign.)

The absolute value of a vector x is defined as the length of the corresponding segment; the formula is analogous to that in the three-dimensional case:

$$|x| = (x_0^2 + x_1^2 + x_2^2 + x_3^2)^{1/2} = \sqrt{(x, x)}.$$

A rotation of a vector space is a linear transformation

$$x'_i = \sum_j \alpha_i^j x_j, \qquad i, j = 0, 1, 2, 3, \tag{1}$$

which leaves the absolute values of the vectors invariant, i.e. $|x| = |x'|$ for any x.

The scalar product of vectors is clearly invariant under a rotation, since

$$(xy) = \frac{1}{4}\{(x+y)^2 - (x-y)^2\} = \frac{1}{4}\{|x+y|^2 - |x-y|^2\}.$$

Under the rotation (1), the basis vectors $e_0(1,0,0,0), \cdots, e_3(0,0,0,1)$ are mapped onto the vectors

$$(\alpha_0^0, \alpha_1^0, \alpha_2^0, \alpha_3^0), \cdots, (\alpha_0^3, \alpha_1^3, \alpha_2^3, \alpha_3^3).$$

Hence, in view of the invariance of the scalar product, we see that the matrix (α_i^j) is orthogonal:

$$\alpha_0^i\alpha_0^j + \alpha_1^i\alpha_1^j + \alpha_2^i\alpha_2^j + \alpha_3^i\alpha_3^j = \begin{cases} 0 & \text{for} \quad i \neq j, \\ 1 & \text{for} \quad i = j. \end{cases} \tag{2}$$

The determinant of the transformation (1) is

$$\Delta = \begin{vmatrix} \alpha_0^0 & \alpha_1^0 & \alpha_2^0 & \alpha_3^0 \\ \alpha_0^1 & \alpha_1^1 & \alpha_2^1 & \alpha_3^1 \\ \alpha_0^2 & \alpha_1^2 & \alpha_2^2 & \alpha_3^2 \\ \alpha_0^3 & \alpha_1^3 & \alpha_2^3 & \alpha_3^3 \end{vmatrix} = \pm 1.$$

To verify this it suffices to square Δ and use the orthogonality conditions (2).

Let us call a rotation proper if $\Delta = +1$, and improper if $\Delta = -1$. Let A be any rotation. Then

$$(Ax, Ay, Az) = \pm (x, y, z),$$
$$(Aa, Ab, Ac, Ad) = \pm (abcd).$$

where the plus sign is valid if the rotation is proper, the minus sign if it is improper. The second of these formulas is a simple corollary of the multiplication theorem for determinants, since

$$(Aa, Ab, Ac, Ad) = (abcd)\,\Delta.$$

The first formula follows from the second, since for any vector u we have, on the one hand,

$$(Ax, Ay, Az, Au) = (Ax, Ay, Az)\,Au,$$

and on the other

$$(Ax, Ay, Az, Au) = \Delta(xyzu) = \Delta A\,(xyz)\,Au.$$

An elliptic space may be defined as a complete three-dimensional Riemannian manifold of constant curvature, homeomorphic to projective space. The euclidean four-sphere is also a three-dimensional Riemannian manifold with constant curvature, and is therefore locally isometric to elliptic space. It turns out that, if we start from some point and progressively superimpose the sphere isometrically on an elliptic space of the same curvature, the sphere will cover the space twice; the points of the sphere that coincide in this construction are pairs of antipodal points. We may therefore interpret elliptic space as a three-sphere in which antipodal points have been identified. We shall constantly use this easily visualized model of elliptic space. The curvature of the space will be assumed equal to $+1$, and the sphere is therefore the unit sphere.

Since any motion of an elliptic space is an isometric mapping and the only isometric mappings of the sphere into itself are rotations (proper or improper), a motion of elliptic space is represented on our model as a rotation of the sphere about its center. Furthermore, the straight lines of elliptic space, as geodesics on the spherical model, are great circles, and the planes are intersections of hyperplanes through the center with the sphere.

We introduce a coordinate system in elliptic space by associating with each point the cartesian coordinates of the corresponding point on the unit sphere. These are known as Weierstrass coordinates in elliptic space. Since the isometry of elliptic space onto the sphere is not single valued, the Weierstrass coordinates are defined only up to sign. Instead of the entire space, we shall often consider only the part remaining after deletion of a plane, e.g. $x_0 = 0$; the indeterminacy may then be eliminated by imposing the additional condition $x_0 > 0$.

A plane in elliptic space is defined in Weierstrass coordinates by a linear homogeneous equation:

$$a_0x_0 + a_1x_1 + a_2x_2 + a_3x_3 = 0,$$

where the a_i are constants not all zero. A straight line is defined by an independent system of two such equations:

$$a_0x_0 + a_1x_1 + a_2x_2 + a_3x_3 = 0,$$
$$b_0x_0 + b_1x_1 + b_2x_2 + b_3x_3 = 0.$$

These equations are easily expressed in vector notation, using the scalar product:

(3) $$(ax) = 0$$

and

(4) $$(ax) = 0, \qquad (bx) = 0.$$

Starting from the above representation of a straight line, we can derive its parametric representation: $x = \rho(c + dt)$, where c and d are constant vectors, t the parameter, and ρ a normalizing factor determined by the condition $x^2 = 1$. As c and d we can take any two distinct solutions of equations (4). Similarly, the plane (3) can be expressed in parametric form: $x = \rho(c + du + ev)$, where c, d, e are three independent solutions of equation (3), u and v are parameters, and ρ is a normalizing factor determined as before.

Let us express the line element of elliptic space in Weierstrass coordinates. Since the Weierstrass coordinates are equal (up to sign) to the cartesian coordinates of the corresponding point on the sphere, the line element of elliptic space in Weierstrass coordinates is simply the line element of the sphere in cartesian coordinates:

$$ds^2 = dx_0^2 + dx_1^2 + dx_2^2 + dx_3^2.$$

We shall also need the projective model of elliptic space. To construct this model, complete the euclidean space $x_0 = 1$ by ideal elements. Denote the resulting projective space by P_0. Now associate with each point (x_i) of elliptic space the point of P_0 with homogeneous coordinates $\bar{x}_i$ defined by $\bar{x} = x/(xe_0)$, where e_0 is the unit basis vector $(1, 0, 0, 0)$. It is obvious that this mapping is one-to-one. Geometrically, it can be construed as a projection of the unit sphere from its center onto the hyperplane $x_0 = 1$.

This mapping of elliptic space onto projective space has many remarkable properties. It is geodesic, in the sense that straight lines of elliptic space go into straight lines of projective space. The motions of elliptic space correspond to projective mappings that preserve the imaginary oval space $\bar{x}^2 = 0$. This follows from the invariance of the scalar product (x, x) under rotations.

We may thus interpret elliptic space as a projective space in which the role of motions is played by projective mappings which map the imaginary oval space $\bar{x}^2 = 0$ (the "absolute") into itself.

Moreover, if the plane $x_0 = 0$ is removed from elliptic space, the remaining part is mapped one-to-one onto the euclidean space $x_0 = 1$. This euclidean space is an osculating space of elliptic space at the point $(1,0,0,0)$. Indeed, the line element of the euclidean space is $d\bar{s}^2 = d\bar{x}^2$. Since $(xe_0) = x_0 = 1$ at the point $(1,0,0,0)$ and this product is stationary there, it follows that

$$d\bar{s}^2 - ds^2 = 0, \qquad \delta(d\bar{s}^2 - ds^2) = 0.$$

And this means that the euclidean space is indeed an osculating space of elliptic space at this point.

The definition of arc length in elliptic space is the same as in euclidean space: the limit of the lengths of polygons regularly inscribed in the curve, when their sides tend to zero. The length of the side of a polygon with endpoints (t) and $(t + \Delta t)$ on the curve $x = x(t)$ is defined as

$$\cong |x(t+\Delta t) - x(t)|. \tag{5}$$

Hence, just as in the euclidean case, we get the formula

$$s = \int \sqrt{x'^2}\, dt.$$

The natural parametrization of a curve is that in which the parameter is arc length s.

Let γ be a curve in elliptic space and $x = x(s)$ its natural parametrization. Let us study the properties of the vector $\gamma = x'(s)$.

First, τ is a unit vector, since

$$s = \int \sqrt{\tau^2}\, ds,$$

and therefore $\tau^2 - 1$.

Now consider the straight line defined parametrically by $\rho x = x(s) + t\tau(s)$ (parameter t). This line is tangent to the curve γ at the point (s). Indeed, it passes through the point (s) $(t = 0)$. Let us estimate the distance of the point $(s + \Delta s)$ on the curve from the line. It cannot exceed the distance of the point from the point $x(s) + \Delta s t(s)$ on the line. Now the distance between these points is

$$\cong |x(s+\Delta s) - x(s) - \tau(s)\Delta s|,$$

which is of the same order of magnitude as Δs^2. Hence the line is indeed a tangent. We shall therefore call τ the unit tangent vector of the curve.

Let γ' and γ'' be curves issuing from the same point P, and $x = x_1(s)$ and $x = x_2(\sigma)$ their natural parametrizations. Let us find the angle between the curves at the point P. By definition, the angle between curves is the angle between the corresponding curves in the osculating euclidean space at P.

Let $P \equiv (1,0,0,0)$. The curves corresponding to γ' and γ'' in the osculating euclidean space at P are defined by the equations

$$y = x_1/(x_1 e_0), \qquad y = x_2/(x_2 e_0).$$

Since $(x_1 e_0)$ and $(x_2 e_0)$ are stationary, the unit tangent vectors of these curves at P are τ' and τ'', and so $\cos\vartheta = (\tau'\tau'')$. Since scalar multiplication is invariant under motions of space, this formula is valid for any point P.

We now calculate the curvature k of a curve $\gamma : x = x(s)$ at an arbitrary point. By definition, the curvature of γ at P is the curvature of the corresponding curve in the osculating euclidean space. Let $P \equiv (1,0,0,0)$. The corresponding curve in the osculating euclidean space is defined by the equation

$$y = x(s)/(x(s)e_0).$$

The curvature of this curve is of course

$$k^2 = (y'^2 y''^2 - (y'y'')^2)/(y'^2)^3.$$

Expressing y in this formula in terms of x and using the relations

$$x^2 = 1, \quad xx' = 0, \quad x'^2 = 1, \quad xx'' = -1, \quad x = e_0,$$

we get $k^2 = x''^2 - 1$. Since this formula is invariant under motions, it is valid not only for the point $(1,0,0,0)$ but for any point.

We now determine the osculating plane of the curve γ at the point (s). To this end, consider the plane defined by the following parametric equations:

$$\rho x = x(s) + ux'(s) + vx''(s)$$

(parameters u, v).

This plane obviously passes through the point (s) on the curve $(u = v = 0)$. The distance of the point $(s + \Delta s)$ on the curve from this plane is not greater than its distance from the point

$$\frac{1}{\rho}\left(x(s) + \Delta s x'(s) + \frac{\Delta s^2}{2} x''(s)\right).$$

The latter distance is

$$\simeq \left| x(s+\Delta s) - x(s) - \Delta s x'(s) - \frac{\Delta s^2}{2} x''(s) \right|$$

whose order of magnitude is at least that of Δs^3. Hence this is indeed the osculating plane.

The straight line $\rho x = x(s) + \lambda x''(s)$ (parameter λ) is the principal normal to the curve. Indeed, it lies in the osculating plane. Furthermore, $\rho^2 = 1 - 2\lambda + o(\lambda^2)$, and therefore $x'_\lambda = x''(s) + x$ when $\lambda = 0$, which implies $x'_\lambda x' = 0$; that is, the vector x'_λ is perpendicular to the tangent vector of the curve; hence the line is indeed the principal normal.

The unit vector defined by the condition $x'' + x = \lambda\nu$ will be called the unit vector of the principal normal of the curve. The factor λ has a simple meaning: it is the curvature of the curve. To verify this, square both sides of the equality:

$$\lambda^2 = x''^2 - 1 = k^2.$$

Note that, up to a term of order Δs^2, the vector $x(s + \Delta s)$ has the convenient representation

$$x(s + \Delta s) = x + \Delta s \tau + \frac{\Delta s^2}{2}(k\nu - x).$$

The vectors x, τ, ν are mutually perpendicular.

Let F be a surface in elliptic space, defined by the equation $x = x(u, v)$. A tangent vector to this space at a point P is a tangent vector to any curve on the surface issuing from P. In particular, x_u and x_v are tangent vectors.

The plane defined by the parametric equation

$$\rho x = x + \alpha x_u + \beta x_v \qquad (\text{parameters } \alpha, \beta)$$

is the tangent plane to the surface, as is easily verified by the same reasoning as for the tangent to a curve.

It follows from the properties of the vector triple and mixed products of vectors that the vector $n = (x x_u x_v)$ is perpendicular to all tangent vectors $(\lambda x_u + \mu x_v)$ to the surface at the point (u, v). We shall therefore call n the normal vector of the surface. The unit normal vector will be denoted by ξ.

As in the euclidean case, one defines two fundamental quadratic forms for a surface in elliptic space: the first fundamental form (the line element of the surface)

$$ds^2 = dx^2 = E du^2 + 2F du dv + G dv^2,$$

and the second fundamental form

$$-(dx d\xi) = (d^2x, \xi) = L du^2 + 2M du dv + N dv^2.$$

Just as in the euclidean case, the length of curves, angle between curves, geodesic curvature of a curve, and Gauss (intrinsic) curvature of a surface can be expressed in terms of the first fundamental form.

Let us consider the second fundamental form in greater detail. Consider a curve on the surface:

$$u = u(s), \qquad v = v(s),$$

where s is arc length. Let us find its curvature. We have $x''_{ss} = k\nu - x$, where k is the curvature of the curve and ν the unit vector of the principal normal. Multiplying this equality by ξ and observing that the vectors x and ξ are perpendicular, we see that $(x''_{ss}\xi) = k\cos\vartheta$, where ϑ is the angle between the principal normal of the curve and the normal to the surface.

On the other hand, since the vectors x_u and x_v are perpendicular to ξ,

$$(x''_{ss}\xi) = (x_{uu}\xi)\,u'^2 + 2\,(x_{uv}\xi)\,u'v' + (x_{vv}\xi)\,v'^2 = Lu'^2 + 2Mu'v' + Nv'^2.$$

Hence

$$(*) \qquad k\cos\vartheta = \frac{L\,du^2 + 2M\,du\,dv + N\,dv^2}{E\,du^2 + 2F\,du\,dv + G\,dv^2}.$$

The right-hand side of this formula has a simple geometrical interpretation: it is the curvature of the normal section of the surface ($\vartheta = 0$).

The principal curvatures of the surface at a point are defined as the extremum values of the normal curvatures at the point. Standard techniques yield the usual equation for them:

$$\begin{vmatrix} L - kE & M - kF \\ M - kF & N - kG \end{vmatrix} = 0,$$

which implies the following expression for the product of the principal curvatures:

$$k_1 k_2 = (LN - M^2)/(EG - F^2).$$

The product of the principal curvatures is called the extrinsic curvature of the surface. In elliptic space, the extrinsic curvature K_e of a surface is different from the intrinsic curvature, which is expressed in the usual manner in terms of the coefficients of the first fundamental form and its derivatives.

Since we have a formula (*) for the normal curvature of a surface, we could define asymptotic lines and lines of curvature just as in euclidean space, and prove the relevant theorems. We shall not do this here, since these theorems are almost never used in the sequel.

We now wish to show that the fundamental characterization of geodesics on surfaces in euclidean space—the fact that their principal normal coincides with the normal to the surface—is also valid for geodesics on surfaces in elliptic space.

Let γ be a geodesic on a surface F in elliptic space, and P a point on F. Without loss of generality, we may assume that $P \equiv (1,0,0,0)$. Construct the osculating euclidean space at P. Let $\bar{F}$ be the surface and $\bar{\gamma}$ the curve corresponding to F and γ in the osculating space. Obviously, $\bar{\gamma}$ has zero geodesic curvature at P on $\bar{F}$. Therefore the principal normal of $\bar{\gamma}$ at P coincides with the normal to the surface

$$(\bar{\nu}, \bar{x}_u) = 0, \qquad (\bar{\nu}, \bar{x}_v) = 0.$$

In this equality, substitute

$$\bar{x} = x/(xe_0), \qquad \bar{\nu} = \bar{x}''/\bar{k}$$

and note that $x = e_0$ at P; we get

$$(\nu x_u) = 0, \qquad (\nu x_v) = 0,$$

where $\nu = (x'' + x)/k$ is the unit vector of the principal normal to γ at P. Since moreover $x\nu = 0$, the vectors ν and ξ either coincide or are opposite in direction. This proves our assertion.

Let F be a surface in elliptic space. The vectors x, x_u, x_v, ξ form a basis at the point (x), since $(xx_ux_v\xi) \neq 0$. Hence any vector at this point is a linear combination of $x, \cdots, \xi$. In particular,

$$\begin{aligned}
x_{uu} &= A_{11}^1 x_u + A_{11}^2 x_v + C_{11}x + \lambda_{11}\xi,\\
x_{uv} &= A_{12}^1 x_u + A_{12}^2 x_v + C_{12}x + \lambda_{12}\xi,\\
x_{vv} &= A_{22}^1 x_u + A_{22}^2 x_v + C_{22}x + \lambda_{22}\xi,\\
\xi_u &= B_1^1 x_u + B_1^2 x_v + D_1 x + H_1\xi,\\
\xi_v &= B_2^1 x_u + B_2^2 x_v + D_2 x + H_2\xi.
\end{aligned}$$

The coefficients of these formulas can be expressed in terms of the coefficients of the first and second fundamental forms of the surface. Indeed, taking the scalar products of the formulas with ξ, we get

$$\lambda_{11} = L, \quad \lambda_{12} = M, \quad \lambda_{22} = N, \quad H_1 = H_2 = 0.$$

Multiplying the equalities by x and noting that

$$x^2 = 1, \quad x_{uu}x = -E, \quad x_{uv}x = -F, \quad x_{vv}x = -G, \quad x\xi_u = x\xi_v = 0,$$

we obtain

$$C_{11} = -E, \quad C_{12} = -F, \quad C_{22} = -G, \quad D_1 = D_2 = 0.$$

To find the coefficients A_{11}^1 and A_{11}^2 we multiply the first equality by x_u and x_v. Noting that $x_{uu}x_u = \frac{1}{2}E_u$ and $x_{uu}x_v = F_u - \frac{1}{2}E_v$, we get

$$\frac{1}{2}E_u = EA_{11}^1 + FA_{11}^2,$$

$$F_u - \frac{1}{2}E_v = FA_{11}^1 + GA_{11}^2.$$

This is none other than the system of equations for the Christoffel symbols of the second kind, Γ_{11}^1 and Γ_{11}^2. Therefore

$$A_{11}^1 = \Gamma_{11}^1, \qquad A_{11}^2 = \Gamma_{11}^2.$$

Similarly,

$$A_{12}^1 = \Gamma_{12}^1, \quad A_{12}^2 = \Gamma_{12}^2, \quad A_{22}^1 = \Gamma_{22}^1, \quad A_{22}^2 = \Gamma_{22}^2.$$

To find the coefficients B_2^1 and B_2^2, multiply the last of the above equalities by x_u and x_v; this gives

$$-M = B_2^1 E + B_2^2 F, \qquad -N = B_2^1 F + B_2^2 G.$$

Now this is precisely the system of equations for the coefficients of the Weingarten equations for surfaces in euclidean space. An analogous conclusion holds for the coefficients B_1^1 and B_1^2.

Thus surfaces in elliptic space satisfy the Gauss-Weingarten equations

$$\begin{aligned}
x_{uu} &= \Gamma_{11}^1 x_u + \Gamma_{11}^2 x_v - Ex + L\xi,\\
x_{uv} &= \Gamma_{12}^1 x_u + \Gamma_{12}^2 x_v - Fx + M\xi,\\
x_{vv} &= \Gamma_{22}^1 x_u + \Gamma_{22}^2 x_v - Gx + N\xi,\\
\xi_u &= B_1^1 x_u + B_1^2 x_v,\\
\xi_v &= B_2^1 x_u + B_2^2 x_v,
\end{aligned}$$

where the coefficients Γ and B have the same expressions in terms of E, F, G, L, M, N as the corresponding coefficients for surfaces in euclidean space.

The conditions satisfied by the coefficients of the first and second fundamental forms are again obtained just as in the euclidean case. To elucidate: starting from the identities

$$\begin{aligned}
(\xi_u)_v - (\xi_v)_u &= 0,\\
(x_{uu})_v - (x_{uv})_u &= 0,\\
(x_{vv})_u - (x_{uv})_v &= 0
\end{aligned}$$

we replace the expressions in parentheses by their values according to the Gauss-Weingarten equations, and repeat this substitution after formal differentiation. The result is

$$\alpha_{11}x_u + \alpha_{12}x_v + \alpha_{10}x + \alpha_1\xi = 0,$$
$$\alpha_{21}x_u + \alpha_{22}x_v + \alpha_{20}x + \alpha_2\xi = 0,$$
$$\alpha_{31}x_u + \alpha_{32}x_v + \alpha_{30}x + \alpha_3\xi = 0.$$

Since x_u, x_v, x and ξ are independent vectors, $\alpha_{ij} = 0$ and $\alpha_i = 0$.

Note that had the Gauss equations (the first three equations in the above system) not contained terms in x, the same expressions would have been obtained for α_{11} and α_{12}. Now, since in euclidean space $\alpha_{11} = 0$ and $\alpha_{12} = 0$ are the Peterson-Codazzi relations, it follows that the coefficients of the fundamental forms also satisfy the Peterson-Codazzi relations in surfaces of elliptic space.

Omitting the detailed analysis, we mention that of the remaining ten relations only one is new. It determines the relation between the intrinsic (Gauss) and extrinsic curvatures of the surface:

$$(LN - M^2)/(EG - F^2) = K_i - 1.$$

In conclusion, note that if two quadratic forms

$$Edu^2 + 2Fdudv + Gdv^2, \qquad Ldu^2 + 2Mdudv + Ndv^2$$

are given, where the first is positive definite, then the above three relations for the coefficients of the forms constitute a sufficient condition for the existence of a surface of which these are the first and second fundamental forms, respectively. This surface is uniquely determined, up to a rigid motion.

§2. Convex bodies and convex surfaces in elliptic space

The bulk of this chapter is devoted to the study of isometric general convex surfaces in elliptic and euclidean spaces. We shall therefore consider certain properties of convex bodies and convex surfaces in an elliptic space.

As we indicated in §1, the unit three-sphere admits a two-to-one locally isometric mapping onto an elliptic space with unit curvature. This mapping is unique up to a rigid motion.

A body in elliptic space is said to be convex if it is the image of a convex body of the unit sphere under the above mapping. This definition is invariant with respect to motions in the elliptic space and is independent of the specific mapping of the sphere.

As for convex bodies on the sphere in four-space, these can be defined as the intersections of the sphere with convex cones whose apex is at the center of the sphere. Since a convex cone has a supporting plane at the apex, a convex body on the sphere lies in one of the hemispheres defined by this plane. Moreover, if the apex of the cone is a strictly conic point, in the sense that it has a supporting plane which has no other points in common with the cone, then the convex body defined by the cone lies in the interior of the hemisphere, i.e. has no points in common with its boundary.

We now turn to the projective model of elliptic space. If the supporting plane of the cone is taken as the plane $x_0 = 0$ and the cone is stipulated to lie in the halfspace $x_0 > 0$, then a convex body of elliptic space on $x_0 = 1$ has the appearance of a euclidean convex body and can be completed by ideal points. The following remarks will clarify this statement.

At the apex of a convex four-dimensional cone there are four possible cases:

a) There exists a supporting plane which has no points other than the apex in common with the cone; we then call the apex a strictly conic point.

b) There exists a supporting plane whose intersection with the cone is a straight line.

c) There exists a supporting plane whose intersection with the cone is a two-dimensional plane.

d) The entire supporting plane is contained in the cone.

In case a), a convex body in the projective model $x_0 = 1$ is a bounded convex body. In case b) it is a euclidean convex cylinder with finite span, completed by a point at infinity. In case c) the convex body consists of a lamina between two parallel planes, completed by the line at infinity in which these planes intersect. Finally, in case d) the convex body is the entire space.

In cases b), c) and d) the convex bodies are in a certain sense degenerate; their surfaces are easily constructed and are of little interest here. In the sequel, therefore, we shall confine the discussion to the convex bodies (and their surfaces) obtained in case a), i.e. when the apex of the projecting cone is strictly conic.

To summarize: we shall consider convex bodies of elliptic space represented in the spherical model by convex bodies in the interior of a hemisphere, and in the projective model by bounded euclidean convex bodies.

Discussion of convex bodies in elliptic space is made more convenient by removing from the space a certain plane not cutting the body. In the spherical model, the role of this plane can be played by the intersection of a supporting plane of the projecting cone with the sphere.

The corresponding plane in the projective model is the plane at infinity. When this is done, convex bodies in the spherical model are those in the open hemisphere $x^2 = 1$, $x_0 > 0$, and in the projective model they are bounded convex bodies in the euclidean space $x_0 = 1$.

When the plane $x_0 = 0$ has been removed from elliptic space, the straight lines and planes in the remaining part of the space possess the properties of euclidean straight lines and planes. We can thus define various concepts for convex bodies in elliptic space exactly as for convex bodies in euclidean space and prove the relevant theorems. This applies in particular to supporting planes and tangent cones.

As in the euclidean case, a convex surface in elliptic space is defined as a domain on the boundary of a convex body.

The entire boundary of a convex body is called a complete convex surface.

The distance between two points on a convex surface is defined as the infimum of the lengths of curves connecting the points on the surface. A curve whose length is equal to the distance between its endpoints is called a segment (shortest join).

Any two points on a complete convex surface in elliptic space can be joined by a segment. An incomplete convex surface (surface with boundary) may contain points which cannot be connected by a segment, but each point has a neighborhood in which any two points can be connected by a segment on the surface.

Convex surfaces in euclidean space satisfy Busemann's theorem, which states that the length of any curve connecting two points of a complete convex surface within the convex body bounded by the surface, and not lying entirely on the surface, is greater than the distance between the points on the surface (Chapter II, §1). In elliptic space, Busemann's theorem is not valid, at least in this formulation. Counterexamples are easily found. However, the theorem is valid locally, i.e. in a sufficiently small neighborhood of each point on the surface.

To verify this, use the projective model of elliptic space. Here a convex body is represented by a bounded euclidean convex body, its surface by a closed convex surface. Let δ denote the infimum of the lengths of the curves (in the metric of elliptic space) connecting the surface to the ideal plane. Now let the distance s between the points A and B be less than δ. We claim that any curve connecting A and B outside the body bounded by the surface has length greater than s. Suppose this is false, so that there exist such curves of length at most s.

Consider all curves of length at most s connecting A and B outside the surface. They clearly do not cut the ideal plane. One of these curves has minimal length. We may assume that it does not lie entirely on the surface. For if its length is less than s, it cannot lie on the surface because the distance between its endpoints is s. But if its length is s, then, by assumption, there exist curves of length s connecting A and B and not lying entirely on the surface. On the other hand, every component of this shortest curve outside the surface is a straight-line segment with endpoints on the surface, and therefore must lie on the surface. This contradiction proves our assertion.

Busemann's theorem enables us to extend Liberman's theorem on geodesics to convex surfaces in elliptic space, as follows.

Let γ be a geodesic on a convex surface (i.e. a curve which is locally a segment). Let O be a point in the interior of the body bounded by the surface. Connect O to all points of the geodesic by straight-line segments inside the body. Now, if the conic surface generated by these segments is developed on an elliptic plane, γ becomes a curve $\bar{\gamma}$ convex toward the "outside" of the domain covered by the cone. Indeed, otherwise $\bar{\gamma}$ would contain arbitrarily close points which can be connected by a small straight-line segment δ outside the domain covered by the developed cone. Now if γ_δ is the curve on the projecting cone of the geodesic γ that corresponds to the segment δ, then it is shorter than the corresponding subarc of the geodesic, which contradicts Busemann's theorem.

This fact has many corollaries, which we shall not list here, remarking only that it yields the existence of right and left semitangents at each point of a geodesic, and these semitangents are continuous from the right and left, respectively.

Let S be a point on a convex surface in elliptic space and γ a geodesic issuing from S. Let X be a point on γ close to S, and connect it to S by a straight-line segment. Let s be the distance between S and X along the geodesic, δ the extrinsic distance (i.e. distance in space) between the points, and ϑ the angle between the semitangent to γ at S and the straight-line segment SX. Then $s/\delta \to 1$ and $\vartheta \to 0$ as $X \to S$. This property of geodesics is a simple corollary of the existence and continuity of semitangents.

The image of the tangent cone of a convex surface under the geodesic mapping of elliptic space onto euclidean space (§1) is the tangent cone of the corresponding convex surface. Hence the mapping also preserves the property of a curve to have a semitangent. Since the semitangents of curves issuing from a point on a convex surface in euclidean space are the generators of its tangent cone, the same holds for convex surface in elliptic space. In particular, the semitangents of the geodesics issuing from a point on a convex surface in elliptic space are the generators of the tangent cone.

Another problem is that of the extrinsic curvature[1)] of convex surfaces in elliptic space. In euclidean space, the extrinsic curvature of a set M on a convex surface is defined as the area of the spherical image of the set. In elliptic space this definition causes difficulties, since we do not have the concept of parallel transport needed to take the normals to a single point.

Aleksandrov [**10**] overcomes this difficulty in the following way. Divide the set M into finitely many subsets m_k. Let P_k be a point in m_k, and construct the osculating euclidean space at this point. The osculating space is mapped geodesically onto a space of constant curvature (as in §1). Let $\omega(m_k)$ be the extrinsic curvature of the set corresponding to m_k on the convex surface in the osculating space. One proves that if the diameters of the sets m_k tend to zero, then $\sum \omega(m_k)$ has a well-defined limit which is independent of the partitions of M into subsets m_k. This limit is defined to be the extrinsic curvature of the surface on the set M.

Aleksandrov then proceeds to prove various properties of the extrinsic curvature of convex surfaces in elliptic space; in particular, he shows that it is completely additive on the ring of Borel sets.

When the convex surface is regular, the above definition of extrinsic curvature yields the following formula:

$$\omega(M) = \iint_M k_1 k_2 \, ds,$$

where k_1 and k_2 are the principal curvatures and ds is the area element of the surface.

Let F be a convex surface in elliptic space, O a point on F and X a point on F close to O. Connect X and O by a segment γ and let t denote the semitangent to the segment at O. Let OY be a straight-line segment on t, of length equal to the length s of the segment γ. We wish to estimate the length δ of the straight-line segment XY and to determine its direction.

The following statements hold for convex surfaces in euclidean space (Chapter II, §1):

[1)] That is to say, integral curvature.

1. The direction of XY, as a point on the unit sphere, is in the convex hull of the spherical image of γ.

2. $\delta/s \to 0$ as $X \to O$.

3. The angle ϑ between the straight-line segments OY and YX tends to $\pi/2$ as $X \to O$.

We shall now show that analogous statements hold for convex surfaces in elliptic space.

Consider the osculating euclidean space at the point O, which is a geodesic image of the elliptic space. Duplicate the above construction in this space, retaining the same notation with a bar added.

In the euclidean space, the first statement is proved first for polyhedra, as follows. The faces containing the geodesic γ are rotated about their edges until their planes coincide with the plane of the first "side" of the geodesic (which is a polygonal line). The geodesic γ then becomes a straight-line segment of length s whose direction is that of the first side; the endpoint of the geodesic (the point X) describes a smooth curve $\tilde{\gamma}$ consisting of circular arcs, which is an orthogonal trajectory of the family of supporting planes to the surface along γ. Thus all tangents of $\tilde{\gamma}$ are exterior normals to the surface along γ, and this implies property 1.

There are obviously no obstacles to performing this transformation of a geodesic polygon on a polyhedron in elliptic space. The trajectory $\tilde{\gamma}$ of the point X is mapped onto a smooth curve in the osculating euclidean space, which cuts the supporting planes of the polyhedron $\bar{F}$ along $\bar{\gamma}$ at angles which can be made arbitrarily close to $\pi/2$ provided s is small. It follows that the direction of the segment $\overline{XY}$ is in an ϵ'-neighborhood of the convex hull of the spherical image of $\bar{\gamma}$, where $\epsilon' \to 0$ as $s \to 0$.

The passage to an arbitrary convex surface, by approximating the surface with polyhedra, is essentially the same as in euclidean space. The final result can be formulated as follows.

The direction of the segment $\overline{XY}$ (the image of XY in the osculating euclidean space at O), as a point of the unit sphere, lies in an $\epsilon(s)$-neighborhood of the convex hull of the spherical image of $\bar{\gamma}$ (the image of γ), where $\epsilon(s) \to 0$ as $s \to 0$.

We now prove that the angle $\bar{\vartheta}$ between the segments $O\overline{X}$ and $\overline{XY}$ tends to $\pi/2$ as $s \to 0$. Suppose that this is false, and there is a sequence of points X_k such that $|\pi/2 - \vartheta_k| > \alpha > 0$. Without loss of generality, we may assume that the semitangents to γ_k at O converge to a generator t_0 of the tangent cone at O. Construct a geodesic sector V with apex at O, whose interior contains the direction t_0, which is so small that the spherical image of its image $\overline{V}$ on $\overline{F}$ is contained in a small neighborhood of the spherical image of $\bar{t}_0$, which is generated by the normals to the supporting planes through $\bar{t}_0$.

The geodesics γ_k are inside V, by the inclusion property. Therefore the directions of the segments $\overline{X}_k\overline{Y}_k$ are contained in an arbitrarily small neighborhood of the spherical image of the sector $\overline{V}$; hence the angle between the segments $O\overline{X}_k$ and $\overline{X}_k\overline{Y}_k$ is arbitrarily close to $\pi/2$, a contradiction.

It now follows from the properties of the osculating euclidean space that the angle ϑ between OX and XY also tends to $\pi/2$ as $s\to 0$.

Finally, let us prove that $\delta/s\to 0$ as $s\to 0$. If this is false, there exists a sequence of points X_k such that $s_k\to 0$ but $\delta_k/s_k > m > 0$. We repeat the same construction as above for this sequence, and take a sector V with small angle at the vertex. Without loss of generality, we may assume that the straight lines OX_k converge to some straight-line generator t_0' of the tangent cone. Since $\delta_k/s_k > m$, while the segments X_kY_k and X_kO are "almost perpendicular" (as we have just proved), it follows that the angle between OX_k and OY_k is always greater than some $\beta > 0$. Hence the generators t_0 and t_0' also form an angle greater than β. But this is impossible if the angle of the sector V is chosen sufficiently small. This contradiction proves the assertion.

Let R be a two-dimensional metric manifold, i.e. a two-dimensional manifold which is a metric space. The metric ρ of the manifold is said to be intrinsic if the distance $\rho(X, Y)$ between any two points X, Y of R is equal to the infimum of the lengths of curves connecting the points.

Let X be a point of R and γ_1 and γ_2 curves issuing from X. Take arbitrary points X_1 and X_2 on these curves, and construct a plane triangle with sides $\rho(X,X_1)$, $\rho(X,X_2)$ and $\rho(X_1,X_2)$. Let $\alpha(X_1,X_2)$ be the angle of the triangle opposite the side $\rho(X_1,X_2)$. The angle between the curves γ_1 and γ_2 at X is defined as $\underline{\lim}_{X_1,X_2\to X}\alpha(X_1,X_2)$. In this sense, the angle between any two curves at their common point always exists.

A manifold R with intrinsic metric is called a manifold of curvature at least K if the sum of the angles of any geodesic triangle is not less than the sum of angles of a triangle with the same sides on a plane in a space of constant curvature K. Manifolds of curvature at least K were introduced by Aleksandrov [11]. We shall recall the basic results of Aleksandrov's theory that have a bearing on our discussion.

If a manifold of curvature at least K is Riemannian, its Gauss curvature is at least K. Conversely, every Riemannian manifold with this property is a manifold of curvature at least K.

Any convex surface in an elliptic space of curvature K is a manifold of curvature at least K.

Let γ_1 and γ_2 be segments issuing from a point X on a manifold of curvature at least K. The angle between the segments at X is defined as $\underline{\lim}\, \alpha(X_1, X_2)$ as $X_1, X_2 \to X$. It turns out that $\alpha(X_1, X_2)$ always has a limit for segments, and so henceforth the angle between arbitrary curves γ_1 and γ_2 will be understood as the limit of the angle $\alpha(X_1, X_2)$ as $X_1, X_2 \to X$. In this sense the angle between curves does not always exist.

A necessary and sufficient condition for curves to form a definite angle is that each of them form a definite angle with itself (which is of course zero). Such curves are said to have a definite direction. A curve on a convex surface in elliptic space has a definite direction at a given point if and only if it has a semitangent at that point.

Given a manifold of curvature at least K, we define intrinsic curvature as the additive function defined for open geodesic triangles by $\alpha + \beta + \gamma - \pi$, where α, β, γ are the angles of the triangle. If the manifold of curvature at least K is Riemannian, the intrinsic curvature thus defined is identical with the integral curvature $\int\int K_i d\sigma$, where K_i is the Gauss curvature of the manifold and $d\sigma$ its area element.

Area in a manifold of curvature at least K is defined as follows. Let G be a domain in the manifold and Δ any triangulation of G. For each triangle $\delta \in \Delta$, construct a plane triangle with the same sides, and let $s(\delta)$ be its area. Then the area of G is defined as

$$\limsup \sum_{\Delta} \sigma(\delta)$$

as the diameters of the triangles δ tend to zero. For Riemannian manifolds this yields the usual area concept.

Gauss's theorem on the relation between intrinsic and extrinsic curvature can be extended to convex surfaces in a space of constant curvature K: If G is any domain on the convex surface, $\varphi(G)$ its extrinsic curvature, $\omega(G)$ its intrinsic curvature and $\sigma(G)$ its area, then

$$\omega(G) = \varphi(G) + K\sigma(G).$$

Let X be a point on a manifold of curvature at least K, and let γ and γ' be two curves issuing from X (which is their only common point) and having definite directions at X. These curves divide any neighborhood of the point X into two "sectors." Let V be one of them. Let γ_k be segments issuing from X inside V, indexed consecutively from γ to γ'. Let $\alpha_1, \cdots, \alpha_n$ be the angles between consecutive segments. The number

$$\vartheta(\gamma, \gamma') = \sup \sum_k \alpha_k,$$

where the supremum extends over all systems of segments satisfying the above conditions, is called the angle of the sector between γ and γ' (the angle as viewed from V). The sector angle possesses the usual properties: If curves γ_1, γ_2 and γ_3 issuing from X define three sectors V_{12}, V_{23} and V_{13}, where $V_{12} + V_{23} = V_{13}$, then $\vartheta(\gamma_1, \gamma_2) + \vartheta(\gamma_2, \gamma_3) = \vartheta(\gamma_1, \gamma_3)$. For a convex surface in a space of curvature K, the sector angle coincides with the angle between the semitangents of γ and γ' on the surface obtained by developing the tangent cone at X on a plane.

Let γ be a simple curve on a manifold of curvature at least K with definite directions at its endpoints. Fix some direction on the curve γ, defining thereby a right and left semineighborhood. Connect the endpoints of γ by a simple geodesic polygon Γ in the right semineighborhood of γ. Let α_k be the angles between consecutive sides of the polygon, α and β the angles between the initial and terminal sides, respectively, and γ; all angles are measured as viewed from the domain bounded by γ and Γ. The right integral geodesic curvature (i.g.c.) is defined as

$$\lim_{\Gamma \to \gamma} (2\pi - \alpha - \beta) + \sum_k (\pi - \alpha_k).$$

The left i.g.c. is defined similarly.

The i.g.c. has the following property. If C is a point dividing the curve γ into curves γ_1 and γ_2 which have definite directions at C, then

$$\psi(\gamma) = \psi(\gamma_1) + \psi(\gamma_2) + \pi - \vartheta,$$

where ψ denotes the (right or left) i.g.c. and ϑ the angle between the curves γ_1 and γ_2 at C (from the right or left).

If the manifold is Riemannian and the curve regular, the i.g.c. is precisely the integral of the usual geodesic curvature with respect to arc length along the curve.

There is a simple relation between the right and left i.g.c. of a curve and the intrinsic curvature of the surface along the curve: The sum of the right and left i.g.c. is equal to the intrinsic curvature of the surface on the set of points of the curve.

The Gauss-Bonnet theorem extends to manifolds of curvature at least K: If G is a domain homeomorphic to a disk, bounded by a closed curve γ, $\omega(G)$ is the curvature of G and $\psi(\gamma)$ the i.g.c. of γ toward G, then $\omega(G)+\psi(\gamma)=2\pi$.

Let R_1 and R_2 be two manifolds of curvature at least K, and γ_1 and γ_2 subarcs of their boundaries which can be mapped isometrically onto each other. Assume that the sum of i.g.c. of any two corresponding subarcs of γ_1 and γ_2 is nonnegative. Then there exists a manifold R of curvature at least K which is the union of two domains, isometric to R_1 and R_2 respectively, whose common boundary in R is a curve γ isometric to γ_1 and γ_2 (gluing theorem).

As shown above, every convex surface in an elliptic space of curvature K is a manifold of curvature at least K. It is natural to ask whether every manifold of curvature at least K is isometric to some convex surface in an elliptic space.

The most general result in this respect is due to Aleksandrov. He proved that any complete manifold of curvature at least $K>0$ is isometric to some closed convex surface in an elliptic space of curvature K. A manifold of curvature at least K, homeomorphic to a disk and having nonnegative i.g.c. on any subarc of its boundary, is isometric to a convex cap.

In Chapter VI we shall study the general problem of isometric embedding of a two-dimensional Riemannian manifold in a three-dimensional Riemannian manifold. Applied to elliptic space in all theorems of Aleksandrov, the results of Chapter VI guarantee the existence of a regular embedding, provided the metric of the original manifold is sufficiently regular and the curvature of the manifold strictly greater than that of the space.

§3. Mappings of congruent figures

In this section we shall define and examine certain special mappings of congruent figures in elliptic space onto congruent figures in euclidean space, and vice versa. The results will be applied to the study of pairs of isometric surfaces in the next section.

Let R be an elliptic space with curvature $K=1$. Introduce Weierstrass coordinates x_i in R. Remove the plane $x_0=0$ and denote the remaining part of the space by R_0. In R_0 we can impose the additional requirement $x_0>0$, so that the Weierstrass coordinates become unique. The usual mapping of a point of elliptic space in R_0 with Weierstrass coordinates x_i onto the point of euclidean four-space with the same cartesian coordinates is now an isometry, mapping R_0 onto the unit hemisphere $x^2 = 1$, $x_0 > 0$.

Let F be some figure in the domain R_0 of elliptic space, which is mapped by a motion A of elliptic space onto a congruent figure AF in R_0. Map each point $x \in F$ onto the point of the euclidean space E_0 ($x_0 = 0$) given by the formula

$$Tx = \frac{x - e_0(xe_0)}{e_0(x + Ax)}.$$

The point Tx is indeed in E_0, since $e_0 Tx = 0$.

As the point x describes the figure F, the corresponding point Tx of E_0 describes a certain figure TF.

LEMMA 1. *T is a geodesic mapping.*

PROOF. We first show that T is a homeomorphism.

Since F and AF are both in R_0, we have $e_0(x + Ax) > 0$, so that T is a continuous mapping. We now show that the images of different points x and y under T are different. Suppose that $Tx = Ty$. Then clearly $y = \lambda x + \mu e_0$, and so

$$Ty = \frac{x - e_0(xe_0)}{e_0(x + Ax) + \frac{\mu}{\lambda} e_0(e_0 + Ae_0)}.$$

Without loss of generality we may assume that $Ae_0 \neq -e_0$, since A and $-A$ define the same motion in elliptic space. Therefore $e_0(e_0 + Ae_0) > 0$, and hence $Ty = Tx$ can hold in only two cases: $\mu = 0$ and $x - e_0(e_0x) = 0$. In the first case $y = \lambda x$, and since $x^2 = y^2 = 1$ and $x_0, y_0 > 0$, we must have $x = y$. In the second case $x = e_0$ and $y = e_0(\lambda + \mu)$, which again implies $x = y$. Since T is also continuous, it is a homeomorphism.

We now prove that T is a geodesic mapping. It will suffice to show that the inverse mapping T^{-1} is geodesic. Consider an arbitrary plane in E_0, defined by the equation $(a, y) + b = 0$, where a is a vector and b a scalar. Setting $y = Tx$ in the equation, we get the equation of its image in R under the mapping T^{-1}. Multiplying $(A, Tx) + b = 0$ by $e_0(x + Ax)$, we get an equation which is linear in x and is therefore the equation of a plane. Since the homeomorphism T^{-1} maps planes of E_0 onto planes of R, it is a geodesic mapping, and hence so is T. This proves the lemma.

Let F and F' be two congruent figures in the domain R_0 of elliptic space. Let A be a motion of elliptic space which maps F onto F'. By Lemma 1, the mapping T' of F' into the euclidean space E_0, defined by

$$T'x' = \frac{x' - e_0(x'e_0)}{e_0\,(x' + A^{-1}x')}$$

is a geodesic mapping.

LEMMA 2. *The mapping Ω of the figure TF onto the figure $T'F'$ in the euclidean space E_0, under which each point Tx is mapped onto TAx, is a motion, and so the figures TF and $T'F'$ are congruent.*

PROOF. First note that the mapping Ω depends continuously on A. Moreover, if A is the identity, then so is Ω. It will therefore suffice to prove that Ω is an isometry for any A.

Let x and y be two arbitrary points of F. We claim that

$$(Tx - Ty)^2 = (T'Ax - T'Ay)^2.$$

Since the formulas for Tx and $T'Ax$ are homogeneous of degree zero in x, while $e_0(x + Ax) > 0$ on F, it follows that the coordinates of x can be normalized by the condition $e_0(x + Ax) = 1$. Setting $x - y = z$, we get

$$(Tx - Ty)^2 = (z - e_0\,(ze_0)\,)^2 = z^2 - (ze_0)^2,$$
$$(T'Ax - T'Ay)^2 = (Az - e_0\,(Az,\ e_0)\,)^2 = (Az)^2 - (e_0,\ Az)^2.$$

Now $(Az)^2 = z^2$; by normalization $e_0(z + Az) = 0$, and so $(ze_0)^2 = (e_0, Az)^2$; thus

$$(Tx - Ty)^2 = (T'Ax - T'Ay)^2.$$

This proves the lemma.

Let Φ be an arbitrary figure in the euclidean space E_0 and B any motion of E_0. We define a mapping S of Φ into elliptic space by

$$Sx = \rho(2x + e_0(1 + (Bx)^2 - x^2)),$$

where ρ is a normalizing factor defined by $(Sx)^2 = 1$. It is easy to see that

$$1/\rho^2 = 1 + 2(x^2 + (Bx)^2) + (x^2 - (Bx)^2)^2 \geqq 1.$$

LEMMA 3. *S is a geodesic mapping.*

PROOF. We first show that S is a homeomorphism. It is sufficient to prove that S is continuous and one-to-one.

The continuity of S is obvious from $1/\rho^2 \geqq 1$.

Now suppose that x and y are two points such that $Sx = Sy$. Then, since x and y are perpendicular to e_0, it follows that x and y are parallel. Let τ be a unit vector parallel to x and y. Then

$$x = \lambda\tau, \qquad y = \mu\tau,$$
$$Sx = \rho(2\lambda\tau + e_0(1 - \lambda^2 + (Bx)^2)),$$
$$Sy = \rho(2\mu\tau + e_0(1 - \mu^2 + (By)^2)).$$

Since $Sx = Sy$ and $\tau \perp e_0$,

$$\frac{1 - \lambda^2 + (Bx)^2}{\lambda} = \frac{1 - \mu^2 + (By)^2}{\mu}.$$

Decomposing B into a rotation B^* and a translation C, we get

$$Bx = \lambda\tau^* + C, \quad By = \mu\tau^* + C, \quad \tau^* = B^*\tau.$$

The preceding equality now becomes

$$2\tau^*C + \frac{1 + C^2}{\lambda} = 2\tau^*C + \frac{1 + C^2}{\mu}.$$

Hence $\lambda = \mu$, and so $x = y$. This contradiction proves that S is one-to-one and hence a homeomorphism.

We now show that S is a geodesic mapping. Again, it will suffice to show that S^{-1} maps planes of R onto planes of E_0.

A plane in R is defined by an equation $ax = 0$. Substituting Sx for x in this equation, we get the equation of its image in E_0 under S^{-1}:

$$a(2x + e_0(1 + (Bx)^2 - x^2)) = 0.$$

It is not hard to see that this equation is linear in x. In fact,

$$(Bx)^2 = (B^*x + C)^2 = x^2 + 2CB^*x + C^2,$$

and the equation becomes

$$a(2x + e_0(1 + 2CB^*x + C^2)) = 0.$$

Thus the mapping S^{-1} takes planes of R to planes of E_0 and is therefore a geodesic mapping. Hence its inverse S is also geodesic, and the lemma is proved.

Now let Φ and Φ' be congruent figures in the euclidean space E_0 and B a motion mapping Φ onto Φ'. By Lemma 3, the mapping S' of Φ' into elliptic space defined by

$$S'x' = \rho(2x' + e_0(1 + (B^{-1}x')^2 - x'^2))$$

is a geodesic mapping.

LEMMA 4. *The mapping $\bar{\Omega}$ of the figure $S\Phi$ onto the figure $S'\Phi'$ in elliptic space which takes each point Sx onto $S'Bx$ is a motion, and so the figures $S\Phi$ and $S'\Phi'$ are congruent.*

PROOF. Since the mapping $\overline{\Omega}$ depends continuously on B and is the identity whenever B is the identity, it will suffice to prove that it is an isometry for any B. This will be proved if we can show that for any x and y in E_0

$$(Sx - Sy)^2 = (S'Bx - S'By)^2.$$

Denote

$$\alpha(x) = 2x + e_0(1 + (Bx)^2 - x^2),$$
$$\beta(x) = 2Bx + e_0(1 - (Bx)^2 + x^2).$$

We claim that $\alpha(x)\alpha(y) = \beta(x)\beta(y)$ for any x and y in E_0.

Since x and y are perpendicular to e_0,

$$\alpha(x)\alpha(y) = 4xy + (1 + (Bx)^2 - x^2)(1 + (By)^2 - y^2),$$
$$\beta(x)\beta(y) = 4(Bx)(By) + (1 - (Bx)^2 + x^2)(1 - (By)^2 + y^2).$$

If the motion B is now decomposed into a rotation B^* and a translation C, we get

$$(Bx)^2 = x^2 + 2C(B^*x) + C^2, \quad (By)^2 = y^2 + 2C(B^*y) + C^2,$$
$$(Bx)(By) = xy + C(B^*x + B^*y) + C^2.$$

Substituting these expressions into $\alpha(x)$, $\alpha(y)$ and $\beta(x)$, $\beta(y)$, we get $\alpha(x)\alpha(y) = \beta(x)\beta(y)$.

Now for any z (in particular, for $z = x$ and $z = y$) we have

$$(Sz)^2 = 1, \ (S'Bz)^2 = 1, \ Sz = \frac{\alpha(z)}{\sqrt{(\alpha(z))^2}}, \qquad S'Bz = \frac{\beta(z)}{\sqrt{(\beta(z))^2}}$$

Hence $(Sx - Sy)^2 = (S'Bx - S'By)^2$, and the proof is complete.

Note that the correspondences between pairs of congruent figures in elliptic and euclidean space established in Lemmas 2 and 4 are mutually inverse. That is to say, if we start from a pair of congruent figures F and F' in elliptic space, construct a pair of congruent figures Φ and Φ' in euclidean space by Lemma 2, and then construct the corresponding pair of congruent figures in elliptic space by Lemma 4, we come back to F and F'.

A deformation of a figure F at $t = 0$ is called an infinitesimal motion if the distance between any two points of F is stationary at $t = 0$. Now the distance between points x and y is stationary if and only if the function $(x - y)^2$ [2)] is stationary; hence the field of velocities $\zeta = dx/dt$

[2)] In euclidean space, the expression $(x - y)^2$ denotes the square of the distance d between the points x and y; in elliptic space it is $4\sin^2(d/2)$.

of an infinitesimal motion satisfies the condition

$$(x-y)(\zeta(x)-\zeta(y))=0.$$

Conversely, any deformation of the figure whose velocity field satisfies this condition is an infinitesimal motion.

LEMMA 5. *Let $\zeta(x)$ be the velocity field of an infinitesimal motion of a figure F in the elliptic space R. With each point*

$$Tx=(x-e_0(xe_0))/(e_0x)$$

of the euclidean space E_0, associate a vector

$$z(Tx)=(\zeta-e_0(e_0\zeta))/(e_0x).$$

Then $z(Tx)$ is the velocity field of an infinitesimal motion of TF in E_0.

PROOF. We first prove that the mapping A of R onto itself which takes each point $u=\rho(x+\epsilon\zeta)$ into $Au=\rho(x-\epsilon\zeta)$ is a motion.

The mapping A depends continuously on ϵ and is the identity for $\epsilon=0$. Hence we need only prove that A is an isometry for any ϵ.

Since $x^2=1$ we have $x\zeta=0$, and consequently the normalizing factors ρ in the expressions for u and Au are equal and $1/\rho^2=x^2+\epsilon^2\zeta^2$. Furthermore, since

$$x\zeta(x)=y\zeta(y)=0, \qquad (x-y)(\zeta(x)-\zeta(y))=0.$$

it follows that

$$(x+\varepsilon\zeta(x))(y+\varepsilon\zeta(y))=(x-\varepsilon\zeta(x))(y-\varepsilon\zeta(y)).$$

Recalling now that $u^2=Au^2=1$, we see that $(u-v)^2=(au-Av)^2$ for any u and v. But this means that A is a motion in the space R.

By Lemma 1, the mapping of E_0 which takes each point

$$w=(u-e_0(e_0u))/e_0(u+Au)$$

onto the point

$$Bw=(Au-e_0(e_0Au))/e_0(u+Au)$$

is a motion. Hence

$$z=\lim_{\varepsilon\to 0}\frac{w-Bw}{2\varepsilon}=\frac{\zeta-e_0(\zeta e_0)}{(e_0x)},$$

referred to the point

$$Tx=(x-e_0(xe_0))/(e_0x)$$

is the velocity field of an infinitesimal motion. This proves the lemma.

LEMMA 6. *Let $z(x)$ be the velocity field of an infinitesimal motion of a figure Φ in the euclidean space E_0. Map each point*

$$Sx = \rho(x + e_0), \qquad x \subset \Phi,$$

of the elliptic space R onto the vector

$$\zeta(S(x)) = \frac{z - e_0(xz)}{\sqrt{1 + x^2}}.$$

Then ζ is the field of an infinitesimal motion in R.

PROOF. We first prove that the mapping B of E_0 onto itself under which each point $u = x + \epsilon z$ goes into the point $Bu = x - \epsilon z$ is a motion. Indeed, it is continuous in ϵ, the identity for $\epsilon = 0$, and for any x and y

$$(x - y \pm \epsilon(z(x) - z(y)))^2 = (x - y)^2 + \epsilon^2(z(x) - z(y))^2,$$

i.e., for any u and v,

$$(u - v)^2 = (Bu - Bv)^2.$$

By Lemma 4, the motion B of euclidean space corresponds to a motion A of elliptic space which takes the point

$$w = \rho(2u + e_0(1 + (Bu)^2 - u^2))$$

to the point

$$Aw = \rho(2Bu + e_0(1 - (Bu)^2 + u^2)).$$

It follows that

$$\xi = \lim_{\varepsilon \to 0} \frac{w - Aw}{4\varepsilon} = \frac{z - e_0(xz)}{\sqrt{1 + x^2}},$$

referred to the point

$$Sx = \frac{x + e_0}{\sqrt{1 + x^2}}$$

is the field of velocities of an infinitesimal motion in elliptic space. Q.E.D.

The correspondences between figures in elliptic and euclidean spaces and their infinitesimal motions, as established in Lemmas 5 and 6, are mutually inverse.

Consider two straight lines g' and g'' in the elliptic space R, given by equations

$$x' = \rho(a' + \lambda\tau'), \qquad x'' = \rho(a'' + \lambda\tau''),$$

where

$$a'^2 = a''^2 = 1, \quad a'\tau' = a''\tau'' = 0, \quad \tau'^2 = \tau''^2 \neq 0.$$

It is readily seen that for any λ and μ

$$(x'(\lambda) - x'(\mu))^2 = (x''(\lambda) - x''(\mu))^2$$

and so the mapping of g' onto g'' which associates points corresponding to the same values of λ is a motion.

By Lemma 2 the equations

$$y' = (x' - e_0(x'e_0))/e_0(x' + x''), \quad y'' = (x'' - e_0(x''e_0))/e_0(x' + x'')$$

define two congruent figures h' and h'' in the euclidean space E_0. By Lemma 1 they are straight lines.

Since the normalizing factors ρ in the equations of g' and g'' are equal, the right-hand sides of the equations of h' and h'' are linear rational functions of λ.

The mapping of the straight line g' onto the straight line h', defined by equal parameters λ, is a homeomoephism. Hence the derivative y'_λ cannot vanish identically. And since $y'(\lambda)$ is a linear rational function of λ, it follows that $y'_\lambda \neq 0$ for all λ. The same holds for y''_λ.

By Lemma 2, the mapping of h' onto h'' defined by equal parameters λ is a motion. Hence $|y'_\lambda| = |y''_\lambda|$.

Now let h' and h'' be two straight lines in the euclidean space E_0, defined by the equations

$$y' = b' + \mu t', \qquad y'' = b'' + \mu t'',$$

where $|t'| = |t''| \neq 0$. Obviously the mapping of h' onto h'' defined by equal parameters μ is a motion.

By Lemma 4 the equations

$$x' = \rho(2y' + e_0(1 - y'^2 + y''^2)), \quad x'' = \rho(2y'' + e_0(1 - y''^2 + y'^2))$$

define two congruent figures g' and g'' in the elliptic space R. By Lemma 3 they must be straight lines.

Since the functions

$$2y' + e_0(1 - y'^2 + y''^2), \qquad 2y'' + e_0(1 - y''^2 + y'^2)$$

are linear in μ with nonzero coefficients, the derivatives x'_μ and x''_μ never vanish.

Since the mapping of g' onto g'' defined by equal parameters μ is a motion, it follows that $|x'_\mu| = |x''_\mu|$.

Analogous conclusions hold for planes, as we shall now see.

Let α' and α'' be two planes in elliptic space, defined by equations

$$x' = \rho(a' + \lambda\tau'_1 + \mu\tau'_2), \qquad x'' = \rho(a'' + \lambda\tau''_1 + \mu\tau''_2),$$

where

$$a'^2 = a''^2 = 1, \quad a'\tau_1' = a''\tau_1'', \quad a'\tau_2' = a''\tau_2'',$$
$$\tau_1'^2 = \tau_1''^2, \quad \tau_2'^2 = \tau_2''^2, \quad \tau_1'\tau_2' = \tau_1''\tau_2'', \qquad (a'\tau_1'\tau_2') \neq 0 \quad (a''\tau_1''\tau_2'') \neq 0.$$

It is easy to see that the mapping of α' onto α'' defined by equal parameters λ and μ is a motion.

By Lemmas 1 and 2 the equations

$$y' = \frac{x' - e_0\,(x'e_0)}{e_0\,(x' + x'')}, \qquad y'' = \frac{x'' - e_0\,(x''e_0)}{e_0\,(x' + x'')}$$

define two planes β' and β'' in euclidean space, and the mapping of one of these planes onto the other defined by equal λ and μ is a motion.

Since $y'(\lambda, \mu)$ is a linear rational function of λ and μ, while the mapping of α' onto β' (defined by equal parameters) is a homeomorphism, the vectors y_λ' and y_μ' are independent and nonzero for any λ and μ. The same holds for the vectors y_λ'' and y_μ''.

The correspondence between the planes β' and β'' defined by equal parameters λ and μ is an isometry (Lemma 2), and so

$$y_\lambda'^2 = y_\lambda''^2, \qquad y_\lambda' y_\mu' = y_\lambda'' y_\mu'', \qquad y_\mu'^2 = y_\mu''^2.$$

Now consider two planes β' and β'' in euclidean space, defined by equations

$$y' = b' + \lambda t_1' + \mu t_2', \qquad y'' = b'' + \lambda t_1'' + \mu t_2'',$$

where

$$t_1'^2 = t_1''^2, \quad t_2'^2 = t_2''^2, \quad t_1' t_2' = t_1'' t_2'', \quad t_1'^2 t_2'^2 - (t_1' t_2')^2 \neq 0.$$

It is clear that the mapping of β' onto β'' defined by equal parameters λ, μ is a motion. By Lemmas 3 and 4, the equations

$$x' = \rho(2y' + e_0(1 - y'^2 + y''^2)), \quad x'' = \rho(2y'' + e_0(1 - y''^2 + y'^2))$$

define two planes α' and α'' in elliptic space, and the mapping defined here by equal parameters λ and μ is again a motion.

It is easily seen that the functions

$$2y' + e_0(1 - y'^2 + y''^2)), \quad 2y'' + e_0(1 - y''^2 + y'^2)$$

are linear in λ and μ, and the coefficients of λ and μ in each of these functions are independent, by virtue of the independence of the pairs of vectors τ_1', τ_2' and τ_1'', τ_2''. It follows that

$$(x' x_\lambda' x_\mu') \neq 0, \qquad (x'' x_\lambda'' x_\mu'') \neq 0.$$

Since the correspondence between the planes α' and α'' defined by equal parameters λ and μ is an isometry, we have

$$x_\lambda'^2 = x_\lambda''^2, \quad x_\mu' x_\lambda' = x_\lambda'' x_\mu'', \quad x_\mu'^2 = x_\mu''^2.$$

§4. Isometric surfaces

The mapping considered in the preceding section which takes pairs of congruent figures in elliptic space into pairs of congruent figures in euclidean space, and vice versa, is easily extended to congruence in the small; in particular, to isometric surfaces. This is the goal of the present section.

Let F' and F'' be two isometric regular surfaces in the domain R_0 ($x_0 > 0$) of the elliptic space R with curvature $K = 1$. Introduce a coordinate net u, v on each surface such that points corresponding to each other under the isometry are given by the same parameter values u, v. Let $x' = x'(u,v)$ and $x'' = x''(u,v)$ be the equations of these surfaces, where $(x' x_u' x_v') \neq 0$ and $(x'' x_u'' x_v'') \neq 0$.

Consider the two surfaces in the euclidean space E_0 ($x_0 = 0$) defined by the equations $y = y'(u,v)$ and $y = y''(u,v)$, where

$$y' = \frac{x' - e_0(x' e_0)}{e_0(x' + x'')}, \qquad y'' = \frac{x'' - e_0(x'' e_0)}{e_0(x' + x'')}.$$

THEOREM 1. *The surfaces Φ' and Φ'' are regular, have no singularities and are isometric. They are congruent if and only if F' and F'' are congruent.*

PROOF. Let A' and A'' be two points on F' and F'' which correspond under the isometry. Without loss of generality we may assume that they are given by parameter values $u = v = 0$. Let α' and α'' be the tangent planes at A' and A'', respectively. These planes can be defined by the equations $x = \bar{x}'(u,v)$ and $x = \bar{x}''(u,v)$, where

$$\bar{x}' = \rho\left(x'(0,0) + u x_u'(0,0) + v x_v'(0,0)\right),$$
$$\bar{x}'' = \rho\left(x''(0,0) + u x_u''(0,0) + v x_v''(0,0)\right)$$

The vectors in the right-hand sides of these expressions for $\bar{x}'$ and $\bar{x}''$ obviously satisfy the conditions

$$x'^2 = x''^2 = 1, \quad x' x_u' = x' x_v' = 0, \quad x'' x_u'' = x'' x_v'' = 0,$$
$$x_u'^2 = x_u''^2, \quad x_u' x_v' = x_u'' x_v'', \quad x_v'^2 = x_v''^2.$$

Therefore, according to §3, the equations

$$y = \bar{y}'(u,v), \qquad y = \bar{y}''(u,v),$$

where

$$\bar{y}' = \frac{\bar{x}' - e_0(\bar{x}'e_0)}{e_0(\bar{x}' + \bar{x}'')}, \qquad \bar{y}'' = \frac{\bar{x}'' - e_0(\bar{x}''e_0)}{e_0(\bar{x}' + \bar{x}'')},$$

define two planes β' and β'' in the euclidean space E_0. Define a mapping of α^i onto β^i by taking each point (u, v) of α^i to the point of β^i with the same coordinates (u, v); this mapping is a nondegenerate projective (geodesic) mapping (§3). The mapping of α' onto α'' defined by equal parameters u, v is a motion.

Since the distance between corresponding points of the surface F^i and its tangent plane α^i is of order at least u^2+v^2 in the neighborhood of A^i, it follows that the distance between corresponding points of β^i and Φ^i near the point B^i corresponding to A^i is also of order at least u^2+v^2.

It follows that each of the figures Φ^i is indeed a surface (necessarily nondegenerate), β^i is a tangent plane to Φ^i, and the mapping of the surface Φ^i onto the plane β^i defined by equal parameters u, v is an isometry at the point B^i.

Since the mapping of β' onto β'' defined by equal parameters u, v is also an isometry, the mapping of Φ' onto Φ'' defined in the same way is an isometry at $u=v=0$. But the point $u=v=0$ was chosen arbitrarily. Hence the surfaces Φ' and Φ'' are locally isometric.

We know (§3) that the figures F' and F'' are congruent if and only if Φ' and Φ'' are congruent. This completes the proof of Theorem 1.

Let Φ' and Φ'' be two isometric regular surfaces in the euclidean space E_0, parametrized in such a way that points corresponding under the isometry have the same coordinates u, v. Let $y=y'(u,v)$ and $y=y''(u,v)$ be the equations of the surfaces, where

$$y'_u \times y'_v \neq 0, \qquad y''_u \times y''_v \neq 0.$$

Consider the surfaces F' and F'' defined in the elliptic space R by the equations

$$x = x'(u,v), \qquad x = x''(u,v),$$

where

$$x' = \rho(2y' + e_0(1 - y'^2 + y''^2)), \quad x'' = \rho(2y'' + e_0(1 - y''^2 + y'^2)).$$

THEOREM 2. *The surfaces F' and F'' are regular, have no singularities and are isometric. They are congruent if and only if Φ' and Φ'' are congruent.*

The proof is analogous to that of Theorem 1 and will be omitted.

Note that if we start from a pair of isometric surfaces F' and F'' in elliptic space, construct the corresponding pair of isometric surfaces Φ' and Φ'' in euclidean space (Theorem 1), and then use the latter to construct a pair of isometric surfaces in the elliptic space (Theorem 2), we come back to F' and F''.

By Theorem 1, each pair of isometric surfaces F' and F'' in the domain R_0 of the elliptic space R goes into a pair of isometric surfaces Φ' and Φ'' in the euclidean space E_0. It turns out that in certain cases this transformation preserves the convexity of the surfaces F' and F'':

THEOREM 3. *Under the assumptions of Theorem* 1, *if the surfaces F' and F'' in elliptic space are locally convex and have the same orientation, and moreover each of the surfaces F' and F'' is visible from the point e_0 $(1,0,0,0)$ from within, then the corresponding surfaces Φ' and Φ'' in euclidean space are also locally convex.*

A surface is said to be locally convex if any point has a neighborhood which is a convex surface. A sufficient condition for a surface to be locally convex is that its second fundamental form be semidefinite.

PROOF OF THEOREM 3. Let y' be an arbitrary point on the surface Φ', and $y' + \Delta y'$ a point close to y'. Let the corresponding points on F' and F'' be $x', x' + \Delta x'$ and $x'', x'' + \Delta x''$, respectively. Connect the points x' and $x' + \Delta x'$ by a segment γ' on the surface F', the points x'' and $x'' + \Delta x''$ by a segment γ'' on the surface F''. Then by §1

$$x' + \Delta x' = \rho\left(\left(1 + \frac{\Delta s^2}{2}\right) x' + \Delta s \tau' + \frac{\Delta s^2}{2} k' \nu' + \dots\right),$$

$$x'' + \Delta x'' = \rho\left(\left(1 + \frac{\Delta s^2}{2}\right) x'' + \Delta s \tau'' + \frac{\Delta s^2}{2} k'' \nu'' + \dots\right),$$

where τ' and τ'' are the unit tangent vectors of the geodesics γ' and γ'' at the points x' and x''; ν' and ν'' are the unit vectors of the principal normals to the curves; k' and k'' are the curvatures of γ' and γ''; Δs is the distance between the points x^i and $x^i + \Delta x^i$ on the surface F^i. Terms of order higher than Δs^2 are omitted. Substitute these expressions in the formula for y'. Then

$$y' + \Delta y' = \frac{x' + \varepsilon' - e_0\,(e_0 x' + e_0 \varepsilon')}{e_0 x' + e_0 x'' + (e_0 \varepsilon' + e_0 \varepsilon'')} + O\,(\Delta s^3),$$

where

$$\varepsilon' = \Delta s \tau' + \frac{\Delta s^2}{2} k' \nu', \quad \varepsilon'' = \Delta s \tau'' + \frac{\Delta s^2}{2} k'' \nu''.$$

Using the expression for $y' + \Delta y'$ up to terms of order Δs^2, we get

$$\Delta y' = \frac{\Delta s}{e_0 (x' + x'')} \{- y' (e_0 \tau' + e_0 \tau'') + \tau' - e_0 (e_0 \tau')\}$$
$$+ \Delta s^2 \frac{e_0 (\tau' + \tau'')}{e_0 (x' + x'')} \{y' (e_0 \tau' + e_0 \tau'') - \tau' + e_0 (e_0 \tau')\}$$
$$+ \frac{\Delta s^2}{2e_0 (x' + x'')} \{- (k' e_0 \nu' + k'' e_0 \nu'') + k' \nu' - k' e_0 (e_0 \nu')\} + O (\Delta s^3).$$

Multiply $\Delta y'$ by the unit vector n' of the surface Φ' at the point y'. The result must be a quantity of order no lower than Δs^2, and hence

$$n' \{ - y' (e_0 \tau' + e_0 \tau'') + \tau' - e_0 (e_0 \tau') \} = 0.$$

Thus

$$n' \Delta y' = \frac{\Delta s^2 k'}{2e_0 (x' + x'')} (- y' (e_0 \nu') + \nu') n'$$
$$- \frac{\Delta s^2 k''}{2e_0 (x' + x'')} (e_0 \nu'') (y' n') + O (\Delta s^3).$$

Since both expressions

$$\frac{\Delta s^2 k'}{2e_0 (x' + x'')}, \qquad \frac{\Delta s^2 k''}{2e_0 (x' + x'')}$$

are positive, the proof that Φ' is locally convex will be complete if we can show that the expressions

$$(*) \qquad (- y' (e_0 \nu') + \nu') n', \qquad - (e_0 \nu'') (y' n')$$

have the same sign, irrespective of the choice of the points y' and $y' + \Delta y'$ on the surface Φ'.

The proof that this is indeed the case is rather tedious, and we therefore set it apart as a special lemma, which we state now and prove later.

Let α' and α'' be two planes in the elliptic space R, mapped onto each other by an isometry; let β' and β'' be the corresponding planes in the euclidean space E_0 (§3). Let a_0 and b_0 be two corresponding points on the planes α' and α'', a_3 and b_3 corresponding unit normals at these points, and c_0 and c_3 a point and a normal on the plane β'.

LEMMA 1. *If the point e_0 lies on the same side of each plane α' and α'', then the expressions*

$$A = c_3 (- (e_0 a_3) c_0 + a_3), \qquad B = - (e_0 b_3) (c_0 c_3)$$

have the same sign. The signs of these expressions are invariant under a motion of the planes α' and α''.[3]

[3] The correspondence between the sides of two planes is established in the usual manner, via the isometry.

The isometry of the surfaces F' and F'' induces a natural isometry of their tangent planes α' and α'' at the points x' and x''. The fact that the expressions (*) have the same sign now follows from Lemma 1. The fact that Φ'' is locally convex is proved in a similar manner. This completes the proof of Theorem 3.

PROOF OF LEMMA 1. Let a_1 and a_2 denote perpendicular unit vectors at the point a_0 in the plane α', b_1 and b_2 the corresponding vectors in the plane α'', and c_1 and c_2 the corresponding vectors in the plane β'. Then

$$a_3 = (a_0 a_1 a_2), \quad b_3 = (b_0 b_1 b_2), \quad c_3 = (e_0 c_1 c_2).$$

To find an expression for the vectors c_1 and c_2, consider the equation of the plane β':

$$y = (x_1 - e_0(x_1 e_0))/e_0(x_1 + x_2),$$

where x_1 and x_2 are corresponding points of α' and α''. Differentiating y in the directions of c_1 and c_2 at the point c_0, we get

$$c_1 = \frac{1}{\lambda_0^2}(a_1\lambda_0 - a_0\lambda_1) + e_0\,(*),$$

$$c_2 = \frac{1}{\lambda_0^2}(a_2\lambda_0 - a_0\lambda_2) + e_0\,(*),$$

where we have used the abbreviations

$$\lambda_0 = e_0(a_0 + b_0), \quad \lambda_1 = e_0(a_1 + b_1), \quad \lambda_2 = e_0(a_2 + b_2).$$

Substituting these expressions for c_1 and c_2 in $c_3 = (e_0 c_1 c_2)$, we get

$$c_3 = \frac{1}{\lambda_0^4}(e_0,\ a_1\lambda_0 - a_0\lambda_1,\ a_2\lambda_0 - a_0\lambda_2)$$
$$= \frac{1}{\lambda_0^3}\{\lambda_0 (e_0 a_1 a_2) - \lambda_1 (e_0 a_0 a_2) - \lambda_2 (e_0 a_1 a_0)\}.$$

Now consider the expressions A and B:

$$B = -(e_0 b_3)(c_0 c_3), \quad (e_0 b_3) = (e_0 b_0 b_1 b_2).$$

Since $c_0 = a_0/\lambda_0 + e_0(*)$, it follows that

$$(c_0 c_3) = -\frac{1}{\lambda_0^3}(e_0 a_0 a_1 a_2).$$

By the assumptions of the lemma, both planes α' and α'' are visible from the same side from the point e_0. In analytical terms, this means that the expressions $(e_0 b_0 b_1 b_2)$ and $(e_0 a_0 a_1 a_2)$ have the same sign. Hence

$$B = \frac{1}{\lambda_0^3}(e_0 a_0 a_1 a_2)(e_0 b_0 b_1 b_2) > 0.$$

Now consider $A = c_3(a_3 - (e_0a_3)c_0)$. First,

$$-(e_0a_3)(c_0c_3) = \frac{1}{\lambda_0^3}(e_0a_0a_1a_2)^2.$$

Furthermore,

$$(c_3a_3) = \frac{1}{\lambda_0^3}\{\lambda_0(e_0a_1a_2)(a_0a_1a_2) - \lambda_1(e_0a_0a_2)(a_0a_1a_2) - \lambda_2(e_0a_1a_0)(a_0a_1a_2)\}.$$

Applying the vector identities of §1 to each of the scalar products in the curly brackets, we get

$$(a_0a_1a_2)(e_0a_1a_2) = (e_0a_0),$$
$$(a_0a_1a_2)(e_0a_0a_2) = -(e_0a_1),$$
$$(a_0a_1a_2)(e_0a_1a_0) = -(e_0a_2).$$

Thus

$$(c_3a_3) = \frac{1}{\lambda_0^3}(\lambda_0(e_0a_0) + \lambda_1(e_0a_1) + \lambda_2(e_0a_2)).$$

Substituting the full expressions for λ_0, λ_1 and λ_2, we get

$$(c_3a_3) = \frac{1}{\lambda_0^3}\{(e_0a_0)^2 + (e_0a_1)^2 + (e_0a_2)^2 + (e_0a_0)(e_0b_0) + (e_0a_1)(e_0b_1) + (e_0a_2)(e_0b_2)\}.$$

Hence, in view of the fact that

$$2|(e_0a_i)(e_0b_i)| \leqq (e_0a_i)^2 + (e_0b_i)^2,$$

we get

$$(c_3a_3) \geqslant \frac{1}{2\lambda_0^3}\{(e_0a_0)^2 + (e_0a_1)^2 + (e_0a_2)^2 - (e_0b_0)^2 - (e_0b_1)^2 - (e_0b_2)^2\}.$$

And since

$$(e_0a_0)^2 + (e_0a_1)^2 + (e_0a_2)^2 + (e_0a_3)^2 = (e_0b_0)^2 + (e_0b_1)^2 + (e_0b_2)^2 + (e_0b_3)^2,$$

it follows that

$$(c_3a_3) \geqslant \frac{1}{2\lambda_0^3}((e_0b_3)^2 - (e_0a_3)^2) = \frac{1}{2\lambda_0^3}((e_0b_0b_1b_2)^2 - (e_0a_0a_1a_2)^2).$$

In view of the expression for $-(e_0a_3)(c_0c_3)$ and the lower estimate for (c_3a_3), we get

$$A \geqslant \frac{1}{2\lambda_0^3}((e_0b_0b_1b_2)^2 + (e_0a_0a_1a_2)^2) > 0.$$

This completes the proof of the lemma.

By Theorem 2, each pair of isometric surfaces Φ' and Φ'' in euclidean space goes into a pair of isometric surfaces F' and F'' in elliptic space. This result can be extended by the following theorem.

THEOREM 4. *Under the assumptions of Theorem 2, if the surfaces Φ' and Φ'' of the euclidean space E_0 are locally convex, identically oriented and visible from the point $(1,0,0,0)$ from within, then the corresponding surfaces F' and F'' of elliptic space are locally convex.*

PROOF. Let x' be an arbitrary point on the surface F' and $x' + \Delta x'$ a point on F' close to x'. The corresponding points on the surfaces Φ' and Φ'' are $y', y' + \Delta y'$ and $y'', y'' + \Delta y''$. Join the points y' and $y' + \Delta y'$ by a segment γ' on Φ' and the points y'' and $y'' + \Delta y''$ by a segment γ'' on Φ''. Then

$$\Delta y' = \Delta s \tau' + \frac{\Delta s^2}{2} k' \nu' + \dots$$

$$\Delta y'' = \Delta s \tau'' + \frac{\Delta s^2}{2} k'' \nu'' + \dots,$$

where τ' and τ'' are the unit tangent vectors of the geodesics γ' and γ'' at the points y' and y'', k' and k'' their curvatures, and ν' and ν'' their principal normals.

Expressing x' in terms of y' and y'' as

$$x' = \rho(2y' + e_0(1 - y'^2 + y''^2)),$$

we get the following expression for $\Delta x'$:

$$\Delta x' = \frac{\Delta \rho}{\rho} x' + \rho \Delta \Omega + \Delta \rho \Delta \Omega,$$

where

$$\Delta \Omega = 2 \Delta s \{\tau' + e_0(- y'\tau' + y''\tau'')\} + \\ + \Delta s^2 \{k'\nu' + e_0(- k'\nu' y' + k''\nu'' y'')\} + O(\Delta s^3).$$

Multiply $\Delta x'$ by the unit vector n' of the normal to F' at x'. Since $n'x' = 0$, while $\Delta x' n'$ must be of order at least Δs^2, we get

$$n'(\tau' + e_0(- y'\tau' + y''\tau'')) = 0$$

and therefore

$$\Delta x' n' = \Delta s^2 k'(n'\nu' - (n'e_0)(\nu' y')) + \Delta s^2 k''(n'e_0)(\nu'' y'') + \dots.$$

Terms of orders higher than Δs^2 have been omitted.

To complete the proof it will suffice to show that the expressions $n'\nu' - (e_0 n')(\nu' y')$ and $(e_0 n')(\nu'' y'')$ have the same sign. Just as in the proof of Theorem 3, this involves computations which are easier to perform with more symmetric notation for the vectors. We therefore formulate a lemma for isometric planes, leaving the proof till later.

Let β' and β'' be two isometric planes in the euclidean space E_0. By the results of §3, the corresponding planes α' and α'' in the elliptic space R are also isometric.

LEMMA 2. *Let a_0 and b_0 be two corresponding points on the planes β' and β'', a_3 and b_3 the corresponding unit normals to the planes at these points, and c_0 and c_3 a point and a normal on the plane α'. Then, if the origin lies on opposite sides of the planes β' and β'', the expressions*

$$A = (c_3 a_3) - (e_0 c_3)(a_0 a_3), \qquad B = (e_0 c_3)(b_0 b_3)$$

are different from zero and have the same signs.

With this lemma the proof of Theorem 4 is easily completed. For the isometry between the surfaces Φ' and Φ'' induces a natural isometry of their tangent planes β' and β'' at the points y' and y''. The corresponding planes in elliptic space are tangent planes of F' and F'': α' is tangent to F' at x'. The expressions $n'\nu' - (e_0 n')(\nu' y')$ and $(e_0 n')(\nu'' y'')$ are precisely the quantities A and B of the lemma, and hence they have equal signs. Thus the surface F' is locally convex. The proof that F'' is locally convex is analogous. This completes the proof of Theorem 4.

PROOF OF LEMMA 2. Let a_1 and a_2 denote perpendicular unit vectors at the point a_0 in the plane β', b_1 and b_2 the corresponding unit vectors in the plane β'', and c_1 and c_2 the corresponding vectors in the plane α'. Then

$$a_3 = (e_0 a_1 a_2), \quad b_3 = (e_0 b_1 b_2), \quad c_3 = (c_0 c_1 c_2).$$

The equation of the plane α' is

$$x' = \rho(2y' + e_0(1 - y'^2 + y''^2)).$$

Differentiating x' in the corresponding directions, we get the following expressions for c_1 and c_2:

$$c_1 = \frac{\rho'}{\rho} c_0 + 2\rho(a_1 + e_0(-a_0 a_1 + b_0 b_1)),$$

$$c_2 = \frac{\rho'}{\rho} c_0 + 2\rho(a_2 + e_0(-a_0 a_2 + b_0 b_2)).$$

Hence

$$c_3 = 4\rho^2 (c_0,\ a_1 + e_0(-a_0 a_1 + b_0 b_1),\ a_2 + e_0(-a_0 a_2 + b_0 b_2)).$$

Consider the expression $B = (e_0 c_3)(b_0 b_3)$. Using the above expression for c_3 and

$$c_0 = \rho(2a_0 + e_0(1 - a_0^2 + b_0^2)),$$

we get

$$(e_0 c_3) = 8\rho^3(a_0 a_1 a_2 e_0) = -8\rho^3(a_0 a_3),$$

and hence $B = -8\rho^3(a_0 a_3)(b_0 b_3)$.

Since the origin lies on opposite sides of the planes β' and β'', the expressions $(a_0 a_3)$ and $(b_0 b_3)$ have different signs. Therefore $B > 0$.

Now consider $A = (a_3 c_3) - (e_0 c_3)(a_0 a_3)$. We have

$$-(e_0 c_3)(a_0 a_3) = 8\rho^3(e_0 a_0 a_1 a_2)^2.$$

Applying the vector identity of §1 to the right-hand side, we get

$$(e_0 c_3)(a_0 a_3) = 8\rho^3(a_0^2 - (a_0 a_1)^2 - (a_0 a_2)^2).$$

Similarly,

$$\begin{aligned}(a_3 c_3) = 4\rho^3 (e_0 a_1 a_2)(2a_0 + e_0(1 - a_0^2 + b_0^2),\\ a_1 + e_0(-a_0 a_1 + b_0 b_1), \quad a_2 + e_0(-a_0 a_2 + b_0 b_2))\\ = 4\rho^3 \{1 - a_0^2 + b_0^2 + 2(a_0 a_1)^2 + 2(a_0 a_2)^2 - 2(a_0 a_1)(b_0 b_1) - 2(a_0 a_2)(b_0 b_2)\}.\end{aligned}$$

Hence the following expression holds for A:

$$A = 4\rho^3(1 + a_0^2 + b_0^2 - 2(a_0 a_1)(b_0 b_1) - 2(a_0 a_2)(b_0 b_2)).$$

Since

$$2|(a_0 a_1)(b_0 b_1)| \leqslant (a_0 a_1)^2 + (b_0 b_1)^2,$$
$$2|(a_0 a_2)(b_0 b_2)| \leqslant (a_0 a_2)^2 + (b_0 b_2)^2,$$

we get

$$A \geqslant 4\rho^3(1 + a_0^2 + b_0^2 - (a_0 a_1)^2 - (a_0 a_2)^2 - (b_0 b_1)^2 - (b_0 b_2)^2).$$

And since

$$a_0^2 = (a_0 a_1)^2 + (a_0 a_2)^2 + (a_0 a_3)^2, \quad b_0^2 = (b_0 b_1)^2 + (b_0 b_2)^2 + (b_0 b_3)^2,$$

it follows that

$$A \geqslant 4\rho^3(1 + (a_0 a_3)^2 + (b_0 b_3)^2) > 0.$$

The lemma is proved.

§5. Infinitesimal bendings of surfaces in elliptic space

In this section we shall establish a one-to-one correspondence between surfaces in euclidean and elliptic space and their infinitesimal bendings. Thus the problem of infinitesimal bendings of surfaces in elliptic space reduces to the analogous problem in euclidean space.

Let τ be the velocity field of an infinitesimal deformation of a surface F in elliptic space. The deformation is called an infinitesimal bending if the lengths of all curves on the surface are stationary. This implies that the line element of the surface must be stationary. Hence, as in the euclidean case, we derive the equation of an infinitesimal bending by differentiating the line element with respect to the deformation parameter: $\delta dx^2/\delta t = 0$, or $dxd\tau = 0$, where $\tau = \delta x/\delta t$ is the velocity field of the deformation. This is precisely the form of the equation of an infinitesimal bending for a surface in euclidean space.

Let Φ_1 and Φ_2 be two isometric surfaces in euclidean space, close together, and defined by equations in cartesian coordinates: $y_1 = y_1(u, v)$ and $y_2 = y_2(u, v)$, where points that correspond under the isometry are determined by the same parameter values u, v.

We know that the vector field $z = y_1 - y_2$ is the field of an infinitesimal bending of the surface Φ defined by $y = y_1 + y_2$. Indeed,

$$dydz = dy_1^2 - dy_2^2 = 0$$

since the line elements of the surfaces Φ_1 and Φ_2 are equal.

Now consider a surface $\Phi: y = y(u, v)$ and let z be the field of an infinitesimal bending. Then, for sufficiently small λ, the equations $y_1 = y + \lambda z$ and $y_2 = y - \lambda z$ define two isometric surfaces, since

$$dy_1^2 = dy^2 + \lambda^2 dz^2, \qquad dy_2^2 = dy^2 + \lambda^2 dz^2.$$

Both these results can be extended to surfaces in elliptic space. Let F_1 and F_2 be two isometric surfaces in elliptic space, close together, and defined by equations in Weierstrass coordinates: $x = x_1(u, v)$ and $x = x_2(u, v)$, where corresponding points are determined by equal parameters. Then the vector field $\zeta = \rho(x_1 - x_2)$ is the field of an infinitesimal bending of the surface F: $x = \rho(x_1 + x_2)$.

Indeed,

$$dx = d\rho\,(x_1 + x_2) + \rho\,(dx_1 + dx_2),$$
$$d\zeta = d\rho\,(x_1 - x_2) + \rho\,(dx_1 - dx_2).$$

Hence, in view of the fact that

$$x_1^2 = x_2^2 = 1, \quad x_1 dx_1 = x_2 dx_2 = 0, \quad dx_1^2 = dx_2^2,$$

we get $dxd\zeta = 0$, i.e. ζ is the field of an infinitesimal bending of F.

Now let $F: x = x(u, v)$ be a surface in elliptic space and ζ the field of an infinitesimal bending on F. Then for small λ, the equations

$$x_1 = \rho(x + \lambda\zeta), \qquad x_2 = \rho(x - \lambda\zeta)$$

define two isometric surfaces F_1 and F_2. Indeed,

$$dx_1 = d\rho\,(x + \lambda\zeta) + \rho\,(dx + \lambda\, d\zeta),$$
$$dx_2 = d\rho\,(x - \lambda\zeta) + \rho\,(dx - \lambda\, d\zeta).$$

In view of the fact that

$$x\zeta = 0, \quad d(x\zeta) = x d\zeta + \zeta dx = 0, \quad dx d\zeta = 0,$$

we get $dx_1^2 = dx_2^2$, so that F_1 and F_2 are isometric.

Now any field of an infinitesimal bending of a surface essentially determines an infinitesimally close isometric surface. On the other hand, in §4 we established a correspondence between pairs of isometric surfaces in elliptic and euclidean space. Hence there is a similar correspondence between infinitesimal bendings of surfaces in elliptic and euclidean space. This statement is made precise in the following two theorems:

THEOREM 1. *If ζ is the field of an infinitesimal bending of the surface $F: x = x(u, v)$ in the elliptic space R, then*

$$z = (\zeta - e_0(\zeta e_0)) / (e_0 x)$$

is the field of an infinitesimal bending of the surface Φ:

$$y = (x - e_0(x e_0)) / (e_0 x)$$

in the euclidean space E_0. The field z is trivial if and only if ζ is trivial.

THEOREM 2. *If z is the field of an infinitesimal bending of the surface $\Phi: y = y(u, v)$ in the euclidean space E_0, then*

$$\zeta = \frac{z - e_0\,(yz)}{\sqrt{1 + y^2}}$$

is the field of an infinitesimal bending of the surface F:

$$x = \frac{y + e_0}{\sqrt{1 + y^2}}$$

in the elliptic space R. The field ζ is trivial if and only if z is trivial.

PROOF OF THEOREM 1. As we demonstrated above, for small λ the equations

$$x_1 = \rho(x + \lambda\zeta), \qquad x_2 = \rho(x - \lambda\zeta)$$

define two isometric surfaces F_1 and F_2 in the elliptic space R. By Theorem 1 of §4, the surfaces Φ_1 and Φ_2 defined in the euclidean space E_0 by the equations

$$y_1 = \frac{x_1 - e_0(x_1 e_0)}{e_0(x_1 + x_2)}, \qquad y_2 = \frac{x_2 - e_0(x_2 e_0)}{e_0(x_1 + x_2)},$$

are isometric. For small λ they are close together. But then the vector field $z = y_1 - y_2$ is the field of an infinitesimal bending of the surface Φ defined in euclidean space by $y = y_1 + y_2$. Obviously the field $z = (y_1 - y_2)/\lambda$ is also the field of an infinitesimal bending of the surface Φ.

Now let $\lambda = 0$. The limit surface Φ is defined by the equation

$$y = (x - e_0(xe_0))/(e_0 x),$$

and the field of the corresponding infinitesimal bending by

$$z = (\zeta - e_0(\zeta e_0))/(e_0 x).$$

By Lemma 5 of §3, if the field ζ is trivial (i.e. the field of an infinitesimal motion), then z is also trivial.

Before completing the proof and showing that triviality of z implies that of ζ, we prove Theorem 2.

PROOF OF THEOREM 2. Since z is the field of an infinitesimal bending of the surface $\Phi: y = y(u, v)$, it follows that for small λ the equations $y_1 = \frac{1}{2}y + \lambda z$ and $y_2 = \frac{1}{2}y - \lambda z$ define two isometric surfaces Φ_1 and Φ_2 in the euclidean space E_0. By Theorem 2 of §4, these correspond to isometric surfaces F_1 and F_2 in the elliptic space R defined by the equations

$$x_1 = \rho(2y_1 + e_0(1 - y_1^2 + y_2^2)),$$
$$x_2 = \rho(2y_2 + e_0(1 - y_2^2 + y_1^2)).$$

For sufficiently small λ the surfaces F_1 and F_2 are close together. Hence, if F is the surface defined in euclidean space by the equation $x = \rho(x_1 + x_2)$, then $\zeta = \rho(x_1 - x_2)$ is the field of an infinitesimal bending of F. Thus $\zeta = \rho(x_1 - x_2)/\lambda$ is also the field of an infinitesimal bending of F.

Letting $\lambda \to 0$, we get a surface F defined by the equation

$$x = \frac{y + e_0}{\sqrt{1 + y^2}},$$

and an infinitesimal bending whose field is

$$\zeta = \frac{z - e_0(yz)}{\sqrt{1 + y^2}}.$$

By Lemma 6 of §3, if the field z is trivial then so is ζ.

Now a direct check will show that the correspondences between surfaces and infinitesimal bendings in elliptic and euclidean spaces established in these two theorems are mutually inverse. Hence the field ζ of Theorem 2 is trivial if and only if z is trivial, and the field z of Theorem 1 is trivial if and only if ζ is trivial.

This completes the proof of both theorems.

Thanks to Theorem 1, many results relating to infinitesimal bendings of surfaces in euclidean space can be carried over to surfaces in elliptic space.

Note that the correspondence between surfaces in elliptic and euclidean space defined in Theorems 1 and 2 is independent of their infinitesimal bendings; it is defined by a geodesic mapping of one space onto the other (§1).

When an elliptic space is mapped geodesically onto a euclidean space (the projective model of elliptic space), convex surfaces in elliptic space go into convex surfaces of euclidean space.

Since closed convex surfaces in euclidean space which contain no flat pieces are rigid, i.e. admit no nontrivial infinitesimal bendings, it follows that convex surfaces in elliptic space which contain no flat pieces are also rigid. To prove this assertion we need only apply Theorem 1.

In the general case, if a closed (not necessarily convex) surface F in elliptic space is mapped geodesically onto a rigid surface in euclidean space, then F is also rigid.

EXAMPLES. A surface F in elliptic space is said to be *of type* T if its convex part lies entirely on its convex hull (recall that the *convex hull* of F is the smallest convex surface containing F). Our geodesic mapping maps any surface of type T onto a surface of type T in euclidean space. Aleksandrov has proved that analytic surfaces of type T in euclidean space are rigid [4]. By Theorem 1, this implies that analytic surfaces of type T in elliptic space are also rigid.

It is known that if a convex surface in euclidean space, containing no flat pieces, is cut at most once by any ray issuing from a point O separated from the surface by a plane, and the distances of the boundary points of the surface from O are stationary, then the surface is rigid. A similar theorem holds in elliptic space. That the distances of boundary points from O remain stationary under the geodesic mapping into euclidean space follows from Lemma 5 of §3, applied to straight-line segments joining the boundary points to O. It follows, in particular, that convex caps are rigid in the class of caps in elliptic space.

I. N. Vekua has proved that if the curvature of the curves bounding a strictly convex surface is stationary, then the surface is rigid. This theorem carries over to strictly convex surfaces in elliptic space. This also follows from Theorem 1. For a proof, observe that if the curvature of the curves bounding the surface is stationary under an infinitesimal bending, then the tangent circles to these curves perform infinitesimal motions; hence, by Lemma 5 of §3, the curves corresponding to these circles in euclidean space also perform motions. Since geodesic mappings preserve order of contact, the curvature of the boundary of the image surface in euclidean space is also stationary. The rigidity of the surface in elliptic space now follows from Vekua's theorem [**69**].

Cohn-Vossen [**28**] constructed examples of nonrigid closed surfaces in euclidean space. By Theorem 2, these examples carry over to elliptic space.

Efimov has proved the existence of surfaces which are rigid "in the small," i.e. rigid in an arbitrarily small neighborhood of each point [**33**]. Theorem 2 implies that such surfaces also exist in elliptic space.

Sauer proved that, given an infinitesimal bending of a surface F in euclidean space, one can find explicit expressions for infinitesimal bendings of any image of F under a projective mapping. Now projective transformations of figures correspond to each other under the geodesic mapping of elliptic space onto euclidean space, and hence this result of Sauer carries over to surfaces in elliptic space.

In euclidean space, all infinitesimal bendings of second-order surfaces are known. Now our geodesic mapping of euclidean space into elliptic space maps second-order surfaces onto second-order surfaces. Hence Theorem 2 yields all infinitesimal bendings of second-order surfaces in elliptic space.

General convex surfaces in spaces of constant curvature satisfy a theorem analogous to Aleksandrov's theorem on surfaces of euclidean space (Chapter IV, §1): *A field z is a bending field for a surface F if and only if it satisfies a Lipschitz condition in every compact domain on the surface and $dx dz = 0$ almost everywhere on the surface.*

The proof of this theorem is essentially the same as the proof of the euclidean version.

Theorems 1 and 2, which were proved for regular surfaces and regular bending fields, are easily extended to general convex surfaces. Indeed, a Lipschitz condition for z implies a Lipschitz condition for ζ, and if $dx dz = 0$ almost everywhere on F, then $dr d\zeta = 0$ almost everywhere on Φ.

It follows from Theorems 1, 2 and Theorem 1 of §8 in Chapter IV that if a regular surface with positive extrinsic curvature is in class $C^{n+\alpha}$ ($n \geqq 2$, $0 < \alpha < 1$), then its bending fields are in class $C^{n+\alpha'}$, $0 < \alpha' < \alpha$.

To prove this it suffices to note that the geodesic mapping of elliptic space onto euclidean space maps $C^{n+\alpha}$ surfaces onto $C^{n+\alpha}$ surfaces.

§6. Monotypy of general convex surfaces in elliptic space

Theorem 1 of §4, which deals with transformations of regular isometric surfaces in elliptic space, can be extended to general convex surfaces. One can thus prove various monotypy theorems in elliptic space as corollaries of the corresponding theorems for convex surfaces in euclidean space.

We shall say that a plane α cutting a convex surface F cuts its edges if at every ridge point of F lying in α the edge of the tangent dihedral angle cuts the plane α (i.e. does not lie in α). The following lemma holds.

LEMMA 1. *Almost all planes which cut a convex surface F in elliptic space cut its edges.*

PROOF. First note that it is sufficient to prove the lemma for convex surfaces in euclidean space, for we need only go over to the surface in euclidean space corresponding to F under the geodesic mapping of elliptic space onto euclidean space.

It is also clear that we need only consider small pieces of the surface, and hence we may assume without loss of generality that F can be projected in one-to-one fashion onto the xy-plane in the square $K: 0 \leqq x \leqq 1,\ 0 \leqq y \leqq 1$.

Let σ and ω be small numbers such that $\omega \ll \sigma$. Divide the square K into small squares k by straight lines $x = m'/2^n$, $y = m''/2^n$, $m'\, m'' < 2^n$. Denote the domain on F projected into the square k by F_k. Let K_ω denote the union of all squares k which contain projections of conic points of F at which the curvature exceeds ω. Divide the part of the square K lying outside K_ω into even smaller squares k', and let K_σ denote the union of all squares k' containing projections of ridge points of F at which the exterior plane angle is greater than σ.

If ω and the squares k' are sufficiently small, the directions of the edges in $F_{k'}$ with angles greater than σ differ by arbitrarily small amounts; otherwise, the domain $F - F_{K_\omega}$ would contain conic points at which the curvature exceeds ω, which is impossible.

It follows that the measure of the set of planes cutting the domain $F_{k'}$ and containing at least one edge with angle greater than σ is at most $c\epsilon\delta$, where δ is the side of k', c a constant depending only on the greatest angle between the supporting planes of F and the xy-plane, and ϵ tends to zero together with ω and δ.

The number $\sum_{k'\subset K_\sigma}\delta\sigma$ is bounded by a constant M which depends only on the integral mean curvature of the surface (Chapter III, §6). Hence the measure μ_σ of the set of planes cutting $F - F_{K_\omega}$ and containing at least one edge with exterior angle greater than σ is at most $cM\epsilon/\delta$. And since ϵ is arbitrarily small, it follows that $\mu_\sigma = 0$.

Letting the sides of the squares k tend to zero, we see that the measure of the set of planes cutting the surface F and containing edges with angles greater than σ is zero.

Now let $\sigma \to 0$; it follows that almost all planes cutting the surface cut its edges. This completes the proof of the lemma.

Let F_1 and F_2 be two isometric, identically oriented convex surfaces in the domain R_0 of the elliptic space R, which are cut in at most one point by any ray issuing from the point $e_0(1,0,0,0)$ and are visible from this point from within. Let Φ_1 and Φ_2 denote the corresponding surfaces defined in the euclidean space E_0 by equations

$$(*)\qquad y_1 = \frac{x_1 - e_0\,(x_1 e_0)}{e_0\,(x_1 + x_2)}, \qquad y_2 = \frac{x_2 - e_0\,(x_2 e_0)}{e_0\,(x_1 + x_2)},$$

where x_1 and x_2 are corresponding points of F_1 and F_2.

Since F_1 and F_2 are cut at most once by any ray issuing from e_0, the same holds for Φ_1 and Φ_2 relative to the origin. Indeed, any straight line in R through e_0 has the equation $x = \rho(e_0 + \lambda a)$ (parameter λ), and any straight line in E_0 through the origin has the equation $y = \mu b$ (parameter μ). It is easily seen that if $x_1 = \rho(e_0 + \lambda a)$, regarded as an equation for λ, has at most one solution for any a, then the equation

$$(x_1 - e_0(x_1 e_0))/e_0(x_1 + x_2) = \mu b,$$

regarded as an equation for μ, also has at most one solution for any b. This proves our statement for the surface Φ_1. The same holds for the surface Φ_2.

Let F_1 and F_2 be regular surfaces. Then, as we proved in §4, the surfaces Φ_1 and Φ_2 are locally convex. We claim that each of these surfaces turns its interior side to the point O, i.e. is visible from O from within.

Let a_0 denote an arbitrary point on F_1, a_1 and a_2 unit tangent vectors at this point, and a_3 the exterior normal there. The corresponding vectors for the surfaces F_2 and Φ_1 will be denoted by b_0, b_1, b_2, b_3 and c_0, c_1, c_2, c_3, respectively. By results of §4,

$$c_0 = \frac{1}{\lambda_0}\,(a_0 - e_0\,(a_0 e_0)), \qquad c_3 = \frac{1}{\lambda_0^4}\,(e_0,\ \lambda_0 a_1 - a_0\lambda_1,\ a_2\lambda_0 - a_0\lambda_2).$$

Hence

$$(c_0c_3) = \frac{1}{\lambda_0^3}(e_0a_0a_1a_2) = -\frac{1}{\lambda_0^3}(a_3e_0).$$

Since F_1 is visible from e_0 from within, we have $(e_0a_3) < 0$, and therefore $c_0c_3 > 0$. But this means that the surface Φ_1 is visible from O from within. Analogous arguments prove the statement for Φ_2.

Now suppose that F_1 and F_2 are two isometric, identically oriented dihedral angles and A_1 and A_2 corresponding points on their edges (the edges themselves need not correspond under the isometry). Consider the line on F_1 corresponding to the edge of F_2, and the line on F_2 corresponding to the edge of F_1. These lines, together with the edges themselves, divide each of the dihedral angles F_1 and F_2 into four plane angles with common vertex A_1 and A_2, respectively.

By (*), the dihedral angles F_1 and F_2 correspond to certain tetrahedral angles Φ_1 and Φ_2 in euclidean space (which may degenerate into dihedral angles if the edges of F_1 and F_2 correspond under the isometry). We claim that Φ_1 and Φ_2 are convex and visible from O from within.

The proof is simple. Approximate the angles F_1 and F_2 by regular isometric cylindrical surfaces $\widetilde{F}_1$ and $\widetilde{F}_2$. The corresponding convex surfaces $\tilde{\Phi}_1$ and $\tilde{\Phi}_2$ in euclidean space are visible from O from within. In the limit, when $\widetilde{F}_1 \to F_1$ and $\widetilde{F}_2 \to F_2$, the surfaces $\tilde{\Phi}_1$ and $\tilde{\Phi}_2$ must converge to convex surfaces visible from O from within. But the limits of $\tilde{\Phi}_1$ and $\tilde{\Phi}_2$ are the tetrahedral angles Φ_1 and Φ_2, respectively.

Let F_1 and F_2 be two isometric convex cones. We claim that the corresponding surfaces Φ_1 and Φ_2 in euclidean space E_0 are convex cones, each visible from the origin O from within.

It is obvious that Φ_1 and Φ_2 are cones, since straight lines are mapped onto straight lines by the isometric mapping (*) (§3). Approximate the cones F_1 and F_2 by isometric cones $\widetilde{F}_1$ and $\widetilde{F}_2$ with regular surfaces. Their images in euclidean space are regular convex cones visible from O from within. We now draw the desired conclusion by letting $\widetilde{F}_1 \to F_1$ and $\widetilde{F}_2 \to F_2$.

Now consider the case of general convex surfaces F_1 and F_2 in the neighborhood of conic points A_1 and A_2 that correspond under the isometry. We claim that the surfaces Φ_1 and Φ_2 are locally convex at the corresponding points B_1 and B_2.

Construct the tangent cones V_1 and V_2 to F_1 and F_2 at the points A_1 and A_2. The isometry of surfaces induces an isometry of the cones. Let x_1 be a point on F_1 close to A_1, and x_2 the corresponding point on F_2. Join A_1 and x_1 by a segment, and let $\bar{x}_1$ be a point on the semitangent at A_1 such that the length Δs of the straight-line segment

$A_1\bar{x}_1$ is equal to the distance between A_1 and x_1. As was shown in §1, $x_1 = \bar{x}_1 + \epsilon_1 \Delta s$, where $\epsilon_1 \to 0$ as $x_1 \to A_1$. Carrying out the corresponding construction on F_2, we get $x_2 = \bar{x}_2 + \epsilon_2 \Delta s$.

Substituting these expressions for x_1 and x_2 into (*), we get

$$y_1 = \frac{\bar{x}_1 - e_0(\bar{x}_1 e_0)}{e_0(\bar{x}_1 + \bar{x}_2)} + \bar{\varepsilon}_1 \Delta s, \quad y_2 = \frac{\bar{x}_2 - e_0(\bar{x}_2 e_0)}{e_0(\bar{x}_1 + \bar{x}_2)} + \bar{\varepsilon}_2 \Delta s.$$

Since the first terms of these expressions represent points of the cones corresponding to V_1 and V_2, while the second terms have order of magnitude smaller than Δs, these cones are the tangent cones to Φ_1 and Φ_2 at B_1 and B_2, respectively.

It follows that the surfaces Φ_1 and Φ_2 are locally convex at points corresponding to conic points of F_1 and F_2. The same conclusion obviously holds for ridge points A_1 and A_2 if the directions of the edges do not correspond under the isometry.

If the edges at A_1 and A_2 correspond under the isometry, we can still state that any plane section of the surface Φ_1 at the point B_1 is locally convex, provided the edge of the tangent dihedral angle at B_1 is not in the plane cutting the surface.

Let us say that the surface Φ_1 is weakly convex at a point B and turns its interior side to O if there exists a plane β through B such that any sphere touching β at B and lying on the side of β opposite to O is a local support, i.e. a sufficiently small neighborhood of B on the surface lies outside the sphere.

We claim that the surface Φ_1 is weakly convex at any smooth point B_1 and turns its interior side to O. To prove this, we first note that the points A_1 and A_2 of F_1 and F_2 that correspond to B_1 must be smooth.

To simplify the exposition, we shall denote the points A_1 and A_2 on F_1 and F_2 by their radius vectors x_1 and x_2. Let $x_1 + \Delta x_1$ be a point on F_1 close to x_1, and $x_2 + \Delta x_2$ the corresponding point on F_2. Connect the points x_1 and $x_1 + \Delta x_1$ by a segment γ_1 on F_1, and x_2 and $x_2 + \Delta x_2$ by the corresponding segment γ_2 on F_2.

In view of the properties of segments on a convex surface studied in §1, we have the following resolutions of the vectors $x_1 + \Delta x_1$ and $x_2 + \Delta x_2$:

$$x_1 + \Delta x_1 = \rho\left(\left(1 + \frac{\Delta s^2}{2}\right) x_1 + \Delta s \tau_1 + \delta_1 \nu_1 + \varepsilon_1\right),$$

$$x_2 + \Delta x_2 = \rho\left(\left(1 + \frac{\Delta s^2}{2}\right) x_2 + \Delta s \tau_2 + \delta_2 \nu_2 + \varepsilon_2\right),$$

where τ_1 and τ_2 are the unit tangent vectors to the geodesics at the points x_1 and x_2; ν_1 and ν_2 are the interior unit normals to the surfaces

F_1 and F_2 at these points; δ_1 and δ_2 are the distances of $x_1+\Delta x_1$ and $x_2+\Delta x_2$ from the tangent planes to F_1 and F_2 at the points x_1 and x_2, and ϵ_1/δ_1 and ϵ_2/δ_2 tend to zero together with Δs (the length of the segments γ_1 and γ_2).

Substituting these expressions into (*), we get

$$\begin{aligned}\Delta y_1 = {} & \frac{\Delta s}{e_0(x_1+x_2)}\{-y_1(e_0\tau_1+e_0\tau_2)+\tau_1-e_0(e_0\tau_1)\} \\ & + \frac{\Delta s^2 e_0(\tau_1+\tau_2)}{e_0(x_1+x_2)}\{y_1(e_0\tau_1+e_0\tau_2)-\tau_1+e_0(e_0\tau_1)\} \\ & + \frac{1}{e_0(x_1+x_2)}\{-(\delta_1 e_0\nu_1+\delta_2 e_0\nu_2)y_1+\delta_1\nu_1-\delta_1 e_0(e_0\nu_1)\} \\ & + \varepsilon'\delta_1+\varepsilon''\delta_2+\varepsilon\Delta s^2,\end{aligned}$$

where ϵ' and ϵ'' tend to zero with Δs.

In order to verify that Φ_1 is weakly convex at B_1, it is sufficient to show that if n denotes the normal to Φ_1 at B_1, then $(n\Delta y_1)$ has the same sign, up to a quantity of order $\epsilon\Delta s^2$, for any point $x_1+\Delta x_1$ sufficiently close to x_1, i.e. for sufficiently small Δs.

Multiplying Δy_1 by the normal n, we must get a quantity of lower order of magnitude than Δs. Hence, since $\delta_1/\Delta s\to 0$ and $\delta_2/\Delta s\to 0$ as $\Delta s\to 0$, we see that

$$n\{-y_1(e_0\tau_1+e_0\tau_2)+\tau_1-e_0(e_0\tau_1)\}=0.$$

And the product $(n\Delta y_1)$ becomes

$$(n\Delta y_1)=\frac{\delta_1}{e_0(x_1+x_2)}(-y_1(e_0\nu_1)+\nu_1)n+\frac{\delta_2}{e_0(x_1+x_2)}(e_0\nu_2)(y_1 n)+\varepsilon\Delta s^2.$$

As we established in §4, the expressions $(-y_1(e_0\nu_1)+\nu_1)n$ and $-(e_0\nu_2)(y_1 n)$ are not zero and have equal signs. Hence Φ_1 is weakly convex at B_1.

The fact that Φ_1 turns its interior side to O at B_1 follows from the truth of this statement for regular surfaces F_1 and F_2. Indeed, let $\widetilde{F}_1$ and $\widetilde{F}_2$ be two isometric regular surfaces tangent to F_1 and F_2 at the points A_1 and A_2, turning their interior sides to e_0. Then the surface $\tilde{\Phi}_1$ is convex and turns its interior side to O. Now since the expressions

$$(-y_1(e_0\nu_1)+\nu_1)n, \quad -(e_0\nu_2)(y_1 n)$$

are the same for $\widetilde{F}_1$ and $\widetilde{F}_2$ as for F_1 and F_2, it follows that the surface Φ_1, like $\tilde{\Phi}_1$, turns its interior side to O at B_1.

The local convexity of the surface Φ_2 at smooth points is proved in analogous fashion.

Map the domain R_0 of elliptic space onto the euclidean space E_0, taking each point x with Weierstrass coordinates x_0, x_1, x_2, x_3 onto the point with cartesian coordinates x_1/x_0, x_2/x_0, x_3/x_0. The images of F_1 and F_2 in the euclidean space E_0 are surfaces which are cut at most once by any ray issuing from the origin and turn their interior sides to the origin. The correspondence between the points of F_1 and Φ_1, F_2 and Φ_2 defined by (*) is in this case simply projection from the origin.

We claim that the part Φ' of the surface Φ_1 (Φ_2) projected from the origin by a convex cone V is a convex surface. To avoid the need for new notation, we shall assume that $\Phi' \equiv \Phi_1$. To prove that Φ_1 is convex, it will clearly suffice to show that any plane α cutting the cone V cuts the surface Φ_1 in a curve which is convex away from the apex of the cone.

Suppose that the plane α cuts the edges of the surface F_1. Then this surface is at least weakly convex at all points of the curve γ along which α cuts Φ_1.

Let $\bar{\gamma}$ denote the convex hull of the curve γ. $\bar{\gamma}$ is divided into two parts by its points of contact with the generators of the cone V that lie in the plane α. Let γ' be the part farthest from the point O. We must prove that γ coincides with γ'.

If this is not so, the part of γ' not in γ consists of straight-line segments with endpoints on γ. Let g be one of these segments and $\tilde{\gamma}'$ the part of γ' corresponding to g under central projection from O.

Let P and Q be points on the continuation of g, on different sides of g; describe a circular arc through P and Q, of sufficiently large radius, so that the domain bounded by the arc and the straight line PQ contains the point O. Now deform the circular arc in such a way that its center moves perpendicular to g, the arc itself always passing through P and Q. It is obvious that at some stage of this deformation the interior side of the arc will touch the curve $\tilde{\gamma}'$. But this is absurd, since Φ_1 is weakly convex along the curve γ. This contradiction proves that the curve γ is convex.

We have assumed that the plane α cuts the edges of the surface F_1. We now relax this restriction, and let α be an arbitrary plane. By Lemma 1, an arbitrarily small displacement of the surface F_1 will ensure that α cuts the edge of F_1. And since Φ_1 depends continuously on the position of F_1, it follows that any plane α cuts Φ_1 in a convex curve. This proves that F_1 is convex.

The proof that Φ_2 is convex is similar.

We now prove that the surfaces Φ_1 and Φ_2 are isometric. Let $\bar{\gamma}_1$ be an arbitrary rectifiable curve on Φ_1. Since the complete angle of the cone projecting this curve from O is finite, the corresponding curve γ_1 on F_1 is also rectifiable. (The image Φ_1 of F_1 in E_0 is obtained by projection from O.)

The curve on F_2 corresponding to γ_1 under the isometry is a rectifiable curve γ_2. Hence the curve $\bar{\gamma}_2$ on Φ_2 corresponding to $\bar{\gamma}_1$ is rectifiable. The proof that Φ_1 and Φ_2 are isometric will be completed by showing that $\bar{\gamma}_1$ and $\bar{\gamma}_2$ have equal lengths.

Since γ_1 is rectifiable, the vector valued function $x_1(s)$, where s is arc length along γ_1, satisfies a Lipschitz condition. For the same reason, the vector valued function $x_2(s)$ satisfies a Lipschitz condition. It follows that the vector valued functions $y_1(s)$ and $y_2(s)$ defined by (*) also satisfy Lipschitz conditions, and the lengths of the curves $\bar{\gamma}_1$ and $\bar{\gamma}_2$ are given by the integrals

$$\int \left| \frac{dy_1}{ds} \right| ds, \qquad \int \left| \frac{dy_2}{ds} \right| ds.$$

It will therefore suffice to show that $|y_1'(s)| = |y_2'(s)|$ for almost all s.

Suppose that the derivatives $x_1'(s)$ and $x_2'(s)$ of the curves γ_1 and γ_2 exist at the corresponding points $x_1(s)$ and $x_2(s)$. We shall show that then the derivatives $y_1'(s)$ and $y_2'(s)$ exist, and $|y_1'(s)| = |y_2'(s)|$.

The existence of $x_1'(s)$ and $x_2'(s)$ implies that γ_1 and γ_2 have tangents t_1 and t_2 at the points x_1 and x_2. Map the curves γ_1 and γ_2 onto the tangents t_1 and t_2, taking each point $x_i(s)$ onto the point $\bar{x}_i(s)$ defined on t_i by equal arc length. The isometry between the curves γ_1 and γ_2 then induces an isometry between the tangents t_1 and t_2.

Let $\bar{x}_1(s)$ and $\bar{x}_2(s)$ be the points on the tangents corresponding to arc length s. The equations

$$\bar{y}_1 = \frac{\bar{x}_1 - e_0(\bar{x}_1 e_0)}{e_0(\bar{x}_1 + \bar{x}_2)}, \qquad \bar{y}_2 = \frac{\bar{x}_2 - e_0(\bar{x}_2 e_0)}{e_0(\bar{x}_1 + \bar{x}_2)}$$

define two straight lines in the euclidean space E_0 (§3), and moreover the correspondence between these lines defined by equal parameters is an isometry. Hence $|\bar{y}_1'(s)| = |\bar{y}_2'(s)|$.

Now, using the equalities

$$x_1(s) = \bar{x}_1(s), \quad x_2(s) = \bar{x}_2(s), \quad x_1'(s) = \bar{x}_1'(s), \quad x_2'(s) = \bar{x}_2'(s),$$

and comparing the expressions for the derivatives $y_1'(s)$ and $\bar{y}_1'(s)$, $y_2'(s)$ and $\bar{y}_2'(s)$ in terms of x_1, x_1', $\bar{x}_1'$, x_2, x_2', $\bar{x}_2'$, we get $y_1'(s) = \bar{y}_1'(s)$ and $y_2'(s) = \bar{y}_2'(s)$. Thus $|y_1'(s)| = |y_2'(s)|$ almost everywhere, and the curves

$\bar{\gamma}_1$, $\bar{\gamma}_2$ have equal lengths.

Since the curve $\bar{\gamma}_1$ was arbitrary, the surfaces Φ_1 and Φ_2 are isometric.

THEOREM 1. *Isometric closed convex surfaces in elliptic space are congruent.*

PROOF. Let F_1 and F_2 be two isometric closed convex surfaces in the elliptic space R. Without loss of generality we may assume that the surfaces are identically oriented, lie in the domain R_0 of the space, and contain the point e_0 in their interior.

This position of the surfaces may be ensured as follows. If the surfaces have opposite orientation, one of them is replaced by its mirror image. Then both surfaces are moved (by a suitable rigid motion) to the domain R_0.

Now, if one of the surfaces, F_1 say, does not contain the point e_0, connect it to e_0 by a shortest straight-line segment g in R_0. Let A_1 be the endpoint of this segment on the surface. Consider two supporting planes to the surface in R_0, perpendicular to the line g. They intersect along some straight line h on the boundary of R_0. Now rotate the surface F_1 about h until the points A_1 and e_0 coincide. The surface obviously remains in R_0. An arbitrarily small motion will now bring the surface to a position in which e_0 lies within it.

As we proved above, a convex cone V with apex at O cuts out a convex domain from each of the surfaces Φ_1 and Φ_2. Hence Φ_1 and Φ_2 are closed convex surfaces.

The surfaces Φ_1 and Φ_2 are isometric. By the monotypy theorem for general convex surfaces (Chapter III, §6), Φ_1 and Φ_2 are congruent. Hence, by Lemma 4 of §3, the surfaces F_1 and F_2 are congruent. This proves the theorem.

THEOREM 2. *Isometric convex caps in elliptic space are congruent.*

Indeed, if F_1 and F_2 are isometric convex caps, they can be completed to isometric closed convex surfaces by adding to each surface its reflection in the plane of its boundary. By Theorem 1 these surfaces are congruent, and hence so are the caps.

THEOREM 3. *Let F_1 and F_2 be two isometric convex surfaces in the domain R_0 of elliptic space, each of which is visible from the point e_0 from within (from without), and the point e_0 is not in the convex hulls of the surfaces. If boundary points of F_1 and F_2 that correspond under the isometry lie at equal distances from e_0, the surfaces F_1 and F_2 are congruent.*

This theorem, like the theorem of §4, reduces to the corresponding theorem for surfaces in euclidean space. Let us consider the case in which F_1 and F_2 are visible from e_0 from within. The other case is treated by suitable modification of the auxiliary arguments presented throughout this section.

The surfaces Φ_1 and Φ_2 in euclidean space corresponding to F_1 and F_2 are isometric, locally convex, and turn their interior side to the point O. We shall show that corresponding boundary points of Φ_1 and Φ_2 lie at equal distances from O.

The isometry of the surfaces F_1 and F_2 induces a natural isometry of the straight-line segments connecting e_0 to corresponding boundary points. By formulas (*), the straight-line segments connecting e_0 to the boundaries of F_1 and F_2 go into equal straight-line segments connecting the origin O to the corresponding points on the boundaries of Φ_1 and Φ_2 (§3).

The point O lies outside the convex hulls of Φ_1 and Φ_2. This is because formulas (*), applied to straight-line segments, amount to projection of the segments onto E_0 followed by dilatation.

By the theorem of Aleksandrov and Sen′kin (Chapter III, §7) the surfaces Φ_1 and Φ_2 are congruent. Hence F_1 and F_2 are also congruent. Q.E.D.

§7. Regularity of convex surfaces with regular metric

The question to be considered in this section may be formulated as follows. To what degree does the regularity of the intrinsic metric of a convex surface, i.e. the regularity of the coefficients $E(u,v)$, $F(u,v)$, $G(u,v)$ of its line element, imply the regularity of the surface itself—the regularity of $x_0(u,v)$, $x_1(u,v)$, $x_2(u,v)$, $x_3(u,v)$?

Let F be a regular surface in an elliptic space, and ζ one of the Weierstrass coordinates of an arbitrary point on F. We shall derive an equation for ζ when the line element of the surface is

$$ds^2 = E du^2 + 2F\,du dv + G dv^2.$$

To this end, we appeal to the Gauss equations:

$$x_{uu} = \Gamma^1_{11} x_u + \Gamma^2_{11} x_v - Ex + L\xi,$$

$$x_{uv} = \Gamma^1_{12} x_u + \Gamma^2_{12} x_v - Fx + M\xi,$$

$$x_{vv} = \Gamma^1_{22} x_u + \Gamma^2_{22} x_v - Gx + N\xi.$$

Multiplying these equations by the unit basis vector e along the ζ-axis, we get

$$(*)\qquad \begin{aligned} \zeta_{uu} &= \Gamma_{11}^1\zeta_u + \Gamma_{11}^2\zeta_v - E\zeta + L(\xi e), \\ \zeta_{uv} &= \Gamma_{12}^1\zeta_u + \Gamma_{12}^2\zeta_v - F\zeta + M(\xi e), \\ \zeta_{vv} &= \Gamma_{22}^1\zeta_u + \Gamma_{22}^2\zeta_v - G\zeta + N(\xi e). \end{aligned}$$

Transfer all terms except the last from the right-hand to the left-hand sides of the equations, multiply the first equation by the third and subtract the square of the second equation. The result is

$$(\zeta_{uu} - \Gamma_{11}^1\zeta_u - \Gamma_{11}^2\zeta_v + E\zeta)(\zeta_{vv} - \Gamma_{22}^1\zeta_u - \Gamma_{22}^2\zeta_v + G\zeta) - (\zeta_{uv} - \Gamma_{12}^1\zeta_u - \Gamma_{12}^2\zeta_v + F\zeta)^2 = (LN - M^2)(e\xi)^2$$

Let us express $(\xi e)^2$ in terms of ζ. We have

$$\xi = \frac{(xx_ux_v)}{|xx_ux_v|}, \qquad (\xi e)^2 = \frac{(exx_ux_v)^2}{(xx_ux_v)^2}.$$

Applying the vector identities of §1 to the numerator and denominator of $(\xi e)^2$, we get

$$(xx_ux_v)^2 = \begin{vmatrix} 1 & 0 & 0 \\ 0 & E & F \\ 0 & F & G \end{vmatrix} = EG - F^2,$$

$$(exx_ux_v)^2 = \begin{vmatrix} 1 & \zeta & \zeta_u & \zeta_v \\ \zeta & 1 & 0 & 0 \\ \zeta_u & 0 & E & F \\ \zeta_v & 0 & F & G \end{vmatrix} = (1-\zeta^2)(EG - F^2) - (E\zeta_v^2 - 2F\zeta_u\zeta_v + G\zeta_u^2).$$

Now, observing that

$$(LN - M^2)/(EG - F^2) = K - 1,$$

where K is the Gauss curvature of the surface, we obtain the equation for ζ in the following form:

$$\begin{aligned} &(\zeta_{uu} - \Gamma_{11}^1\zeta_u - \Gamma_{11}^2\zeta_v + E\zeta)(\zeta_{vv} - \Gamma_{22}^1\zeta_u - \Gamma_{22}^2\zeta_v + G\zeta) \\ &\qquad - (\zeta_{uv} - \Gamma_{12}^1\zeta_u - \Gamma_{12}^2\zeta_v + F\zeta)^2 \\ &\qquad = (K-1)\{(1-\zeta^2)(EG - F^2) - (E\zeta_v^2 - 2F\zeta_u\zeta_v + G\zeta_u^2)\}. \end{aligned}$$

The coefficients of this equation are expressed in terms of the coefficients E, F, G of the line element and their derivatives.

We now introduce a new function by the substitution $\zeta = \sin z$. The geometrical meaning of z is easily seen: Up to sign, z is the distance of a point on the surface from the plane $\zeta = 0$. Noting that

$$z_u = \frac{\zeta_u}{\sqrt{1-\zeta^2}}, \quad z_v = \frac{\zeta_v}{\sqrt{1-\zeta^2}}, \quad \tan z = \frac{\zeta}{\sqrt{1-\zeta^2}},$$

$$\frac{\zeta_{uu}}{\sqrt{1-\zeta^2}} = z_{uu} - z_u^2 \tan z, \quad \frac{\zeta_{uv}}{\sqrt{1-\zeta^2}} = z_{uv} - z_u z_v \tan z,$$

$$\frac{\zeta_{vv}}{\sqrt{1-\zeta^2}} = z_{vv} - z_v^2 \tan z,$$

we get the following equation for z:

$$\begin{aligned}\{z_{uu} - \Gamma_{11}^1 z_u - \Gamma_{11}^2 z_v + (E - z_u^2)\tan z\}\{z_{vv} - \Gamma_{22}^1 z_u - \Gamma_{22}^2 z_v^2 + (G - z_v^2)\tan z\} \\ - \{z_{uv} - \Gamma_{12}^1 z_u - \Gamma_{12}^2 z_v + (F - z_u z_v)\tan z\}^2 \\ = (K-1)\{(E - z_u^2)(G - z_v^2) - (F - z_u z_v)^2\}.\end{aligned}$$

Note that the only difference between this equation and the bending equation for a surface in euclidean space is the presence of the last terms in curly brackets on the left-hand side and the -1 appearing in the factor $K-1$ on the right-hand side.

If we substitute z for ζ in (*), we get the following expressions for the coefficients L, M, N of the second fundamental form of the surface:

$$L = \frac{1}{Q}\{z_{uu} - \Gamma_{11}^1 z_u - \Gamma_{11}^2 z_v + (E - z_u^2)\tan z\},$$

$$M = \frac{1}{Q}\{z_{uv} - \Gamma_{12}^1 z_u - \Gamma_{12}^2 z_v + (F - z_u z_v)\tan z\},$$

$$N = \frac{1}{Q}\{z_{vv} - \Gamma_{22}^1 z_u - \Gamma_{22}^2 z_v + (G - z_v^2)\tan z\},$$

where

$$Q^2 = \frac{(E - z_u^2)(G - z_v^2) - (F - z_u z_v)^2}{EG - F^2}.$$

Now assume that the parametrization u, v of the surface is semi-geodesic, and so the line element of the surface is $ds^2 = du^2 + c^2 dv^2$. Then the Christoffel symbols are

$$\Gamma_{11}^1 = 0, \ \Gamma_{12}^1 = 0, \ \Gamma_{22}^1 = -cc_u, \qquad \Gamma_{11}^2 = 0. \ \Gamma_{12}^2 = c_u/c, \ \Gamma_{22}^2 = c_v/c.$$

The bending equation becomes

$$\begin{aligned}(r + (1-p^2)\tan z)\left(t + cc_u p - \frac{c_v}{c} q + (c^2 - q^2)\tan z\right) \\ - \left(s - \frac{c_u}{c} q - pq\tan z\right)^2 = (K-1)(c^2 - c^2p^2 - q^2).\end{aligned}$$

The coefficients L, M, N are

$$L = c\,\frac{r + (1-p^2)\tan z}{(c^2 - c^2p^2 - q^2)^{1/2}},$$

$$M = c\,\frac{s - \frac{c_u}{c}\,q - pq\tan z}{(c^2 - c^2p^2 - q^2)^{1/2}},$$

$$N = c\,\frac{t + cc_u p - \frac{c_v}{c}\,q + (c^2 - q^2)\tan z}{(c^2 - c^2p^2 - q^2)^{1/2}},$$

where r, s, t, p, q are the Monge symbols for the derivatives of z.

Lemma 1. *Let F be a regular convex cap in an elliptic space. Assume that the Gauss curvature of the cap is everywhere greater than that of the space. Then the normal curvature of the cap at a point x has an estimate which depends only on the intrinsic metric of the cap and the distance $h(x)$ of x from the base plane of the cap.*

Proof. Let x be an arbitrary point on the cap and γ an arbitrary geodesic issuing from x. Let $\kappa_\gamma(x)$ denote the normal curvature of the cap at x in the direction of γ. Consider the function $w_\gamma(x) = h(x)\kappa_\gamma(x)$.

The function $w_\gamma(x)$ is nonnegative and vanishes on the boundary of the cap. It assumes a maximum at some interior point x_0 on the cap, for some geodesic γ_0 issuing from x_0. Let w_0 denote this maximum.

Let $\bar{\gamma}_0$ be a geodesic issuing from x_0 perpendicular to γ_0; introduce a semigeodesic coordinate net u, v in the neighborhood of x_0, taking the geodesics perpendicular to $\bar{\gamma}_0$ as the u-curves. The parameters u, v will be arc length along the geodesics γ_0 and $\bar{\gamma}_0$, measured from x_0.

Let $\gamma(x)$ denote the u-curve passing through a point x close to x_0, and consider the function $w(x) = w_{\gamma(x)}(x)$. This function obviously has a maximum w_0 at x_0.

Set up the bending equation for F, with the tangent plane to the cap at x_0 as the plane $z = 0$. To fix ideas, we shall assume that the cap is in the halfspace $z > 0$. Setting

$$\alpha = (1-p^2)\tan z,\quad \beta = -\frac{c_u}{c}\,q - pq\tan z,$$

$$\gamma = cc_u p - \frac{c_v}{c}\,q + (c^2 - q^2)\tan z,\quad \delta = \left(\frac{c_{uu}}{c} + 1\right)(c^2 - c^2p^2 - q^2),$$

we write the bending equation as

$$(r+\alpha)(t+\gamma) - (s+\beta)^2 + \delta = 0.$$

We first express the normal curvature $\kappa_{\gamma(x)}$ in terms of z, using the formula for the coefficient L of the second fundamental form:

$$\kappa_{\gamma(x)} = c\,(r + (1 - p^2)\tan z)/(c^2 - c^2p^2 - q^2)^{1/2}.$$

Hence

$$w = hc\,(r + (1 - p^2)\tan z)/(c^2 - c^2p^2 - q^2)^{1/2}.$$

Solving this equation for r, we get

$$r = \frac{w}{h}\left(1 - p^2 - \frac{q^2}{c^2}\right)^{1/2} - (1 - p^2)\tan z.$$

We now find the first and second derivatives of this expression for r with respect to u and v at the point x_0, using the following information. By our choice of coordinate system we have

$$c = 1, \quad c_u = c_v = c_{uv} = c_{vv} = 0, \quad c_{uu} = -K,$$

where K is the Gauss curvature of the surface. The derivatives p and q vanish, since the plane $z = 0$ is tangent to the surface. The derivative s also vanishes, since $M = 0$ (the directions of the coordinate curves at x_0 are the principal directions). Finally, $w_u = 0$ and $w_v = 0$, since w assumes a maximum at x_0. Hence

$$r_u = \left(\frac{1}{h}\right)_u w, \quad r_v = \left(\frac{1}{h}\right)_v w,$$

$$r_{uu} = \frac{w_{uu}}{h} + w\left(\left(\frac{1}{h}\right)_{uu} - \frac{r^2}{h}\right) - r,$$

$$r_{vv} = \frac{w_{vv}}{h} + w\left(\left(\frac{1}{h}\right)_{vv} - \frac{t^2}{h}\right) - t.$$

Differentiating the bending equation at x_0 with respect to u, we get

$$rt_u + tr_u - K_u = 0.$$

Hence $t_u = (K_u - tr_u)/r$.

Differentiating the bending equation twice with respect to u, we get

$$r_{uu}t + t_{uu}r + 2r_ut_u - 2s_u^2 - r^2 + 2K^2 - K - 1 - K_{uu} = 0.$$

Now substitute the above values of r_u, r_v, t_u, $s_u = r_v$, r_{uu} and $t_{uu} = r_{vv}$ into this equation, and replace w by its value at x_0, which is rh:

$$(**) \qquad \begin{aligned} &\frac{1}{h}(w_{uu}t + w_{vv}r) - h(K-1)(r^2+t^2) - 2r^2h^2\left(\left(\frac{1}{h}\right)_u^2 + \left(\frac{1}{h}\right)_v^2\right) \\ &\quad + h(K-1)\left(\left(\frac{1}{h}\right)_{uu} + \left(\frac{1}{h}\right)_{vv}\right) \\ &\quad + 2\left(\frac{1}{h}\right)_u hK_u - r^2 + 2K^2 - 3K + 1 - K_{uu} = 0. \end{aligned}$$

To estimate the second derivatives h_{uu} and h_{vv}, we use the formulas for L and N with the base plane as the plane $z=0$. Then the principal curvatures at x_0, which are equal to r and t, are expressed in terms of h by

$$r=-\frac{h_{uu}+(1-h_u^2)\tanh}{(1-h_u^2-h_v^2)^{1/2}},\qquad t=-\frac{h_{vv}+(1-h_v^2)\tanh}{(1-h_u^2-h_v^2)^{1/2}}.$$

Hence we get expressions for h_{uu} and h_{vv} and substitute them into (**).

Since the first three terms in the left-hand side of (**) are nonpositive, this gives an inequality

$$-r^2h^2+Arh+B\geqq 0,$$

where A and B are certain bounded functions of the Gauss curvature K and its derivatives, h and its first derivatives h_u and h_v. Since $rh=w$, we thus get

$$-w_0^2+Aw_0+B\geqq 0.$$

This implies the existence of an upper limit $\bar{w}_0$ for w_0. The normal curvature of the cap at a point whose distance from the base is h cannot exceed $\bar{w}_0/h$. This proves the lemma.

Let F be a convex surface in an elliptic space. The specific extrinsic curvature of F on a set M is the quotient $\omega(M)/s(M)$, where $\omega(M)$ is the extrinsic curvature of the surface on M and $s(M)$ is the area of the surface there. We shall say that the specific curvature of the surface is bounded above (below) if there exists a positive constant c such that, for any Borel set M on the surface, $\omega(M)/s(M)\leqq(\geqq)c$.

We are already familiar with Aleksandrov's theorems on surfaces of bounded specific curvature in a euclidean space (Chapter II, §3):

1) *Convex surfaces with specific curvature bounded above contain neither ridge points nor conic points.*

2) *Convex surfaces with specific curvature bounded below are strictly convex, i.e. they can contain no straight-line segments.*

We wish to extend these theorems to convex surfaces of bounded specific curvature in elliptic space.

Let R_ρ denote a sphere of radius $\rho<\pi/2$ about $e_0(1,0,0,0)$ in an elliptic space R. Let F be any convex surface in the interior of R_ρ.

Map the elliptic space R geodesically onto the euclidean space $x_0=1$, associating with a point x of elliptic space the point $x/(xe_0)$ of euclidean space. If elliptic space is interpreted in terms of the three-sphere, this mapping is simply a projection of the sphere from its center onto the tangent hyperplane at e_0. Under this mapping, a convex surface F in R goes into a convex surface $\bar{F}$ in euclidean space.

Denote the extrinsic curvature and area of F over a set M by $\omega(M)$ and $S(M)$, the curvature and area of $\overline{F}$ on the corresponding set $\overline{M}$ by $\bar{\omega}(\overline{M})$ and $\overline{S}(\overline{M})$.

LEMMA 2. *There exist positive constants A_1, B_1 and A_2, B_2 depending only on ρ such that, for any open set M on the surface F,*

$$A_1 \leqslant \frac{S(M)}{\overline{S}(\overline{M})} \leqslant B_1, \qquad A_2 \leqslant \frac{\omega(M)}{\bar{\omega}(\overline{M})} \leqslant B_2.$$

PROOF. Without loss of generality, we may assume that the surface F is closed, since any convex surface may be completed to a closed surface. Take a sufficiently dense net of points on the surface F and construct their convex hull, which is a convex polyhedron P. Define a one-to-one correspondence between the points of F and P by projection from a point O in the interior of F. As the density of the net on F increases, the polyhedron P converges to F, and

$$S(M_P) \to S(M), \qquad \omega(M_P) \to \omega(M),$$

where M_P denotes the image of the set M on P under this projection.

The geodesic mapping takes the polyhedron P into a euclidean polyhedron $\overline{P}$. $\overline{P}$ is in one-to-one correspondence with $\overline{F}$ under projection from the point $\overline{O}$—the image of O under the geodesic mapping—and moreover, when $P \to F$ and $\overline{P} \to \overline{F}$, then

$$\overline{S}(\overline{M}_P) \to \overline{S}(M), \qquad \bar{\omega}(\overline{M}_P) \to \bar{\omega}(\overline{M}).$$

It follows that the lemma need be proved only for polyhedra. Moreover, since area and curvature are additive set functions, it will suffice to prove the first inequality for sets lying in a single face and the second for the curvature at a single vertex.

To do this, we appeal to the projective model of elliptic space in the euclidean space $x_0 = 1$. Here the polyhedron P is represented by a euclidean polyhedron $\overline{P}$. Now, when restricted to a bounded region of the space, the scaling factor determining the transition from distances in the euclidean metric to distances in the elliptic metric lies between positive limits, and hence the same holds for areas. This proves the first inequality.

The second is proved in a similar manner. Since the scaling factor defining the transition from angles between planes in euclidean space to angles between the corresponding planes in elliptic space lies between positive limits, the same holds for the curvature at a vertex of the polyhedron. Q.E.D.

Together with the theorems of Aleksandrov cited above, Lemma 2 implies the following assertions.

1. *Convex surfaces in elliptic space whose specific curvature is bounded above contain neither ridge points nor conic points.*

2. *Convex surfaces in elliptic space whose specific curvature is bounded below are strictly convex, i.e. they contain no straight-line segments.*

We shall say that the metric of a surface is regular (k times differentiable) if there exists a coordinate net u, v on the surface with respect to which the coefficients E, F, G of the line element of the surface are regular (k times differentiable) functions of u, v.

Regularity of a surface implies regularity of its metric. More precisely, if the surface is k times differentiable, then its metric is at least $k-1$ times differentiable. The converse is generally false: a surface may have a regular, even analytic metric without being regular itself. However:

THEOREM 1. *If a convex surface F in elliptic space has a regular (k times differentiable, $k \geqq 5$) metric and its Gauss curvature is greater than the curvature of the space, then the surface itself is regular (at least $k-1$ times differentiable). If the metric of the surface is analytic, then the surface is analytic.*

PROOF. In a sufficiently small neighborhood G of an arbitrary point x on F, the Gauss curvature K_i satisfies the inequality $1 < a \leqq K_i \leqq b$. Since the intrinsic curvature of F differs from its extrinsic curvature by the magnitude of its area

$$\omega(M) = \iint_M K_i ds - S(M),$$

it follows that the specific curvature of the surface in G is bounded between positive limits, i.e.

$$a - 1 \leqslant \frac{\dot{\omega}(M)}{S(M)} \leqslant b - 1.$$

Hence the surface F is smooth and strictly convex.

We shall prove that F is regular in a neighborhood of an arbitrary point x. We first consider the case of an analytic metric.

Cut off a small cap ω containing x from the surface F by a plane α; assume that ω is so small that any plane β cutting F in ω must cut off a cap. That this is possible follows from the fact that F is smooth and strictly convex.

Since the boundary γ of the cap ω has nonnegative i.g.c. over any subarc, it can be approximated by an analytic curve γ^* with everywhere positive geodesic curvature. This curve bounds a certain domain ω^* on ω.

Let Ω be the closed manifold formed by "gluing together" two copies of the surface ω^* (identifying corresponding boundary points). This manifold has an analytic metric everywhere except along the "glued" curve. In order to avoid this, the two copies of ω^* can be glued together not directly but through a narrow, regular strip, so chosen that the passage from the strip to ω^* is regular in the sense of the intrinsic metric. Moreover, since the geodesic curvature of the boundary of ω^* is positive, the strip can be so chosen that its Gauss curvature is everywhere greater than 1 (the curvature of the space).

It will be proved in Chapter VI that there exists a regular closed convex surface Ω^* in elliptic space isometric to Ω. Let $\tilde{\omega}^*$ be the domain on this surface corresponding to ω^*.

It follows from the monotypy theorem for closed convex surfaces (§6) that when the width of the strip tends to zero and $\gamma^* \to \gamma$, the surface Ω^* tends to the closed convex surface formed by the cap ω and its reflection in the plane of the boundary, and so $\tilde{\omega}^* \to \omega$.

Displace the plane α in the direction of the cap ω into a new position α'; let ω' and $\tilde{\omega}'$ denote the caps cut from the surfaces ω and $\tilde{\omega}^*$, respectively, by the plane α'.

Let $\widetilde{G}$ be a closed domain on $\tilde{\omega}'$ whose distance from the plane α' is greater than $h_0 > 0$. In this domain we can find upper estimates, which depend only on the intrinsic metric of F, for the normal curvatures of $\tilde{\omega}'$.

An estimate for the normal curvatures is equivalent to an estimate for the first and second derivatives of the coordinates x_i of a point on the surface. Hence, using the bending equation, which is an elliptic equation for x_i, we can establish estimates for the derivatives of x_i of all orders.

Now that intrinsic estimates are available for all derivatives of x_i, we can conclude that the surface F is infinitely differentiable in the domain $G = \lim \widetilde{G}$. Together with the ellipticity of the bending equation, this implies that the coordinates of x_i are analytic functions of u and v, i.e. the surface F is analytic.

Now assume that the metric of F is k times differentiable. Approximate the metric of F in ω by an analytic metric, and the curve γ^* by an analytic curve. This yields an analytic manifold with intrinsically convex boundary. By Aleksandrov's theorem, this manifold is isometric to a cap, which is analytic by what we have just proved. Now since we can find a priori estimates for the kth derivatives of the coordinates of a point on this cap, it follows that the limit cap, which by the monotypy theorem for convex caps is a domain on F, is $k-1$ times differentiable. Even more: the $(k-1)$th derivatives of the coordinates of its points satisfy a Hölder condition. This completes the proof of Theorem 1.

Just as in the euclidean case (Chapter II, §10), Theorem 1 can be sharpened by using Heinz's estimates for the second derivatives of the solution of the Darboux equation. Dubrovin [**30**], using Theorem 1, has proved that if the metric of the surface is in class C^n, $n \geqq 2$, then the surface is in class $C^{n-1+\nu}$, $0<\nu<1$.

The following two theorems are corollaries of Theorem 1 and Aleksandrov's theorem on the realizability of manifolds of curvature at least K in an elliptic space of curvature K.

THEOREM 2. *A closed two-dimensional manifold of curvature greater than K, with regular metric, is isometric to a regular closed convex surface in an elliptic space of curvature K. If the metric of the manifold is k times differentiable, $k \geqq 2$, then the surface is $k-1$ times differentiable. If the metric is analytic, the surface is analytic.*

THEOREM 3. *Every point of a two-dimensional manifold of curvature greater than K, with regular metric, has a neighborhood isometric to a regular convex cap in an elliptic space of curvature K. If the metric of the manifold is $k \geqq 2$ times differentiable, then the cap is $k-1$ times differentiable. If the metric is analytic, the cap is analytic.*

In conclusion, we remark that, as in the euclidean case, the theorem on regularity of convex surfaces with regular metric makes the synthetic methods of Aleksandrov's theory of general convex surfaces (including the gluing theorem) applicable to various problems in classical surface theory which are usually confined to sufficiently regular objects.

§8. Regularity of convex surfaces with regular metric in a hyperbolic space

In this section we discuss the regularity of convex surfaces with regular metric in a hyperbolic (Lobačevskiĭ) space. We have two reasons for doing this. First, formal application of the method presented in §7 to surfaces in a hyperbolic space yields incomplete results [**58**]. Estimates for the second derivatives at interior points of the cap can be established by this method only on the assumption that the Gauss curvature of the surface is positive, while a more natural condition would be that the extrinsic curvature is positive. Second, the result will be used in the next chapter in solving the embedding problem for two-dimensional Riemannian manifolds in a three-dimensional Riemannian space.

THEOREM 1. *If a convex surface in a hyperbolic space has a regular* (k *times differentiable*, $k \geqq 5$) *metric and its Gauss curvature is greater than the curvature of the space, then the surface itself is regular* (*at least* $k-1$ *times differentiable*). *If the metric is analytic, the surface is analytic.*

The proof of Theorem 1 resembles that of the analogous theorem for euclidean space (Chapter II, §10). We shall therefore pay special attention to those parts of the proof requiring essential modifications. The remainder of the proof, which is more or less a repetition of the reasoning in the euclidean case, will only be outlined.

We first prove the theorem for an analytic metric. Let F be a surface with analytic metric and O an arbitrary point on F.

In euclidean space, a convex surface with regular metric and positive curvature is smooth and strictly convex. An analogous statement holds for spaces of constant curvature, provided the Gauss curvature of the surface is greater than the curvature of the space. The proof for elliptic space was given in §7. It carries over word for word to the case of convex surfaces in hyperbolic space.

Since the surface F is smooth and strictly convex, there exists a unique supporting plane at the point O—the tangent plane to the surface. It has no points in common with the surface other than O. A plane perpendicular to the normal to the surface at O cuts off a small cap ω from the surface F. To prove that F is analytic, it will now suffice to prove the following two propositions:

1. There exists an analytic convex cap $\tilde{\omega}$ isometric to ω.
2. Isometric convex caps in a hyperbolic space are congruent.

The proof of the second proposition follows the same arguments as those applied in the euclidean case in [**55**]. This was the method employed by Danelič in [**29**] to prove the monotypy of convex caps in hyperbolic spaces.

To construct a surface realizing the metric given by a line element we must solve a boundary-value problem for the bending equation (Darboux equation). Thus the proof of the first proposition reduces to solution of this boundary-value problem. The derivation of the Darboux equation in hyperbolic space is the same as in elliptic space; we shall therefore omit the details and present the final result.

Let F be a regular surface with line element $ds^2 = Edu^2 + 2Fdudv + Gdv^2$, which can be projected in one-to-one fashion onto a plane σ, and $z(u,v)$ the distance of a point (u,v) on the surface from the plane, with suitable sign. Then, if the curvature K_0 of the hyperbolic space is -1, the function $z(u,v)$ satisfies the equation

$$(*) \qquad (r+\alpha)(t+\gamma) - (s+\beta)^2 + \delta = 0,$$

where

$$\alpha = -\Gamma_{11}^1 z_u - \Gamma_{11}^2 z_v - (E - z_u^2)\tanh z,$$
$$\beta = -\Gamma_{12}^1 z_u - \Gamma_{12}^2 z_v - (F - z_u z_v)\tanh z,$$
$$\gamma = -\Gamma_{22}^1 z_u - \Gamma_{22}^2 z_v - (G - z_v^2)\tanh z,$$
$$\delta = (K+1)\{(E - z_u^2)(G - z_v^2) - (F - z_u z_v)^2\};$$

Γ_{ij}^k are the Christoffel symbols of the surface, K is its Gauss curvature, and r, s, t are the second derivatives of $z(u,v)$.

Just as in the euclidean case, $r+\alpha$, $s+\beta$, and $t+\gamma$ are equal to the coefficients L, M, N of the second fundamental form, up to a multiplicative factor: if we set

$$Q^2 = \frac{(E - z_u^2)(G - z_v^2) - (F - z_u z_v)^2}{EG - F^2},$$

then

$$L = \frac{1}{Q}(r+\alpha), \quad M = \frac{1}{Q}(s+\beta), \quad N = \frac{1}{Q}(t+\gamma).$$

As in the case of euclidean caps, we construct a convex analytic curve (i.e. an analytic curve with positive geodesic curvature) approximating the boundary of ω, and let ω' denote the domain on F bounded by this curve. We claim that there exists an analytic domain $\tilde{\omega}'$ isometric to ω'. To this end it is sufficient to solve the Dirichlet problem for equation (*) in the domain G' of the u, v plane corresponding to ω', with boundary condition $z=0$. The existence of a solution will follow from Bernšteĭn's theorem if we can find a priori estimates for the posited solution and its first and second derivatives.

There is no difficulty in estimating the absolute value of the solution and its first derivatives p and q. In fact, $|z|$ cannot exceed the intrinsic diameter of the domain ω', and

$$|p| \leqslant \max_{\omega'} \frac{1}{\sqrt{E}}, \quad |q| \leqslant \max_{\omega'} \frac{1}{\sqrt{G}}.$$

Estimates for the second derivatives of the solution $z(u,v)$ are first derived on the boundary of G'. To this end, we introduce a semigeodesic coordinate system u, v in ω' near its boundary, with the boundary itself as the curve $u=0$. In this system the line element of the surface is $ds^2=du^2+c^2dv^2$, and the coefficients of equation (*) are

$$\alpha=-(1-p^2)\tanh z,\ \beta=pq\tanh z-\frac{c_u}{c}q,$$

$$\gamma=-(c^2-\dot{q}^2)\tanh z+cc_up-\frac{c_v}{c}q,$$

$$\delta=-(K+1)(c^2-c^2p^2-q^2).$$

Since the new coordinates u, v are defined by purely intrinsic means, it follows that derivation of a priori estimates for the second derivatives of z in either of the two coordinate systems is equivalent.

In the semigeodesic coordinates along the boundary of $\tilde{\omega}'$, we have $t=z_{vv}=0$. It will therefore suffice to find estimates for s and r. As in the euclidean case, we begin with s.

Define a domain ω_0': $u\leqq u_0$ along the boundary of $\tilde{\omega}'$ by the conditions $|p|<1$ and $|q|<2$. u_0 is clearly bounded below, irrespective of the form of the analytic cap $\tilde{\omega}'$ isometric to ω'. We now consider a surface Φ defined in rectangular cartesian coordinates u, v by the equation $\zeta=\zeta(u,v)$, where $\zeta(u,v)$ is a v-periodic function equal to $q(u,v)$ over one period. The surface Φ is, as it were, an infinite tape whose projection lies in the strip $0\leqq u\leqq u_0$ with one edge on the axis $u=0$. Let σ be a plane through the axis $u=0$ which forms the minimum angle with the u, v plane, such that the surface Φ lies beneath the plane σ. There are two possibilities: the plane σ either supports Φ along its edge $u=u_0$, or is tangent to Φ at some point.

In the first case,

$$\zeta_u\leqslant\frac{1}{u_0}\max_v|q(u_0,v)|,$$

and so the derivative s satisfies the following estimate along the boundary of the cap:

$$s\leqslant\frac{1}{u_0}\max_v|q(u_0,v)|<\frac{2}{u_0}.$$

Similarly, we derive a lower estimate for s:

$$-\frac{1}{u_0}\max_v|q(u_0,v)|\leqslant s.$$

Consequently $|s|<2/u_0$.

Now consider the second case: the plane σ is tangent to Φ at some point. Differentiating equation (*) with respect to v, we get

$$(**)\qquad q_{uu}(t+\gamma)-2q_{uv}(s+\beta)+q_{vv}(r+\alpha)-\alpha_v(t+\gamma)\\ -2\beta_v(s+\beta)+\gamma_v(r+\alpha)+\delta_v=0.$$

At the point of contact of σ and Φ we have

$$q_{uu}(t+\gamma)-2q_{uv}(s+\beta)+q_{vv}(r+\alpha)\leqq 0,$$

while $q_v=t=0$. Therefore equations (*) and (**) imply

$$(***)\qquad \alpha_v\gamma-2\beta_v(s+\beta)+\gamma_v\frac{(s+\beta)^2-\delta}{\gamma}+\delta_v\geqslant 0.$$

Now α_v, β_v, γ_v and δ_v are linear functions of the second derivatives of z; in fact, only of s and r (since $t=0$). Thus the function $\alpha_v\gamma-2\beta_v(s+\beta)+\delta_v$ has order of magnitude at most s^2 for large s.

With regard to the remaining term (the third term in the inequality), it is essential to note that $\gamma_v=cc_us+\cdots$, where terms not involving s have been omitted. Since $c_u=-\kappa_0$ on the boundary of the cap $\tilde{\omega}'$, where κ_0 is the geodesic curvature of the boundary, it follows that for small u_0 we may assume that $c_u<-\kappa_0/2$. Then the function

$$\gamma_v((s+\beta)^2-\delta)/\gamma$$

is of order $-\kappa_0s^3/4\gamma$ for large s. At the point of contact $\gamma>0$, since $t+\gamma>0$ and $t=0$. Now multiply inequality (***) by γ:

$$-\kappa_0s^3/4+O(s^2)\geqq 0,$$

where $O(s^2)$ denotes a quadratic polynomial in s with bounded coefficients. This implies the existence of an a priori estimate for $q_u=s$ at the point of contact, and hence also on the boundary of $\tilde{\omega}'$, where it cannot be greater.

It remains to derive an estimate for r. This can be done using the equation (*) itself, provided the coefficient γ can be bounded away from zero. Indeed, on the boundary of the cap,

$$r=-\alpha+((s+\beta)^2-\delta)/\gamma.$$

Now, on the boundary of the cap $\tilde{\omega}'$,

$$\gamma=cc_up,\quad c=1,\quad c_u=-\kappa_0,$$

and hence to estimate γ we need only find a lower estimate for p. This is done exactly as in the euclidean case. We have thus proved the existence of a priori estimates for the second derivatives on the boundary of the cap $\tilde{\omega}'$. The existence of estimates for the second derivatives on the entire surface of the cap will be proved later. Assuming that these have been found, we proceed with the proof of the theorem for analytic caps.

Since we have a priori estimates for the solution of equation (*), we can assert that there exists an analytic cap $\tilde{\omega}'$ isometric to ω'. Now let the domain ω' approach ω. Without loss of generality, we may assume that the cap $\tilde{\omega}'$ then converges to a cap $\tilde{\omega}$ isometric to ω.

It remains to prove that the cap $\tilde{\omega}'$ remains analytic as $\omega' \to \omega$. This follows from the fact that we can derive uniform a priori estimates for all the derivatives of the function z on a set of points of $\tilde{\omega}'$ whose distance from the plane of the boundary is greater than $\epsilon > 0$. This will be done later.

Thus in order to complete the proof it remains to establish a priori estimates for

1) the derivatives r, s and t on the entire cap $\tilde{\omega}'$, on the assumption that estimates are available on the boundary of the cap, and

2) the derivatives of z at a certain distance from the plane of the boundary of $\tilde{\omega}'$, without any assumptions on the geometry of the boundary.

The proof of the theorem for a metric which is k times differentiable utilizes analytic approximation of the metric of the cap and a limit procedure, as in the euclidean case.

To estimate the second derivatives of the solution z of equation (*), it suffices to estimate the normal curvature k_n of the cap $\tilde{\omega}'$. Indeed,

$$k_n = \frac{L\,du^2 + 2M\,du\,dv + N\,dv^2}{E\,du^2 + 2F\,du\,dv + G\,dv^2}.$$

If $|k_n| \leqq C$, then

$$|L| \leqslant CE, \quad |N| \leqslant CG, \quad |M| \leqslant \frac{C}{2}(E + 2|F| + G).$$

Hence, using the expressions obtained above for L, M, N, we get estimates for the second derivatives r, s, t.

An estimate for the normal curvature of $\tilde{\omega}'$ will first be derived on the boundary of the cap. Since we already have estimates there for the derivatives, it will suffice to establish a lower estimate for

$$Q = \left(\frac{(E - z_u^2)(G - z_v^2) - (F - z_u z_v)^2}{EG - F^2}\right)^{1/2},$$

which figures in the denominator of the coefficients L, M, N. This expression has a simple geometrical meaning: it is the cosine of the angle between the normal to the cap and the perpendicular dropped onto the base plane. Along the boundary of the cap, this is the cosine of the angle between the tangent plane to the cap and its base plane.

In §7 of Chapter II, we derived a lower estimate for Q in the euclidean case. The derivation can be repeated almost word for word in hyperbolic space, and we shall therefore assume that the required estimate has been established, hence also an upper estimate for the normal curvature of $\tilde{\omega}'$ along its boundary.

We must now find an estimate for the normal curvature on the surface of the entire cap $\tilde{\omega}'$, given an estimate on the boundary. Let $\vartheta(X)$ denote the smallest angle formed by the normal to the cap at the point X and the perpendicular dropped from X to the base plane of the cap. Denote the maximum normal curvature at X by $\bar{\kappa}(X)$, and consider the function

$$\bar{w}(X) = \bar{\kappa}(X)/(\cos\vartheta(X))^{\mu},$$

where μ is a positive constant, to be defined later. It is obvious that, in order to estimate the normal curvature of the cap, it is sufficient to estimate the function $\bar{w}(X)$.

The function $\bar{w}(X)$ has an absolute maximum at some point X_0 on the surface of the cap $\tilde{\omega}'$. There are two possible cases:

1. X_0 is on the boundary of the cap.
2. X_0 is an interior point of the cap.

In the first case, an estimate for $\bar{w}$ is easily obtained, since we have an upper estimate for the normal curvature on the boundary of the cap and an upper estimate for $\cos\vartheta$.

Thus it remains to estimate $\bar{w}$ in the second case, when the point X_0 at which $\bar{w}$ assumes its maximum is an interior point.

Introduce semigeodesic coordinates u, v in a neighborhood of X_0 on $\tilde{\omega}'$, taking a geodesic through X_0 in the direction of maximal normal curvature as the u-curve, and a perpendicular geodesic as the v-curve. The parameters u and v are arc length along these geodesics. The line element of the cap becomes $ds^2 = du^2 + c^2dv^2$. At the point X_0 we have $c = 1$, $c_u = c_v = 0$, $c_{uu} = -K$, $c_{uv} = 0$, $c_{vv} = 0$, where K is the Gauss curvature of the cap at X_0.

Now let $\kappa(X)$ denote the normal curvature of the cap at a point X in the u direction; consider the function

$$w(X) = \kappa(X)/(\cos\vartheta(X))^{\mu}.$$

Since $\kappa(X) \leqq \bar{\kappa}(X)$, while $\kappa(X_0) = \bar{\kappa}(X_0)$, the function $w(X)$ also has a maximum at X_0, and all we need is an estimate for $w(X)$ at the point X_0.

Subsequent manipulations will be somewhat simpler if the plane $z=0$ is not the base of the cap, as hitherto, but the tangent plane of the cap at X_0. The bending equation, though of the same form, becomes much simpler at the point X_0, since there $z=0$, $p=0$, $q=0$, and so $\alpha=\beta=\gamma=0$.

Since

$$L=\frac{1}{Q}(r+\alpha), \quad \alpha=-(1-p^2)\tanh z, \quad Q=\left(1-p^2-\frac{q^2}{c^2}\right)^{1/2}$$

and $E=1$, we get

$$\varkappa(X)=\frac{r-(1-p^2)\tanh z}{\left(1-p^2-\frac{q^2}{c^2}\right)^{1/2}}.$$

Note that the normal curvature in the v direction is

$$\varkappa'(X)=\frac{t+\gamma}{c^2\left(1-p^2-\frac{q^2}{c^2}\right)^{1/2}}.$$

And since $\kappa' \leqq \kappa$, $\alpha=\gamma=0$, $c=1$ at the point X_0, we have $t \leqq r$ there. Moreover, the coefficient M vanishes at X_0, and therefore $s=0$.

We now consider the bending equation

$$(*) \qquad (r+\alpha)(t+\gamma)-(s+\beta)^2+\delta=0,$$

$$\alpha=-(1-p^2)\tanh z, \quad \beta=pq\tanh z-\frac{c_u}{c}q,$$

$$\gamma=-(c^2-q^2)\tanh z+cc_u p-\frac{c_v}{c}q, \qquad \delta=-(1+K)(c^2-c^2p^2-q^2).$$

At the point X_0 we have $c=1$, $c_u=1$, $c_u=c_v=0$, $c_{uv}=c_{vv}=0$, $c_{uu}=-K$, $z=0$, $p=q=0$, $s=0$. Hence, at the same point,

$$\begin{aligned} &\alpha=0, \qquad \beta=0, \qquad \gamma=0, \qquad \delta=-(1+K),\\ &\alpha_u=0, \qquad \beta_u=0, \qquad \gamma_u=0, \qquad \delta_u=-K_u,\\ &\alpha_{uu}=-r, \quad \beta_{uu}=0, \quad \gamma_{uu}=-r(1+2K),\\ &\delta_{uu}=-K_{uu}+2K(1+K)-2r^2(1+K).\end{aligned}$$

The bending equation at X_0 becomes $rt+\delta=0$.

Differentiating equation (*) with respect to u at X_0, we get $r_u t+t_u r+\delta_u=0$. Hence, again at the point X_0,

$$t_u=-(\delta_u+r_u t)/r.$$

Differentiating equation (*) twice with respect to u at X_0, we get

$$\begin{aligned}(r_{uu}+\alpha_{uu})(t+\gamma)-2(s_{uu}+\beta_{uu})(s+\beta)+(t_{uu}+\gamma_{uu})(r+\alpha)\\ +2(r_u+\alpha_u)(t_u+\gamma_u)-2(s_u+\beta_u)^2+\delta_{uu}=0.\end{aligned}$$

Substituting the values of the derivatives of α, β, γ and δ at X_0, we get

$$r_{uu}t + r_{vv}r + 2r_u t_u - 2s_u^2 + \alpha_{uu}t + \gamma_{uu}r + \delta_{uu} = 0.$$

For brevity, we set $\lambda = 1/(\cos\vartheta)^\mu$. Then

$$w = \lambda \frac{r - (1-p^2)\tanh z}{\left(1 - p^2 - \frac{q^2}{c^2}\right)^{1/2}},$$

and hence

$$r = \frac{w}{\lambda}\left(1 - p^2 - \frac{q^2}{c^2}\right)^{1/2} + (1-p^2)\tanh z.$$

Since $w_u = w_v = 0$ at X_0, we have there

$$r_u = w\left(\frac{1}{\lambda}\right)_u, \quad r_v = w\left(\frac{1}{\lambda}\right)_v,$$

$$r_{uu} = \frac{w_{uu}}{\lambda} + w\left(\frac{1}{\lambda}\right)_{uu} - \frac{w}{\lambda}r^2 + r,$$

$$r_{vv} = \frac{w_{vv}}{\lambda} + w\left(\frac{1}{\lambda}\right)_{vv} - \frac{w}{\lambda}t^2 + t.$$

We now substitute the expressions for the derivatives r_u, r_v, r_{uu} and r_{vv} into the equation obtained by twice differentiating (*) with respect to u. After a few simple manipulations, we get

$$\left(\frac{w_{uu}}{\lambda} + w\left(\frac{1}{\lambda}\right)_{uu}\right)t + \left(\frac{w_{vv}}{\lambda} + w\left(\frac{1}{\lambda}\right)_{vv}\right)r - \frac{2w^2t}{r}\left(\frac{1}{\lambda}\right)_u^2 - 2w^2\left(\frac{1}{\lambda}\right)_v^2$$
$$+ \frac{2w}{r}K_u\left(\frac{1}{\lambda}\right)_u + Ar^2 + 2Br + C = 0,$$

where A, B, C are bounded functions which depend only on the curvature K and its derivatives of order up to 2. Noting that

$$\left(\frac{1}{\lambda}\right)_{uu} - 2\lambda\left(\frac{1}{\lambda}\right)_u^2 = -\frac{\lambda_{uu}}{\lambda^2}, \qquad \left(\frac{1}{\lambda}\right)_{vv} - 2\lambda\left(\frac{1}{\lambda}\right)_v^2 = -\frac{\lambda_{vv}}{\lambda^2},$$

we can bring our equation to the form

$$\frac{1}{\lambda}(w_{uu}t + w_{vv}r) - (1+K)\frac{\lambda_{uu}}{\lambda} - r^2\frac{\lambda_{vv}}{\lambda} + 2K_u\frac{\lambda_u}{\lambda} + O(r^2) = 0,$$

where $O(r^2)$ is a quadratic trinomial in r with bounded coefficients.

Since w has a maximum at X_0, so that $w_{uu} \leqq 0$ and $w_{vv} \leqq 0$, it follows that

$$-(1+K)\frac{\lambda_{uu}}{\lambda} - r^2\frac{\lambda_{vv}}{\lambda} + 2K_u\frac{\lambda_u}{\lambda} + O(r^2) \geqslant 0.$$

Henceforth we shall refer to this as the *fundamental inequality*.

We now turn to the derivatives of λ. If h denotes the distance of an arbitrary point of the cap $\tilde{\omega}'$ from its base plane, then

$$\lambda = \frac{1}{\Delta^{\mu}}, \qquad \Delta = 1 - h_u^2 - \frac{h_v^2}{c^2}.$$

Hence it is clear that the required derivatives of λ can be expressed in terms of the derivatives of h of order up to 3. We therefore begin with a calculation of the latter.

Since the coefficients L, M, N of the second fundamental form of the surface are independent of the choice of the plane $z = 0$, we must have

$$\frac{r - (1 - p^2)\tanh z}{\left(1 - p^2 - \frac{q^2}{c^2}\right)^{1/2}} = - \frac{h_{uu} - (1 - h_u^2)\tanh h}{\left(1 - h_u^2 - \frac{h_v^2}{c^2}\right)^{1/2}},$$

$$\frac{s + pq\tanh z - \frac{c_u}{c} q}{\left(1 - p^2 - \frac{q^2}{c^2}\right)^{1/2}} = - \frac{h_{uv} + h_u h_v \tanh h - \frac{c_u}{c} h_v}{\left(1 - h_u^2 - \frac{h_v^2}{c^2}\right)^{1/2}}.$$

$$\frac{t - (c^2 - q^2)\tanh z + c c_u p - \frac{c_v}{c} q}{\left(1 - p^2 - \frac{q^2}{c^2}\right)^{1/2}} = - \frac{h_{vv} - (c^2 - h_v^2)\tanh h + c c_u h_u - \frac{c_v}{c} h_v}{\left(1 - h_u^2 - \frac{h_v^2}{c^2}\right)^{1/2}}.$$

Hence we obtain the following expressions for the second derivatives of h at X_0:

$$h_{uu} = (1 - h_u^2)\tanh h - r(1 - h_u^2 - h_v^2)^{1/2},$$
$$h_{uv} = - h_u h_v \tanh h - s(1 - h_u^2 - h_v^2)^{1/2},$$
$$h_{vv} = (1 - h_v^2)\tanh h - t(1 - h_u^2 - h_v^2)^{1/2}.$$

Since $s = 0$ and $rt = 1 + K$ at the point X_0, it follows that for large r

$$h_{uu} = -r\Delta^{1/2} + \cdots, \quad h_{uv} = 0 + \cdots, \quad h_{vv} = 0 + \cdots,$$

where terms of order $O(1)$ are omitted.

Differentiating $1/\lambda$ with respect to u and v at the point X_0, we get

$$\left(\frac{1}{\lambda}\right)_u = 2\mu\Delta^{\mu - 1/2} h_u r + \ldots, \qquad \left(\frac{1}{\lambda}\right)_v = 0 + \ldots,$$

writing only the principal terms in r.

We now calculate the principal part (in r) of the third derivatives h_{uuu} and h_{uuv}:

$$h_{uuu} = - r_u \Delta^{1/2} + r\Delta^{-1/2} h_u h_{uu} + \ldots,$$

$$r_u = w\left(\frac{1}{\lambda}\right)_u = 2\mu\lambda h_u \Delta^{\mu - 1/2} r^2 + \ldots.$$

Consequently

$$h_{uuu} = -(2\mu+1)h_u r^2 + \cdots,$$

where terms of order lower than r^2 are omitted.

Similarly,

$$h_{uuv} = -r_v \Delta^{1/2} + \cdots, \quad r_v = w(1/\lambda)_v = 0 + \cdots.$$

Thus $h_{uuv} = 0 + \cdots$, omitting terms of order lower than r^2.

We can now determine the principal part of the derivative λ_{uu}. We have

$$\lambda_{uu} = \left(\frac{1}{\Delta^{\mu}}\right)_{uu} = \mu(\mu+1)\Delta^{-\mu-2}\Delta_u^2 - \mu\Delta^{-\mu-1}\Delta_{uu}$$
$$= 4\mu(\mu+1)\Delta^{-\mu-2}(h_u h_{uu})^2 + 2\mu\Delta^{-\mu-1}(h_{uu}^2 + h_u h_{uuu}) + \ldots$$

Hence

$$\lambda_{uu} = 2\mu\Delta^{-\mu} r^2 + 2\mu\Delta^{-\mu-1} h_u^2 r^2 + \cdots,$$

omitting terms of order lower than r^2.

Now consider the derivative λ_{vv}. We first calculate the corresponding derivatives of h. Note that at the point X_0

$$h_{uu} = -\Delta^{1/2} r + O(1),$$
$$h_{uv} = -h_u h_v \tanh h,$$
$$h_{vv} = (1 - h_v^2)\tanh h + O\left(\frac{1}{r}\right).$$

Now

$$h_{vvu} = ((1-h_v^2)\tanh h)_u - t_u \Delta^{1/2} + t\Delta^{-1/2} h_u h_{uu} + K h_u + O\left(\frac{1}{r}\right),$$
$$-t_u \Delta^{1/2} = 2\mu h_u (1+K) + O\left(\frac{1}{r}\right),$$
$$t\Delta^{-1/2} h_u h_{uu} = -(1+K) h_u + O\left(\frac{1}{r}\right),$$

$$((1-h_v^2)\tanh h)_u = -2h_v h_{uv} \tanh h + (1-h_v^2)\frac{h_u}{\cosh^2 h} =$$
$$= 2h_u h_v^2 \tanh^2 h + (1-h_v^2)(1-\tanh^2 h) h_u.$$

Hence

$$h_u h_{vvu} = 2h_u^2 h_v^2 \tanh^2 h + (1-h_v^2)(1-\tanh^2 h) h_u^2$$
$$+ 2\mu h_u^2 (1+K) - h_u^2 + O\left(\frac{1}{r}\right).$$

Thus for sufficiently large μ

$$h_u h_{vvu} = (*)^2 + O(1/r),$$

where $(*)^2$ denotes a nonnegative function.

We now calculate $h_v h_{vvv}$:

$$h_{vvv} = \left((1-h_v^2)\tanh h\right)_v - t_v \Delta^{1/2} + O\left(\frac{1}{r}\right),$$

$$t_v = -\frac{1}{r}(\delta_v + r_v t) + O\left(\frac{1}{r}\right),$$

$$\left((1-h_v^2)\tanh h\right)_v = -2h_v(1-h_v^2)\tanh^2 h + (1-h_v^2)(1-\tanh^2 h)\, h_v + O\left(\frac{1}{r}\right).$$

Hence

$$h_v h_{vvv} = h_v^2(1-h_v^2)(1-3\tanh^2 h) + O(1/r).$$

And since the height of the cap $\tilde{\omega}'$ is small, we have

$$h_v h_{vvv} = (*)^2 + O(1/r).$$

Now, for the derivative λ_{vv}, we have

$$\begin{aligned}\lambda_{vv} &= \mu(\mu+1)\Delta^{-\mu-2}\Delta_v^2 - \mu\Delta^{-\mu-1}\Delta_{vv}\\ &= 4\mu(\mu+1)\Delta^{-\mu-2}(h_u h_{uv} + h_v h_{vv})^2\\ &\qquad + 2\mu\Delta^{-\mu-1}(h_{uv}^2 + h_{vv}^2 + h_u h_{vvu} + h_v h_{vvv}).\end{aligned}$$

Hence, using the expressions for $h_u h_{vvu}$ and $h_v h_{vvv}$, we see that

$$\lambda_{vv} = (*)^2 + O(1/r).$$

We now substitute our expressions for the derivatives of λ into the fundamental inequality; the result is:

$$-2\mu(1+K)r^2 + Ar^2 + O(r) \geqq 0,$$

where A is bounded above by a constant A_0 which depends on the curvature of the surface and its derivatives, and $O(r)$ is of order r. Multiplying the inequality by λ^2, we get

$$-2\mu(K+1)w^2 + Aw^2 + O(w) \geqq 0.$$

Now assume that, apart from the conditions imposed in the course of the foregoing arguments, μ satisfies the inequality $-2\mu(K+1) + A_0 < 1$. Our inequality then implies $-w^2 + O(w) \geqq 0$. Hence w cannot be arbitrarily large, and this proves the existence of an estimate w_0 for w.

The estimate for w is at the same time an estimate for the normal curvature, since

$$\kappa(x) = w(\cos\vartheta)^\mu \leqq w.$$

The a priori estimates we have established for the normal curvatures complete the proof of the statement that there exists an analytic cap $\tilde{\omega}'$ isometric to the domain ω on the cap ω. Now let the domain ω'

approach ω. Then, because of the monotypy of caps in the class of caps, the surface $\tilde{\omega}'$ converges to a convex cap $\tilde{\omega}$ congruent to ω. Without loss of generality, we may assume that this cap is ω itself.

In order to prove that $\tilde{\omega}$ is analytic, it suffices to establish in a neighborhood of each point uniform estimates for the derivatives of order up to 4 of the functions z defining the surfaces $\tilde{\omega}'$ converging to $\tilde{\omega}$. In fact, as in the euclidean case, this will imply that the limit cap $\tilde{\omega}$ is thrice differentiable. Hence, by Bernšteĭn's theorem on the analyticity of solutions of elliptic equations, applied to the bending equation, $\tilde{\omega}$ is analytic. As in the euclidean case, estimates for the third and fourth derivatives, given estimates for the second derivatives, are derived by general arguments applied to the bending equation. Thus our problem is to establish estimates for the first and second derivatives of the functions z defining the caps $\tilde{\omega}'$. For the first derivatives, this is trivial. To estimate the second derivatives, we first consider the normal curvature.

Move the base plane σ of the cap ω to a position σ'' in which it cuts off from ω a cap ω'' whose boundary is not on the plane σ. If $\tilde{\omega}'$ is sufficiently close to $\tilde{\omega} \equiv \omega$, the plane σ'' will cut off from $\tilde{\omega}'$ a cap $\tilde{\omega}''$ such that the angles between the tangent planes along its boundary and the plane σ'' are less than $\pi/2 - \epsilon$, $\epsilon > 0$. This property is essential for the sequel. To avoid introducing new notation, we shall assume that the caps $\tilde{\omega}'$ themselves have this property. Thus we assume that the tangent planes to $\tilde{\omega}'$ form angles less than $\pi/2 - \epsilon$ with the base plane σ, where ϵ is a positive number, the same for all caps $\tilde{\omega}'$ sufficiently near ω.

Retaining the previous notation, consider the auxiliary function

$$\bar{w}(X) = h(X)\bar{\kappa}(X)/(\cos\vartheta(X))^{\mu},$$

where h is the distance of the point X on the cap $\tilde{\omega}'$ from the base plane σ. This function is nonnegative, vanishes on the boundary of the cap, and therefore assumes an absolute maximum at some interior point X_0. Introducing a semigeodesic coordinate system in a neighborhood of X_0 on the surface of the cap, consider the function

$$w(X) = h(X)\kappa(X)/(\cos\vartheta(X))^{\mu}.$$

This function also assumes a maximum at X_0. Setting $\lambda = h/(\cos\vartheta)^{\mu}$, we repeat the foregoing arguments word for word, up to the proof of the fundamental inequality

$$-(1+K)\frac{\lambda_{uu}}{\lambda} - r^2\frac{\lambda_{vv}}{\lambda} + 2K_u\frac{\lambda_u}{\lambda} + O(r^2) \geqslant 0.$$

We now find expressions for h_{uu}, h_{uv}, h_{vv}, h_{uuu} and h_{uuv}. We have

$$\lambda_{uu} = 2h\mu\Delta^{-\mu}r^2 + 2h\mu\Delta^{-\mu-1}h_u^2r^2 + \cdots,$$

where only the principal part of λ_{uu} in r has been written.

We consider λ_{vv} in greater detail. We have

$$\lambda_{vv} = \left(\frac{n}{\Lambda^{\mu}}\right) = \frac{h_{vv}}{\Lambda^{\mu}} + 2h_v(-\mu)\Delta^{-\mu-1}\Delta_v + h\left(\frac{1}{\Delta}\right)_v .$$

At the point X_0 we have $h_{vv} = (1-h_v^2)\tanh h - t\Delta^{1/2}$, while $t = (K+1)/r$. Consequently

$$h_{vv}/\Delta^{\mu} = (*)^2 + O(1/r).$$

Now, $\Delta_v|_{X_0} = -h_u h_{uv} - h_v h_{vv}$. Therefore

$$2h_v(-\mu)\Delta^{-\mu-1}\Delta_v = 2\mu\Delta^{-\mu-1}(1-h_u^2-h_v^2)h_v^2\tanh h + O\left(\frac{1}{r}\right)$$

$$= (*)^2 + O\left(\frac{1}{r}\right).$$

As for $(1/\Delta^{\mu})_{vv}$, we have already shown that at X_0

$$(1/\Delta^{\mu})_{vv} = (*)^2 + O(1/r).$$

Thus $\lambda_{vv} = (*)^2 + O(1/r)$.

Now substitute our expressions for λ_{uu} and λ_{vv} in the fundamental inequality, and replace r by w:

$$-2\mu(K+1)w^2 + Aw^2 + O(w) \geqq 0.$$

This implies the existence of the required estimate: $w \leqq w_0$.

Now consider the neighborhood of a point $\widetilde{X}$ whose distance from the base plane σ is at least $h > 0$. The normal curvatures of the cap in this neighborhood satisfy the estimate

$$\kappa \leqq w_0(\cos\vartheta)^{\mu}/h \leqq w_0/ \quad ,$$

Q.E.D.

The following theorem is a corollary of Theorem 1 and Aleksandrov's theorem on the realizability of a manifold of curvature at least K, homeomorphic to a sphere, by a closed convex surface in a hyperbolic space of curvature K.

THEOREM 2. *A regular metric with curvature everywhere greater than K, defined on a sphere, is realizable by a regular convex surface in a hyperbolic space of curvature K. If the metric is in class C^n, $n \geqq 5$, then the surface is in class C^{n-1}. If the metric is analytic, the surface is analytic.*

CHAPTER VI

CONVEX SURFACES IN RIEMANNIAN SPACE

In 1916 Hermann Weyl posed the following problem [**71**].

Consider a Riemannian metric defined on a sphere or a manifold homeomorphic to a sphere by the line element

$$ds^2 = g_{ij} du^i du^j.$$

Assume that the Gauss curvature of the manifold (defined by the usual formula of differential geometry in terms of the coefficients g_{ij}) is positive. Do there exist a closed surface in euclidean space and a parametrization u^i (system of parametrizations) of the surface in which the line element of the surface coincides with the given element ds^2? Equivalently, can the given two-dimensional Riemannian manifold be isometrically embedded in euclidean three-space?

In principle, the problem was solved by Weyl himself in the cited paper. At any rate, he indicated the basic steps of the solution. This solution was later completed by Lewy [**40**] (see also [**49**]). He proved that a two-dimensional closed Riemannian manifold homeomorphic to a sphere, with analytic metric and positive Gauss curvature, can be isometrically embedded in euclidean three-space as a closed analytic surface.

A new solution to Weyl's problem follows from Aleksandrov's theorems (a metric of positive curvature defined on a manifold homeomorphic to a sphere is realizable by a convex surface; a convex surface with regular metric is regular; Chapter II, §10). Other theorems of Aleksandrov (a metric with Gauss curvature greater than K is realizable by a convex surface in a space of curvature K; a convex surface with regular metric is regular) yield a solution to Weyl's problem for spaces of constant curvature (Chapter V, §§7 and 8).

Motivated by these results, the author posed and solved the problem of isometric embedding of a Riemannian manifold homeomorphic to a sphere in general Riemannian three-space [**59, 54**]. The main subject of this chapter is a proof of this result.

Also discussed in this chapter is the problem of infinitesimal bendings of regular closed convex surfaces in Riemannian three-space. We prove the existence of a priori estimates for the normal curvatures of a closed convex surface, which depend only on the metrics of the surface and the space.

In §11 we apply our theorem on isometric embeddability to solution of a problem of Cohn-Vossen on isometric transformations of a punctured convex surface in euclidean space.

§1. Surfaces in Riemannian spaces

Much of the theory of curves and surfaces in euclidean space can be carried over to Riemannian three-space. In this section we shall recall the parts of the theory of curves and surfaces in Riemannian space needed for the sequel. A more detailed account may be found in Cartan's book [**26**].

Let R be a Riemannian space with line element

$$ds^2 = g_{ij}\,dx^i dx^j$$

and consider a regular surface $F: x^i = x^i(u^1, u^2)$ in R, where u^1, u^2 are the coordinates on the surface. Denoting the point of the surface with coordinates u^1; u^2 by $x(u^1, u^2)$, we write the equations of the surface in abbreviated form: $x = x(u^1, u^2)$.

Just as in the euclidean case, one associates two quadratic differential forms with the surface:

$$\mathrm{I} = (dx)^2 = \tilde{g}_{ij}\,du^i\,du^j,$$
$$\mathrm{II} = -dx \cdot Dn = \lambda_{ij}\,du^i\,du^j,$$

where D is the symbol for absolute differentiation in the space and n the unit normal vector of the surface.[1)] The first fundamental form is the line element of the surface.

The normal curvature in a given direction on a surface in Riemannian three-space may be defined as the normal curvature of the surface in the osculating euclidean space. It is given by the usual formula,

$$\varkappa = \frac{\lambda_{ij}\,du^i\,du^j}{\tilde{g}_{ij}\,du^i\,du^j}.$$

[1)] By the absolute differential Da of a vector a we mean the principal part of the difference between the vector $a(x + dx)$ (transported parallel to itself to the point x) and the vector $a(x)$.

The principal directions on a surface in a Riemannian space are defined as the directions in which the normal curvatures assume their extremum values. They are expressed in the usual manner in terms of the coefficients of the first and second fundamental forms. Absolute differentiation along a principal direction is given by the Rodrigues formula, $Dn = -\kappa dx$, where κ is the normal curvature in this direction.

Extrinsic (total) curvature K_e is defined as the curvature of the surface in the osculating euclidean space. It is again given by the usual formula,

$$K_e = \frac{\lambda_{11}\lambda_{22} - \lambda_{12}^2}{\tilde{g}_{11}\tilde{g}_{22} - \tilde{g}_{12}^2}.$$

In contradistinction to the euclidean case, the extrinsic curvature of a surface in a Riemannian space is not equal to its intrinsic (Gauss) curvature K_i; the difference is the sectional curvature K_R of the space in a plane section tangent to the surface: $K_e = K_i - K_R$.

Equations analogous to the Gauss and Weingarten equations for surfaces in euclidean space are valid for surfaces in a Riemannian space.

The tangent vectors to the surface at a point x,

$$x_1 = \partial x/\partial u^1, \qquad x_2 = \partial x/\partial u^2$$

and the unit normal vector n form a basis. Therefore any vector at x can be expressed as a linear combination of the vectors x_1, x_2 and n. In particular,

$$(*) \qquad \begin{aligned} D_i x_j &= A_{ij}^k x_k + a_{ij} n, \\ D_i n &= B_i^k x_k + b_i n, \end{aligned}$$

where D_i denotes absolute differentiation with respect to u^i.

The coefficients of these equations are found just as in the euclidean case. First, calculating the scalar product of the first equation and n, and recalling that

$$n^2 = 1, \qquad nx_j = 0,$$

$$nD_i x_j = D_i(nx_j) - x_j D_i n = \lambda_{ij},$$

we get $a_{ij} = \lambda_{ij}$.

Then, taking the scalar product of the first equation and the vector x_α $(\alpha = 1, 2)$, and noting that

$$x_k x_\alpha = \tilde{g}_{k\alpha}, \qquad D_i x_j = D_j x_i,$$

we get

$$\tilde{\Gamma}_{i\alpha j} = A_{ij}^k \tilde{g}_{k\alpha},$$

where $\tilde{\Gamma}_{i\alpha j}$ are the Christoffel symbols of the first kind for the surface:

$$\tilde{\Gamma}_{i\alpha j}=\frac{1}{2}\left(\frac{\partial \tilde{g}_{i\alpha}}{\partial u^j}+\frac{\partial \tilde{g}_{\alpha j}}{\partial u^i}-\frac{\partial \tilde{g}_{ij}}{\partial u^\alpha}\right).$$

Multiplying the equality by $\tilde{g}^{\alpha\beta}$ and summing over γ, we get

$$A_{ij}^{\beta}=\tilde{\Gamma}_{ij}^{\beta},$$

where $\tilde{\Gamma}_{ij}^{\beta}$ are the Christoffel symbols of the second kind.

Multiplying the second equation of (*) by n, we get $b_i=0$.

Now multiply the second equation of (*) by x_α, and note that $x_\alpha D_i n = -\lambda_{i\alpha}$. The result is

$$-\lambda_{i\alpha}=B_i^k \tilde{g}_{k\alpha}.$$

If we now multiply this equality by $\tilde{g}^{\alpha\beta}$ and sum over α, we get

$$B_i^\beta=-\lambda_{i\alpha}\tilde{g}^{\alpha\beta}.$$

We thus have the following differential equations for surfaces in a Riemannian space:

$$(**)\qquad \begin{aligned} D_i x_j &= \tilde{\Gamma}_{ij}^k x_k+\lambda_{ij} n,\\ D_i n &= -\lambda_{i\alpha}\tilde{g}^{\alpha k} x_k. \end{aligned}$$

The coefficients of these equations are expressed in terms of the coefficients of the first and second fundamental forms exactly as in the euclidean case. This is not surprising, for equations (**) are simply the Gauss and Weingarten equations for the surface in the osculating euclidean space.

The fundamental equations of surface theory in euclidean space are the Gauss formula [the *theorema egregium*] and the Peterson-Codazzi equations.[2)] Analogous equations can be found for surfaces in a Riemannian space. We shall call these the fundamental equations of surface theory in Riemannian space.

Let us calculate the expression

$$\omega_{ij\alpha}=(D_iD_j-D_jD_i)x_\alpha.$$

To do this, we introduce semigeodesic coordinates v^i in a neighborhood of the surface. The coordinates of an arbitrary point of space near the surface will be the signed distance (v^3) of the point from the surface

[2)] The Latvian geometer K. M. Peterson (1828-81) apparently discovered the Codazzi equations a few years before Codazzi and Mainardi.

and the coordinates u^1, u^2 of the foot of the geodesic perpendicular dropped to the surface (v^1, v^2). In this parametrization, the vectors x_1, x_2, n coincide with the local basis vectors e_1, e_2, e_3 defined by the parametrization of the space.

We have

$$D_j e_\alpha = \Gamma^k_{j\alpha} e_k,$$

$$D_i D_j e_\alpha = \left(\frac{\partial \Gamma^\beta_{j\alpha}}{\partial v^i} + \Gamma^k_{j\alpha}\Gamma^\beta_{ik}\right) e_\beta.$$

Hence

$$\omega_{ij\alpha} = (D_i D_j - D_j D_i)\, e_\alpha = \left(\frac{\partial \Gamma^\beta_{j\alpha}}{\partial v^i} - \frac{\partial \Gamma^\beta_{i\alpha}}{\partial v^j} + \Gamma^k_{j\alpha}\Gamma^\beta_{ik} - \Gamma^k_{i\alpha}\Gamma^\beta_{jk}\right) e_\beta.$$

But the expression in parentheses is precisely the component $R^{\cdot\beta}_{\alpha\cdot ij}$ of the mixed Riemann-Christoffel tensor. Thus

$$\omega_{ij\alpha} = R^{\cdot\beta}_{\alpha\cdot ij} e_\beta.$$

On the other hand, $\omega_{ij\alpha}$ can be calculated by using the Gauss equations: Replace $D_j x_\alpha$ and $D_i x_\alpha$ in the expression

$$\omega_{ij\alpha} = D_i(D_j x_\alpha) - D_j(D_i x_\alpha)$$

by their values from the Gauss equations, perform the differentiations D_i and D_j formally, and repeat the substitution; we get

$$\omega_{ij\alpha} = P^\beta_{\alpha ij} x_\beta + Q_{ij\alpha} n,$$

where

$$P^\beta_{\alpha ij} = \frac{\partial \tilde{\Gamma}^\beta_{j\alpha}}{\partial u^i} - \frac{\partial \tilde{\Gamma}^\beta_{i\alpha}}{\partial u^j} + \tilde{\Gamma}^s_{j\alpha}\tilde{\Gamma}^\beta_{is} - \tilde{\Gamma}^s_{i\alpha}\tilde{\Gamma}^\beta_{js} + (\lambda_{i\alpha}\lambda_{jk} - \lambda_{j\alpha}\lambda_{ik})\, \tilde{g}^{k\beta},$$

$$Q_{ij\alpha} = \frac{\partial \lambda_{j\alpha}}{\partial u^i} - \frac{\partial \lambda_{i\alpha}}{\partial u^j} + \tilde{\Gamma}^s_{j\alpha}\lambda_{is} - \tilde{\Gamma}^s_{i\alpha}\lambda_{js}.$$

The first four terms of $P^\beta_{\alpha ij}$ are obviously the component $\tilde{R}^{\cdot\beta}_{\alpha\cdot ij}$ of the mixed Riemann-Christoffel tensor of the surface.

Setting $i, j, \alpha, \beta < 3$, let us compare the coefficients of $e_\beta = x_\beta$ in the two expressions for $\omega_{ij\alpha}$. This gives

$$R^{\cdot\beta}_{\alpha\cdot ij} = \tilde{R}^{\cdot\beta}_{\alpha\cdot ij} + (\lambda_{i\alpha}\lambda_{jk} - \lambda_{j\alpha}\lambda_{ik})\, \tilde{g}^{k\beta}.$$

Since $g_{ij} = \tilde{g}_{ij}$ and $g_{i3} = 0$ for $i, j < 3$, this equality easily implies

$$R_{\alpha\beta ij} = \tilde{R}_{\alpha\beta ij} + (\lambda_{i\alpha}\lambda_{j\beta} - \lambda_{j\alpha}\lambda_{i\beta}).$$

All these equalities, which are not identities, are equivalent to the equality

$$R_{1212} = \widetilde{R}_{1212} + (\lambda_{11}\lambda_{22} - \lambda_{12}^2).$$

Dividing by $\widetilde{g}_{11}\widetilde{g}_{22} - \widetilde{g}_{12}^2 = g_{11}g_{22} - g_{12}^2$, we get

$$(***) \qquad -K_R = -K_i + K_e,$$

where K_i is the Gauss curvature of the surface, K_e its extrinsic curvature, and K_R the sectional curvature of the space in a plane section tangent to the surface.

In the euclidean case, formula (***) is simply the well-known formula of Gauss for the total curvature of the surface in terms of the coefficients of the first fundamental form and their derivatives.

Comparing coefficients of $e_3 = n$ in the two expressions for $\omega_{ij\alpha}$, we get

$$R_{\alpha\cdot ij}^{\cdot 3} = \frac{\partial \lambda_{j\alpha}}{\partial u^i} - \frac{\partial \lambda_{i\alpha}}{\partial u^j} + \widetilde{\Gamma}_{j\alpha}^s \lambda_{is} - \widetilde{\Gamma}_{i\alpha}^s \lambda_{js}.$$

Only two of these equations are different. Recalling that $g_{23} = g_{31} = 0$, $g_{33} = 1$, and therefore $R_{\alpha\cdot ij}^{\cdot 3} = R_{\alpha 3ij}$, we can rewrite these equations as

$$R_{1312} = \frac{\partial \lambda_{12}}{\partial u^1} - \frac{\partial \lambda_{11}}{\partial u^2} + \widetilde{\Gamma}_{12}^s \lambda_{1s} - \widetilde{\Gamma}_{11}^s \lambda_{2s},$$

$$R_{2312} = \frac{\partial \lambda_{22}}{\partial u^1} - \frac{\partial \lambda_{12}}{\partial u^2} + \widetilde{\Gamma}_{22}^s \lambda_{1s} - \widetilde{\Gamma}_{12}^s \lambda_{2s}.$$

In the euclidean case these are the Peterson-Codazzi equations.

Other relations can be derived in a similar manner from the expression $(D_iD_j - D_jD_i)n$, but we shall not do this, since they will not be needed.

At many points of the forthcoming exposition, we shall employ a semigeodesic parametrization of space in the neighborhood of the surface. Let us study it in greater detail.

As indicated above, in semigeodesic coordinates we have $g_{13} = g_{23} = 0$, $g_{33} = 1$.

We now compute the second fundamental form of the basis surface (the surface $v^3 = 0$). By definition, $\mathrm{II} = -dx \cdot Dn$. But

$$dx = e_i du^i, \quad i < 3, \qquad dn = \Gamma_{3k}^j du^k e_j, \quad k < 3.$$

Hence

$$\mathrm{II} = -\Gamma_{3k}^j g_{ij}\, du^i\, du^k = -\Gamma_{3ik}\, du^i\, du^k,$$

$$\Gamma_{3ik} = \frac{1}{2}\left(\frac{\partial g_{3i}}{\partial u^k} + \frac{\partial g_{ik}}{\partial v^3} - \frac{\partial g_{3k}}{\partial u^i}\right).$$

Since $g_{3i} = g_{3k} = 0$ for $i, k < 3$, it follows that

$$\Gamma_{3ik} = \frac{1}{2}\frac{\partial g_{ik}}{\partial v^3}.$$

Thus the second fundamental form of the surface is

$$\mathrm{II} = -\frac{1}{2}\frac{\partial g_{ik}}{\partial v^3}\, du^i\, du^k$$

and hence the coefficients of the second fundamental form are

$$\lambda_{ij} = -\frac{1}{2}\frac{\partial g_{ij}}{\partial v^3}.$$

Note that if the extrinsic curvature of the surface is positive, the form II is definite. We may assume that it is positive definite; it is sufficient to choose the direction of the v^3-axis in a suitable manner.

§2. A priori estimates for normal curvature

In the sequel we shall need estimates for the normal curvatures of a closed surface in a Riemannian space under certain conditions, estimates which depend only on the metrics of the surface and of the space. The fact that such estimates can be established is decisive for our study of isometric embedding of a Riemannian manifold homeomorphic to a sphere in Riemannian three-space. This section is devoted to a derivation of these estimates.

Let F be a surface in a Riemannian space, X_0 a point on F. Assume that X_0 is not an umbilical point, i.e. the principal normal curvatures of the surface at X_0 are different. In this case the role of the coordinate curves on the surface in the neighborhood of X_0 may be played by the lines of curvature. We take the parameters u^1 and u^2 to be arc length along the coordinate curves through X_0.

Introduce a semigeodesic parametrization (v^i) of the space in the neighborhood of X_0, with F as the basis surface. Thus the coordinates v^i of a point in space are: v^3, the signed distance from the basis surface F; and u^1, u^2, the coordinates of the foot of the geodesic perpendicular dropped to the surface F. The parametrization (v^i) is orthogonal along the surface F.

Let us calculate the Christoffel symbols of the first kind for the space along the surface F.

On the surface F,

$$\frac{\partial g_{12}}{\partial v^3} = -2\lambda_{12} = 0, \qquad \frac{\partial g_{i3}}{\partial v^j} = 0,$$

and hence $\Gamma_{ijk} = 0$ when all three indices i, j, k are different.

When two consecutive indices are equal,

$$\Gamma_{iik} = \Gamma_{kii} = \frac{1}{2}\frac{\partial g_{ii}}{\partial u^k}.$$

If the first and third indices are equal, then

$$\Gamma_{lkl} = -\frac{1}{2}\frac{\partial g_{ll}}{\partial u^k}.$$

For future reference, we list the Christoffel symbols in a table:

$$\begin{array}{lll}
\Gamma_{111} = \frac{1}{2}\frac{\partial g_{11}}{\partial u^1} & \Gamma_{211} = \frac{1}{2}\frac{\partial g_{11}}{\partial u^2} & \Gamma_{311} = -\lambda_{11} \\
\Gamma_{112} = \frac{1}{2}\frac{\partial g_{11}}{\partial u^2} & \Gamma_{212} = -\frac{1}{2}\frac{\partial g_{22}}{\partial u^1} & \Gamma_{312} = 0 \\
\Gamma_{113} = -\lambda_{11} & \Gamma_{213} = 0 & \Gamma_{313} = 0 \\
\Gamma_{121} = -\frac{1}{2}\frac{\partial g_{11}}{\partial u^2} & \Gamma_{221} = \frac{1}{2}\frac{\partial g_{22}}{\partial u^1} & \Gamma_{321} = 0 \\
\Gamma_{122} = \frac{1}{2}\frac{\partial g_{22}}{\partial u^1} & \Gamma_{222} = \frac{1}{2}\frac{\partial g_{22}}{\partial u^2} & \Gamma_{322} = -\lambda_{22} \\
\Gamma_{123} = 0 & \Gamma_{223} = -\lambda_{22} & \Gamma_{323} = 0 \\
\Gamma_{131} = \lambda_{11} & \Gamma_{231} = 0 & \Gamma_{331} = 0 \\
\Gamma_{132} = 0 & \Gamma_{232} = \lambda_{22} & \Gamma_{332} = 0 \\
\Gamma_{133} = 0 & \Gamma_{233} = 0 & \Gamma_{333} = 0
\end{array}$$

Since the parametrization of the space is orthogonal along the surface F, we have $g^{ij} = 0$ for $i \neq j$, and $g^{ii} = 1/g_{ii}$. Consequently it is easy to calculate the Christoffel symbols of the second kind:

$$\Gamma_{ij}^k = \Gamma_{i\alpha j} g^{\alpha k} = \Gamma_{ikj}/g_{kk}.$$

The Christoffel symbols of the second kind are also listed in a table:

$$\begin{array}{lll}
\Gamma_{11}^1 = \frac{1}{2g_{11}}\frac{\partial g_{11}}{\partial u^1} & \Gamma_{21}^1 = \frac{1}{2g_{11}}\frac{\partial g_{11}}{\partial u^2} & \Gamma_{31}^1 = -\frac{1}{g_{11}}\lambda_{11} \\
\Gamma_{12}^1 = \frac{1}{2g_{11}}\frac{\partial g_{11}}{\partial u^2} & \Gamma_{22}^1 = -\frac{1}{2g_{11}}\frac{\partial g_{22}}{\partial u^1} & \Gamma_{32}^1 = 0 \\
\Gamma_{13}^1 = -\frac{1}{g_{11}}\lambda_{11} & \Gamma_{23}^1 = 0 & \Gamma_{33}^1 = 0 \\
\Gamma_{11}^2 = -\frac{1}{2g_{22}}\frac{\partial g_{11}}{\partial u^2} & \Gamma_{21}^2 = \frac{1}{2g_{22}}\frac{\partial g_{22}}{\partial u^1} & \Gamma_{31}^2 = 0 \\
\Gamma_{12}^2 = \frac{1}{2g_{22}}\frac{\partial g_{22}}{\partial u^1} & \Gamma_{22}^2 = \frac{1}{2g_{22}}\frac{\partial g_{22}}{\partial u^2} & \Gamma_{32}^2 = -\frac{1}{g_{22}}\lambda_{22} \\
\Gamma_{13}^2 = 0 & \Gamma_{23}^2 = -\frac{1}{g_{22}}\lambda_{22} & \Gamma_{33}^2 = 0 \\
\Gamma_{11}^3 = \lambda_{11} & \Gamma_{21}^3 = 0 & \Gamma_{31}^3 = 0 \\
\Gamma_{12}^3 = 0 & \Gamma_{22}^3 = \lambda_{22} & \Gamma_{32}^3 = 0 \\
\Gamma_{13}^3 = 0 & \Gamma_{23}^3 = 0 & \Gamma_{33}^3 = 0
\end{array}$$

Let us estimate the derivatives of g_{11} and g_{22} at the point where the normal curvature is a maximum. Suppose that the normal curvature assumes its absolute maximum at a point X_0. As before, introduce a semigeodesic parametrization (v) in the neighborhood of this point. Assume that the maximum normal curvature is assumed in the u^1-direction $(du^2=0)$.

The principal curvatures of the surface at X_0 are

$$\varkappa_1 = \frac{\lambda_{11}}{g_{11}} = \lambda_{11}, \qquad \varkappa_2 = \frac{\lambda_{22}}{g_{22}} = \lambda_{22}.$$

The Peterson-Codazzi equations are

$$R_{1312} = -\frac{\partial \lambda_{11}}{\partial u^2} + \frac{1}{2}\frac{\partial g_{11}}{\partial u^2}\left(\frac{\lambda_{11}}{g_{11}} + \frac{\lambda_{22}}{g_{22}}\right),$$

$$R_{2312} = \frac{\partial \lambda_{22}}{\partial u^1} - \frac{1}{2}\frac{\partial g_{22}}{\partial u^1}\left(\frac{\lambda_{11}}{g_{11}} + \frac{\lambda_{22}}{g_{22}}\right).$$

Replace λ_{ii} in these formulas by the corresponding principal curvatures κ_i:

$$(*)\qquad \begin{aligned} R_{1312} &= -g_{11}\frac{\partial \varkappa_1}{\partial u^2} + \frac{1}{2}\frac{\partial g_{11}}{\partial u^2}(-\varkappa_1 + \varkappa_2), \\ R_{2312} &= g_{22}\frac{\partial \varkappa_2}{\partial u^1} - \frac{1}{2}\frac{\partial g_{22}}{\partial u^1}(\varkappa_1 - \varkappa_2). \end{aligned}$$

Let us estimate the derivatives $\partial g_{11}/\partial u^2$ and $\partial g_{22}/\partial u^1$ at the point X_0, on the assumption that $\kappa_1 \gg \sqrt{K_e}$.

Solving equations (*) for $\partial g_{11}/\partial u^2$ and $\partial g_{22}/\partial u^1$, we get

$$\frac{\partial g_{11}}{\partial u^2} = \frac{2\left(R_{1312} + g_{11}\dfrac{\partial \varkappa_1}{\partial u^2}\right)}{-\varkappa_1 + \varkappa_2}, \qquad \frac{\partial g_{22}}{\partial u^1} = -\frac{2\left(R_{2312} - g_{22}\dfrac{\partial \varkappa_2}{\partial u^1}\right)}{-\varkappa_1 + \varkappa_2}.$$

Since $\partial \kappa_1/\partial u^2 = 0$ at X_0, we have

$$\partial g_{11}/\partial u^2 = 2R_{1312}/(-\kappa_1 + \kappa_2).$$

Since R_{1312} is a component of a tensor, it is independent of the orientation of the surface at the point. That is, we may assume that R_{1312} is given in normal Riemann coordinates $\bar{v}^i$, related to our coordinates v^i at the point X_0 by the conditions $\partial \bar{v}^i/\partial v^j = \delta^i_j$. Hence $\partial g_{11}/\partial u^2$ is of first order in $1/\kappa_1$, i.e.

$$\partial g_{11}/\partial u^2 = O(1/\kappa_1).$$

It is a little more difficult to derive a similar estimate for $\partial g_{22}/\partial u^1$. Since $\kappa_1\kappa_2 = K_e = K_i - K_R$, where K_i is the Gauss curvature of the surface and K_R the sectional curvature of the space in a plane section tangent to the surface, it follows that

$$\frac{\partial g_{22}}{\partial u^1} = \frac{2\left(R_{2312} - \frac{g_{22}}{\varkappa_1}\frac{\partial}{\partial u^1}(K_l - K_R)\right)}{-\varkappa_1 + \varkappa_2}, \qquad \frac{\partial K_l}{\partial u^1} = O(1).$$

Since

$$K_R = -R_{1212}/g_{11}g_{22},$$

it follows that, at the point X_0,

$$\frac{\partial K_R}{\partial u^1} = -\frac{\partial}{\partial u^1}R_{1212} - R_{1212}\frac{\partial g_{22}}{\partial u^1}.$$

We now consider the covariant derivative $R_{ijrs,\alpha}$ of the Riemann tensor. We have

$$R_{1212,1} = \partial R_{1212}/\partial u^1 - R_{i212}\Gamma^i_{11} - \cdots.$$

Since we can assume that the components of the Riemann tensor and its covariant derivative are computed in normal Riemann coordinates, it follows that they are $O(1)$. Therefore the derivative $\partial R_{1212}/\partial u^1$ is of first order in Γ^k_{ij}. Thus

$$\frac{\partial R_{1212}}{\partial u^1} = \varkappa_1 O(1) + \frac{\partial g_{22}}{\partial u^1}O(1) + O(1).$$

Thus our expression for $\partial g_{22}/\partial u^1$ becomes

$$\frac{\partial g_{22}}{\partial u^1} = \frac{O(1) + \frac{1}{\varkappa_1}\frac{\partial g_{22}}{\partial u^1}O(1)}{-\varkappa_1 + \varkappa_2}.$$

Hence it is clear that for large κ_1

$$\partial g_{22}/\partial u^1 = O(1/\kappa^1).$$

We now determine the order of magnitude (relative to κ_1) of

$$\partial R_{1312}/\partial u^2, \quad \partial R_{2312}/\partial u^1, \quad \partial^2 R_{1212}/(\partial u^1)^2.$$

Again using the covariant derivative of the Riemann tensor, we get

$$R_{1312,2} = \partial R_{1312}/\partial u^2 - R_{k312}\Gamma^k_{12} - \cdots.$$

Since the components of the Riemann tensor and its covariant derivative are $O(1)$, while the coefficients Γ^k_{ij} are of at most first order in $1/\kappa^1$, except for three:

$$\Gamma^1_{13} = \Gamma^1_{31} = -\kappa_1, \quad \Gamma^3_{11} = \kappa_1,$$

the above equality implies

$$\partial R_{1312}/\partial u^2 = O(1).$$

The derivative $\partial R_{2312}/\partial u^1$ is treated similarly:

$$R_{2312,1} = \partial R_{2312}/\partial u^1 - R_{k312}\Gamma^k_{21} - \cdots.$$

Hence

$$\partial R_{2312}/\partial u^1 = \kappa_1(R_{1212} - R_{2323}) + O(1).$$

Finally, to estimate $\partial^2 R_{1212}/(\partial u^1)^2$ we use the second covariant derivative of the Riemann tensor:

$$R_{ijrs,\alpha\beta} = \partial R_{ijrs,\alpha}/\partial u^\beta - R_{kjrs,\alpha}\Gamma^k_{i\beta} - \cdots.$$

Hence

$$\partial R_{ijrs,\alpha}/\partial u^\beta = O(\kappa_1).$$

Differentiate the equality

$$R_{1212,1} = \partial R_{1212}/\partial u_1 - R_{k212}\Gamma^k_{11} - \cdots$$

with respect to u^1. The derivative of the left-hand side is at most $O(\kappa_1)$. The derivative of the right-hand side contains $\partial^2 R_{1212}/(\partial u^1)^2$ and two groups of terms of type $\Gamma(\partial R/\partial u^1)$ and $R(\partial\Gamma/\partial u^1)$.

Let us now consider the second group of terms, each of which contains either $\partial\Gamma^s_{11}/\partial u^1$ or $\partial\Gamma^s_{21}/\partial u^1$.

In view of the relations

$$\frac{\partial g_{11}}{\partial u^1} = \frac{\partial g_{22}}{\partial u^2} = 0, \quad \frac{\partial g_{11}}{\partial u^2} = O\left(\frac{1}{\varkappa_1}\right), \quad \frac{\partial g_{22}}{\partial u^1} = O\left(\frac{1}{\varkappa_1}\right),$$

$$\frac{\partial \lambda_{11}}{\partial u^1} = \frac{\partial}{\partial u^1}(\varkappa_1 g_{11}) = 0, \quad \frac{\partial \lambda_{22}}{\partial u^1} = \frac{\partial}{\partial u^1}\left(g_{22}\frac{K_e}{\varkappa_1}\right) = O(1),$$

we see that all derivatives $\partial\Gamma^s_{11}/\partial u^1$ and $\partial\Gamma^s_{12}/\partial u^1$ are at most $O(1)$, except perhaps three:

$$\frac{\partial\Gamma^1_{12}}{\partial u^1}, \quad \frac{\partial\Gamma^2_{11}}{\partial u^1} \quad \text{and} \quad \frac{\partial\Gamma^2_{12}}{\partial u^1} = \frac{\partial^2 g_{22}}{(\partial u^1)^2}.$$

Fortunately, the components of the Riemann tensor multiplying the first two of these derivatives are zero (the first or second pair involves equal indices).

It follows that the second group of terms can be expressed as

$$-R\,\partial\Gamma/\partial u^1 = -R_{1212}\partial^2 g_{22}/(\partial u^1)^2 + O(1).$$

Now consider the second group of terms, $-\Gamma(\partial R/\partial u^1)$. Since all the derivatives $\partial R/\partial u^1$ are at most $O(\kappa_1)$, and all the Γ are at most $O(1/\kappa_1)$, except

$$\Gamma^1_{13} = \Gamma^1_{31} = -\kappa_1, \quad \Gamma^3_{11} = \kappa_1,$$

it follows that this group of terms can be expressed as

$$-2\kappa_1 \partial R_{3212}/\partial u^1 + O(1).$$

But

$$R_{3212,1} = \partial R_{3212}/\partial u^1 - R_{k212}\Gamma^k_{31} - \cdots.$$

Hence, by the same arguments for the order of magnitude of the Γ, we get

$$\partial R_{3212}/\partial u^1 = -\kappa_1(R_{1212} - R_{3232}) + O(1).$$

The second group of terms is therefore

$$-\Gamma\,\partial R/\partial u^1 = 2\kappa_1^2(R_{1212} - R_{3232}) + O(\kappa_1).$$

We now estimate $\partial^2 R_{1212}/(\partial u^1)^2$:

$$\frac{\partial^2 R_{1212}}{(\partial u^1)^2} = -2\varkappa_1^2(R_{1212} - R_{3232}) + R_{1212}\frac{\partial^2 g_{22}}{(\partial u^1)^2} + O(\varkappa_1).$$

We shall show that, under certain conditions, an estimate can be established for the normal curvatures of a closed surface F in a Riemannian space R which depends only on the metrics of the surface and the space.

Assume that the normal curvature κ_1 assumes an absolute maximum at a point X_0, and moreover $\kappa_1 \gg \sqrt{K_e}$. In the neighborhood of X_0, introduce a coordinate system of lines of curvature, with the parameters u^1, u^2 of arc length on the coordinate curves passing through X_0. Introduce a semigeodesic parametrization of the space in the neighborhood of X_0, taking the surface F as basis surface, as before.

By the Gauss formula for the intrinsic (Gauss) curvature of the surface F at X_0, we have

$$K_i = -\frac{1}{2}\frac{\partial^2 g_{11}}{(\partial u^2)^2} - \frac{1}{2}\frac{\partial^2 g_{22}}{(\partial u^1)^2} + \frac{1}{4}\left(\frac{\partial g_{11}}{\partial u^2}\right)^2 + \frac{1}{4}\left(\frac{\partial g_{22}}{\partial u^1}\right)^2.$$

As we proved above,

$$\frac{\partial g_{11}}{\partial u^2} = O\left(\frac{1}{\varkappa_1}\right), \qquad \frac{\partial g_{22}}{\partial u^1} = O\left(\frac{1}{\varkappa_1}\right).$$

We now calculate the second derivatives $\partial^2 g_{11}/(\partial u^2)^2$ and $\partial^2 g_{22}/(\partial u^1)^2$, up to $O(1/\kappa_1)$. We have

$$\frac{\partial g_{11}}{\partial u^2} = 2\frac{R_{1312} + g_{11}\dfrac{\partial \varkappa_1}{\partial u^2}}{-\varkappa_1 + \varkappa_2}.$$

Differentiating this formula with respect to u^2 and noting that

$$\frac{\partial \varkappa_1}{\partial u^2}=0, \qquad \frac{\partial R_{1312}}{\partial u^2}=O(1),$$

$$\varkappa_2=\frac{K_e}{\varkappa_1}=\frac{1}{\varkappa_1}(K_i-K_R)=\frac{K_i}{\varkappa_1}+\frac{R_{1212}}{\varkappa_1 g_{11}g_{22}},$$

$$\frac{\partial K_i}{\partial u^2}=O(1), \qquad \frac{\partial R_{1212}}{\partial u^2}=O(1),$$

we see that

$$\frac{\partial^2 g_{11}}{(\partial u^2)^2}=O\left(\frac{1}{\varkappa_1}\right)+\frac{2\dfrac{\partial^2 \varkappa_1}{(\partial u^2)^2}}{-\varkappa_1+\varkappa_2}.$$

Now consider $\partial^2 g_{22}/(\partial u^1)^2$. We have

$$\frac{\partial g_{22}}{\partial u^1}=2\frac{R_{2312}-g_{22}\dfrac{\partial \varkappa_2}{\partial u^1}}{-\varkappa_1+\varkappa_2}.$$

Differentiating with respect to u^1 and noting that

$$\frac{\partial R_{2312}}{\partial u^1}=-\varkappa_1(R_{1212}-R_{2323})+O(1), \qquad \frac{\partial \varkappa_2}{\partial u^1}=O(1),$$

we get

$$\frac{\partial^2 g_{22}}{(\partial u^1)^2}=2\frac{\varkappa_1(R_{1212}-R_{2323})-\dfrac{\partial^2 \varkappa_2}{(\partial u^1)^2}}{-\varkappa_1+\varkappa_2}+O\left(\frac{1}{\varkappa_1}\right).$$

We now calculate $\partial^2\kappa_2/(\partial u^1)^2$. We have

$$\frac{\partial^2 \varkappa_2}{(\partial u^1)^2}=\frac{1}{\varkappa_1}\frac{\partial^2 K_e}{(\partial u^1)^2}-\frac{K_e}{\varkappa_1^2}\frac{\partial^2 \varkappa_1}{(\partial u^1)^2},$$

$$\frac{\partial^2 K_e}{(\partial u^1)^2}=\frac{\partial^2 K_i}{(\partial u^1)^2}+\frac{\partial^2}{(\partial u^1)^2}\left(\frac{R_{1212}}{g_{12}g_{22}}\right)$$
$$=\frac{\partial^2 K_i}{(\partial u^1)^2}+\frac{\partial^2 R_{1212}}{(\partial u^1)^2}-R_{1212}\frac{\partial^2 g_{22}}{(\partial u^1)^2}+O(1).$$

Recalling that

$$\frac{\partial^2 R_{1212}}{(\partial u^1)^2}=-2\varkappa_1^2(R_{1212}-R_{3232})+R_{1212}\frac{\partial^2 g_{22}}{(\partial u^1)^2}+O(\varkappa_1),$$

we see that

$$\frac{\partial^2 K_e}{(\partial u^1)^2}=\frac{\partial^2 K_i}{(\partial u^1)^2}-2\varkappa_1^2(R_{1212}-R_{3232})+O(\varkappa_1).$$

The derivative $\partial^2 K_i/(\partial u^1)^2$ is $O(1)$, since the geodesic curvature of the u^1-curve ($u^2=0$) at X_0 has the same order of magnitude as $\partial g_{11}/\partial u^2$ and $\partial g_{22}/\partial u^1$.

Substituting this expression for $\partial^2 K_e/(\partial u^1)^2$ into $\partial^2\kappa_2/(\partial u^1)^2$ and the latter into $\partial^2 g_{22}/(\partial u^1)^2$, we get

$$\frac{\partial^2 g_{22}}{(\partial u^1)^2} = 2\,\frac{2\varkappa_1(R_{1212}-R_{3232}) + \dfrac{K_e}{\varkappa_1^2}\dfrac{\partial^2\varkappa_1}{(\partial u^1)^2}}{-\varkappa_1+\varkappa_2} + O\left(\frac{1}{\varkappa_1}\right).$$

Substituting the expression for $\partial^2 g_{11}/(\partial u^2)^2$ and $\partial^2 g_{22}/(\partial u^1)^2$ into the Gauss formula, we get

$$K_i = \frac{\dfrac{\partial^2\varkappa_1}{(\partial u^2)^2} + \dfrac{K_e}{\varkappa_1^2}\dfrac{\partial^2\varkappa_1}{(\partial u^1)^2}}{\varkappa_1-\varkappa_2} + 3\,\frac{R_{1212}-R_{2323}}{1-\dfrac{\varkappa_2}{\varkappa_1}} + O\left(\frac{1}{\varkappa_1}\right).$$

Assume that $K_i > 3(R_{1212} - R_{2323})$ at the point X_0. Then κ_1 cannot be arbitrarily large. Indeed, the first term on the right-hand side of the equality is nonpositive, since κ_1 has a maximum at X_0, and so

$$\partial^2\kappa_1/(\partial u^2)^2 \leqq 0, \qquad \partial^2\kappa_1/(\partial u^1)^2 \leqq 0.$$

For sufficiently large κ_1,

$$K_i > 3\,\frac{R_{1212}-R_{2323}}{1-\dfrac{\varkappa_2}{\varkappa_1}},$$

and $O(1/\kappa_1)$ is arbitrarily small. Thus for sufficiently large κ_1 the above equality fails.

Since the component R_{1212} at the point X_0 is the sectional curvature of the space in a plane section tangent to the surface while R_{2323} is the sectional curvature of the space in a plane section perpendicular to the surface, we arrive at the following theorem.

THEOREM. *Let F be a closed regular (four times continuously differentiable) surface in a regular (four times continuously differentiable) Riemannian space R. Assume that at each point of the surface*

$$K_i > K_R, \quad K_i - 3(K_R - \overline{K}_R) > 0,$$

where K_i is the Gauss curvature of the surface, K_R the sectional curvature of the space in a plane section tangent to the surface, and $\overline{K}_R$ the sectional curvature in any plane section perpendicular to the surface.

Then the normal curvature of the surface satisfies an estimate which depends only on the metrics of the surface and the space.

This theorem has a more precise formulation. Suppose that the surface F is situated in a region G of the Riemannian space R. Let G be covered by a finite family of neighborhoods Ω, each with its own parametrization (v^i), and let $g_{ij}dv^idv^j$ be the line element of the space in Ω.

Let the surface F be covered by a finite number of neighborhoods ω, each with its own parametrization (u^i), and let $\tilde{g}_{ij}du^idu^j$ be the line element of the surface in ω.

Assume that at each point of the surface

$$K_i - K_R > \delta, \quad K_i - 3(K_R - \overline{K}_R) > \delta, \quad \delta > 0.$$

Then, if the absolute values of g_{ij}, $\tilde{g}_{ij}$ and their derivatives of order up to four are bounded by a constant M in all the neighborhoods Ω and ω, and the discriminants of the forms $g_{ij}dv^idv^j$ and $\tilde{g}_{ij}du^idu^j$ are at least $1/M$ in these neighborhoods, the normal curvatures of F are bounded by a constant $c(M, \delta)$ which depends only on M and δ.

§3. A priori estimates for the derivatives of space coordinates with respect to surface coordinates

Let F be a two-dimensional Riemannian manifold isometrically embedded in a three-dimensional Riemannian space R. Suppose that the space is covered by a system of parametrized neighborhoods Ω, and the surface F by a system of parametrized neighborhoods ω. The space coordinates v^i of a point on F are then functions of the surface coordinates u^i.

In this section we shall prove that, given a priori estimates for the normal curvatures of the surface (§2), estimates can be established for the derivatives of v^i of arbitrary order with respect to the variables u^i, and these estimates depend only on the metrics of the surface and the space.

Let $x(u^i)$ be an arbitrary point on F. We have

$$\frac{\partial x}{\partial u^i} = \frac{\partial x}{\partial v^\alpha}\frac{\partial v^\alpha}{\partial u^i} = e_\alpha \frac{\partial v^\alpha}{\partial u^i}.$$

Squaring this equality, we get

$$\tilde{g}_{ij} = g_{\alpha\beta}\frac{\partial v^\alpha}{\partial u^i}\frac{\partial v^\beta}{\partial u^j},$$

where $\tilde{g}_{ij}$ and $g_{\alpha\beta}$ are the coefficients of the surface and space line elements, respectively.

Since the form $g_{\alpha\beta}\xi^\alpha\xi^\beta$ is positive definite, there exists $\epsilon > 0$ such that

$$g_{\alpha\beta}\xi^\alpha\xi^\beta \geqslant \varepsilon\sum_\alpha(\xi^\alpha)^2 \geqslant \varepsilon(\xi^\alpha)^2.$$

The number ϵ depends only on the parametrization of the space. Thus

$$\epsilon(\partial v^\alpha/\partial u^i) \leqq \tilde{g}_{ii},$$

whence

$$\left|\frac{\partial v^\alpha}{\partial u^i}\right| \leqq \sqrt{\frac{\tilde{g}_{ii}}{\varepsilon}}.$$

This proves the existence of estimates for the first derivatives.

We now proceed to the second derivatives of v^α with respect to u^i. The first step is to estimate the coefficients of the second fundamental form of the surface.

Let κ_0 be the upper limit of the normal curvatures of the surface. Then, for any ξ^i,

$$|\lambda_{ij}\xi^i\xi^j|/\tilde{g}_{ij}\xi^i\xi^j \leqq \kappa_0.$$

It follows that

$$|\lambda_{11}| \leqq \kappa_0\tilde{g}_{11}, \qquad |\lambda_{22}| \leqq \kappa_0\tilde{g}_{22}.$$

And since

$$\lambda_{11}\lambda_{22} - \lambda_{12}^2 = K_e\tilde{g},$$

$$K_e = K_i - K_R, \qquad \tilde{g} = \tilde{g}_{11}\tilde{g}_{22} - g_{12}^2,$$

it follows that

$$|\lambda_{12}|^2 \leqslant \varkappa_0^2\tilde{g}_{11}\tilde{g}_{22} + K_e\tilde{g}.$$

We now appeal to the Gauss equations (§1):

$$D_j x_i = \tilde{\Gamma}^\alpha_{ij} x_\alpha + \lambda_{ij} n,$$

$$x_i = \frac{\partial x^i}{\partial v^\lambda}\frac{\partial v^\lambda}{\partial u^i} = e_\lambda\frac{\partial v^\lambda}{\partial u^i},$$

$$D_j x_i = e_\lambda\frac{\partial^2 v^\lambda}{\partial u^i\,\partial u^j} + \frac{\partial v^\lambda}{\partial u^i}\Gamma^\mu_{\lambda j} e_\mu.$$

Setting $n = \xi^\lambda e_\lambda$, we get

$$e_\lambda\frac{\partial^2 v^\lambda}{\partial u^i\,\partial u^j} = \tilde{\Gamma}^\alpha_{ij}\frac{\partial v^\lambda}{\partial u^\alpha}e_\lambda - \Gamma^\lambda_{j\mu}\frac{\partial v^\mu}{\partial u^i}e_\lambda + \lambda_{ij}\xi^\lambda e_\lambda.$$

Hence

$$\frac{\partial^2 v}{\partial u^i\,\partial u^j} = \tilde{\Gamma}^\alpha_{ij}\frac{\partial v^\lambda}{\partial u^\alpha} - \Gamma^\lambda_{j\mu}\frac{\partial v^\mu}{\partial u^i} + \lambda_{ij}\xi^\lambda.$$

We already have estimates for the first derivatives $\partial v^\alpha/\partial u^\beta$ and the coefficients of the second fundamental form; thus to derive estimates for the second derivatives $\partial^2 v^\lambda/\partial u^i \partial u^j$ we need only estimate ξ^λ. This is easily done: we have $n = \xi^\alpha e_\alpha$, so that $g_{\alpha\beta}\xi^\alpha\xi^\beta = 1$. And since the form $g_{\alpha\beta}\xi^\alpha\xi^\beta$ is positive definite, it is now easy to find estimates for ξ^α.

Each of the cartesian coordinates v^i of a point on the surface in euclidean space realizing the line element $\tilde{g}_{ij}du^i du^j$ satisfies a second-order partial differential equation (the Monge-Ampère equation) with coefficients which depend only on $\tilde{g}_{ij}$ and its derivatives.

We shall now derive a system of differential equations for the space coordinates of a point of a surface in a Riemannian space (the so-called Darboux equation) which will play an analogous role in our investigations. The Darboux equation plays an essential role in the realization of abstractly prescribed metrics by surfaces in euclidean space.

As before, let e_i be the basis vectors at a point x_i of the space. We have

$$x_k = \partial x/\partial u^k = e_i \partial v^i/\partial u^k.$$

Let e_i^* denote the vector at the point x such that $e_i^* e_k = \delta_{ik}$, where δ_{ik} is the Kronecker symbol.

We have two expressions for the vector $D_i x_j$:

$$D_i x_j = \tilde{\Gamma}^k_{ij} x_k + \lambda_{ij} n, \quad D_i x_j = e_k \frac{\partial^2 v^k}{\partial u^i \partial u^j} + \frac{\partial v^k}{\partial u^j} \Gamma^\alpha_{ks} e_\alpha \frac{\partial v^s}{\partial u^i}.$$

The first is one of the Gauss equations, while the second is obtained by direct differentiation.

Consider the scalar product of these expressions with e_β^*:

$$e_\beta^* D_i x_j = \tilde{\Gamma}^k_{ij} \frac{\partial v^\beta}{\partial u^k} + \lambda_{ij} (n e_\beta^*),$$

$$e_\beta^* D_i x_j = \frac{\partial^2 v^\beta}{\partial u^i \partial u^j} + \Gamma^\beta_{ks} \frac{\partial v^k}{\partial u^i} \frac{\partial v^s}{\partial u^j}.$$

Comparing the right-hand sides of these equalities, using the abbreviations

$$\frac{\partial^2 v^\beta}{\partial u^i \partial u^j} = v^\beta_{ij}, \quad \Gamma^\beta_{ks} \frac{\partial v^k}{\partial u^i} \frac{\partial v^s}{\partial u^j} - \tilde{\Gamma}^k_{ij} \frac{\partial v^\beta}{\partial u^k} = A^\beta_{ij},$$

we get

(*) $$v^k_{ij} + A^k_{ij} = \lambda_{ij} (n e_k^*).$$

These equations imply

$$(v^k_{11} + A^k_{11})(v^k_{22} + A^k_{22}) - (v^k_{12} + A^k_{12})^2 = (\lambda_{11}\lambda_{22} - \lambda^2_{12})^2 (n e_k^*)^2.$$

Let us express the product $(n e_k^*)$ in terms of $g_{\alpha\beta}$, $\tilde{g}_{ik}$ and the derivatives $\partial v^\alpha/\partial u^\beta$. We have

$$e_k^* = \frac{e_i \times e_j}{\sqrt{g}}, \qquad n = \frac{x_\alpha \times x_\beta}{\sqrt{\tilde{g}}},$$

where $i \neq j$, $i,j \neq k$; $\alpha = 1$ and $\beta = 2$, or $\alpha = 2$ and $\beta = 1$. Hence

$$(ne_k^*) = \frac{1}{\sqrt{g\tilde{g}}} \begin{vmatrix} e_i x_\alpha & e_i x_\beta \\ e_j x_\alpha & e_j x_\beta \end{vmatrix} = \frac{1}{\sqrt{g\tilde{g}}} \begin{vmatrix} g_{ki} \dfrac{\partial v^k}{\partial u^\alpha} & g_{si} \dfrac{\partial v^s}{\partial u^\beta} \\ g_{kj} \dfrac{\partial v^k}{\partial u^\alpha} & g_{sj} \dfrac{\partial v^s}{\partial u^\beta} \end{vmatrix}.$$

This gives (ne_k^*) in terms of g_{ij}, $\tilde{g}_{ij}$ and the derivatives $\partial v^\lambda / \partial u^\mu$. Note, moreover, that

$$\lambda_{11}\lambda_{22} - \lambda_{12}^2 = \tilde{g}K_e = \tilde{g}(K_i - K_R),$$

so that the product may also be expressed in terms of g_{ij}, $\tilde{g}_{ij}$ and their derivatives.

The final result can be stated as follows: *The space coordinates v^i of a point on the surface, as functions of the intrinsic coordinates u^i on the surface, satisfy a system of differential equations*

$$(v_{11}^k + A_{11}^k)(v_{22}^k + A_{22}^k) - (v_{12}^k + A_{12}^k)^2 = a^k.$$

The coefficients A_{ij}^k and a^k are polynomials in the first derivatives $\partial v^\alpha / \partial u^\beta$, with coefficients involving g_{ij}, $\tilde{g}_{ij}$ and their derivatives.

If the extrinsic curvature of the space $K_e = K_i - K_R$ is positive and $ne_k^* \neq 0$, then $a^k > 0$ (the system is elliptic).

Let X_0 be an arbitrary point on a surface F in a three-dimensional Riemannian space. For sufficiently small δ, depending only on the curvature of the space, the intrinsic curvature of the surface, and the supremum of the normal curvatures in the neighborhood of X_0, one can introduce normal coordinates v^i and u^i in the space and on the surface, respectively, such that

1. $|g_{ij} - \delta_{ij}| < \epsilon$,
2. $|\tilde{g}_{ij} - \delta_{ij}| < \epsilon$,
3. $\partial v^i / \partial u^j = \delta_i^j$ at the point X_0 $(i, j = 1, 2)$, and
4. the vector e_3^* defined by the equation $e_i e_3^* = \delta_{i3}$ satisfies the inequality $(ne_3^*) > 1 - \epsilon$.

Here δ_{ij} and δ_i^j are Kronecker symbols, g_{ij} and $\tilde{g}_{ij}$ are the coefficients of the line elements of the space and the surface, and e_i are the basis vectors.

We claim that, in a $\delta/4$-neighborhood of the point X_0, estimates can be given for the third derivatives $\partial^3 v^\alpha / \partial u^i \partial u^j \partial u^k$ which depend only on the metrics of the space and the surface, i.e. g_{ij}, $\tilde{g}_{ij}$ and their derivatives, the supremum of the normal curvatures, and the number δ.

To prove this statement we shall apply the method used in Chapter II (§11) to establish the existence of estimates for the third derivatives of the solution of an elliptic equation

$$F(r,s,t,p,q,z,x,y)=0$$

given estimates for the solution itself and its first and second derivatives.

In fact, the functions v^k satisfy the equation

$$(v^3_{11}+A^3_{11})(v^3_{22}+A^3_{22})-(v^3_{12}+A^3_{12})^2=a^3. \tag{**}$$

This equation involves the second derivatives of the function v^3 alone. The coefficients A^3_{ij} and a^3 depend only on the first derivatives. We have $a^3>0$, since $K_i-K_R>0$.

If the functions v^1 and v^2 are regarded as known functions in equation (**), it becomes a second-order elliptic equation for v^3. To this equation we apply the above-mentioned method.

To see that the method is indeed applicable to equation (**), it suffices to observe that the third derivatives of the functions v^1 and v^2 can be expressed as linear combinations of the corresponding derivatives of the function v^3, with bounded coefficients. To see this, note that

$$\frac{\partial^2 v^\alpha}{\partial u^i\,\partial u^j}+A^\alpha_{ij}=\lambda_{ij}\,(n\overset{*}{e}_\alpha).$$

Differentiating this equality with respect to u^k, we get

$$\frac{\partial^3 v^\alpha}{\partial u^i\,\partial u^j\,\partial u^k}+\dots=\frac{\partial\lambda_{ij}}{\partial u^k}\,(n\overset{*}{e}_\alpha),$$

where terms involving only the first and second derivatives of v have been omitted. Hence

$$\frac{\partial^3 v^\alpha}{\partial u^i\,\partial u^j\,\partial u^k}=\frac{(n\overset{*}{e}_\alpha)}{(n\overset{*}{e}_3)}\,\frac{\partial^3 v^3}{\partial u^i\,\partial u^j\,\partial u^k}+\dots$$

The estimates in §11 of Chapter II for the third derivatives of the solution of the equation $F(r,s,\cdots)=0$ depend on the supremum of the absolute values of the solution and its first two derivatives, the supremum of the absolute values of the derivatives of F of order up to three, the infimum of the discriminant of the equation $4F_rF_t-F_s^2$, and the distance from the point at which the estimation is being carried out to the boundary of the domain in which the solution exists.

Accordingly, we get the following result for equation (**).

In the $\delta/4$-neighborhood of the point X_0, the third derivatives $\partial^3v^3/\partial u^i\partial u^j\partial u^k$ are bounded by a constant which depends only on the derivatives of g_{ij} and $\tilde{g}_{ij}$ of order up to five, the infimum of the extrinsic curvature K_i-K_R of the surface, the supremum of the normal curvatures of the surface, and the number δ.

The third derivatives of the other functions v^1 and v^2 satisfy obvious estimates in terms of the corresponding third derivatives of v^3.

Estimation of the fourth and higher-order derivatives of the functions v^j can be based on the following theorem of Schauder [67] for linear elliptic equations.

In a bounded domain G of the (x_1, x_2)-plane, consider the linear elliptic partial differential equation

$$a_{11}(x_1, x_2)\, u_{x_1x_1} + 2a_{12}(x_1, x_2)\, u_{x_1x_2} + a_{22}(x_1, x_2)\, u_{x_2x_2} = f(x_1, x_2),$$

where $a_{11}a_{22} - a_{12}^2 = 1$. Assume that in the closed domain $\overline{G}$ the coefficients a_{ij} and the free term f satisfy a Hölder condition with exponent $\alpha + \epsilon$ $(0 < \alpha < 1,\ \epsilon > 0)$ and Hölder coefficient M.

Then, if the second derivatives of the solution $u(x_1, x_2)$ satisfy a Hölder condition with exponent α, estimates can be determined for the suprema of the absolute values of the first and second derivatives of $u(x_1, x_2)$ in a domain B whose closure is contained in G, and for the minimum Hölder coefficients of the second derivatives of $u(x_1, x_2)$ in B for exponent α. These estimates depend only on M, the maximum modulus of $u(x_1, x_2)$ in G, and the distance of the domain B from the boundary of G.

The function $v^3(u^1, u^2)$ satisfies the equation

$$(**) \qquad \left(\frac{\partial^2 v^3}{(\partial u^1)^2} + A_{11}^3\right)\left(\frac{\partial^2 v^3}{(\partial u^2)^2} + A_{22}^3\right) - \left(\frac{\partial^2 v^3}{\partial u^1\, \partial u^2} + A_{12}^3\right) = a^3.$$

Differentiating this equation with respect to u^k, we get an equation of the form

$$A_{11}\frac{\partial^2 v}{(\partial u^1)^2} + 2A_{12}\frac{\partial^2 v}{\partial u^1\, \partial u^2} + A_{22}\frac{\partial^2 v}{(\partial u^2)^2} = A,$$

for $\partial v^3/\partial u^k = v$, where A_{ij} and A involve only the first and second derivatives of the functions v^j, and $A_{11}A_{22} - A_{12}^2 = a^3$.

Since the third derivatives of v^j are already bounded in the $\delta/4$-neighborhood of X_0, the coefficients A_{ij} and A satisfy a Hölder condition with exponent $\alpha' < 1$. We can therefore estimate the Hölder coefficients of the second derivatives of v, i.e. of the third derivatives of v^3, for exponent $\alpha'' < \alpha'$ in the $\delta/5$-neighborhood of X_0. Estimates for the Hölder coefficients for the third derivatives of v^1 and v^2 are then obtained by expressing them in terms of the third derivatives of v^3.

Similarly, differentiating equation (**) twice with respect to u^k and u^l and applying Schauder's theorem, we get estimates for the fourth derivatives and their Hölder coefficients for exponent $\alpha''' < \alpha''$ in the $\delta/6$-neighborhood of X_0.

If the coefficients of equation (**) are sufficiently regular, this procedure can be prolonged arbitrarily. To be precise, if g_{ij} and $\tilde{g}_{ij}$ are k times differentiable ($k \geqq 5$) and their kth derivatives satisfy Hölder conditions with exponent α, estimates can be established for the kth derivatives of the functions v^i and their Hölder coefficients relative to exponent α, in the $\delta/(k-2)$-neighborhood of X_0.

The results of this section can be summarized as follows.

THEOREM. *Let R be a three-dimensional Riemannian space covered by a system of domains Ω with parametrization v^i, and let F be a closed two-dimensional Riemannian manifold covered by a system of domains ω with parametrization u^i. Let the manifold F be isometrically and $k+\alpha$ times differentiably ($k \geq 5$, $0 < \alpha < 1$) embedded in R, as a surface $\tilde{F}$ with positive extrinsic curvature.*

Then there exist estimates for the derivatives of the space coordinates v^i of a point on the surface and their Hölder coefficients relative to exponent α, which depend only on the supremum of the absolute values of the coefficients g_{ij} and $\tilde{g}_{ij}$ of the surface and space line elements, their derivatives of order up to k, the Hölder coefficients of these derivatives relative to exponent α, the infimum of the discriminants of the forms $g_{ij}dv^idv^j$, $\tilde{g}du^idu^j$, the supremum of the normal curvatures of the surface $\tilde{F}$, and the infimum of its extrinsic curvature K_e.

§4. Infinitesimal bendings of surfaces in Riemannian spaces

Let F be a surface in a Riemannian space R; consider a continuous deformation of F which takes it at time t into a surface F_t. This deformation is said to be an *infinitesimal bending* if the lengths of curves on the surface are stationary at $t=0$.

An infinitesimal bending of a surface F determines a vector field

$$\xi = dx/dt|_{t=0},$$

where $x(t)$ is the transform on F_t of a point x of the surface F. This vector field will be called the field of the infinitesimal bending.

Closed convex surfaces in euclidean space admit only trivial infinitesimal bendings, i.e. bendings whose field is the velocity field of a rigid motion of the surface, which thus have the representation

$$\xi(x) = r(x) \times a + b,$$

where $r(x)$ is the radius vector of a point x on the surface, and a and b are constant vectors. It follows that if we consider only deformations whose fields satisfy the conditions $\xi = 0$ and $d\xi = 0$ at a point x_0, then

$\xi(x) \equiv 0$. A proof of the corresponding theorem for surfaces in Riemannian space is the goal of this section.

Let F be a surface in a three-dimensional Riemannian space R. Introduce a semigeodesic parametrization v^i in a neighborhood of F in R (§1). The metric tensors g_{ij} and $\tilde{g}_{ij}$ of the space and the surface then satisfy the relations

$$g_{13} = 0, \quad g_{23} = 0, \quad g_{33} = 1$$

throughout the parametrized neighborhood, and $g_{ij}|_F = \tilde{g}_{ij}$ for $i, j < 3$ on the surface F.

The second fundamental form of the surface is

$$\lambda_{ij}\, du^i du^j = -\tfrac{1}{2}\partial g_{ij}/\partial v^3|_F\, du^i du^j.$$

Consider an infinitesimal bending, under which the transform F_t of the surface F at time t is a surface defined by equations

$$v^i = v^i(u^1, u^2, t).$$

To determine the line element ds_t^2 of the surface F_t, we must substitute $v^i = v^i(u^1, u^2, t)$ in the expression for the line element of the surface. This gives

$$g_{ij}\, dv^i dv^j = ds_t^2.$$

Compute the total differential of both sides of this equality with respect to t. Since the deformation is an infinitesimal bending, $d(ds_t^2)/dt = 0$. Denoting the contravariant coordinates of the vector $\xi = dx/dt$ by $\xi^i = \partial v^i/\partial t$, we get

$$\left(\frac{\partial g_{ij}}{\partial v^k}\xi^k\, dv^i\, dv^j + g_{ij}\, d\xi^i\, dv^j + g_{ij}\, d\xi^j\, dv^i\right)_{t=0} = 0.$$

Hence, observing that $v^1 = u^1$, $v^2 = u^2$, $v^3 = 0$, $g_{13} = g_{23} = 0$ and $g_{33} = 1$ on the surface F, we get

$$\frac{\partial g_{ij}}{\partial v^k}\xi^k\, du^i\, du^j + g_{ij}\, d\xi^i\, du^j + g_{ij}\, d\xi^j\, du^i = 0,$$

where the summation indices i, j assume the values 1, 2 only.

The left-hand side of this equality is a quadratic form in du^1, du^2. Since it vanishes, so must its coefficients. We thus get three equations for the functions ξ^i:

$$g_{\alpha j}\frac{\partial \xi^\alpha}{\partial u^i} + g_{i\alpha}\frac{\partial \xi^\alpha}{\partial u^j} + \xi^k\frac{\partial g_{ij}}{\partial v^k} = 0. \tag{1}$$

We now replace the contravariant coordinates of the vector ξ in these equations by the covariant coordinates $\xi_i = g_{i\alpha}\xi^\alpha$ and note that $g_{i3} = 0$ for $i = 1, 2$; this yields

$$\frac{\partial \xi_j}{\partial u^i} + \frac{\partial \xi_i}{\partial u^j} - \left(\frac{\partial g_{\alpha j}}{\partial u^i} + \frac{\partial g_{i\alpha}}{\partial u^j} - \frac{\partial g_{ij}}{\partial u^\alpha}\right) g^{\alpha k}\xi_k = 0.$$

That is,

$$\frac{\partial \xi_j}{\partial u^i} + \frac{\partial \xi_i}{\partial u^j} - 2\Gamma^k_{ij}\xi_k = 0, \qquad i, j = 1, 2. \tag{2}$$

This equation can be written in a more compact form:

$$D_i \xi_j + D_j \xi_i = 0, \qquad i, j = 1, 2,$$

where D_i denotes covariant differentiation.

Now for $i, j, k < 3$ the Christoffel symbols Γ^k_{ij} of the space coincide with the corresponding Christoffel symbols $\tilde{\Gamma}^k_{ij}$ of the surface, and

$$\Gamma^3_{ij} = -\tfrac{1}{2}\partial g_{ij}/\partial v^3 = \lambda_{ij}, \qquad i, j = 1, 2,$$

so we can write equations (2) as

$$\frac{\partial \xi_i}{\partial u^j} + \frac{\partial \xi_j}{\partial u^i} - 2\tilde{\Gamma}^k_{ij} - 2\lambda_{ij}\xi_3 = 0, \tag{3}$$

where the indices i, j, k assume the values 1, 2 only.

Now let F be a closed surface homeomorphic to a sphere, with positive extrinsic curvature. We now introduce an isothermal-conjugate parametrization on F; that is, a parametrization in which the second fundamental form of the surface is

$$\text{II} = \nu((du^1)^2 + (du^2)^2).$$

Let us see how this is done.

Since the extrinsic curvature of F is positive, the quadratic form $\text{II} = \lambda_{ij}du^i du^j$ is positive definite. In the metric defined by the quadratic form II, the surface F is a Riemannian manifold homeomorphic to a sphere. If the coefficients λ_{ij} are sufficiently regular, there exists a conformal mapping of F onto the unit sphere, which is regular to the same order as the λ_{ij}. Introduce stereographic coordinates on the sphere. Now define the coordinates u^1, u^2 of a point on F as the coordinates of the corresponding point on the sphere under the above conformal mapping. It is obvious that the coordinate system thus defined on the surface F is an isothermal-conjugate system.

In isothermal coordinates, we can eliminate ξ_3 by term-by-term subtraction of the two equations of (3) corresponding to $i = j = 1$ and $i = j = 2$; this gives the following final form of the equations of an infinitesimal bending on the surface:

$$\begin{aligned} &\frac{\partial \xi_1}{\partial u^1} - \frac{\partial \xi_2}{\partial u^2} - (\tilde{\Gamma}^{\alpha}_{11} - \tilde{\Gamma}^{\alpha}_{22})\, \xi_\alpha = 0, \\ &\frac{\partial \xi_1}{\partial u^2} + \frac{\partial \xi_2}{\partial u^1} - 2\tilde{\Gamma}^{\alpha}_{12} \xi_\alpha = 0. \end{aligned} \tag{4}$$

Our system of isothermal-conjugate coordinates u^1, u^2 on the surface has a singularity at one point, which we denote by X_0. The image of this point under the above-mentioned conformal mapping is the pole P_0 of the sphere from which it is projected in order to define stereographic coordinates.

We now wish to study the behavior of the coefficients of the equations for an infinitesimal bending near the point X_0, i.e. when $(u^1)^2 + (u^2)^2 = \rho^2 \to \infty$.

Let $\bar{u}^i$ be isothermal-conjugate coordinates on the surface, defined by replacing the point P_0 on the sphere by its antipodal point $\bar{P}_0$. Without loss of generality, we may assume that the relation between the coordinates $\bar{u}^i$ and u^i is given by formulas

$$\bar{u}^i = u^i/\rho^2, \qquad \rho^2 = (u^1)^2 + (u^2)^2.$$

Let $\bar{\Gamma}^{\alpha}_{ij}$ denote the Christoffel symbols of the surface in the coordinate system $\bar{u}^i$. Obviously, in a neighborhood of the point X_0 $(\bar{u}^1 = \bar{u}^2 = 0)$ the coefficients $\bar{\Gamma}^{\alpha}_{ij}$ are regular, and hence bounded functions.

Define a euclidean metric on the surface F, given in the coordinate system $\bar{u}$ by the quadratic form

$$ds_0^2 = (d\bar{u}^1)^2 + (d\bar{u}^2)^2.$$

The difference between the Christoffel symbols of the second kind defined by the forms ds^2 and ds_0^2 is a tensor A^{α}_{ij}. The components of this tensor in the coordinate system $\bar{u}$ are the quantities $\bar{\Gamma}^{\alpha}_{ij}$, since the Christoffel symbols for ds_0^2 vanish.

Let us estimate the components of the tensor A^{α}_{ij} near the point X_0 in the coordinate system u^i. We have

$$A^k_{ij} = \bar{\Gamma}^{\gamma}_{\alpha\beta} \frac{\partial \bar{u}^\alpha}{\partial u^i} \frac{\partial \bar{u}^\beta}{\partial u^j} \frac{\partial u^k}{\partial \bar{u}^\gamma}.$$

A direct check easily shows that

$$\left| \frac{\partial \bar{u}^\lambda}{\partial u^\mu} \right| \leqslant \frac{1}{\rho^2}, \qquad \left| \frac{\partial u^\lambda}{\partial \bar{u}^\mu} \right| \leqslant \rho^2.$$

Hence the components A^k_{ij} are $O(1/\rho^2)$ near the point X_0.

It is now easy to determine the behavior of the Christoffel symbols of the surface near X_0 in the coordinates u^i, since they are equal to the Christoffel symbols for the plane line element ds_0^2 in the coordinates u^i, up to quantities of order $O(1/\rho^2)$. We have

$$ds_0^2 = ((du^1)^2 + (du^2)^2)/\rho^4.$$

A direct computation of the Christoffel symbols of the plane gives

$$\Gamma_{11}^1 = -2(\ln\rho)_{u^1}, \qquad \Gamma_{11}^2 = 2(\ln\rho)_{u^2},$$
$$\Gamma_{12}^1 = -2(\ln\rho)_{u^2}, \qquad \Gamma_{12}^2 = -2(\ln\rho)_{u^1},$$
$$\Gamma_{22}^1 = 2(\ln\rho)_{u^1}, \qquad \Gamma_{22}^2 = -2(\ln\rho)_{u^2},$$

and the corresponding Christoffel symbols $\tilde\Gamma_{ij}^k$ of the surface are equal to these up to quantities of order $O(1/\rho^2)$.

If we now write the equations of an infinitesimal bending as

$$\frac{\partial\xi_1}{\partial u^1} - \frac{\partial\xi_2}{\partial u^2} = a\xi_1 + b\xi_2,$$
$$\frac{\partial\xi_1}{\partial u^2} + \frac{\partial\xi_2}{\partial u^1} = c\xi_1 + d\xi_2,$$

we get the following expressions for the coefficients a, b, c, d near the point X_0:

$$a = -4(\ln\rho)_{u^1} + O\left(\frac{1}{\rho^2}\right), \qquad c = -4(\ln\rho)_{u^2} + O\left(\frac{1}{\rho^2}\right),$$
$$b = 4(\ln\rho)_{u^2} + O\left(\frac{1}{\rho^2}\right), \qquad d = -4(\ln\rho)_{u^1} + O\left(\frac{1}{\rho^2}\right).$$

We now transform the equations of the infinitesimal bending by the substitution $\lambda_1 = \vartheta\xi_1$, $\lambda_2 = \vartheta\xi_2$. The equations for λ_1 and λ_2 are

$$\frac{\partial\lambda_1}{\partial u^1} - \frac{\partial\lambda_2}{\partial u^2} = \left(a + \frac{\vartheta_{u^1}}{\vartheta}\right)\lambda_1 + \left(b - \frac{\vartheta_{u^2}}{\vartheta}\right)\lambda_2,$$
$$\frac{\partial\lambda_1}{\partial u^2} + \frac{d\lambda_2}{\partial u^1} = \left(c + \frac{\vartheta_{u^2}}{\vartheta}\right)\lambda_1 + \left(d + \frac{\vartheta_{u^1}}{\vartheta}\right)\lambda_2.$$

Hence it is clear that if we take $\vartheta = \rho^4$, the coefficients of the equations for λ_1 and λ_2 decrease as $1/\rho^2$ when $\rho \to \infty$.

We now prove the following theorem.

THEOREM. *A closed surface in a Riemann space, homeomorphic to a sphere, with positive extrinsic curvature, anchored at one point together with a pencil of directions at one point, is rigid; i.e. it admits no non-trivial infinitesimal bendings.*

In analytical language, the theorem states that if the vector field ξ of an infinitesimal bending of the surface satisfies the condition $\xi = 0$, $D\xi = 0$ at some point X_0 of the surface, then it is identically zero.

We define two systems of isothermal-conjugate coordinates u^i and $\bar{u}^i$ on the surface, related by the formulas $\bar{u}^i = u^i/\rho^2$, such that the $\bar{u}$-coordinates of the point X_0 are $(0, 0)$. A detailed description of these coordinates was given above.

By assumption, $\bar{\xi}_i = 0$ and $\partial\bar{\xi}_i/\partial\bar{u}^j = 0$. But the $\bar{\xi}_i$ satisfy the system of equations

$$\frac{\partial\bar{\xi}_i}{\partial\bar{u}^j} + \frac{\partial\bar{\xi}_j}{\partial\bar{u}^i} - 2\Gamma^{\alpha}_{ij}\bar{\xi}_{\alpha} = 0, \qquad i, j = 1, 2.$$

Differentiating this system with respect to $\bar{u}^k$, we get

$$\frac{\partial^2\bar{\xi}_i}{\partial\bar{u}^j\,\partial\bar{u}^k} + \frac{\partial^2\bar{\xi}_j}{\partial\bar{u}^i\,\partial\bar{u}^k} = 0, \qquad i, j, k = 1, 2,$$

at the point X_0, which implies that all second derivatives of the functions $\bar{\xi}_1$ and $\bar{\xi}_2$ vanish.

We now examine the behavior of ξ_1 and ξ_2 near the point X_0, i.e. as $\rho \to \infty$. We have

$$\xi_i = \bar{\xi}_{\alpha}\,\partial\bar{u}^{\alpha}/\partial u^i.$$

And since $\bar{\xi}_{\alpha}$ decreases as $1/\rho^3$ as a function of u^1, u^2 when $\rho \to \infty$, while $\partial\bar{u}^{\alpha}/\partial u^i$ decreases as $1/\rho^2$, it follows that near X_0 the functions ξ_i are of the order of $1/\rho^5$.

Hence the functions $\lambda_i = \rho^4\xi_i$ decrease as $1/\rho$ when $\rho \to \infty$. But the functions λ_i satisfy the equations

$$\text{(5)} \qquad \begin{aligned} \frac{\partial\lambda_1}{\partial u^1} - \frac{\partial\lambda_2}{\partial u^2} &= a\lambda_1 + b\lambda_2, \\ \frac{\partial\lambda_1}{\partial u^2} + \frac{\partial\lambda_2}{\partial u^1} &= c\lambda_1 + d\lambda_2, \end{aligned}$$

whose coefficients a, b, c, d decrease as $1/\rho^2$ when $\rho \to \infty$. It will thus suffice to show that any solution λ_i of the system (5) which decreases as $1/\rho$ when $\rho \to \infty$ is identically zero.

Following I. N. Vekua [**68**], we express the equations of (5) in the form

$$\text{(6)} \qquad \partial\lambda/\partial\bar{z} = A\lambda + B\bar{\lambda},$$

where

$$\lambda = \lambda_1 + i\lambda_2, \qquad \overline{\lambda} = \lambda_1 - i\lambda_2,$$
$$A = \frac{1}{4}(a + d + ic - ib), \qquad B = \frac{1}{4}(a - d + ic + ib),$$
$$\frac{\partial}{\partial \overline{z}} = \frac{1}{2}\left(\frac{\partial}{\partial u^1} + i\frac{\partial}{\partial u^2}\right).$$

The solution of equation (6) can be expressed as $\lambda(z) = \varphi(z)e^{\omega(z)}$, where $\varphi(z)$ is an analytic function of the complex variable $z = u^1 + iu^2$, and

$$\omega(z) = -\frac{1}{\pi}\int_E\int \frac{A + B\dfrac{\overline{\lambda}}{\lambda}}{t - z}\, dE_t,$$

where the domain of integration is the entire complex t-plane.

Since $\omega(z)$ is bounded, while $\lambda(z) \to 0$ as $z \to \infty$, the analytic function $\varphi(z)$ tends to zero as $z \to \infty$, and is hence identically zero. But then $\lambda(z) = 0$, i.e. $\lambda_1 = 0$, $\lambda_2 = 0$.

This implies that ξ_1 and ξ_2 are also zero. And since

$$\partial \xi_i / \partial u^i = \tilde{\Gamma}^{\alpha}_{ii}\xi_\alpha - \lambda_{ii}\xi_3 = 0,$$

while $\lambda_{ii} \neq 0$, it follows that $\xi_3 = 0$. Q.E.D.

§5. On solutions of a certain elliptic system of differential equations

In the next section we shall study the problem of isometrically embedding two-dimensional closed manifolds which are "almost" embeddable (in a sense to be made precise) in a Riemannian space. This problem will be reduced to consideration of a system of linear differential equations:

$$\frac{\partial u}{\partial x} - \frac{\partial v}{\partial y} = au + bv + f,$$
$$\frac{\partial u}{\partial y} + \frac{\partial v}{\partial x} = cu + dv + g.$$

In this section we shall study the solutions of this system, considered in the entire xy-plane. All the requisite auxiliary material may be found in Vekua [**68, 69**].

Let $U(z)$ be a continuous complex function of the complex variable $z = x + iy$ in a domain G. If $U(z)$ has in G continuous derivatives with respect to x and y, we set

$$\frac{\partial U}{\partial z} = \frac{1}{2}\frac{\partial U}{\partial x} + \frac{i}{2}\frac{\partial U}{\partial y}.$$

It is readily seen that if T is a domain in G bounded by a rectifiable contour L, then

$$\iint_T \frac{\partial U}{\partial \bar{z}}\, dT = \frac{1}{2i} \int_L U(t)\, dt.$$

Using this formula, we can introduce a new definition of the operation $\partial/\partial\bar{z}$ which does not require the existence of the derivatives:

$$\frac{\partial U}{\partial \bar{z}} = \lim_{L \to z} \frac{1}{2i\,|T|} \int_L U(t)\, dt, \tag{1}$$

where $|T|$ is the area of T and $L \to z$ means that the contour L bounding T contracts to the point z, its length decreasing to zero.

It is obvious that, if $U(z)$ is continuously differentiable with respect to x and y, this definition is equivalent to the original one. The converse is in general false: the existence of $\partial U/\partial\bar{z}$ in the sense of the second definition does not imply the existence of continuous derivatives $\partial U/\partial x$ and $\partial U/\partial y$.

The function $U(z)$ is said to be in class $C_{\bar{z}}$ if it has a continuous derivative $\partial U/\partial\bar{z}$ in the sense of (1).

An important example of a function of class $C_{\bar{z}}$ is

$$U(z) = -\frac{1}{\pi} \iint_T \frac{f(\xi, \eta)\, d\xi\, d\eta}{t - z} \qquad (t = \xi + i\eta),$$

where $f(x, y)$ is integrable and continuous in T. Its derivative is

$$\partial U/\partial\bar{z} = f(x, y).$$

This special function of class $C_{\bar{z}}$ is in a certain sense the most general, as is evidenced by the following theorem.

Any function of class $C_{\bar{z}}$ in a domain T bounded by a rectifiable contour L has the representation

$$U(z) = \frac{1}{2\pi i} \int_L \frac{U(t)\, dt}{t - z} - \frac{1}{\pi} \iint_T \frac{\partial U}{\partial \bar{t}} \frac{d\xi\, d\eta}{t - z}, \qquad z \in T.$$

Note that the first integral in the right-hand side is an analytic function in T. As we mentioned above, a function of class $C_{\bar{z}}$ may not have derivatives in x and y, but it satisfies a Hölder condition with any positive exponent $\alpha < 1$. More precisely, if $U(z) \in C_{\bar{z}}(G)$ and T is a closed domain contained in G, then for any $z_1, z_2 \in T$ and $\alpha < 1$

$$\frac{|U(z_1) - U(z_2)|}{|z_1 - z_2|^{\alpha}} \leqslant c\left(\|U\| + \left\|\frac{\partial U}{\partial \bar{z}}\right\|\right),$$

where $\| \ \|$ denotes the maximum modulus and c is a function depending only on G, T and α.

Consider the system of equations

$$\frac{\partial u}{\partial x} - \frac{\partial v}{\partial y} = au + bv,$$

$$\frac{\partial u}{\partial y} - \frac{\partial v}{\partial x} = cu + dv.$$

Setting

$$U(z) = u + iv, \quad A = \frac{1}{4}(a + d + ic - ib). \quad B = \frac{1}{4}(a - d + ic + ib),$$

we rewrite the system as

$$\partial U / \partial \bar{z} = AU + B\overline{U}. \tag{2}$$

If the operator $\partial / \partial \bar{z}$ is taken in the sense of definition (1), equation (2) is in general not equivalent to the original system.

The functions of class $C_{\bar{z}}$ that satisfy equation (2) possess many remarkable properties. We mention two which will be used later:

1. A function of class $C_{\bar{z}}$ which satisfies equation (2) has isolated zeros.
2. Any solution $U(z)$ of equation (2) in the domain T has the representation $U(z) = \varphi(z) e^{\omega(z)}$, where $\varphi(z)$ is an analytic function in T and

$$\omega(z) = -\frac{1}{\pi} \iint_T \frac{A + B\frac{\overline{U}}{U}}{t - z} \, dT_t.$$

Consider the system of differential equations

$$\frac{\partial u}{\partial x} - \frac{\partial v}{\partial y} = au + bv + f,$$

$$\frac{\partial u}{\partial y} + \frac{\partial v}{\partial x} = cu + dv + g,$$

whose coefficients are regular functions in the entire xy-plane which decrease at infinity as $1/\rho^2$, $\rho^2 = x^2 + y^2$. This system may be written in the following compact form:

$$\partial U / \partial \bar{z} = AU + B\overline{U} + F, \tag{3}$$

where

$$U = u + iv, \qquad \overline{U} = u - iv,$$

$$A = \frac{1}{4}(a + d + ic - ib), \qquad B = \frac{1}{4}(a - d + ic + ib),$$

$$F = f + ig, \qquad \frac{\partial}{\partial \bar{z}} = \frac{1}{2}\left(\frac{\partial}{\partial x} + i\frac{\partial}{\partial y}\right).$$

Let U be a solution of equation (3) with continuous derivatives, which vanishes at infinity. In any bounded domain T whose boundary is a rectifiable contour L, the function $U(z)$ $(z = x + iy)$ has the representation

$$U(z) = \frac{1}{2\pi i}\int_L \frac{U(t)\,dt}{t-z} - \frac{1}{\pi}\iint_T \frac{\partial U}{\partial \bar t}\,\frac{dT_t}{t-z}.$$

It follows that a solution of equation (3) satisfies the integral equation

$$U(z) + \frac{1}{\pi}\iint_T \frac{A(t)\,U(t) + B(t)\,\overline{U}(t)}{t-z}\,dT = G(z),$$

where

$$G(z) = \Phi(z) - \frac{1}{\pi}\iint_T \frac{F(t)}{t-z}\,dT,$$

and Φ is a function holomorphic in T and continuous in $T + L$, defined by a Cauchy integral

$$\Phi(z) = \frac{1}{2\pi i}\int_L \frac{U(t)}{t-z}\,dt.$$

Let T be a disk of radius R. As $R \to \infty$, all the integrals converge, and $\Phi(z) \to 0$. Hence we get the following integral equation for $U(z)$:

$$U(z) + \frac{1}{\pi}\iint_E \frac{\Omega(U)}{t-z}\,dE = G(z),$$

where

$$\Omega(U) = AU + B\overline{U}, \qquad G(z) = -\frac{1}{\pi}\iint_E \frac{F(t)\,dE}{t-z},$$

and the integration extends over the entire plane.

Note that since $F(t)$ decreases as $1/|t|^2$ when $t \to \infty$, it follows that $G(z)$ decreases as $1/|z|$ when $z \to \infty$.

Let B^0 denote the space of functions continuous in the entire z-plane which vanish at infinity. We norm this space in the usual way: $\|U\| = \max_E |U|$.

We claim that the integral equation (4) has a unique solution in the space B^0 for any function $G(z) \in B^0$.

Let S denote the operator which maps each function $U \in B^0$ onto the function V given by

$$V = \frac{1}{\pi}\iint_E \frac{\Omega(U)}{t-z}\,dE.$$

Then equation (4) becomes $U+SU=G$.

To prove that this equation is solvable in B^0 for any $G\in B^0$, it suffices to show that

1. $SU\in B^0$ for any $U\in B^0$.
2. The operator S is compact.
3. If $U+SU=0$, then $U=0$.

The first property follows in an obvious manner from the fact that U is a bounded function, while A and B decrease as $1/|t|^2$ when $t\to\infty$.

To prove that S is a compact operator, it is sufficient to show that it maps any bounded set of functions U into a set of functions V which converge uniformly to zero as $z\to\infty$ and are equicontinuous in any bounded domain.

The fact that the functions V converge uniformly to zero as $z\to\infty$ follows from the estimate $|\Omega(U)|\leqq\|U\|c\mu$, where

$$\mu(z)=\begin{cases}1 & \text{for} \quad |z|\leqslant 1,\\ \dfrac{1}{|z|^2} & \text{for} \quad |z|\geqslant 1,\end{cases}$$

and c is a constant depending only on A and B.

To prove that the functions V are equicontinuous, we set

$$V_T=\frac{1}{\pi}\iint_T\frac{\Omega(U)}{t-z}\,dE, \qquad V_{E-T}=\frac{1}{\pi}\iint_{E-T}\frac{\Omega(U)}{t-z}\,dE,$$

where T is the disk $x^2+y^2\leq R^2$. For sufficiently large R, the norms $|V_{E-T}|$ are bounded by ϵ uniformly for all U in the given bounded set. The functions $V_T(z)$ satisfy a Hölder condition with any exponent α, $0<\alpha<1$, in the disk $x^2+y^2<R_1^2<R^2$, with a coefficient $c(\alpha)$ which depends only on R_1 and $\|\Omega(U)\|$; we may therefore assume that the coefficient $c(\alpha)$ is the same for all functions U. Hence, if z_1 and z_2 are in the disk $x^2+y^2<R_1^2$, $|z_1-z_2|<\delta$ and δ is sufficiently small, then $|V(z_1)-V(z_2)|<\epsilon$ for all the functions V. This proves that the functions V are equicontinuous.

Finally, we prove that the only solution of the equation $U+SU=0$ in B^0 is $U=0$.

A continuous solution U of the equation

$$U+\frac{1}{\pi}\iint_E\frac{\Omega(U)}{t-z}\,dE=0$$

is a solution of the equation

$$\partial U/\partial\bar{z}=AU+B\bar{U},$$

if the derivative $\partial U/\partial \bar{z}$ is interpreted in the generalized sense.

Any solution of the equation

$$\partial U/\partial \bar{z} = AU + B\bar{U}$$

has the representation

$$U(z) = \varphi(z)e^{\omega(z)},$$

where $\varphi(z)$ is an analytic function and

$$\omega(z) = -\frac{1}{\pi}\iint_B \frac{A + B\dfrac{\bar{U}}{U}}{t - z}\, dE.$$

Since $\omega(z)$ is bounded and $U(z) \to 0$ as $z \to \infty$, it follows that $\varphi(z) \to 0$ as $z \to \infty$. Hence $\varphi(z) \equiv 0$, and hence also $U(z) \equiv 0$.

This implies the third property of the operator S, and thus equation (4) is solvable.

It is known [**68**] that any generalized solution of equation (3) is regular if its coefficients are regular. Precisely, if the coefficients are n times differentiable and their nth derivatives satisfy a Hölder condition in any bounded domain of the plane, then the solution $U(z)$ is $n+1$ times differentiable and its $(n-1)$th derivatives satisfy a Hölder condition with the same exponent, in any bounded domain.

We state the final result as a lemma.

LEMMA 1. *The system of linear equations*

$$\frac{\partial u}{\partial x} - \frac{\partial v}{\partial y} = au + bv + f,$$

$$\frac{\partial u}{\partial y} + \frac{\partial v}{\partial x} = cu + dv + g,$$

where $a, b, \cdots, g$ are regular functions which decrease as $1/(x^2+y^2)$ when $x^2+y^2 \to \infty$, has a unique solution, which vanishes at infinity.

If $a, \cdots, g$ are n times differentiable and their nth derivatives satisfy a Hölder condition in any bounded domain of the plane, then the solution is $n+1$ times differentiable and its $(n+1)$th derivatives satisfy a Hölder condition with the same exponent, in any bounded domain.

We now estimate the norm of the solution of equation (4). Since the operator S is compact and the homogeneous equation $U + SU = 0$ has no nontrivial solution in B^0, the operator $(1+S)^{-1}$ is bounded in B^0. Hence the norm of the solution $U(z)$ of the equation $U + SU = F$ satisfies the estimate $\|U\| \leq K\|F\|$, where K is a constant depending only on the functions A and B (but not on F).

We now estimate the derivatives of $U(z)$ in a disk T of radius R. To this end, consider the following problem. Suppose that the function $U(z)$ is defined in a disk T' of radius R' by

$$U(z)=\iint_{T'}\frac{\mu(t)}{t-z}\,dT',$$

where $\mu(t)$ is continuous and satisfies a Hölder condition with exponent α in T'. Let us estimate the derivatives of U and their Hölder expressions for exponent α in a disk T'' of radius $R''<R'$.

A direct check shows that

$$U(z)=\frac{\partial}{\partial z}\left(-2\iint_{T'}\mu(t)\ln|t-z|\,dT'\right).$$

Thus it is sufficient to estimate the second derivatives of the function of z in parentheses in this equality.

Let $h(z)$ be a function having derivatives of order up to n in the domain G; assume that its nth derivatives satisfy a Hölder condition with exponent α. Let $\|h\|^{G}_{n,\alpha}$ denote the maximum of the function h, its derivatives of order up to n, and its Hölder expressions with exponent α, in the domain G.

A theorem of potential theory states that

$$\left\|\iint_{T'}\mu(t)\ln|t-z|\,dT'\right\|^{T''}_{2,\alpha}\leqslant c\|\mu\|^{T'}_{\alpha},$$

where c is a constant independent of the function μ. In our case this implies

$$\|U(z)\|^{T''}_{1,\alpha}\leqq c\|\mu\|^{T'}_{\alpha}.$$

If nothing is assumed of the function $\mu(t)$ besides continuity in a closed domain T', then the function

$$U(z)=\iint_{T'}\frac{\mu(t)}{t-z}\,dT'$$

and its Hölder expressions satisfy the estimate

$$\|U\|^{T''}_{\alpha}\leqq c\|\mu\|^{T'},$$

where the constant c is independent of the function μ.

We now consider estimates for the derivatives of the solution of the equation

$$\partial U/\partial\bar z=AU+B\overline{U}+F.$$

As was shown above, throughout the z-plane we have $\|U\| \leq K\|F\|$, where K depends only on A and B.

We estimate the derivatives of U and their Hölder expressions in a disk T of radius R. In a disk T_0 containing T, $U(z)$ satisfies the equation

$$U(z) + \frac{1}{\pi}\iint_{T_0} \frac{\Omega(U)}{t-z}\, dT_0 = G(z),$$

where

$$G(z_0) = \Phi(z) - \frac{1}{\pi}\iint_{T_0} \frac{F(t)}{t-z}\, dT_0,$$

and Φ is an analytic function defined by the Cauchy integral

$$\Phi(z) = \frac{1}{2\pi i}\int_{L_0} \frac{U(t)}{t-z}\, dt.$$

It follows that, in a disk T_1 contained in T_0 and containing T,

$$\|U\|_{1,\alpha}^{T_1} \leqslant c(\|U\|^{T_0} + \|F\|^{T_0}).$$

At the same time, in a disk T_2 contained in T_0 and containing T_1,

$$\|U\|_{1,\alpha}^{T_2} \leqslant c_1(\|U\|^{T_0} + \|F\|_{\alpha}^{T_0}),$$

where the constants c and c_1 depend on $\|A\|_{\alpha}^{T_0}$, $\|B\|_{\alpha}^{T_0}$.

Differentiating the equation for $U(z)$ with respect to x and setting $\partial U(z)/\partial x = U_1(z)$, we get

$$\partial U_1/\partial \bar{z} = AU_1 + B\bar{U}_1 + F_1,$$

where

$$F_1 = \frac{\partial F}{\partial x} + U\frac{\partial A}{\partial x} + \bar{U}\frac{\partial B}{\partial x}.$$

Using the integral equation for the function U_1 in a disk T_3 containing T and contained in T_2, we get the estimate

$$\|U_1\|_{1,\alpha}^{T_3} \leqslant c_2(\|U\|^{T_0} + \|F\|_{1,\alpha}^{T_0}),$$

where c_2 depends on $\|A\|_{1,\alpha}^{T_0}$, $\|B\|_{1,\alpha}^{T_0}$. An estimate for $\partial U/\partial y$ is derived in the same way.

Differentiating the equation for $U(z)$ twice and going over to the integral equation for the second derivative, we get a similar estimate:

$$\|U_2\|_{1,\alpha}^{T_0} \leqslant c_3(\|U\|^{T_0} + \|F\|_{2,\alpha}^{T_0}).$$

The constant c_3 depends on $\|A\|_{2,\alpha}^{T_0}$, $\|B\|_{2,\alpha}^{T_0}$.

This procedure for successive estimation of the derivatives of U can be continued up to estimates for the nth derivatives, provided A, B and F have $(n-1)$th derivatives satisfying a Hölder condition.

Since $\|U\|^{T_0} \leqq \|U\|^E$, and $\|U\|^E$ has been estimated in terms of $\|F\|^E$, it follows that all derivatives of $U(z)$ can finally be estimated in terms of $\|F\|^E$ and $\|F\|^{T_0}_{n-1,\alpha}$. We state the final result as a lemma.

LEMMA 2. *Let $U(z)$ be a solution, vanishing at infinity, of the equation*

$$\partial U/\partial \bar{z} = AU + B\bar{U} + F,$$

whose coefficients decrease as $1/|z|^2$ when $z \to \infty$ and have nth derivatives in any bounded domain of the plane which satisfy a Hölder condition with exponent α.

Then the following estimate is valid in a disk T contained in the interior of the disk T_0:

$$\| U \|^T_{n+1,\alpha} \leqslant c\left(\| F \|^E + \| F \|^{T_0}_{n,\alpha}\right).$$

The constant c depends on the disks T and T_0 and the coefficients A and B, but not on F.

§6. Embedding of manifolds infinitesimally close to embeddable manifolds

Let F be a closed surface homeomorphic to a sphere, with positive extrinsic curvature K_e, in a Riemannian space R. Suppose this surface is subjected to a regular deformation, under which its transform at time t is a surface F_t with line element

$$ds_t^2 = ds^2 + t\,d\sigma_t^2,$$

where ds^2 is the line element of the original surface. As $t \to 0$, the form $d\sigma_t^2$ converges to some limit $d\sigma^2$ which is uniquely determined by the deformation of the surface.

In this section we shall discuss the converse problem: to determine a deformation of the surface which induces a given variation of the line element, up to quantities of higher than first order in t, i.e. a deformation which corresponds in the above manner to a given form $d\sigma^2$.

As in the preceding section, introduce an isothermal-conjugate coordinate net u^i on the surface F, in such a way that $(u^1)^2 + (u^2)^2 = \infty$ at a given point X_0 on the surface. The space near the surface is provided with a semigeodesic parametrization v^i.

Let F_t be a regular deformation of the surface F, $v^i(u^1, u^2, t)$ the point which is the transform of a point (u^1, u^2) on F at time t. The line element

ds_t^2 of the surface F_t is found by substituting $v^i = v^i(u_1, u_2, t)$ in the line element $g_{ij}dv^i dv^j$ of the space:

$$(*) \qquad g_{ij}dv^i dv^j = ds_t^2.$$

In order that the deformation F_t be a solution to our problem, i.e. induce the given variation $d\sigma^2$ of the line element, the total derivative of the left-hand side of (*) with respect to t must be equal to $d\sigma^2$ at time $t=0$. This gives the equation

$$(**) \qquad \frac{\partial g_{ij}}{\partial v^k}\xi^k\, du^i\, du^j + g_{ij}\, d\xi^i\, du^j + g_{ij}\, d\xi^j\, du^i = d\sigma^2,$$

where ξ^k are the components of the displacement velocity of the points of F at time $t=0$,

$$\xi^k = \partial v^k / \partial t.$$

Our problem will be solved if we can find a regular vector field ξ satisfying equation (**). In fact, as is easily seen, the deformation of F which takes a point (u^1, u^2) to the point of space with coordinates

$$v^1 = u^1 + t\xi^1, \quad v^2 = u^2 + t\xi^2, \quad v^3 = t\xi^3$$

at time t induces the given variation of the surface line element.

The left-hand side of equation (**) coincides with that of the equation of an infinitesimal bending of the surface F. We can thus employ all the arguments applied in the preceding section to the equation of an infinitesimal bending. In particular, equation (**) can be brought to an equivalent form

$$g_{\alpha j}\frac{\partial \xi^\alpha}{\partial u^i} + g_{i\alpha}\frac{\partial \xi^\alpha}{\partial u^j} + \xi^k\frac{\partial g_{ij}}{\partial v^k} = \sigma_{ij}, \qquad i,\ j = 1,\ 2,$$

where σ_{ij} are the coefficients of the quadratic form $d\sigma^2$. Hence, replacing the contravariant components of the vector ξ by its covariant components $\xi_i = g_{i\alpha}\xi^\alpha$, we get

$$\frac{\partial \xi_i}{\partial u^j} + \frac{\partial \xi_j}{\partial u^i} - 2\Gamma_{ij}^k \xi_k = \sigma_{ij}, \qquad i,\ j = 1,\ 2,$$

or

$$(***) \qquad \frac{\partial \xi_i}{\partial u^j} + \frac{\partial \xi_j}{\partial u^i} - 2\tilde{\Gamma}_{ij}^k \xi_k - 2\nu\delta_{ij}\xi_3 = \sigma_{ij},$$

where $\tilde{\Gamma}_{ij}^k$ are the Christoffel symbols of the surface and $\nu\delta_{ij}$ the coefficients of the second fundamental form. In particular, we get two equations for ξ_1 and ξ_2:

$$\frac{\partial \xi_1}{\partial u^1} - \frac{\partial \xi_2}{\partial u^2} - (\tilde{\Gamma}^{\alpha}_{11} - \tilde{\Gamma}^{\alpha}_{22}) \xi_{\alpha} = \frac{\sigma_{11} - \sigma_{22}}{2},$$

$$\frac{\partial \xi_1}{\partial u^2} + \frac{\partial \xi_2}{\partial u^1} - 2\tilde{\Gamma}^{\alpha}_{12}\xi_{\alpha} = \sigma_{12}.$$

On the other hand, upon differentiation of equation (***) with respect to u^i and u^j, followed by termwise addition and subtraction, we obtain

$$-\Delta(\nu\xi_3) = \frac{\partial^2}{(\partial u^2)^2}\left(\tilde{\Gamma}^k_{11}\xi_k + \frac{\sigma_{11}}{2}\right) - 2\frac{\partial^2}{\partial u^1 \partial u^2}\left(\tilde{\Gamma}^k_{12}\xi_k + \frac{\sigma_{12}}{2}\right) + \frac{\partial^2}{(\partial u^1)^2}\left(\tilde{\Gamma}^k_{22}\xi_k + \frac{\sigma_{22}}{2}\right).$$

We have derived an equation for the velocity field of a deformation of the surface under which the line element undergoes the given variation $d\sigma^2$. We have already stated that determination of the field ξ is sufficient to determine the required deformation.

We now determine this field, on the assumption that $d\sigma^2 = 0$ and $Dd\sigma^2 = 0$ at the origin X_0, i.e. the line element is stationary at the point X_0.

The components ξ_1 and ξ_2 of the velocity of the deformation satisfy the system of equations

$$\frac{\partial \xi_1}{\partial u^1} - \frac{\partial \xi_2}{\partial u^2} = (\tilde{\Gamma}^{\alpha}_{11} - \Gamma^{\alpha}_{22}) \xi_{\alpha} + \frac{\sigma_{11} - \sigma_{22}}{2},$$

$$\frac{\partial \xi_1}{\partial u^2} + \frac{\partial \xi_2}{\partial u^1} = 2\tilde{\Gamma}^{\alpha}_{12}\xi_{\alpha} + \sigma_{12}.$$

We shall solve these equations for the unknowns $\lambda_i = \vartheta\xi_i$, where ϑ is a positive and sufficiently regular function equal to ρ^4 outside the unit disk. The equations then become

$$\frac{\partial \lambda_1}{\partial u^1} - \frac{\partial \lambda_2}{\partial u^2} = a\lambda_1 + b\lambda_2 + \vartheta\frac{\sigma_{11} - \sigma_{22}}{2},$$

$$\frac{\partial \lambda_1}{\partial u^2} + \frac{\partial \lambda_2}{\partial u^1} = c\lambda_1 + d\lambda_2 + \vartheta\sigma_{12},$$

where the coefficients a, b, c, d decrease as $1/\rho^2$ when $\rho \to \infty$ (§4). Let us study the behavior of the free terms as $\rho \to \infty$.

By assumption, $d\sigma^2 = 0$, $Dd\sigma^2 = 0$ at the point X_0. This means that, in terms of the coordinates $\bar{u}$, we have $\bar{\sigma}_{ij} = O(\bar{\rho}^2)$ at X_0. And since

$$\sigma_{ij} = \bar{\sigma}_{\alpha\beta}\frac{\partial \bar{u}^{\alpha}}{\partial u^i}\frac{\partial \bar{u}^{\beta}}{\partial u^j},$$

$$\left|\frac{\partial \bar{u}^{\lambda}}{\partial u^{\mu}}\right| < \frac{1}{\rho^2}, \qquad \rho\bar{\rho} = 1,$$

it follows that $\sigma_{ij} = O(1/\rho^6)$ near X_0. Hence the free terms of the equations decrease as $1/\rho^2$ when $\rho \to \infty$.

According to Lemma 1 of §5, there exists a solution λ_1, λ_2 of the system which decreases as $1/\rho$ when $\rho \to \infty$. This solution is unique. ξ_1 and ξ_2 may now be determined from λ_1 and λ_2, and ξ_3 found from any equation of the system

$$\frac{\partial \xi_i}{\partial u^j} + \frac{\partial \xi_j}{\partial u^i} - 2\tilde{\Gamma}^k_{ij}\xi_k - 2\nu\delta_{ij}\xi_3 = \sigma_{ij}$$

for $i = j$. We have thus found a field ξ, regular on the entire surface (except perhaps at X_0), which induces the given variation $d\sigma^2$ of the line element.

We now determine the behavior of the field ξ near the point X_0. To this end, we go over to the coordinates $\bar{u}$. Then, except at X_0, the components $\bar{\xi}_1$ and $\bar{\xi}_2$ of the field ξ satisfy the equations

$$\frac{\partial \bar{\xi}_1}{\partial \bar{u}^1} - \frac{\partial \bar{\xi}_2}{\partial \bar{u}^2} - (\bar{\Gamma}^\alpha_{11} - \bar{\Gamma}^\alpha_{22})\,\bar{\xi}_\alpha = \frac{\bar{\sigma}_{11} - \bar{\sigma}_{22}}{2},$$

$$\frac{\partial \bar{\xi}_1}{\partial \bar{u}^2} + \frac{\partial \bar{\xi}_2}{\partial \bar{u}^1} - 2\bar{\Gamma}^\alpha_{12}\bar{\xi}_\alpha = \bar{\sigma}_{12},$$

whose coefficients $\bar{u}_1$ and $\bar{u}_2$ are regular near X_0.

The functions $\bar{\xi}_1$ and $\bar{\xi}_2$ are continuous near X_0. Indeed,

$$\bar{\xi}_i = \xi_\alpha \, \partial u^\alpha / \partial \bar{u}^i.$$

Near the point X_0 we have $\xi_\alpha = O(\bar{\rho}^5)$, while $\partial u^\alpha / \partial \bar{u}^i = O(1/\bar{\rho}^2)$. Hence it is clear that $\bar{\xi}_i$ $(i = 1, 2)$ tends to zero with $\bar{\rho}^3$ when $\bar{\rho} \to 0$.

Set

$$z = \bar{u}^1 + i\bar{u}^2, \qquad U(z) = \bar{\xi}_1 + i\bar{\xi}_2.$$

Then the system of equations for $\bar{\xi}_1$ and $\bar{\xi}_2$ can be expressed in complex form:

$$\partial U / \partial \bar{z} = AU + B\bar{U} + F,$$

where the functions A, B, F have known expressions in terms of the coefficients of the system. In the neighborhood of X_0, the functions A, B, F, U are continuous and tend to zero. Hence $\partial U/\partial \bar{z}$ is continuous. It follows that if we set $U(0) = 0$ this function will satisfy the equation in a neighborhood of X_0 $(z = 0)$ including X_0 itself. Now the coefficients of the equation are regular functions, and hence $U(z)$ is a regular function. Since $\bar{\xi}_1$ and $\bar{\xi}_2$ are regular, so is $\bar{\xi}_3$, which can be expressed in terms of $\bar{\xi}_1$, $\bar{\xi}_2$ and their first derivatives.

We have thus proved that there exists a regular field ξ which induces the given variation $d\sigma^2$ of the line element. Since the functions $\bar{\xi}_1$ and $\bar{\xi}_2$ are $O(\bar{\rho}^3)$ near the point X_0, so that $\bar{\xi}_3 = O(\bar{\rho}^2)$, it follows that $\xi = 0$, $D(\xi) = 0$ at the point X_0. The degree of regularity of the field ξ depends on that of the coefficients A, B, F (§5).

The final result can be formulated as the following lemma.

LEMMA. *Let $d\sigma^2$ be a given variation of the line element of a closed surface F, homeomorphic to a sphere, with positive extrinsic curvature, such that $d\sigma^2 = 0$, $Dd\sigma^2 = 0$ at a point X_0.*

Then there exists a vector field ξ which induces the variation $d\sigma^2$ of the line element of the surface and satisfies the condition $\xi = 0$, $D\xi = 0$ at the point X_0.

The field ξ is regular. Precisely, if the metric of the surface is n times differentiable and its nth derivatives satisfy a Hölder condition with exponent α, and $d\sigma^2$ is $n-1$ times differentiable and its $(n-1)$th derivatives satisfy a Hölder condition with exponent α, then the field ξ is n times differentiable and its nth derivatives satisfy a Hölder condition with the same exponent.

We now estimate the norm of the velocity field of the deformation.

Let f be some function (scalar or vector valued) defined for points of the surface. Given any parametrization u^i, the function f becomes a function of the variables u^i. We shall say that the function f is in class $H_{n,\alpha}$ if there exists an $n+\alpha$ times differentiable parametrization of the neighborhood of any point X on the surface such that f, as a function of the surface parameters in the neighborhood of X, is $n+\alpha$ times differentiable, i.e. its nth derivatives exist and satisfy a Hölder condition with exponent α.

We shall say that a quadratic form $d\omega^2$ defined on the surface is in class $H_{n,\alpha}$ if its coefficients are $n+\alpha$ times differentiable in some $n+\alpha$ times differentiable parametrization of the surface.

Note that if the metrics of the space and the surface with positive extrinsic curvature are $n+\alpha$ times differentiable, then the isothermal-conjugate parametrization of the surface and the corresponding semi-geodesic parametrization of the space are $n+\alpha$ times differentiable.

The functions of class $H_{n,\alpha}$ and the quadratic differential forms of class $H_{n,\alpha}$ form linear spaces, which can be normed in the following way.

Let $f \in H_{n,\alpha}$. Let $f = f_1(u^1, u^2)$ in the isothermal-conjugate parametrization u^i, and $f = f_2(\bar{u}^1, \bar{u}^2)$ in the isothermal-conjugate parametrization $\bar{u}^i$. Let $\| f_1 \|_{n,\alpha}$ denote the maximum of the function f_1, its derivatives of order up to n, and the Hölder expressions for the nth derivatives, in the disk $(u^1)^2+(u^2)^2\leqq 1$, and $\|f_2\|_{n,\alpha}$ the analogous maximum for the function f_2 in the disk $(\bar{u}^1)^2+(\bar{u}^2)^2\leqq 1$. The norm in $H_{n,\alpha}$ is defined by

$$\|f\|_{n,\alpha}=\max(\|f_1\|_{n,\alpha},\|f_2\|_{n,\alpha}).$$

The norm of the space $H_{n,\alpha}$ of quadratic forms is defined similarly, using the coefficients of the forms.

Our problem is now to estimate the norm of the vector field ξ of a deformation which induces the given variation $d\sigma^2$ of the line element, in terms of the norm of $d\sigma^2$.

First, by Lemma 2 of §5,

$$|\xi_i|<c\|d\sigma^2\|,\qquad i=1,2.$$

Hence $|\bar{\xi}_i|<\bar{c}\|d\sigma^2\|$, $i=1,2$, since within the unit disk $\bar{\rho}\leqq 1$ we have

$$|\partial u^\alpha/\partial\bar{u}^\beta|\leqq\bar{\rho}^{-2},$$

and

$$\bar{\xi}_i=\xi_\alpha\,\partial u^\alpha/\partial\bar{u}^i.$$

Higher-order derivatives of ξ_i $(i=1,2)$ in the disk $\rho\leqq 1$ are estimated by the method of §5.

The result is

$$\|\xi_i\|_{k,\alpha}\leqq c_1\|d\sigma^2\|_{k-1,\alpha},\quad i=1,2.$$

Similarly, for the functions $\bar{\xi}_i$ in the disk $\bar{\rho}\leqq 1$,

$$\|\bar{\xi}_i\|_{k,\alpha}\leqq c_2\|d\sigma^2\|_{k-1,\alpha}.$$

Now consider the component $\xi_3=\bar{\xi}_3$. As was shown above, it satisfies the equation

$$-\Delta(\nu\xi_3)=\frac{\partial^2}{(\partial u^2)^2}\left(\tilde{\Gamma}_{11}^k\xi_k+\frac{\sigma_{11}}{2}\right)+\cdots.$$

If we regard the functions ξ_1, ξ_2 as known (they have been estimated), this is a Poisson equation for $\nu\xi_3$. We may assume that an estimate for $|\xi_3|$ is known. Estimates for the derivatives of ξ_3 are established by the now familiar method of §5. We get the following estimate in the disk $\rho\leqq 1$:

$$\|\xi_3\|_{k,\alpha}\leqq c_3\|d\sigma^2\|_{k,\alpha}.$$

Our primary interest is in the case when the form $d\sigma^2$ is $a\,d\lambda\,d\mu$, where a, λ and μ are certain functions of u^1 and u^2. It is easily seen that the third derivatives of λ and μ do not appear in the expression $\Delta(\nu\xi_3)$. In this case, therefore, one obtains a better estimate for the norm of ξ_3:

$$\|\xi_3\|_{k,\alpha} \leqq c_3 \|a\|_{k,\alpha} \|d\lambda d\mu\|_{k-1,\alpha}.$$

We have found a vector field ξ corresponding to the given variation $d\sigma^2$ of the line element, such that $\xi=0$ and $D\xi=0$ at the point X_0. We now wish to find a field ξ such that $\xi=a$ at X_0, where a is a given nonzero vector.

Construct a regular vector field η on the surface F, which satisfies the following conditions at X_0:

$$\eta=a, \qquad d\sigma_\eta^2=0, \qquad Dd\sigma_\eta^2=0,$$

where

$$d\sigma_\eta^2=\left(g_{\alpha j}\frac{\partial\eta^\alpha}{\partial u^i}+g_{i\alpha}\frac{\partial\eta^\alpha}{\partial u^j}+\eta^k\frac{\partial g_{ij}}{\partial v^k}\right)du^i\,du^j.$$

It is obvious that this construction involves a good deal of freedom.

Let ξ be the velocity field of a deformation which induces the variation $d\sigma^2-d\sigma_\eta^2$ of the line element, and satisfies the conditions $\xi=0$ and $D\xi=0$ at the point X_0.

Now consider the field $\zeta=\xi+\eta$. The corresponding variation of the line element of the surface is $d\sigma^2$, but now $\zeta=a$ at the point X_0. Obviously

$$\|\zeta\|_{k,\alpha} \leqq c(\|d\sigma^2\|_{k,\alpha}+\|\eta\|_{k,\alpha}).$$

As was shown in §4, a closed surface in a Riemannian space, homeomorphic to a sphere, with positive extrinsic curvature, anchored to some point X_0 together with a pencil of directions on the surface at X_0, is rigid. In analytical language, this means that the equation of an infinitesimal bending

$$g_{\alpha j}\frac{\partial\xi^\alpha}{\partial u^i}+g_{i\alpha}\frac{\partial\xi^\alpha}{\partial u^j}+\xi^k\frac{\partial g_{ij}}{\partial v^k}=0$$

has no nontrivial solutions such that $\xi=0$ and $D\xi=0$ at the point X_0.

In euclidean space, this equation has nontrivial solutions if no restrictions are imposed at the point X_0. They are all of the form $\xi=a\times r+b$, where a and b are constant vectors and r the radius vector of a point on the surface. These solutions constitute the field of velocities of a rigid motion of the surface, and are said to be trivial.

In this connection, it is natural to ask whether the equation of an infinitesimal bending in Riemannian space has nontrivial solutions if no restrictions are imposed on the solution, and moreover how many such solutions exist. We shall now show that such solutions exist, and the degree of freedom involved is the same as in the euclidean case.

Let a and b be any two vectors at the point X_0 on the surface. Construct a vector field η on the surface such that $\eta = a$, $D\eta = b \times dx$ at the point X_0 (this field is by no means unique).

Now introduce the quadratic form $d\sigma_\eta^2$ for the field η, as in the foregoing arguments. We claim that this form satisfies the conditions $d\sigma_\eta^2 = 0$, $Dd\sigma_\eta^2 = 0$ at the point X_0.

The simplest way to see this is as follows. Construct the osculating euclidean space at X_0. In this euclidean space the conditions $\eta = a$ and $D\eta = b \times dx$ mean that in the neighborhood of X_0 the field η is the velocity field of a rigid motion of the surface, up to second-order quantities. It follows that in the euclidean space $d\sigma_\eta^2$ and $D(d\sigma_\eta^2)$ vanish at the point X_0. But this means that in the Riemannian space we have $d\sigma_\eta^2 = 0$ and $D(d\sigma_\eta^2) = 0$ at X_0.

Now construct the velocity field ξ of a deformation of the surface which induces the variation $d\sigma_\eta^2$ of the line element and satisfies the conditions $\xi = 0$ and $D\xi = 0$ at X_0. The vector field $\zeta = \eta + \xi$ obviously satisfies the equation of an infinitesimal bending, and $\zeta = a$, $D\zeta = b \times dx$ at the point X_0.

The fields ζ thus constructed exhaust all possible infinitesimal bendings of the surface. In fact, there obviously exist six linearly independent solutions of type ζ. There cannot be more, for, given any seven solutions, we can always find a linear combination ζ^* such that $\zeta^* = 0$, $D\zeta^* = 0$ at the point X_0. And then, by the rigidity theorem (§4), $\zeta^* \equiv 0$. The set of fields ζ may contain trivial ones, in the sense that the deformations that they generate leave stationary not only the distance between points on the surface but also their extrinsic distances.

§7. Isometric embedding of almost embeddable manifolds

The next three sections are devoted to the main result of this chapter. We shall prove that, under certain assumptions, a closed Riemannian manifold homeomorphic to a sphere admits an isometric embedding in a given three-dimensional Riemannian space R, i.e. the Riemannian space contains a surface isometric to the given manifold.

The proof of the theorem proceeds in three steps. First, we shall construct a continuous family of manifolds M_t which contains the given manifold M and an embeddable manifold M_0.

Second, we shall prove that if a manifold M_t of the family is embeddable, then the manifolds of the family in its "neighborhood" are also embeddable.

The third and final step is to prove that if each of the manifolds M_{t_n}

of the family is embeddable and $t_n \to t^*$, then the manifold M_{t^*} is embeddable.

It is clear that this will imply that the given manifold is embeddable. The present section presents the first two steps of the proof.

Let M be a closed Riemannian manifold homeomorphic to a sphere, with positive extrinsic curvature, A_0 a point on the manifold, and P_0 a pencil of directions at this point. Let R be a Riemannian space, X_0 a point in R, α a plane section at X_0 and U_0 a pencil of directions in α with center X_0. The pencils of directions P_0 and U_0 are isometric; fix some isometric mapping between the directions in the pencils P_0 and U_0 (i.e. a mapping preserving the angles between the directions).

We shall assume that the metric of M is sufficiently regular (at least five times differentiable). It can then be isometrically embedded in a euclidean space as a regular (at least four times differentiable) surface F (Chapter II, §10). Let A be the point on this surface corresponding to A_0.

Let γ be a geodesic issuing from the point X_0 in R, perpendicular to the plane section α, in either of the two possible directions. Fix some order of succession of directions in U_0, which forms a right-handed screw with the direction of the geodesic γ at X_0. Without loss of generality, we may assume that the corresponding order in the pencil P on F at the point A also forms a right-handed screw with the interior normal to the surface. Otherwise we need only consider the mirror image of the surface.

Let O_0 be a point on the geodesic γ, at a distance δ from X_0. Describe a sphere ω_0 of radius δ about O_0 in R. If the space R is sufficiently regular, the sphere ω_0 will have a sufficiently regular metric. For sufficiently small δ, its Gauss curvature will be strictly positive. The sphere ω_0 may therefore be isometrically embedded in euclidean space as a regular surface ω. Let X be the point on this surface corresponding to X_0.

The isometric correspondence of the directions in the pencils P_0 and U_0 induces a natural isometric correspondence of the directions on the surfaces F and ω at the points A and X, respectively.

Identify the points A and X and the corresponding directions of the pencils on the surfaces F and ω. Without loss of generality, we may assume that the surfaces then lie on the same side of their common tangent plane at the point $O \equiv A \equiv X$. Otherwise we need only reflect one of them in the tangent plane.

For sufficiently small δ, the normal curvature of the surface ω is arbitrarily large. Since this is fairly obvious, we confine ourselves to an outline of the proof.

The normal curvature of a closed convex surface in euclidean space satisfies the inequality

$$\varkappa \leqslant \sqrt{K_1 + \frac{K_{11}}{K_2}},$$

where K_1 is the maximum and K_2 the minimum Gauss curvature, and K_{11} is the maximum modulus of the second derivative of the Gauss curvature with respect to arc length along a geodesic (Chapter II, §8).

Introduce normal Riemannian coordinates x^i in a neighborhood of the point O_0 in R. The line element of the space is then

$$ds^2 = \delta_{ij} dx^i dx^j + h_{ij} dx^i dx^j,$$

where the h_{ij} are functions which vanish at O_0 together with their first derivatives. The line element of the sphere is obtained by setting

$$x^1 = \delta \cos u \cos v, \quad x^2 = \delta \cos u \sin v, \quad x^3 = \delta \sin u.$$

We get

$$ds_\omega^2 = \delta^2 (du^2 + \cos^2 u \, dv^2 + \delta^2(\cdots)).$$

Since the derivatives of h_{ij} are bounded by a constant, it is not hard to derive estimates for the quantities K_1, K_2 and K_{11} appearing in the inequality for the normal curvature. The result is:

$$\kappa \leqq (1 + \epsilon(\delta))/\delta,$$

where ϵ tends to zero with δ. Since the Gauss curvature of the surface is

$$K = (1 + \epsilon_1(\delta))/\delta^2,$$

the normal curvature satisfies the inequality

$$\kappa \geqq (1 + \epsilon_2(\delta))/\delta,$$

where $\epsilon_2(\delta)$ tends to zero with δ.

We now assume δ to be so small that the minimum normal curvature of ω is greater than the maximum normal curvature of F.

Let H_0 and H_1 be the supporting functions of the surfaces ω and F. Let F_t be the surface with supporting function

$$H = (1 - t) H_0 + t H_1.$$

Establish a homeomorphism between the surfaces F_t by projection from a point S on the interior normal to ω at O.

The surfaces F_t are closed convex surfaces. We claim that if the Gauss curvature of the surface F is greater than K_0, the same holds for any surface F_t.

Let Z_0, Z_1 and Z_t be cylinders projecting the surfaces ω, F and F_t in some direction g. The normal curvatures of these cylinders at points with parallel normals satisfy the relation

$$1/\kappa_t = (1-t)/\kappa_0 + t/\kappa_1.$$

Since $\kappa_0 > \kappa_1$, it follows that $\kappa_t > \kappa_1$. Hence the Gauss curvatures of the surfaces F_t and F at points with parallel normals satisfy the inequality $K_t \geqq K_1$. Since $K_1 \geqq K_0$, this implies that $K_t \geqq K_0$.

We now define a family of metrics on F, stipulating that the distance between two arbitrary points X and Y be equal to the distance between the corresponding points on the surface F_t. We thus obtain a continuous family of Riemannian manifolds M_t. The manifold M_1 coincides with the given manifold M, while the manifold M_0 is known to be isometrically embeddable in R.

Let u^i be any coordinate system on the surface F. Let ds_t^2 be the line element of the manifold M_t in these coordinates. We claim that for any t the metric $d\sigma_t^2 = ds_t^2 - ds_0^2$ satisfies the condition $d\sigma_t^2 = 0$, $Dd\sigma_t^2 = 0$ at the point O. This follows from the fact that, under the correspondence set up by projection from the point S, the tangent plane to each of the surfaces F_t is an osculating euclidean space.

Assume that a manifold M_{t_0} is isometrically embeddable in the Riemannian space R as a regular surface Φ_{t_0}, and the following three conditions hold:

1. The point O of M_{t_0} corresponds under the isometry to the point X_0 of the space.
2. The surface Φ_{t_0} is tangent to the plane section α at X_0 and the interior normal to Φ_{t_0} is a semitangent to the geodesic γ at X_0.
3. Corresponding directions in the plane section α and on the surface Φ_{t_0} coincide.

The third condition needs explanation. The pencil of directions on F at O is mapped isometrically onto the pencil of directions in α. The pencils of directions on the manifolds M_t at O are mapped onto each other isometrically by the homeomorphism of the manifolds; this mapping is an isometry of the pencils at the point O. Thus the pencils of directions in α and on the M_t at O are mapped isometrically onto each other. This is the correspondence mentioned in condition 3.

Our problem is now to prove that when t is close to t_0 the manifold

M_t is also isometrically embeddable in R as a regular surface Φ_t satisfying conditions 1, 2 and 3.

Assuming that the surface Φ_t indeed exists and is close to Φ_{t_0}, we shall derive its equation. To this end we first define a coordinate system v^i in the neighborhood of Φ_{t_0}, which has no singularities and is the same for the entire neighborhood. This can be done, for example, as follows.

Let $\overline{\Phi}_{t_0}$ be a surface in the euclidean space E isometric to Φ_{t_0}. Map a neighborhood of the surface $\overline{\Phi}_{t_0}$ onto a neighborhood of the surface Φ_{t_0} in the Riemannian space, as follows. An arbitrary point $\overline{X}$ of the surface $\overline{\Phi}_{t_0}$ is mapped onto the point on Φ_{t_0} corresponding to it under isometry. Construct the normal to $\overline{\Phi}_{t_0}$ at $\overline{X}$ and a geodesic normal to Φ_{t_0} at X. Now define the mapping for points on these normals by associating points at equal distances from $\overline{X}$ and X, respectively, either both inside or both outside the surfaces. The coordinate system v^i in the Riemannian space induced by the cartesian coordinate system of euclidean space under this mapping obviously has no singularities. Moreover, the line element of the space along the surface Φ_{t_0} in this coordinate system has the form $\delta_{ij}\,dv^i dv^j$.

We now derive an equation for the surface Φ_t. Let X be an arbitrary point of Φ_{t_0}, $v_0^i(X)$ its coordinates, and $v^i(X)$ the coordinates of the corresponding point of Φ_t. Set $\xi^i(X) = v^i(X) - v_0^i(X)$. Substituting the functions $v^i(X)$ into the line element of the space, we get the line element of the surface Φ_t:

$$(*) \qquad g_{ij}\,dv^i dv^j = ds_t^2.$$

By Taylor's formula with remainder,

$$g_{ij} = g_{ij}^0 + \frac{\partial g_{ij}^0}{\partial v^k}\xi^k + c_{ij\alpha\beta}\xi^\alpha\xi^\beta.$$

Substituting this expression for g_{ij} into equation (*) and noting that $dv^i = dv_0^i + d\xi^i$, we get

$$(**) \qquad g_{ij}^0\,d\xi^i\,dv_0^j + g_{ij}^0\,dv_0^i\,d\xi^j + \xi^k\frac{\partial g_{ij}^0}{\partial v^k}\,dv_0^i\,dv_0^j = d\sigma_t^2 + \Omega,$$

where

$$d\sigma_t^2 = ds_t^2 - ds_{t_0}^2.$$

and

$$\Omega = a_{ij}\xi^i\xi^j + a'_{ij}\xi^i d\xi^j + a''_{ij}d\xi^i d\xi^j.$$

Note that if we equate the left-hand side of equation (**) to zero, we get the equation of an infinitesimal bending of the surface. If the right-hand side is regarded as a known quadratic form defined on the surface, we get the equation of the velocity field of a deformation inducing the given variation of the line element of the surface.

We now prove that the surface Φ_t exists for t close to t_0.

Denote

$$L\xi = \overset{0}{g}_{ij}\, d\xi^i\, dv_0^j + \overset{0}{g}_{ij}\, dv_0^i\, d\xi^j + \xi^k \frac{\partial \overset{0}{g}_{ij}}{\partial v^k}\, dv_0^i\, dv_0^j.$$

The equation of the surface Φ_t may then be expressed as

$$L\xi = d\sigma_t^2 + \Omega. \qquad (***)$$

We construct a sequence of vector fields ξ_n as follows. The field ξ_1 is a solution of the equation $L\xi = d\sigma_t^2$ such that $\xi_1 = 0$, $D\xi_1 = 0$ at the point X_0. The existence of this field is guaranteed by the lemma of §6. Now substitute the components of the field ξ_1 into Ω and define a field ξ_2 as a solution of the equation

$$L(\xi) = d\sigma_t^2 + \Omega(\xi_1),$$

such that $\xi_2 = 0$ and $D\xi_2 = 0$ at the point X_0. The existence of ξ_2 follows from the same lemma, since if $\xi_1 = 0$, $D\xi_1 = 0$ at the point X_0, then $\Omega(\xi_1) = 0$ and $D\Omega(\xi_1) = 0$ at the point. The vector fields $\xi_3, \xi_4, \cdots$ are defined similarly.

We claim that if $|t - t_0|$ is sufficiently small, the sequence of fields ξ_n converges to a solution of the equation $L\xi = d\sigma_t^2 + \Omega(\xi)$, which satisfies the condition $\xi = 0$, $D\xi = 0$ at the point X_0.

We first prove that the sequence is bounded.

Estimates were established in §6 for the norm of the velocity of deformation of a surface inducing a given variation of its metric. In other words, we have estimates for the norms of the solutions to the equation $L\xi = d\sigma^2$. Applying the results to the form $d\sigma^2 = d\sigma_t^2 + \Omega(\xi_1)$ and taking into consideration the expression of Ω in terms of ξ_1, we get

$$|\xi|_{k,\alpha} \leqq c\|d\sigma_t\|_{k,\alpha} + c(\epsilon)\|\xi_1\|_{k,\alpha}^2.$$

Obviously, as $\epsilon \to 0$ the coefficient $c(\epsilon)$ surely cannot increase, while $\|d\sigma_t\|_{k,\alpha} \to 0$ as $t \to t_0$. Hence for sufficiently small ϵ and $|t - t_0|$ we have

$$2c\|d\sigma_t^2\| < \epsilon, \quad 2\epsilon c(\epsilon) < 1.$$

If ϵ and t are chosen in this way, then, as is easily seen, for all n

$$\|\xi_n\|_{k,\alpha} < \epsilon,$$

and this proves that ξ_n is bounded.

We now show that for sufficiently small $|t-t_0|$ the successive approximations ξ_n converge. We have

$$L\xi_{n+1}=d\sigma_t^2+\Omega(\xi_n),\qquad L\xi_n=d\sigma_t^2+\Omega(\xi_{n-1}).$$

Subtracting the second equation from the first term by term, we get

$$L(\xi_{n-1}-\xi_n)=\Omega(\xi_n)-\Omega(\xi_{n-1}),$$

whence, again in view of the specific dependence of Ω on ξ, we get

$$\|\xi_{n+1}-\xi_n\|_{k,\alpha}\leqq\bar{c}(\epsilon)\,\|\xi_n-\xi_{n-1}\|_{k,\alpha},$$

where the constant $\bar{c}(\epsilon)$ can be made arbitrarily small if ϵ is sufficiently small. Take ϵ so small that $\bar{c}(\epsilon)<1$. Then the sequence ξ_n clearly converges. If $k\geqq 1$, then $\xi\equiv\lim\xi_n$ satisfies the conditions $\xi=0$ and $D\xi=0$ at the point X_0. Consequently the surface Φ_t defined by the equations $v^i=v_0^i+\xi^i$ satisfies conditions 1, 2 and 3 as to its position relative to the two-dimensional element at X_0.

We now show that if $|t-t_0|$ and $\|\xi\|_{k,\alpha}$ are sufficiently small the surface Φ_t is unique. Suppose that there are two surfaces Φ_t and $\tilde{\Phi}_t$. Then

$$L\xi=d\sigma_t^2+\Omega(\xi),\qquad L\tilde{\xi}=d\sigma_t^2+\Omega(\tilde{\xi}).$$

Subtracting the second equation from the first term by term, and noting that

$$\|\Omega(\xi)-\Omega(\tilde{\xi})\|_{k,\alpha}\leqq\bar{c}\|\xi-\tilde{\xi}\|_{k,\alpha},$$

we get

$$\|\xi-\tilde{\xi}\|_{k,\alpha}\leqq\bar{c}\|\xi-\tilde{\xi}\|_{k,\alpha}.$$

If $|t-t_0|$, $\|\xi\|_{k,\alpha}$, $\|\tilde{\xi}\|_{k,\alpha}$ are all sufficiently small, then $\bar{c}$ is small, e.g., $\bar{c}<1$, and the inequality can hold only if $\xi-\tilde{\xi}\equiv 0$. We have thus proved the following lemma.

Lemma 1. *If the manifold M_{t_0} is isometrically embeddable in R, then any manifold M_t close to M_{t_0} is also embeddable; this embedding is unique for surfaces near M_{t_0}.*

We now consider a problem which, though resembling the preceding problem from the analytical viewpoint, has a different geometric interpretation. We have proved the isometric embeddability of a manifold M_t close to M_{t_0}, retaining the same boundary condition at the point X_0; in particular, the surface Φ_t contains the point X_0 of the space.

We shall now assume that the manifold M_t is isometric to M_{t_0}, but the boundary condition at X_0 depends continuously on the parameter t. This condition can be formulated as follows.

1′. $\xi = \xi(X_0, t)$ defines a regular curve in the space, issuing from X_0 and perpendicular to the surface Φ_{t_0}.

2′. For any t close to t_0, the condition $\xi = \eta$, $D\xi = D\eta$ must hold at X_0, where $\eta(X, t)$ is a given regular vector valued function of X and t such that

$$L\eta - \Omega(\eta) = 0, \qquad D(L\eta - \Omega(\eta)) = 0,$$
$$\|\eta\|_{k,\alpha} \to 0 \quad \text{as} \quad t \to t_0$$

at the point X_0.

We claim that for sufficiently small $|t - t_0|$ there exists a solution $\xi(X, t)$ of the embedding equation $L\xi - \Omega(\xi) = 0$ which depends continuously on t and satisfies conditions 1′ and 2′. The solution $\xi(X, t)$ defines a continuous bending of the surface Φ_{t_0} under which the plane section at X_0 moves along a certain curve in the required way.

For the proof we replace ξ by a vector valued function ζ defined by $\xi - \eta = \zeta$. Since $\xi^i = v^i - v_0^i$, it follows that $\zeta^i = v^i - v_0^i - \eta^i$. Set $v_0^i - \eta^i = v_\eta^i$. Then $\zeta^i = v^i - v_\eta^i$. The equation for ζ can be set up directly, in the same way as for ξ. However, now the role of v_0^i is played by v_η^i; the equation has exactly the same form:

$$L_\eta \zeta = d\sigma_\eta^2 + \Omega_\eta(\zeta),$$

where

$$L_\eta \zeta = g_{ij}^\eta \, d\zeta^i \, dv_\eta^j + g_{ij}^\eta \, dv_\eta^i \, d\zeta^j + \zeta^k \frac{\partial g_{ij}^\eta}{\partial v^k} \, dv_\eta^i dv_\eta^j,$$

$$\Omega_\eta(\zeta) = a_{ij}^\eta \zeta^i \zeta^j + a_{ij}^{\prime\eta} \zeta^i \, d\zeta^j + a_{ij}^{\prime\prime\eta} \, d\zeta^i \, d\zeta^j,$$

and

$$d\sigma_\eta^2 = g_{ij}^\eta \, dv_\eta^i \, dv_\eta^j.$$

When $X = X_0$, we have $\zeta = 0$ and $D\zeta = 0$.

This equation is also solved by successive approximations. We construct a sequence of vector fields ζ_k as follows. ζ_1 is a solution of the equation $L_\eta \zeta = d\sigma_\eta^2$ such that $\zeta_1 = 0$, $D\zeta_1 = 0$ at X_0. This field exists, for it is the velocity field of a deformation of the surface Φ_η: $v^i = v_0^i + \eta^i$, inducing the variation $d\sigma_\eta^2$ of the line element. Note that by virtue of the conditions imposed on η we have $d\sigma_\eta^2 = 0$ and $Dd\sigma_\eta^2 = 0$ for $X = X_0$.

Each field is now defined in terms of the preceding one: ζ_n satisfies the equation

$$L_\eta \zeta_n = d\sigma_\eta^2 + \Omega_\eta(\zeta_{n-1}).$$

To prove that the sequence is bounded and convergent, we apply the same reasoning as used for ξ_n, with one reservation. The constant c in the norm estimate

$$\| \zeta_{n+1} - \zeta_n \|_{k,\alpha} \leqq \bar{c}_n \| \zeta_n - \zeta_{n-1} \|_{k,\alpha},$$

as in all other estimates of this type, depends on the field η. Therefore, to prove that the sequence ζ_n is bounded and convergent, we must impose an additional condition: $\| \eta \|_{k,\alpha} \to 0$ as $t \to t_0$.

We show that if this condition holds, then suitable estimates, implying that the successive approximations converge, may indeed be established. To this end it is clearly sufficient to show that the solution of the equation $L_\eta \zeta = d\sigma^2$ satisfies the estimate

$$\| \zeta \|_{k,\alpha} \leqq c \| d\sigma^2 \|_{k,\alpha},$$

where the constant c is the same for all η if $|t - t_0|$ is small.

We have

$$L\zeta = d\sigma^2 + (L\zeta - L_\eta \zeta).$$

Hence

$$\| \zeta \|_{k,\alpha} \leqq c' \| d\sigma^2 \|_{k,\alpha} + \epsilon(t) \| \zeta \|_{k,\alpha}.$$

Now $\epsilon(t) \to 0$ as $t \to t_0$, since $\| \eta \|_{k,\alpha} \to 0$. Therefore

$$\| \zeta \|_{k,\alpha} \leqslant \frac{c}{1 - \varepsilon(t)} \| d\sigma^2 \|_{k,\alpha}.$$

Thus, if $\| \eta \|_{k,\alpha} \to 0$ as $t \to t_0$, the norm of the solution of the equation $L_\eta \zeta = d\sigma^2$ satisfying the conditions $\zeta = 0$ and $D\zeta = 0$ at X_0, satisfies the estimate

$$\| \zeta \|_{k,\alpha} \leqq c \| d\sigma^2 \|_{k,\alpha},$$

where the constant c is independent of $d\sigma^2$ and is the same for all η if $|t - t_0|$ is sufficiently small.

Hence, if $|t - t_0|$ is small, the successive approximations ζ_n converge to a solution of the equation $L_\eta \zeta = d\sigma_\eta^2 + \Omega_\eta(\zeta)$, and this solution satisfies conditions 1′ and 2′ at the point X_0. The solution is locally unique.

To conclude this section, we construct an example of the vector field $\eta(X, t)$.

Let γ be a regular curve issuing from a point X_0 on the surface Φ_{t_0}, perpendicular to Φ_{t_0} at X_0. Define a mapping of a small neighborhood of the surface Φ_{t_0} into the euclidean space E such that:

1. the image of the surface Φ_{t_0} is a surface $\bar{\Phi}_{t_0}$ isometric to Φ_{t_0};
2. the space E is an osculating space for R along the curve γ.

It is clear that such a mapping is easily constructed. For it is well known how to construct a mapping satisfying the second condition, and we have already constructed a mapping satisfying the first condition. Thus we need only consider a regular combination of these two mappings.

Introduce a coordinate net v^i in the neighborhood of Φ_{t_0} in R, corresponding to the cartesian coordinates in E. Now let the surface $\overline{\Phi}_{t_0}$ be subjected to a parallel displacement along the curve $\overline{\gamma}$ (the image of γ) such that the point v moves along the curve. Let $\eta^i(X,t)$ denote the variation at time t of the coordinates of the point of $\overline{\Phi}_{t_0}$ corresponding to the point X of Φ_{t_0} under the isometry.

Let ds^2 be the line element of the manifold M_{t_0}, $\bar{g}_{ij}\,dv^i dv^j$ the line element of the space E, and $g_{ij}\,dv^i dv^j$ the line element of the space R. Then, on the surface $\overline{\Phi}_t$ in E obtained by the displacement of $\overline{\Phi}_{t_0}$, we have

$$\bar{g}_{ij}\,dv^i dv^j|_{\overline{\Phi}_t} - ds^2 \equiv 0.$$

Thus the line element of the corresponding surface Φ_t in the Riemannian space (defined by the same equations in v^i) satisfies the equations

$$g_{ij}\,dv^i dv^j|_{\Phi_t} - ds^2 = 0, \quad D(g_{ij}\,dv^i dv^j|_{\Phi_t} - ds^2) = 0$$

at the point $X_0(t)$ on the curve γ.

It is clear that $\|\eta\|_{k,\alpha} \to 0$ as $t \to t_0$.

We have thus proved the following lemma.

LEMMA 2. *Let F be a closed surface with positive extrinsic and intrinsic curvature, γ a regular curve issuing from a point X_0 on F and perpendicular to F.*

Then there exists a continuous bending of the surface F which subjects the plane section at X_0 to a parallel displacement along γ; this bending is unique for surfaces close to F.

§8. Partial solution of the isometric embedding problem

In this section we shall prove that, under certain assumptions concerning the curvature of the manifold, a two-dimensional closed manifold M can be isometrically embedded in a given three-dimensional Riemannian space R as a certain surface F.

We shall show that this embedding admits the same degree of freedom as an embedding in euclidean space. Under certain natural additional assumptions, the embedding is unique.

We shall also study the question of the regularity of the embedding, as a function of the regularity of the manifold M and the space R.

Let R be a three-dimensional Riemannian space with regular (four times continuously differentiable) metric. Let K_R^* be a nonnegative number with the following property: For any regular (four times continuously differentiable) closed surface F with regular (four times continuously differentiable) metric and Gauss curvature greater than K_R^*, the normal curvatures are bounded by a constant which depends only on the metrics of the space and the surface.

The existence of the number K_R^* was proved in §2. To be precise, let X be a point of the Riemannian space R, and K_1 the maximum and K_2 the minimum sectional curvature in a plane section at X; then

$$K_R^* = \max_R \max\{K_1, 3(K_1 - K_2)\}.$$

Theorem 1. *Let R be a three-dimensional Riemannian space with regular (k times differentiable, $k \geqq 6$) metric, and G a compact domain in R containing a ball of radius d about a point O. Let M be a two-dimensional closed Riemannian manifold with regular (k times differentiable) metric ds^2, intrinsic diameter d and Gauss curvature greater than K_R^*.*

Then the manifold M admits an isometric embedding into R as a regular ($k-2$ times differentiable) surface Φ satisfying the following condition.

A given point S of the surface coincides with the point O of the space. A pencil of directions α_S at S coincides with a given pencil of directions α_0 at O. A given order of succession of directions in α_0 forms a right-handed screw with the interior normal to Φ.

To prove this theorem, we construct a continuous family of manifolds M_t, as in the preceding section, containing the given manifold $M \equiv M_1$ and a manifold M_0 which is known to be embeddable (a sphere of small radius tangent to the plane section spanned by the pencil α_0). The method by which the manifolds M_t are constructed will ensure that their metrics are $k-2$ times differentiable.

By the lemma of the preceding section, if the manifold M_{t_0} is embeddable, then all manifolds sufficiently close to it are embeddable; moreover, if the surface Φ_{t_0} isometric to M_{t_0} is $k-2+\alpha$ times differentiable ($0 < \alpha < 1$), then for t close to t_0 the surfaces Φ_t are also $k-2+\alpha$ times differentiable.

Thus the set ω of all t such that the manifold M_t is embeddable is open. We now show that this set is also closed.

In this connection we make two observations concerning the manifolds M_t. First, the Gauss curvature of each manifold is greater than K_R^*, since the minimum Gauss curvature of the manifold $M \equiv M_1$ is less than the minimum Gauss curvature of any M_t, by construction. Second, the intrinsic diameter of any manifold M_t is less than that of M. Let us prove this statement.

By construction, the manifolds M and M_t can be embedded in euclidean space as surfaces F and F_t, with F_t contained in the interior of F. Let X_t and Y_t be two points on F_t. Let X and Y be the points on F at which the exterior normals to F_t at X_t and Y_t cut F. Join X_t and Y_t by a segment γ_t on F_t, X and Y by a segment γ on F. By Busemann's theorem the length of γ is at least that of γ_t. Hence the intrinsic diameter of F_t cannot exceed that of F, and is therefore at most d.

Let $t_1, t_2, \cdots$ be a sequence of elements of ω converging to t_0. We want to prove that t_0 is in ω, i.e. M_{t_0} is embeddable.

Let M_{t_n} be embedded in R as a surface F_n. Since the diameter of F_n is less than d and passes through O, it is contained in the domain G. Furthermore, since the Gauss curvature of F_n is greater than K_R^*, it follows that the derivatives of the space coordinates with respect to the surface coordinates of order up to $k-2$, and also the Hölder relations for the $(k-2)$th derivatives, satisfy estimates which depend only on the metric of the space (in G) and the metric of the surface (§§2 and 3). These estimates may obviously be assumed uniform in t_n (i.e. the same for all surfaces F_n).

It follows, first, that the sequence F_n contains a convergent subsequence, and second, that the limit surface F_0 is $k-2+\alpha$ times differentiable. The surface F_0 is obviously isometric to M_{t_0} and touches the plane section spanned by the pencil of directions α_0 at O in the required manner.

The set ω is thus closed. Since ω is both closed and open, it must be the entire interval $0 \leqq t \leqq 1$. In particular, $1 \in \omega$, i.e. M is embeddable. Q.E.D.

COROLLARY. *If the manifold M and the space R have analytic metrics, the embedding of M into R is analytic, i.e. the surface Φ whose existence is ensured by the theorem is analytic.*

Indeed, by the theorem, the space coordinates are infinitely differentiable with respect to the surface coordinates. Since, moreover, the space coordinates satisfy an elliptic system of differential equations (§3), they are analytic functions of the surface coordinates, as required.

THEOREM 2. *The isometric embedding of the manifold M into the space R whose existence is ensured by Theorem 1 is unique.*

Suppose that, besides the surface Φ whose existence was proved in

Theorem 1, there exists another surface Φ' isometric to M and touching the plane section spanned by the pencil α_0 at the point O.

The manifold M_1 of the family M_t admits a realization Φ'. Hence, by the lemma of §6, manifolds close to M_1 admit realizations Φ_t' close to Φ'.

By continuously decreasing t, we can construct a continuous family of surfaces Φ_t', at least for t near 1. We claim that this continuous family can be extended down to $t=0$.

Suppose that the continuity breaks off at $t=t_0$. Let Φ_{t_0}' be the limit surface of Φ_t' as $t\to t_0$ from above. Then the family Φ_t' $(t_0<t<1)$ is continuous, and by the lemma of §6 it can be extended down to t_0. We thus construct a continuous family of surfaces Φ_t' $(0\leqq t\leqq 1)$ which contains the surface Φ' (for $t=1$). Similarly, starting from the surface Φ, we construct a family of surfaces Φ_t $(0\leqq t\leqq 1)$ containing the surface Φ (for $t=1$).

Suppose that the surfaces Φ_{t_0} and Φ_{t_0}' coincide. Then, since for t close to t_0 the surfaces Φ_t and Φ_t' are unique in the class of surfaces close to $\Phi_{t_0}\equiv\Phi_{t_0}'$, they must also coincide. Hence the surfaces Φ and Φ' coincide, which is impossible. Thus the surfaces Φ_0 and Φ_0' are also different.

Let Q be a point near O on the normal to the pencil α_0, outside the surfaces Φ and Φ'. If Q is sufficiently close to O, it will also lie outside the surfaces Φ_t and Φ_t'.

Deform the space R in a small neighborhood of the point Q in such a way that it remains regular and becomes euclidean near Q. This is easily done as follows.

Introduce normal Riemannian coordinates in the neighborhood of Q. The line element of the space is

$$ds^2=\delta_{ij}\,dv^i dv^j+h_{ij}\,dv^i dv^j,$$

where the h_{ij} are functions which vanish with their first and second derivatives at the point Q. Now define a function $\mu(\rho)$, where ρ is the distance from Q, by the following conditions:

1) $\mu(\rho)$ is k times differentiable;
2) $\mu(\rho)=0$ for $\rho<\delta/2$;
3) $\mu(\rho)=1$ for $\rho<\delta$.

Now define the metric of the space in the neighborhood of Q by the line element

$$ds^2=\delta_{ij}\,dv^i dv^j+\mu(\rho)\,h_{ij}\,dv^i dv^j.$$

It is clear that the space remains regular to the same degree (the metric is k times differentiable), and it is euclidean in the $\delta/2$-neighborhood of Q.

Without loss of generality, we may assume that the Gauss curvatures of the surfaces Φ_0 and Φ_0' are greater than the quantity K_G^* for the deformed space, and their intrinsic diameters are small.

Now displace the pencil α_0 parallel to itself along the normal at Q. This corresponds to the continuous bending of the surfaces Φ_0 and Φ_0' established in the lemma of §7. Thus we again obtain two continuous families Φ_τ and Φ_τ'.

Since the intrinsic diameters of the surfaces Φ_0 and Φ_0' may be assumed arbitrarily small, it follows that when α_0 reaches the point Q the surfaces Φ_τ and Φ_τ' lie entirely within the euclidean neighborhood of Q, and therefore coincide. Now, reasoning as before, we conclude that the surfaces Φ_0 and Φ_0' coincide, which is impossible. This completes the proof of the theorem.

THEOREM 3. *Let Φ and Φ' be two isometric closed surfaces in a domain G of the Riemannian space. Assume that these surfaces have Gauss curvature greater than K_G^* and intrinsic diameters less than δ. Let G contain a curve connecting two points of Φ and Φ' which correspond to each other under the isometry and lie at distances greater than δ from the boundary of G.*

Then there exists a continuous bending of Φ into Φ'.

This theorem follows in an obvious manner from Theorem 2.

Theorem 1 (isometric embedding of closed Riemannian manifolds in Riemannian spaces) yields various theorems on embedding of not necessarily closed manifolds.

THEOREM 4. *Let G be a compact domain in a Riemannian space, containing a ball of radius $2d$. Let M be a two-dimensional Riemannian manifold, homeomorphic to a disk, such that*

1) *the intrinsic diameter of M is less than d;*
2) *the Gauss curvature of M is everywhere greater than K_G^*;*
3) *the geodesic curvature of the boundary of M is everywhere positive.*

Then there exists an isometric embedding of M into G as a surface Φ, such that if the metrics of the space and the manifold are k times differentiable ($k \geqq 6$), the surface Φ is at least $k-2$ times differentiable.

If the metrics M and G are analytic, then Φ is an analytic surface.

Take two copies of the manifold M and "glue" them together along their boundaries, identifying points that correspond under isometry. The result is a closed manifold homeomorphic to a sphere. Unfortunately, it need not be regular, since there may be singularities along the common boundary of the copies of M.

For this reason we glue the manifolds M together not directly but through an intermediate regular strip $\overline{M}$ of width δ. There is no difficulty in defining the metric on $\overline{M}$ in such a way that its contact with the metric of the copies of M is regular (k times differentiable) and its Gauss curvature is greater than K_G^*.

A practical way to define the metric on the strip is as follows. Introduce semigeodesic coordinates u, v along the boundary of M in such a way that the u-curves are geodesics perpendicular to the boundary. The parameter v is arc length along the boundary. The line element of M along the boundary is

$$ds^2 = du^2 + \bar{g}\,dv^2.$$

The metric is continued to the strip beyond the boundary of M in the same form. The choice of the function g is governed by symmetry conditions on the straight line $u = \delta/2$, the required regularity of the contact with the function $\bar{g}(u, v)$ along the curve $u = 0$, and the condition

$$(\sqrt{g})_{uu}/\sqrt{g} > K_G^*,$$

which holds for $\bar{g}(u, v)$.

By Theorem 1, the manifold M constructed by this gluing procedure is isometrically embeddable into G as a $k-2$ times differentiable surface $\tilde{\Phi}$. This surface contains a domain isometric to M. The proof is complete.

THEOREM 5. *Let R be a three-dimensional Riemannian space and M a two-dimensional Riemannian manifold. Let X be an arbitrary point of R, Y an arbitrary point on M.*

Then, if the Gauss curvature of M at Y is greater than K_X^*, a sufficiently small neighborhood of the point Y on M is isometrically embeddable as a surface Φ in a given neighborhood of X in R.

If the metrics of the manifold and the space are k times differentiable ($k \geqq 6$), then Φ is $k-2$ times differentiable.

If the metrics of the manifold and the space are analytic, then Φ is an analytic surface.

This theorem follows from Theorem 4, applied to a geodesic disk of sufficiently small radius about the point Y.

§9. More about estimates for normal curvature on a closed convex surface

One of the central points of the proof of the isometric embedding theorem for two-dimensional Riemannian manifolds homeomorphic to a sphere is the derivation of a priori estimates for the normal curvatures of a closed convex surface, homeomorphic to a sphere, in a Riemannian space. The estimates obtained in §2 do not furnish a complete solution to the problem. In a certain sense, a complete solution requires estimates of the normal curvatures of a surface under the sole assumption of positive extrinsic curvature, i.e. the Gauss curvature of the surface is everywhere greater than the sectional curvature of the space in the plane sections tangent to the surface.

In this section we shall solve the problem of estimates for the normal curvatures of a convex surface, in a form which yields the complete solution of the embedding problem. We shall prove the following theorem.

THEOREM. *If the extrinsic curvature of a closed regular surface, homeomorphic to a sphere, in a complete Riemannian space with nonpositive curvature, is strictly positive, then the normal curvatures of the surface are bounded by a constant which depends only on the metrics of the surface and the space.*

Before we proceed to the proof, a remark is in order concerning the statement of the theorem. It is well known that in euclidean space positive extrinsic curvature is a sufficient condition for a surface to be homeomorphic to a sphere. In Riemannian spaces of nonpositive curvature this is in general false, as is easily seen from the following example.

In a hyperbolic space of curvature $K_R = -1$, let A_1A_2 be a straight-line segment; construct three planes perpendicular to the segment: σ_1 and σ_2 through the endpoints, σ_0 through the midpoint. Let R' denote the region of the space bounded by the planes σ_1 and σ_2. Set up a pointwise correspondence between σ_1 and σ_2, associating points which are symmetric with respect to the plane σ_0. This correspondence is clearly an isometry. Now identify corresponding points of the planes σ_1 and σ_2; then R' becomes a complete Riemannian space of curvature -1. The straight-line segment A_1A_2 is a closed geodesic γ in this space.

Let F be the locus of the points of R' whose distance from γ is ρ. F is obviously homeomorphic to a torus. By the symmetry of our construction, F admits a transitive group of motions into itself, and consequently has constant Gauss curvature. Since the total curvature of a surface homeomorphic to a torus is zero, F has zero Gauss curvature and so its extrinsic curvature is unity, i.e. positive.

Thus the requirement that the surface be homeomorphic to a sphere does not follow from the fact that the extrinsic curvature is positive. The foregoing example will convince the reader that the requirement is at the same time essential. Indeed, the surface F is analytic and its extrinsic curvature is unity. However, no estimate can be established for the normal curvatures of F in terms of the metrics of the space and the surface, since as ρ decreases the surface F is always locally isometric to a euclidean plane and the maximum normal curvature increases as $1/\rho$.

We now outline the proof of the theorem.

Without loss of generality we may assume that the space is simply connected. Otherwise we go over to the universal covering space. It is well known that a simply connected Riemannian space with non-positive curvature is homeomorphic to a euclidean space. Any two of its points can be connected by a unique geodesic, which is a segment (shortest join). As in euclidean space, a closed surface with positive extrinsic curvature bounds a convex body, and is thus a convex surface in the usual sense.

Let F be a closed convex surface with regular metric and positive extrinsic curvature; we wish to establish estimates for its normal curvatures.

We shall first show that the interior of the body bounded by F contains a point O whose distance from the surface is bounded below by a constant ρ_0, which depends only on the metrics of the surface and the space.

To prove this, take any point O_0 inside the surface F. Let K_0 and K_1 be the minimum and maximum extrinsic curvatures of F, $K_0 > 0$, and d its intrinsic diameter. Consider the set of all convex surfaces containing O_0 in their interior, with extrinsic curvatures bounded between K_0 and K_1, and diameter at most d.

If our assertion is false, then for any n we can find a surface F_n in the set whose interior contains no ball of radius $1/n$. The sequence F_n contains a convergent subsequence. Its limit F_0 is obviously degenerate; it is either a convex domain on a completely geodesic surface, a subarc of a geodesic, or a point. In the first case, F_0 contains points with zero extrinsic curvature; but this is absurd, for the extrinsic curvatures of the surfaces F_n are bounded below by the positive number K_0. In the second case, for sufficiently large n the surfaces F_n contain points, near the endpoints of the geodesic segment F_0, at which the extrinsic curvature is arbitrarily large, which is impossible. The same applies when F_0 degenerates to a point. Thus, in all cases we have a contradiction, and the assertion is proved.

We now introduce polar geodesic coordinates in the space R, with center at O, and define a function $\bar{w}(X)$ on the surface by

$$\bar{w}(X) = \bar{\kappa}(X)/(\cos\varphi(X))^{\mu},$$

where $\bar{\kappa}(X)$ is the maximum normal curvature of the surface at X,

$\varphi(X)$ is the angle that the geodesic ray issuing from O and passing through X forms with the exterior normal to the surface, and μ is some positive constant.

We claim that $\varphi < \pi/2 - \epsilon$, where ϵ is a positive constant depending only on ρ_1 and ρ_0, the maximum and minimum distances of the points of the surface from the pole O (in the space metric). Indeed, let X be an arbitrary point on F. Connect it by segments (shortest joins) to the points of a ball of radius ρ_0 about O. These segments lie in the interior of F and fill out a solid cone with apex at X. Any geodesic issuing from X and forming an angle smaller than some constant α with the geodesic XO is in the cone, and hence within the body bounded by the surface F. The angle α can be estimated from below in terms of the distance OX, the radius ρ_0, and quantities depending on the metric of the space. Obviously $\varphi \leqq \pi/2 - \alpha$. This proves our assertion.

Since $\cos\varphi \geqq \sin\alpha > 0$, the function $\bar{w}(X)$ attains an absolute maximum at some point X_0 on F. For suitable choice of the constant μ the quantity $\bar{w}(X_0)$ is bounded by a number w_0 which depends only on the metrics of the surface and the space. With this estimate available, we get an estimate for the normal curvatures κ at any point of the surface and in any direction; namely, $\kappa \leqq w_0$.

In §3 we derived a system of equations for the space coordinates of a point on a surface with given line element ds^2, as functions of the surface coordinates u^i (the analog of the Darboux equation). In this section these equations will be extensively used, and therefore we recall some of the notation.

Let R be a Riemannian space, referred to a coordinate system v^i. Let F be a surface in R, and u^1, u^2 a coordinate net on F. The space coordinates v^i of a point on the surface are certain functions of the surface coordinates u^i:

$$v^i = v^i(u^1, u^2), \qquad i = 1, 2, 3.$$

Let e_i be the basis vectors at a point x of the space. Then

$$x_k = \partial x/\partial u^k = e_i \partial v^i/\partial u^k.$$

Let e_i^* denote the vectors at x defined by $e_i^* e_k = \delta_{ik}$. Set

$$\frac{\partial^2 v^\beta}{\partial u^i\,\partial u^j} = v_{ij}^\beta, \qquad \Gamma_{rs}^\beta \frac{\partial v^r}{\partial u^i}\frac{\partial v^s}{\partial u^j} - \tilde{\Gamma}_{ij}^k \frac{\partial v^\beta}{\partial u^k} = A_{ij}^\beta,$$

where Γ_{ij}^k are the Christoffel symbols of the space and $\tilde{\Gamma}_{ij}^k$ the Christoffel symbols of the surface. Denoting the coefficients of the second fundamental form of the surface by λ_{ij} $(i, j - 1, 2)$, we now get the following formulas:

$$v_{ij}^k + A_{ij}^k = \lambda_{ij}(ne_k^*) \qquad i,j = 1,2,$$

where n is the unit normal vector of the surface.

The scalar product (ne_k^*) is given by

$$(ne_k^*) = \frac{1}{\sqrt{g\bar{g}}} \begin{vmatrix} g_{i\beta}\dfrac{\partial v^\beta}{\partial u^1}, & g_{i\alpha}\dfrac{\partial v^\alpha}{\partial u^2} \\ g_{j\beta}\dfrac{\partial v^\beta}{\partial u^1}, & g_{j\alpha}\dfrac{\partial v^\alpha}{\partial u^2} \end{vmatrix},$$

where $\bar{g}$ and g are the discriminants of the quadratic forms $d\tilde{s}^2$ and ds^2 (the line elements of the surface and the space), and the indices i, j are distinct from each other and from k.

Finally, let K_i denote the Gauss curvature of the surface, K_R the sectional curvature of the space in a plane section tangent to the surface, and $K_e = K_i - K_R$ the extrinsic curvature of the surface.

We can now write down a system of equations for the space coordinates v^i of a point on the surface as functions of the intrinsic coordinates u^1 and u^2 on the surface:

$$(v_{11}^k + A_{11}^k)(v_{22}^k + A_{22}^k) - (v_{12}^k + A_{12}^k)^2 = K_e\bar{g}(ne_k^*)^2.$$

In our outline of the proof of the theorem, we defined the auxiliary function

$$\bar{w}(X) = \bar{\kappa}(X)/(\cos\varphi(X))^\mu.$$

To abbreviate the following arguments we set $\bar{w} = \sigma\bar{\kappa}$.

Let X_0 be a point on F at which the function $\bar{w}$ assumes its maximum. Define a semigeodesic parametrization u^1, u^2 of the surface in a neighborhood of X_0. The curve $u^1 = 0$ is a geodesic γ through X_0 in the direction of minimum normal curvature. The curves $u^2 = \text{const}$ are geodesics perpendicular to γ; the curves $u^1 = \text{const}$ are the orthogonal trajectories of the latter. The parameters u^1 and u^2 are arc length along the coordinate curves through X_0. In this parametrization the line element of the surface is

$$d\tilde{s}^2 = (du^1)^2 + c^2(du^2)^2.$$

At the point X_0 we have $c = 1$, $c_u = c_v = 0$, $c_{uv} = c_{vv} = 0$, $c_{uu} = -K_i$. All the Christoffel symbols $\tilde{\Gamma}_{ij}^k$ for this parametrization of the surface vanish at X_0.

Now let $\kappa(X)$ denote the normal curvature of the surface at a point X in the direction of the u^1-curve; consider the function $w = \sigma\kappa$. Since

$\kappa(X) \leqq \bar{\kappa}(X)$, with equality at X_0, it follows that w also assumes a maximum at X_0. In order to estimate $\bar{w}(X)$, therefore, it suffices to estimate $w(X_0)$.

We define a semigeodesic parametrization of the space in a neighborhood of X_0 as follows. As the coordinate surface $v^3=0$ we take a geodesic surface Φ through X_0, tangent to the surface F. Define a semigeodesic parametrization v^1, v^2 on the surface Φ, based on two geodesics issuing from X_0 in the principal directions of the surface F. Now define the coordinates v^1, v^2, v^3 of a point of the space near X_0 to be the signed distance of the point from Φ (v^3) and the coordinates v^1, v^2 of the foot of the geodesic perpendicular dropped from the point onto the surface Φ. Using the formulas for the Christoffel symbols in a parametrization of this type (§2), we easily see that all the Christoffel symbols vanish at the point X_0.

In this parametrization of the space, e_3 is a unit vector and is equal to e_3^*. Therefore (ne_3^*) is the cosine of the angle between the surface F and the coordinate surface $v^3=\text{const}$. It follows that (ne_3^*) is the ratio of discriminants of the two forms $d\tilde{s}^2-(dv^3)^2$ and $d\tilde{s}^2$, i.e.

$$(ne_3^*)^2 = 1-\left(\frac{\partial v^3}{\partial u^1}\right)^2-\frac{1}{c^2}\left(\frac{\partial v^3}{\partial u^2}\right)^2.$$

To underline the analogy with the arguments of Chapter V (§8) for caps in a hyperbolic space, we set

$$A_{11}^3=\alpha,\quad A_{12}^3=\beta,\quad A_{22}^3=\gamma,\quad -K_e\tilde{g}(ne_3^*)^2=\delta.$$

The first and second derivatives of the function $v^3(u^1,u^2)$ will be denoted by p, q, r, s, t. Then the bending equation for $k=3$ is similar in form to the Darboux equation:

$$(*)\qquad (r+\alpha)(t+\gamma)-(s+\beta)^2+\delta=0.$$

As we indicated above, the Christoffel symbols of the surface vanish at X_0. Hence at the point X_0 we have $\alpha=\beta=\gamma=0$ and $\delta=-K_e$, and equation (*) at this point is $rt-s^2=K_e$. The coefficients of the second fundamental form of the surface at X_0 are $\lambda_{11}=r$, $\lambda_{12}=s$ and $\lambda_{22}=t$. Since the coordinate curves on the surface at X_0 are conjugate, we have $\lambda_{12}=0$. Therefore, at X_0,

$$s=0,\quad t=K_e/r=O(1/r).$$

Differentiating equation (*) with respect to u^1 at X_0, we get

$$(r_1+\alpha_1)t+(t_1+\gamma_1)r+\delta_1=0.$$

Hence

$$t_1 + \gamma_1 = -(\delta_1 + (r_1 + \alpha_1)t)/r.$$

Since the Christoffel symbols of both surface and space vanish at X_0, the derivatives α_1, β_1 and γ_1 are of order $O(1)$, while $\delta_1 = (K_e)_1$. Therefore

$$t_1 + \gamma_1 = -t(r_1 + \alpha_1)/r + O(1/r).$$

Now differentiate equation (*) twice with respect to u^1 at X_0. We get

$$(r_{11} + \alpha_{11})(t + \gamma) + (t_{11} + \gamma_{11})(r + \alpha) + 2(r_1 + \alpha_1)(t_1 + \gamma_1) \\ - 2(s + \beta)(s_{11} + \beta_{11}) - 2(s_1 + \beta_1)^2 + \delta_{11} = 0.$$

Noting that $t_{11} = r_{22}$, $s_{11} = r_{12}$, $s = 0$ and $\alpha = \beta = \gamma = 0$, and substituting the expression for $t_1 + \gamma_1$, we have

$$(r_{11} + \alpha_{11})t + (r_{22} + \gamma_{22})r - 2(r_1 + \alpha_1)\left(\frac{t}{r}(r_1 + \alpha_1) + O\left(\frac{1}{r}\right)\right) \\ - 2(r_2 + \beta_1)^2 + \delta_{11} = 0.$$

As we mentioned above, the first derivatives of α, β and γ at the point X_0 are of order $O(1)$, since the Christoffel symbols vanish there. For the same reason, the second derivatives of α, β and γ are of order at most $O(r)$. Indeed, these derivatives are linear in the derivatives v_{ij}^k, while the latter are of first order in λ_{ij}, i.e. at most $O(r)$. In view of this we can rewrite our equation as follows:

$$(r_{11} + \alpha_{11})t + (r_{22} + \alpha_{22})t - 2(r_1 + \alpha_1)\frac{t}{r} - 2(r_2 + \alpha_2)^2 + O\left(\frac{r_1}{r}\right) \\ + O(r_2) + O(r^2) = 0.$$

Into this equation we insert the auxiliary function $w = \sigma\kappa$. Since $\kappa = \lambda_{11}$, while

$$r + \alpha = \lambda_{11}(n\overset{*}{e}_3), \quad (n\overset{*}{e}_3) = \left(1 - p^2 - \frac{q^2}{c^2}\right)^{\frac{1}{2}}$$

it follows that

$$r + \alpha = \frac{w}{\sigma}\Delta^{\frac{1}{2}}, \quad \Delta = 1 - p^2 - \frac{q^2}{c^2}.$$

The function w assumes a maximum at X_0, and so $w_1 = 0$ and $w_2 = 0$. Hence, at the point X_0,

$$(r + \alpha)_1 = w\left(\frac{1}{\sigma}\right)_1, \qquad (r + \alpha)_2 = w\left(\frac{1}{\sigma}\right)_2,$$

$$(r + \alpha)_{11} = \frac{w_{11}}{\sigma} + w\left(\left(\frac{1}{\sigma}\right)_{11} + \frac{r^2}{\sigma}\right), \qquad (r + \alpha)_{22} = \frac{w_{22}}{\sigma} + w\left(\left(\frac{1}{\sigma}\right)_{22} + \frac{t^2}{\sigma}\right).$$

Substituting these expressions for the derivatives of $r + \alpha$ into the above equality, we get

$$\frac{1}{\sigma}(w_{11}t+w_{22}r)+wt\left(\frac{1}{\sigma}\right)_{11}+wr\left(\frac{1}{\sigma}\right)_{22}-2w^2\left(\frac{1}{\sigma}\right)_1^2\frac{t}{r}-2w^2\left(\frac{1}{\sigma}\right)_2^2$$
$$+O\left(\frac{w}{r}\left(\frac{1}{\sigma}\right)_1\right)+O\left(w\left(\frac{1}{\sigma}\right)_2\right)+O(r^2)=0$$

After regrouping terms and substituting w for σr, we can rewrite this equality as

$$\frac{1}{\sigma}(w_{11}t+w_{22}r)-\frac{K_e\sigma_{11}}{\sigma}-\frac{r^2\sigma_{22}}{\sigma}+O\left(\frac{\sigma}{r}\left(\frac{1}{\sigma}\right)_1\right)+O\left(\sigma r\left(\frac{1}{\sigma}\right)_2\right)+O(r^2)=0.$$

Since the function w has a maximum at X_0, it follows that $w_{11}t+w_{22}r\leqq 0$ at this point. Hence

$$-\frac{K_e\sigma_{11}}{\sigma}-\frac{r^2\sigma_{22}}{\sigma}+O\left(\frac{\sigma}{r}\left(\frac{1}{\sigma}\right)_1\right)+O\left(\sigma r\left(\frac{1}{\sigma}\right)_2\right)+O(r^2)\geqslant 0.$$

Henceforth we call this the fundamental inequality.

Let us calculate the derivatives of σ. Define a polar geodesic coordinate system in a neighborhood of the point X_0, with pole O. To do this, take a sphere S_0 through X_0, with center O, and define a semigeodesic parametrization in the neighborhood of X_0 on the sphere, based on two geodesics through X_0 in the principal directions of S_0. The coordinates of a point X near X_0 are now defined as the distance v^3 from O and the coordinates v^1, v^2 of the intersection of the geodesic ray OX with the surface S_0. Since we shall have to deal with two systems of space coordinates, all quantities referred to the coordinate system introduced previously will be indicated by a bar. Sometimes the coordinate v^3 in the polar system will be denoted by h.

Now the coefficients of the second fundamental form of the surface in the parametrization u^1, u^2 are preserved (up to sign) when we go from one space parametrization to another; hence

$$(**)\qquad \frac{\bar{v}_{ij}^k+\bar{A}_{ij}^k}{(n\bar{e}_k^*)}=-\frac{v_{ij}^k+A_{ij}^k}{(ne_k^*)}$$

The minus sign in the right-hand side is due to the choice of directions $v^3>0$ and $\bar{v}^3>0$ in our coordinate systems.

Now consider the derivatives of the function

$$\sigma=1/\Delta^\mu,\quad\text{where}\quad \Delta=(1-h_1^2-h_2^2/c^2).$$

Note that in the polar geodesic system

$$\cos^2\varphi=1-h_1^2-h_2^2/c^2.$$

We have

$$(1/\sigma)_1=-2\mu\Delta^{\mu-1}(h_1h_{11}+h_2h_{12}).$$

Setting $k=3$ in (**), we get

$$h_{11}=O(r), \quad h_{12}=O(1), \quad h_{22}=O(1).$$

Therefore $(1/\sigma)_1=O(r)$. Similarly,

$$(1/\sigma)_2=-2\mu(h_1h_{12}+h_2h_{22})=O(1).$$

Now consider the second derivatives σ_{11} and σ_{22}. We begin with σ_{11}:

$$\sigma_{11}=\mu(\mu+1)\Delta^{-\mu-2}\Delta_1^2-\mu\Delta^{-\mu-1}\Delta_{11},$$
$$\Delta_1=-2(h_1h_{11}+h_2h_{12}).$$

As was shown above, h_{11} is of order $O(r)$, and h_{12} is $O(1)$. To find the principal part of h_{11} with respect to r we use (**) with $k=3$, $i=j=1$. At the point X_0,

$$r=-(h_{11}+A_{11}^3)/\Delta^{1/2}.$$

Hence

$$h_{11}=-\Delta^{1/2}r+O(1).$$

Consequently

$$\Delta_1=2h_1\Delta^{1/2}r+O(1).$$

We now calculate Δ_{11} at X_0:

$$\Delta_{11}=-2(h_{11}^2+h_{12}^2+h_1h_{111}+h_2h_{112})+O(1).$$

The derivative h_{111} is found by differentiating (**) for $i=j=1$, $k=3$. Note that

$$\bar{A}_{11}^3=O(1), \quad (n\bar{\overset{*}{e}}_3)=O(1), \quad (A_{11}^3)_1=O(r),$$
$$(n\overset{*}{e}_3)_1=-\bar{\Delta}^{1/2}(h_1h_{11}+h_2h_{12})=h_1r+O(1).$$

Therefore

$$h_{111}=-h_1r^2-\Delta^{1/2}r_1+O(r).$$

But

$$r_1=w\left(\frac{1}{\sigma}\right)_1+O(1),$$
$$\left(\frac{1}{\sigma}\right)_1=-2\mu\Delta^{\mu-1}(h_1h_{11}+h_2h_{12})=2\mu\Delta^{\mu-1/2}h_1r+O(1).$$

Hence

$$r_1=2\mu\Delta^{-1/2}h_1r^2+O(r).$$

Consequently

$$h_{111}=-(1+2\mu)h_1r^2+O(r).$$

The derivative h_{221} is found by differentiating (**) with respect to u^1 for $i=j=2$, $k=3$. The result is

$$h_{221} = -\Delta^{1/2}t_1 + O(r).$$

But

$$t_1 = -K_e r_1/r^2 + O(1).$$

Therefore $h_{221} = O(1)$.

Inserting the values of the derivatives of h into Δ_{11}, we get

$$\Delta_{11} = -2\Delta r^2 + 2(2\mu+1)h_1^2 r^2 + O(r).$$

Using Δ_1 and Δ_{11}, we finally find σ_{11}:

$$\sigma_{11} = 2\mu\Delta^{-\mu}r^2 + 2\Delta^{-\mu-1}h_1^2 r^2 + O(r).$$

The only essential point about the second term of this equality is that it is nonnegative. Hence σ_{11} can be expressed as

$$\sigma_{11} = 2\mu\Delta^{-\mu}r^2 + (*)^2 + O(r).$$

Now consider the derivative σ_{22}. We have

$$\sigma_{22} = \mu(\mu+1)\Delta^{-\mu-2}\Delta_2^2 - \mu\Delta^{-\mu-1}\Delta_{22},$$
$$\Delta_2 = -2(h_1h_{12} + h_2h_{22}).$$

Using (**) at the point X_0, we get

$$h_{12} = -A_{12}^3, \quad h_{22} = -A_{22}^3 + O\left(\frac{1}{r}\right),$$
$$A_{12}^3 = \Gamma_{ij}^2 v_1^i v_2^j, \quad A_{22}^3 = \Gamma_{ij}^3 v_2^i v_2^j.$$

The Christoffel symbols Γ_{ij}^3 have the following values at X_0 (§2):

$$\Gamma_{ij}^3 = 0 \quad \text{for} \quad i \neq j, \quad \Gamma_{11}^3 = a_{11}, \quad \Gamma_{22}^3 = a_{22}, \quad \Gamma_{33}^3 = 0,$$

where a_{11} and a_{22} are the principal curvatures of the spherical surface $v^3 = \text{const}$ at the point X_0. Thus,

$$A_{12}^3 = a_{11}v_1^1 v_2^1 + a_{22}v_1^2 v_2^2, \quad A_{22}^3 = a_{11}(v_2^1)^2 + a_{22}(v_2^2)^2.$$

Suppose that a_{11} is the smaller of the principal curvatures. Then, since

$$(v_2^1)^2 + (v_2^2)^2 + h_2^2 = 1,$$

if follows that

$$A_{22}^3 = a_{11}(1 - h_2^2) + (a_{22} - a_{11})(v_2^2)^2.$$

Consequently, $A_{22}^3 > a_{11}\Delta$. Setting $A_{22}^3 = a$, $A_{12}^3 = b$ for brevity, we can write $\Delta_2 = 2(ah_2 + bh_1)$, where a is bounded below, provided a_{11} does not vanish. Note that a and b do not depend on μ.

We now calculate Δ_{22}:

$$\Delta_{22} = -2(h_{12}^2 + h_{22}^2 + h_1 h_{221} + h_2 h_{222}).$$

Hence

$$\Delta_{22} = -2(h_1 h_{221} + h_2 h_{222}) - (*)^2.$$

The third derivatives of h are found with the aid of (**). At the point X_0,

$$h_{221} = -(A_{22}^3) - t_1 \Delta^{1/2} - t(\Delta^{1/2})_1.$$

Since $(A_{22}^3)_1$ involves the second derivatives of v^k only as v_{12}^k, and these are $O(1)$, we get $(A_{22}^3)_1 = O(1)$, and this estimate is independent of μ. Moreover, $t(\Delta^{1/2})_1 = O(1)$, and this estimate is also independent of μ. Now consider $t_1 \Delta^{1/2}$. We have

$$t_1 = -K_e r_1 / r^2 + O(1), \quad r_1 = 2\mu \Delta^{-1/2} h_1 r^2 + O(r).$$

Hence

$$t_1 \Delta^{1/2} = -2\mu K_e h_1 + O(1).$$

Adding the various terms appearing in h_{221}, we get

$$h_{221} = 2\mu K_e h_1 + O(1),$$

where the estimate $O(1)$ is independent of μ.

The third derivative h_{222} is determined in a similar manner. The final result is

$$h_{222} = O(1) + \mu O(1/r),$$

where $O(1)$ is independent of μ.

We can now write Δ_{22} as

$$\Delta_{22} = -(*)^2 - 4\mu K_e h_1^2 + A h_1 + B h_2 + O(1/r),$$

where A and B satisfy estimates independent of μ.

Finally, we determine σ_{22}:

$$\sigma_{22} = \mu(\mu+1)\Delta^{-\mu-2}\Delta_2^2 - \mu \Delta^{-\mu-1}\Delta_{22}.$$

Substituting the values of Δ_2 and Δ_{22}, we get

$$\sigma_{22} = \frac{1}{\Delta^{\mu}} \left\{ \frac{4\mu^2}{\Delta^2}(ah_2 + bh_1)^2 + \frac{4\mu^2 K_e h_1^2}{\Delta} + \mu A' h_1 + \mu B' h_2 + (*)^2 + O\left(\frac{1}{r}\right) \right\}.$$

Now recall the fundamental inequality, for the point X_0 of F at which w assumes a maximum:

$$\frac{-K_e\sigma_{11}}{\sigma}-\frac{r^2\sigma_{22}}{\sigma}+O\left(\frac{\sigma}{r}\left(\frac{1}{\sigma}\right)_1\right)+O\left(\sigma r\left(\frac{1}{\sigma}\right)_2\right)+O(r^2)\geqslant 0.$$

Substituting the expressions obtained for the derivatives of σ, we get

$$\Delta^{-\mu}\left\{-2\mu K_e r^2-\frac{4\mu^2}{\Delta^2}(ah_2+bh_1)^2-\frac{4\mu^2K_eh_1^2}{\Delta}+\mu A'h_1+\mu B'h_2\right\} + Cr^2+(*)\geqslant 0,$$

where C is a bounded expression independent of μ, and (*) is an expression which increases more slowly than r^2 with r. Divide this inequality by Δ^μ and replace r by the function $w=r\Delta^{-\mu}$. The result is

$$(***)\qquad -2\mu K_e w^2-\frac{4\mu^2w^2}{\Delta^2}\{(ah_2+bh_1)^2+K_e\Delta h_1^2\} +\mu w^2(A'h_1+B'h_2)+C\Delta^\mu w^2+(*)\geqslant 0,$$

where terms of less than second order in w have been omitted.

The quantity $|a|$ is bounded away from zero. This is equivalent to the existence of a positive lower estimate for the normal curvature of the spherical surface S_0 ($v^3=h=\text{const}$) through the point X_0. By construction, the coordinate curves v^1 and v^2 at X_0 on the surface S_0 are parallel to the principal directions. To fix ideas, suppose that the minimum normal curvature of S_0 corresponds to the v^1-direction at X_0. Since the v^1-curve through X_0 is a geodesic on the spherical surface S_0, it follows that the normal curvature a_{11} is the geodesic curvature of the curve $v^3=\text{const}$ on the coordinate surface $v^2=\text{const}$ passing through the point X_0. The line element of this surface is

$$ds^2=(dv^3)^2+\bar{c}^2(dv^1)^2.$$

Since the surface $v^2=\text{const}$ is generated by geodesics of the space R, its extrinsic curvature is nonpositive, and so its Gauss curvature is also nonpositive. Denoting $v^3=u$ and $v^1=v$, we have

$$ds^2=du^2+c^2dv^2.$$

The Gauss curvature is $\bar{K}=-\bar{c}_{uu}/\bar{c}$. Hence, since $\bar{K}\leqq 0$, we have $\bar{c}_{uu}\geqq 0$.

The geodesic curvature of the curve $u=\text{const}$ is given by $K_g=\bar{c}_u/\bar{c}$. Thus we can find a positive lower estimate for the normal curvature $|a_{11}|$ which depends only on such an estimate for $|\bar{c}_u|$. But $\bar{c}_u=1$ at the point $u=0$, and since $\bar{c}_{uu}\geqq 0$, it follows that $\bar{c}_u\geqq 1$ for all u. Thus

$$|a_{11}|=|\bar{c}_u/\bar{c}|\geqq 1/\bar{c}.$$

Thus the existence of a positive lower estimate for $|a_{11}|$, and hence also for $|a|$, is evident.

Now let a_0, b_0, c_0 be positive numbers such that

$$a_0 \leqq |a|, \quad |b| \leqq b_0, \quad K_e\Delta \geqq c_0.$$

Consider the quadratic form

$$\omega = (ah_2 + bh_1)^2 + K_e\Delta h_1^2$$

for $h_1^2 + h_2^2 \geqq 1$. Let ϵ_0 be a positive number such that

$$\epsilon_0 < \tfrac{1}{2}, \quad a_0/2 - b_0\epsilon_0 > a_0/3.$$

Obviously, if $h_1 \leqq \epsilon_0$, then

$$\omega > (a_0/3)^2 + (*)^2,$$

and if $h_1 \geqq \epsilon_0$, then

$$\omega \geqq c_0\epsilon_0^2 + (*)^2.$$

Let ω_0 denote the smaller of the two numbers $c_0\epsilon_0^2$ and $(a_0/3)^2$. Then, in either case, $\omega > \omega_0 > 0$.

Consequently if $h_1^2 + h_2^2 > \epsilon$, then

$$(ah_2 + bh_1)^2 + K_e\Delta h_1^2 > \epsilon^2\omega_0.$$

The conditions defining the number ϵ_0 bound it only from above:

$$\epsilon_0 \leqq \min\{\tfrac{1}{2}, a_0/6b_0\}.$$

Defining constants K_0 and D such that

$$0 < K_0 < K_e, \quad |A'| \leqq D, \quad |B'| \leqq D,$$

we now require that ϵ_0 also satisfy the condition $2\epsilon_0 D \leqq K_0$, which can clearly be done without violating the previous conditions:

$$0 < \varepsilon_0 \leqslant \frac{1}{2}, \qquad \varepsilon_0 \leqslant \frac{1}{6}\frac{a_0}{b_0}.$$

Then, if $h_1^2 + h_2^2 \leqq \epsilon^2$,

$$-K_e + A'h_1 + B'h_2 \leqq 0.$$

We now define the positive constant μ figuring in the expression for w by $w = \kappa/\Delta^\mu$. First we take the number μ so large that when $h_1^2 + h_2^2 \geqq \epsilon$

$$-\frac{4\mu^2}{\Delta^2}(ah_2 + bh_1)^2 - \frac{4\mu^2 K_e h_1^2}{\Delta} + \mu A'h_1 + \mu B'h_2 \leqslant 0.$$

This condition holds if

$$-\mu^2\epsilon^2\omega_0 + 2\mu D\epsilon \leqq 0,$$

i.e. if $\mu > \mu_0 = 2D/\epsilon\omega_0$.

Finally, we require that μ be so large that $\mu K_0 > C+1$.

We now claim that our inequality (***) can be brought to the form $-w^2+(*) \geqq 0$, where (*) denotes a function of less than second order in w. Indeed, at the point X_0 of F where w assumes a maximum, we have either $h_1^2+h_2^2>\epsilon^2$ or $h_1^2+h_2^2 \leqq \epsilon^2$. In the first case,

$$\frac{-4\mu^2w^2}{\Delta^2}\left\{(ah_2+bh_1)^2+K_e\Delta h_1^2\right\}+\mu w^2(A'h_1+B'h_2)\leqslant 0$$
$$-\mu K_e w^2+C\Delta^\mu w^2\leqslant -w^2,$$

and so $-w^2+(*) \geqq 0$. In the second case,

$$-\frac{4\mu^2w^2}{\Delta^2}\{(ah_2+bh_1)^2+K_e\Delta h_1^2\}\leqslant 0,$$
$$-\mu K_e w^2+\mu w^2(A'h_1+B'h_2)\leqslant 0,$$
$$-\mu K_e w^2+C\Delta^\mu w^2\leqslant -w^2$$

and so $-w^2+(*) \geqq 0$ again.

Now, since (*) is of order lower than w^2, it follows from $-w^2+(*) \geqq 0$ that w cannot be arbitrarily large, and it is therefore bounded from above. The estimate for w implies an estimate for the normal curvatures of the surface, since $\kappa \leqq w$.

In conclusion, we remark that the above proof of the theorem on estimation of the normal curvatures carries over without modifications to the case in which, instead of requiring that R be a complete space, we require that the surface F be the boundary of a region homeomorphic to a ball.

§10. Complete solution of the isometric embedding problem

In this section we present the complete solution to the problem of isometrically embedding a two-dimensional Riemannian manifold homeomorphic to a sphere in a three-dimensional Riemannian space. We shall prove the following theorem:

THEOREM 1. *Let R be a complete three-dimensional Riemannian space and M a closed Riemannian manifold, homeomorphic to a sphere, with Gauss curvature everywhere greater than a constant c (positive, negative or zero). If the curvature of R is everywhere less than c, then there exists an isometric embedding of M into R as a regular surface F.*

Moreover, the embedding can be accomplished in such a way that a given plane section M (a point S and a pencil of directions at S) coincides with a given plane section α', isometric to α, in the space R, and the surface F lies on a specified side of α'.

If the metrics of R and M are k times differentiable ($k \geqq 6$), *then F is at least* $k-1$ *times differentiable.*

If the metrics of R and M are analytic, then F is analytic.

We shall first prove Theorem 1 for a space R with nonpositive curvature ($c \leqq 0$), proceeding thence to the general case. The general lines of the proof are the same as in §8. It includes three steps:

1. There exists a continuous family of manifolds M_t, containing the given manifold M and a manifold M_0 known to be embeddable.
2. If the manifold M_t is embeddable, then all manifolds in the family close to M_t are also embeddable.
3. The final step: if each manifold M_{t_n} in the family is embeddable and $t_n \to t^*$, then M_{t^*} is embeddable.

With regard to the proofs of these three statements, note the following. The proof in §8 that an "almost embeddable" manifold M_t is embeddable was essentially based on the fact that the Gauss curvature of M_t is greater than the curvature of the space. Therefore, if we can construct the family M_t in such a way that the Gauss curvatures of the manifolds are greater than c, we may consider step 2 as accomplished.

The proof in step 3 that the limit manifold M_{t^*} is embeddable is based only on the existence of a priori estimates for the normal curvatures of the surfaces F_t isometric to M_t. By the theorem of §9, the existence of such estimates is guaranteed if the Gauss curvatures of the surfaces are greater than the curvature of the space. Thus step 3 requires no further proof if we can construct the manifolds M_t subject to this restriction.

Thus the proof will be complete if we can construct a continuous family of manifolds M_t with Gauss curvatures greater than c, which contains the given manifold M and a manifold M_0 known to be embeddable in R. We proceed to the construction.

Let γ be a semigeodesic issuing from a point O_R in the space R, perpendicular to the plane of the pencil α_R, such that the order of directions in the pencil forms a right-handed screw with the direction of γ at O_R. Let S be a point on γ close to O_R, and ω a spherical surface of radius SO_R about the point S. This surface is regular (at least k times differentiable) and convex. Hence its Gauss curvature is greater than the sectional curvature of the space in a plane section tangent to ω; in particular, it is greater than c. Denote the pencil of directions on ω coinciding with α_R by α_ω.

Now consider a hyperbolic space of curvature c and let $\bar{F}$ and $\bar{\omega}$ be convex regular surfaces in this space, isometric to M and ω, respectively.[3) That this construction is possible follows from Aleksandrov's theorem on the regularity of a convex surface with regular metric in hyperbolic space (Chapter V, §8). Without loss of generality we may assume that the pencils α of the surfaces $\overline{F}$ and $\bar{\omega}$ coincide, and the surfaces themselves lie on the same side of the plane of α. Otherwise we need only reflect one of the surfaces in the plane of α.

Now map the hyperbolic space containing $\overline{F}$ and $\bar{\omega}$ geodesically onto the interior of a euclidean ball (the Cayley-Klein interpretation of hyperbolic space). The surfaces $\overline{F}$ and $\bar{\omega}$ are thereby mapped onto closed convex surfaces in euclidean space in the interior of the ball, with common point O and common tangent plane at O. Denote these surfaces by $\widetilde{F}$ and $\tilde{\omega}$, respectively.

Let H_F and H_ω be the supporting functions of the surfaces $\widetilde{F}$ and $\tilde{\omega}$. Define a family of convex surfaces $\widetilde{F}_t$ by their supporting functions:

$$H_t = tH_F + (1-t)H_\omega, \qquad 0 \leqq t \leqq 1.$$

This family contains the surfaces $\widetilde{F}$ $(t=1)$ and $\tilde{\omega}$ $(t=0)$.

Let Q be a point on the interior normal to $\widetilde{F}$ and $\tilde{\omega}$ at the point O, in the interior of both surfaces (and therefore inside all the surfaces $\widetilde{F}_t$). We now define a family of metrics M_t on the surface $\widetilde{F}$, as follows. Let X and Y be two arbitrary points on $\widetilde{F}$; project X and Y onto the surface $\widetilde{F}_t$, and call the corresponding points X_t and Y_t. Define the distance between X and Y in the metric M_t to be the distance between X_t and Y_t on $\widetilde{F}_t$ in the hyperbolic metric.

Since the surface $\widetilde{F}_t$ is strictly convex for any t $(0 \leqq t \leqq 1)$, the Gauss curvature of the manifold with metric M_t is greater than the curvature of the hyperbolic space, i.e. greater than c. We have thus constructed a continuous family of manifolds M_t containing the given manifold M and an embeddable manifold M_0 (isometric to ω), such that the Gauss curvature of each M_t is everywhere greater than c. This proves Theorem 1 for nonpositive c.

Now consider the general case. First, note that in steps 1 and 2 of the proof the assumption that R has nonpositive curvature is not essential. It is essential only in step 3, where it was used to establish estimates for the normal curvature of a surface F_t isometric to M_t. Thus to prove Theorem 1 it is sufficient to show that the limit M_{t^*} of

---

3) If $c = 0$, consider a euclidean space.

embeddable manifolds is embeddable, without assuming that the space R has nonpositive curvature. We may assume here that $c > 0$, since the case $c \leqq 0$ has already been treated.

Let S be the point in R which is the center of the pencil α'. Let Ω be a ball of radius $\pi/\sqrt{c}$ about S, and construct a covering Riemannian manifold R_0 for Ω. To do this, consider a ball Ω_0 of radius $\pi/\sqrt{c}$ in euclidean space and set up an isometric correspondence between the directions in R at S and the directions in euclidean space at the center S_0 of Ω_0.

Let X be an arbitrary point of the ball Ω_0. Join it to the center S_0 by a straight-line segment. Lay off a subarc of length S_0X on a geodesic issuing from S in the corresponding direction. Denote the endpoint of this arc by X'. This defines a mapping of Ω_0 onto Ω in R. This mapping is locally homeomorphic, since the curvature of R is less than c and therefore the geodesic issuing from S contains no points associated with S at distances less than $\pi/\sqrt{c}$. Define a Riemannian metric in the interior of Ω_0, with the distance between close points X and Y equal to the distance between their images in R. The resulting Riemannian manifold R_0 is a covering manifold for R in Ω. To define an isometric embedding of M into R, it will suffice to define an isometric embedding of M into R_0 and then to map R_0 into R as indicated. We may thus confine our attention to an embedding of M into the covering manifold R_0.

Let M_0 be a manifold in the family M_t which is known to be embeddable. By what we have proved, manifolds M_t close to M_0 are also embeddable. The corresponding surfaces F_t are obtained by continuous deformations from F_0. We claim that each of the surfaces F_t bounds a body V_t homeomorphic to a ball which is cut at most once by any geodesic ray issuing from S_0. This is obvious for F_0, which is a small sphere. Since the family F_t is continuous, the property can fail to be valid only if for some t the arc of a geodesic ray connecting a point A on F_t to S_0 is tangent to the surface at A. But this is impossible: this arc would lie at least partly outside the body bounded by F_t, since the latter is strictly convex.

Now all the surfaces F_t lie within a ball of radius $\pi/\sqrt{c'}$, $c' > c$, about S_0. Indeed, the Gauss curvature of the surface F_t is strictly greater than c, and hence greater than some $c' > c$, since the domain of the parameter t is closed. The intrinsic diameter of a surface with Gauss curvature greater than $c' > 0$ is less than $\pi/\sqrt{c'}$. And since the surface F_t contains the point S_0, it is contained in a ball of radius $\pi/\sqrt{c'}$ about S_0.

Let t^* denote the supremum of all parameter values t such that the manifold M_t is isometrically embeddable in R_0 for any $t < t^*$. We claim that the manifold M_{t^*} is embeddable. To prove this, it suffices to show that for small $|t-t^*|$ there exists an estimate for the normal curvatures of F_t which is independent of t. Suppose that the manifold M_{t^*} is not embeddable in R_0 as a regular surface, and so there can be no estimates for the normal curvatures of surfaces F_t (with t close to t^*) independent of t. This means that for any $n>0$ we can find a surface F_{t_n} such that $|t-t_n|<1/n$ and the maximum normal curvature is greater than n. Suppose this maximum is assumed on F_{t_n} at a point P_n. Without loss of generality we may assume that the sequence of points P_n converges to a point P_0.

Let G_0 be a ball about P_0, with radius so small that any two of its points can be connected by a unique segment (shortest join) in R_0. Let ω_n be the intersection of F_{t_n} and G_0. ω_n is clearly a convex surface: it is a domain on the boundary of the intersection of V_{t_n} and G_0, which is a convex body. Without loss of generality we may assume that the sequence of convex surfaces ω_n converges to some general convex surface ω_0. This surface has positive Gauss curvature. Hence, as in the euclidean case, it follows that it is strictly convex, i.e. contains no geodesic segments of R_0.

Join an arbitrary point A on the surface ω_0 to the point P_0. Let $\vartheta(A)$ be the angle between the interior normal to ω_0 at A and the direction of the segment AP_0. Since ω_0 is strictly convex, $\vartheta(A)<\pi/2$, and for all points A whose distance from P_0 is at least $\epsilon>0$ we have $\vartheta(A)<\pi/2-\eta(\epsilon)$, where $\eta(\epsilon)>0$.

P_0 cannot be a conic point of ω_0. Hence there exists a geodesic γ_0 through P_0 tangent to the surface ω_0. Let γ' be a semigeodesic issuing from P_0, within the body whose boundary contains ω_0, such that the angle between γ' and the geodesic γ_0 at P_0 is less than $\eta(\epsilon)/2$. Let P' be a point on γ'. Let P' be sufficiently close to P_0; replace the latter by P' in the definition of the angle $\vartheta(A)$. Then, for points A whose distance from P_0 is greater than ϵ, this angle will still be less than $\pi/2-\eta(\epsilon)$. Since ω_n converges to ω_0, it follows that for sufficiently large n the angle $\vartheta(A)$ defined for the surface ω_n and point P' will also be less than $\pi/2-\eta(\epsilon)$ for $AP_0>\epsilon$, and if $A\equiv P_n$ the angle will be greater than $\pi/2-\eta(\epsilon)/2$.

We now define the following function on the surface ω_n:

$$\bar{w}(X)=\bar{\kappa}(X)/(\cos\vartheta(X))^{\mu},$$

where $\bar{\kappa}(X)$ is the maximum normal curvature at the point S, $\vartheta(X)$ is the angle between the interior normal to the surface and the direction of the semigeodesic XP' at X, and μ is a positive constant. For ω_n sufficiently close to ω_0, any point X such that $XP_0 > \epsilon$ satisfies the inequality $\bar{w}(X) < \bar{w}(P_n)$. Indeed, by the definition of P_n we have $\bar{\kappa}(X) \leqq \bar{\kappa}(P_n)$. Furthermore, $\vartheta(X) < \pi/2 - \eta(\epsilon)$, whereas $\vartheta(P_n) > \pi/2 - \eta(\epsilon)/2$. Hence $\vartheta(X) < \vartheta(P_n)$, and so $\bar{w}(X) < \bar{w}(P_n)$.

Since $\bar{w}(X) < \bar{w}(P_n)$ for $XP_0 > \epsilon$, the maximum of the function $\bar{w}$ is assumed at a point X_0 whose distance from P_0 is at most ϵ, and hence whose distance from P' is at most $\epsilon + P_0P'$.

Now define a polar geodesic coordinate system in the neighborhood of the point P', and repeat word for word the proof of the theorem on estimates in §9. In this proof the fact that the curvature of the space is nonpositive was used only in demonstrating that the extrinsic curvature of the spherical surface $v^3 = \text{const}$ through the point X_0 is positive. This now holds independently of the curvature of the space, since the radius $P'X_0$ of this sphere is small.

The estimate for $\bar{w}$ is also an estimate for $\bar{\kappa}(P_n)$. But by assumption $\bar{\kappa}(P_n) > n$, and we have a contradiction. Thus the manifold M_{t^*} is embeddable, and therefore so is the given manifold M. This completes the proof of Theorem 1.

Remark. It is clear from the proof of Theorem 1 that the completeness of the space R is inessential. It is sufficient that the ball of radius d (the intrinsic diameter of the manifold M) about the point S be compact. Equivalently, the distance of the point S from the boundary of R must be greater than d.

Dubrovin has proved that the part of Theorem 1 relating to regular embedding may be improved by using Heinz's results on estimates for the solutions of elliptic Monge-Ampere equations. Namely, if the metrics of R and M are in class C^n, $n \geqq 3$, then the surface F whose existence is proved in Theorem 1 is in class $C^{n-1+\nu}$, $0 < \nu < 1$.

Our complete solution of the problem of estimates for the normal curvatures yields sharpened versions of other theorems of §8.

Theorem 2. *The surface F whose existence is asserted in Theorem 1 is uniquely determined by the plane section α'.*

Theorem 3. *If F_1 and F_2 are two regular isometric surfaces homeomorphic to a sphere, with Gauss curvature greater than c, in a complete Riemannian space whose sectional curvature is everywhere less than c, then F_1 can be continuously bent into F_2.*

§11. Isometric mappings of a punctured space into a euclidean space

Let F be a closed regular convex surface in a euclidean space. The surface F' obtained from F by deleting a finite number of points $P_1, \cdots, P_n$ is said to be punctured at these points. In this section we consider the following question: Do there exist nontrivial isometric mappings of a punctured space?

Let F be a closed regular convex surface with positive curvature, punctured at a single point S. We claim that this surface admits no nontrivial isometric mappings. Indeed, let F_1 be a regular surface isometric to F. Let $\overline{F}_1$ be the convex hull of F_1. There are two possibilities: the point S_1 corresponding to S lies either on $\overline{F}_1$ or inside $\overline{F}_1$. We first consider the second case.

Since S_1 is inside $\overline{F}_1$, it follows that $\overline{F}_1$ is a smooth convex surface. Every point at which $\overline{F}_1$ is strictly convex is a point of F_1. And since the total curvature of $\overline{F}_1$, equal to 4π, is concentrated on the set of points of strict convexity, it follows that all points of F_1 must be in $\overline{F}_1$. But points close to S_1 must lie within $\overline{F}_1$. This is a contradiction.

Now consider the second case: S_1 is on the convex hull $\overline{F}_1$. Assume that the curvature of the surface $\overline{F}_1$ at S_1 is zero. Then, reasoning as before, we conclude that the entire surface F_1 must lie on $\overline{F}_1$ and it is therefore a convex surface. Therefore, if we complete the surfaces F and F_1 by the points S and S_1, respectively, we get two isometric closed convex surfaces, which we know are congruent. Therefore F and F_1 are also congruent.

Now suppose that the convex hull $\overline{F}_1$ has nonzero curvature at S_1, and hence S_1 is a conic point of $\overline{F}_1$. Let α_0 be a supporting plane of $\overline{F}_1$ through the point S_1, such that S_1 is the only common point of α_0 and $\overline{F}_1$. The existence of such a plane follows from the fact that S_1 is a conic point.

Now let α be an arbitrary plane parallel to α_0, cutting the surface F_1. Let G be the set of all points on F which correspond under the isometry to points on F_1 lying in the halfspace defined by α which does not contain S_1. Let G' be a connected component of G and G_1' the corresponding domain on F_1. The domain G_1' is a convex surface whose boundary is in the plane α (Chapter IX, §5).

Now let the plane α approach α_0. Then the surface G_1' becomes a closed surface punctured at S_1. Indeed, since S_1 is the only common point of $\overline{F}_l$ and the plane α_0, the boundary of the surface G_1', which lies in the plane α, contracts to the point S_1 as $\alpha \to \alpha_0$. Hence as $\alpha \to \alpha_0$ the domain G' extends through the entire surface F, and F_1 is a closed convex surface punctured at S_1. We now conclude, as before, that the surfaces F and F_1 are congruent.

Thus, *a closed convex surface punctured at one point admits no nontrivial isometric mappings.*

The situation is different when the surface is punctured at two points. For example, consider a sphere punctured at two antipodal points S and S'. Divide the sphere into n congruent lunes by meridians m_k connecting S and S'. Form a closed surface from each lune by gluing the two sides together. The result is a set of spindle-shaped surfaces of revolution with constant curvature. Combine these surfaces in such a way that the glued lines coincide, and cut them along the glued lines. Now glue the lunes together again, in the order in which they appeared on the surface of the sphere, and delete the vertices. The result is a surface isometric to the sphere punctured at the points S and S'. Since the number n in this construction is arbitrary, there exist at least countably many nontrivial isometric images of the sphere punctured at two points.

A similar construction is valid for any closed surface of revolution punctured at the poles.

The following question arises naturally. Let F be an arbitrary closed convex surface punctured at two arbitrary points. Does F admit nontrivial isometric mappings? The answer to this question is in the affirmative, as the following theorem shows.

THEOREM. *Any regular closed convex surface with positive curvature, punctured at any two points, admits at least a countably infinite number of nontrivial isometric mappings in the class of regular surfaces.*

The proof will employ a specially constructed Riemannian space of nonpositive curvature.

Introduce polar coordinates ρ, ϑ in the plane and construct a function $\varphi(\rho)$ such that

1. $\varphi(\rho)$ is sufficiently regular,
2. $\varphi(\rho) = 0$ for $\rho \geqq \epsilon > 0$,
3. $\varphi(\rho) < 0$ for $\rho < \epsilon$.

Such a function is readily constructed.

Now consider the differential equation

$$g'' + \alpha\varphi(\rho)g = 0,$$

where α is a positive parameter. Let $g(\rho,\alpha)$ denote the solution of this equation satisfying the following conditions at $\rho=0$:

$$g(0,\alpha)=0, \qquad g'(0,\alpha)=1.$$

The function $g(\rho,\alpha)$ is obviously nonnegative, vanishes only at $\rho=0$, and is linear for $\rho \geqq \epsilon$.

Define a metric in the plane with polar coordinates by the line element

$$ds^2=d\rho^2+g^2d\vartheta^2.$$

The resulting Riemannian manifold has Gauss curvature $K=-g''/g=\alpha\varphi \leqq 0$. The integral curvature of the manifold is

$$\omega=\int\int \alpha\varphi g\, d\rho\, d\vartheta=-\int\int g''\, d\rho\, d\vartheta=-2\pi(g'(\varepsilon,\ \alpha)-1).$$

If $\alpha=0$, then $g'(\epsilon,\alpha)=1$, and hence $\omega=0$. For sufficiently large α it is obvious that $g'(\epsilon,\alpha)$ is arbitrarily large, and therefore ω is arbitrarily large in absolute value. Choose the parameter α in such a way that $\omega=-2\pi(n-1)$, where n is a positive integer.

Now introduce cylindrical coordinates ρ, ϑ, h in euclidean space, and define a metric by the line element

$$ds^2=d\rho^2+g^2d\vartheta^2+dh^2.$$

Denote the Riemannian space thus defined by R. What can we say about the curvature of R?

R is symmetric with respect to any surface $h=\text{const}$. Indeed, the mapping of R onto itself defined by taking the point with coordinates ρ, ϑ, $h+\zeta$ onto the point with coordinates ρ, ϑ, $h-\zeta$ is an isometry. It follows that any surface $h=\text{const}$ is completely geodesic. Hence the sectional curvature of the space in any plane section $dh=0$ coincides with the Gauss curvature of the surface $h=\text{const}$. Now the line element of a surface $h=\text{const}$ is $d\rho^2+g^2d\vartheta^2$, and thus its Gauss curvature is $K=-g''/g=\alpha\varphi\leqq 0$. Thus the sectional curvature of the space R in a plane section $dh=0$ is nonpositive, and it vanishes for $\rho\geqq\epsilon$.

Similarly, the space R is symmetric with respect to any surface $\vartheta=\text{const}$. Consequently these surfaces are completely geodesic and the sectional curvature of the space R in plane sections $d\vartheta=0$ is equal to the Gauss curvature of the surfaces. The surface $\vartheta=\text{const}$ has a euclidean line element $d\rho^2+dh^2$, and therefore its Gauss curvature is zero. Thus the sectional curvature of R in plane sections $d\vartheta=0$ is zero.

Finally, consider the sectional curvature of R in plane sections $d\rho=0$. To this end, note that any surface $\rho=\text{const}$ is isometric to a euclidean cylinder and therefore has zero Gauss curvature. In addition, this surface has nonzero extrinsic curvature, since the geodesics h on the surface are lines of curvature, owing to the symmetry of the space with respect to the planes $\vartheta = \text{const}$. Since the extrinsic and intrinsic curvatures of the surface $\rho = \text{const}$ vanish, and therefore are equal, it follows that the sectional curvature of the space R in plane sections tangent to this surface (i.e. $d\rho = 0$) is zero.

Since the space R is symmetric relative to the surfaces $h = \text{const}$ and $\vartheta = \text{const}$, it follows that the directions of the h- and ϑ-coordinate curves, and therefore also the ρ-coordinate curves, are the principal directions of the Riemann indicatrix. Thus the sectional curvature of the space R in plane sections corresponding to the principal directions of the Riemann indicatrix is nonpositive. It follows that it is nonpositive in any direction.

Since the sectional curvature of the space in the principal directions $d\rho = 0$, $d\vartheta = 0$, $dh = 0$ vanishes for $\rho \geqq \epsilon$, it follows that the space R is locally euclidean in the region R_0 defined by the condition $\rho > \epsilon$.

Our Riemannian space R is thus a space of nonpositive curvature, and is locally euclidean in the region R_0.

We now construct a special locally euclidean space E. Introduce cartesian coordinates x, y, z and cut the space along the halfplane $y = 0$, $x \geqq 0$. Take n copies of spaces $E_1, \cdots, E_n$ cut in this way, and glue them together along the cuts in such a way that the cuts of E_i and E_{i+1} adjacent to the halfspaces $y > 0$ and $y < 0$, respectively, are glued together. The result is a locally euclidean space E with a singularity along the z-axis. Let E_0 denote the region in E consisting of all points whose distance from the z-axis is greater than ϵ'.

It is obvious that the locally euclidean spaces R_0 and E_0 are isometric if the geodesic curvature of the edge $\rho = \epsilon$ of the surface $h = \text{const}$ is $1/\epsilon'$. And since the geodesic curvature of the edge $\rho = \epsilon$ of the surface $h = \text{const}$ is $k_g = g'(\epsilon, \alpha)/g(\epsilon, \alpha)$, we have the following condition for isometry of the spaces R_0 and E_0:

$$g'(\epsilon, \alpha)/g(\epsilon, \alpha) = 1/\epsilon'.$$

We claim that ϵ' tends to zero with ϵ. In fact,

$$g'(\epsilon, \alpha) = n, \quad g'(\rho, \alpha) \leqq n, \quad g(0, \alpha) = 0.$$

Hence $g(\epsilon, \alpha) \leqq n\epsilon$ and

$$g'(\epsilon, \alpha)/g(\epsilon, \alpha) \geqq 1/\epsilon.$$

Consequently $\epsilon' < \epsilon$, and so ϵ' indeed tends to zero as $\epsilon \to 0$.

We can now prove the theorem.

Let F be a regular closed convex surface with positive curvature in euclidean space, S and S' two arbitrary points on F. Let R be the Riemannian space of nonpositive curvature constructed above. By the theorem on isometric embedding of a two-dimensional Riemannian manifold into a three-dimensional Riemannian space of nonpositive curvature, the surface F can be isometrically embedded in R, in such a way that the point S is mapped onto a given point A of the space and the pencil of directions α_F at S on F is mapped onto a given pencil of directions isometric to α_F at A, and the order of the directions in the pencil forms a right-handed screw with the direction of the interior normal to the surface.

Let t_1 and t_2 be two perpendicular directions at the point S on F. Take a point A on the s-axis ($\rho = 0$) of the space R, and let t_1' and t_2' be two perpendicular directions at A. By the above-mentioned theorem, F can be isometrically embedded in R in such a way that S is mapped onto A, the directions t_1 and t_2 are mapped onto t_1' and t_2', and the directions t_1', t_2' form a right-handed screw with the direction t_3' of the interior normal to the surface. Let F' be the surface in R onto which F is mapped.

Let α_1, α_2, α_3 denote the cosines of the angles between the direction of the s-axis in R and the directions t_1', t_2', t_3', respectively. The surface F' obtained by isometrically embedding F into R cuts the s-axis of the space at two points, if $\alpha_3 \neq 0$. One of these points is A, by construction; denote the other by B. Let $S(B)$ be the point on F that corresponds to B by the isometry of F onto F'. By virtue of the symmetry of the space relative to the geodesic s, the position of the point $S(B)$ depends only on the angles between the direction of s at A and the directions t_1', t_2', t_3', i.e. on α_1, α_2, α_3.

We claim that the directions t_1' and t_2' can be so chosen that the s-axis cuts the surface F' in a point B corresponding under the isometry to the point S' on F, i.e. $S(B) \equiv S'$.

As we mentioned above, the position of $S(B)$ on F is uniquely determined by the numbers α_1, α_2, α_3. We can thus define a mapping T of the set of directions $(\alpha_1, \alpha_2, \alpha_3)$ onto the set of points of F:

$$(\alpha_1, \alpha_2, \alpha_3) \rightarrow S(B).$$

This mapping is continuous. Let G^* denote the closed domain, homeomorphic to a disk, defined in the set of directions $(\alpha_1, \alpha_2, \alpha_3)$ by the condition $\alpha_3 \geqq \epsilon^* > 0$.

Since the normal curvatures of the surface F' satisfy an estimate which is independent of the choice of the directions t_1' and t_2', the image of the boundary of G^* under the mapping T is for sufficiently small ϵ^* a closed curve separating the points S and S'. It follows that G^* contains a direction corresponding to the point S' under the mapping T. But this means that the surface F can be isometrically embedded in R in such a way that S and S' are mapped onto points on the s-axis of the space, S coinciding with A. Henceforth we shall assume that F has been embedded in R in this way.

The space R depends on the parameter ϵ. Let $\epsilon \to 0$. Then R becomes a locally euclidean space E, and F' becomes a surface F_0' which is regular everywhere except at two points S_0 and S_0' on the axis of E. Every ray issuing from a point O in E on the segment S_0S_0' cuts F_0' in one point.

Map the locally euclidean space E onto the euclidean space E_0 from which E is obtained, by associating with a point X of E the geometrically equal point of E_0. This mapping is locally isometric. It takes the surface F_0' into a surface F_0'' isometric to F_0', and therefore isometric to F. Since the space E was formed by combining n copies of the space E_0, the surface F_0'' is cut in n points (not necessarily geometrically distinct) by every ray issuing from an interior point of the segment S_0S_0'.

We have thus proved the existence of a nontrivial isometric mapping of the surface F punctured at two points S and S'. Because of the above-mentioned geometrical property (n points of intersection with a ray), this mapping depends essentially on the integer parameter n. Hence there exist at least a countably infinite number of different surfaces isometric to the surface F punctured at two points. Q.E.D.

§12. Rigidity of closed surfaces, not homeomorphic to a sphere, in a Riemannian space

Consider a continuous deformation of a surface F in a Riemannian space, such that the transform of the surface at time t is a surface F_t. As usual, this deformation is called an infinitesimal bending if the lengths of all curves on the surface are stationary at $t = 0$. With any infinitesimal bending one associates the vector field

$$\xi = dx(t)/dt|_{t=0},$$

where $x(t)$ is the point of F_t into which the deformation takes a point x on F. This vector field is known as the bending field.

It was proved in §3 that any closed surface in a Riemannian space, homeomorphic to a sphere, with positive extrinsic curvature, anchored at one point together with a pencil of directions at this point, is rigid in the sense that any bending field as described above is identically zero.

In this section we shall prove analogous theorems for closed surfaces with positive extrinsic curvature, not homeomorphic to a sphere. In this connection we first show that in a Riemannian space there may exist surfaces with positive extrinsic curvature homeomorphic to any closed oriented two-dimensional manifold. As we know, such manifolds are characterized by their genus p. A manifold of genus 1 is homeomorphic to a torus; a manifold of genus $p > 1$ is homeomorphic to a sphere with p handles. A manifold of genus p can be constructed by identifying parallel and oppositely oriented sides of a regular polygon with $2(p+1)$ sides.

In §9 we constructed a closed surface homeomorphic to a torus, with positive extrinsic curvature. Let us recall this construction. Let A_1A_2 be a straight-line segment in a hyperbolic space and construct three planes perpendicular to the segment: σ_1 and σ_2 through the endpoints, and σ through the midpoint. Let R' denote the region of the space enclosed between the planes σ_1 and σ_2. If we identify points of the planes σ_1 and σ_2 which are symmetric relative to the plane σ, the region R' becomes a complete Riemannian space of constant negative curvature. The segment A_1A_2 is a closed geodesic γ in this space. Let F denote the locus of the points of R equidistant from γ. F is obviously homeomorphic to a torus. It admits a transitive group of motions into itself, induced by motions in R, and therefore has constant (zero) Gauss curvature. Since the curvature of the space R is negative, the extrinsic curvature of the surface F, which is equal to the difference between its Gauss curvature and the sectional curvature of the space, must be positive.

We now construct a closed surface with positive extrinsic curvature, of genus $p > 1$. Consider a regular $2(p+1)$-gon on a hyperbolic plane. The interior angles of this polygon depend on its dimensions and may assume any values from 0 to $\pi - \pi(p+1)$. Hence, if $p > 1$, there exists a polygon with interior angles equal to $\pi/(p-1)$. Now identify opposite sides of this polygon; we get a closed manifold of genus p with constant negative curvature, which we denote by M.

Now consider the three-dimensional manifold $M^* = M \times g$, the topological product of M and a straight line g. We define a Riemannian metric in M^* as follows.

Let P be an arbitrary point on M. In a neighborhood of P the metric of M is given by a quadratic form

$$ds^2 = g_{ij}\,dv^i dv^j, \quad i,j = 1,2.$$

Map the manifold M in the neighborhood of P isometrically onto the plane σ in the hyperbolic space R'. Then in the neighborhood of the point P' corresponding to P the line element of σ is also defined by the form $g_{ij}\,dv^i dv^j$. Introduce a semigeodesic coordinate system in R' in the neighborhood of the perpendicular to σ at P', with the geodesics perpendicular to σ as v^3-curves and the planes parallel to σ as surfaces $v^3 = \text{const}$. The coordinates of a point are its signed distance from the plane σ (v^3) and the coordinates v^1, v^2 of the foot of a perpendicular dropped on the plane σ. The line element of the space in this system is

$$ds_1^2 = \varphi(v^3) g_{ij}\,dv^i dv^j + (dv^3)^2.$$

Now define a metric in M^* in the neighborhood of the straight line g by the line element ds_1^2; then M^* becomes a Riemannian space R^* locally isometric to a hyperbolic space. Any surface $v^3 = \text{const} \neq 0$ in the space R^*, as a surface equidistant from a plane in a hyperbolic space, has positive extrinsic curvature. We have thus proved the existence of closed surfaces of genus $p > 1$ with everywhere positive extrinsic curvature in a Riemannian space, and hence the question of infinitesimal bendings of such surfaces is meaningful.

We now state the two theorems to be proved in this section.

THEOREM 1. *A closed surface in a Riemannian space, homeomorphic to a torus, with positive extrinsic curvature, anchored at one point, is rigid, i.e. any bending field of the surface which vanishes at a point vanishes identically.*

THEOREM 2. *A closed surface in a Riemannian space, of genus $p > 1$, with positive extrinsic curvature, is rigid, i.e. admits no bending field other than the identically zero field.*

Before proceeding to the proofs, we note that the condition in Theorem 1 that the surface be anchored is essential. Indeed, a surface homeomorphic to a torus which is not anchored generally admits nontrivial bending fields. To show this, take a surface homeomorphic to a torus in the space of constant curvature constructed above. This surface admits a motion into itself. The velocity field of this motion is a bending field. It is indeed trivial, in the sense that it is the velocity field of a motion of the entire space. However, by an arbitrarily small deformation of the metric of the space in a neighborhood of the surface (but not on the surface itself) we get an inhomogeneous space which no longer admits motions. And here a motion of the surface into itself involves modification of its extrinsic form, i.e. modification of the extrinsic distances between its points.

PROOF OF THEOREM 1. Let F be a surface in a Riemannian space R, homeomorphic to a torus, with positive extrinsic curvature. Define some parametrization u^1, u^2 on the surface F and define a semigeodesic parametrization of the space near F, the coordinates of a point being the signed distance of the point from F (v^3) and the coordinates u^1, u^2 of the foot of a geodesic perpendicular dropped on the surface (v^1, v^2).

If the surface parametrization is so chosen that the second fundamental form $\lambda_{ij}\,du^i du^j$ is of isothermal type ($\lambda_{11} = \lambda_{22}$, $\lambda_{12} = 0$), the equations of an infinitesimal bending assume a particularly simple form. Namely, the covariant components ξ_1 and ξ_2 of the vector of a bending field satisfy the system

$$(*) \qquad \frac{\partial \xi_1}{\partial u^1} - \frac{\partial \xi_2}{\partial u^2} - (\tilde{\Gamma}^\alpha_{11} - \tilde{\Gamma}^\alpha_{22})\,\xi_\alpha = 0, \qquad \frac{\partial \xi_1}{\partial u^2} + \frac{\partial \xi_2}{\partial u^1} - 2\tilde{\Gamma}^\alpha_{12}\xi_\alpha = 0,$$

where $\tilde{\Gamma}^\alpha_{ij}$ are the Christoffel symbols of the surface. These equations were derived in §3. To make the system (*) available for the proof of Theorem 1, we must first define an isothermal-conjugate coordinate system on the surface F.

Map the surface F onto a torus of zero curvature. (A torus of this type is obtained by identifying opposite sides of a rectangle Δ on a euclidean plane.) Introduce cartesian coordinates u^1, u^2 on the plane of the rectangle, with the coordinate axes on the sides of the rectangle, and define the coordinates of a point on F as the coordinates of the corresponding point of the rectangle; then the first and second fundamental forms of the surface are

$$g_{ij}\,du^i du^j, \qquad \lambda_{ij}\,du^i du^j,$$

where g_{ij} and λ_{ij} are functions periodic in u^1 and u^2, with periods equal to the sides of the rectangle. The universal covering surface $\widetilde{F}$ of F has the same fundamental forms, but these are defined over the entire (u^1, u^2)-plane.

Let M denote a two-dimensional Riemannian manifold whose metric is defined by the second fundamental form of the surface $\widetilde{F}$, i.e. the form $\lambda_{ij}\,du^i du^j$ in the (u^1, u^2)-plane. Since the extrinsic curvature of the surface F is positive, so that the form $\lambda_{ij}\,du^i du^j$ is definite (we may assume without loss of generality that it is positive definite), this form indeed defines a Riemannian metric.

Map the manifold M onto the (u^1, u^2)-plane with its natural metric, associating points with equal coordinates u^1, u^2. This mapping is quasiconformal. It takes an infinitesimal disk of M into an infinitesimal ellipse on the plane with uniformly bounded ratio of semiaxes. It follows that M can be mapped conformally onto the plane. Let S be a conformal mapping of M onto the plane, and let u and v be the cartesian coordinates of the image of a point (u^1, u^2) on the manifold under S. If we define the coordinates of a point on M as the coordinates of its image under the conformal mapping S, the line element of M is of isothermal type:

$$ds^2 = \lambda(du^2 + dv^2).$$

Set $\tilde{z} = u^1 + iu^2$ and $z = u + iv$. Then the mapping S is defined by a certain function s of the complex variable $\tilde{z}$ with values in the complex z-plane: $z = s(\tilde{z})$. Let α_1 and α_2 be two nonzero complex numbers, representing the vertices of the rectangle Δ on the u^1- and u^2-axes, respectively.

The mapping S has the following property. There exist complex numbers ω_1 and ω_2 such that for any $\tilde{z}$ and integers m and n

$$s(\tilde{z} + m\alpha_1 + n\alpha_2) = s(\tilde{z}) + m\omega_1 + n\omega_2.$$

To prove this assertion for any m and n, it is sufficient to verify it for $m = 1$, $n = 0$, and $m = 0$, $n = 1$. Consider the case $m = 1$, $n = 0$.

The mapping of M onto itself which takes a point $\tilde{z}$ onto the point $\tilde{z} + \alpha_1$ is an isometry by virtue of the periodicity of the coefficients of the form $\lambda_{ij}du^i du^j$—the line element of M. Hence the mapping of the z-plane onto itself under which a point $s(\tilde{z})$ goes into $s(\tilde{z} + \alpha_1)$ is conformal. But the only conformal mappings of a euclidean plane onto itself are linear mappings. Thus

$$s(\tilde{z} + \alpha_1) = cs(\tilde{z}) + \omega_1.$$

Since this mapping can have no fixed points (the mapping $\tilde{z} \to \tilde{z} + \alpha_1$ has no fixed points), the constant c is equal to unity, and so $s(\tilde{z} + \alpha_1) = s(\tilde{z}) + \omega_1$. Similarly we prove the existence of ω_2 such that $s(\tilde{z} + \alpha_2) = s(\tilde{z}) + \omega_2$. It now follows immediately that for any integers m and n

$$s(\tilde{z} + m\alpha_1 + n\alpha_2) = s(\tilde{z}) + m\omega_1 + n\omega_2.$$

We now introduce u and v as coordinates on the universal covering surface $\tilde{F}$. The first and second fundamental forms of the surface are then defined in the u, v-plane, and the second fundamental form is of isothermal type: $\lambda(du^2 + dv^2)$. The coefficients of both forms are doubly periodic functions of the complex variable $z = u + iv$ with periods ω_1 and ω_2.

REMARK. Let Δ' be the curvilinear quadrangle in the z-plane into which the mapping S maps the rectangle Δ. The above property of S implies that Δ' becomes a torus of zero curvature if corresponding points of its opposite sides are identified. Thus we have proved in passing that any Riemannian manifold homeomorphic to a torus can be mapped conformally onto a torus of zero curvature.

We can extend the bending field ξ of the surface F to the universal cover $\widetilde{F}$, associating with each point of $\widetilde{F}$ the vector of the bending field of F at the geometrically identical point. In coordinates u, v on $\widetilde{F}$, the components ξ_1 and ξ_2 of the bending field are also doubly periodic functions of the complex variable $z=u+iv$.

Now consider the equations of the bending field:

$$\frac{\partial \xi_1}{\partial u}-\frac{\partial \xi_2}{\partial v}=a\xi_1+b\xi_2,$$

$$\frac{\partial \xi_1}{\partial v}+\frac{\partial \xi_2}{\partial u}=c\xi_1+d\xi_2.$$

The coefficients $a,\cdots,d$ of this system are functions of the Christoffel symbols of the surface, and are therefore doubly periodic functions of z.

Setting

$$\zeta=\xi_1+i\xi_2,\quad \bar{\zeta}=\xi_1-i\xi_2,$$

$$A=\frac{1}{4}(a+d+ic-ib),\quad B=\frac{1}{4}(a-d+ic+ib),$$

$$\frac{\partial}{\partial \bar{z}}=\frac{1}{2}\left(\frac{\partial}{\partial u}+i\,\frac{\partial}{\partial v}\right),$$

we can rewrite the equations of the bending field as follows:

$$\partial\zeta/\partial\bar{z}=A\zeta+B\bar{\zeta}.$$

Now let K_R be a disk of radius R about the point $z=0$. Within this disk the function ζ has the representation $\zeta=\varphi(z)e^{\omega(z)}$, where $\varphi(z)$ is an analytic function and

$$\omega(z)=\iint_{K_R}\frac{C(t)\,ds}{t-z},\quad C(t)=-\frac{1}{\pi}\left(A(t)+B(t)\frac{\bar{\zeta}(t)}{\zeta(t)}\right),$$

where the integration is performed over the area of the disk K_R.

Let $q<1$ be some fixed number. Let us estimate $|\omega(z)|$ for z such that $|z|\leq qR$. Let $K(z)$ be the disk about z internally tangent to the disk K_R. Then

$$\omega(z)=\iint_{K(z)}\frac{C(t)\,ds}{t-z}+\iint_{K_R-K(z)}\frac{C(t)\,ds}{t-z}.$$

Since $C(t)$, as a periodic function, is bounded, and since when $|z| \leqq qR$ in the domain $K_R - K(z)$ we have $|t-z| \geqq (1-q)R$, it follows that

$$\left| \iint_{K_R - K(z)} \frac{C(t)\, ds}{t-z} \right| \leqslant c_1 |z|,$$

where c_1 is a constant.

Let Δ_0 be the parallelogram with vertices $z + \frac{1}{2}(\pm\omega_1, \pm\omega_2)$. Let Δ_{mn} denote the parallelogram congruent to Δ_0 and parallel to it, with center at the point $z + m\omega_1 + n\omega_2$, where m and n are arbitrary integers. Now divide the disk $K(z)$ into two domains: $K_1(z)$ consists of those parallelograms Δ_{mn} entirely contained in the disk $K(z)$, and $K_2(z) = K(z) - K_1(z)$. Obviously

$$\left| \iint_{K_2(z)} \frac{C(t)\, ds}{t-z} \right| < c_2,$$

where c_2 is a constant. It remains to estimate

$$\omega_1(z) = \iint_{K_1(z)} \frac{C(t)}{t-z}\, ds.$$

Making use of the periodicity of $C(t)$, we can reduce the integration in $\omega_1(z)$ to the basic parallelogram Δ_0. Then

$$\omega_1(z) = \iint_{\Delta_0} \left(\sum_{m,n} \frac{1}{t - z + \omega_{mn}} \right) C(t)\, ds, \quad \omega_{mn} = m\omega_1 + n\omega_2.$$

By a substitution of variables in the integral, we can make z vanish. Thus

$$\omega_1(z) = \iint_{\Delta_0} \left(\sum_{m,n} \frac{1}{t - \omega_{mn}} \right) C_1(t)\, ds.$$

The sum appearing in this expression contains the terms $1/(t - \omega_{mn})$ and $1/(t + \omega_{mn})$ in pairs. Hence

$$\omega_1(z) = \frac{1}{2} \iint_{\Delta_0} \left(\sum_{m,n} \frac{1}{t^2 - \omega_{mn}^2} \right) t C_1(t)\, ds.$$

Consider the difference

$$\iint_{\Delta_{mn}} \frac{ds_\tau}{t^2 - \omega_{mn}^2} - \iint_{\Delta_{mn}} \frac{ds_\tau}{t^2 - \tau^2}.$$

For sufficiently large $|\omega_{mn}|$,

$$\left|\frac{1}{t^2-\omega_{mn}^2}-\frac{1}{t^2-\tau^2}\right|\leqslant c_3\left|\frac{1}{\tau^3}\right|,$$

where c_3 is a constant. Therefore

$$\omega_1=\iint\limits_{\Delta_0}\left(\iint\limits_{K_1(z)}\frac{ds_\tau}{t^2-\tau^2}\right)tC_1(t)\,ds_t+O(1).$$

The inner integration may be performed over the disk $K(z)$, since the concomitant modification of the integral part of ω_1 can be included in the $O(1)$ term. But

$$\iint\limits_{K(z)}\frac{ds_\tau}{t^2-\tau^2}=\int\oint\frac{d\rho\,d\tau}{(t^2-\tau^2)\,\tau},$$

where the inner integration in the right-hand side is performed with respect to the complex variable τ along a circle of radius ρ, and the outer integration with respect to ρ from zero to the radius of the disk $K(z)$. Since

$$\oint\frac{d\tau}{(t^2-\tau^2)\,\tau}=0$$

for $|t|<\rho$, we have

$$\left|\iint\limits_{K(z)}\frac{ds_\tau}{t^2-\tau^2}\right|\leqslant c_1,$$

where c_1 is a constant. This implies that $|\omega_1|$ is bounded by a constant, and therefore $|\omega(z)|\leqq c_0|z|$ for large z.

Now take a sequence of disks K_n of radius n and construct the corresponding sequences of functions ω_n and φ_n:

$$\zeta(z)=\varphi_n(z)e^{\omega_n(z)}.$$

Since $\zeta(z)$, as a periodic function, is bounded, it follows that the sequence φ_n contains a convergent subsequence. The limit function $\varphi(z)$ is analytic in the entire z-plane.

By the assumptions of the theorem, the function ζ vanishes at some point z_0 (the anchoring point of the surface). Since ζ is periodic, it vanishes at all points $z_{mn}=z_0+\omega_{mn}$. But the functions $\varphi_n(z)$ must vanish at these points, and hence so does the limit function $\varphi(z)$.

Since the function $\varphi(z)$ cannot increase more rapidly than $e^{c_0|z|}$ and it vanishes at the points z_{mn}, it follows from a standard theorem that it must vanish identically. Hence, since the functions $\omega_n(z)$ are uniformly bounded in any bounded region of the z-plane, we conclude that $|\zeta(z)|$

is arbitrarily small in any bounded region of the z-plane. Hence $\zeta(z)$ is identically zero.

Thus the components ξ_1 and ξ_2 of our bending field vanish. Now the third component ξ_3 must vanish together with ξ_1 and ξ_2, since

$$2\,\partial\xi_1/\partial u - \Gamma_{11}^k\xi_k - 2\lambda\xi_3 = 0.$$

This completes the proof of Theorem 1.

We now outline the proof of Theorem 2.

Let F be a closed surface of genus p with positive extrinsic curvature, in a Riemannian space R. Let F_0 be a closed two-dimensional manifold of genus p with constant negative Gauss curvature; suppose that we have some mapping of F onto F_0. This mapping induces a mapping of the universal covering surface $\widetilde{F}$ of F onto a hyperbolic plane. Define a metric on $\widetilde{F}$ using the second fundamental form of F, and denote the resulting Riemannian manifold by M.

The mapping of $\widetilde{F}$ onto the hyperbolic plane induces a quasiconformal mapping of the manifold M. Now map the hyperbolic plane conformally onto a unit disk in the euclidean plane; we get a quasiconformal mapping of M. Hence there exists a conformal mapping of the manifold M onto the unit disk. Denote this mapping by S.

Let M_0 be the domain on the manifold M corresponding to the surface F. The manifold M admits a discrete group G of isometries g_k. The images $M_k = g_k M_0$ under these isometries cover the entire manifold M.

The isometric mappings g_k of M into itself generate a transformation $h_k = S g_k S^{-1}$ of the unit disk into itself. We claim that this transformation is a Möbius transformation. Indeed, since the mapping g_k is an isometry, while S is a conformal mapping, it follows that h_k is a conformal mapping of the disk into itself. But every conformal mapping of the disk into itself is a Möbius transformation.

The mapping S of M onto the unit disk is at the same time a mapping of $\widetilde{F}$ onto the disk. If we define the coordinates of a point on $\widetilde{F}$ as the cartesian coordinates u, v in the disk, the second fundamental form of the surface $\widetilde{F}$ is of isothermal type, and the isometric mappings of $\widetilde{F}$ into itself are represented by Möbius transformations in the complex variable $z = u + iv$.

Retaining the notation introduced above, we consider the equation of the bending field in its complex form:

$$\partial\zeta/\partial\bar{z} = A\zeta + B\bar{\zeta}.$$

Within a disk $|z| < \rho < 1$ the solution of this equation can be expressed as $\zeta = \varphi(z) e^{\omega(z)}$.

Now suppose that we have somehow proved that the functions $\varphi(z)$ are uniformly bounded as $\rho \to 1$. The proof is then completed as follows.

Construct a sequence of disks K_n of radius $\rho = 1 - 1/n$ and define the function φ_n for each disk. Since the functions φ_n are uniformly bounded, we can extract a convergent subsequence. Let φ be the limit of this subsequence. Consider the existence and distribution of the zeros of φ. Since they are the same as the zeros of ζ, it will suffice to consider the latter.

In each domain $\Delta_k = S(M_k)$, the function ζ has at least one zero. Indeed, otherwise we could define on the surface F a continuous, nowhere vanishing tangent vector field (the projection of the bending field on the surface). But such a field can be defined only on a surface homeomorphic to a torus, while our surface is of genus $p > 1$.

Let z_k be the zeros of ζ, hence also of φ. Consider the series

$$U = \sum_k (1 - |z_k|).$$

We claim that it is divergent. In fact, the points A_k on the hyperbolic plane corresponding to the zeros z_k of ζ are uniformly distributed. Hence the distribution density of the roots z_k increases together with $1/(1-\rho^2)^2$ as one approaches the circumference of the disk $|z| = 1$. In other words, the area element ds of the disk contains $\approx c_0 ds/(1-\rho^2)^2$ zeros. Hence we can write

$$U = \iint_{|z|<1} \frac{c_0 (1-\rho)\, ds}{(1-\rho^2)^2} + O(1),$$

where $O(1)$ is a bounded quantity. Since the integral in the right-hand side of this equality is divergent, so is the series U.

Since the series U is divergent and the function φ bounded, it follows that φ is identically zero. This easily implies that ζ vanishes identically, and together with it the bending field.

Thus it remains to prove that the function φ is bounded in the entire disk $|z| < 1$. We now indicate an approach to the proof which apparently yields the desired result. [See the Appendix, p. 656.]

Cover the entire disk $|z| < 1$ by the domains $\Delta_k = S(M_k)$. Let Δ_0 be the domain containing the point $z = 0$. We compare the values of the function $C(t)$ at corresponding points of the domains Δ_0 and Δ'. The correspondence implied here is that defined by the Möbius transformation h_k which maps the disk $|z| \leqq 1$ into itself and the domain Δ_0 into Δ'. Let z be an arbitrary point of Δ_0, and z' the corresponding point of Δ'. Regard the coordinates u, v of z as the coordinates of the point z'. This coordinate transformation is accompanied by a corresponding transformation of the covariant components ξ_α of the bending field. For the complex component ζ, this transformation is given by

$$\zeta' = \zeta d\bar{z}/d\bar{z}'.$$

It is clear from the equation of an infinitesimal bending in complex form that

$$C(z) = -(1/\pi\zeta)\partial\zeta/\partial\bar{z}.$$

Hence, using the relation between ζ and ζ' under the coordinate transformation, we get

$$C(z') = C(z) - \frac{1}{\pi}\frac{d^2\bar{z}}{d\bar{z}'^2}\Big/\frac{d\bar{z}}{d\bar{z}'}.$$

Since the function $C(z)$ is bounded in Δ_0, the function ω can be expressed as

$$\omega = \iint\limits_{K_\rho} \frac{D(t)\,ds}{t-z} + O(1),$$

where D is a function defined in Δ' by the equality

$$D(z') = -\pi\frac{d^2\bar{z}}{(d\bar{z}')^2}\Big/\frac{d\bar{z}}{d\bar{z}'}.$$

Here z' is an arbitrary point of Δ', and z is the corresponding point of Δ_0 under the mapping h_k of Δ_0 into Δ'.

Now let E denote the function defined in Δ' by the equality

$$E(z') = d\bar{z}/d\bar{z}'.$$

Then the function φ is bounded if and only if the function

$$\Psi = E(z)\,e^{-\pi\iint\limits_{K_\rho}\frac{D(t)}{t-z}ds}$$

is bounded.

The essential point here is that the function Ψ is independent of the specific form of the surface F. It is uniquely determined by the group of transformations h_k. The latter depends only on the topology of the surface F, i.e. its genus p. It can be shown that the group of transformations h_k coincides with the group of transformations of an automorphic function in the disk $|z| < 1$.

CHAPTER VII

CONVEX SURFACES WITH GIVEN SPHERICAL IMAGE

All the basic problems considered hitherto in this book have involved two or more objects related by isometric mappings: a manifold with given metric and realization of this metric as a surface (embedding problem); two isometric surfaces (monotypy problem); a continuous family of isometric surfaces (in the bending problem). In the regular case we have accordingly discussed manifolds and surfaces with common line element ds^2, i.e. common first fundamental form. Specification of this form in the nonregular case amounts to specification of an intrinsic metric.

This chapter is devoted to problems of geometry in the large relating to surfaces with given third fundamental form (in the regular case) or given spherical image (in the general case). A typical problem of this kind is the congruence of two closed convex surfaces for which the principal radii of curvature R_1 and R_2 at points with parallel and identically oriented normals n satisfy a relation

$$(*) \qquad f(R_1, R_2, n) = \varphi(n).$$

The exposition begins with a solution of the classical problems of Christoffel and Minkowski. We then formulate and solve the general problem of existence and uniqueness of a closed convex surface satisfying equation (*). The analogous problem is formulated and solved for convex polyhedra. At the end of the chapter we consider the stability of solutions of certain problems of geometry in the large, and other problems.

§1. Christoffel's problem

Christoffel's problem concerns the existence and uniqueness of a closed convex surface F such that the sum of the principal radii of curvature at a point with unit normal vector n is a given function $f(n)$. This problem was essentially solved by Christoffel himself, who constructed the surface (i.e. determined its equation) on the basis of the given function f. Here we shall present Blaschke's solution [**22**].

Let F be a closed convex surface, X an arbitrary point of space distinct from the origin O. Let α be a supporting plane to F with exterior normal OX. Let Y be the point on the surface lying in the plane α. Then the function

$$P(X) = \overrightarrow{OX} \cdot \overrightarrow{OY}$$

of the coordinates x_1, x_2, x_3 of X is known as the *supporting function* of the surface F. We consider a few properties of the supporting function.

The supporting function is obviously a positive homogeneous function of first degree. Therefore, if it is differentiable, Euler's theorem shows that

$$x_1 P_1 + x_2 P_2 + x_3 P_3 = P,$$

where P_1, P_2, P_3 denote the first derivatives of P with respect to x_1, x_2, x_3, respectively.

If the convex surface contracts to a point, the supporting function is a linear function. Indeed, if the coordinates of this point are a_1, a_2, a_3, then

$$P = a_1 x_1 + a_2 x_2 + a_3 x_3.$$

If the surface F contains a surface F_1 in its interior, the supporting functions P and P_1 satisfy the inequality

$$P_1(x_1, x_2, x_3) \leqq P(x_1, x_2, x_3),$$

and equality holds if and only if the surfaces have a common supporting plane with exterior normal (x_1, x_2, x_3).

The last two properties imply that the supporting function is convex:

$$P(\lambda x' + (1-\lambda) x'') \leqslant \lambda P(x') + (1-\lambda) P(x''), \qquad 0 \leqslant \lambda \leqslant 1.$$

Indeed, together with the supporting function P of the surface F, consider the supporting function P' of a point A. We may assume that the point A lies on F in such a position that, for the given x', x'' and λ,

$$P(\lambda x' + (1-\lambda) x'') = P'(\lambda x' + (1-\lambda) x'').$$

For this to be true, the exterior normal to F at A must have the direction $\lambda x' + (1-\lambda) x''$. Since $P' \leqq P$, while P' obviously has the required property:

$$P'(\lambda x' + (1-\lambda) x'') \leqslant \lambda P'(x') + (1-\lambda) P'(x''),$$

it follows that

$$P(\lambda x' + (1-\lambda) x'') \leqslant \lambda P(x') + (1-\lambda) P(x'').$$

This proves our assertion.

The supporting function of a convex polyhedron is piecewise linear.

To be precise, if ω is the spherical image of a vertex A of the polyhedron and V the polyhedral angle projecting the domain ω from the origin, then the supporting function of the polyhedron is linear inside the angle V and coincides with the supporting function of the vertex A.

If the surface F is translated along a vector a, the supporting function is modified by addition of a linear term $a_1x_1+a_2x_2+a_3x_3$, where a_1, a_2, a_3 are the components of the vector a.

Continuity of the supporting function follows directly from its definition.

Since the supporting function is a positive homogeneous function of degree 1, it is uniquely determined by its values on the unit sphere about the origin. In this interpretation it has a simple geometric meaning: the value of $P(n)$ at a point of the sphere with vector n is the signed distance from the origin O to the supporting plane with exterior normal n; if the origin lies within the surface, this is simply the distance.

The supporting function of a convex curve on the plane is defined in analogous fashion, and it possesses the above properties. Let γ be a closed convex curve with positive curvature. We wish to express the radius of curvature of γ in terms of the supporting function p defined on the unit circle. The argument of the function p is arc length σ on the circle. Let $r(s)$ be the radius vector of the point on the curve corresponding to arc length s. Then $p=rn$. Differentiating this equality with respect to σ and noting that $r'n=0$, we get $p'=r\tau=q$, where τ is the unit vector tangent to the curve at the point with normal n. Note that the derivative p' has a simple geometric meaning: it is equal to the length q of the segment on the tangent to γ whose endpoints are the point of contact and the foot of a perpendicular dropped on the tangent from the origin O (Figure 32).

Differentiate the equality $p=rn$ twice with respect to σ. Noting that $ds/d\sigma=R$ (the radius of curvature of γ) and $r'_\sigma=\tau R$, $n''=-n$, we get

$$p''=R\tau^2-rn=R-p.$$

Hence for the radius of curvature we have $R=p''+p$.

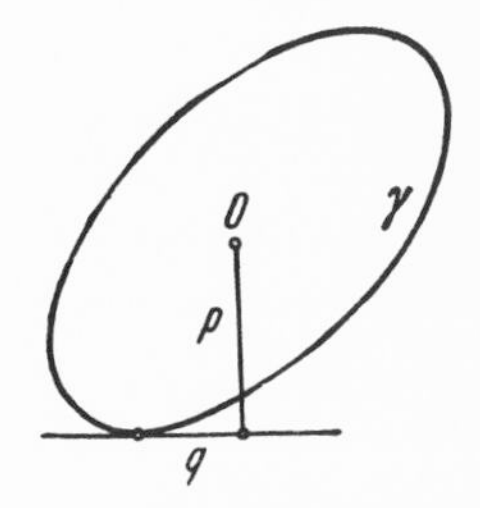

FIGURE 32

Now consider a regular, twice differentiable closed convex surface. Let $P(a_1,a_2,a_3)$ be the value of its supporting function at the point with radius vector $a(a_1,a_2,a_3)$. The equation of the tangent plane to the surface at the point with exterior normal a is clearly

$$a_1 x_1 + a_2 x_2 + a_3 x_3 = P(a_1, a_2, a_3).$$

Set

$$\Phi(x_1, x_2, x_3, a_1, a_2, a_3) \equiv a_1 x_1 + a_2 x_2 + a_3 x_3 - P(a_1, a_2, a_3).$$

Let $\bar{x}$ be a point on F and $\bar{a}$ the exterior normal at $\bar{x}$. Then, when $\bar{x}$ is fixed and a varies, the function $\Phi(\bar{x}, a)$ has fixed sign, and when $a = \bar{a}$ we have $\Phi(\bar{x}, \bar{a}) = 0$. Thus for $x = \bar{x}$ and $a = \bar{a}$,

$$\frac{\partial \Phi}{\partial a_1} = \frac{\partial \Phi}{\partial a_2} = \frac{\partial \Phi}{\partial a_3} = 0.$$

This yields the following formulas for the coordinates of the point of contact:

$$x_1 = \frac{\partial P}{\partial a_1}, \quad x_2 = \frac{\partial P}{\partial a_2}, \quad x_3 = \frac{\partial P}{\partial a_3},$$

where a_1, a_2, a_3 are the coordinates of the exterior normal vector at the point of contact.

We now express the principal radii of curvature of a convex surface as functions of its supporting function. Following Blaschkė, we consider a translation from a given point on the surface to an infinitesimally close point in a principal direction on the surface. By the Rodrigues formula we have $dx_i + R d\xi_i = 0$, where ξ_i are the components of the exterior normal vector to the surface and R is the radius of curvature in the direction of the translation. Since $x_i = P_i$, it follows that

$$\sum_k P_{ik} \, d\xi_k + R \, d\xi_i = 0, \qquad i = 1, 2, 3.$$

Regarding these equalities as a system of linear equations for $d\xi_i$, we get the compatibility condition

$$\begin{vmatrix} P_{11} + R; & P_{12}, & P_{13} \\ P_{21}, & P_{22} + R, & P_{23} \\ P_{31}, & P_{32}, & P_{33} + R \end{vmatrix} = 0$$

Two of the roots R of this equation are the principal radii of curvature of the surface. The third root is zero. Indeed, by virtue of the homogeneity of P we have

$$\sum_k P_{ik} a_k = 0, \qquad i = 1, 2, 3.$$

This implies that the determinant $|P_{ik}|$ vanishes, which means that one of the roots of the equation for R is zero.

From the equation determining the principal radii of curvature R_1 and R_2, we get

$$-(R_1 + R_2) = P_{11} + P_{22} + P_{33},$$

where the supporting function is first formally differentiated and the coordinates of the unit normal vector ξ_1, ξ_2, ξ_3 then inserted as arguments.

We now turn to Christoffel's problem. We wish to determine the surface F, given the sum of principal radii of curvature $f = R_1 + R_2$.

Let U_k be a homogeneous polynomial of degree k satisfying the Laplace equation

$$\Delta U_k = \frac{\partial^2 U_k}{\partial \xi_1^2} + \frac{\partial^2 U_k}{\partial \xi_2^2} + \frac{\partial^2 U_k}{\partial \xi_3^2} = 0.$$

It is well known that one can construct a complete orthonormal system consisting of functions U_k defined on the unit sphere about the origin (spherical harmonics). Expand the supporting function P of the surface F on the unit sphere in a series in the spherical harmonic functions U_k:

$$P = \sum_k c_k U_k (\xi_1, \xi_2, \xi_3).$$

The supporting function P is then clearly given in the entire space by the formula

$$P(x_1, x_2, x_3) = \sum_k c_k \frac{U_k(x_1, x_2, x_3)}{r^{k-1}} \qquad r^2 = x_1^2 + x_2^2 + x_3^2.$$

We now compute $P_{11} + P_{22} + P_{33}$. Noting that

$$\Delta U_k = 0, \quad x_1 \frac{\partial U_k}{\partial x_1} + x_2 \frac{\partial U_k}{\partial x_2} + x_3 \frac{\partial U_k}{\partial x_3} = k U_k,$$

we get

$$P_{11} + P_{22} + P_{33} = -\sum_k c_k \frac{(k-1)(k+2)}{r^{k+1}} U_k.$$

Now expand the given function $f(n) = R_1 + R_2$ in a series in the spherical harmonics:

$$f = \sum_k d_k U_k.$$

Since $P_{11} + P_{22} + P_{33} = -(R_1 + R_2)$ on the unit sphere, it follows that

$$(k-1)(k+2) c_k = d_k.$$

Hence

$$c_k = \frac{d_k}{(k-1)(k+2)}$$

Consequently

$$P = \sum_k \frac{d_k}{(k-1)(k+2)} U_k, \tag{*}$$

where the d_k are the expansion coefficients of f with respect to the spherical harmonics.

For $k=1$ the equality $(k-1)(k+2)c_k = d_k$ implies that $d_1 = 0$. It follows that

$$\int_\omega f U_1 \, d\omega = 0,$$

integrating over the unit sphere. Since the function U_1 can be chosen arbitrarily,

$$\int_\omega f \xi_i \, d\omega = 0, \qquad i = 1, 2, 3.$$

Or, in vector notation,

$$\int_\omega f n \, d\omega = 0.$$

Since $d_1 = 0$, the coefficient c_1 in the expansion of P remains undetermined. This corresponds to the fact that the sum of principal radii of curvature determines the surface F up to translation.

The above integral condition for the function f is necessary but not sufficient for the existence of a closed *convex* surface with the given sum f of principal radii of curvature. This was observed by Aleksandrov. It is natural to ask whether sufficient conditions can be found. We shall now show that they can. The result itself is of no little interest, but the main point here is the method for estimating the principal radii of curvature, which will be applied later in more complicated problems (see, for example, §5).

The idea underlying the derivation of sufficient conditions is as follows. Let P be a function on the unit sphere, defined in terms of the given function f by formula (*). If C is a sufficiently large positive number, the function $P+C$ is the supporting function of some closed convex surface (we assume that P is twice differentiable). If the smaller of the principal radii of curvature of the surface is greater than C, then the surface parallel to it at a distance C is convex, and it is defined by the supporting function P. Thus the problem is to estimate the minimum radius of curvature of a convex surface.

Let Φ be a convex surface defined by the supporting function $h(\xi) = P(\xi) + C$, and $F(\xi)$ the sum of its principal radii of curvature at the point with normal ξ. We define a function $w(\xi, t)$ of a point ξ and a direction t at ξ, on the unit sphere Ω about the origin, by $h(\xi) + h_{ss}(\xi)$, where h is differentiated with respect to arc length on the great circle $K(\xi, t)$ through the point ξ in the direction t. The function $w(\xi, t)$ is the radius of curvature of a cylinder tangent to the surface Φ, with generators perpendicular to the plane of the circle $K(\xi, t)$ at the point with normal ξ. For fixed ξ, the maximum and minimum of $w(\xi, t)$ are attained when the direction t is parallel to a principal direction of the surface Φ, and the maximum and minimum values of w are precisely the principal radii of curvature of the surface.

The function $w(\xi, t)$ assumes an absolute minimum at some point y_0, for some direction t_0 at this point. Define geographical coordinates φ, ϑ on the unit sphere Ω, with the great circle through ξ_0 in the direction t_0 as equator and the great circle perpendicular to it through ξ_0 as the zero meridian. In this coordinate system, the sum of principal radii of curvature of the surface Φ at the point with normal $\xi(\varphi, \vartheta)$ is

$$F(\xi) = 2h + h_{\vartheta\vartheta} + \frac{1}{\cos^2 \vartheta} h_{\varphi\varphi} - h_\vartheta \tan \vartheta.$$

In a neighborhood of the point ξ_0, the function $\bar{w}(\xi) = h + h_{\vartheta\vartheta}$ satisfies the inequality $\bar{w}(\xi) \geqq w(\xi_0, t_0)$, with equality at the point ξ_0. Hence the function $\bar{w}(\xi)$ assumes a minimum at ξ_0, and this minimum is $w(\xi_0, t_0)$.

Since the function $\bar{w}(\xi)$ has a minimum at ξ_0, the following conditions must hold there:

$$\begin{aligned} \bar{w}_\vartheta &= h_\vartheta + h_{\vartheta\vartheta\vartheta} = 0, \\ \bar{w}_\varphi &= h_\varphi + h_{\vartheta\vartheta\varphi} = 0, \\ \bar{w}_{\vartheta\vartheta} &= h_{\vartheta\vartheta} + h_{\vartheta\vartheta\vartheta\vartheta} \geqslant 0, \\ \bar{w}_{\varphi\varphi} &= h_{\varphi\varphi} + h_{\vartheta\vartheta\varphi\varphi} \geqslant 0. \end{aligned}$$

Without loss of generality, we may assume that the point of Φ whose neighborhood is under consideration (the point with normal ξ_0) is the origin. Then, obviously, $h = h_\vartheta = h_\varphi = 0$. Hence, by the conditions $\bar{w}_\vartheta = \bar{w}_\varphi = 0$ for a minimum, we get $h_{\vartheta\vartheta\vartheta} = h_{\vartheta\vartheta\varphi} = 0$.

Differentiating $F(\xi)$ twice with respect to ϑ at the point ξ_0, we get

$$F_{\vartheta\vartheta} = h_{\vartheta\vartheta\vartheta\vartheta} + h_{\varphi\varphi\vartheta\vartheta} + 2h_{\varphi\varphi}.$$

Now at the same point ξ_0 we have $F = h_{\vartheta\vartheta} + h_{\varphi\varphi}$, and so

$$F - F_{\vartheta\vartheta} = -(h_{\vartheta\vartheta} + h_{\vartheta\vartheta\vartheta\vartheta}) - (h_{\varphi\varphi} + h_{\vartheta\vartheta\varphi\varphi}) + 2h_{\vartheta\vartheta}.$$

The first two terms on the right-hand side are nonpositive, because of the minimum condition $\bar{w}_{\vartheta\vartheta} \geqq 0$, $\bar{w}_{\varphi\varphi} \geqq 0$. $h_{\vartheta\vartheta}$ is equal to $w(\xi_0, t_0)$, i.e. the minimum radius of curvature of Φ, at the point ξ_0. Consequently

$$2R_{\min}|_{\Phi} \geqq F - F_{\vartheta\vartheta}.$$

Obviously $F = f + 2C$. Therefore

$$R_{\min}|_{\Phi} \geqq C + (f - f_{\vartheta\vartheta})/2.$$

Now assume that the given function f satisfies the condition $f - f_{ss} \geqq 0$ on the unit sphere; the differentiation is with respect to arc length s along any great circle. Then

$$R_{\min}|_{\Phi} \geqq C.$$

Thus the surface parallel to Φ at a distance C is convex. This surface has supporting function P defined by the given function f in accordance with (*). This proves the following:

THEOREM. *The following conditions are sufficient*[1] *for the existence of a closed convex surface with given sum $f(\xi)$ of principal radii of curvature:*

$$f(\xi) \geqslant 0, \quad f - f_{ss} \geqslant 0, \quad \int_{\Omega} \xi f(\xi)\, d\omega = 0.$$

§2. Minkowski's problem

Let F be a closed twice differentiable convex surface with positive Gauss curvature. Define a function $K(n)$ of the unit vector n on the unit sphere Ω, whose value for a point n on the sphere is the Gauss curvature of F at the point with exterior normal n.

Now let $K(n)$ be an arbitrary positive continuous function defined on Ω. Does there exist a closed convex surface F with Gauss curvature $K(n)$ at the point with exterior normal n, and is it unique? This is Minkowski's problem, which was solved by Minkowski himself [**44**]. We shall present Aleksandrov's solution [**3**]. It begins with a consideration of the corresponding problem for convex polyhedra. The general arguments used here will also be applied in other proofs of this chapter.

[1] [These conditions are not necessary. At the time of writing the author was apparently not aware of W. J. Firey's derivation of necessary and sufficient conditions, first for regular surfaces (*The determination of convex bodies from their mean values of curvature functions*, Mathematika **14** (1967), 1-13; MR **36** #788), and later for the general case (*Christoffel's problem for general convex bodies*, ibid. **15** (1968), 7-21; MR **37** #5822). This remark is due to H. Busemann (MR **39** #6222).]

LEMMA 1. *Let $n_1, n_2, \cdots, n_k$ be a k-tuple of noncoplanar unit vectors and $F_1, F_2, \cdots, F_k$ positive numbers such that*

$$(*) \qquad \sum_i n_i F_i = 0.$$

Then there exists a closed convex polyhedron P, unique up to translation, whose faces have exterior normals n_i and areas F_i.

The proof of this lemma will be based on Aleksandrov's "Mapping Lemma." We first consider some properties of polyhedra satisfying the assumptions of Lemma 1, which will enable us to apply the Mapping Lemma.

First note that the exterior normals n_i of a polvhedron satisfying condition (*) cannot point into the same halfspace. Indeed, suppose that the vectors n_i all point into, say, the halfspace $z > 0$. Then, since the vectors n_i are not coplanar, the vector $\sum n_i F_i$ is a nonzero vector pointing into the same halfspace. Thus condition (*) cannot hold.

For any k-tuple of vectors n_i satisfying condition (*), there exists a closed convex polyhedron whose faces have exterior normals n_i. To prove this, take rays l_i issuing from the center of a sphere in the directions n_i. They cut the sphere in points A_i. Let α_i be the tangent planes to the sphere at the points A_i, and let E_i denote the halfspaces [defined by these planes] in which the sphere lies. The intersection of the half-spaces E_i is a convex polyhedron P, which is not a priori bounded. Its faces lie in the planes α_i and have exterior normals n_i. We now show that P must be bounded. For were P unbounded, it would contain a ray l not intersecting any of the planes α_i. Then the rays l_i would form with l angles at least $\pi/2$. The vectors n_i would then all point into the same halfspace, which is impossible.

Now consider the set of all closed convex polyhedra with exterior normals n_i such that the areas of their faces F_i satisfy the condition $0 < a < F_i < b$. We claim that the polyhedra of this set are uniformly bounded, i.e. they are all contained within a ball of sufficiently large radius which depends only on the numbers a, b and the vectors n_i.

We first observe that, whatever the direction l, at least one of the vectors n_i forms an angle less than $\pi/2 - \epsilon$ with l, where $\epsilon > 0$ depends only on the vectors n_i. Indeed, otherwise we could find a sequence of directions l_s, each forming an angle greater than $\pi/2 - 1/s$ with all the vectors n_i. Without loss of generality we may assume that the sequence l_s converges to some direction l. The angle between this direction and the vectors n_i is at least $\pi/2$, and hence the vectors n_i all point into a halfspace bounded by a plane perpendicular to l, which is impossible.

We can now prove that the set of polyhedra with exterior normals n_i and face areas F_i such that $0 < a < F_i < b$ is bounded. Suppose that the set is not bounded. Then there exist polyhedra P_s of arbitrarily large diameters satisfying these conditions. Assume, for example, that the diameter of P_s is greater than s. The interior of P_s contains two points A_s and B_s separated by a distance greater than s. Project the polyhedron P_s onto a plane α perpendicular to the segment A_sB_s. The projection is a convex polygon $\overline{P}_s$. Since the areas of the faces of P_s are at least a and the angle between A_sB_s and one of the vectors n_i is at most $\pi/2 - \epsilon$, the area of the polygon $\overline{P}_s$ is clearly bounded below by some number σ depending only on ϵ and a.

The interior of the polygon $\overline{P}_s$ contains two points $\overline{C}_s$ and $\overline{D}_s$, the distance between which is at least some number δ depending only on σ, hence only on ϵ and a. Let C_s and D_s be the points in the polyhedron P_s whose projections on the plane α are $\overline{C}_s$ and $\overline{D}_s$. Let β be a plane parallel to the segments A_sB_s on C_sD_s. Project the polyhedron P_s onto the plane β. The projection is a quadrangle whose vertices are the projections of A_s, B_s, C_s, D_s. The area of this quadrangle is obviously not less than $s\delta$, and can therefore be made arbitrarily large by increasing s. Hence the area of the projection of the polyhedron is arbitrarily large, and thus the area of the polyhedron P_s itself is also arbitrarily large. But the latter cannot exceed bk, where k is the number of vectors n_i. This contradiction proves our assertion.

As mentioned above, the proof of Lemma 1 is based on the Mapping Lemma, which we now state.

MAPPING LEMMA. *Let A and B be two manifolds of the same dimension n. Let φ be a mapping of A into B with the following properties.*

1. *Each component of B contains images of points of A.*
2. *φ is one-to-one.*
3. *φ is continuous.*
4. *If b_m are points of B which are images of points a_n of A and converge to a point b, then A contains a point a whose image is b and which is a limit point of the points a_n.*

Then $\varphi(A) = B$, i.e. φ maps A onto B.

Let us define the manifolds A and B to which the Mapping Lemma will be applied. The points of A are equivalence classes of congruent and parallel polyhedra with normals n_i, i.e. the polyhedra of each class are translates of each other. Since each polyhedron is determined by the signed distances of its k faces from the origin, the manifold A is of dimension $k-3$.

The points of B are the points of euclidean k-space with coordinates F_i such that $F_i > 0$ and $\sum_i n_i F_i = 0$. B is also of dimension $k-3$.

The mapping φ takes each equivalence class of polyhedra onto the point of space with coordinates $F_1, F_2, \cdots, F_k$ equal to the areas of the faces of a representative polyhedron of the class.

We have to show that the assumptions of the Mapping Lemma hold for A, B and φ as defined. Condition 1 holds because the manifold B, as the intersection of convex sets, is convex and therefore connected. Moreover, as was shown above, there exists a polyhedron with exterior normals n_i. Consequently each component (here there is only one!) contains the image of a point of A.

We must now show that φ is one-to-one. Indeed, this means that two closed convex polyhedra whose faces have given directions and areas are congruent. We have the following theorem of Aleksandrov [**3**].

If all pairs of parallel faces of two closed convex polyhedra are such that neither face can be made to lie within the other by a translation, then the polyhedra are congruent and parallel.

Now since the areas of parallel faces of our polyhedra are equal, it is clear that the assumption of Aleksandrov's theorem holds. Hence the polyhedron is uniquely determined by the directions and areas of its faces, up to translation. This means that the mapping φ of A into B is one-to-one, and the second condition of the Mapping Lemma holds.

The third condition, which requires the mapping φ to be continuous, implies essentially that the areas of the faces must be continuous under continuous deformations which preserve their directions. This is obvious.

Finally, we prove that the fourth condition of the lemma holds. Let $b_s(F_1^s, \cdots, F_k^s)$ be a sequence of points of B, converging to a point $b_0(F_1^0, \cdots, F_k^0)$, P^s a closed polyhedron with exterior normals $n_1, \cdots, n_k$ and faces with areas $F_1^s, \cdots, F_k^s$. The fourth condition requires that there exist a polyhedron P^0 with exterior normals $n_1, \cdots, n_k$ and areas of faces $F_1^0, \cdots, F_k^0$ which is the limit of a sequence of polyhedra congruent to P^s. Now, since $F_i^0 > 0$, there exist positive numbers a and b such that $a < F_i^s < b$ for all i and s. Therefore, as was shown previously, the polyhedra P^s are uniformly bounded. Without loss of generality we may assume that the sequence P^s converges to a polyhedron P^0. Now the faces of the polyhedron P^0 obviously have exterior normals n_i and areas F_i^0, so that the fourth condition of the Mapping Lemma also holds.

By the Mapping Lemma, each point of B is the image of some point of A. This means that for any k-tuple of unit vectors n_i and positive numbers F_i such that $\sum n_i F_i = 0$ there exists a closed convex polyhedron with exterior normals n_i and faces of area F_i. This completes the proof of Lemma 1.

Theorem. *Let Ω be the unit sphere about the origin O. Let $K(n)$ be a given continuous positive function defined for the points n on the sphere, which satisfies the condition*

$$(**) \qquad \int_{\Omega} \frac{n\, d\omega}{K(n)} = 0,$$

where $d\omega$ is the area element on Ω and the integration extends over the entire sphere.

Then there exists a closed convex surface Φ, unique up to translation, whose Gauss curvature at the point with exterior normal n is $K(n)$.

In this section we shall prove only the existence part of the theorem. Uniqueness will follow in §4 from a more general theorem.

Proof. Divide the sphere Ω into small domains ω_α by meridians and parallels. Set

$$\int_{\omega_\alpha} \frac{n\, d\omega}{K(n)} = n_\alpha F_\alpha,$$

where n_α is a unit vector and F_α a positive number. In view of condition (**), for vectors n_α and numbers F_α satisfying the assumptions of Lemma 1 we have $\sum n_\alpha F_\alpha = 0$. Hence there exists a closed convex polyhedron P with exterior normals n_α and faces of area F_α. The area of the projection of P onto a plane perpendicular to a unit vector τ is

$$S_\tau = \frac{1}{2} \int_{\Omega} \frac{|n\tau|}{K(n)}\, d\omega > \varepsilon > 0,$$

where ϵ is a constant.

Construct a sequence of polyhedra P_s corresponding to progressively finer partitions of the sphere into domains ω_α. The polyhedra P_s are uniformly bounded. Indeed, the areas of the polyhedra P_s are bounded above, for they do not exceed

$$\int_{\Omega} \frac{d\omega}{K(n)},$$

and as was shown above their projections on any plane have areas bounded below by $\epsilon > 0$. It now suffices to repeat the reasoning of the analogous part of Lemma 1.

Since the polyhedra P_s are uniformly bounded, we may assume without loss of generality that they converge to a closed convex surface Φ. We claim that this surface is smooth and strictly convex. Assume that Φ has a conic point A. Without loss of generality, we may assume that for large s the polyhedra P_s all lie within the surface Φ in the neighborhood of A. Let G be the spherical image of the point A of Φ. Since A is a conic point, G contains a small disk g whose circumference does not cut the boundary of G. As $s \to \infty$, the total area of the faces of the polyhedra P_s whose normals lie in g tends to zero. But this is impossible, since

$$\int_g \frac{d\omega}{K} > \sum{}' F_\alpha > \bar{\varepsilon}(g) > 0,$$

where $\bar{\epsilon}$ is a constant which depends only on g and is the same for all the polyhedra.

If A is a ridge point, we employ the construction applied in §3 of Chapter II to prove that a surface with bounded specific curvature is smooth. To prove that Φ is strictly convex, we use the construction used to prove that a surface with positive specific curvature is strictly convex (ibid.). Both these proofs are omitted here.

Since the surface Φ is smooth and strictly convex, there exists for any unit vector n exactly one point with exterior normal n. We claim that the Gauss curvature of the surface at this point is $K(n)$. The Gauss curvature at a point $A(n)$ is defined here as the limit of the ratio of the curvature to area of a domain containing the point, when the domain contracts arbitrarily to the point $A(n)$.

Let G be an arbitrary domain on the unit sphere Ω. Let G' be a domain, contained in G together with its boundary, whose area ω' differs from that of G by at most $\epsilon > 0$. Let σ and σ' denote the total area of those faces of the polyhedron P_s whose spherical images lie in the domains G and G', respectively. Let σ_0 and σ_0' denote the areas of the domains on the surface Φ whose spherical images are G and G', respectively. For sufficiently large s, since Φ is smooth and strictly convex, we have $\sigma_0 > \sigma'$ and $\sigma > \sigma_0'$. Hence, letting $\epsilon \to 0$, we conclude that the domain G_Φ on Φ whose spherical image is G has area

$$S = \int_G \frac{d\omega}{K(n)}.$$

Hence the specific curvature of G_Φ is

$$\omega \Big/ \int_G \frac{d\omega}{K(n)}.$$

Letting the domain G contract to the point $A(n)$, and recalling that the function $K(n)$ is continuous, we see that the Gauss curvature of the surface Φ at the point with exterior normal n is indeed $K(n)$. Q.E.D.

§3. Regularity of a convex surface whose Gauss curvature is positive and a regular function of the normal

Both Minkowski's own solution of his problem and that presented in §2 are incomplete, if the function $K(n)$ is not only continuous but also regular to some degree. In such cases we would like to know whether a regular surface exists with the given Gauss curvature $K(n)$. In this section we shall prove that a convex (not necessarily closed) surface with regular $K(n)$ is regular. Combined with the result of §2, this furnishes a complete solution to Minkowski's problem.

The basic idea of the proof is as follows. Let F be a convex surface whose Gauss curvature at a point with exterior normal n is $K(n)$, where $K(n)$ is a regular function. Let Ω be the spherical image of F, X_0 an interior point of Ω and ω a disk about X_0 on the unit sphere, contained in the domain Ω. Let F_ω denote the part of the surface F whose spherical image is the disk ω.

First we show that there exists a regular convex surface F' possessing the following properties.

1. The convex surface F' has spherical image ω.

2. At each point with normal n, the surface F' has the same Gauss curvature as F_ω, i.e. $K(n)$.

3. The supporting functions $H(n)$ and $H'(n)$ of F_ω and F' coincide on the boundary of the disk ω.

We then prove that the surfaces F_ω and F' coincide. Since F' is regular, F has the same degree of regularity in a neighborhood of any point X_0, and hence everywhere, as required.

Analytically speaking, the construction of the convex surface F' amounts to solution of a Dirichlet problem for the equation

$$rt - s^2 = \varphi(x, y) > 0. \tag{1}$$

We therefore begin our account with the solution of this problem.

Lemma 1. *Let $\varphi(x,y)$ be a function analytic in the disk $x^2+y^2 \le R^2$, and ψ any function analytic on the circumference. Then equation* (1) *has two solutions in the disk which are equal to ψ on the circumference. If these solutions are $z_1(x,y)$ and $z_2(x,y)$, then the surfaces defined by the equations $z=z_1(x,y)$ and $z=z_2(x,y)$ are convex, one of them toward $z>0$ and the other toward $z<0$.*

The proof of this lemma is based on Bernšteǐn's theorem on the solvability of the Dirichlet problem for elliptic partial differential equations. This theorem states that the Dirichlet problem for an elliptic equation $f(r,s,t,p,q,z,x,y)=0$, $f'_z \geqq 0$ in a domain G bounded by an analytic contour $\widetilde{C}$ is solvable if, given the contour data, one can establish a priori upper estimates for the absolute values of the solution and its first and second derivatives in $G+\widetilde{C}$. To obtain estimates for the absolute values of the solution of equation (1) and its derivatives, we shall employ the arguments applied by Bernšteǐn to a particular case of the equation $\varphi(x,y)=\text{const}$ in [**21**].

Let $z(x,y)$ be a solution of equation (1). The surface $z=z(x,y)$ is obviously convex, since its Gauss curvature is positive. Assume that this surface is convex toward $z<0$, i.e. the derivatives r and t are positive. This being so, the function $z(x,y)$ assumes its maximum on the circle $x^2+y^2=R^2$, and therefore the maximum cannot exceed the maximum modulus of the function φ.

The minimum of the function $z(x,y)$ is easily estimated. To this end, consider the difference $z-k(x^2+y^2)/2=\bar{z}$, where k is a positive constant such that $k^2>\varphi(x,y)$ in the closed disk. It is readily seen that the function $\bar{z}(x,y)$ satisfies the equation

$$A\bar{z}_{xx}-2B\bar{z}_{xy}+C\bar{z}_{yy}=\varphi-k^2,$$

where

$$A=\frac{z_{yy}+k}{2},\quad B=\frac{z_{xy}}{2},\quad C=\frac{z_{xx}+k}{2}.$$

Hence the quadratic form $d^2\bar{z}$ cannot be positive definite at any point of the disk, and so $\bar{z}$ assumes its minimum on the boundary of the disk. Hence the minimum of $z(x,y)$ can be estimated in terms of k and the minimum of ψ.

We now estimate the maximum moduli of the derivatives p and q of z. First, note that since the surface $z=z(x,y)$ is convex, p and q assume their maximum and minimum on the circumference, and their maximum moduli need be estimated only there.

Bernšteǐn showed that there exists a plane $\sigma: z=ax+by+c$ through the tangent to the contour γ bounding the surface $z=z(x,y)$, at any point M, such that no point of γ lies below the plane σ, and the coefficients a, b, c of its equation are bounded in absolute value by a number which depends only on the maximum modulus of ψ and its derivatives of order up to three.

Estimates for the maximum moduli of p and q on the circle $x^2+y^2=R^2$

will clearly follow if we can estimate the derivative of z with respect to the interior normal to the circle. Without loss of generality, we shall establish such an estimate at the point $M(-R,0)$. At this point the derivative of z with respect to the normal is simply p. Since the surface $z=z(x,y)$ is convex toward $z<0$, the maximum of p can be estimated in an obvious way in terms of the maximum modulus of ψ. It is therefore sufficient to estimate the minimum of p. Consider the function

$$\tilde{z}=z-\frac{k}{2}(x^2+y^2-R^2)-ax-by-c.$$

This function has the following properties. On the circle $x^2+y^2=R^2$, we have $\tilde{z}\geqq 0$, and moreover $\tilde{z}=0$ at M. At no point of the disk $x^2+y^2\leqq R^2$ is the form $d^2\tilde{z}$ positive definite. It follows that the minimum of $\tilde{z}$ is assumed on the circumference, and hence $\tilde{z}\geqq 0$ in the disk. But then $\partial\tilde{z}/\partial x\geqq 0$ at M. Since a, b, c are bounded in absolute value by a number which depends only on the maximum moduli of ψ and its derivatives, we have the required estimate for the minimum of p.

In conclusion, note that the moduli of the first derivatives of $z(x,y)$ at an interior point of the disk $x^2+y^2\leqq R^2$ can be shown to satisfy estimates which depend only on the maximum modulus of ψ and the distance of the point from the circumference. More precisely, if m_0 is the maximum modulus of ψ and d the distance of X from the circumference, then the derivatives p and q are bounded in absolute value by $(m_0+kR^2/2)/d$.

It remains to derive estimates for the absolute values of the second derivatives. We first establish these estimates on the circle $x^2+y^2=R^2$.

In polar coordinates ρ, ϑ, equation (1) is

$$z_{\rho\rho}\left(\frac{1}{\rho^2}z_{\vartheta\vartheta}+\frac{1}{\rho}z_\rho\right)-\left(\frac{1}{\rho}z_{\rho\vartheta}-\frac{1}{\rho^2}z_\vartheta\right)^2=\varphi.$$

Differentiating this equation with respect to ϑ and denoting z_ϑ by z_1, we get

$$z_{1\rho\rho}\left(\frac{1}{\rho^2}z_{\vartheta\vartheta}+\frac{1}{\rho}z_\rho\right)-2\left(\frac{1}{\rho}z_{1\rho\vartheta}-\frac{1}{\rho^2}z_{1\vartheta}\right)\left(\frac{1}{\rho}z_{\rho\vartheta}+\frac{1}{\rho^2}z_\vartheta\right)$$
$$+\left(\frac{1}{\rho^2}z_{1\vartheta\vartheta}+\frac{1}{\rho}z_{1\rho}\right)z_{\rho\rho}=\varphi_\vartheta.$$

Hence, if μ is any constant, the function $z_2=z_1-\mu z$ satisfies the equation

$$(z_2)_{\rho\rho}\left(\frac{1}{\rho^2}z_{\vartheta\vartheta}+\frac{1}{\rho}z_\rho\right)-2\left(\frac{1}{\rho}(z_2)_{\rho\vartheta}-\frac{1}{\rho^2}(z_2)_\vartheta\right)\left(\frac{1}{\rho}z_{\rho\vartheta}-\frac{1}{\rho^2}z_\vartheta\right)$$
$$+\left(\frac{1}{\rho^2}(z_2)_{\vartheta\vartheta}+\frac{1}{\rho}(z_2)_\rho\right)z_{\rho\rho}=\varphi_\vartheta-\mu\varphi.$$

Let μ be so large that $\varphi_\vartheta - \mu\varphi < 0$ in the disk $x^2 + y^2 \leqq R^2$. Since

$$(z_2)_{xx}(z_2)_{yy} - (z_2)^2_{xy} = (z_2)_{\rho\rho}\left(\frac{1}{\rho^2}(z_2)_{\vartheta\vartheta} + \frac{1}{\rho}(z_2)_\rho\right) - \left(\frac{1}{\rho}(z_2)_{\rho\vartheta} - \frac{1}{\rho^2}(z_2)_\vartheta\right)^2 \leqslant 0,$$

the form $d^2 z_2$ cannot be positive definite; we can thus obtain an estimate for $|z_{2\rho}| = |z_{\rho\vartheta} - \mu z_\rho|$, hence also for $|z_{\rho\vartheta}|$, which depends only on the maximum modulus of ψ and its derivatives of order up to three with respect to ϑ.

To establish an estimate for $|z_{\rho\rho}|$, we first find a lower bound for the positive expression $z_{\vartheta\vartheta}/\rho^2 + z_\rho/\rho$.

Suppose that we wish to find a lower bound at the point $M(\vartheta = 0)$. In [21], Bernšteĭn proved that there exists a function

$$z_0 = a_{11}x^2 + 2a_{12}xy + a_{22}y^2 + a_1 x + a_2 y + a_0,$$

possessing the following properties:

1. $a_{11} > 0,\ a_{22} > 0,\ a_{11}a_{22} - a_{12}^2 = k^2$.
2. At the point M,

$$z_0 = \psi, \quad (z_0)_\vartheta = \psi_\vartheta, \quad (z_0)_{\vartheta\vartheta} = \psi_{\vartheta\vartheta}, \quad (z_0)_{\vartheta\vartheta\vartheta} = \psi_{\vartheta\vartheta\vartheta}.$$

3. At points of the circle other than M, $z_0 > \psi$.
4. At the point M,

$$\frac{1}{\rho^2}(z_0)_{\vartheta\vartheta} + \frac{1}{\rho}(z_0)_\rho > 2\alpha_0 > 0,$$

where α_0 depends only on the maximum moduli of the derivatives of ψ of order up to four.

Let k satisfy the condition $k^2 < \varphi(x, y)$ in the disk $x^2 + y^2 \leqq R^2$. Then the function $z_0 - z$ cannot be negative in the disk, and hence $(z_0 - z)_\rho \leqq 0$ at M; that is, $(z_0)_\rho \leqq z_\rho$. Hence at the point M the quantity $z_{\vartheta\vartheta}/\rho^2 + z_\rho/\rho$ cannot be smaller than

$$\frac{1}{\rho^2}(z_0)_{\vartheta\vartheta} + \frac{1}{\rho}(z_0)_\rho > 2\alpha_0 > 0.$$

We thus have estimates for the second derivatives of $z(x, y)$ on the circumference of the disk $x^2 + y^2 \leqq R^2$. To establish estimates in the interior, we consider the function $w = \lambda r$, where $\lambda > 0$ is some function of the variables x, y. Suppose that w assumes a maximum at a point P inside the disk. At P we have $w_x = 0$ and $w_y = 0$, whence

$$r_x = w\left(\frac{1}{\lambda}\right)_x, \qquad r_y = w\left(\frac{1}{\lambda}\right)_y,$$

$$r_{xx} = \frac{w_{xx}}{\lambda} + w\left(\frac{1}{\lambda}\right)_{xx}, \quad r_{xy} = \frac{w_{xy}}{\lambda} + w\left(\frac{1}{\lambda}\right)_{xy}, \quad r_{yy} = \frac{w_{yy}}{\lambda} + w\left(\frac{1}{\lambda}\right)_{yy}.$$

Differentiating equation (1) with respect to x, we get

$$r_x t + t_x r - 2ss_x = \varphi_x, \tag{2}$$

$$(r_{xx}t - 2r_{xy}s + r_{yy}r) + 2(r_x t_x - r_y^2) = \varphi_{xx}. \tag{3}$$

Using equation (2) and the expressions for the derivatives r_x and r_y at P, we can express $r_x t_x - r_y^2$ as follows:

$$r_x t_x - r_y^2 = -\left(\frac{\lambda_x}{\lambda}\right)\varphi_x + 2rs\left(\frac{\lambda_x}{\lambda}\right)\left(\frac{\lambda_y}{\lambda}\right) - rt\left(\frac{\lambda_x}{\lambda}\right)^2 - r^2\left(\frac{\lambda_y}{\lambda}\right)^2.$$

Now substitute into equation (3) our values for the second derivatives of r at P and the expression for $r_x t_x - r_y^2$:

$$\frac{1}{2}(tw_{xx} - 2sw_{xy} + rw_{yy}) + \lambda r\left\{t\left(\frac{1}{\lambda}\right)_{xx} - 2s\left(\frac{1}{\lambda}\right)_{xy} + r\left(\frac{1}{\lambda}\right)_{yy}\right\}$$
$$+ 2\left\{-\varphi_x\left(\frac{\lambda_x}{\lambda}\right) + 2rs\left(\frac{\lambda_x}{\lambda}\right)\left(\frac{\lambda_y}{\lambda}\right) - rt\left(\frac{\lambda_x}{\lambda}\right)^2 - r^2\left(\frac{\lambda_y}{\lambda}\right)^2\right\} = \varphi_{xx}.$$

Noting that

$$\left(\lambda\left(\frac{1}{\lambda}\right)_{xx} - 2\left(\frac{\lambda_x}{\lambda}\right)^2\right) rt = -\left(\frac{\lambda_{xx}}{\lambda}\right) rt,$$
$$-2\left(\lambda\left(\frac{1}{\lambda}\right)_{yy} - 2\left(\frac{\lambda_x}{\lambda}\right)\left(\frac{\lambda_y}{\lambda}\right)\right) rs = 2\left(\frac{\lambda_{xy}}{\lambda}\right) rs,$$
$$\left(\lambda\left(\frac{1}{\lambda}\right)_{yy} - 2\left(\frac{\lambda_y}{\lambda}\right)^2\right) r^2 = -\left(\frac{\lambda_{yy}}{\lambda}\right) r^2,$$

we transform the last equality as follows:

$$\frac{1}{\lambda}(tw_{xx} - 2sw_{xy} + rw_{yy}) - \frac{\lambda_{xx}}{\lambda} rt + 2\frac{\lambda_{xy}}{\lambda} rs$$
$$- \frac{\lambda_{yy}}{\lambda} r^2 - 2\varphi_x\left(\frac{\lambda_x}{\lambda}\right) = \varphi_{xx}.$$

Replacing rt by $s^2 + \varphi$ in this equality, and noting that

$$\varphi_{xx} + 2\varphi_x\left(\frac{\lambda_x}{\lambda}\right) + \varphi\frac{\lambda_{xx}}{\lambda} = \frac{1}{\lambda}(\varphi\lambda)_{xx},$$

we get

$$(tw_{xx} - 2sw_{xy} + rw_{yy}) - \lambda_{xx}s^2 + 2\lambda_{xy}rs - \lambda_{yy}r^2 = (\varphi\lambda)_{xx}.$$

Since the function $tw_{xx} - 2sw_{xy} + rw_{yy}$ is nonpositive at the point P (where w assumes a maximum), it follows that, at the same point,

$$\lambda_{xx}s^2 - 2\lambda_{xy}rs + \lambda_{yy}r^2 \leqq -(\varphi\lambda)_{xx}. \tag{4}$$

Setting $\lambda = 1 + y^2/2$, we see that $r^2 \leqq \{\varphi(1 + y^2/2)\}_{xx}|_{(P)}$ at the point P, and now we can estimate the derivative r throughout the disk $x^2+y^2 \leqq R^2$, inasmuch as an estimate on the circumference is known. An estimate for the derivative t is established in similar fashion. Finally, an estimate for s is derived from equation (1).

We have thus proved that a priori estimates can be established for the function $z(x,y)$ and its first and second derivatives; this implies the existence of a solution of equation (1) in the disk $x^2+y^2 \leqq R^2$ for arbitrary analytic boundary values on the circumference. Lemma 1 is proved.

LEMMA 2. *Let $\varphi(x,y)$ be a function k times differentiable $(k \geqq 3)$ in the disk $x^2+y^2 \leqq R^2$, and ψ a continuous function on the circumference of the disk. Then there exists a $k+1$ times differentiable solution $z(x,y)$ of equation* (1) *in the disk which coincides with ψ on the circumference.*

PROOF. We shall establish estimates for the maximum moduli of the solution of (1) and its first and second derivatives at an interior point of any domain G bounded by a convex contour $\bar{\gamma}$, without using the derivatives of the function ψ.

Using the same arguments as in the case of a disk G, the following estimate for $|z|$ is easily established:

$$|z| \leqq \|\psi\| + D^2\|\varphi\|/8,$$

where $\|\varphi\|$ and $\|\psi\|$ are the maximum moduli of φ and ψ in G and on its boundary $\bar{\gamma}$ and D is the diameter of the domain G.

Using the convexity of the surface $z=z(x,y)$, estimates for the absolute values of the derivatives p and q are readily obtained: If m_0 denotes the maximum modulus of the function $z(x,y)$ in the domain G, then, at a point M whose distance from the boundary of G is at least δ,

$$|p| \leqq 2m_0/\delta, \qquad |q| \leqq 2m_0/\delta.$$

We now establish estimates for the absolute values of the second derivatives at an arbitrary point $\bar{M}(\bar{x},\bar{y})$ in G. Let M be the point of the surface $z=z(x,y)$ whose projection is the point $\bar{M}$ in the domain G. Let $z=z_0(x,y)$ be a plane which separates M from the boundary γ of the surface $z=z(x,y)$ such that $z_0(x,y)-z(x,y) \leqq 0$ at every point $(x.y)$ of the domain whose distance from $\bar{\gamma}$ is at most $\delta/2$. Set

$$\Delta(\bar{M}) \equiv \sup(z_0(\bar{x},\bar{y}) - z(\bar{x},\bar{y})),$$

where the sup extends over all planes with these properties. We claim that $\Delta(\bar{M})$ can be bounded below by a constant which depends only on δ, m_0 (the maximum modulus of $z(x,y)$), and n_0 (the minimum of the function $\varphi(x,y)$).

Indeed, were this false, there would exist a number $\delta_0 > 0$, sequences of functions $z_k(x,y)$, $\varphi_k(x,y)$ and points $\overline{M}_k$ with the following properties:

1. In the domain G we have $|z_k| \leqq m_0$, $|\varphi_k| \geqq n_0$ and $r_k t_k - s_k^2 = \varphi_k$.
2. The distance of $\overline{M}_k$ from $\overline{\gamma}$ is at least δ_0, and $\Delta_k(\overline{M}_k) < 1/k$.

Without loss of generality we may assume that the sequence $z_k(x,y)$ converges to a function $z^*(x,y)$, and the sequence $\overline{M}_k$ to a point $\overline{M}^*$. Let $G_{(\delta_0/2)}$ be the set of points of G whose distance from $\overline{\gamma}$ is at least $\delta_0/2$. The surfaces defined by the equations $z = z_k(x,y)$ in $G_{(\delta_0/2)}$ are convex and their specific curvatures are bounded away from zero. Hence the same holds for the limit surface $z = z^*(x,y)$. On the other hand, it is clear that the tangent plane to this surface at the point M^* whose projection is $\overline{M}^*$ contains a point P of the surface whose projection $\overline{P}$ on the xy-plane lies at a distance at most $\delta/2$ from $\overline{\gamma}$. It follows that the limit surface contains a straight-line segment, and this is impossible (Chapter II, §3). This proves our assertion.

Let $z = z_0(x,y) = ax + by + c$ be the equation of a plane such that $z_0 - z \leqq 0$ in $G_{(\delta/2)}$ and $z_0 - z = \Delta(\overline{M})$ at the point $\overline{M}$. Consider the function $w = \lambda r = (z_0 - z) re^p$. It has a maximum in $G_{(\delta/2)}$. At the maximum point equation (4) holds; substituting $\lambda = (z_0 - z)e^p$ and simplifying, this equation becomes

$$e^p (z_0 - z) \varphi r^2 + (-2(p-a) e^p \varphi + 3(z_0 - z) e^p \varphi_x - 2e^p \varphi) r \\ - 2\varphi_x e^p (p - a) + e^p \varphi_{xx} (z_0 - z) \leqslant 0.$$

Hence w is bounded by some number w_0 which depends only on the maximum modulus of $z(x,y)$, the maximum modulus of $\varphi(x,y)$ and its derivatives, and the distance δ of the point $\overline{M}$ from the boundary of G. Thus $w = \Delta e^p r \leqq w_0$ at the point $\overline{M}$, and this yields an estimate for r. An estimate for t is established in a similar manner, and the estimate for s then follows from equation (1).

Now that we have estimates for the absolute values of $z(x,y)$ and its first and second derivatives within the domain G, we can establish estimates for the third and fourth derivatives of $z(x,y)$ which depend on the upper estimates for the absolute values of the first and second derivatives of z, the distance of $\overline{M}$ from the boundary of G, and the suprema of the absolute values of the functions φ, $1/\varphi$ and the derivatives of φ of order up to three (Chapter II, §11).

Construct a sequence of analytic functions φ_m which converge uniformly with their derivatives of order up to three in the disk $x^2 + y^2 \leq R^2$, and a sequence of analytic functions ψ_m which converge to the function ψ on the circumference $x^2+y^2=R^2$. We know that there exists an analytic function z_m satisfying the equation $rt-s^2=\varphi_m$ in the disk and equal to ψ_m on the circumference, such that the surface F_m defined by the equation $z=z_m(x,y)$ is convex toward $z<0$.

Since the functions $z_m(x,y)$ are uniformly bounded, we can extract from the sequence F_m a convergent subsequence. Without loss of generality, we may assume that the sequence F_m itself converges, say to a surface F whose equation is $z=z(x,y)$.

It is clear that an estimate can be found for the absolute values of the first derivatives of z_m in the disk $x^2+y^2\leqq R_1^2<R^2$, which depends only on m. It follows that the Gauss curvature of the surface F_m at a point whose projection lies in the disk $x^2+y^2\leqq R_1^2$ given by

$$\frac{rt-s^2}{(1+p^2+q^2)^2}$$

is bounded both above and below by positive constants independent of m. Consequently the specific curvature of any domain $G\subset F$ whose projection lies within the disk $x^2+y^2\leqq R_1^2$ is bounded between the same constants. Hence, by Aleksandrov's theorem, the surface F is smooth and strictly convex.

Since the limit surface F is strictly convex, the quantities $\Delta^m(\overline{M})$ appearing in the estimates for the second derivatives of $z_m(x,y)$ can be estimated from below by a positive constant which is independent of m and of the position of the point $\overline{M}$ in the disk $x^2+y^2\leqq R_1^2$. But this means that estimates independent of m can be found for the derivatives of $z_m(x,y)$ of order up to four in the disk $x^2+y^2\leqq R_1^2$. Hence the limit function $z(x,y)$ is at least thrice differentiable.

Since the function $\varphi(x,y)$ is by assumption k times differentiable, the function $z(x,y)$ (the solution of the equation $rt-s^2=\varphi$) is at least $k+1$ times differentiable. This proves Lemma 2.

THEOREM 1. *If the Gauss curvature of a convex surface is everywhere positive and regular* (m *times differentiable*, $m\geqq 3$) *as a function of the exterior normal, then the surface itself is regular* (*at least* $m+1$ *times differentiable*). *If the Gauss curvature is an analytic function of the normal, then the surface is analytic.*

PROOF. Let F be a convex surface with regular Gauss curvature $K(n)$. By Aleksandrov's theorem (Chapter II, §3), F is smooth and strictly convex.

Define rectangular cartesian coordinates x, y, z in space, with the origin O at some interior point of the convex body whose boundary contains the surface F. Let $H(x,y,z)$ be the supporting function of F.

Aleksandrov has shown [7] that the function H has a second differential almost everywhere on the unit sphere. Moreover, if the second differential of H exists at a point n on the unit sphere, then the Gauss curvature at the point of F with normal n (i.e. the limit of the ratio of the area of the spherical image to the area of the domain) is given by the formula

$$K(n) = \frac{1}{\begin{vmatrix} H_{xx} & H_{xy} \\ H_{xy} & H_{yy} \end{vmatrix} + \begin{vmatrix} H_{yy} & H_{yz} \\ H_{yz} & H_{zz} \end{vmatrix} + \begin{vmatrix} H_{zz} & H_{xz} \\ H_{xz} & H_{zz} \end{vmatrix}},$$

where the derivatives $H_{xx}, \cdots, H_{zz}$ refer to the point n on the unit sphere.

Let Ω be the spherical image of the surface F, and n_0 an arbitrary point on Ω. Let the z-axis be a straight line in the direction of the vector n_0, and consider the bounded region F_1 on F whose spherical image Ω_1 lies inside a spherical segment smaller than the hemisphere with pole n_0. Consider the function $h(x,y) = H(x,y,1)$, where $H(x,y,1)$ is the supporting function of the surface F_1. The function $h(x,y)$ is defined in the domain G_1 obtained by central projection of the spherical image of F_1 onto the plane $z=1$, followed by orthogonal projection onto the xy-plane. Since H is a convex function, so is h. If the function H has a second differential at the point $(x,y,1)$, then the Gauss curvature of F_1 at the point with exterior normal in the direction $(x,y,1)$ is

$$K = \frac{1}{(1+x^2+y^2)\left(h_{xx}h_{yy} - h_{xy}^2\right)}.$$

Consider the surface $\overline{F}_1$ defined in rectangular coordinates x, y, z by the equation $z = h(x,y)$. The surface $\overline{F}_1$ is convex. We claim that if the specific curvature of the surface F_1 is bounded, then so is that of $\overline{F}_1$. To prove this, we set up a pointwise correspondence between F_1 and $\overline{F}_1$, mapping the point (x,y,z) of $\overline{F}_1$ onto the point of F_1 at which the exterior normal has direction $(x,y,1)$.

Assume that the surface F_1 is twice differentiable. Let G be an arbitrary domain on F_1 and $\overline{G}$ the corresponding domain on $\overline{F}_1$. The specific curvatures of the domains G and $\overline{G}$ on the surfaces F_1 and $\overline{F}_1$ are

$$K_G = \frac{\sigma}{\iint (1+x^2+y^2)\left(h_{xx}h_{yy} - h_{xy}^2\right) d\sigma}, \qquad K_{\overline{G}} = \frac{\iint \frac{h_{xx}h_{yy} - h_{xy}^2}{(1+p^2+q^2)^2}\, dS}{S},$$

respectively, where σ is the area of the spherical image of the domain G and S the area of the domain $\overline{G}$. The definition of the correspondence between F_1 and $\overline{F}_1$ implies that the ratio σ/S is bounded between positive constants which are independent of the specific form of G. The specific curvature K_G is also bounded between positive constants independent of G; the derivatives p and q are uniformly bounded in a domain G_1 containing $\overline{G}$, since the surface F_1 is bounded (while p and q are always the abscissa and ordinate of some point of F_1). All this implies that $K_{\overline{G}}$ is also bounded between positive constants which are independent of the choice of $\overline{G}$. To prove this without assuming that F_1 is regular, construct a sequence of analytic convex surfaces converging to $\overline{F}_1$, determine the spherical images of the domains G on them, and pass to the limit.

Let n_0 be an arbitrary point of the domain Ω (the spherical image of F) and ω a small segment of the unit sphere with pole n_0, lying within the domain Ω. Define rectangular coordinates in space, with a point O on the exterior normal n_0 as origin and the z-axis along n_0. Let H be the supporting function of the surface F, and $h(x,y) = H(x,y,1)$. Project the segment ω onto the xy-plane. The projection is a disk $x^2+y^2 \leq R^2$. The function $h(x,y)$ has a second differential almost everywhere in this disk, and satisfies the equation

$$rt - s^2 = \frac{1}{K\,(1 + x^2 + y^2)},$$

where K is the Gauss curvature of F at the point with normal $(x,y,1)$.

Construct a regular function $\bar{h}(x,y)$ satisfying this equation in the disk and assuming the values $h(x,y)$ on its circumference. As was shown above, the surfaces defined by the equations $z = h(x,y)$ and $z = \bar{h}(x,y)$ have bounded specific curvature.

Consider the function $\tilde{h}(x,y) = h(x,y) - \bar{h}(x,y)$. It vanishes on the circumference of the disk $x^2+y^2 \leq R^2$. We claim that it must vanish throughout the disk. Suppose that, say, $\tilde{h} < 0$ at some point of the disk. Since $h(x,y)$ is a convex function, it has a second differential almost everywhere. Let M denote the set of points of the disk at which the second differential does not exist. Let $\widetilde{F}$ denote the surface defined by the equation $z = \tilde{h}(x,y)$. The boundary of this surface is the circle $x^2+y^2 = R^2$ in the xy-plane.

As was shown in §5 of Chapter II, there exists a convex surface $\Phi: \bar{z} = \varphi(x,y)$ whose boundary lies in the xy-plane within the circle $x^2+y^2 = R^2$, such that the surface is convex toward $z < 0$ and satisfies the following conditions:

1. The surface Φ has positive lower curvature at every smooth point.
2. If the projection of a smooth point X of Φ lies in a given set M of zero measure, the upper curvature of the surface at this point is infinite.

Consider the surface Φ_λ defined by the equation $z = \lambda\varphi(x,y)$. This surface also satisfies conditions 1 and 2. For a suitable value of the parameter λ, the surfaces $\widetilde{F}$ and Φ_λ have a common smooth point X, and the surface $\widetilde{F}$ lies above Φ_λ. Let Φ_λ' denote the surface defined by $z = \bar{h}(x,y) + \lambda\varphi(x,y)$.

The surface Φ_λ' and the surface $\overline{F}_1$ defined by the equation $z = h(x,y)$ also have a common point X', which lies on the same vertical straight line as X, and $\overline{F}_1$ lies above Φ_λ'. The surface Φ_λ' also satisfies conditions 1 and 2. Because of conditions 1 and 2 and the relative position of $\overline{F}_1$ and Φ_λ', the projection of the point X' cannot lie in the set M on the xy-plane, since then the lower curvature of the surface $\overline{F}_1$ at X' would be positive and the upper curvature infinite (Chapter II, §5). Thus the surface $\overline{F}_1$ must be twice differentiable at the point X'.

Let $\overline{X}$ be the projection of the points X and X' on the xy-plane. At this point

$$h_{xx}h_{yy} - h_{xy}^2 = \bar{h}_{xx}\bar{h}_{yy} - \bar{h}_{xy}^2 > 0.$$

Hence the quadratic form

$$(h_{xx} - \bar{h}_{xx})\xi^2 + 2(h_{xy} - \bar{h}_{xy})\xi\eta + (h_{yy} - \bar{h}_{yy})d\eta^2$$

is either indefinite or identically zero. On the other hand, by condition 1 the second differential

$$d^2(h - \bar{h}) = (h_{xx} - \bar{h}_{xx})dx^2 + 2(h_{xy} - \bar{h}_{xy})dx\,dy + (h_{yy} - \bar{h}_{yy})dy^2$$

must be a positive definite form. This is a contradiction, and so $h \equiv \bar{h}$. Consequently the function $h(x,y)$ is regular, and hence so is $H(x,y,z)$. This completes the proof of Theorem 1.

THEOREM 2. *Let Ω be the unit sphere about the origin O, and $K(n)$ a regular (k times differentiable, $k \geqq 3$) function defined for points n on the sphere. Assume that*

$$\int\limits_{\Omega} \frac{n\,d\omega}{K(n)} = 0,$$

where $d\omega$ is the area element on Ω and the integration extends over the entire sphere.

Then there exists a regular (at least $k+1$ times differentiable) surface

F whose Gauss curvature at the point with exterior normal n is $K(n)$. The surface F is uniquely determined up to translation.

This theorem is a corollary of the theorem of §2 and Theorem 1 of this section.

§4. Uniqueness theorems for convex surfaces with given function of principal radii of curvature

Let $f(R_1, R_2, n)$ be a function defined for unit vectors n and positive numbers R_1 and R_2 $(R_1 \leqq R_2)$, strictly monotone in R_1 and R_2, i.e. $f(R_1', R_2) > f(R_1, R_2)$ for $R_1' > R_1$, and $f(R_1, R_2') > f(R_1, R_2)$ for $R_2' > R_2$.

In this section we shall prove uniqueness theorems for surfaces with equal functions f of the principal radii of curvature and unit normal vector n.

Let F' and F'' be two strictly convex surfaces with a common point P and a common exterior normal n_0 at P. Let H' and H'' be the values of the supporting functions of these surfaces on the unit sphere. We say that the surfaces F' and F'' are in strong internal contact at the point P if for all n sufficiently close to n_0

$$|H'(n) - H''(n)| > c|n - n_0|^2, \quad c > 0, \quad n \neq n_0.$$

We say that two convex surfaces admit a strong internal contact if one of them can be brought by a translation to a position in which it is in strong internal contact with the other surface at some common point.

THEOREM 1. *Let F' and F'' be strictly convex surfaces, with a common spherical image ω lying together with its boundary on the unit hemisphere $x_1^2 + x_2^2 + x_3^2 = 1$, $x_3 > 0$.*

Then, if the supporting functions of the surfaces coincide on the boundary of the spherical image ω, the surfaces either coincide or admit a strong internal contact.

PROOF. Let $H'(n)$ and $H''(n)$ be the supporting functions of the surfaces F' and F''. If $H'(n) = H''(n)$ for all n in ω, the surfaces F' and F'' clearly coincide. If F' and F'' do not coincide, there exists a vector n_0 such that $H'(n_0) \neq H''(n_0)$; to fix ideas, let $H'(n_0) > H''(n_0)$.

Let $\bar{\omega}$ denote the maximal domain on the unit sphere, containing n_0, in which $H'(n) > H''(n_0)$. Consider the vector-valued function

$$\varphi(n) = n/(H'(n) - H''(n))$$

in the domain $\bar{\omega}$. The unbounded surface Φ defined by the equation

$r = \varphi(n)$ lies within a certain cone $x_3^2 = k(x_1^2 + x_2^2)$, $k > 0$ in the halfspace $x_3 > 0$. Cut the surface Φ by a plane $\alpha : z = c = \text{const}$. Let $\bar{\Phi}$ be the part of the surface below the plane α.

For sufficiently small ϵ, $\bar{\Phi}$ lies in the interior of the paraboloid defined by the equation $x_3 = \epsilon^2(x_1^2 + x_2^2)$. If the paraboloid is subjected to an affine transformation "flattening it out" onto the plane α, then at some stage of the transformation it will touch the surface $\bar{\Phi}$ at a point X_0. Let σ be the tangent plane to the paraboloid at X_0. It is also a tangent plane to the surface $\bar{\Phi}$ and therefore cannot contain the origin.

Let $r = \psi(n) = n/(an)$ be the equation of the plane σ (a is a constant vector). If n_0 is the unit vector directed to the point X_0, then for all n sufficiently close to n_0 we obviously have

$$|\varphi(n) - \psi(n)| \geqq c_1 |n - n_0|^2,$$

where c_1 is a positive constant, and equality holds only for $n = n_0$. Substituting φ and ψ into this inequality, and noting that $H'(n_0) \neq H''(n_0)$ and $(an_0) \neq 0$, we get

$$|(an) + H''(n_0) - H'(n_0)| \geqq c_2 |n - n_0|^2.$$

The term an can be included in either of the functions H' or H'', by simply translating the corresponding surface along the vector a $(-a)$. In the new position,

$$|H''(n) - H'(n)| \geqq c_2 |n - n_0|^2,$$

i.e. the surfaces F' and F'' admit a strong internal contact. Theorem 1 is proved.

The following theorem is a corollary of Theorem 1.

THEOREM 2. *Let $f(R_1, R_2, n)$ be an arbitrary function of the positive variables R_1, R_2, $R_1 \leqq R_2$ and the unit vector n, monotone in R_1 and R_2. If F' and F'' are twice differentiable convex surfaces satisfying the assumptions of Theorem* 1, *with equal values of the function f for the principal radii of curvature and unit exterior normal, then the surfaces are identical.*

Indeed, otherwise Theorem 1 states that the surfaces must admit a strong internal contact at some point with normal n. The second differential of the function $H = H'(x_1, x_2, x_3) - H''(x_1, x_2, x_3)$ at this point is a nonnegative form, not identically zero. We claim that this is impossible. Let R_1', R_2' be the principal radii of curvature of the surface F' at the point with normal n, and R_1'', R_2'' the principal radii of curvature of F'' at the point with the same exterior normal. Since the function

f is monotone, one of the following conditions must hold: 1. $R_1' = R_1''$, $R_2' = R_2''$. 2. $R_1' < R_1''$, $R_2' > R_2''$. 3. $R_1' > R_1''$, $R_2' < R_2''$.

Since the principal radii of curvature of a surface are the eigenvalues of the second differential of the supporting function (§1), it follows that in any of these three cases the form d^2H is either indefinite or identically zero. This contradiction proves Theorem 2.

THEOREM 3. *Let F' and F'' be two closed convex surfaces and $f(R_1, R_2, n)$ a function of the positive variables R_1, R_2 and the unit vector n, strictly monotone in R_1, R_2. Then, if*

$$f(R_1', R_2', n) = f(R_1'', R_2'', n), \quad R_1' \leqq R_2', \quad R_1'' \leqq R_2'',$$

at points of F' and F'' with parallel and identically oriented exterior normals, where R_1', R_2' and R_1'', R_2'' are the principal radii of curvature of the surfaces, then the surfaces are congruent and parallel, i.e. they are translates of each other.

This theorem was first proved by Aleksandrov for analytic surfaces F', F'' and any function f satisfying the monotonicity condition [**13**]. Aleksandrov himself then extended the proof to piecewise analytic surfaces and an analytic function f [**14**]. The present author weakened the requirement of analyticity of the surfaces and the function f to three-fold differentiability [**60**]. Recently Aleksandrov has weakened the differentiability conditions in a series of papers [**15**]. However, as yet the theorem has not been proved for the natural class of twice differentiable surfaces, with monotonicity the sole condition imposed on f.

Here we shall prove Theorem 3 for thrice differentiable surfaces, with monotonicity of the function f replaced by the slightly stronger condition

$$\partial f / \partial R_1 > 0, \qquad \partial f / \partial R_2 > 0.$$

In addition, in order to simplify the presentation we shall assume that

$$f(R_1, R_2, n) \equiv \tilde{f}(R_1 + R_2, R_1 R_2, n),$$

where $\tilde{f}$ is regular to the same degree as f, i.e. thrice differentiable.

LEMMA 1. *Let F' and F'' be twice differentiable convex surfaces for which the function $f(R_1, R_2, n)$ assumes equal values at points with the same normal n. Let $H'(x, y, z)$ and $H''(x, y, z)$ be the supporting functions of the surfaces. Set $h(x, y) = H'(x, y, 1) - H''(x, y, 1)$.*

Then the surface Φ defined by the equation $z = h(x, y)$ can have only hyperbolic points and flat points.

PROOF. Let $P(x^0, y^0, z^0)$ be an elliptic or parabolic point of the surface Φ. Let $z = ax + by + c$ be the equation of the tangent plane to Φ at P. Without loss of generality we may assume that $h_{xx} > 0$ at P. Then there exists a positive constant k^2 such that, for any fixed $\epsilon^2 > 0$ and points $Q(x, y, z)$ on Φ sufficiently close to P,

$$h(x,y) - (ax + by + c) \geqq k^2(x - x^0)^2 - \epsilon^2(y - y^0)^2.$$

Replace x, y by homogeneous variables, setting $x = x_1/x_3$, $y = x_2/x_3$. Then

$$H'(x_1, x_2, x_3) - H''(x_1, x_2, x_3) - (ax_1 + bx_2 + cx_3) \geqslant k^2 (x_1 x_3^0 - x_1^0 x_3)^2 - \varepsilon^2 (x_2 x_3^0 - x_2^0 x_3)^2.$$

Hence the second differential at the point (x_1^0, x_2^0, x_3^0) satisfies the inequality

$$d^2(H' - H'') \geqq k^2(x_3^0 dx_1 - x_1^0 dx_3)^2 - \epsilon^2(x_3^0 dx_1 - x_2^0 dx_3)^2.$$

Since ϵ^2 is arbitrarily small, we have

$$d^2(H' - H'') \geqq k^2(x_3^0 dx_1 - x_1^0 dx_3)^2,$$

at this point, i.e. the second differential $d^2(H' - H'')$ is neither identically zero nor an indefinite form. This contradiction[2)] proves Lemma 1.

LEMMA 2. *If the function $h_1 = \partial h/\partial x$ has an extremum at some point P of the surface Φ, then P is a flat point of the surface.*

In fact, by the extremum conditions at P we have $h_{11} = 0$ and $h_{12} = 0$. The point P cannot be hyperbolic, since $h_{11}h_{22} - h_{12}^2 = 0$. Hence it must be flat.

As was shown in §1, the principal radii of curvature of a convex surface with supporting function H are the nonzero roots of the equation

$$\begin{vmatrix} H_{11} + R & H_{12} & H_{13} \\ H_{21} & H_{22} + R & H_{23} \\ H_{31} & H_{32} & H_{33} + R \end{vmatrix} = 0.$$

This gives the following expressions for the sum and product of the principal radii of curvature:

---

2) It was shown at the end of the proof of Theorem 2 that the form $d^2(H' - H'')$ is either indefinite or identically zero.

$$R_1 + R_2 = H_{11} + H_{22} + H_{33},$$

$$R_1R_2 = \begin{vmatrix} H_{11} & H_{12} \\ H_{21} & H_{22} \end{vmatrix} + \begin{vmatrix} H_{22} & H_{23} \\ H_{32} & H_{33} \end{vmatrix} + \begin{vmatrix} H_{33} & H_{31} \\ H_{13} & H_{11} \end{vmatrix}.$$

If we set

$$H(x, y, z) = zh\left(\frac{x}{z}, \frac{y}{z}\right), \qquad h(\alpha, \beta) = H(\alpha, \beta, 1)$$

the sum and product of the principal radii of curvature at the point with normal in direction $(x, y, 1)$ are

$$R_1 + R_2 = (h_{11}(1 + x^2) + 2h_{12}xy + h_{22}(1 + y^2))(1 + x^2 + y^2)^{1/2},$$
$$R_1R_2 = (h_{11}h_{22} - h_{12}^2)(1 + x^2 + y^2).$$

(The subscripts in the right-hand sides indicate differentiation with respect to the corresponding arguments.)

Let φ denote the common value of the function f for the surfaces F' and F''. If we replace $R_1 + R_2$ and R_1R_2 in the function f by their expressions in terms of h, we get a differential equation for the function h:

(*) $$\psi(h_{11}, h_{12}, h_{22}, x, y) = \varphi(x, y).$$

We claim that this equation is elliptic.

Indeed, the type of the equation is invariant under a substitution of variables, in particular, if the coordinate system x, y, z is so chosen that the x- and y-directions coincide with the principal directions at the given point on the surface. Then, at this point,

$$R_1 + R_2 = h_{11} + h_{22}, \qquad R_1R_2 = h_{11}h_{22}.$$

Consequently $h_{11} = R_1$, $h_{22} = R_2$. Therefore

$$\frac{\partial\psi}{\partial h_{11}} = f_1 + f_2R_2 = \frac{\partial f}{\partial R_1}$$
$$\frac{\partial\psi}{\partial h_{22}} = f_1 + f_2R_1 = \frac{\partial f}{\partial R_2},$$
$$\frac{\partial\psi}{\partial h_{12}} = 0,$$

where f_1 and f_2 are the derivatives of f with respect to $R_1 + R_2$ and R_1R_2. The ellipticity condition

$$\frac{\partial\psi}{\partial h_{11}}\frac{\partial\psi}{\partial h_{22}} - \frac{1}{4}\left(\frac{\partial\psi}{\partial h_{12}}\right)^2 > 0$$

is clearly satisfied at the point, since by assumption $\partial f/\partial R_1 > 0$ and

$\partial f/\partial R_2 > 0$. Hence the equation is indeed elliptic.

Set

$$h' = H'(x, y, 1), \qquad h'' = H''(x, y, 1).$$

We claim that the difference $h = h' - h''$ satisfies a linear elliptic equation

$$(**) \qquad Ah_{11} + Bh_{12} + Ch_{22} = 0.$$

Indeed,

$$\psi(h'_{11}, h'_{12}, h'_{22}, x, y) - \psi(h''_{11}, h''_{12}, h''_{22}, x, y) = 0.$$

Using the Hadamard lemma, we indeed get an equation of type (**) for h. We must show that it is elliptic. This will be done for a neighborhood of the point P at which $h_{11} = h_{12} = h_{22} = 0$. At this point

$$A = \partial\psi/\partial h_{11}, \qquad B = \partial\psi/\partial h_{12}, \qquad C = \partial\psi/\partial h_{22},$$

and the ellipticity of equation (**) at the point P, hence also in a neighborhood of P, follows from that of equation (*).

Proof of Theorem 3. Let $\alpha(\alpha_1, \alpha_2, \alpha_3)$ be an arbitrary but fixed point of the unit sphere Ω about the origin. Let H_α denote the derivative of the function $H = H' - H''$ in the direction $(\alpha_1, \alpha_2, \alpha_3)$; H_α, regarded as a function defined on the sphere Ω, assumes a maximum at some point. There are two possible cases:

1. For each α, H_α assumes its maximum at α or at the antipodal point $-\alpha$.

2. For some α, H_α assumes its maximum neither at α nor at $-\alpha$.

Consider the first case. We first show that if H_α has a maximum at the point α (or $-\alpha$), then $d^2H = 0$ at this point. Let the direction of the z-axis be α. Setting $x/z = \xi$, $y/z = \eta$, we have $H_z = h - \xi h_1 - \eta h_2$. By Lemma 1, the surface Φ defined by the equation $\zeta = h(\xi, \eta)$ in cartesian coordinates ξ, η, ζ can have only hyperbolic or flat points. If $\xi = \eta = 0$ is a flat point, then $h_{11} = h_{12} = h_{22} = 0$ there. It is readily seen that the second derivatives H_{ij} are linear and homogeneous in the derivatives h_{kl}. Hence the derivatives H_{ij} all vanish if the corresponding point of Φ is flat. We claim that the point $\xi = \eta = 0$ cannot be hyperbolic. Indeed, in a neighborhood of this point

$$h = c + a\xi + b\eta + A\xi^2 + 2B\xi\eta + C\eta^2 + \cdots,$$

where higher order terms have been omitted, and

$$H_z = c - (A\xi^2 + 2B\xi\eta + C\eta^2) + \cdots.$$

Since the form in parentheses is indefinite, H_z cannot have a maximum at the point $\xi=\eta=0$. Thus $\xi=\eta=0$ cannot be a hyperbolic point.

Let N be the set of all points α on the unit sphere Ω such that H_α or $H_{-\alpha}$ assumes a maximum at α. N is obviously a closed set, and so its complement $M=\Omega-N$ is open. The set N has interior points. This is obvious if $N\equiv\Omega$. If not, then M is not empty. The set M^* antipodal to M consists entirely of interior points and is a subset of N.

Let P be an interior point of N. Define the coordinate system so that P is on the hemisphere $z>0$. In a neighborhood of the point P, we have $d^2H\equiv 0$, and so H is a linear function. By subjecting one of the surfaces to a translation, we can ensure that $H\equiv 0$ in a neighborhood of P. Hence the function $h(x,y)=H(x,y,1)$ will vanish in a neighborhood of a point $\bar{P}$ in the xy-plane. By the uniqueness theorem for equation (**), it follows that $h(x,y)\equiv 0$ throughout the xy-plane.

Consider the surface $\bar{\Phi}$ defined by the equation $z=\bar{h}(x,y)=H'(x,y,-1)-H''(x,y,-1)$. Since $h\equiv 0$, we have $\bar{h}\to 0$ as $x^2+y^2\to\infty$. Since $\bar{\Phi}$ has nonpositive curvature, it follows that $\bar{h}\equiv 0$. But $h\equiv 0$ and $\bar{h}\equiv 0$ imply that $H'-H''\equiv 0$, i.e. the surfaces F' and F'' coincide.

Now consider the second case. Suppose that α is a point such that H_α does not assume a maximum at α or $-\alpha$. Define the coordinates x, y, z so that the x-axis passes through α and $-\alpha$, while $z>0$ at the point where H_α assumes a maximum. We again consider the surface Φ defined by $z=h(x,y)=H'(x,y,1)-H''(x,y,1)$. Since the function $h_1=\partial h/\partial x$ assumes a maximum at some point P, we have $h_{11}=h_{12}=0$ at this point. Hence also $h_{22}=0$ (the point cannot be hyperbolic). Without loss of generality, we may assume that $h=h_1=h_2=0$ at the point P. This can always be guaranteed by a suitable translation of one of the surfaces F' or F''. If all points in a neighborhood of P are flat points of Φ, it follows as in the previous case that F' and F'' are congruent. Otherwise there exists a sequence of hyperbolic points P_k converging to P.

Let α_k be the tangent plane to Φ at the point P_k. For sufficiently large k, this plane approaches the tangent plane at P, i.e. the xy-plane. Without loss of generality we may assume that the slopes of the plane α_k in the x and y directions are at most $1/k$ in absolute value.

Since P_k is a hyperbolic point, the neighborhood of P_k on Φ is divided by the plane α_k into four sectors V_1, V_2, V_3, V_4, two of them, say V_1 and V_3, above the plane α_k, and the other two (V_2 and V_4) below it. Though the sectors V_1, V_3 and V_2, V_4 lie on one side of α_k, they belong to different components G_k of the partition of Φ induced by the plane α_k. Indeed, suppose that points A_1 and A_3 of the sectors V_1 and V_3 can be connected by a curve γ lying above the plane α_k. Connect A_1 and A_3

to P_k by curves γ_1 and γ_3 in the sectors V_1 and V_3. The closed contour $\gamma+\gamma_1+\gamma_3$ encloses one of the components G_k, containing either V_2 or V_4. But then the plane α_k cuts off a "hummock" from this component. Since the surface Φ has nonpositive curvature, this is impossible.

Now as $x \to +\infty$ and $x \to -\infty$ the function $h_1(x,0)$ converges to a negative limit (the maximum of h_1 is zero). Hence, for sufficiently large positive x, all points $(x,0)$ of the surface Φ lie in one component of the partition of Φ by α_k. Denote this component by G^+. For sufficiently large positive x the points $(-x,0)$ all lie in a component G^-. Moreover, for every fixed y and large x the points (x,y) are in G^+, the points $(-x,y)$ in G^-.

Let G_1, G_2, G_3, G_4 be the components of the partition of Φ by α_k containing the sectors V_1, V_2, V_3, V_4, respectively. One of these components, say G_1, is neither G^+ nor G^-. Since G_1 is not bounded, it goes to infinity in at least one of the directions $y>0$ or $y<0$, say in the direction $y>0$.

Cut the surface Φ by a plane $y=\text{const}>0$. Denote the intersections of the components G^+, G^- and G_1 with this plane by g^+, g^- and g_1, respectively. Since g^+ and g^- contain curves going to infinity in the directions $x>0$ and $x<0$, it follows that g_1 consists of bounded curves with endpoints on the plane α_k. Consequently g_1 contains at least one point for which $|h_1|<1/k$. Letting $P_k \to P$, we see that in one of the halfspaces, say $y>0$, each straight line $y=\text{const}$ contains a point at which $h=0$ and $h_1=0$.

Let M denote the set of all points of the xy-plane at which $h=0$ and $h_1=0$. As we have already proved, $h_{11}=h_{12}=h_{22}=0$ at these points. Every straight line $y=\text{const}>0$ contains points of M, and the intersection of each such line with M is either a single point or a segment. Otherwise the intersection would contain points with $x_1>0$, which is impossible (the maximum of h_1 is zero). If one of the intersections contains a segment, then the function h is linear along this segment, since the derivatives h_{ij} vanish there. By a suitable translation of one of the surfaces we can ensure that $h=h_1=h_2=0$ along the segment. By the uniqueness theorem for equation (**) it follows that $h\equiv 0$. Reasoning as before, we now conclude that the surfaces F' and F'' are identical.

If each intersection $y=\text{const}>0$ contains only one point $P(y)$ of the set M, this point is a continuous function of y. Otherwise, some intersection for which $P(y)$ is discontinuous would have to contain at least two points of M, hence an entire segment. We claim that at each point $P(y)$ we have $h_2=0$. Indeed, if $h_2\neq 0$ at $P(y)$, then M is a smooth curve in the neighborhood of this point which can be projected one-to-one onto the x- and y-axes. Since $h_1 = 0$ along this curve, its tangents must be parallel to the y-axis at each point near $P(y)$. Hence the curve contains segments parallel to the y-axis; but this contradicts the fact that $P(y)$ is single-valued. Thus $h_2 = 0$ at each point $P(y)$. Since $P(y)$ is continuous in y, the point P is the limit of points in the xy-plane at which $h = h_1 = h_2 = 0$. By the uniqueness theorem for equation (**) it follows that $h \equiv 0$. As before, this implies that the surfaces F' and F'' coincide. Theorem 3 is proved.

In view of the fact that Theorem 3 was proved under certain differentiability assumptions on the function f, the following theorem is not devoid of interest.

THEOREM 4. *Let F be a twice differentiable convex surface and $f(R_1, R_2)$ any function monotone in both variables in the domain $R_1 > 0$, $R_2 > 0$, $R_1 \leqq R_2$.*

Then, if the principal radii of curvature R_1 and R_2 of the function F satisfy the condition $f(R_1, R_2) = \text{const}$, *the surface F is a sphere.*

PROOF. Introduce arbitrary cartesian coordinates x, y, z. Let $H(x,y,z)$ be the supporting function of the surface F. Set

$$h(x,y) = H(x,y,1) - H(x,y,-1).$$

As in Lemma 1, it can be proved that the surface Φ defined by the equation $z = h(x,y)$ can have only hyperbolic or flat points.

We claim that the function $h(x,y)$ remains bounded as $x^2 + y^2 \to \infty$. Otherwise there exists a sequence of points (x_k, y_k) such that $x_k^2 + y_k^2 \to \infty$ and $h(x_k, y_k) \to \infty$. Without loss of generality we may assume that as $k \to \infty$

$$\frac{x_k}{\sqrt{1 + x_k^2 + y_k^2}} \to \xi, \qquad \frac{y_k}{\sqrt{1 + x_k^2 + y_k^2}} \to \eta.$$

This can always be achieved by extracting a suitable subsequence of (x_k, y_k).

Set

$$x' = \frac{x}{\sqrt{1 + x^2 + y^2}}, \quad y' = \frac{y}{\sqrt{1 + x^2 + y^2}}, \quad z' = \frac{1}{\sqrt{1 + x^2 + y^2}}.$$

Let P be the point on the unit sphere with coordinates ξ, η, 0. As $k \to \infty$,

$$h(x_k, y_k) = \frac{H(x'_k, y'_k, z'_k) - H(x'_k, y'_k, -z'_k)}{z'_k} \to 2H_z(\xi, \eta, 0) < \infty.$$

This contradiction shows that $h(x,y)$ is indeed bounded.

A well-known theorem of Bernšteĭn [20] states that the only surfaces Φ containing only hyperbolic or flat points, for which the function h is bounded, are the surfaces $z = \text{const}$.

By a translation of the coordinate systems we can make $h(x,y)$ identically zero. Then the xy-plane is a plane of symmetry for the surface F. Since the coordinate systems xyz was arbitrary, the surface F has planes of symmetry in all directions, and hence it is a sphere. Q.E.D.

In conclusion, we mention that Theorem 3 has been proved for arbitrary differentiability properties of the function f, on the assumption that f is an even function of n: $f(R_1, R_2, n) = f(R_1, R_2, -n)$ (Medjanik [42]).

§5. Existence of a convex surface with given function of principal radii of curvature

Let F be a twice differentiable closed convex surface with positive Gauss curvature, and $f(R_1, R_2)$ a function defined in the domain $R_1 > 0$, $R_2 > 0$, $R_1 \geqq R_2$. Let $R_1(n)$ and $R_2(n)$ $(R_1(n) \geqq R_2(n))$ be the principal radii of curvature of F at the point with exterior normal n. The surface F and function f define a certain function $\varphi(n) = f(R_1(n), R_2(n))$ of the unit vector n. A natural problem is to find sufficient conditions on the given functions f and φ for the existence of a closed convex surface such that

$$f(R_1, R_2) = \varphi(n). \tag{1}$$

The problems studied in §§1 and 2 are special cases of this general problem. Christoffel's problem corresponds to $f(R_1, R_2) = R_1 + R_2$, Minkowski's problem to $f(R_1, R_2) = R_1 R_2$.

In this section we shall consider the general problem. In its analytical interpretation the problem reduces to the solvability of a certain nonlinear equation for the supporting function of the surface; this equation is derived from equation (1). The main idea of the method of solution is to embed the equation (surface) in a continuous family of equations (surfaces), one of which has a known solution. This presupposes the availability of a priori estimates of the solution together with its first and second derivatives, or, equivalently, estimates of the principal radii of curvature of the surface. We therefore begin our discussion with the derivation of a priori estimates for the principal radii of curvature of a closed convex surface satisfying equation (1). Both functions f and φ are assumed to be twice differentiable, and f strictly monotone, i.e.

$$\partial f / \partial R_1 > 0, \qquad \partial f / \partial R_2 > 0. \tag{2}$$

THEOREM 1. *Let P_1 be a point on F at which R_1 assumes a maximum, P_2 a point at which R_2 assumes a minimum. Then*

$$(R_2 - R_1)\frac{\partial f}{\partial R_2} + \frac{\partial^2 f}{\partial R_2^2}\frac{\varphi_s^2}{(\partial f/\partial R_2)^2} \geqslant \varphi_{ss}$$

at P_1, and

$$(R_1 - R_2)\frac{\partial f}{\partial R_1} + \frac{\partial^2 f}{\partial R_1^2}\frac{\varphi_s^2}{(\partial f/\partial R_1)^2} \leqslant \varphi_{ss}$$

at P_2. Here φ is differentiated with respect to arc length s along a great circle on the unit sphere, in the corresponding principal direction.

PROOF. Assume that $R_1 > R_2$ at the point P_1. Then, in a neighborhood of this point,

$$f(R_1, R_2) \equiv g(R_1 R_2, R_1 + R_2),$$

where g is also a twice differentiable function. Thus in a neighborhood of P_1 equation (1) becomes

$$g(R_1 R_2, R_1 + R_2) = \varphi(n). \tag{3}$$

Take the point P_1 on F as the origin, and the principal directions at P_1 as the directions of the x- and y-axes. To fix ideas, let the x-axis correspond to the principal direction with radius of curvature R_1. Let $H(x,y,z)$ denote the supporting function of the surface. Then for $x = y = 0$, $z = 1$ we have

$$H = 0, \quad H_x = 0, \quad H_y = 0.$$

Thus, setting $h(x,y) = H(x,y,1)$, we have

$$h = 0, \quad h_x = 0, \quad h_y = 0.$$

In §4 we derived expressions for the sum and product of the principal radii of curvature:

$$\begin{aligned} R_1 R_2 &= (rt - s^2)(1 + x^2 + y^2) \\ R_1 + R_2 &= ((1 + x^2) r + 2xys + (1 + y^2) t)\sqrt{1 + x^2 + y^2}, \end{aligned} \tag{4}$$

where r, s, t are the usual symbols for the second derivatives of h. Note that at the point P_1 itself ($x = y = 0$) we have $s = 0$ owing to the specific choice of coordinate system, and so $r = R_1$, $t = R_2$.

By projection from the point P_1 onto the unit sphere $x^2 + y^2 + z^2 = 1$, the cartesian coordinate system on the plane $z = 1$ induces a curvilinear system on the sphere. The coordinate curves $x = \text{const}$ and $y = \text{const}$ on the sphere are great circles. Let Z be a cylinder projecting the surface

F onto the plane of a great circle $y = \text{const}$. The radius of curvature R of this cylinder along its line of contact with the surface is bounded between the principal radii of curvature of the surface, i.e.

$$R_2 \leqq R \leqq R_1. \tag{5}$$

The radius of curvature R is given by the usual formula (see §1): $R = p + p_{ss}$, where p is the value of the supporting function H on the unit circle $y = \text{const}$, and the differentiation is with respect to arc length along this circle. The radius of curvature R is a certain function $w(x, y)$ of the coordinates x, y. Noting that

$$p = \frac{1}{\sqrt{1 + x^2 + y^2}} h(x, y),$$

we get the following formula for $w(x, y)$:

$$w = r \frac{(1 + x^2 + y^2)^{3/2}}{1 + y^2}.$$

Since the direction $y = 0$ at the point $(0, 0)$ corresponds to the principal direction with radius R_1, it follows that $w(0, 0) = R_1$ and hence, by (5), the function $w(x, y)$ has a maximum at the point $(0, 0)$. At this point, therefore,

$$w_x = r_x = 0, \qquad w_y = r_y = 0,$$
$$w_{xx} = r_{xx} + 3r \leqslant 0, \qquad w_{yy} = r_{yy} + r \leqslant 0.$$

Differentiating (3) with respect to x at $(0, 0)$, we get

$$g_1 (r_x t + r t_x) + g_2 (r_x + t_x) = \varphi_x.$$
$$g_1 (r_{xx} + 2 r_x t_x + r t_{xx} + 4rt) + g_2 (r_{xx} + t_{xx} + 3r + t)$$
$$+ g_{11} (r_x t + r t_x)^2 + 2 g_{12} (r_x t + r t_x)(r_x + t_x) + g_{22} (r_x + t_x)^2 = \varphi_{xx}.$$

Since $r_x = 0$, $r = R_1$, $t = R_2$ at $(0, 0)$, the first equality implies

$$t_x = \frac{\varphi_x}{g_1 R_1 + g_2} = \varphi_x \Big/ \frac{\partial f}{\partial R_2}.$$

The second equality then becomes

$$w_{xx} \frac{\partial f}{\partial R_1} + w_{yy} \frac{\partial f}{\partial R_2} + (R_2 - R_1) \frac{\partial f}{\partial R_2} + \frac{\partial^2 f}{\partial R_2^2} \frac{\varphi_x^2}{(\partial f / \partial R_2)^2} = \varphi_{xx}.$$

Hence, since $w_{xx} \leqq 0$, $w_{yy} \leqq 0$, we get the inequality

$$(R_2 - R_1) \frac{\partial f}{\partial R_2} + \frac{\partial^2 f}{\partial R_2^2} \frac{\varphi_x^2}{(\partial f / \partial R_2)^2} \geqslant \varphi_{xx}.$$

It is readily seen that

$$\partial x/\partial s = \frac{1+x^2+y^2}{\sqrt{1+y^2}}.$$

Therefore $\varphi_x = \varphi_s$ and $\varphi_{xx} = \varphi_{ss}$ at the point $(0,0)$, and this gives the required inequality.

At the beginning of the proof we assumed that $R_1 > R_2$ at the point P_1. But our inequality is clearly valid for $R_1 = R_2$ as well: in this case it represents a trivial fact—the function φ has a maximum at P_1. The proof of the second assertion of the theorem, concerning the minimum of R_2, is analogous, and will therefore be omitted.

We consider two examples. Let $f(R_1, R_2) = R_1 R_2$. Then our inequality is $(R_2 - R_1) R_1 \geqq \varphi_{ss}$. Hence, since $R_1 R_2 = \varphi$, we get $R_1^2 \leqq \varphi - \varphi_{ss}$. This estimate plays an essential role in some solutions of Minkowski's problem (Miranda [**46**], Nirenberg [**49**]).

Now let $f(R_1, R_2) = R_1 + R_2$. Then $R_1 - R_2 < \varphi_{ss}$ at the point P_2 at which R_2 assumes a minimum. Hence we obtain a lower estimate for R_2:

$$2R_2 \geqq \varphi - \varphi_{ss}.$$

We used this estimate in our solution of Christoffel's problem in §1.

Theorem 4 of §4, which states that a closed convex surface for which $f(R_1, R_2) = \text{const}$ is a sphere, is a corollary of Theorem 1. Indeed, if the surface is not a sphere, then $R_1 > R_2$ at the point P_1. On the other hand, by Theorem 1, $(R_2 - R_1)\partial f/\partial R_2 \geqq 0$ at this point, i.e. $R_2 \geqq R_1$, which is absurd. However, this proof cannot replace that of §4, since it assumes that the surface is four times differentiable and the function f twice differentiable.

Theorem 1 yields general conditions for the existence of a priori estimates of the radii of curvature of a closed convex surface satisfying equation (1). Let us say that functions f and φ satisfy condition (*) if

$$(*) \qquad \varlimsup_{\substack{R_2 = R_2(R_1, n) \\ R_1 \to \infty}} \left\{ (R_2 - R_1)\frac{\partial f}{\partial R_2} + \frac{\partial^2 f}{\partial R_2^2}\frac{\varphi_s^2}{(\partial f/\partial R_2)^2} \right\} < \varphi_{ss}$$

for any point n on the unit sphere and any direction of differentiation (with respect to s) at this point. We shall say that these functions satisfy condition (**) if

$$(**) \qquad \varliminf_{\substack{R_1 = R_1(R_2, n) \\ R_2 \to 0}} \left\{ (R_1 - R_2)\frac{\partial f}{\partial R_1} + \frac{\partial^2 f}{\partial R_1^2}\frac{\varphi_s^2}{(\partial f/\partial R_1)^2} \right\} > \varphi_{ss}.$$

Under these conditions, $R_1(R_2, n)$ and $R_2(R_1, n)$ are solutions of equation (1) for R_1 and R_2, respectively.

Theorem 2. *If the functions f and φ satisfy condition* (*), *then the radii of normal curvature of a closed convex surface F satisfying equation* (1) *satisfy an a priori upper estimate. If f and φ satisfy condition* (**), *then the radii of normal curvature satisfy a positive lower estimate. These estimates depend only on f and φ.*

Theorem 2 follows directly from Theorem 1.

Remark. This method for proving the existence of a priori estimates is also applicable to the more general case, in which the function f depends also on n and equation (1) is $f(R_1, R_2, n) = \varphi(n)$.

We now consider the existence of a closed convex surface F satisfying equation (1). Since the statement of the final result (Theorem 3) is rather complicated, we shall postpone it to the end of the section and begin with the existence proof, subjecting f and φ to suitable restrictions as the need arises. We first assume that the function f is defined for all positive R_1 and R_2, symmetric ($f(R_1, R_2) = f(R_2, R_1)$), and strictly increasing:

$$\partial f/\partial R_1 > 0, \qquad \partial f/\partial R_2 > 0.$$

The function φ is assumed to be even ($\varphi(n) = \varphi(-n)$). Both functions f and φ are first assumed to be analytic (later we shall relax this condition to two-fold differentiability).

By Theorem 3 of §4, equation (1) determines the surface F uniquely up to translation. Let F^* be a surface symmetric to F with respect to a point O. Since the function $\varphi(n)$ is even, F^* also satisfies equation (1). Hence F and F^* are translates of each other, and this means that F must have a center of symmetry.

Let the surface F be transformed by an infinitesimal centrally-symmetric deformation into a surface F_t with supporting function $H_t = H + tZ$, where H is the supporting function of F. If the function f is stationary for the surface F_t at $t = 0$, i.e. $\partial f(R_1(t), R_2(t))/\partial t = 0$, then $Z \equiv 0$. This assertion can be proved using the same arguments as in the proof of Theorem 3 of §4. We shall not present the proof, since we need the result only for the case in which F and the deformation are analytic, and in this form it has in effect been proved by Aleksandrov [**13**].

Let $H(n)$ be the supporting function of the surface F satisfying equation (1), on the unit sphere ω. If we take the center of the surface F as the origin O, then $H(n)$ is an even function ($H(n) = H(-n)$). It satisfies an elliptic equation

$$\Phi(H_{11}, H_{12}, \cdots, H, n) = 0,$$

derived from (1). The corresponding variational equation is

$$\frac{\partial \Phi}{\partial H_{11}} p_{11} + \frac{\partial \Phi}{\partial H_{12}} p_{12} + \ \dots \ + \frac{\partial \Phi}{\partial H} p \equiv L(p) = 0. \tag{6}$$

As was shown above, the equation $L(p) = 0$ has no nontrivial solutions in the class of even functions p $(p(n) = p(-n))$.

The linear operator L induces a Riemannian metric on the unit sphere ω, with line element

$$ds^2 = \frac{\partial \Phi}{\partial H_{11}} du_1^2 + \frac{\partial \Phi}{\partial H_{12}} du_1 du_2 + \frac{\partial \Phi}{\partial H_{22}} du_2^2,$$

where u_1, u_2 are curvilinear coordinates on ω. Since F is symmetric (and H even), the mapping of ω onto itself defined by symmetry with respect to its center is an isometry in the metric ds^2.

The problem of reducing equation (6) on the sphere ω to canonical form

$$\Delta p + Ap_1 + Bp_2 + Cp = 0, \tag{7}$$

where Δ is Beltrami's differential parameter of second order, is known to involve a conformal mapping of a manifold M with metric ds^2 onto the sphere. In view of the symmetry properties of the metric of M, we may assume that this conformal mapping preserves the symmetry.

We shall construct a surface satisfying equation (1) by embedding a surface for which the solution is known in a continuous family. Consider the following continuous family of functions $\varphi_\lambda(n)$, which contains $\varphi(n)$:

$$\varphi_\lambda(n) = \lambda\varphi + (1-\lambda)f(1,1), \qquad 0 \leqq \lambda \leqq 1.$$

Our problem is trivially solvable for $\lambda = 0$: the corresponding surface is a unit sphere. Assume that the problem is solvable for a function φ_{λ_0}. We shall show that it is then solvable for any function φ_λ if λ is sufficiently close to λ_0.

The equation for the supporting function $P_\lambda(n)$ of the surface F_λ can be expressed as

$$L(\overline{P}) + R(\overline{P}) + (\lambda - \lambda_0)(\varphi(n) - f(1,1)) = 0, \tag{8}$$

where $\overline{P} = P_\lambda - P_{\lambda_0}$, L is the linear elliptic operator (6) for the surface F_{λ_0}, and R is a quadratic polynomial in the function $\overline{P}$ and its first and second derivatives.

We now go from (8) to an integral equation, using the even fundamental solution of the equation $\Delta\overline{P} = 0$ with a logarithmic singularity at two antipodal points. We then obtain

$$\overline{P}+\Omega\overline{P}=A, \tag{9}$$

where

$$\Omega\overline{P}=\int_{\omega} K(n, n')\,\overline{P}\,d\omega, \qquad A=\int_{\omega} K_1(n, n')\,R'\,d\omega.$$

The kernels K and K' are even in both variables. As $|n-n'|\to 0$ we have

$$K\sim 1/|n-n'|, \qquad K_1\sim \ln|n-n'|.$$

We solve equation (9) by successive approximations. Let $C'_{2,\alpha}$ denote the space of twice differentiable even (in n) functions on the unit sphere whose second derivatives satisfy a Hölder condition with exponent $\alpha>0$. The operator Ω is compact in this space. Since the homogeneous equation $\overline{P}+\Omega\overline{P}=0$ has no nontrivial solutions in $C'_{2,\alpha}$, equation (9) has a unique solution in $C'_{2,\alpha}$ for any A in $C'_{2,\alpha}$.

We now define successive approximations by the recurrence relations

$$\overline{P}_k+\Omega\overline{P}_k=A(\overline{P}_{k-1}).$$

The initial approximation is $\overline{P}_0(n)\equiv 0$. The successive approximations converge if λ is sufficiently close to λ_0, and furnish a solution of our problem for these values of λ. Thanks to the analyticity and ellipticity of the original equation, the solution is analytic.

We now show that, under suitable assumptions on the functions f and φ, the problem is solvable for any λ. It will obviously suffice to guarantee the existence of a priori estimates for the function $P_\lambda(n)$ and its first and second derivatives. Since the surface F_λ is closed and convex, this is equivalent to the existence of a priori estimates for the principal radii of curvature. But conditions for this are furnished by Theorem 2.

Assume that, differentiating φ with respect to arc length on a great circle, we get $a\leqq\varphi'^2\leqq b$ and $A\leqq\varphi''^2\leqq B$. Then, by Theorem 2, a sufficient condition for the existence of a priori estimates of the principal radii of curvature of the surface F_λ is that, for arbitrary constants α,β,λ such that $a\leqq\alpha\leqq b$, $A\leqq\beta\leqq B$ and $0\leqq\lambda\leqq 1$,

$$(***)\qquad \begin{aligned} &\overline{\lim_{\substack{R_2=R_2(R_1;\,n)\\ R_1\to\infty}}}\left\{(R_2-R_1)\frac{\partial f}{\partial R_2}+\frac{\partial^2 f}{\partial R_2^2}\frac{\lambda^2\alpha}{(\partial f/\partial R_2)^2}\right\}<\beta\lambda\\ &\lim_{\substack{R_1=\overline{R_1}(R_2,\,n)\\ R_2\to 0}}\left\{(R_1-R_2)\frac{\partial f}{\partial R_1}+\frac{\partial^2 f}{\partial R_1^2}\frac{\lambda^2\alpha}{(\partial f/\partial R_1)^2}\right\}>\beta\lambda.\end{aligned}$$

Once the problem has been solved for analytic functions f and φ, the analyticity condition can be weakened to two-fold differentiability. To do this one approximates the functions f and φ by analytic functions, and subjects the corresponding solutions to a limit procedure. That the limit surface is regular follows from the existence of a priori estimates for its radii of curvature and from the regularity of its equation. The final result is the following theorem.

THEOREM 3. *Let $f(R_1, R_2)$ be a twice differentiable strictly increasing function of both variables. Then, for any even twice differentiable function φ, if condition (***) holds, there exists a closed convex surface satisfying the equation $f(R_1, R_2) = \varphi(n)$, where R_1 and R_2 are the principal radii of curvature of the surface and n is the unit exterior normal.*

In conclusion, we remark that a similar result can be proved in a more general case, when the function f is also an even function of n. The equation is $f(R_1, R_2, n) = \varphi(n)$.

§6. Convex polyhedra with given values of a monotone function on their faces

A function σ is defined for convex space polygons by associating a number with each polygon Q. Examples of such functions are the area and the perimeter of a polygon. Let P be a convex polyhedron with exterior normals n_k. With each vector n_k we can then associate a number σ_k, the value of the function σ on the face with normal n_k. The following question is natural. Under what conditions does there exist a convex polyhedron with given exterior normals n_k and given values σ_k of the function σ on its faces? The present section is devoted to a solution of this problem, which is an analog of the corresponding problem for convex surfaces, considered in §5.

Let P_1 and P_2 be two unbounded convex polyhedra. We shall say that P_1 and P_2 coincide at infinity if the planes of their unbounded faces coincide, or, equivalently, if the polyhedra coincide outside a ball of sufficiently large radius. If polyhedra P_1 and P_2 coincide at infinity, they have the same spherical image.

Let $A_1, \cdots, A_n$ be any finite n-tuple of interior points of the spherical image $\omega(P)$ of an unbounded convex polyhedron P. It is easy to see that there exists a polyhedron (indeed, more than one) P' which coincides with P at infinity such that the spherical images of the bounded faces are the points A_k.

Let σ be a function defined on a set of bounded convex polygons in space, satisfying the following conditions:

1. The function σ is continuous and nonnegative, and vanishes if and only if the polygon is degenerate (segment or point).

2. If a polygon Q_1 is included in a polygon Q_2, then $\sigma(Q_1) < \sigma(Q_2)$.

3. If a polygon Q_2 is obtained from a polygon Q_1 by a translation in the direction $z > 0$, then $\sigma(Q_1) \leqq \sigma(Q_2)$.

An example of a function with these properties is the area of a polygon. In the sequel, a function σ satisfying conditions 1, 2 and 3 will be called a monotone function.

THEOREM 1. *Let P be an unbounded convex polyhedron which can be projected in one-to-one fashion onto the xy-plane, and is convex toward $z < 0$; $n_1, \cdots, n_m$ are arbitrary unit vectors pointing into the spherical image $\omega(P)$ of P; $\sigma_1, \cdots, \sigma_m$ are arbitrary positive numbers. Let σ be a monotone function defined for convex polygons with planes in the directions n_k. Let Ω_P be the collection of all unbounded convex polyhedra coinciding with P at infinity, whose bounded faces α_k have exterior normals n_k.*

If Ω_P contains a polyhedron P_0 such that $\sigma(\alpha_i) \leqq \sigma_i$, $i = 1, \cdots, m$, and moreover

$$(*) \qquad \sum_i \sigma(\alpha_i) \geqslant \sum_i \sigma_i,$$

for any polyhedron in Ω_P lying above a plane $z = ax + by + c$, then there exists a convex polyhedron in Ω_P on whose bounded faces the values of the function σ are equal to the given numbers σ_i.

PROOF. Let Q be an arbitrary polyhedron in Ω_P whose bounded faces satisfy the condition

$$(**) \qquad \sigma(\alpha_i) \leqq \sigma_i, \quad i = 1, \cdots, m.$$

The set Ω_Q of polyhedra Q is not empty (it contains P_0), and it is closed (this follows from the continuity of σ).

Let c_i be the z-coordinate of the point at which the z-axis cuts the face α_i of Q. Set $\varphi(Q) = c_1 + \cdots + c_m$. By condition (*) the function φ is bounded on the set Ω_Q, and hence it attains an absolute maximum (the set Ω_Q is closed and the function φ continuous).

Let $\overline{Q}$ be a polyhedron for which φ is a maximum. We claim that for this polyhedron $\sigma(\alpha_i) = \sigma_i$ $(i = 1, \cdots, m)$. Indeed, if $\sigma(\alpha_j) < \sigma_j$ for some $i = j$, then this inequality will still hold if we subject the plane of the face α_j to a small displacement in the direction $z > 0$. The values of σ on the other faces can only decrease under this displacement, and so the inequalities $\sigma(\alpha_k) \leqq \sigma_k$ continue to hold irrespective of the magnitude of the displacement.

Thus our transformation of the polyhedron $\overline{Q}$ leaves it in the set Ω_Q, but at the same time it increases the value of $\varphi(\overline{Q})$, which is impossible by the definition of $\overline{Q}$ (φ has an absolute maximum on $\overline{Q}$). This completes the proof.

REMARK. A practical construction of the polyhedron whose existence is proved in Theorem 1 is as follows. Displace the plane of the face α_1 of the polyhedron P_0 in the direction $z > 0$, to get a polyhedron for which $\sigma(\alpha_1) = \sigma_1$. The values of the function σ on the other faces are not increased thereby. Now displace the plane of the face α_2, and so on, till α_m, and then again displace the face α_1, etc. The result is a monotone sequence of polyhedra converging to a polyhedron with the desired properties.

We now consider the uniqueness of the polyhedron given by Theorem 1.

Suppose that there exist two polyhedra P' and P'' coinciding at infinity on whose bounded and identically oriented faces the function σ has the same values.

Let α_i' and α_i'' denote corresponding (identically oriented) bounded faces of P' and P''. The z-axis cuts the planes of these faces at points with coordinates c_i' and c_i'', respectively. Since the polyhedra P' and P'' are different, at least one of the differences $c_i' - c_i''$ is not zero. Without loss of generality we may assume that

$$\max_i (c_i' - c_i'') = \delta > 0.$$

Otherwise we simply interchange the roles of P' and P''.

Let S' and S'' denote the sets of all faces of P' and P'', respectively, such that $c' - c'' = \delta$. Displace the polyhedron P'' toward $z > 0$ for the distance δ. The polyhedron P'' will then lie within P' and the planes of corresponding faces of S' and S'' coincide. Thus after the displacement the corresponding faces in S' and S'' are such that $c' - c'' = 0$, while for the other faces (including the unbounded ones) we have $c' - c'' < 0$.

We claim that any face α' in S' can be adjacent only to other faces of S'. In fact, let $\bar{\alpha}_i'$ be the faces adjacent to α' on the polyhedron P', and $\bar{\alpha}_i''$ the corresponding faces of P''. If our assertion is false, there must be at least one pair of corresponding faces $\bar{\alpha}_i'$ and $\bar{\alpha}_i''$ such that $\bar{c}_i' - \bar{c}_i'' < 0$. Since the planes of the faces α' and α'' coincide, while $\bar{c}_i' - \bar{c}_i'' \leqq 0$ for each pair of faces $\bar{\alpha}_i'$ and $\bar{\alpha}_i''$ and strict inequality holds in at least one case, it follows that the face α'' lies strictly inside the face α'. But this is impossible because of the monotonicity of σ, which, by assumption, assumes identical values on corresponding bounded faces of P' and P''.

Since the faces adjacent to a face α' in S' are also in S', it follows that all faces of P' must be in S'. But the unbounded faces cannot be in S', and we have a contradiction. Thus the polyhedra P' and P'' coincide, and we have proved

THEOREM 2. *If two unbounded convex polyhedra can be projected in one-to-one fashion onto the xy-plane and are both convex toward $z < 0$, and if the planes of their unbounded faces coincide while a monotone function assumes equal values on their bounded and identically oriented faces, then the polyhedra are identical.*

We now consider the analogous problem for closed convex polyhedra.

Let α be an arbitrary plane and ω_α a function defined on closed convex polygons in planes parallel to α, which satisfies the following conditions:

1'. The function ω_α is positive and continuous.

2'. The function ω_α is invariant under translation. That is to say, if the polygons P and Q can be made to coincide by a translation, then the corresponding values of ω_α are equal.

3'. The function ω_α is strictly monotone, in the sense that if Q is included in P, then $\omega_\alpha(Q) < \omega_\alpha(P)$.

4'. If a polygon P is varied in such a way that its area $S(P)$ tends to infinity, then $\omega_\alpha(P) \to \infty$. If the polygon P degenerates (to a segment), then $\omega_\alpha(P) \to 0$.

THEOREM 3. *Let $\alpha_1, \cdots, \alpha_n$ $(n \geqq 3)$ be an n-tuple of planes not parallel to one straight line, and $\omega_1, \cdots, \omega_n$ an n-tuple of functions defined on polygons parallel to the planes $\alpha_1, \cdots, \alpha_n$, respectively, which satisfy conditions 1' to 4'.*

Then for any positive numbers $\varphi_1, \cdots, \varphi_n$ there exists a closed convex polyhedron with $2n$ faces parallel to the planes $\alpha_1, \cdots, \alpha_n$, on which the functions ω_k assume the values $\varphi_1, \cdots, \varphi_n$. This polyhedron is centrally symmetric and is defined uniquely up to translation.

We first prove that the polyhedron (if it exists) has a center of symmetry and is unique. Let P be a polyhedron satisfying the conditions of the theorem. Since it has $2n$ faces and no polyhedron can have more than two faces in a given direction, it follows that P has exactly two faces in each direction α_k. Construct a polyhedron P^* symmetric to P with respect to an arbitrary point O. Set up a correspondence between pairs of parallel faces of P and P^* with identically oriented exterior normals. The values of the functions ω on corresponding faces of P and P^* are equal, and hence neither face can be made to lie within the other by a translation. By a well-known theorem of Aleksandrov [**3**], the polyhedra P and P^* are congruent and translates of each other. This implies that P is centrally symmetric. Similarly one proves that P is uniquely determined.

We now prove the existence of the polyhedron. An arbitrary n-tuple of positive numbers $\varphi_1, \cdots, \varphi_n$ will be called an abstract polyhedron. If there exists a closed convex polyhedron with $2n$ faces parallel to the planes $\alpha_1, \cdots, \alpha_n$ on which the functions ω_k assume the values φ_k, we shall call it a realization of the abstract polyhedron $(\varphi_1, \cdots, \varphi_n)$. We define two manifolds. The elements of the manifold M_1 are the closed convex symmetric polyhedra with $2n$ faces parallel to the planes $\alpha_1, \cdots, \alpha_n$. Each such polyhedron is determined by n positive numbers, the values of its supporting function in the directions perpendicular to the planes α_k. It can be represented by a point in euclidean n-space, with coordinates equal to these values of the supporting function. The second manifold M_2 consists of the abstract polyhedra; it can be interpreted as the interior of the first "quadrant" of the euclidean n-space with cartesian coordinates $\varphi_1, \cdots, \varphi_n$.

Each polyhedron M_1 corresponds in a natural way to a certain abstract polyhedron in M_2. We claim that this correspondence is a homeomorphism, so that each abstract polyhedron has a geometric realization. To prove this assertion we apply Aleksandrov's Mapping Lemma (see §2) to the manifolds M_1, M_2. The assumptions of the Mapping Lemma hold. Indeed, the manifolds M_1 and M_2 are of equal dimensions (n). M_2 is connected, since it is a convex set. There exist realizable polyhedra. The images of different points of M_1 in M_2 are different by virtue of the uniqueness property proved above (polyhedra which are translates of each other are identified). It remains to prove that an abstract polyhedron which is the limit of realizable polyhedra is realizable.

Let $P_1, P_2, \cdots$ be a sequence of polyhedra of M_1, and $P_1', P_2', \cdots$ the sequence of corresponding abstract polyhedra, which converge to an abstract polyhedron P. We must show that P is realizable. Let $h_1^k, \cdots, h_n^k$ denote the supporting numbers [i.e., the values of the supporting function] of the polyhedron P_k in the directions perpendicular to the planes $\alpha_1, \cdots, \alpha_n$. Without loss of generality we may assume that each sequence H_s $(h_s^1, h_s^2, \cdots)$, $s = 1, \cdots, n$, converges to a (finite or infinite) limit.

We claim that each sequence H_s converges to a finite nonzero limit. Indeed, the surface areas of the polyhedra P_k are uniformly bounded. Therefore, were some sequence H_s to diverge to infinity, then for sufficiently large k the polyhedron P_k would lie in a cylinder of arbitrarily small radius. At least one of the faces of this polyhedron would have a diameter of the same order of magnitude as the diameter of the cylinder. Hence the function ω would converge to zero for this face as $k \to \infty$, and this is impossible.

Further, were some sequence H_s to converge to zero, then for sufficiently large k the polyhedron P_k would lie between two arbitrarily close parallel planes. Hence it would have a face degenerating to a segment as $k \to \infty$, and again the function ω would tend to zero, which is impossible.

Thus all the sequences H_s have positive limits. This implies that the polyhedra P_k converge to a realization of the limit polyhedron of the sequence P'_k. Aleksandrov's Mapping Lemma now implies that the manifolds M_1 and M_2 are homeomorphic, and hence all abstract polyhedra are realizable. This proves the theorem.

The following theorem is a corollary of Theorem 3.

THEOREM 4. *Let* $\alpha_1, \cdots, \alpha_n$ $(n \geqq 3)$ *be an n-tuple of planes not parallel to one straight line,* $f(s,p,\nu_k)$ *a function of the positive variables s, p and unit vectors* ν_k *perpendicular to the planes* α_k*, such that*

1) *the function f is positive, continuous, strictly increasing in s, p, and even in* ν, *i.e.* $f(s,p,\nu) = f(s,p,-\nu)$;

2) $f(s,p,\nu) \to \infty$ *as* $s \to \infty$; $f(s,p,\nu) \to 0$ *as* $s \to 0$.,

Then for any even function $\varphi(\nu_k)$ *there exists a closed convex polyhedron with* $2n$ *faces parallel to the planes* α_k *such that the areas* s_k, *perimeters* p_k *and exterior normals* ν_k *of the faces satisfy the conditions* $f(s_k, p_k, \nu_k) = \varphi(\nu_k)$.

This polyhedron is centrally symmetric and is unique up to translation.

§7. Convex polyhedra with vertices on given rays and with given values of a monotone function on their polyhedral angles

In the preceding section we considered convex polyhedra whose faces can be paired off so that corresponding faces are translates of each other. If the space is completed by ideal elements (elements at infinity), this means that the planes of corresponding faces intersect along straight lines lying in one plane. In this section we shall consider a problem which is in a sense dual to that of §6. We shall now pair off the vertices in such a way that the straight lines connecting corresponding vertices pass through a single point (in particular, are parallel). Instead of functions on the faces of the polyhedron we shall consider functions of the polyhedral angles at the vertices.

Let V be a convex polyhedral angle and A its apex. If a ray can be drawn from A within V parallel to the z-axis in the direction $z > 0$, we shall say that the angle V is convex toward $z < 0$.

We shall consider functions ϑ defined on polyhedral angles convex toward $z < 0$, satisfying the following conditions:

1. The function ϑ is continuous, nonnegative, and vanishes identically if and only if the angle degenerates into a dihedral angle or a plane.
2. If V_1 and V_2 are angles with a common apex and V_1 is contained in V_2, then $\vartheta(V_1) \geqq \vartheta(V_2)$, with equality if and only if the angles are identical.
3. If the angle V_2 is obtained from V_1 by a displacement toward $z < 0$, then $\vartheta(V_1) \leqq \vartheta(V_2)$.

An example is the curvature of the angle—the area of its spherical image. In the sequel any function ϑ satisfying conditions 1, 2 and 3 will be called a monotone function of the angle.

Theorem 1. *Let γ be a closed polygonal line in space, which can be projected in one-to-one fashion by straight lines parallel to the z-axis onto a convex polygonal line $\bar{\gamma}$ bounding a polygon G in the xy-plane, so that the vertices of γ correspond to the vertices of $\bar{\gamma}$. Let $g_1, \cdots, g_n$ be any n-tuple of straight lines parallel to the z-axis and cutting the polygon G, $\vartheta_1, \cdots, \vartheta_n$ arbitrary positive numbers, and ϑ a monotone function defined on polyhedral angles convex toward $z < 0$ with apices on the straight lines g_k. Let Ω_P denote the collection of all polyhedra P with boundary γ which can be projected in one-to-one fashion onto the xy-plane, are convex toward $z < 0$, and have vertices on the straight lines g_k.*

If Ω_P contains a polyhedron P_0 for whose polyhedral angles with apices at the points A_k

$$(*) \qquad \vartheta(A_k) \leqq \vartheta_k, \quad k = 1, \cdots, n,$$

and moreover

$$(**) \qquad \sum_k \vartheta(A_k) \geqslant \sum_k \vartheta_k$$

for any polyhedron in Ω_P cutting a plane $z =$ const, then Ω_P contains a polyhedron for whose polyhedral angles the function ϑ assumes the values ϑ_k, i.e. $\vartheta(A_k) = \vartheta_k$, $k = 1, \cdots, n$.

Proof. Let Ω_Q denote the set of all polyhedra Q in Ω_P which satisfy condition (*). The set Ω_Q is not empty (since it contains at least P_0); it is bounded, because of condition (**); it is closed, since the function ϑ is continuous.

Let c_k be the z-coordinate of the vertex A_k of a polyhedron Q in Ω_Q. Consider the function $\varphi(Q) = c_1 + \cdots + c_n$.

Since the set Ω_Q is closed and bounded and the function φ continuous, it assumes an absolute minimum for some polyhedron $\overline{Q}$ in Ω_Q. We claim that $\overline{Q}$ is the required polyhedron.

Suppose the assertion false. Then $\vartheta(A_k) < \vartheta_k$ for some vertex A_k of the polyhedron $\overline{Q}$. Displace the vertex A_k along the straight line g_k, in the direction $z < 0$, for a small distance δ; call the new point A_k'. Now construct the smallest polyhedron containing the polygonal line γ and the points $A_1, \cdots, A_k', \cdots, A_n$. That part of this polyhedron convex toward $z < 0$ is a convex polyhedron $\overline{Q}'$ with boundary γ, in the set Ω_P.

If δ is sufficiently small, each vertex A_i $(i \neq k)$ of the polyhedron $\overline{Q}$ coincides with the corresponding vertex of the polyhedron $\overline{Q}'$, and the polyhedral angle of $\overline{Q}'$ contains the polyhedral angle of $\overline{Q}$. If the polyhedral angle of $\overline{Q}$ at A_i $(i \neq k)$ degenerates to a dihedral angle or a plane, the same holds for the angle at the corresponding vertex of $\overline{Q}'$.

Since the function ϑ is continuous, we have $\vartheta(A_k') < \vartheta_k$ for sufficiently small δ. At the other vertices of the polyhedron $\overline{Q}'$ the inequality $\vartheta(A_i') \leqq \vartheta_i$ continues to hold, because of the way the angles are modified in going from $\overline{Q}$ to $\overline{Q}'$.

Thus for a sufficiently small displacement δ of the vertex A_k the new polyhedron $\overline{Q}'$ is also in Ω_Q. In the transformation from $\overline{Q}$ to $\overline{Q}'$ the vertices may be displaced only toward $z < 0$, and one of them (A_k) is certainly displaced; hence $\varphi(\overline{Q}') < \varphi(\overline{Q})$, which contradicts the definition of $\overline{Q}$ as the polyhedron on which φ assumes an absolute minimum. This proves the theorem.

Let P be an unbounded polyhedron which is not a prism. Let S be some interior point of the polyhedron, and consider all rays issuing from S and not cutting the polyhedron. These rays generate a certain polyhedral angle, which we call the limit angle of the polyhedron. The limit angle may degenerate into a plane angle or even a straight line. By the definition of the limit angle, it is unique up to translation (its position depends on the choice of the point S).

We shall now prove a theorem analogous to Theorem 1 for unbounded convex polyhedra. Instead of requiring that the polyhedron pass through a given contour, we shall require it to have a given limit angle.

THEOREM 2. *Let V be a polyhedral angle which can be projected in one-to-one fashion onto the xy-plane and is convex toward $z < 0$, $g_1, \cdots, g_n$ an n-tuple of straight lines parallel to the z-axis, $\vartheta_1, \cdots, \vartheta_n$ arbitrary* *positive numbers, and ϑ a monotone function defined for polyhedral angles convex toward $z < 0$ with vertices on the straight lines g_k. Let Ω_P denote the set of all polyhedra having limit angles which are translates of V and vertices A_k on the straight lines g_k.*

If the set Ω_P contains a polyhedron such that $\vartheta(A_k) \leqq \vartheta_k$, $k = 1, \cdots, n$, and moreover $\sum_k \vartheta(A_k) > \sum_k \vartheta_k$ for every polyhedron in Ω_P which cuts a plane $z = ax + by + c$, then Ω_P contains a polyhedron for whose polyhedral angles the function ϑ assumes the given values ϑ_k.

The proof is analogous to that of Theorem 1.

Let Ω_Q denote the set of all polyhedra in Ω_P such that $\vartheta(A_k) \leqq \vartheta_k$, $k = 1, \cdots, n$. The set Ω_Q is not empty, is bounded and closed. (The set is bounded in the sense that all the polyhedra in Ω_Q lie above the plane $z = ax + by + c$.) The function φ, defined as before, assumes an absolute maximum on the set Ω_Q for some polyhedron $\overline{Q}$. This is precisely the polyhedron whose existence we want to prove. Indeed, suppose that at some vertex A_k of $\overline{Q}$ we have $\vartheta(A_k) < \vartheta_k$. Displace this vertex for a small distance δ toward $z < 0$, and denote the new point by A_k'. Let $\overline{V}$ be the limit angle of $\overline{Q}$. Let $\overline{Q}'$ denote the smallest polyhedron containing the points $A_1, \cdots, A_k', \cdots, A_n$ and the angle $\overline{V}$. This polyhedron is in Ω_P. If δ is sufficiently small, it is also in Ω_Q, but $\varphi(\overline{Q}')$ is clearly smaller than $\varphi(\overline{Q})$. This proves the theorem.

The next theorem relates to uniqueness of polyhedra with vertices on given straight lines, on whose polyhedral angles a monotone function assumes given values.

THEOREM 3. *Let P' and P'' be convex polyhedra with common boundary γ, both of which can be projected in one-to-one fashion onto the xy-plane and are convex toward $z < 0$. Assume that their vertices can be paired off so that corresponding vertices lie on the same straight line parallel to the z-axis, and moreover a certain monotone function ϑ assumes equal values on the polyhedral angles of P' and P'' at corresponding vertices. Then the polyhedra P' and P'' are identical.*

PROOF. Let A_k' and A_k'' be corresponding vertices, c_k' and c_k'' their z-coordinates. If the polyhedra are different, the differences $c_k' - c_k''$ cannot all be zero. Without loss of generality we may assume that

$$\max_{(k)}(c_k' - c_k'') = \delta > 0.$$

Displace the polyhedron P'' toward $z > 0$ for a distance δ. Let S' and S'' be the sets of vertices of the polyhedra P' and P'' that coincide after this displacement. Call two vertices of a polyhedron adjacent if they are endpoints of the same edge. We claim that all vertices of P' adjacent to vertices in S' are also in S'.

Indeed, let A' be a vertex in S'. If there are vertices of P' adjacent to A' but not in S', then, since $c_k' - c_k'' \leqq 0$, with equality if and only if the corresponding vertex is in S', it follows that after the above displacement the polyhedral angle at the vertex A'' lies within the polyhedral angle at P' at A'. This is impossible by the monotonicity of the function ϑ.

Since all vertices adjacent to vertices of S' are in S', it follows that all vertices of P' are in S'. On the other hand, the vertices in the boundary of P' are certainly not in S', since $\delta > 0$. This contradiction proves the theorem.

THEOREM 4. *Let P' and P'' be two unbounded convex polyhedra which can be projected in one-to-one fashion onto the xy-plane, are convex toward $z < 0$, and have a common limit angle V. Assume that the vertices can be paired off so that corresponding vertices lie on the same straight line parallel to the z-axis, and a monotone function ϑ assumes equal values on the polyhedral angles at corresponding vertices.*

Then either the polyhedra coincide, or one can be obtained from the other by a translation in the direction of the z-axis. The latter case is impossible if the function ϑ is strictly increasing for translation toward $z < 0$.

The proof is a word-for-word repetition of that of Theorem 3. The fact that all vertices of P' are in S' implies that the polyhedron P'' can be made to coincide with P' by a translation in the direction of the z-axis. If the function ϑ is strictly increasing for translation toward $z < 0$, then the polyhedra P' and P'' simply coincide, since the function must assume different values on polyhedral angles which are translates of each other but not identical.

We now consider the case of closed convex polyhedra. Let ϑ be a function defined on polyhedral angles V containing the origin O, which satisfies the following conditions.

1. The function $\vartheta(V)$ is continuous and nonnegative, and vanishes if and only if the angle degenerates to a dihedral angle or plane.

2. If V_1 and V_2 have a common apex and V_1 lies within V_2 (but is not identical with it), then $\vartheta(V_1) > \vartheta(V_2)$.

3. If A is the apex of the angle V_1 and the angle V_2 is obtained by a displacement of the angle V_1 in the direction OA, then $\vartheta(V_1) \leqq \vartheta(V_2)$.

For brevity we shall call the function ϑ a monotone function. A simple example is the curvature of a polyhedral angle—the area of its spherical image.

THEOREM 5. *Let $g_1, \cdots, g_n$ be an n-tuple of rays issuing from the origin O which are not contained in a single halfspace, ϑ a monotone function defined on polyhedral angles with apices on the rays g_k, and $\vartheta_1, \cdots, \vartheta_n$ arbitrary positive numbers. Assume that the following conditions hold:*

(a) *There exists a closed convex polyhedron with vertices on the rays g_k and polyhedral angles V_k at these vertices such that $\vartheta(V_k) \geqq \vartheta_k$, $k = 1, \cdots, n$.*

(b) *There exists $\epsilon > 0$ such that, for every closed convex polyhedron with vertices on the g_k, there is at least one polyhedral angle V_k for which $\vartheta(V_k) < \vartheta_k$, provided that at least one vertex of the polyhedron is at a distance less than ϵ from O.*

Then there exists a closed convex polyhedron with vertices on the rays g_k on whose polyhedral angles the function ϑ assumes the given values ϑ_k, i.e. $\vartheta(V_k) = \vartheta_k$, $k = 1, \cdots, n$.

Either this polyhedron is unique, or all such polyhedra are transforms of each other under a similarity mapping with center of similitude O. The latter case is impossible if the function ϑ is strictly monotone for displacement along a ray issuing from O.

PROOF. Let Ω_P denote the set of closed convex polyhedra with vertices on the rays g_k and polyhedral angles V_k such that $\vartheta(V_k) \geqq \vartheta_k$, $k = 1, \cdots, n$. This set is not empty, for by assumption it contains at least one polyhedron.

For polyhedra P in the set Ω_P, define the function

$$\varphi(P) = \sum_k |OA_k|,$$

where A_k are the vertices of the polyhedron P. φ, being continuous, has an absolute minimum on the set Ω_P, by virtue of the last assumption of the theorem, according to which the distances of the vertices of polyhedra in Ω_P from the origin cannot be less than ϵ.

We claim that the polyhedron P for which φ assumes its minimum is the required polyhedron, i.e. $\vartheta(V_k) = \vartheta_k$ for all k.

If this is false, then $\vartheta(V_m) > \vartheta_m$ for some $k = m$, Displace the vertex A_m for a small distance δ in the direction $A_m O$, and call the new point A'_m. Since ϑ is continuous, it follows that if δ is small the inequality $\vartheta(V_m) > \vartheta_m$ will still hold in a polyhedron P' with vertex A'_m. The inequality $\vartheta(V_k) \geqq \vartheta_k$ at the other vertices can only be strengthened, since the polyhedral angles at these vertices cannot increase. Thus the polyhedron P' is in Ω_P. But this is impossible, since $\varphi(P') < \varphi(P)$. This proves existence.

As to uniqueness, suppose that there are two polyhedra P' and P'' with vertices on the rays g_k and polyhedral angles at these vertices such that $\vartheta(V_k') = \vartheta(V_k'')$, $k = 1, \cdots, n$. Set

$$\max_k \frac{|OA_k''|}{|OA_k'|} = m.$$

Without loss of generality we may assume that $m > 1$. Expand P' by a similarity transformation with center O and ratio of similitude m; call the new polyhedron $\overline{P}'$. $\overline{P}'$ contains P'', and they have at least one common vertex. If P' and P'' are not similar, there must be at least one common vertex of $\overline{P}'$ and P'' at which the polyhedral angle V'' of P'' is smaller than the polyhedral angle $\overline{V}'$ of $\overline{P}'$, and so $\vartheta(\overline{V}') < \vartheta(V'')$. Now, since ϑ is monotone we have $\vartheta(V') \leqq \vartheta(\overline{V}')$; hence $\vartheta(V') < \vartheta(V'')$, which is a contradiction. Thus the polyhedra P' and P'' are indeed similar, with center of similitude O.

If the function ϑ is strictly monotone for displacement along a ray issuing from O, the similar polyhedra P' and P'' must coincide, since otherwise the function ϑ would assume different values for polyhedral angles with vertices on the same ray through O. This completes the proof of Theorem 5.

Special cases of Theorems 4 and 5, with the function $\vartheta(V)$ equal to the curvature of the angle V, were first proved by Aleksandrov [6] with the aid of the Mapping Lemma.

§8. Uniqueness of a convex surface with given intrinsic metric and metric of spherical image

The three fundamental quadratic forms of the theory of regular surfaces are

$$\mathrm{I} = dr^2, \quad \mathrm{II} = -dr\,dn, \quad \mathrm{III} = dn^2,$$

where r is the radius vector of a point on the surface and n the unit vector of the exterior normal. The first fundamental form defines the metric of the surface; if it is known, one can determine the distance between any two points on the surface. The third fundamental form defines the metric on the spherical image of the surface; once it is known, one can determine the angle between the normals to the surface at any two points. In this section we shall study the uniqueness problem for a surface with given intrinsic metric and metric of spherical image.

A well-known theorem of Bonnet states that a regular surface is uniquely determined, up to motions and reflections, by its first and second fundamental forms. If the mean curvature of the surface is nowhere zero, then the first and third fundamental forms uniquely determine the second fundamental form. Thus a surface whose mean curvature never vanishes is uniquely determined by its first and third fundamental forms. This theorem generalizes to arbitrary (not necessary regular) convex surfaces.

Let F_1 and F_2 be convex surfaces whose points and supporting planes can be put in a one-to-one correspondence satisfying the following conditions.

1) If P_1 and Q_1 are arbitrary points of F_1, P_2 and Q_2 the corresponding points of F_2, then the intrinsic distance between P_1 and Q_1 is equal to the intrinsic distance between P_2 and Q_2.

2) If π_1 is a supporting plane to F_1 at a point P_1, π_2 the corresponding supporting plane to F_2 and P_2 the point on F_2 corresponding to P_1, then P_2 lies in the plane π_2.

3) If π_1 and σ_1 are any two supporting planes to F_1, π_2 and σ_2 the corresponding supporting planes to F_2, then the angle between the exterior normals to π_1 and σ_1 is equal to the angle between the exterior normals to π_2 and σ_2.

We shall assume that neither of the surfaces F_1, F_2 is a plane domain (the case in which one of them is a plane domain is not interesting). Then the exterior normal to the supporting plane can be defined as a ray perpendicular to the supporting plane and pointing into the halfspace not containing points of the surface. We shall prove the following theorem.

THEOREM. *A convex surface is uniquely determined, up to motions and reflections, by the intrinsic distance between each two points and the angle between the exterior normals to each two supporting planes. In other words, two convex surfaces whose points and supporting planes can be put in a one-to-one correspondence satisfying conditions* 1), 2) *and* 3) *are either congruent or mirror images of each other.*

Without loss of generality, we may assume that neither of the surfaces is a plane domain. (Otherwise the other surface is also a plane domain and the statement of the theorem is a trivial consequence of the isometry between the domains.)

We proceed by way of several auxiliary propositions.

A) Suppose that the points and supporting planes of F_1 and F_2 can be put in a one-to-one correspondence satisfying conditions 1), 2) and 3). Then, by a suitable motion or a motion plus a reflection, the surface F_2 can be brought to a position in which the exterior normals to its supporting planes are parallel and identically oriented with the exterior normals to the corresponding supporting planes of F_1.

The surface F_1 has at least two, generally three supporting planes whose exterior normals are not parallel. By condition 3), a suitable motion or motion plus reflection will bring F_2 to a position in which the exterior normals to these two or three supporting planes of F_1 are parallel to and identically oriented with the exterior normals to the corresponding supporting planes of F_2. Since the direction of the exterior normal to any supporting plane of the surface is uniquely determined by the angles that it forms with the above-mentioned two or three exterior normals, it follows from condition 3) that the exterior normals to any pair of corresponding supporting planes of F_1 and F_2 are parallel and identically oriented.

Throughout the sequel we shall assume that corresponding supporting planes of the surfaces F_1 and F_2 are parallel and their exterior normals identically oriented.

Let P_1 be a point on F_1, γ_1 a small contour enclosing P_1, and g_1 a directed straight line through P_1 pointing into the convex body whose boundary contains the surface F_1. Let P_2, γ_2 and g_2 be the corresponding point, contour and straight line for the surface F_2. We shall say that the surfaces F_1 are identically or oppositely oriented according to whether points describing the contours γ_1 and γ_2 in accordance with the surface isometry encircle the straight lines g_1 and g_2 in identical or opposite senses.

B) The tangent cones at corresponding points of the surfaces F_1 and F_2 are translates of each other.

This follows from the fact that the exterior normals to corresponding supporting planes are parallel and identically oriented.

C) If P_1 is a ridge point of F_1, a direction τ_1 along the edge g_1 of the dihedral angle V_1 at P_1 (i.e. the tangent cone at P_1) corresponds to a direction τ_2 along the edge g_2 of the dihedral angle V_2 (the tangent cone at P_2 on F_2). If the surfaces F_1 and F_2 are identically oriented, then $\tau_1 = \tau_2$; if they are oppositely oriented, then $\tau_1 = -\tau_2$.

In fact, the dihedral angles V_1 and V_2 are translates of each other. Assume that the assertion is false, so that there are geodesics γ_1 and

γ_2 issuing from P_1 and P_2 on F_1 and F_2, whose semitangents at P_1 and P_2 lie in nonparallel faces of the angles V_1 and V_2. Let P_{1k} and P_{2k} ($k=1,2,\cdots$) be two sequences of smooth points on γ_1 and γ_2 converging to P_1 and P_2, respectively. Let π_{1k} and π_{2k} be the corresponding supporting planes at the points P_{1k} and P_{2k}. Since the semitangents to γ_1 and γ_2 are continuous at P_1 and P_2, the sequences π_{1k} and π_{2k} converge; their limits are necessarily nonparallel faces of the angles V_1 and V_2. This is impossible, since $\pi_{1k} \| \pi_{2k}$ for any k. This proves the proposition.

By proposition B) it follows that if the surfaces F_1 and F_2 are identically oriented, all other pairs of corresponding directions at P_1 and P_2 must coincide. If the surfaces are oppositely oriented, then rays perpendicular to the edges g_1, g_2 of the dihedral angles V_1, V_2 and lying in parallel faces of these angles define corresponding directions at the points P_1, P_2.

D) Let P_1 be a smooth point of F_1, π_1 the supporting plane at P_1. Let the intersection of F_1 and π_1 be a straight-line segment Δ_1, of which P_1 is an interior point. The point P_2 corresponding to P_1 on F_2 possesses the same properties. A direction τ_1 along the segment Δ_1 corresponds to a direction τ_2 along Δ_2. The segments Δ_1 and Δ_2 are parallel. If the surfaces F_1 and F_2 are identically oriented, then $\tau_1 = \tau_2$. If they are oppositely oriented, then $\tau_1 = -\tau_2$.

It is obvious that the directions on the surfaces F_1 and F_2 at P_1 and P_2 lying in the segments Δ_1 and Δ_2 correspond to each other. The remainder of the proposition is proved in the same way as proposition C). Namely, assume that the proposition is false. Let γ_1 and γ_2 be geodesics issuing from P_1 and P_2 whose semitangents at these points do not coincide with the segments Δ_1 and Δ_2. As in the proof of C), construct sequences of supporting planes π_{1k} and π_{2k}. They cut the planes π_1 and π_2 in parallel straight lines g_{1k} and g_{2k}. As $k \to \infty$, these lines converge to the straight lines g_1 and g_2 containing Δ_1 and Δ_2. Thus $\Delta_1 \| \Delta_2$. Subject the surface F_2 to a translation so that the points P_1 and P_2 coincide. Then the planes π_1, π_2 and lines g_1, g_2 also coincide. The semitangents to the geodesics γ_1 and γ_2 lie in different components of the partition of the plane $\pi_1 \equiv \pi_2$ by the straight line $g_1 \equiv g_2$. Let $\overline{P}_{1k}$ and $\overline{P}_{2k}$ be the projections of P_{1k} and P_{2k} on the plane $\pi_1 \equiv \pi_2$. The straight-line segment connecting them cuts the segment $\Delta_1 \equiv \Delta_2$ in a point $\overline{P}$. The plane π_{1k} cuts the segment $\overline{P}_{1k}\overline{P}$; the plane π_{2k} cuts the segment $\overline{P}_{2k}\overline{P}$. Thus the point $\overline{P}$ lies between the parallel planes π_{1k} and π_{2k}. If the points P_{1k} and P_{2k} are sufficiently close to P_1 and P_2, respectively, then the interior normal to the plane $\pi_1 \equiv \pi_2$ at $\overline{P}$ does not cut the planes π_{1k} and π_{2k}.

But this implies that π_{1k} and π_{2k} are perpendicular to the plane $\pi_1 \equiv \pi_2$, and this is impossible since P_1 is a smooth point. This contradiction proves the assertion.

If F_1 and F_2 are identically oriented, it follows that any two corresponding directions at P_1 and P_2 coincide. If they are oppositely oriented, corresponding directions perpendicular to the segments Δ_1 and Δ_2 coincide.

E) Let the surfaces F_1 and F_2 be identically oriented, P_1 a smooth point on F_1 and P_{1k}, P_{2k} $(k = 1, 2, \cdots)$ sequences of corresponding points on F_1 and F_2 which converge to P_1 and P_2; assume, moreover, that corresponding directions at the points P_{1k} and P_{2k} are parallel (not necessarily identical). Then corresponding directions at the points P_1 and P_2 are also parallel.

In fact, connect the points P_{1k} by segments (shortest joins) γ_{1k} to a point Q_1 close to P_1; connect the points P_{2k} by the corresponding segments γ_{2k} to the point Q_2 corresponding to Q_1. A sequence k_j of values of k can be so chosen that the curves γ_{1k_j} and the semitangents to the curves γ_{1k_j} and γ_{2k_j} at the points P_{1k_j} and P_{2k_j} converge. Then the limit rays are semitangents to the limit segments. As limits of parallel convergent sequences of semitangents, they must be parallel. Since P_1 and P_2 are smooth points and the surfaces F_1 and F_2 are identically oriented, it follows that all corresponding directions at the points P_1 and P_2 are parallel.

F) Let P_1 be a smooth point on the surface F_1, π_1 the supporting plane at P_1; let P_1 be the only point common to π_1 and F_1. Then, if the surfaces F_1 and F_2 are identically oriented, corresponding directions at the points P_1 and P_2 (where P_2 is the point corresponding to P_1) are parallel.

If there exists a sequence of ridge points on F_1 converging to P_1, this follows from propositions C) and E). If P_1 has a neighborhood which contains no ridge points of F_1, let π_2' be a plane parallel to π_2, near π_2, which cuts the surface F_2 in a curve γ_2' containing only smooth points of the surface (this is always possible, since there are no ridge points in the neighborhood of P_2 and the set of conic points is at most countable). The curve γ_1' corresponding to γ_2' is also smooth. Let Q_1 be the point on γ_1' farthest from the plane π_1. Corresponding directions at Q_1 and Q_2 are parallel, since there is a pair of parallel corresponding directions at these points (the directions of γ_1' and γ_2') and the surfaces F_1 and F_2 are identically oriented. It now suffices to apply proposition E).

G) Let F_1 and F_2 be identically oriented, P_1 a smooth point, and assume that the intersection of the plane π_1 and the surface F_1 is a plane region ω_1. Then corresponding directions at the points P_1 and P_2 (P_2 corresponding to P_1) are parallel.

If P_1 is on the boundary of ω_1, there exists a sequence of points of the types considered in propositions C), D) and F) which converges to P_1, and it then suffices to apply proposition E). Let P_1 be an interior point of ω_1. The boundary of ω_1 contains a smooth point or a ridge point, and for such points the assertion has been proved. Since the orientations of F_1 and F_2 are identical and ω_1 and ω_2 are congruent and lie in parallel planes, the assertion is valid in this case too.

H) Let P_1 be a point on F_1 such that: 1) there exists a supporting plane π_1 at P_1 which contains no other point of the surface F_1; 2) the interior normal to π_1 points into the body whose boundary contains F_1. Then the surfaces F_1 and F_2 are identically oriented.

In fact, make the points P_1 and P_2 coincide by a translation of one of the surfaces. Let γ_1 be a small contour on F_1 enclosing P_1. If the surfaces F_1 and F_2 are oppositely oriented, then the contours γ_1 and γ_2 (where γ_2 is the contour corresponding to γ_1) must contain a pair of corresponding points Q_1 and Q_2 whose projections $\overline{Q}_1$ and $\overline{Q}_2$ on the plane $\pi_1 \equiv \pi_2$ lie on a straight line containing the point $P_1 \equiv P_2$, but on different sides of the latter point. Let π_1' and π_2' be corresponding supporting planes at the points Q_1 and Q_2. They cut the segments $\overline{Q}_1 P_1$ and $\overline{Q}_2 P_2$, respectively. If the contour γ_1 is small, then the interior normal to the plane π_1 at P_1 does not cut the planes π_1' and π_2', but the point P_1 lies between these planes. Therefore the planes π_1' and π_2' are perpendicular to π_1, and this is impossible if Q_1 and Q_2 are sufficiently close to P_1 and P_2, i.e. if the contour γ_1 is sufficiently small.

I) Let the surfaces F_1 and F_2 be identically oriented. Let P_1 and Q_1 be arbitrary points on the surface F_1 and γ_1 a geodesic connecting them. Since the right semitangent to the geodesic is continuous from the right, and the left semitangent continuous from the left, the set of points of the curve γ_1 (without its endpoints) for which the corresponding points on F_2 have the same directions is closed. The same statement holds when the corresponding points have opposite directions to those on F_1. But γ_1 is a connected set and hence this can be true only if one of these sets is empty.

J) Let F_1 and F_2 be oppositely oriented. Each surface F_i $(i = 1, 2)$ is part of some closed convex surface $\Phi_i = F_i + \overline{F}_i$. Construct the minimum convex body containing the set $\overline{F}_i$. Denote its surface by ψ_i. Since the surface F_i contains no points of the type considered in proposition H), it follows that $F_i = \psi_i - \overline{F}_i$. Therefore each supporting plane of F_i cuts

F_i either in a system of straight-line open segments lying on one straight line in the supporting plane, or in a system of plane domains bounded by straight-line segments. It was shown in D) that in the first case corresponding segments Δ_1 and Δ_2 are parallel, and they contain corresponding directions, which are opposite. Arguments similar to those in proposition D) show that the boundary segments in the second case also have this property. Since the plane domains ω_1 and ω_2 common to the surfaces F_1 and F_2 and the corresponding supporting planes are parallel, and one is the mirror image of the other, and since moreover corresponding directions along the boundary segments of ω_1 and ω_2 are opposite, it follows that all the boundary segments of ω_1 are parallel.

K) Enlarge the family of straight-line generators of the surfaces F_1 and F_2 at the points of their plane domains (see proposition J)) by adding straight lines parallel to the boundary segments of these domains. There is now one straight-line generator through each point on F_1 (or F_2). The straight-line generators through corresponding points of F_1 and F_2 are parallel. The directions perpendicular to the generators of F_1 correspond to those perpendicular to the corresponding generators of F_2, and these directions coincide (see D) and J)).

We claim that all straight-line generators of the surface F_1 are parallel. Indeed, the orthogonal trajectories of the families of generators of F_1 and F_2 at corresponding points have parallel semitangents by propositions D) and J). Therefore, if r_1 and r_1' (r_2 and r_2') denote the radius vectors of the intersections of corresponding orthogonal trajectories with two generators g_1 and g_1' (g_2 and g_2'), then $r_1 - r_1' = r_2 - r_2'$. Moreover, $dr_1 = -dr_2$, $dr_1' = -dr_2'$ (see D) and J)). Hence $dr_1 = dr_1'$ and $dr_2 = dr_2'$, i.e. the generators g_i and g_i' are parallel. By reflection of the surface F_2 in a plane perpendicular to its generators, we can alter the orientation of the surface. We may therefore assume that the surfaces F_1 and F_2 are identically oriented.

Proof of the Theorem. By proposition I) the unit tangent vectors of the right semitangents of corresponding geodesics must satisfy either $\tau_1 = \tau_2$ or $\tau_1 = -\tau_2$, and the same equality must hold for all points on the geodesics. Integrating these equalities, we get

$$|r_1(P_1) - r_1(Q_1)| = |r_2(P_2) - r_2(Q_2)|,$$

i.e. the extrinsic distances between corresponding pairs of points are equal. Consequently the original surfaces are either congruent or mirror images of each other.

§9. Convex surfaces with almost umbilical points

In geometry as in the theory of differential equations, one often considers three problems: existence, uniqueness and stability of the solution. The main point in the last-named problem is to determine how the solution varies when the initial data vary slightly, in some sense. Solution of this problem presupposes a quantitative estimate of the variation in the solution as a function of a given variation in the initial data. The qualitative solution is usually inherent in the corresponding uniqueness problem.

In the last two sections of this chapter we shall consider some questions concerning the stability problem.

A well-known theorem of differential geometry states that the sphere is the only convex surface all of whose points are umbilical. In other words, if the ratio R_1/R_2 of principal radii of curvature of a surface is equal to 1 at each point, then the surface is a sphere. No assumption is made here concerning equality of the principal radii of curvature themselves. It is natural to inquire whether a surface all of whose points are "almost umbilical" is "almost a sphere". The answer is given in the following

THEOREM. *If the ratio of principal radii of curvature of a closed convex surface is close to unity, then the surface is close to a sphere. More precisely, if the principal radii of curvature satisfy the condition* $1-\epsilon < R_1/R_2 < 1+\epsilon$, *then for sufficiently small* ϵ *the surface lies between two concentric spheres of radii* r_1 *and* r_2 *such that*

$$1 - c\epsilon < r_1/r_2 < 1 + c\epsilon, \quad c < 32\pi.$$

To prove this theorem we need three lemmas.

LEMMA 1. *Let* ϑ *be the angle between two conjugate diameters of an ellipse with semiaxes* a, b. *Then*

$$|\tan\vartheta| \geqslant \frac{2\frac{b}{a}}{1-\left(\frac{b}{a}\right)^2}. \tag{1}$$

Indeed, the slopes k and k' of the conjugate diameters of an ellipse with semiaxes a, b satisfy the relation $kk' = -b^2/a^2$. Since k and k' have opposite signs,

$$|k' - k| \geqq 2\sqrt{-k'k}.$$

Hence

$$|\tan\vartheta| = \left|\frac{k'-k}{1+k'k}\right| \geqslant \frac{2\dfrac{b}{a}}{\left|1-\dfrac{b^2}{a^2}\right|},$$

as required.

Let γ be a closed strictly convex curve, O a point in the interior of the domain bounded by γ. Let t be the tangent to γ at an arbitrary point X, p the distance from O to the tangent t,[3] and q the distance between X and the foot Y of a perpendicular dropped from O to the tangent t. We know (see §1) that $|p'| = q$, where p is differentiated with respect to the angle through which the tangent turns as X describes the curve.

LEMMA 2. *If the quotient q/p is small for all tangents to the curve, then the curve is "almost" a circle. Precisely, if $q/p < \epsilon$, then for sufficiently small ϵ the curve lies between two concentric circles of radii ρ_1, ρ_2 about O such that*

$$1 - 2\pi\epsilon < \rho_1/\rho_2 < 1 + 2\pi\epsilon.$$

Indeed, by assumption,

$$-\epsilon < p'/p < \epsilon. \tag{2}$$

Let ρ_1 and ρ_2 be the maximum and minimum distance of the points of the curve from O. Obviously the curve lies between concentric circles of radii ρ_1, ρ_2 about O. Let X_1 and X_2 be the points of the curve corresponding to the distances ρ_1 and ρ_2. The angle through which the curve turns over one of the arcs X_1X_2 is at most π. Integrating inequality (2) along this curve, we get

$$-\pi\epsilon < \ln(\rho_1/\rho_2) < \pi\epsilon.$$

Hence $e^{-\pi\epsilon} < \rho_1/\rho_2 < e^{\pi\epsilon}$. For sufficiently small ϵ we get

$$1 - 2\pi\epsilon < \rho_1/\rho_2 < 1 + 2\pi\epsilon,$$

as required.

Let F be a closed convex surface and α the tangent plane to F at some point P. Let $\alpha(h)$ denote the plane parallel to α which cuts the surface F at a distance h from α. Let $l(h)$ be the length of the curve $\gamma(h)$ in which the plane cuts the surface.

[3] Also called the "supporting distance" of the curve γ at the point X.

LEMMA 3. *If $l(h) \leqq l(h_0)$ for all $h \leqq h_0$, then the function $l(h)$ is monotone increasing in the interval $0 \leqq h \leqq h_0$.*

Indeed, let $p(h, \vartheta)$ be the supporting distance of the curve $\gamma(h)$ in the plane $\alpha(h)$. For fixed ϑ the function $p(h, \vartheta)$ is convex. Since

$$l(h) = \int_0^{2\pi} p(h, \vartheta)\, d\vartheta,$$

it follows that $l(h)$ is also a convex function. It has a maximum for $h = h_0$, and vanishes for $h = 0$. Hence it must be a monotone function of h. Q.E.D.

PROOF OF THE THEOREM. Let α and α' be two parallel tangent planes of F at a minimum distance from each other, and P and P' their points of contact with the surface. The straight line PP' is obviously normal to the surface at P and P'. Let H denote the distance between the planes α and α', i.e. the length of the segment PP'. We consider plane sections of F by planes parallel to α and α'. As before, let $\alpha(h)$ denote the plane parallel to α at a distance h, $\gamma(h)$ the curve in which it cuts the surface and $l(h)$ the length of this curve.

We claim that each curve $\gamma(h)$ lies between two concentric circles with center on the straight line PP' and radii ρ_1, ρ_2 such that

$$(*) \qquad 1 - c\epsilon < \rho_1/\rho_2 < 1 + c\epsilon,$$

where $c = 16\pi$.

We first show that condition (*) holds for sections $\alpha(h)$ close to α, i.e. for small h. Let β be a plane through the straight line PP'. Project the surface F onto this plane. The projection is a convex domain $\bar{F}$ bounded by some curve $\bar{\gamma}_\beta$. Let $\bar{l}$ be the length of the arc PQ on the curve $\bar{\gamma}_\beta$ (see Figure 33). For sufficiently small h we have $\bar{l} < 2OQ$ irrespective of the choice of the plane β. This follows directly from the fact that F is smooth at P.

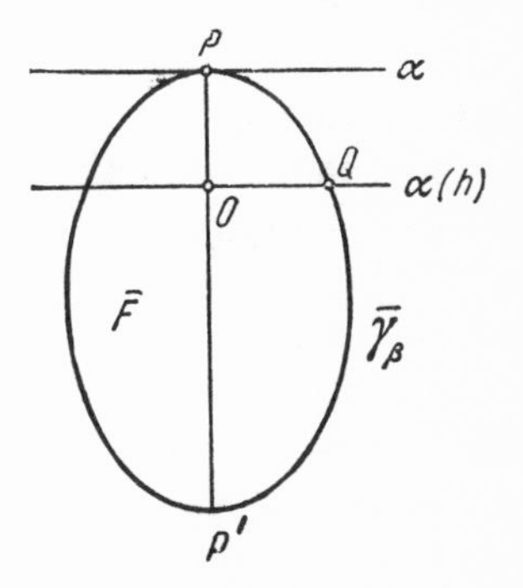

FIGURE 33

Now consider the curve $\gamma(h)$ in which the plane $\alpha(h)$ cuts F. In this section, the segment OQ is a supporting distance p for the tangent perpendicular to the plane β. Let us estimate the length of the corresponding segment q (Figure 32).

Let γ_β be the curve on F whose projection on the plane β is the curve $\bar{\gamma}_\beta$. At each point the directions of the curves γ_β and $\gamma(h)$ define conjugate diameters of the Dupin indicatrix of the surface F. The Dupin indicatrix is an ellipse with semiaxes $\sqrt{R_1}$, $\sqrt{R_2}$, where R_1 and R_2 are the principal radii of curvature. By Lemma 1 the curve γ_β cuts the curves $\gamma(h)$ at an angle ϑ such that

$$|\tan\vartheta| \geqslant \frac{2\sqrt{\frac{R_1}{R_2}}}{1-\frac{R_1}{R_2}}, \qquad R_1 \leqslant R_2.$$

Since $|1 - R_1/R_2| < \epsilon$, for sufficiently small ϵ we have $|\tan\vartheta| > 1/\epsilon$.

It is now easy to estimate q in terms of $p = OQ$:

$$q \leqq \bar{l}/\min|\tan\vartheta| < 2p\epsilon.$$

By Lemma 2 this implies that for small ϵ the curve $\gamma(h)$ lies between two concentric circles with center on PP' and radii ρ_1, ρ_2 satisfying condition (*).

Suppose that $l(h)$ assumes a maximum for $h = h_0$. Let h_2 denote the greatest number $\leqq h_0$ such that for any $h < h_2$ the section $\alpha(h)$ satisfies condition (*). We claim that $h_2 = h_0$. Suppose that $h_2 < h_0$. Then, since condition (*) involves a strict inequality, it cannot hold in $\alpha(h_2)$, since otherwise, by continuity, it would hold for h sufficiently close to h_2 and greater than h_2, which is impossible by the definition of h_2. We shall now show that condition (*) holds in $\alpha(h_2)$, and this will prove that $h_2 = h_0$.

At any rate, a weakened version of (*) surely holds in $\alpha(h_2)$:

$$1 - c\epsilon \leqq \rho_1/\rho_2 \leqq 1 + c\epsilon.$$

Since the curve $\gamma(h_2)$ contains a disk of radius ρ_1 and is contained in a disk of radius ρ_2, both with center on PP', it follows that the supporting distance p of the curve and its length $l(h_2)$ satisfy the inequalities

$$\rho_1 \leqq p \leqq \rho_2, \qquad 2\pi\rho_1 \leqq l(h_2) \leqq 2\pi\rho_2.$$

Let us estimate q for the curve $\gamma(h_2)$, in the case of an arbitrary plane β. We have

$$q \leqq \bar{l}_2/\min\tan\vartheta \leqq \epsilon\bar{l}_2,$$

where $\bar{l}_2$ is the length of the arc PQ_2 of the curve $\bar{\gamma}$ (Figure 34). Each point of the curve $\bar{\gamma}$ on the arc PQ_2 lies at a distance at most $l(h_2)/2$ from the straight line PP'. Indeed, for $h \leqq h_2$ each curve $\gamma(h)$ encircles the line PP' and its length is at most $l(h_2)$, since $l(h)$ is monotone. Hence each point of the arc PQ_2 of $\bar{\gamma}$ lies at a distance at most $l(h_2)/2$ from PP'. The polygonal line $PABO_2$, with sides $AB = BO_2 = l(h_2)/2$, envelopes the arc PQ_2 of the curve $\bar{\gamma}$, and its length is therefore at least that of the arc PQ_2. Hence $\bar{l}_2 \leqq l(h_2)+h_2$. Consequently $q \leqq (l(h_2)+h_2)\epsilon$.

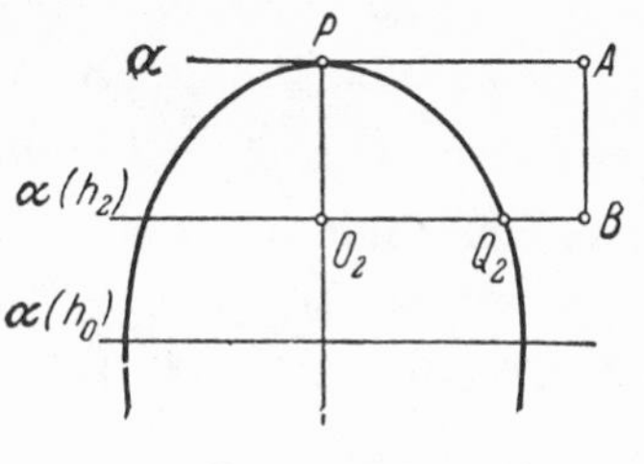

FIGURE 34

We now show that $h_2 < l(h_2)$. We first show that $H \leqq l(h_0)$. Indeed, assume that $l(h_0) < H - \delta$, $\delta > 0$. Then the distance of any point on the surface from the straight line PP' is less than $(H-\delta)/2$. Hence the surface must lie between two planes parallel to the straight line PP' and separated by a distance smaller than $H - \delta < H$. But this contradicts the definition of H. Thus $H \leqq l(h_0)$. Since the function $l(h)$ is convex and monotone increasing (Lemma 3),

$$l(h_2)/h_2 \geqq l(h_0)/h_0 > l(h_0)/H \geqq 1.$$

Hence $h_2 < l(h_2)$, and so $q < 2l(h_2)$.

Since

$$\rho_1 \leqq p, \quad l(h_2) \leqq 2\pi\rho_2, \quad \rho_2/\rho_1 \leqq 1 + 4\pi\epsilon,$$

it follows that $l(h_2) < 2\pi p(1 + 4\pi\epsilon)$. Consequently, $q < 4\pi p(1 + 4\pi\epsilon)$.

By Lemma 2, this implies that the curve $\gamma(h_2)$ lies between concentric circles with radii ρ_1 and ρ_2 such that

$$1 - 8\pi\epsilon(1 + 4\pi\epsilon) < \rho_1/\rho_2 < 1 + 8\pi\epsilon(1 + 4\pi\epsilon).$$

Obviously $8\pi\epsilon(1 + 4\pi\epsilon) < 16\pi\epsilon$ for sufficiently small ϵ, and thus condition (*) holds in the section $\alpha(h_2)$, contrary to assumption.

It follows that condition (*) holds for all $h \leqq h_0$. Exchanging the roles of the planes α and α', we conclude that condition (*) holds for the other sections $\alpha(h)$ of the surface F.

Now let $\bar{\alpha}$ and $\bar{\alpha}'$ be two planes tangent to F, both parallel to the straight line PP'. Let $\bar{\alpha}_0$ be a plane parallel to $\bar{\alpha}$ and $\bar{\alpha}'$, which cuts F in a curve of maximum length. Without loss of generality we may assume that the straight line PP' lies either between the planes $\bar{\alpha}$ and $\bar{\alpha}_0$ or on the plane α_0.

Now repeat the previous reasoning word for word, replacing α by $\bar{\alpha}$, PP' by the normal to F at its point of contact with $\bar{\alpha}$, and $\alpha(h_0)$ by the plane α_0. We conclude that the section of F by a plane parallel to $\bar{\alpha}$ through the line PP' lies between two concentric circles of radii ρ_1 and ρ_2 such that

$$1 - 16\pi\epsilon < \rho_1/\rho_2 < 1 + 16\pi\epsilon.$$

Combining our two results for plane sections perpendicular to PP' and plane sections through PP', we see that the surface F itself lies between two concentric spheres of radii r_1 and r_2 such that

$$1 - 32\pi\epsilon < r_1/r_2 < 1 + 32\pi\epsilon.$$

REMARK. The fact that the surface is closed is not essential; this assumption was made merely to simplify the proof.

§10. Stability of the solutions of Minkowski's and Christoffel's problems

In §2 we considered Minkowski's problem, which deals with the existence of a closed convex surface with given Gauss curvature. It was proved that if a positive continuous function $K(n)$ defined on the sphere satisfies the condition

$$\int_{\Omega} \frac{n}{K(n)}\, d\omega = 0,$$

then there exists a closed convex surface with Gauss curvature $K(n)$ at each point with normal n. This surface is unique up to translation.

The stability problem here is to estimate the difference between surfaces whose Gauss curvatures at points with parallel and identically oriented exterior normals differ by small amounts. This problem was solved by Volkov [**70**]. Volkov's solution applies to Minkowski's problem in a more general setting than that considered above.

The point is that specification of the Gauss curvature of a regular convex surface is equivalent to specification, for any set E on the sphere Ω, of the area $S(E)$ of that part of the surface whose spherical image is E. The set function $S(E)$ is known as the surface function. It retains its meaning for general convex surfaces, and is completely additive on the ring of Borel sets. The general problem of Minkowski concerns the existence and uniqueness of a closed convex surface whose surface function is a given completely additive set function $S(E)$ defined on the sphere Ω. Aleksandrov solved the problem in this formulation in [**16**]. Volkov considered the stability of the solution, proving the following theorem.

THEOREM 1. *If the surface functions of two closed convex surfaces* Φ_0 *and* Φ_1 *satisfy the condition*

$$|S(E)/S_0(E) - 1| \leqq \epsilon,$$

then the surface Φ_1 *can be brought by a translation into a* $\delta(\epsilon)$*-neighborhood of the surface* Φ_0, *and the surface* Φ_0 *into a* $\delta(\epsilon)$*-neighborhood of* Φ_1.

In other words, in a suitable relative position of the surfaces, their supporting functions on the unit sphere satisfy the inequality

$$|H_1(n) - H_0(n)| \leqq \delta(\epsilon).$$

Here $\delta(\epsilon) \leqq c_0\epsilon^{1/5}(1 + c_1(\epsilon))$, *where* c_0 *and* c_1 *are constants which depend on the maximum and positive minimum areas of the projections of* Φ_0 *and* Φ_1 *on arbitrary planes, and* $c_1(\epsilon) \to 0$ *as* $\epsilon \to 0$.

Volkov's proof of this theorem is rather intricate, and we shall therefore confine ourselves to a sketch of the proof, omitting details.

We first show that the radii of the circumscribed and inscribed spheres of the surfaces Φ_i may be estimated above and below, respectively, by the maximum area b and positive minimum area a of the areas of the projections of the surfaces on all possible planes. In fact, let d be the diameter of the surface. Then there are points C and D on the surface the distance between which is d. The area of the projection of the surface on a plane perpendicular to the straight-line segment CD is at least a. Hence there exist points A and B within the surface, the distance between whose projections on the plane is at least $\sqrt{a/\pi}$. The area of the projection of the surface on a plane parallel to the segments AB and CD is at least $\frac{1}{2}d\sqrt{a/\pi} \leqq b$. Thus we have an estimate for d and hence for the radius of the circumscribed sphere.

We now prove the existence of a positive lower estimate for the radius of the inscribed sphere. Let A_1 and A_2 be two points inside the surface, whose distance from each other is at least $\sqrt{a/\pi}$. Since the area of the projection of the surface on any plane is at least a, while the diameter of the surface has already been bounded, it follows that there is a point A_3 inside the surface such that the angles A_i of the triangle $A_1A_2A_3$ satisfy the inequalities $\epsilon' \leqq A_i \leqq \pi - \epsilon'$, where ϵ' can be estimated from below in terms of a and b. To find such a point, it suffices to consider the projection of the surface on a plane parallel to the segment A_1A_2. We can now find a point A_4 inside the surface such that the angles between all the edges of the tetrahedron $A_1A_2A_3A_4$ issuing from the vertex A_4 and the plane of the base $A_1A_2A_3$ lie between ϵ'' and $\pi - \epsilon''$, where $\epsilon'' > 0$ depends only on a and b. To find a point A_4 it suffices to consider the projection of the surface on a plane perpendicular to the plane of the triangle $A_1A_2A_3$. Having these estimates for the segment A_1A_2 and the angles of the tetrahedron, we can now find an estimate for the radius of the sphere inscribed in the tetrahedron, which depends only on a and b.

Let T_0 and T_1 be the convex bodies bounded by the surfaces Φ_0 and Φ_1, $\widetilde{T}_0$ and $\widetilde{T}_1$ bodies of unit volume similar to T_0 and T_1, $\tilde{\Phi}_0$ and $\tilde{\Phi}_1$ the surfaces of the bodies T_0 and T_1. The difference $\delta(\epsilon)$ is first estimated for the surfaces $\tilde{\Phi}_0$ and $\tilde{\Phi}_1$. One proves an estimate of the form

$$\tilde{\delta} \leq c(\widetilde{V}_{12} - 1)^{1/5}, \tag{1}$$

where $\widetilde{V}_{12}$ is the mixed volume of the bodies T_0 and T_1. We recall its definition.

Let r_0 and r_1 be the radius vectors of points P_0 and P_1 of the convex bodies T_0 and T_1. Consider the vector $r = \lambda r_0 + \mu r_1$. When the points P_0 and P_1 independently trace out the bodies T_0 and T_1, the end of the vector r traces out a certain convex body $T = \lambda T_0 + \mu T_1$. The volume of this body is a homogeneous polynomial of degree three in λ and μ (see Bonnesen and Fenchel [23]):

$$V = \sum C_3^k \lambda^k \mu^{k-1} V_{k,k-1}$$

(C_3^k are the binomial coefficients). The coefficients $V_{k,k-1}$ are called the mixed volumes; we have $V_{0,3} = V_0$, $V_{3,0} = V_1$.

The estimate (1) is the pivot of the proof of Theorem 1. We shall outline its derivation. Without loss of generality we may assume that the centers of gravity of the bodies $\widetilde{T}_0$ and $\widetilde{T}_1$ are at the origin O. Let α_i be the supporting plane of the body $\widetilde{T}_1$ with exterior normal n. Let $\alpha_i(v)$ be a plane parallel to α_i which cuts off a volume v from the body $\widetilde{T}_i$, and $F_i(v)$ the area of the section of $\widetilde{T}_i$ by the plane $\alpha_i(v)$. It is easy to see that the distance of O (the center of gravitv) to the plane α_i, i.e. the supporting distance, is

$$\widetilde{H}_l(n) = \int_0^1 (1-v)\frac{dv}{F_l(v)}.$$

Hence

$$|\widetilde{H}_1(n) - \widetilde{H}_0(n)| \leqslant \int_0^1 \left|\frac{1}{F_1} - \frac{1}{F_0}\right| dv = \psi(n). \tag{2}$$

Thus to estimate the difference $\tilde{\delta}$ for the surfaces $\tilde{\Phi}_0$ and $\tilde{\Phi}_1$ it suffices to estimate ψ.

In his proof of the Minkowski inequality for mixed volumes, Brunn obtained the inequality

$$f(t) = V((1-t)\tilde{T}_0 + t\tilde{T}_1)$$
$$\geqslant \int_0^1 ((1-t)F_0^{1/2} + tF_1^{1/2})\left(\frac{1-t}{F_0} + \frac{t}{F_1}\right) dv \equiv \varphi(t).$$

Since $f(0) = \varphi(0)$, it follows that $f'(0) \geqq \varphi'(0)$. Hence

$$3(\tilde{V}_{12} - 1) \geqslant \int_0^1 \left[\frac{F_0}{F_1} + 2\left(\frac{F_1}{F_0}\right)^{1/2} - 3\right] dv = I. \tag{3}$$

To get the estimate (1) we need only estimate ψ in terms of I. Set $\alpha = (F_0/F_1)^{1/2}$. Then

$$I = \int_0^1 \left(\alpha^2 + \frac{2}{\alpha} - 3\right) dv.$$

Let δ be a number such that $0 < \delta < \frac{1}{2}$. Divide the integration interval $(0,1)$ into three sets E_0, E_1, E_2: E_0 contains all points at which $|\alpha - 1| < \delta$; E_1 all points at which $\alpha > 1 + \delta$; E_2 all points at which $\alpha < 1 - \delta$. The integrals over E_0, E_1, E_2 are shown to satisfy the following estimates:

$$\psi_0 \leqq c'\delta, \quad \psi_1 \leqq c''(mE_1)^{1/3}, \quad \psi_2 \leqq c'''(mE_2)^{1/3},$$

where mE_1 and mE_2 denote the measures of E_1 and E_2. On the other hand,

$$I > \bar{c}'\delta^2(mE_1), \qquad I > \bar{c}''\delta^2(mE_2).$$

Hence

$$\psi \leqq c_1(1/\delta^2)^{1/3} + c_0\delta.$$

Finding the minimum of the right-hand side with respect to δ, we get

$$\psi \leqq \tilde{c} I^{1/5}. \tag{4}$$

The estimate (1) now follows easily from (2), (3) and (4).

To go from (1) to an estimate of $\tilde{\delta}$ in terms of ϵ, one proves that $\tilde{V}_{12} - \tilde{V}_i \leqq c_i\epsilon$. This estimate follows from an integral representation of the differences $\tilde{V}_{12} - \tilde{V}_i$ in terms of the supporting and surface functions.

To estimate the difference for the original surfaces Φ_0 and Φ_1, subject the surface Φ_0 to a similarity transformation with ratio λ such that the surfaces $\lambda\Phi_0$ and Φ_1 bound equal volumes. One proves that λ differs from unity by a quantity of first order in ϵ. The difference $\delta(\epsilon)$ for the surfaces Φ_0 and Φ_1 is then estimated by the sum of the differences for Φ_0, $\lambda\Phi_0$ and $\lambda\Phi_0$, Φ_1. This completes our sketch of the proof of Theorem 1.

We now consider the stability of the solution to Christoffel's problem. This problem is comparatively simple, since an explicit expression is available for the supporting function H of the surface in terms of the sum of principal radii of curvature $S = R_1 + R_2$. This expression was derived in §1. Other analytic expressions of H in terms of S are available. One of these is due to Weingarten (see [**22**], German p. 232):

$$H(n) = \frac{1}{4\pi} \int\limits_{\Omega} S(\nu)(1 - n\nu)\ln(1 - n\nu)\, d\omega_\nu,$$

where n and ν are unit vectors and the integration is with respect to area on the unit sphere Ω.

Let $S(\nu)$ and $S'(\nu)$ be the sums of principal radii of curvature of two closed convex surfaces F and F' at points with exterior normal ν. Then, by Weingarten's formula,

$$H(n) = \frac{1}{4\pi} \int\limits_{\Omega} S(\nu)(1 - n\nu)\ln(1 - n\nu)\, d\omega_\nu,$$

$$H'(n) = \frac{1}{4\pi} \int\limits_{\Omega} S'(\nu)(1 - n\nu)\ln(1 - n\nu)\, d\omega_\nu.$$

Assume that for any ν the functions S and S' differ by at most $\epsilon > 0$. Then

$$|H(n) - H'(n)| \leqslant \frac{1}{4\pi} \int\limits_{\Omega} |S(\nu) - S'(\nu)|(1 - n\nu)\ln(1 - n\nu)\, d\omega_\nu$$

$$\leqslant \frac{\varepsilon}{4\pi} \int\limits_{\Omega} (1 - n\nu)|\ln(1 - n\nu)|\, d\omega_\nu = \left(\ln 2 - \frac{1}{4}\right)\varepsilon.$$

THEOREM 2. *If the sums $S(\nu)$ and $S'(\nu)$ of the principal radii of curvature of two closed convex surfaces at points with parallel and identically oriented exterior normals differ by at most ϵ,*

$$|S(\nu) - S'(\nu)| \leqq \epsilon,$$

then the surfaces can be brought to a position in which their supporting functions $H(n)$ and $H'(n)$, considered on the unit sphere, differ by at most $(\ln 2 - \frac{1}{4})\epsilon$:

$$|H(n) - H'(n)| \leqq (\ln 2 - \tfrac{1}{4})\epsilon.$$

Chapter VIII

Geometric Theory of Monge-Ampère Equations of Elliptic Type

In their analytic interpretation, many problems of geometry in the large reduce to existence and uniqueness problems for differential equations. Examples are the isometric embedding problem, which reduces to investigation of the Darboux equation; the problem of infinitesimal bendings, which involves a certain system of linear equations; the existence and uniqueness of a convex surface with given function of principal radii of curvature; and so on. Analytically speaking, many geometric problems lead to Monge-Ampère equations. In such situations, results on the existence and uniqueness of various geometric objects may be treated as theorems on the solvability of an equation and the uniqueness of the solution.

In §§6 and 7 of Chapter VII we proved very general theorems on the existence and uniqueness of a convex polyhedron on whose faces or polyhedral angles given monotone functions assumed prescribed values. A limit procedure applied to these theorems, for a special class of monotone functions, yields existence theorems for the general Monge-Ampère equation. These solutions are first obtained in a generalized sense; one then proves that they are regular, if the coefficients of the equations are sufficiently regular. As a result we obtain general theorems on the solvability of boundary-value problems for elliptic Monge-Ampère equations in the usual (classical) formulation. The present chapter is devoted to a systematic exposition of the relevant results.

§1. Convex polyhedra with given generalized area of faces and given generalized curvature at the vertices

In §§6 and 7 we proved general theorems on the existence of convex polyhedra with prescribed values of a monotone function on the faces or on the polyhedral angles at the vertices. The simplest examples of such functions are the area of a face and the curvature at a vertex. In this section we consider natural generalizations of these two functions and accordingly specialize our existence theorems for polyhedra.

Let $\sigma'(x,y,\zeta,p,q)$ be any positive continuous function, nondecreasing in ζ, and α an arbitrary polygon in the plane $z=px+qy+\zeta$. Set

$$\sigma(\alpha)=\int\int_{(\alpha)} \sigma'(x, y, \zeta, p, q)\,dx\,dy.$$

Any function thus defined for polyhedra will be called a generalized area function. It reduces to the ordinary area if we put

$$\sigma' = \sqrt{1+p^2+q^2}.$$

Obviously the fact that σ' is positive and nondecreasing in ζ guarantees the monotonicity of σ in the sense of §6 of Chapter VII.

Our first goal is to prove existence theorems when the function defined on the faces of the polygon is generalized area. We wish to reformulate the assumptions of the theorem so that they apply to the function σ' defining the generalized area. Before we can do this, some observations are necessary.

Let G be an arbitrary convex polygon in the p,q-plane. With each point (p,q) of the polygon we associate a unit vector n with components

$$p/\sqrt{1+p^2+q^2}, \quad q/\sqrt{1+p^2+q^2}, \quad -1/\sqrt{1+p^2+q^2}.$$

The endpoints of the vectors n on the unit sphere trace out a convex polygon whose vertices correspond to those of G. Geometrically, this mapping of the (p,q)-plane onto the unit sphere is simply a projection of the plane $\zeta=-1$ in the p, q, ζ space onto the unit sphere $p^2+q^2+\zeta^2=1$ from its center.

Hitherto we have characterized the position of the plane of the face of a polyhedron $z=px+qy+\zeta$ by its exterior normal n and supporting number h. We shall now characterize it by the direction ratios[1] p, q of the plane, and the free term ζ, which we shall call the *elevation* of the plane.

In this new setting we can prove the existence of a polyhedron with unbounded faces lying in given planes as follows. If the points (p_k,q_k) of the pq-plane are the vertices of a convex polygon, then for any ζ_k there always exists a polyhedron whose unbounded faces lie in planes with direction ratios p_k, q_k and elevation ζ_k. This polyhedron is the boundary of the body defined by the intersection of halfspaces $z\geqq p_kx+q_ky+\zeta_k$. We now reinterpret Theorem 1, §6 of Chapter VII.

[1] [The direction ratios of this plane are in fact $(p,q,-1)$. The term will be used in this restricted sense throughout the sequel.]

Let G be a convex polygon in the pq-plane, $\bar{A}_k(\bar{p}_k, \bar{q}_k)$ its vertices, $A_k(p_k, q_k)$ points in the interior of G and σ_k positive numbers. What conditions must be imposed on the function σ' defining the generalized area σ in order that there exist a polyhedron, convex toward $z<0$, whose unbounded faces lie in the planes $\bar{\alpha}_k: z=\bar{p}_k x+\bar{q}_k y+\bar{\zeta}_k$, and whose bounded faces α_k have direction ratios p_k, q_k and conditional areas σ_k? It is obvious that we need only ensure that the assumptions of Theorem 1 in §6 of Chapter VII are satisfied.

The intersection of the halfspaces $z \geqq \bar{p}_k x+\bar{q}_k y+\bar{\zeta}_k$ is an unbounded convex polyhedron $\bar{P}$. It has no bounded faces and its unbounded faces lie in the planes $\bar{\alpha}_k$, each of which contains exactly one face. However, we may assume that the polyhedron $\bar{P}$ has *degenerate* bounded faces α_k in the directions p_k, q_k. Since the generalized area of a degenerate face is zero, the polyhedron $\bar{P}$ satisfies the condition $\sigma(\alpha_k) < \sigma_k$, $k=1, \cdots, n$.

Now let P be a polyhedron which coincides with $\bar{P}$ at infinity, i.e. its unbounded faces lie in the same planes as those of $\bar{P}$; assume that it lies above a plane $z=p_0 x+q_0 y+c$, where (p_0, q_0) is a point in the interior of G. Then the generalized area of its bounded faces α_k is

$$\sigma_P=\sum_k \iint_{(\alpha_k)} \sigma'(x, y, \zeta_k, p_k, q_k)\, dx\, dy.$$

We wish to find conditions for the function σ' under which σ_P is greater than $\sum \sigma_k$ for sufficiently large c.

Set

$$\bar{\sigma}'(x, y)=\varliminf_{\substack{(p, q) \in G \\ \zeta \to \infty}} \sigma'(x, y, \zeta, p, q).$$

Then

$$\sigma_P \geqslant \iint_{(\Sigma \alpha_k)} \bar{\sigma}'(x, y)\, dx\, dy.$$

If $c \rightarrow \infty$, then all the elevations ζ_k tend to infinity and the domain of integration (with respect to x, y) becomes the entire xy-plane. Therefore

$$\varliminf_{c \to \infty} \sigma_P \geqslant \iint_{(xy)} \bar{\sigma}'(x, y)\, dx\, dy,$$

where the domain of integration in the right-hand side is the entire xy-plane.

Hence the second condition of Theorem 1 (§6 of Chapter VII) will hold if

$$\iint_{(xy)} \bar{\sigma}'(x, y)\, dx\, dy > \sum_k \sigma_k.$$

We have proved the following theorem.

Theorem 1. *Let G be a convex polygon in the pq-plane, $\bar{p}_k$, $\bar{q}_k$ the coordinates of its vertices, $\bar{\zeta}_k$ arbitrary numbers, (p_k, q_k) arbitrary points inside the polygon G, σ_k arbitrary positive numbers, and σ the generalized area defined by a function σ' such that*

$$\iint_{(xy)} \bar{\sigma}'(x, y)\, dx\, dy > \sum_k \sigma_k,$$

where

$$\bar{\sigma}'(x, y) = \varliminf_{\substack{(p, q) \in G \\ \zeta \to \infty}} \sigma'(x, y, \zeta, p, q).$$

Then there exists a unique polyhedron, convex toward $z < 0$, whose unbounded faces lie in the planes $z = \bar{p}_k x + \bar{q}_k y + \bar{\zeta}_k$ and whose bounded faces have directions p_k, q_k and generalized areas σ_k.

Let $\vartheta'(x, y, p, q)$ be any positive continuous function, nonincreasing in z, V an arbitrary polyhedral angle convex toward $z < 0$, with vertex (x, y, z), and α an arbitrary supporting plane of V with direction ratios p, q. Set

$$\vartheta(V) = \iint_{(V)} \vartheta'(x, y, z, p, q)\, dp\, dq,$$

where the integration extends over all supporting planes α. The fact that the function ϑ' is positive and nonincreasing in z guarantees that ϑ is a monotone function of angles V, in the sense of §7 of Chapter VII. We shall call this function generalized curvature. Ordinary curvature is the special case in which

$$\vartheta' = 1/\sqrt{1 + p^2 + q^2}.$$

Let us consider the existence of a polyhedron with given boundary, vertices on given straight lines, and given generalized curvatures at the vertices. In other words, we wish to reformulate the assumptions of Theorem 1 in §7 of Chapter VII, in a form directly applicable to the function ϑ' defining the conditional curvature.

Thus, suppose that a closed polygonal line γ can be uniquely projected onto the xy-plane as a convex polygonal line bounding a polygon G. Let $g_1, \cdots, g_n$ be straight lines parallel to the z-axis cutting the polygon

G, and $\vartheta_1, \cdots, \vartheta_n$ positive numbers. What conditions must be imposed on the function ϑ' in order that there exist a polyhedron with boundary γ, which can be projected in one-to-one fashion onto the xy-plane, is convex toward $z < 0$, has vertices on the lines g_k and generalized curvatures ϑ_k at these vertices?

Consider the convex hull of γ (the smallest convex polyhedron containing γ). That part of the convex hull which is convex toward $z < 0$ is a polyhedron with boundary γ, containing no interior vertices, but we may assume that the polyhedron has *degenerate* vertices A_k on the lines g_k. Hence $\vartheta(A_k) \leqq \vartheta_k$, $k = 1, \cdots, n$, for this polyhedron, so that the first condition of Theorem 1 in §7 of Chapter VII is always satisfied, whatever the function ϑ'.

Now consider the second condition. The sum of generalized curvatures over all vertices of a polyhedron P with boundary γ and vertices on the straight lines g_k is

$$\vartheta_P = \sum_k \iint_{(A_k)} \vartheta'(x_k, y_k, z_k, p, q)\, dp\, dq.$$

Set

$$\bar{\vartheta}'(p, q) = \varliminf_{\substack{(x, y) \in G \\ z \to -\infty}} \vartheta'(x, y, z, p, q).$$

Then

$$\vartheta_P \geqslant \iint_{(\Sigma A_k)} \bar{\vartheta}'(p, q)\, dp\, dq.$$

Suppose that the polyhedron P cuts a plane $z = -c$. Then, as $c \to \infty$, the domain of integration with respect to p, q in the above inequality extends over the entire pq-plane. Hence

$$\lim_{c \to \infty} \vartheta_P \geqslant \iint_{(pq)} \vartheta'(p, q)\, dp\, dq,$$

where the domain of integration is the entire pq-plane.

Consequently the second condition of Theorem 1 in §7 of Chapter VII will hold if

$$\iint_{(pq)} \bar{\vartheta}(p, q)\, dp\, dq > \sum_k \vartheta_k,$$

and we have proved the following theorem.

Theorem 2. *Let γ be a closed polygonal line which can be projected in one-to-one fashion onto the xy-plane as a convex polygonal line bounding a polygon G. Let $g_1, \cdots, g_n$ be straight lines parallel to the z-axis cutting* *the polygon G, $\vartheta_1, \cdots, \vartheta_n$ arbitrary positive numbers, and ϑ the generalized curvature defined by a function ϑ' such that*

$$\int\int_{(pq)} \bar{\vartheta}'(p, q)\, dp\, dq > \sum_k \vartheta_k,$$

where

$$\bar{\vartheta}'(p, q) = \overline{\lim_{\substack{(x, y) \in G \\ z \to -\infty}}} \vartheta'(x, y, z, p, q).$$

Then there exists a unique polyhedron with boundary γ, which can be projected in one-to-one fashion onto the xy-plane, is convex toward $z < 0$, and has vertices on the straight lines g_k and generalized curvatures ϑ_k at the vertices.

We now determine conditions on the function ϑ' for the existence of an unbounded polyhedron with a given limit polyhedral angle V convex toward $z < 0$ whose projection covers the entire xy-plane; the vertices of the polyhedron are to lie on straight lines g_k parallel to the z-axis and the generalized curvatures at the vertices are ϑ_k.

The second condition of Theorem 4 in §7 of Chapter VII will hold if

$$\int\int_{(V)} \bar{\vartheta}'(p, q)\, dp\, dq > \sum_k \vartheta_k;$$

where the domain of integration is the generalized spherical image of the angle V, i.e. the set of p, q-values which are direction ratios of its supporting planes.

Consider the first condition of Theorem 4 in §7 of Chapter VII. We must prove that there exists a convex polyhedron with limit angle V and vertices A_k on the lines g_k such that $\vartheta(A_k) \leq \vartheta_k$, $k = 1, \cdots, n$.

Introduce the function

$$\tilde{\vartheta}'(p, q) = \overline{\lim_{\substack{(x, y) \in g_k \\ z \to \infty}}} \vartheta'(x, y, z, p, q).$$

Let

$$(*) \qquad \int\int_{(V)} \tilde{\vartheta}'(p, q)\, dp\, dq < \sum_k \vartheta_k.$$

We claim that this is a sufficient condition for the existence of the required polyhedron, i.e., a polyhedron with limit angle V and vertices A_k on the lines g_k, such that $\vartheta(A_k) \leq \vartheta_k$, $k = 1, \cdots, n$.

Without loss of generality, we may assume that the apex of the limit angle V is the origin and the z-axis is one of the lines g_k, say g_0.

By slightly enlarging the function $\tilde{\vartheta}'$ (if necessary), we may assume that it is positive, defined throughout the pq-plane, and satisfies condition (*). The new function will also be denoted by $\tilde{\vartheta}'$. Define a number $\mu < 1$ by the condition

$$\int\int_{(pq)} \tilde{\vartheta}'\, dp\, dq = \mu \sum \vartheta_k.$$

Let α be a plane cutting the angle V, such that the intersection is a polygon γ enclosing all the straight lines g_k. By Theorem 2 there exists a polyhedron P_ϵ with boundary γ, vertices on the lines g_k and generalized curvatures (defined by the function ϑ') equal to $(1-\epsilon)\mu\vartheta_k$ at the vertices.

Assume that for some ϵ the vertex A_0 (which is on the z-axis) does not lie above the point O. Let P be the convex hull of the vertices A_k of P_ϵ and the angle V. P is a polyhedron whose only vertices are A_k, and the polyhedral angle of P at A_k contains the corresponding polyhedral angle of P_ϵ. Since the function ϑ' is monotone, we see that, for the polyhedron P,

$$\vartheta(A_k) \leqq (1-\epsilon)\mu\vartheta_k < \vartheta_k.$$

Thus, if for some $\epsilon > 0$ there exists a polyhedron P_ϵ with vertex A_0 on the negative z-axis, we can also construct a polyhedron P with limit angle V, and vertices A_k on the lines g_k, such that $\vartheta(A_k) \leqq \vartheta_k$.

Now assume that the vertex A_0 of P_ϵ remains on the positive z-axis as $\epsilon \to 0$. Letting $\epsilon \to 0$, we get a polyhedron P_0 with boundary γ, vertices on the lines g_k, and generalized curvatures at the vertices (as defined by the function $\tilde{\vartheta}'$) equal to $\mu\vartheta_k$. Hence

$$\int\int_{(P_0)} \tilde{\vartheta}'\, dp\, dq = \mu \sum_k \vartheta_k,$$

which is impossible, since

$$\int\int_{(P_0)} \tilde{\vartheta}'\, dp\, dq < \int\int_{(pq)} \tilde{\vartheta}'\, dp\, dq = \mu \sum_k \vartheta_k.$$

(The domain of integration in the left-hand side is the generalized spherical image of the polyhedron P_0.) Thus we have proved

THEOREM 3. *Let V be a polyhedral angle whose projection covers the entire xy-plane, convex toward $z < 0$; let $g_1, \cdots, g_n$ be straight lines parallel to the z-axis, $\vartheta_1, \cdots, \vartheta_n$ positive numbers and ϑ the generalized curvature defined by a function ϑ' satisfying the following conditions:*

1. $$\iint_{(V)} \bar{\vartheta}'(p, q)\,dp\,dq > \sum_k \vartheta_k,$$

$$\bar{\vartheta}'(p, q) = \varliminf_{\substack{(x, y) \in g_k \\ z \to -\infty}} \vartheta'(x, y, z, p, q).$$

2. $$\iint_{(V)} \tilde{\vartheta}'(p, q)\,dp\,dq < \sum_k \vartheta_k,$$

$$\tilde{\vartheta}'(p, q) = \varlimsup_{\substack{(x, y) \in g_k \\ z \to \infty}} \vartheta'(x, y, z, p, q).$$

Then there exists an unbounded convex polyhedron which can be projected in one-to-one fashion onto the xy-plane, convex toward $z < 0$, *with limit angle* V, *whose vertices lie on the straight lines* g_k, *and with generalized curvatures* ϑ_k *at the vertices.*

This polyhedron is either unique or unique up to translation in the direction of the z-axis. The polyhedron is unique if the function ϑ' *is strictly monotone in* z.

THEOREM 3a. *If the function* ϑ' *defining the generalized curvature* ϑ *depends only on p and q, then conditions* 1 *and* 2 *of Theorem* 3 *must be replaced by the condition*

$$\iint_{(V)} \vartheta'(p, q)\,dp\,dq = \sum_k \vartheta_k.$$

To prove this theorem, replace the function ϑ' by the function

$$\vartheta'^{*} = (1 + \epsilon f(z + c))\,\vartheta'(p, q),$$

where $f(z)$ is a decreasing function of z, equal to 1 for $z = -\infty$ and -1 for $z = \infty$. By Theorem 3, there exists a polyhedron P^* with generalized curvatures ϑ^* equal to ϑ_k at its vertices, and limit angle V. By suitable choice of the parameter c we can make P^* pass through the origin. Now let $\epsilon \to 0$.

We now consider the existence of a closed convex polyhedron with vertices on given rays, with given generalized curvatures at the vertices.

Let ϑ' be a positive continuous function of the point A and unit vector n, which is monotone with respect to displacement of A in the direction OA, i.e. $\vartheta'(A, n) \leqq \vartheta'(A', n)$ if $|OA| < |OA'|$.

Now let V be an angle with vertex A, containing the point O. Define a function ϑ on such angles V by

$$\vartheta(V) = \iint_{\omega(V)} \vartheta'(A, n)\,dn,$$

where dn is the area element on the spherical image of V and the domain of integration is the spherical image $\omega(V)$ of the angle V. The function $\vartheta(V)$ is monotone in the sense of §7 of Chapter VII. We shall call it a generalized curvature function. It coincides with ordinary curvature when $\vartheta' \equiv 1$.

Theorem 4. *Let $g_1, \cdots, g_m$ be rays issuing from the point O, not all contained in a single halfspace, and ϑ the generalized curvature defined by a function ϑ' such that $\vartheta'(A,n) \to \infty$ as $|OA| \to \infty$; let $\vartheta_1, \cdots, \vartheta_m$ be positive numbers such that*

$$\sum_k \vartheta_k > \int\int_{\omega_0} \vartheta'(O, n)\, dn,$$

where the domain of integration is the unit sphere about the point O.

Assume that $\vartheta(V) < \sum' \vartheta_k$ for any convex polyhedral angle V with apex O; here the summation extends over all rays lying within the angle V.

Then there exists a closed convex polyhedron with vertices on the rays g_k and generalized curvatures ϑ equal to ϑ_k at the vertices.

Either this polyhedron is unique, or all such polyhedra are similar, with O as center of similitude. The second alternative is impossible if the function $\vartheta'(A,n)$ is strictly monotone with respect to displacement of the point A along the ray OA.

Proof. We shall show that the conditional curvature ϑ, as a function of angles, satisfies the assumptions of Theorem 5 in §7 of Chapter VII.

Describe a sphere of large radius R about the point O. It cuts the rays g_k at points A_k. Since $\vartheta'(A_k n) \to \infty$ as $|OA| \to \infty$, it follows that for sufficiently large R the convex polyhedron with vertices A_k has generalized curvatures greater than ϑ_k at the vertices.

Now assume that at least one of the vertices of the polyhedron P lies at a distance less than ϵ from the point O. We shall show that if ϵ is sufficiently small, the polyhedron P has a vertex such that $\vartheta(V_k) < \vartheta_k$.

If this is not true, we can find a sequence of polyhedra P such that $\vartheta(V_k) \geqq \vartheta_k$, $k = 1, \cdots, n$, whose vertices are arbitrarily close to O. Without loss of generality we may assume that all the vertices of these polyhedra converge, and at least one of them converges to the point O. They cannot all converge to O, since

$$\int\int_{\omega_0} \vartheta'(O, n)\, dn < \sum_k \vartheta_k.$$

Let V denote the convex hull of those rays g_k along which the vertices of the polyhedra P do not converge to O. Obviously there exist polyhedra in the sequence for which the total generalized curvature at the vertices on the rays g_k within the angle V is arbitrarily close to $\vartheta(V)$. But this contradicts the assumption of the theorem that $\vartheta(V) < \sum' \vartheta_k$.

Thus the function ϑ satisfies the assumptions of Theorem 5 in §7 of Chapter VII, and this implies the statement of our theorem.

Theorem 2a. *Let $g_1, \cdots, g_m$ be rays issuing from a point O, not all contained in a single halfspace, ϑ the generalized curvature defined by a function ϑ' which does not depend on the point A (but only on the unit vector n), and $\vartheta_1, \cdots, \vartheta_m$ positive numbers such that*

$$\sum_k \vartheta_k = \int_{\omega_0}\int \vartheta'(n)\, dn.$$

Assume that $\vartheta(V) < \sum' \vartheta_k$ for any convex polyhedral angle V with apex O, where the summation extends over all rays lying within V.

Then there exists a closed convex polyhedron (unique up to similarity with center O) with vertices on the rays g_k and generalized curvatures ϑ_k at the vertices.

Proof. Let ϵ be a small positive number. Define a function $\vartheta'_\epsilon(A, n)$ as follows:

$$\vartheta'_\varepsilon(A, n) = \begin{cases} \left(1 + \frac{1}{2}|OA| - \frac{\varepsilon}{2}\right)\vartheta'(n) & \text{for} \quad |OA| < \varepsilon, \\ \vartheta'(n) & \text{for} \quad \varepsilon \leqslant |OA| \leqslant 1, \\ |OA|\,\vartheta'(n) & \text{for} \quad |OA| > 1. \end{cases}$$

Denote the generalized curvature defined by this function by ϑ_ϵ.

By Theorem 2, there exists a polyhedron P_ϵ with vertices on the rays g_k and generalized curvatures ϑ_ϵ at the vertices equal to ϑ_k. The polyhedron P_ϵ cannot cut only one of the domains $|OA| < \epsilon$ or $|OA| > 1$, since this would contradict the assumption

$$\sum \vartheta_k = \int_{\omega_0}\int \vartheta'(n)\, dn.$$

If we assume that it cuts both these domains for arbitrarily small ϵ, we get a contradiction to the assumption $\vartheta(V) < \sum' \vartheta_k$.

The only remaining possibility is that for sufficiently small ϵ the polyhedron P_ϵ lies in the domain $\epsilon \leqq |OA| \leqq 1$. Since $\vartheta'_\epsilon = \vartheta'$ in this domain, P_ϵ is the required polyhedron.

The uniqueness part of the theorem is obvious, and so the proof is complete.

Theorem 2a was proved by Aleksandrov [6] for the case $\vartheta' \equiv 1$ (so that ϑ is ordinary curvature).

§2. Convex surfaces with given generalized area and given generalized curvature

The definitions of generalized area and generalized curvature are easily extended to arbitrary convex surfaces. By analogy with the case of polyhedra in §1, here again we can consider the existence of a convex surface with given generalized area or generalized curvature. We shall prove existence theorems analogous to those of §1.

Let F be a convex surface which can be projected in one-to-one fashion onto the xy-plane. Let H denote an arbitrary subset of the projection of F on the xy-plane, $H(F)$ the set of points on the surface whose projections are in H, and $\omega(H)$ the set of points in the pq-plane whose coordinates are direction ratios of the supporting planes $z = px + qy + \zeta$ of F on the set $F(H)$. The mapping taking the set H on the xy-plane onto the set $\omega(H)$ on the pq-plane will be called the generalized spherical mapping.

Together with the mapping ω, we also define a mapping $\bar{\omega}$ of the pq-plane onto the xy-plane: If $\bar{H}$ is an arbitrary subset of the generalized spherical image of the surface F, then $\bar{\omega}(\bar{H})$ will denote the set H on the xy-plane such that the generalized spherical image of each point of H intersects $\bar{H}$.

If the surface F is smooth and strictly convex, i.e. each supporting plane has only one point in common with the surface, then ω and $\bar{\omega}$ are mutually inverse homeomorphisms. In the general case, ω and $\bar{\omega}$ map closed sets onto closed sets and Borel sets onto Borel sets.

As in §1, let $\sigma'(x,y,\zeta,p,q)$ be a positive, continuous function, nondecreasing in ζ. For any Borel subset H of the generalized spherical image of F (in the pq-plane) we define a set function $\sigma(H)$ by

$$\sigma(H) = \iint_{\bar{\omega}(H)} \sigma'(x, y, \zeta(x, y), p(x, y), q(x, y))\, dx\, dy,$$

where $p(x,y)$, $q(x,y)$, $\zeta(x,y)$ are the direction ratios and elevation of the supporting plane of F at the point whose projection on the xy-plane is (x,y). The integrand is not a single-valued function when the supporting plane is not unique. However, the measure of the set of such points (x,y) is zero, and so these exceptional points have no effect on the function σ. The function σ will be called generalized area.

For any two pairwise disjoint Borel sets H_1 and H_2, the intersection of the sets $\bar{\omega}(H_1)$ and $\bar{\omega}(H_2)$ has measure zero; hence the additivity properties of the integral imply that the generalized area function is completely additive on the ring of Borel sets.

Let F_n be a sequence of convex surfaces converging to a convex surface F, σ_n and σ the conditional areas of the surfaces F_n and F, defined by some positive continuous function $\sigma'(x, y, \zeta, p, q)$. Then, if H is a closed set,

$$\overline{\lim}\, \sigma_n(H) \leqq \sigma(H)$$

as $n \to \infty$; and if H is an open set,

$$\underline{\lim}\, \sigma_n(H) \geqq \sigma(H)$$

as $n \to \infty$. We prove the first of these assertions; the proof of the second is analogous. Let H be a closed set. We have

$$\sigma(H) = \iint\limits_{\bar{\omega}_F(H)} \sigma'(x, y, \zeta, p, q)\, dx\, dy,$$

$$\sigma_n(H) = \iint\limits_{\bar{\omega}_{F_n}(H)} \sigma'(x, y, \zeta_n, p_n, q_n)\, dx\, dy.$$

Since H is closed, it follows that for sufficiently large n the set $\bar{\omega}_{F_n}(H)$ is contained in an arbitrarily small neighborhood of the set $\bar{\omega}_F(H)$. Hence, since σ' is bounded, it follows that

$$\sigma_n(H) - \sigma(H)$$
$$= \iint\limits_{\bar{\omega}_F(H)} (\sigma'(x, y, \zeta_n, p_n, q_n) - \sigma'(x, y, \zeta, p, q))\, dx\, dy + \varepsilon_n,$$

where $\lim \epsilon_n \leqq 0$. Furthermore, since the set of points on the xy-plane whose generalized spherical image ω_F is not uniquely defined has measure zero,, the integrand in the above expression for $\sigma_n - \sigma$ tends to zero in measure as $n \to \infty$. Hence

$$\overline{\lim}\, \sigma_n(H) \leqq \sigma(H).$$

These convergence properties of generalized area on closed and open sets H imply that the generalized areas of the surfaces F_n converge weakly to the generalized area of F, i.e., for any continuous function f defined on the pq-plane and vanishing in a neighborhood of the boundary of the spherical image of F,

$$\lim_{n\to\infty} \int f\, d\sigma_n = \int f\, d\sigma.$$

Let G be a strictly convex domain in the pq-plane, λ a completely additive set function defined in G. We wish to construct an unbounded convex surface F with generalized spherical image G and generalized area σ equal to λ, i.e. $\sigma(H) = \lambda(H)$ for any Borel subset of G.

To do this, inscribe a polygon $\overline{G}$ in G. Let $\overline{A}_k(\overline{p}_k, \overline{q}_k)$ be its vertices, and ζ_k the values of some continuous function ζ defined on the boundary of G at the vertices of $\overline{G}$. Partition the polygon $\overline{G}$ into small domains $\overline{G}_k$ and choose points (p_k, q_k) in each of them. Set $\lambda(\overline{G}_k) = \sigma_k$. Now construct an unbounded polyhedron P with unbounded faces in the directions $\overline{p}_k$, $\overline{q}_k$ and elevations $\overline{\zeta}_k$, and bounded faces in the directions p_k, q_k with generalized areas σ_k. This can be done, provided the function σ' defining the generalized area σ satisfies the assumptions of Theorem 1 in §1, i.e.

$$\int\int_{(x,\, y)} \overline{\sigma}'(x, y)\, dx\, dy > \lambda(G),$$

where

$$\overline{\sigma}'(x, y) = \lim_{\substack{(p,\, q) \in G \\ \zeta \to \infty}} \sigma'(x, y, \zeta, p, q).$$

Now let the sides of the polygon $\overline{G}$ inscribed in G and the diameters of the domains $\overline{G}_k$ tend to zero. The polyhedra P then remain bounded (in the sense that the elevations ζ of their supporting planes remain bounded), and we can extract a convergent sequence of polyhedra P_n.

Now the generalized areas of the polyhedra P_n obviously converge weakly to the set function λ, and the weak limit (if it exists) is uniquely defined. Hence the convex surface to which the sequence P_n converges has generalized area equal to λ. We thus obtain the following theorem:

THEOREM 1. *Let λ be a completely additive nonnegative set function defined in a strictly convex domain G of the pq-plane. Let σ be the generalized area defined by a function σ' such that*

$$\int\int_{(xy)} \overline{\sigma}'(x, y)\, dx\, dy > \lambda(G),$$

where

$$\overline{\sigma}'(x, y) = \lim_{\substack{(p,\, q) \in \overline{G} \\ \zeta \to \infty}} \sigma'(x, y, \zeta, p, q).$$

Then there exists an unbounded convex surface F with generalized spherical image G and generalized area λ, i.e. $\sigma(H) = \lambda(H)$ for any Borel subset H of G.

Let F be a convex surface which can be projected in one-to-one fashion onto the xy-plane as a domain G, and $\vartheta'(x,y,z,p,q)$ a positive continuous function, nonincreasing in z. For any measurable subset H of G, we define a set function $\vartheta(H)$ by

$$\vartheta(H)=\iint_{\omega(H)} \vartheta'(x(p,q), y(p,q), z(p,q), p, q)\,dp\,dq,$$

where $x(p,q)$, $y(p,q)$, $z(p,q)$ are the coordinates of the point of F whose supporting plane has direction ratios p, q. Again, the integrand is not single valued if the supporting plane has more than one point in common with the surface, but the measure of the set of such supporting planes is zero and so the function ϑ is not affected. The function ϑ will be called the generalized curvature of the surface.

As in the case of generalized area, one can prove the following properties:

1. Generalized curvature is a completely additive nonnegative function on the ring of Borel sets.

2. If a sequence of convex surfaces F_n converges to a convex surface F, then the generalized curvatures of F_n converge weakly to the generalized curvature of F.

Let G be an arbitrary strictly convex set on the xy-plane and $\mu(H)$ a completely additive nonnegative set function defined in G. We wish to construct a convex surface F which can be projected in one-to-one fashion onto G and has generalized curvature ϑ equal to μ.

To solve this problem, take an arbitrary closed contour γ which can be projected in one-to-one fashion onto the boundary of G, and inscribe in γ a polygonal line $\bar{\gamma}$ with small sides. The projection of this polygonal line on the xy-plane is a convex polygonal line bounding a polygon $\bar{G}$. Partition $\bar{G}$ into small domains $\bar{G}_k$ and choose a point A_k in each of them. Set $\vartheta_k=\mu(\bar{G}_k)$.

Now construct a polyhedron P with boundary $\bar{\gamma}$, vertices on straight lines g_k through the points A_k parallel to the z-axis, and generalized curvatures ϑ_k at these vertices. This can be done, provided the function ϑ' defining the generalized curvature satisfies the condition

$$\iint_{(pq)} \bar{\vartheta}'(p,q)\,dp\,dq>\mu(G),$$

where

$$\bar{\vartheta}'(p,q)=\varliminf_{\substack{(x,y)\in G\\ z\to-\infty}} \vartheta'(x,y,z,p,q).$$

Let the sides of the polygonal line $\bar{\gamma}$ and the diameters of the domains $\bar{G}_k$ tend to zero. Since the polyhedra P are uniformly bounded [in the same sense as before], we may assume without loss of generality that they converge to some convex surface F whose projection on the xy-plane is the domain G.

Since the generalized curvatures of the polyhedra P converge weakly to the set function μ, the limit surface F has generalized curvature ϑ equal to μ, and we have proved the following theorem:

THEOREM 2. *Let G be a strictly convex domain in the xy-plane, and μ a completely additive nonnegative set function in G, ϑ the generalized curvature defined by a function ϑ' such that*

$$\int\int_{(pq)} \bar{\vartheta}'(p, q)\, dp\, dq > \mu(G),$$

where

$$\bar{\vartheta}'(p, q) = \varliminf_{\substack{(x, y) \in G \\ z \to -\infty}} \vartheta'(x, y, z, p, q).$$

Then there exists a convex surface F, projected by straight lines parallel to the z-axis onto the domain G, convex toward $z < 0$, whose generalized curvature ϑ is equal to μ, i.e. $\vartheta(H) = \mu(H)$ for any Borel subset H of G.

For ordinary curvature this theorem was first proved by Aleksandrov [**6**] and the author [**61**]; the proof for the case $\vartheta' = \vartheta'(p,q)$ was given by Bakel'man [**19**], and for the general case $\vartheta' = \vartheta'(x,y,z,p,q)$ again by Aleksandrov [**8**].

REMARK. Despite the fact that the surface F is the limit of polyhedra whose boundaries are polygonal lines converging to γ, the latter need not be the boundary of F. Nonetheless, the following theorem (Bakel'man [**19**]) holds:

THEOREM 2a. *If $\vartheta' = \vartheta'(p,q)$, then the set of all convex surfaces with generalized curvature ϑ equal to μ and boundaries lying no higher than a curve γ contains a surface situated above all the others.*

The proof of this theorem, in outline, is as follows. Replace the function μ by a function μ_ϵ which coincides with μ everywhere in G except in an ϵ-neighborhood of the boundary of G, where it vanishes. Using the same construction as in the proof of Theorem 2, construct a surface F_ϵ with generalized curvature ϑ equal to μ_ϵ. The boundary of this surface is the curve γ. It can be shown that the surface F_ϵ lies above the surface F. Letting $\epsilon \to 0$, we get a surface F_0 with the desired properties.

Let F be an unbounded convex surface, not a cylinder, and S a point in the interior of the body bounded by F. The set of rays issuing from

S and not cutting the surface F describes a certain cone V (possibly degenerate). The cone V, and any cone which is a translate of V, is called the limit cone of the surface F.

Let μ be a completely additive set function defined in the xy-plane, nonnegative in a convex domain G and vanishing outside G. Let V be a convex cone which can be projected in one-to-one fashion onto the xy-plane and is convex toward $z < 0$.

We shall construct an unbounded convex surface with limit cone V and generalized curvature ϑ equal to μ. To do this, inscribe in the cone V a polyhedral angle $\overline{V}$ with small plane angles, divide the domain G into small domains G_k and choose a point A_k in each of them. Now construct a convex polyhedron P with limit angle $\overline{V}$, vertices on straight lines g_k parallel to the z-axis through the points A_k, and generalized curvature $\vartheta_k = \mu(G_k)$ at the vertices.

The existence of this polyhedron is assured if the function ϑ' defining the generalized curvature ϑ satisfies the condition

$$\iint_{\omega(V)} \overline{\vartheta}'(p, q)\, dp\, dq > \mu(G), \quad \iint_{\omega(V)} \tilde{\vartheta}'(p, q)\, dp\, dq < \mu(G),$$

where

$$\overline{\vartheta}'(p, q) = \varliminf_{\substack{(x, y) \in G \\ z \to -\infty}} \vartheta'(x, y, z, p, q),$$

$$\tilde{\vartheta}'(p, q) = \varlimsup_{\substack{(x, y) \in G \\ z \to \infty}} \vartheta'(x, y, z, p, q).$$

In view of the conditions imposed on the function ϑ', the polyhedra P are uniformly bounded in the sense that for sufficiently large N they all cut the segment $-N \leqq r \leqq N$ on the z-axis.

Now let the plane angles of $\overline{V}$ and the diameters of the domains G_k tend to zero. We can then extract a convergent sequence from the polyhedra P; the limit of this sequence is a convex surface F with limit cone V and generalized curvature ϑ equal to μ. Thus we have proved

THEOREM 3. *Let μ be a completely additive set function defined in the xy-plane, nonnegative in a convex domain G and vanishing outside this domain. Let V be a convex cone whose projection covers the entire xy-plane, convex toward $z < 0$.*

Then, if the function ϑ' defining the generalized curvature ϑ satisfies the condition

$$(*) \qquad \iint_{\omega(V)} \overline{\vartheta}'(p, q)\, dp\, dq > \mu(G), \quad \iint_{\omega(V)} \tilde{\vartheta}'(p, q)\, dp\, dq < \mu(G),$$

there exists an unbounded convex surface F with limit cone V and generalized curvature ϑ equal to μ.

THEOREM 3a. *If the function ϑ' defining the generalized curvature ϑ depends only on p and q, Theorem 3 is valid when condition (*) is replaced by the condition*

$$\iint_{\omega(V)} \vartheta'(p, q)\, dp\, dq = \mu(G).$$

This theorem is proved using Theorem 3, as in the case of polyhedra (§3).

THEOREM 3b. *Theorem 3a is valid when the domain G is the entire xy-plane* [**19**].

Theorem 3b is obtained by a limit procedure from Theorem 3a.

Let F be a closed convex surface whose interior contains the origin O, and H any measurable point set on the surface. We define the generalized curvature of the surface F on the set H as

$$\vartheta(H) = \iint_{\omega(H)} \vartheta'(A, n)\, dn,$$

where A is a point of H, n the exterior normal to a supporting plane at A, and the domain of integration is the spherical image of the set H.

Just as in the case of generalized area, the generalized curvature thus defined can be proved to have the following properties:

1. Generalized curvature ϑ is a nonnegative, completely additive function on the ring of Borel sets on the surface F.

2. If F_n is a sequence of convex surfaces converging to a convex surface F, then their generalized curvatures, defined by the same function ϑ', converge weakly to the generalized curvature of F.

THEOREM 4. *Let μ be a nonnegative, completely additive set function defined on the unit sphere ω_0 about O, and ϑ the generalized curvature defined by a function ϑ' such that*

$$\vartheta'(A,n) \to \infty \quad as\ |OA| \to \infty,$$

$$\mu(\omega_0) > \iint_{\omega_0} \vartheta'(A, n)\, dn.$$

Then, if $\vartheta(V) < \mu(\omega_0 - V)$ for any (possibly degenerate) cone V with apex at O, where $\omega_0 - V$ is the set of points on ω_0 lying outside the cone V, there exists a closed convex surface F whose generalized curvature $\vartheta(H)$

on any Borel set H is equal to $\mu(\overline{H})$, where $\overline{H}$ is the projection of H on the sphere ω_0 from its center (the point O).

PROOF. Divide the sphere ω_0 into small domains G_k, and let g_k be rays issuing from the center O and cutting the domains G_k. Set $\vartheta_k = \mu(G_k)$. If the diameters of the domains G_k are sufficiently small, then the rays g_k, the generalized curvature ϑ and the numbers ϑ_k satisfy the assumptions of Theorem 2a in §1. Hence there exists a closed convex polyhedron P with vertices on the rays g_k and generalized curvatures ϑ_k at these vertices.

When the maximal diameter δ of the domains G_k tends to zero, the polyhedra P cannot grow without bound, for then their total generalized curvature $\sum \vartheta_k$, which is equal to $\mu(\omega_0)$, would also increase without bound.

Since the polyhedra P are uniformly bounded as $\delta \to 0$, we can extract a convergent sequence. The limit surface F of this sequence contains the point O in its interior and has generalized curvature ϑ equal to μ.

Indeed, assume that O lies on the surface F (it cannot lie outside F, since it is in the interior of each polyhedron P). Let $\overline{V}$ be a cone with apex O, outside the surface F. Since the total generalized curvature at the vertices of P lying inside the cone $\overline{V}$ is greater than some $c_0 > 0$ for sufficiently small δ, it follows that the ordinary total curvature at these vertices is greater than some $c_0' > 0$. Hence O is a conic point of F.

Let V be the tangent cone of the surface F at O. If the polyhedron P is sufficiently close to F, the total generalized curvature at the vertices lying outside the cone V is arbitrarily close to $\vartheta(V)$ and $\mu(\omega_0 - V)$; but this is impossible, for by assumption $\vartheta(V) < \mu(\omega_0 - V)$. This contradiction shows that O must be in the interior of the surface F.

The fact that the generalized curvature ϑ of F coincides with the given set function μ follows from the weak convergence of the generalized curvature of P to that of F. This completes the proof.

THEOREM 4a. *Let the generalized curvature ϑ be defined by a function ϑ' which depends only on n; let μ be a completely additive nonnegative set function on the unit sphere ω_0 about the origin O, such that*

1. $\iint_{\omega_0} \vartheta'(n)\, dn = \mu(\omega_0)$.

2. *For any cone V with apex at O we have $\vartheta(V) < \mu(\omega_0 - V)$, where $\omega_0 - V$ is the set of all points on ω_0 lying outside the cone V.*

Then there exists a closed convex surface F whose generalized curvature

$\vartheta(H)$ on any Borel set H is $\mu(\overline{H})$, where $\overline{H}$ is the projection of H onto the sphere ω_0 from its center O.

We omit the proof of this theorem, which employs Theorem 4 in the same way as the proof of Theorem 2a uses Theorem 2.

§3. Generalized solutions of the Monge-Ampère equation $rt - s^2 = \varphi(x,y,z,p,q)$. Boundary-value problems for the generalized solutions

An analytical interpretation of our existence theorems for convex surfaces with given generalized area and generalized curvature, with the latter satisfying certain regularity conditions, yields the existence of solutions of the Monge-Ampère equation

$$rt - s^2 = \varphi(x,y,z,p,q).$$

We shall use this to solve various boundary-value problems for equations of this type.

Let F be a convex surface which can be projected in one-to-one fashion onto a domain G in the xy-plane and is convex toward $z<0$, ϑ the generalized curvature of the surface defined by a positive continuous function $\vartheta'(x,y,z,p,q)$, and μ a completely additive set function defined by a positive continuous function $\mu'(x,y)$ as follows:

$$\mu(H) = \int_H\int \mu'(x, y)\,dx\,dy.$$

By definition the surface F has generalized curvature ϑ equal to μ if for any Borel subset H of G

$$(*)\qquad \int_{\omega(H)}\int \vartheta'(x, y, z, p, q)\,dp\,dq = \int_{(H)}\int \mu'(x, y)\,dx\,dy.$$

Similarly, the surface F has generalized area σ defined by a function $\sigma'(x,y,\zeta,p,q)$ and equal to

$$\lambda = \int\int \lambda'(p, q)\,dp\,dq,$$

if for any Borel subset H of its generalized spherical image

$$\int_{\bar{\omega}(H)}\int \sigma'(x, y, \zeta, p, q)\,dx\,dy = \int_{(H)}\int \lambda'(p, q)\,dp\,dq.$$

If F is a twice continuously differentiable surface, condition (*) can be put in the equivalent form

$$(**) \qquad \vartheta'(x,y,z,p,q)(rt-s^2)=\mu'(x,y),$$

where r, s, t denote the second derivatives of the function z. Thus, the function $z(x,y)$ defining the surface F satisfies a Monge-Ampère equation (**).

With this in mind, we define a *generalized* solution of the Monge-Ampère equation (**) to be any convex function $z(x,y)$ defining a convex surface with generalized curvature ϑ equal to μ.

Similarly, for a surface with given generalized area we get an equation for the function ζ:

$$\sigma'(\zeta_p,\zeta_q,\zeta,p,q)(\zeta_{pp}\zeta_{qq}-\zeta_{pq}^2)=\lambda'(p,q)$$

and we thus define a generalized solution of this equation as a function ζ defining a surface with the given generalized areas.[2)]

The surface F defined by a generalized solution $z(x,y)$ of equation (**) is a surface of positive and bounded specific curvature. Indeed, let A be an arbitrary point of the domain G and H any neighborhood of A. Since

$$\iint_{\omega(H)} \vartheta'\,dp\,dq=\iint_{(H)} \mu'\,dx\,dy,$$

it follows from the continuity and positivity of the functions ϑ' and μ' that as $H\to A$ the quotient

$$\iint_{\omega(H)} dp\,dq \Big/ \iint_H dx\,dy$$

remains bounded both above and away from zero. This means that F has positive and bounded specific curvature.

A surface of positive bounded specific curvature is smooth and strictly convex (Chapter II, §3). Hence any generalized solution of equation (**) is a smooth, strictly convex function.

The convex surface F is twice differentiable almost everywhere. Since it has bounded specific curvature, it follows from a theorem of Aleksandrov [5] that the function $rt-s^2$ is almost everywhere bounded and

$$\iint_H (rt-s^2)\,dx\,dy=\iint_{\omega(H)} dp\,dq.$$

---

2) The surface F is defined here by an equation in tangential coordinates p, q, ζ.

Consequently, without any a priori assumptions on the regularity of the surface F, equation (**) is satisfied almost everywhere by its generalized solution, and the generalized solution of equation (**) can be defined as a function $z(x,y)$ determining a convex surface of bounded specific curvature on which the equation holds almost everywhere

Analogous conclusions hold for generalized solutions of the Monge-Ampère equation associated with surfaces of given generalized area.

We now state existence theorems for generalized solutions corresponding to the theorems of the preceding section.

THEOREM 1. *In a convex domain G of the xy-plane, consider the Monge-Ampère equation*

$$\vartheta'(x,y,z,p,q)(rt-s^2)=\mu'(x,y)$$

with continuous coefficients satisfying the following conditions:

1. *Both functions ϑ' and μ' are positive, and ϑ' is a nonincreasing function of z.*

2. $$\iint_{(pq)} \bar{\vartheta}'(p,q)\,dp\,dq > \iint_G \mu'(x,y)\,dx\,dy,$$

where

$$\bar{\vartheta}'(p,q)=\varlimsup_{\substack{(x,y)\in G\\ z\to-\infty}} \vartheta'(x,y,z,p,q).$$

Then the equation has a generalized solution in the domain G.

As a corollary, we get the following theorem.

THEOREM 1a. *The Monge-Ampère equation*

$$rt-s^2=\varphi(x,y,z,p,q)>0$$

in a convex domain G has a generalized solution if φ is a nondecreasing function of Z and increases to at most first order in p^2+q^2 as $p^2+q^2\to\infty$.

Now consider the Monge-Ampère equation

$$\vartheta'(x,y,z,p,q)(rt-s^2)=\mu'(x,y)$$

in a strictly convex domain G of the xy-plane. Assume that the functions ϑ' and μ' are continuous and positive, and that ϑ' is nonincreasing in z. We wish to construct a generalized solution $z(x,y)$ of this equation in the domain G, which coincides on the boundary of G with a given continuous function h (the Dirichlet problem).

Let γ denote the continuous curve defined by the function h, whose projection is the boundary of the domain G. The generalized solution constructed in Theorem 1 was the limit of polyhedra P, with boundaries converging to a prescribed continuous curve whose projection is the boundary of G. Without loss of generality we may assume that this curve is γ.

The limit surface F of the polyhedra P may have a boundary $\bar{\gamma}$ different from γ. All that we can say is that $\bar{\gamma}$ does not lie above γ. (As usual, we are assuming that the polyhedra P are convex toward $z < 0$.) Our problem is to impose conditions on the functions ϑ', μ' and the domain G under which the curves $\bar{\gamma}$ and γ necessarily coincide.

Suppose that $\bar{\gamma}$ does not coincide with γ, and let $\bar{Q}$ and Q be different points of these curves with the same projection on the xy-plane (Q lies above $\bar{Q}$). Let Q' be a point on the segment $Q\bar{Q}$, and Q'' a point beyond $\bar{Q}$ on the continuation of the segment, close to $\bar{Q}$. Let β'' be a horizontal plane through the point Q''. Let Z be the cylinder projecting G, and β' a plane through Q' obtained from the supporting plane to Z at Q' by rotation through a small angle α about a horizontal line. (The xy-plane is assumed to be horizontal.) Let K be the convex body bounded by the cylinder Z and the planes β'', β'.

If the polyhedron P is sufficiently close to the surface F, it has a nonempty intersection with K. Let P_K denote this intersection. We wish to find conditions on the domain G and the functions μ', ϑ' such that the equality

$$\iint_{(P_K)} \vartheta' \, dp \, dq = \iint_{(P_K)} \mu' \, dx \, dy$$

cannot hold for sufficiently small α. These conditions will ensure that the curves γ and $\bar{\gamma}$ coincide, thus yielding solvability conditions for our Dirichlet problem.

Let β'_K be the intersection of the plane β' and the body K. Let V denote the cone projecting the domain β'_K from the point $\bar{Q}$. If P is sufficiently close to F, the generalized spherical image of P_K almost covers the generalized spherical image of the cone V; when $P \to F$ the covering will be complete.

Let β''_K be the intersection of the body K and the plane β''. It is obvious that the projection of β''_K on the xy-plane covers the projection of P_K.

Now assume that the domain G and the functions ϑ', μ' satisfy the following condition: For any convex surface Φ (including the polyhedra P), for sufficiently small α

$$\iint_{\omega(V)} \vartheta' \, dp \, dq >> \iint_{\beta_K''} \mu' \, dx \, dy.$$

Then the points Q and $\overline{Q}$ cannot be distinct, and hence we have a solvability condition for the Dirichlet problem. Since this condition is hard to visualize and not easily checked, we shall replace it by simpler, albeit stronger conditions. These conditions involve the curvature of the boundary of G and the rate of decrease of ϑ' as $p^2+q^2 \to \infty$.

Assume that the specific curvature of the boundary of G is bounded away from zero. That is to say, the ratio of the angle between supporting lines at any two close points on the curve to the distance between the points is bounded below by a constant $c_0 > 0$. For brevity we shall simply say that the curve has positive curvature.

Construct a circular cylinder $\widetilde{Z}$, one of whose generators is the straight line $Q\overline{Q}$, such that $\widetilde{Z}$ encloses the cylinder Z. If we base the construction of the cone V not on Z but on the circular cylinder $\widetilde{Z}$, we obtain a cone $\widetilde{V}$ with the same apex, containing V. Hence $\omega(\widetilde{V}) \subset \omega(V)$ and

$$\iint_{\omega(\widetilde{V})} \vartheta' \, dp \, dq \leqslant \iint_{(V)} \vartheta' \, dp \, dq.$$

It is not difficult to study the generalized spherical image $\omega(\widetilde{V})$ on the cone $\widetilde{V}$. Without loss of generality we may assume for the sake of convenience that the straight line $Q\overline{Q}$ is the z-axis and the cylinder $\widetilde{Z}$ is in the halfspace $y \leq 0$. Simple arguments show that $\omega(\widetilde{V})$ contains an isosceles triangle Δ in the pq-plane with vertices

$$\left(\frac{\varepsilon}{\alpha}, 0\right), \quad \left(\frac{c'}{\alpha}, -\frac{c''}{\sqrt{\alpha}}\right), \quad \left(\frac{c'}{\alpha}, \frac{c''}{\sqrt{\alpha}}\right),$$

where ϵ, c' and c'' tend to positive limits as $\alpha \to 0$, and ϵ/c' is arbitrarily small if the segment ratio $Q''\overline{Q}/Q'\overline{Q}$ is sufficiently small.

Now assume that the rate of decrease of the function $\vartheta'(x,y,z,p,q)$ as $p^2+q^2 \to \infty$ is at most that of $(p^2+q^2)^{-k}$, i.e.

$$\vartheta'(x, y, z, p, q) > \frac{c_0}{(p^2+q^2)^k} \quad \text{as} \quad p^2+q^2 \to \infty.$$

Then, since $\omega(\widetilde{V})$ contains the triangle Δ, it follows that when $\alpha \to 0$,

$$\iint_{\omega(\widetilde{V})} \vartheta' \, dp \, dq > \frac{a\sqrt{\alpha}}{(2k-1)(2k-2)} \left(\frac{\alpha}{\varepsilon}\right)^{2k-2},$$

where a is a constant.

Since the specific curvature of the boundary of the domain is bounded away from zero, it follows that when $\alpha \to 0$

$$\iint_{\beta_K''} \mu' \, dx \, dy < b\alpha^{3/2},$$

where b is a constant.

Now let $k \leqq 3/2$. Then for sufficiently small ϵ (depending on the choice of the points Q' and Q'') and $\alpha \to 0$ we have

$$\iint_{\omega(\tilde{V})} \vartheta' \, dp \, dq > \iint_{\beta_K''} \mu' \, dx \, dy.$$

Consequently, when $\alpha \to 0$,

$$\iint_{(P_K)} \vartheta' \, dp \, dq > \iint_{(P_K)} \mu' \, dx \, dy.$$

Thus the limit surface F of the sequence of polyhedra P will be bounded by the curve γ (i.e. the curves $\bar{\gamma}$ and γ will coincide), provided the following two conditions are satisfied:

1. The specific curvature of the boundary of G is bounded away from zero.

2. The rate of decrease of the function ϑ' as $p^2+q^2 \to \infty$ is at most that of $(p^2+q^2)^{-3/2}$, i.e.

$$\vartheta'(x,y,z,p,q) > c/(p^2+q^2)^{3/2} \quad \text{when } p^2+q^2 \to \infty.$$

Theorem 1 now yields the following existence theorem for the Dirichlet problem.

THEOREM 2. *In a convex domain G on the xy-plane, consider the Monge-Ampère equation*

$$\vartheta'(x,y,z,p,q)(rt-s^2) = \mu'(x,y),$$

where the following conditions hold:

1. *The boundary of G has positive curvature.*

2. *The functions μ' and ϑ' are continuous and positive. The function ϑ' is nonincreasing in z and*

$$\iint_{(p,q)} \bar{\vartheta}(p,q) \, dp \, dq > \iint_{G} \mu'(x,y) \, dx \, dy,$$

where

$$\bar{\vartheta}(p,q) = \varliminf_{\substack{(x,y)\in G \\ z \to -\infty}} \vartheta'(x,y,z,p,q).$$

3. *At any point on the boundary of G, for any finite z and* $p^2+q^2 \to \infty$,

$$\vartheta'(x,y,z,p,q) > c/(p^2+q^2)^{3/2} \quad (c>0).$$

Then the Dirichlet problem for this equation is solvable for any continuous function defined on the boundary of G.

As a corollary of Theorems 1a and 2, we get

THEOREM 2a. *The Dirichlet problem for the equation*

$$rt - s^2 = \varphi(x,y,z,p,q)$$

in a convex domain G is solvable for any continuous boundary values, provided the following conditions hold:

1. *The boundary of G has positive curvature.*
2. *The function φ is continuous and positive, nondecreasing in z, and its rate of increase as $p^2+q^2 \to \infty$ is at most that of p^2+q^2.*

Now consider a convex closed curve in the pq-plane, defined by an equation $\psi(p,q)=0$. Does there exist a solution of the equation

$$\vartheta'(x,y,z,p,q)(rt-s^2) = \mu'(x,y)$$

in a bounded convex domain G, with boundary condition $\psi(p,q)=0$? This is the *second boundary-value problem.*

The planes defined by the directions p, q satisfying the equation $\psi(p,q)=0$ and passing through the origin generate a convex cone V. Under certain conditions on the functions ϑ' and μ' there exists a sequence of polyhedra P converging to a convex surface F with generalized curvature ϑ equal to μ (§2). The degree of freedom involved in the choice of the limit angles of these polyhedra is sufficient to allow us to assume that they converge to the cone V.

Let G^* denote the convex domain on the pq-plane bounded by the curve $\psi(p,q)=0$. Assume that the functions μ' and ϑ' satisfy the condition $\mu'/\vartheta' < \infty$ for $(x,y)\in G$ and $(p,q)\in G^*$. Then the limit surface F of the sequence of polyhedra P is smooth.

In fact, the surface F has bounded specific curvature. By Aleksandrov's theorem, a surface of bounded specific curvature can fail to be smooth only along a straight-line segment with endpoints on the boundary of the surface. But the surface F is unbounded. Hence it can fail to be smooth only along an entire straight line. Now an unbounded convex surface containing a straight line must be a cylinder. Since the surface F has nonzero curvature, it cannot be a cylinder. Hence F is smooth.

F is the union of two surfaces: a bounded surface F_1 whose projection on the xy-plane is the domain G, and the remainder F_2, an unbounded surface. F_2 is a developable surface. Through each point of F_2 there is a straight-line generator with stationary tangent plane. Hence the direction ratios p, q of the tangent planes to F_2 satisfy the equation $\psi(p,q)=0$. By continuity this equation also holds for the direction ratios p, q of the tangent planes along the boundary of F_1. Hence the generalized solution in the domain G, defined by the surface F_1, satisfies the boundary condition $\psi(p,q)=0$.

THEOREM 3. *In a bounded convex domain G on the xy-plane, consider the Monge-Ampère equation*

$$\vartheta'(x,y,z,p,q)(rt-s^2)=\mu'(x,y),$$

and let G^* be a bounded convex domain in the pq-plane whose boundary is defined by an equation $\psi(p,q)=0$. Assume that the coefficients ϑ' and μ' of the equation satisfy the following conditions:

1. *ϑ' and μ' are positive and continuous for $(x,y)\in G$, $(p,q)\in G^*$; ϑ' is a nonincreasing function of z and $\mu'/\vartheta'<\infty$.*
2. $\iint_{G^*}\bar{\vartheta}'\,dp\,dq>\iint_G\mu'\,dx\,dy$, $\bar{\vartheta}'=\varliminf_{\substack{(x,y)\in G\\ z\to-\infty}}\vartheta'(x,y,z,p,q)$.
3. $\iint_{G^*}\tilde{\vartheta}'\,dp\,dq<\iint_G\mu\,dx\,dy$, $\tilde{\vartheta}'=\varlimsup_{\substack{(x,y)\in G\\ z\to\infty}}\vartheta'(x,y,z,p,q)$.

Then there exists a generalized solution of the equation $\vartheta'(rt-s^2)=\mu'$ in the domain G, which satisfies the boundary condition $\psi(p,q)=0$ and is convex toward $z<0$.

Now assume that the function ϑ' depends only on p and q. We shall prove that Theorem 3 then holds with conditions 2 and 3 replaced by the condition

$$\iint_{G^*}\vartheta'(p,q)\,dp\,dq=\iint_G\mu'(x,y)\,dx\,dy.$$

Displace the cone V along a segment N toward $z>0$, and denote the new cone by V_N. Let $\overline{V}$ and $\overline{V}_N$ be the regions of space outside V and inside V_N, respectively, whose projections lie in the domain G on the xy-plane. Now replace the function ϑ' by a function ϑ'' equal to ϑ' outside the regions $\overline{V}$ and $\overline{V}_N$, greater than ϑ' in $\overline{V}$ and smaller than ϑ' in $\overline{V}_N$. The function ϑ'' satisfies the conditions of Theorem 3. Thus we have a solution of the second boundary-value problem for the equation $\vartheta''(rt-s^2)=\mu'$.

For sufficiently large N the surface $F_{\vartheta''}$ defining this solution must cut the segment ON on the z-axis connecting the apices of the cones V and V_N. Otherwise it would cut one of the regions $\overline{V}$ or $\overline{V}_N$, and the resulting equality

$$\iint_{G^*} \vartheta'' \, dp \, dq = \iint_{G} \mu' \, dx \, dy$$

would contradict the assumption that

$$\iint_{G^*} \vartheta' \, dp \, dq = \iint_{G} \mu' \, dx \, dy.$$

Since the surface $F_{\vartheta''}$ cuts the segment ON, we may assume without loss of generality that as $\vartheta'' \to \vartheta'$ the surface $F_{\vartheta''}$ converges to some surface $F_{\vartheta'}$. This surface defines a solution of the equation

$$\vartheta'(p,q)(rt - s^2) = \mu'(x,y),$$

which satisfies the required boundary condition.

Theorem 3a. *If the function ϑ' depends only on p and q, the conclusion of Theorem 3 concerning the solvability of the second boundary-value problem for the equation*

$$\vartheta'(p,q)(rt - s^2) = \mu'(x,y)$$

remains valid when conditions 2 and 3 are replaced by the condition

$$\iint_{G^*} \vartheta'(p, q) \, dp \, dq = \iint_{G} \mu'(x, y) \, dx \, dy.$$

In conclusion, we remark that Theorem 3a is also valid when the domain G is the entire xy-plane [**21**].

Let the set function $\mu(H)$ be the integral of a positive continuous function $\mu'(n)$ on the sphere ω_0:

$$\mu(\overline{H}) = \iint_{\overline{H}} \mu'(n) \, dn.$$

Then the surfaces whose existence is ensured by Theorems 4 and 4a of §2 are smooth, strictly convex surfaces with positive bounded Gauss curvature, and they satisfy the differential equation

$$K(A) \frac{|OA|^3 \vartheta'(A, n)}{(\overrightarrow{OA} \cdot n)} = \mu' \left(\frac{\overrightarrow{OA}}{|OA|} \right)$$

almost everywhere, where A is an arbitrary point on the surface, n is the exterior normal and $K(A)$ the Gauss curvature of the surface at A.

In a neighborhood of the point A_0 at which the surface cuts the negative z-axis, this equation becomes

$$rt - s^2 = \varphi(x, y, z, p, q),$$

where

$$\varphi = \frac{\mu'\left(\frac{x}{z}, \frac{y}{z}\right)(1 + p^2 + q^2)^{3/2}\,(px + qy - z)}{\vartheta'\left(\frac{x}{z}, \frac{y}{z}, z\right)(x^2 + y^2 + z^2)^{3/2}},$$

Note that $\mu_z = 0$ at the point A_0, since there $x = 0$, $y = 0$ and $\vartheta'_z \leqq 0$ by virtue of our assumption that the function $\vartheta'(A, n)$ is monotone in A. Therefore $\varphi_z > 0$ in a neighborhood of A_0.

§4. Uniqueness theorems for solutions of the equation $rt - s^2 = \varphi(x, y, z, p, q)$

Our main subject is Monge-Ampère equations with regular coefficients. We shall prove later, under very broad assumptions, that the generalized solutions of these equations are regular functions.

For this reason, the uniqueness theorems to be proved in this section relate to regular solutions and, accordingly, regular surfaces.

THEOREM 1. *In a domain G on the xy-plane, consider the Monge-Ampère equation*

$$rt - s^2 = \varphi(x, y, z, p, q),$$

where φ is a regular function such that $\varphi > 0$ and $\varphi_z \geqq 0$, and $z_1(x, y)$ and $z_2(x, y)$ are regular solutions of this equation.

Then, if these solutions coincide on the boundary of G and are both convex toward $z < 0$, i.e. $d^2 z_1 \geqq 0$, $d^2 z_2 \geqq 0$, then they are identical throughout G.

PROOF. By assumption,

$$r_1 t_1 - s_1^2 = \varphi(x, y, z_1, p_1, q_1),$$
$$r_2 t_2 - s_2^2 = \varphi(x, y, z_2, p_2, q_2).$$

Subtracting these equalities term by term and setting $z_1 - z_2 = u$ and

$$A = \tfrac{1}{2}(t_1 + t_2), \quad B = \tfrac{1}{2}(s_1 + s_2), \quad C = \tfrac{1}{2}(r_1 + r_2),$$

we get

$$(*) \qquad Au_{xx} - 2Bu_{xy} + Cu_{yy} = \varphi_p u_x + \varphi_q u_y + \varphi_z u,$$

where the derivatives of φ in the right-hand side are computed at z, p, q between z_1, p_1, q_1 and z_2, p_2, q_2.

Since the forms

$$r_1\xi^2 - 2s_1\xi\eta + t_1\eta^2, \quad r_2\xi^2 - 2s_2\xi\eta + t_2\eta^2$$

are positive definite, their arithmetic mean

$$C\xi^2 - 2B\xi\eta + A\eta^2$$

is also a positive definite form, and hence $AC - B^2 > 0$.

Thus our equation (*) for the function u is elliptic. But since $u = 0$ on the boundary of G and $\varphi_z \geqq 0$, it follows from a standard theorem that u cannot have positive maximum and negative minimum in G. Hence $u = 0$ in G, i.e. the solutions z_1 and z_2 coincide in G. Q.E.D.

REMARK. Theorem 1 can be proved for generalized solutions under any of the following assumptions:

1. The function φ is continuous, positive, and strictly increasing in z.
2. The function φ is continuous, positive, nondecreasing in z and satisfies a Lipschitz condition in p, q (see §12).
3. The function φ is continuous, positive, and has the form $\varphi = \varphi_1(x,y)\varphi_2(p,q)$.

Variant 3 is due to Bakel′man [**19**].

THEOREM 2. *Let G be a domain on the xy-plane whose boundary is a convex curve with bounded specific curvature, $\psi(p,q) = 0$ the equation of a closed convex curve in the pq-plane.*

If solutions z_1 and z_2 of the equation

$$rt - s^2 = \varphi(x,y,z,p,q), \quad \varphi > 0, \quad \varphi_z \geqq 0,$$

satisfy the condition $\psi(p,q) = 0$ on the boundary of G and both are convex toward $z < 0$, then they coincide up to an additive constant. If the function φ is strictly increasing in z, the solutions are identical.

PROOF. Let α be an arbitrary tangent plane of the surface $F_1 : z = z_1(x,y)$. Let E_α denote the halfspace defined by α which contains the surface F_1. Since the generalized spherical image of F_1 is a convex domain, the complete surface of the unbounded body defined as the intersection of all halfspaces E_α is an unbounded convex surface $\overline{F}_1$ with the same spherical image as F_1.

Hence the surface $\overline{F}_1$ has bounded specific curvature. Since it is certainly not a cylinder (it contains F_1) it must be smooth. Therefore F_1, as a domain on $\overline{F}_1$, is smooth up to its boundary. Similarly the surface $F_2 : z = z_2(x,y)$ is also smooth up to its boundary.

Consider the difference $u = z_1(x,y) - z_2(x,y)$. The function u assumes a maximum at some point A in the domain G or on its boundary. If A is an interior point it is obvious that $du=0$ at A.

Let A be a boundary point of G. Since $du=0$ at A in the direction of the boundary of G, the tangents to the curves bounding the surfaces F_1 and F_2 at the points corresponding to A are parallel. And since the surfaces F_1 and F_2 have the same generalized spherical image (the convex domain bounded by the curve $\psi(p,q)=0$) it follows that the tangent planes to F_1 and F_2 at these points are parallel, and so $p_1=p_2$ and $q_1=q_2$ at A, i.e. $du=0$ in this case too.

Now the boundary of G has bounded specific curvature: hence, irrespective of whether A is an interior or boundary point of G, there exists a disk whose circumference passes through A and which lies within the domain G. By a theorem of Aleksandrov [**9**], a solution u of equation (*) which has a maximum at a boundary point of a disk in which the solution is defined and is stationary at this point must be a constant. Obviously, if $\varphi_z>0$ this constant must be zero. Q.E.D.

THEOREM 3. *Let z_1 and z_2 be two solutions of the equation*

$$rt-s^2=\varphi_1(x, y)\varphi_2(p, q), \quad \varphi_1>0, \quad \varphi_2>0,$$
$$\iint_{(xy)}\varphi_1\,dx\,dy<\infty,$$

defined over the entire xy-plane, such that the unbounded convex surfaces $F_1: z=z_1(x,y)$ and $F_2: z=z_2(x,y)$ are convex toward $z<0$ and have the same generalized spherical image. Then the solutions z_1 and z_2 are identical up to an additive constant.

This theorem will be proved by Aleksandrov's method [**6**].

Without loss of generality we may assume that the surfaces F_1 and F_2 have nonempty intersection, since if $z(x,y)$ is a solution then any function $z(x,y)+\text{const}$ is also a solution. Moreover, we may assume that F_1 and F_2 have different tangent planes at some common point A, since otherwise $dz_1\equiv dz_2$ and so $z_1-z_2=\text{const}$.

Let G be a component of the set of points in the xy-plane for which $z_1(x,y)<z_2(x,y)$, whose boundary contains the projection of A. Let $\overline{F}_1$ and $\overline{F}_2$ denote the domains on the surfaces F_1 and F_2 whose projection on the xy-plane is the domain G.

Let P_2 be an arbitrary point on the surface $\overline{F}_2$. The tangent plane α_2 at P_2 cuts out a cap from the surface $\overline{F}_1$. Hence there is a point P_1 on the surface $\overline{F}_1$ at which the tangent plane α_1 is parallel to α_2. This implies that the generalized spherical image $\omega(F_2)$ of F_2 is contained in the generalized spherical image $\omega(\overline{F}_1)$ of $\overline{F}_1$.

Let α be the tangent plane to $\overline{F}_2$ at the point A. It also cuts out a cap from $\overline{F}_1$. Hence it follows that the surface $\overline{F}_1$ has not only a tangent plane parallel to α but also tangent planes in arbitrary directions close to that of α. Consequently the image of the point A under the spherical mapping of F_2 is a boundary point of $\omega(\overline{F}_2)$ and an interior point of $\omega(\overline{F}_1)$. Thus the set $\omega(\overline{F}_1)-\omega(\overline{F}_2)$ has positive measure.

Now

$$\iint_{\omega(\overline{F}_2)} \frac{dp\,dq}{\varphi_2(p,q)} = \iint_G \varphi_1(x,y)\,dx\,dy \qquad \iint_{\omega(\overline{F}_1)} \frac{dp\,dq}{\varphi_2(p,q)} = \iint_G \varphi_1(x,y)\,dx\,dy.$$

Subtracting these equalities term by term, we get

$$\iint_{\omega(\overline{F}_1)-\omega(\overline{F}_2)} \frac{dp\,dq}{\varphi_2(p,q)} = 0,$$

but this is impossible, since $1/\varphi_2 > 0$ and the set $\omega(\overline{F}_1)-\omega(\overline{F}_2)$ has positive measure. This completes the proof.

Note that the proof of Theorem 3 does not use the regularity of the solution essentially, and it remains valid for generalized solutions. It is also noteworthy that this method of proof is applicable to Theorem 1 in the case $\varphi=\varphi_1(x,y)\varphi_2(p,q)$, for generalized solutions [**19**].

The existence of a closed convex surface with given generalized curvature yields the solution of a certain Monge-Ampère equation invariantly defined on the sphere (§4). We state a geometric version of the corresponding uniqueness theorem.

THEOREM 4. *A closed convex surface with given generalized curvature ϑ is uniquely defined up to a similarity mapping with center O. If the function $\vartheta'(A,n)$ is strictly increasing with respect to displacement of A along the ray OA, then the surface is uniquely defined.*

PROOF. Let F_1 and F_2 be two closed convex surfaces with equal generalized curvatures on corresponding sets. (The correspondence between points of F_1 and F_2 is defined by projection from the origin O.) Let $A_1(g)$ and $A_2(g)$ denote the points of intersection of the surfaces F_1 and F_2 with an arbitrary ray g issuing from O. If the surfaces F_1 and F_2 are not identical then the quotient $|OA_1(g)|/|OA_2(g)|$, as a function of g, assumes a maximum k for some ray g_0. Without loss of generality we may assume that $k>1$ (otherwise, interchange the roles of the surfaces F_1 and F_2).

Let $\overline{F_2}$ be the image of F_2 under a similarity mapping with ratio of similitude k. The surface F_1 lies within the surface $\overline{F}_2$, and is tangent to it at their common point A_0 on the ray g_0.

Introduce rectangular cartesian coordinates in space, with g_0 as the negative z-axis. In a neighborhood of the point A_0, the surface F_1 will then satisfy an equation

$$rt - s^2 = \varphi(x, y, z, p, q), \quad \varphi > 0, \quad \varphi_z > 0,$$

and the surface $\overline{F}_2$ satisfies the inequality

$$rt - s^2 \leqq \varphi(x, y, z, p, q).$$

Hence the difference $u = z_1(x, y) - z_2(x, y)$ of the functions $z_1(x, y)$ and $z_2(x, y)$ defining the surfaces F_1 and $\overline{F}_2$ in the neighborhood of A_0 satisfies the inequality

$$Au_{xx} - 2Bu_{xy} = Cu_{yy} \geqq \varphi_p u_x + \varphi_q u_y + \varphi_z u.$$

Now at the point $x = y = 0$ we have $u = 0$, $du = 0$ and u assumes a maximum (the surface F_1 lies inside $\overline{F}_2$ and is tangent to it at the point A_0). Hence, by a theorem of Aleksandrov [**9**], $u = 0$ in a neighborhood of $x = y = 0$. But this implies that the surfaces F_1 and $\overline{F}_2$ coincide, so that F_1 and F_2 are indeed similar.

If the function $\vartheta'(A, n)$ is strictly monotone with respect to displacement of A along the ray OA, the similar surfaces F_1 and F_2 must coincide, since otherwise their generalized curvatures on corresponding sets would differ. This completes the proof of Theorem 4.

REMARK. If the function $\vartheta'(A, n)$ defining the generalized curvature depends not on the point A but only on the unit normal n, the proof may be based on the same reasoning as in the proof of Theorem 3, and it remains valid for any continuous function $\vartheta'(n)$ with no assumptions on the regularity of the surfaces. A proof along these lines for the case $\vartheta' \equiv 1$ was given by Aleksandrov in [**6**].

§5. Regular solution of a boundary-value problem for the equation $rt - s^2 = \varphi(x, y, z, p, q)$ with φ regular

This section is devoted to a proof of the following lemma.

LEMMA. *In the disk $G: x^2 + y^2 \leqq R^2$, consider the Monge-Ampère equation*

$$rt - s^2 = \varphi(x, y, z, p, q),$$

where φ is sufficiently regular and satisfies the following conditions: 1) $\varphi > 0$, 2) $\varphi_z \geqq 0$, 3) $\varphi = \text{const}$ *in an ϵ-neighborhood of the circumference*

γ of G. Let h be a sufficiently regular function defined on γ and continuable into the interior of G as a regular function convex toward $z<0$, and such that

$$h_{xx}h_{yy}-h_{xy}^2>\varphi(x,y,m,h_x,h_y),$$

where $m=\max h$ on γ.

Then there exists a regular solution of the equation $rt-s^2=\varphi$ in the disk G, convex toward $z<0$, and equal to h on the circumference γ.

This lemma is of little independent interest, but it is essential for the proof, in the next section, that generalized solutions are regular.

According to Bernšteĭn's theorem [**21**] on the solvability of the Dirichlet problem for nonlinear elliptic equations, and its generalization by Miranda [**45**] to equations with regular coefficients, the proof of the lemma reduces to finding a priori estimates for the maximum moduli of the solution and its first and second derivatives.

Since the solution $z(x,y)$ is assumed convex toward $z<0$, its maximum is assumed on the circumference γ. The same applies to the function h. Therefore $z\leqq \max h=m$. Thus we have an upper estimate for the solution z.

To derive a lower estimate for $z(x,y)$, consider the difference $u=z-h$. We claim that it is nonnegative in the disk G. In fact, for the solution z,

$$rt-s^2=\varphi(x,y,z,p,q),$$

and for the function h

$$h_{xx}h_{yy}-h_{xy}^2>\varphi(x,y,m,h_x,h_y).$$

Hence the following inequality for the function $u=z-h$:

$$(*)\qquad Au_{xx}-2Bu_{xy}+Cu_{yy}<\varphi_p u_x+\varphi_q u_y+\varphi_z(z-m),$$

where

$$A=\tfrac{1}{2}(z_{yy}+h_{yy}),\quad B=\tfrac{1}{2}(z_{xy}+h_{xy}),\quad C=\tfrac{1}{2}(z_{xx}+h_{xx}).$$

Suppose that the function u assumes negative values. Then it has a minimum in the interior of the disk G (for $u=0$ on the circumference), at a point P, say. We have $du=0$, $d^2u\geqq 0$ at the point P. The form $A\xi^2-2B\xi\eta+C\eta^2$ is positive definite, and hence at P

$$Au_{xx}-2Bu_{xy}+Cu_{yy}\geqq 0.$$

Furthermore, since $z\leqq m$ and $\varphi_z>0$, it follows that

$$\varphi_p u_x+\varphi_q u_y+\varphi_z(z-m)=\varphi_z(z-m)\leqq 0$$

at the point P. But these two inequalities imply that

$$Au_{xx} - 2Bu_{xy} + Cu_{yy} \geqq \varphi_p u_x + \varphi_q u_y + \varphi_z(z - m),$$

which contradicts inequality (*). Hence $z - h \geqq 0$ in the disk G. Hence a lower estimate for the solution $z(x,y)$ is $z \geqq \min_G h$.

We now estimate the first derivatives of z. It will suffice to estimate $p^2 + q^2$. Since the function z is convex, the function $p^2 + q^2$ assumes its maximum on the circumference γ of G. Let M be an arbitrary point on g; we shall estimate $p^2 + q^2$ at this point. Define polar coordinates ρ, ϑ in the disk G. Then, on the circumference γ,

$$p^2 + q^2 = z_\rho^2 + \frac{1}{R^2} z_\vartheta^2.$$

Since $z = h$ on γ, the term z_ϑ^2/R^2 satisfies the following estimate at M:

$$\frac{1}{R^2} z_\vartheta^2 \leqslant \frac{1}{R^2} \max_\gamma h_\vartheta^2.$$

Thus it remains to estimate $|z_\rho|$ at the point M.

Since the function z is convex toward $z < 0$,

$$-z_\rho \leqslant \frac{1}{2R}(z(\overline{M}) - z(M)),$$

where $\overline{M}$ is the point on γ antipodal to M. Hence we get a lower estimate for z_ρ:

$$z_\rho \geqslant \frac{1}{2R}(\min_\gamma h - \max_\gamma h).$$

To find an upper estimate for z_ρ at the point M, note that the function u vanishes on the circumference γ (in particular, at the point M) and is nonnegative throughout G. Hence $-u_\rho = -z_\rho + h_\rho \geqq 0$ at M, and we get an upper estimate for $z_\rho : z_\rho \leqq \max_\gamma h_\rho$.

We thus have a priori estimates for the solution $z(x,y)$ and its first derivatives.

To estimate the second derivatives, we consider the original equation in polar coordinates:

$$z_{\rho\rho}\left(\frac{1}{\rho^2} z_{\vartheta\vartheta} + \frac{1}{\rho} z_\rho\right) - \left(\frac{1}{\rho} z_{\rho\vartheta} - \frac{1}{\rho^2} z_\vartheta\right)^2 = \varphi.$$

Since the first derivatives have been estimated, we need only estimate the derivatives $z_{\rho\rho}$, $z_{\rho\vartheta}$, $z_{\vartheta\vartheta}$ to obtain estimates for r, s and t. We do this first on the circumference γ of G.

Since $z = h$ on γ, we have a trivial estimate for $z_{\vartheta\vartheta}$:

$$|z_{\vartheta\vartheta}| \leqq \max_\gamma |h_{\vartheta\vartheta}|.$$

We now estimate the derivative $z_{\rho\vartheta}$.

Differentiating the equation for z with respect to ϑ and setting $z_\vartheta = \zeta$, we get

$$\zeta_{\rho\rho}\left(\frac{1}{\rho^2} z_{\vartheta\vartheta} + \frac{1}{\rho} z_\rho\right) - 2\left(\frac{1}{\rho}\zeta_{\rho\vartheta} - \frac{1}{\rho^2}\zeta_\vartheta\right)\left(\frac{1}{\rho} z_{\rho\vartheta} + \frac{1}{\rho^2} z_\vartheta\right) + \left(\frac{1}{\rho}\zeta_{\vartheta\vartheta} + \frac{1}{\rho}\zeta_\rho\right) z_{\rho\rho} = \frac{d}{d\vartheta}\varphi.$$

Let F_ϑ denote the surface defined by the equation $z = \zeta(x,y)$ in the annulus $(R-\epsilon)^2 \leqq x^2 + y^2 \leqq R^2$. Let γ_ϑ and $\overline{\gamma}_\vartheta$ be the curves bounding this surface (such that the projection of γ_ϑ is γ). Since by assumption $\varphi = \text{const}$ in the annulus $(R-\epsilon)^2 \leqq x^2 + y^2 \leqq R^2$, it follows from the equation for ζ in the annulus that

$$\zeta_{xx}\zeta_{yy} - \zeta_{xy}^2 = \zeta_{\rho\rho}\left(\frac{1}{\rho^2}\zeta_{\vartheta\vartheta} + \frac{1}{\rho}\zeta_\rho\right) - \left(\frac{1}{\rho}\zeta_{\rho\vartheta} - \frac{1}{\rho^2}\zeta_\vartheta\right)^2 \leqslant 0.$$

Thus the surface F_ϑ has nonpositive curvature.

Construct a cone V with boundary γ_ϑ and apex at the point $S(0,0,-N)$ on the z-axis. The number N is assumed to be so large that the cone V is convex toward $z<0$ and the curve $\overline{\gamma}_\vartheta$ lies above the cone. This construction involves no difficulties: To meet the first condition, the point S should be so chosen that it lies below all osculating planes of the curve γ_ϑ. The second condition can be guaranteed in view of the estimates derived above for the first derivatives, hence also for ζ.

Since the surface F_ϑ has nonpositive curvature and its boundary $\gamma_\vartheta + \overline{\gamma}_\vartheta$ lies above the cone V, it follows that the entire surface F_ϑ lies above the cone. Let $z = v(\rho,\vartheta)$ be the equation of the cone V. Since the difference $\zeta - v$ vanishes on the circle γ and is nonnegative in a neighborhood of the circle, we have $-(\zeta - v)_\rho \geqq 0$. Hence

$$z_{\rho\vartheta} = \zeta_\rho \leqq \max_\gamma v_\rho.$$

Without loss of generality we may assume that $z_{\rho\vartheta}$ is positive at the point corresponding to $\max|z_{\rho\vartheta}|$ (otherwise we need only reverse the sense in which the angles ϑ are measured). Hence our estimate for $z_{\rho\vartheta}$ may be regarded as an estimate for its absolute value.

We now turn to $z_{\rho\rho}$. Without loss of generality we may assume that $z_{\rho\rho}$ assumes a maximum at a point M $(\vartheta = 0)$. Examination of the equation for z shows that in order to estimate $z_{\rho\rho}$ it will suffice to find a positive lower bound for the expression $z_{\vartheta\vartheta}/\rho + z_\rho/\rho$.

In the paper cited above, Bernšteĭn proved the existence of a solution

$$z_0 = a_{11}x^2 + 2a_{12}xy + a_{22}y^2 + a_1 x + a_2 y + a_0$$

of the equation

$$rt - s^2 = k_0 = \text{const} > 0,$$

which is convex toward $z < 0$ and satisfies the following conditions at the point M ($\vartheta = 0$) of the circumference γ of the disk G:

1) $z_0 = h,\ (z_0)_\vartheta = h_\vartheta,\ (z_0)_{\vartheta\vartheta} = h_{\vartheta\vartheta},\ (z_0)_{\vartheta\vartheta\vartheta} = h_{\vartheta\vartheta\vartheta}$.

2. At all points of γ other than M, $z_0 > h$.

3. At the point M,

$$\frac{1}{\rho^2}(z_0)_{\vartheta\vartheta} + \frac{1}{\rho}(z_0)_\rho > \alpha_0 > 0,$$

where α_0 is a constant which depends only on k_0 and the maximum moduli of the derivatives of h of order up to four.

Choose k_0 so that $k_0 < \varphi$ for $(x, y) \in G$ and p, q, z satisfying the a priori estimates established above. Then the function $u = z_0 - z$ satisfies the inequality

$$Au_{xx} - 2Bu_{xy} + Cu_{yy} = k_0 - \varphi < 0,$$

where

$$2A = 2a_{22} + t, \quad 2B = 2a_{12} + s, \quad 2C = 2a_{11} + r.$$

Hence there are no points in the disk G at which $d^2u \geqq 0$. Since the function u is nonnegative on the circumference of G, it must be nonnegative throughout the disk.

Since $u = 0$ at M, it follows that, again at M, we have $u_\rho = (z_0 - z)_\rho \leqq 0$, i.e. $(z_0)_\rho \leqq z_\rho$. But this implies that at the point M

$$\frac{1}{\rho^2} z_{\vartheta\vartheta} + \frac{1}{\rho} z_\rho \geqslant \frac{1}{\rho^2}(z_0)_{\vartheta\vartheta} + \frac{1}{\rho}(z_0)_\rho > \alpha_0 > 0,$$

and this yields an estimate for $z_{\rho\rho}$ at M.

We have thus established a priori estimates for the second derivatives of the solution on the circumference γ of the disk G.

Given these estimates for r, s, t, we must now establish corresponding estimates for the entire disk G. We begin with r.

Consider the auxiliary function $w = \lambda r$, where λ is a function, bounded and positive in the closed disk G, which will be specified later. The function r assumes a maximum in the closed disk G. If this maximum w_0 is assumed on the circumference, it can be estimated in terms of the known maximum of r on the circumference, and we then obtain the estimate $r \leqq w_0 / \min \lambda$ throughout G.

Now let w assume its maximum at an interior point A of G. At this point $w_x = w_y = 0$, whence

$$r_x = w\left(\frac{1}{\lambda}\right)_x = -r\frac{\lambda_x}{\lambda}, \qquad r_y = w\left(\frac{1}{\lambda}\right)_y = -r\frac{\lambda_y}{\lambda}.$$

Differentiating, we get the following expressions for the second derivatives of r at A:

$$r_{xx} = \frac{w_{xx}}{\lambda} + w\left(\frac{1}{\lambda}\right)_{xx}, \quad r_{xy} = \frac{w_{xy}}{\lambda} + w\left(\frac{1}{\lambda}\right)_{xy}, \quad r_{yy} = \frac{w_{yy}}{\lambda} + w\left(\frac{1}{\lambda}\right)_{yy}.$$

Differentiating the equation

$$rt - s^2 = \varphi(x, y, z, p, q)$$

with respect to x, we get

$$r_x t - t_x r - 2ss_x = d\varphi/dx.$$

Using this equality, we bring the expression for $r_x t_x - s_x^2$ at the point A to the form

$$r_x t_x - s_x^2 = -\frac{\lambda_x}{\lambda}\frac{d\varphi}{dx} + 2rs\left(\frac{\lambda_x}{\lambda}\right)\left(\frac{\lambda_y}{\lambda}\right) - rt\left(\frac{\lambda_x}{\lambda}\right)^2 - r^2\left(\frac{\lambda_y}{\lambda}\right)^2.$$

Differentiating the equation $rt - s^2$ again with respect to x, we get

$$(r_{xx}t - 2r_{xy}s + r_{yy}r) + 2(r_x t_x - s_x^2) = d^2\varphi/dx^2.$$

Substituting our expression for $r_x t_x - s_x^2$ at A into this expression, we get

$$\frac{1}{\lambda}(tw_{xx} - 2sw_{xy} + rw_{yy}) + \lambda r\left\{t\left(\frac{1}{\lambda}\right)_{xx} - 2s\left(\frac{1}{\lambda}\right)_{xy} + r\left(\frac{1}{\lambda}\right)_{yy}\right\}$$
$$+ 2\left\{-\frac{\lambda_x}{\lambda}\frac{d\varphi}{dx} + 2rs\left(\frac{\lambda_x}{\lambda}\right)\left(\frac{\lambda_y}{\lambda}\right) - rt\left(\frac{\lambda_x}{\lambda}\right)^2 - r^2\left(\frac{\lambda_y}{\lambda}\right)^2\right\} = \frac{d^2\varphi}{dx^2}.$$

Noting that

$$\lambda\left(\frac{1}{\lambda}\right)_{xx} - 2\left(\frac{\lambda_x}{\lambda}\right)^2 = -\frac{\lambda_{xx}}{\lambda}, \quad \lambda\left(\frac{1}{\lambda}\right)_{xy} - 2\left(\frac{\lambda_x}{\lambda}\right)\left(\frac{\lambda_y}{\lambda}\right) = -\frac{\lambda_{xy}}{\lambda},$$
$$\lambda\left(\frac{1}{\lambda}\right)_{yy} - 2\left(\frac{\lambda_y}{\lambda}\right)^2 = -\frac{\lambda_{yy}}{\lambda},$$

we bring the last equality to the form

$$\frac{1}{\lambda}(tw_{xx} - 2sw_{xy} + rw_{yy}) - \frac{\lambda_{xx}}{\lambda}rt + \frac{2\lambda_{xy}}{\lambda}rs - \frac{\lambda_{yy}}{\lambda}r^2 - \frac{2\lambda_x}{\lambda}\frac{d\varphi}{dx} = \frac{d^2\varphi}{dx^2}.$$

Replacing rt by $\varphi + s^2$ and noting that

$$\frac{d^2\varphi}{dx^2} + 2\frac{\lambda_x}{\lambda}\frac{d\varphi}{dx} + \varphi\frac{\lambda_{xx}}{\lambda} = \frac{1}{\lambda}\frac{d^2}{dx^2}(\lambda\varphi),$$

we get

$$(tw_{xx} - 2sw_{xy} + rw_{yy}) - \lambda_{xx}s^2 + 2\lambda_{xy}rs - \lambda_{yy}r^2 = \frac{d^2}{dx^2}(\lambda\varphi).$$

Since the function w has a maximum at A, it follows that

$$tw_{xx} - 2sw_{xy} + rw_{yy} \leqq 0$$

and so

$$d^2(\lambda\varphi)/dx^2 + \lambda_{xx}s^2 - 2\lambda_{xy}rs + \lambda_{yy}r^2 \leqq 0.$$

Now set $\lambda = \exp(\alpha(x^2+y^2)/2)$. Then at A we have

$$(**) \qquad \begin{aligned}(\alpha + 2\alpha^2x^2)\,s^2 + 4\alpha^2xyrs + (\alpha + 2\alpha^2y^2)\,r^2 \\ + (\varphi_{pp}r^2 + 2\varphi_{pq}rs + \varphi_{qq}s^2) + \ldots \leqslant 0,\end{aligned}$$

where terms linear in r and s have been omitted. For sufficiently large α the quadratic form (in r and s)

$$(\varphi_{pp} + \alpha + 2\alpha^2x^2)\,s^2 + 2\,(\varphi_{pq} + 2\alpha^2xy)\,rs + (\varphi_{qq} + \alpha + 2\alpha^2y^2)\,r^2$$

is positive definite, and hence r cannot exceed some quantity r_0 defined by inequality (**). Hence

$$w_0 \leqq r_0 \max_G \lambda$$

and consequently, throughout the disk G,

$$r \leqq r_0 \max\lambda/\min\lambda.$$

An estimate for t is established in analogous fashion. The estimate for s then follows directly from $s^2 = rt - \varphi$.

Since we now have a priori estimates for the solution $z(x,y)$ and its first and second derivatives, the proof of our lemma is complete.

§6. Regularity of generalized solutions of the Monge-Ampère equation $rt - s^2 = \varphi(x,y,z,p,q)$ with φ regular

In this section we shall prove that if the function φ in the equation $rt - s^2 = \varphi(x,y,z,p,q)$ is sufficiently regular, then any generalized solution of the equation is regular. With this result in hand, the theorems of §3 can be applied to solution of classical boundary-value problems for the Monge-Ampère equation.

Let ϵ', ϵ'', ϵ''', ϵ be small positive numbers such that $0 < \epsilon' < \epsilon'' < \epsilon''' < \epsilon$; let λ and μ be two sufficiently regular functions of ρ in the interval $(0,\epsilon)$ such that

$$\begin{aligned} \lambda = 1, \quad & \mu = 0 \quad \text{for} \quad 0 \leqslant \rho < \varepsilon' \\ \lambda = 1, \quad & 0 \leqslant \mu \leqslant 1 \quad \text{for} \quad \varepsilon' \leqslant \rho \leqslant \varepsilon'', \\ 0 \leqslant \lambda \leqslant 1, \quad & \mu = 1 \quad \text{for} \quad \varepsilon'' \leqslant \rho \leqslant \varepsilon''', \\ \lambda = 0, \quad & \mu = 1 \quad \text{for} \quad \varepsilon''' \leqslant \rho \leqslant \varepsilon. \end{aligned}$$

Set

$$\tilde{\varphi} = \lambda(\rho)\varphi + \mu(\rho), \qquad \rho^2 = (x-a)^2 + (y-b)^2.$$

Obviously $0 < \tilde{\varphi} \leqq \varphi + 1$ and $\tilde{\varphi}_z \geqq 0$.

We shall say that a function h defined on the circle $(x-a)^2 + (y-b)^2 = \epsilon^2$ in the xy-plane is in class Ω_R if $-R \leqq h \leqq R$ and the closed curve γ defined by

$$z = h(x,y), \qquad (x-a)^2 + (y-b)^2 = \epsilon^2,$$

is the boundary of a surface convex toward $z < 0$ whose generalized spherical image lies in the interior of the disk $p^2 + q^2 \leqq R^2$ in the pq-plane.

Lemma. *Consider the Monge-Ampère equation $rt - s^2 = \varphi(x,y,z,p,q)$ in some domain on the xy-plane, assuming that the function φ is sufficiently regular and satisfies the conditions $\varphi > 0$ and $\varphi_z \geqq 0$. Let (a,b) be an arbitrary point of the domain. Then for sufficiently small ϵ there exists a regular solution of the equation $rt - s^2 = \tilde{\varphi}$ in the disk $G_\epsilon : (x-a)^2 + (y-b)^2 \leqq \epsilon^2$ which is convex toward $z < 0$ and takes the values h on the circumference of G_ϵ, for any function $h \in \Omega_R$ and arbitrary choice of the numbers ϵ', ϵ'', ϵ''' and functions λ and μ defining $\tilde{\varphi}$.*

Proof. Without loss of generality we may assume that the point (a,b) is the origin. Let $z = z_0(x,y)$ be a surface convex toward $z < 0$, with boundary γ defined by the function h and generalized spherical image contained in the disk $\omega_R : p^2 + q^2 \leqq R^2$. Set

$$\bar{z} = z_0 + \frac{1}{2\sqrt{\varepsilon}}(x^2 + y^2 - \varepsilon^2) \qquad (x^2 + y^2 \leqslant \varepsilon^2).$$

The equation $z = \bar{z}(x,y)$ defines a certain convex surface with boundary γ; if ϵ is sufficiently small, its generalized spherical image is contained in the disk ω_{2R}.

We have

$$\bar{r}\bar{t} - \bar{s}^2 = (r_0 t_0 - s_0^2) + \frac{1}{\sqrt{\varepsilon}}(r_0 + t_0) + \frac{1}{\varepsilon}.$$

Since $r_0 t_0 - s_0^2 \geqq 0$, $r_0 + t_0 \geqq 0$, it follows that $\bar{r}\bar{t} - \bar{s}^2 < 1/\epsilon$.

Let ϵ be so small that the generalized spherical image of the surface $z=\bar{z}(x,y)$ lies in the disk ω_{2R} and moreover, for $x^2+y^2 \leqq \epsilon^2$ and $p^2 + q^2 \leqq 4R^2$,

$$\tilde{\varphi}(x,y,R,p,q) < 1/\epsilon.$$

This ϵ can be chosen independently of the numbers ϵ', ϵ'', ϵ''' and the functions λ and μ, since $\tilde{\varphi} < \varphi + 1$.

Now let $z(x,y)$ be a solution of the equation $rt - s^2 = \tilde{\varphi}$ in the disk G_ϵ, convex toward $z<0$ and assuming the values h on the circumference of the disk. We claim that the function $u = z(x,y) - \bar{z}(x,y)$, which vanishes on the circumference of G_ϵ, is nonnegative inside the disk.

Indeed, suppose that u is negative at some point in the disk. Then it assumes a minimum at an interior point P of the disk. At this point $du=0$, and hence $p=\bar{p}$, $q=\bar{q}$; since $\bar{p}^2+\bar{q}^2<4R^2$, this implies that at P

$$\tilde{\varphi}(x, y, R, p, q) < \frac{1}{\varepsilon} < \bar{r}\bar{t} - \bar{s}^2.$$

As in the proof of the lemma in §5, one establishes the following inequality for the function u:

$$Au_{xx} - 2Bu_{xy} + Cu_{yy} < \tilde{\varphi}_p u_x + \tilde{\varphi}_q u_y + \tilde{\varphi}_z (z-R),$$

but this is impossible, since at P we have $Au_{xx} - 2Bu_{xy} + Cu_{yy} \geqq 0$, $u_x = u_y = 0$ and $\tilde{\varphi}_z(z-R) \leqq 0$, and thus we have a contradiction. Thus the function $z - \bar{z}$ is nonnegative in G.

Reasoning now as in the proof of the lemma in §5, we can establish a priori estimates for the solution $z(x,y)$ and its first and second derivatives. Bernšteĭn's theorem then implies the existence of a solution. This proves the lemma.

Let $z_{\epsilon'}(x,y)$ denote the solution $z(x,y)$ in the disk $x^2+y^2 \leqq \epsilon'^2$. Since $\tilde{\varphi} = \varphi$ in this disk, $z_{\epsilon'}$ is a solution of the original equation $rt-s^2=\varphi$. Now the a priori estimates obtained for the first derivatives of the solution $z(x,y)$ are determined by the function $\bar{z}(x,y)$, which is independent of ϵ', ϵ'', ϵ''', λ and μ. Hence the boundary values h' of the solution $z_{\epsilon'}$ converge to h as $\epsilon' \to \epsilon$.

Let $\bar{z}(x,y)$ be any generalized solution of the equation

$$rt - s^2 = \varphi(x,y,z,p,q)$$

in a domain G on the xy-plane, where φ is a sufficiently regular function such that $\varphi > 0$, $\varphi_z \geqq 0$. Assume that this solution is convex toward $z<0$.

Let $h(x,y)$ be a sufficiently regular convex function close to $\bar{z}(x,y)$. For sufficiently small ϵ there exists a regular solution $z(x,y)$ of the equation

$rt - s^2 = \varphi$ in a disk $G_\epsilon: (x-a)^2 + (y-b)^2 \leqq \epsilon^2$ contained in G, and this solution is convex toward $z < 0$ and takes values arbitrarily close to h on the boundary of the disk. If h is sufficiently close to $\bar{z}$, we may assume that ϵ is independent of the choice of h.

We shall now prove that as $h \to \bar{z}$ the function $z(x,y)$ also converges to $\bar{z}(x,y)$ in the disk G_ϵ. Suppose that this is false. Then, without loss of generality, we may assume that for some function $h(x,y)$ close to $\bar{z}(x,y)$ the function $u(x,y) = z(x,y) - \bar{z}(x,y)$ assumes a maximum m in the interior of the disk G_ϵ, whereas on the circumference $u < m$.

Define an auxiliary function w by $u = \delta w$, where $\delta = 1 - \sigma e^{\lambda x} - \sigma e^{\lambda y}$. For sufficiently small σ and fixed λ this function also assumes a maximum in the interior of G_ϵ. Let M denote the set of points of the disk G_ϵ at which u assumes its maximum, and let M' be the ϵ'-neighborhood of the set M. For sufficiently small ϵ' the functions $p - \bar{p}$ and $q - \bar{q}$ are arbitrarily small in M'.

Let M'' be the subset of M' in which $d^2\bar{z}$ exists and $d^2 w \leqq 0$. At each point $P \in M''$

$$d^2 u = w d^2\delta + 2 d\delta dw + \delta d^2 w < 0,$$

provided ϵ' is sufficiently small. Hence we see that $d^2 z < d^2 \bar{z}$ in M'', and therefore, since $\bar{r}\bar{t} - \bar{s}^2$ is bounded, the derivatives $\bar{r}$, $\bar{t}$ and $\bar{s}$ are uniformly bounded in M''.

Almost everywhere in G_ϵ the function $u(x,y)$ satisfies the equation

$$A u_{xx} - 2B u_{xy} + C u_{yy} = \varphi_p u_x + \varphi_q u_y + \varphi_z u,$$

where

$$A = \tfrac{1}{2}(t + \bar{t}), \quad B = \tfrac{1}{2}(s + \bar{s}), \quad C = \tfrac{1}{2}(r + \bar{r}),$$

and the derivatives φ_p, φ_q, φ_z are computed at certain mean values of the arguments p, q, z and are bounded functions in M'.

Substituting δw into the equation for u, we get

$$\begin{aligned} w(A\delta_{xx} - 2B\delta_{xy} + C\delta_{yy}) &+ 2(A\delta_x w_x - B(\delta_x w_y + w_x \delta_y) + C\delta_y w_y) \\ &+ \delta(A w_{xx} - 2B w_{xy} + C w_{yy}) \\ &= w(\varphi_p \delta_x + \varphi_q \delta_y) + \delta(\varphi_p w_x + \varphi_q w_y) + \delta w \varphi_z. \end{aligned}$$

We now choose the parameters λ, σ, ϵ as follows. Take the number λ so large that the first term on the left-hand side of the equality is much greater in absolute value than the first term on the right-hand side for points in M''. This is possible, for the form $d^2 z$ is strictly positive. Now take σ and ϵ so small that the first term on the left-hand side is much greater in absolute value than the second term on the left-hand side and the second term in the right-hand side for points in M''. That this is possible follows from the fact that A, B, C and φ_p, φ_q are uniformly bounded in M''. Moreover, this can be done in such a way that the first condition still holds.

Now let P be an arbitrary point in M'' (if M'' is not empty). Because of the choice of λ, σ and ϵ', the left-hand side of the equality is smaller than the right-hand side at P. This is a contradiction. It remains to consider the case in which M'' is empty.

Construct the convex hull of the set of points on the surface $z = w(x,y)$ whose projections lie in M'. Let Φ' be the part of the convex hull convex toward $z > 0$, and Φ'' the intersection of Φ' and the surface $z = w$. Φ'' is a closed set. Its measure is zero, since M'' is empty, but its generalized spherical image has positive measure.

Now let $\tilde{z}$ be a regular function close to $\bar{z}$. Construct the function $\tilde{w} = (z - \tilde{z})/\delta$, the surface $\tilde{\Phi}'$, and the set $\tilde{\Phi}''$ on $\tilde{\Phi}'$ as for the function $\bar{z}$. If $\tilde{z}$ is sufficiently close to $\bar{z}$, the set $\tilde{\Phi}''$ lies in an arbitrarily small neighborhood of Φ'', and hence has arbitrarily small measure. However, at the same time,

$$\iint_{(\tilde{\Phi}'')} (\tilde{w}_{xx}\tilde{w}_{yy} - \tilde{w}^2_{xy})\, dx\, dy > \varepsilon_0 > 0.$$

For small ϵ' and $\tilde{z}$ sufficiently close to $\bar{z}$,

$$d^2\tilde{w} = \tilde{w}d^2\delta + 2d\tilde{w}\, d\delta + \delta d^2\tilde{w} > \delta d^2\tilde{w}.$$

Hence

$$\tilde{u}_{xx}\tilde{u}_{yy} - \tilde{u}^2_{xy} > \delta^2(\tilde{w}_{xx}\tilde{w}_{yy} - \tilde{w}^2_{xy}).$$

Consequently

$$\iint_{(\tilde{\Phi}'')} \frac{1}{\delta^2}(\tilde{u}_{xx}\tilde{u}_{yy} - \tilde{u}^2_{xy})\, dx\, dy > \varepsilon_0 > 0.$$

But

$$\tilde{u}_{xx}\tilde{u}_{yy} - \tilde{u}^2_{xy} \leqslant (rt - s^2) + (\tilde{r}\tilde{t} - \tilde{s}^2).$$

Therefore

$$\iint_{(\tilde{\Phi}'')} (rt - s^2)\, dx\, dy + \iint_{(\tilde{\Phi}'')} (\tilde{r}\tilde{t} - \tilde{s}^2)\, dx\, dy > \varepsilon_0 > 0.$$

Since the surfaces $z = z(x,y)$ and $z = \bar{z}(x,y)$ are surfaces of bounded specific curvature, while the set $\tilde{\Phi}''$ is contained in an arbitrarily small neighborhood of Φ'', which is a closed set of zero measure, it follows that for z sufficiently close to $\bar{z}$ each of the integrals

$$\iint\limits_{(\tilde{\Phi}'')} (rt - s^2)\, dx\, dy \quad \text{and} \quad \iint\limits_{(\tilde{\Phi}'')} (\tilde{r}\tilde{t} - \tilde{s}^2)\, dx\, dy$$

is arbitrarily small. This is a contradiction, since their sum is greater than $\epsilon_0 > 0$. Thus the regular solution $z(x,y)$ in the disk G_ϵ converges to the generalized solution $\bar{z}(x,y)$ as $h \to \bar{z}$.

We now show that the second derivatives of a solution z converging to $\bar{z}$ as $h \to \bar{z}$ satisfy estimates on a closed subset H of the interior of G_ϵ, uniformly in $z \to \bar{z}$.

Let P be an arbitrary point of the set H and P' the point on the surface $F: z = z(x,y)$ whose projection is P. Displace the tangent plane to F at P' for a distance $\delta_0 > 0$ in the direction $z > 0$. Let $z = z_0(x,y)$ be the equation of the plane in its new position. Since the surface $\bar{F}$: $z = \bar{z}(x,y)$ is strictly convex, it follows that if F is sufficiently close to $\bar{F}$ and δ_0 sufficiently small, the plane $z = z_0(x,y)$ does not cut the boundary of the surface F, irrespective of the choice of P in H.

Let G_P denote the set of points of G_ϵ at which $z_0 > z$. All such points are in the interior of G_ϵ, and on the boundary of the set G_P we have $z_0 = z$.

We introduce a function w of points and directions (t) in the domain G_P:

$$w_{(t)} = \delta\mu(z_t)\, z_{tt},$$

where $\delta = z_0 - z$ and $\mu(z_t)$ is a positive, bounded function, to be specified below. The function $w_{(t)}$ clearly assumes an absolute maximum at some point P_0 of G_P, for some direction t_0 at P_0. Take this direction as that of the x-axis. Obviously the function $w = \delta\mu(p) r$ also assumes an absolute maximum at P_0.

Let us estimate the derivative s at P_0 in terms of the derivative r. Since the function

$$\delta\mu(p\cos\vartheta + q\sin\vartheta)\,(r\cos^2\vartheta + 2s\sin\vartheta\cos\vartheta + t\sin^2\vartheta),$$

as a function of ϑ, assumes a maximum at $\vartheta = 0$, its derivative with respect to ϑ vanishes at $\vartheta = 0$. Hence $\mu'(p) qr = 2\mu s$. Consequently at P_0

$$|s| \leqq \tfrac{1}{2}|\mu'/\mu|\, |q| r.$$

Denoting $\delta\mu = \lambda$, we see, as was shown in §5, that the following inequality holds at P_0:

$$\lambda_{xx}s^2 - 2\lambda_{xy}rs + \lambda_{yy}r^2 + (\lambda\varphi)_{xx} \leqq 0,$$

where the subscripts stand for total differentiation of the corresponding composite functions.

Setting $\lambda = \delta\mu$ in this inequality and simplifying (with the aid ot the equality $w_x = w_y = 0$ at P_0), we get

$$\delta\mu'' r^2 + \delta\mu(\varphi_{pp} r^2 + 2\varphi_{pq} rs + \varphi_{qq} s^2) + (*) \leqq 0,$$

where (*) denotes a linear function of r and s with bounded coefficients depending on the functions δ, μ, φ and their derivatives.

Since, as we have already proved, $s = \mu' qr/2\mu$ at the point P_0, we have

$$\delta\mu'' r^2 + \delta\mu r^2 \left(\varphi_{pp} + 2\varphi_{pq} q \left(\frac{\mu'}{\mu}\right) + \varphi_{qq} q^2 \left(\frac{\mu'}{\mu}\right)^2\right) + (*) \leqslant 0.$$

Multiplying by $\delta\mu^2$ and substituting $w = \delta\mu r$, we get

$$\mu'' w^2 + \mu w^2 \left(\varphi_{pp} + 2\varphi_{pq} q \left(\frac{\mu'}{\mu}\right) + \varphi_{qq} q^2 \left(\frac{\mu'}{\mu}\right)^2\right) + (*) \leqslant 0,$$

where (*) is linear in w.

The surface $\overline{F}$ is smooth and strictly convex; hence, if F is sufficiently close to $\overline{F}$, the generalized spherical image of the domain F_P on F whose projection is G_P has arbitrarily small diameter for sufficiently small δ_0, and this property holds uniformly in F close to $\overline{F}$ (i.e. z converging to $\bar{z}$).

If the diameter of the generalized spherical image of F_P is small, we can find constants κ and κ_0 such that

$$1/e^{n+1} > \kappa p + \kappa_0 < 1/e^n$$

in the domain G_P, where κ and n may be assumed arbitrarily large.

Now define the function $\mu(p) = -\ln(\kappa p + \kappa_0)$. Then

$$\mu'' > \kappa^2 e^{2n}, \quad n < \mu < n+1, \quad |\mu'| < \kappa e^{n+1}.$$

Hence, if κ and n are sufficiently large, the above inequality for w is impossible for large n. Thus the function w is bounded in G_P by w_P, say. We thus obtain estimates for r, s, t at the point P:

$$r \leqslant \frac{w_P}{s_0\mu(p)}, \quad |s| \leqslant \frac{|\mu' q|\, w_P}{2\delta_0\mu^2(p)}, \quad t \leqslant \frac{w_P}{\delta_0\mu(q)}.$$

Thus for z sufficiently close to $\bar{z}$ the second derivatives of $z(x,y)$ satisfy uniform estimates in a closed subset H of the interior of the disk G_ϵ.

Now that we have estimates for the second derivatives of the solution $z(x,y)$, the general theorem of §11 of Chapter II, valid for any elliptic equation, implies the existence of a priori estimates for the third and fourth derivatives of $z(x,y)$. It follows that the generalized solution $\bar{z}(x,y)$, as the limit of regular solutions $z(x,y)$ in the disk G_ϵ, is at least thrice differentiable and its third derivatives satisfy Lipschitz conditions. Further information on the regularity of $\bar{z}$ follows from the theorem according to which the solutions of an elliptic equation with regular coefficients are regular.

THEOREM. *Any generalized solution of the equation*

$$rt - s^2 = \varphi(x, y, z, p, q), \quad \varphi > 0, \quad \varphi_z \geqq 0,$$

where φ is k times differentiable ($k \geqq 3$), is at least $k+1$ times differentiable.

If φ is analytic, then the solution is analytic.

This theorem was proved by the author [**61**] for $\varphi = \varphi(x, y)$, and subsequently by Bakel′man [**19**] for $\varphi = \varphi_1(x, y)\varphi_2(p, q)$ on the assumption that $d^2\varphi_2 \geqq c(dp^2 + dq^2)$, $c > 0$.

The general case was considered by the author in [**62**].

§7. Strongly elliptic Monge-Ampère equations

The general Monge-Ampère equation

$$rt - s^2 = ar + 2bs + ct + \varphi$$

is said to be *strongly elliptic* if $\varphi > 0$ and the form $a\xi^2 + 2b\xi\eta + c\eta^2$ is nonnegative. In this section we shall prove that any strongly elliptic equation of this type has generalized solutions.

Let $F: z = z(x, y)$ be a convex surface which can be projected in one-to-one fashion onto the xy-plane, and $a(x, y, z, p, q)$ an arbitrary continuous function of the coordinates x, y, z of a point on the surface and the direction ratios p, q of the supporting plane at the point.

Introduce a new system of cartesian coordinates $\bar{x}$, $\bar{y}$ on the xy-plane, and consider the function a in the new variables $\bar{x}$, $\bar{y}$, $\bar{z}$, $\bar{p}$, $\bar{q}$. The Lebesgue-Stieltjes integral

$$h(H) = \int\int_{(H)} \bar{a}\, d\bar{p}\, d\bar{y},$$

or any sum of such integrals for different functions a and coordinate systems $\bar{x}$, $\bar{y}$, will be called a *generalized mean curvature function*.

Any generalized mean curvature function is a completely additive function on the ring of Borel sets.

Let $F_n: z = z_n(x, y)$ be a sequence of convex surfaces converging to the convex surface F. Then the generalized mean curvatures of F_n converge weakly to that of F, i.e., for any continuous function f which vanishes in the neighborhood of the boundary of F.

$$\lim_{F_n \to F} \int\int_{(F_n)} f\, dh_n = \int\int_{(F)} f\, dh.$$

To prove this property of generalized mean curvature it suffices to show that for any closed set H on the xy-plane

$$\overline{\lim_{F_n \to F}}\, h_n(H) \leqslant h(H),$$

and for any open set H

$$\underline{\lim_{F_n \to F}}\, h_n(H) \geqslant h(H),$$

where the generalized curvatures h and h_n of the surfaces F and F_n are computed on the sets whose projections on the xy-plane coincide with H. We prove the first assertion; the proof of the second is analogous.

Without loss of generality we may assume that the coordinate systems x, y and $\bar{x}, \bar{y}$ are the same, and the generalized mean curvature involves a single term

$$h = \int\int a(x, y, z, p, q)\, dp\, dy.$$

We define a correspondence between the xy-plane and the yp-plane as follows. With each point (x, y), associate the set of points (y, p) such that p is the direction ratio of a supporting plane of F at the point $(x, y, z(x, y))$. We call this mapping the generalized semi-spherical mapping.

Let H^* be the generalized semi-spherical image of a set H on F, and H_δ^* the set of all points on the yp-plane having at least two preimages on the xy-plane, the distance between which is at least δ. The set H_δ^* consists of the images of all points whose projections lie on straight-line segments of the surface F parallel to the plane $y = 0$, such that the length of the projection on the xy-plane is at least δ. H_δ^* is a closed set of measure zero, since there are at most countably many points of H_δ^* on any straight line $y = \text{const}$. Let $H_{\delta,\epsilon}^*$ be an open set of measure less than ϵ containing the set H_δ^*.

We write $A \cong B$ to indicate that the quantity B is arbitrarily close to the quantity A. Then, for sufficiently small ϵ,

$$(*) \qquad \begin{aligned} h(H) &\simeq \int\int_{H^* - H_{\delta,\varepsilon}^*} a\, dp\, dy, \\ h_n(H) &\simeq \int\int_{H_n^* - H_{\delta,\varepsilon}^*} a_n\, dp\, dy, \end{aligned}$$

where H_n^* is the image of H_n under the mapping $(x,y) \to (y,p)$ defined by the surface F_n.

Since H is a closed set, so is H^*. If F_n is sufficiently close to F, the set H_n^* is contained in an arbitrarily small neighborhood of H^*. Therefore the measure of the set $H_n^* - H_n$ is arbitrarily small for large n (i.e. when the surface F_n is sufficiently close to F).

Finally we show that a_n, as a function of p and y, converges in measure to a as $n \to \infty$. In fact, for sufficiently large n, $|a_n - a|$ tends to zero together with δ on the set $H^* - H_{\delta,\epsilon}^*$ (since a is continuous in x, y, z, p, q and the set $H_{\delta,\epsilon}^*$ has measure less than ϵ).

The relation (*) now implies that indeed

$$\overline{\lim_{n\to\infty}}\, h_n(H) \leqq h(H).$$

Analogous arguments yield the corresponding inequality for open sets. Thus the generalized mean curvatures of the surfaces F_n converge weakly to the generalized mean curvature of F.

Now let F_n be a sequence of convex surfaces converging to a convex surface F, and let their generalized mean curvatures converge weakly to a completely additive function h. Then h is a generalized mean curvature function on F. This follows from the fact that a sequence of completely additive set functions cannot have two different weak limits.

In order to distinguish the generalized curvature of a surface from the generalized mean curvature, we shall henceforth call it the *generalized total curvature*. In addition, the generalized area of the surface will refer not to its generalized spherical image, as hitherto, but to the projection of the surface on the xy-plane, and we shall define this new area by

$$\sigma(H) = \iint_H \varphi(x, y, z, p, q)\, dx\, dy,$$

where φ is any continuous positive function.

Generalized area is clearly a completely additive function on the ring of Borel sets. It is also obvious that if F_n are convex surfaces converging to a convex surface F, then their generalized areas converge weakly to that of F.

Let G be a convex domain in the xy-plane, $\vartheta'(p,q)$ and $\varphi(x,y,z,p,q)$ continuous positive functions for $(x,y) \in G$ and arbitrary z, p, q. Let $a_i(x,y,z,p,q)$ be a finite system of continuous functions defined in the same domain, and $\bar{x}_i$, $\bar{y}_i$ corresponding cartesian coordinate systems.

We shall consider convex surfaces F which can be projected in one-to-one fashion onto the domain G (the surfaces are open, without boundary), are convex toward $z<0$ and lie below a plane $z=z_0$. We denote this class of surfaces by Ω.

The surfaces in the class Ω will have generalized area σ defined by the function φ, generalized total curvature ϑ defined by the function ϑ' and generalized mean curvature h defined by the functions a_i and the coordinate systems $\bar{x}_i$, $\bar{y}_i$.

Theorem 1. *Let the generalized mean curvature h be nonnegative*[3] *for each convex surface of class Ω, and moreover*

$$(**) \qquad h(G)+\sigma(G)<c<\iint\limits_{(pq)} \vartheta'(p,q)\,dp\,dq.$$

Then Ω contains a convex surface F such that $\vartheta(H)=h(H)+\sigma(H)$ for any Borel set H.

Proof. Let $\bar{F}$ denote the class of surfaces containing a surface $F\in\Omega$ and all surfaces obtained from F by translation in the direction of the z-axis. Denote the set of all such classes by $\bar{\Omega}$.

Let $\bar{F}_1$ and $\bar{F}_2$ be classes defined by the surfaces $z=z_1(x,y)$ and $z=z_2(x,y)$. Then the class $\lambda\bar{F}_1+\mu\bar{F}_2$, λ and $\mu>0$, is defined by the surface $z=\lambda z_1(x,y)+\mu z_2(x,y)$.

To define a distance between two classes $\bar{F}_1$ and $\bar{F}_2$, we choose that surface in each class whose boundary touches the plane $z=z_0$. If these surfaces are defined by functions $z_1(x,y)$ and $z_2(x,y)$, then by definition

$$\rho(\bar{F}_1,\bar{F}_2)=\max_G|z_1(x,y)-z_2(x,y)|\,\delta(x,y),$$

where $\delta(x,y)$ is the distance of the point (x,y) from the boundary of G. Throughout the sequel, the representative of a class $\bar{F}$ will always be the surface F of the class whose boundary touches the plane $z=z_0$.

We now define an operator A on the set of classes $\bar{\Omega}$ as follows. Construct a surface Φ convex toward $z<0$, with generalized total curvature ϑ_Φ equal to $h_F+\sigma_F$, with boundary not higher than the plane $z=z_0$, and lying above any other surface satisfying these conditions. The existence of the surface Φ follows from condition (**) (§4, Theorem 2a). It is unique and its boundary touches the plane $z=z_0$. Now set $A(\bar{F})=\bar{\Phi}$.

---

[3] $h\geqq 0$ if $a_i\geqq 0$ for all i, but this condition is not necessary.

The operator A is continuous. In fact, as $F_n \to F$ the function $h_{F_n} + \sigma_{F_n}$ converges weakly to $h_F + \sigma_F$. Let $\overline{\Phi}_n = A(\overline{F}_n)$. The sequence Φ_n is bounded (this follows from condition (**)). We can therefore extract a convergent subsequence from the sequence Φ_n. Since the generalized total curvatures of the Φ_n converge weakly to $h_F + \sigma_F$, the limit surface Φ has generalized total curvature equal to $h_F + \sigma_F$. Consequently, by the uniqueness theorem, $\overline{\Phi} = A(\overline{F})$.

The set $A(\overline{\Omega})$ is bounded. Take any closed bounded set $\tilde{\Omega}$ in $\overline{\Omega}$ containing $A(\overline{\Omega})$. We claim that $A(\tilde{\Omega})$ is [sequentially] compact. Indeed, let $A(\overline{F}_k)$ be an infinite sequence in $A(\tilde{\Omega})$. The sequence of surfaces F_n contains a convergent subsequence whose limit is some surface F. Since $\tilde{\Omega}$ is closed, we have $\overline{F} \in \tilde{\Omega}$. But then, as we have proved, $A(\overline{F}_k) \to A(F)$. Hence $A(\tilde{\Omega})$ is compact.

Assume that all the surfaces F defining classes in $A(\tilde{\Omega})$ lie above a plane $z = z_1$. Let $\tilde{\Omega}$ be the set of all classes in $\overline{\Omega}$ which contain surfaces lying between the planes $z = z_0$ and $z = z_1$. The set $\tilde{\Omega}$ is convex and closed.

Schauder's fixed-point principle states that any mapping of a convex subset of a normed linear space into a compact subset of itself has a fixed point. Hence Ω contains an element $\overline{F}$ such that $A(\overline{F}) = \overline{F}$. This means that Ω contains a surface F such that $\vartheta_F = h_F + \sigma_F$. This proves the theorem. In another formulation and under different assumptions, it was proved by Bakel'man [**19**].

Let G be a convex domain in the xy-plane. In this domain we consider a strongly elliptic Monge-Ampère equation

$$\vartheta'(rt - s^2) = ar + 2bs + ct + \varphi, \tag{***}$$

where ϑ' depends only on p and q; the other coefficients depend on the five variables x, y, z, p, q. All coefficients are assumed continuous.

We introduce a generalized mean curvature function h, the sum of the following four functions h_i:

$$h_1 = \int\int a\, dp\, dy, \qquad h_2 = \int\int c\, dq\, dx,$$

$$h_3 = \int\int b\, d\bar{p}\, d\bar{y}, \qquad h_4 = -\int\int b\, d\bar{q}\, d\bar{x},$$

where $\bar{x}$ and $\bar{y}$ are related to x and y by

$$\bar{x} = \frac{1}{\sqrt{2}}(x + y), \qquad \bar{y} = \frac{1}{\sqrt{2}}(x - y).$$

It is easy to see that, for a regular surface, this generalized curvature function is equal to

$$\int\int (ar + 2bs + ct)\, dx\, dy,$$

and hence, since the form $a\xi^2 + 2b\xi\eta + c\eta^2$ is nonnegative, it is nonnegative for any convex surface $z = z(x, y)$ which is convex toward $z < 0$. Since any convex surface is the limit of regular surfaces, it follows from the weak convergence property of generalized mean curvatures that $h \geqq 0$ on any Borel set H for an arbitrary convex surface.

Define a generalized total curvature function ϑ by the function ϑ', and a generalized area σ by the function φ. We define a generalized solution of the strongly elliptic Monge-Ampère equation (***) in the domain G as a convex function which defines a surface F convex toward $z < 0$ such that $\vartheta_F = h_F + \sigma_F$.

To formulate an existence theorem for generalized solutions, we need only impose conditions on the coefficients of equation (***) under which the assumptions of Theorem 1 are satisfied.

Set

$$\bar{a}(p, y) = \max_{(q, x, z)} a(x, y, z, p, q),$$

$$\bar{c}(q, x) = \max_{(p, y, z)} c(x, y, z, p, q),$$

$$\bar{b}'(\bar{p}, \bar{y}) = \max_{(\bar{q}, \bar{x}, z)} |b(\bar{x}, \bar{y}, z, \bar{p}, \bar{q})|,$$

$$\bar{b}''(\bar{q}, \bar{x}) = \max_{(\bar{p}, \bar{y}, z)} |b(\bar{x}, \bar{y}, z, \bar{p}, \bar{q})|,$$

$$\bar{\varphi}(x, y) = \max_{(p, q, z)} \varphi(x, y, z, p, q),$$

where the maximum extends over x, y, $\bar{x}$, $\bar{y}$ in G and p, q, $\bar{p}$, $\bar{q}$ from $-\infty$ to $+\infty$ for $z < -N$.

THEOREM 2. *The strongly elliptic equation*

$$\vartheta'(rt - s^2) = ar + 2bs + ct + \varphi$$

has a generalized solution in a convex domain G if its coefficients satisfy the condition

$$\int_{(pq)}\int \vartheta'\, dp\, dq > \int_{(py)}\int \bar{a}\, dp\, dy + \int_{(qx)}\int c\, dq\, dx + \int_{(\bar{p}\bar{y})}\int \bar{b}'\, d\bar{p}\, d\bar{y} + \int_{(\bar{q}\bar{x})}\int \bar{b}''\, d\bar{q}\, d\bar{x} + \int_{G}\int \bar{\varphi}\, dx\, dy,$$

where the limits of integration with respect to p, q, $\bar{p}$, $\bar{q}$ are $-\infty$, $+\infty$ and with respect to x, y, $\bar{x}$, $\bar{y}$ the limits defined by the domain G.

THEOREM 2a. *The strongly elliptic equation*

$$rt - s^2 = ar + 2bs + ct + \varphi$$

has a generalized solution in a convex domain G if its coefficients satisfy the condition

$$a, |b|, c, \varphi \leqq \lambda(p,q)\mu(z),$$

where $\mu(z) \to 0$ as $z \to -\infty$.

To prove Theorem 2a, divide the equation by $\lambda(1+p^2+q^2)$ and use Theorem 2.

THEOREM 2b. *If the coefficients of the strongly elliptic Monge-Ampère equation*

$$rt - s^2 = ar + 2bs + ct + \varphi$$

are bounded, it has a generalized solution in any convex domain G.

THEOREM 2c. *If the coefficients of the strongly elliptic equation*

$$rt - s^2 = ar + 2bs + ct + \varphi$$

satisfy the condition

$$a, |b|, c, \varphi \leqq f(p,q) < \infty,$$

for $(x-\alpha)^2 + (y-\beta)^2 \leqq \epsilon^2$ and $z \leqq N$, then it has a generalized solution in any sufficiently small convex domain G in the disk $(x-\alpha)^2 + (y-\beta)^2 \leqq \epsilon^2$.

To prove Theorem 2c, divide the equation by

$$(1 + f(p,q))(1 + p^2 + q^2),$$

and then use Theorem 2.

THEOREM 2d. *If the coefficients of the strongly elliptic Monge-Ampère equation*

$$\vartheta'(rt - s^2) = ar + 2bs + ct + \varphi$$

satisfy the condition

$$\vartheta' > \frac{\varepsilon}{p^2+q^2}, \quad a, |b|, c < \frac{N}{\sqrt{p^2+q^2}}, \quad \varphi < N,$$

where $p^2 + q^2 \to \infty$ and $z < N$, then there exists a generalized solution in any convex domain G of diameter $D < \pi\epsilon/4N$.

To prove this theorem, divide the equation by $(1+p^2+q^2)^\alpha$ (with sufficiently small α) and use Theorem 2.

§8. First boundary-value problem for generalized solutions of strongly elliptic Monge-Ampère equations

As in the case of the simple Monge-Ampère equation $rt - s^2 = \varphi$ (§3), we shall consider two boundary-value problems for the strongly elliptic Monge-Ampère equation.

The first boundary-value problem is to construct a generalized solution of the equation in a given domain G, with given continuous values on the boundary of G. In this section we shall determine a condition for the solvability of this boundary-value problem and consider the question of uniqueness.

In a convex domain G on the xy-plane, consider the strongly elliptic Monge-Ampère equation

$$\vartheta'(rt - s^2) = ar + 2bs + ct + \varphi, \quad \vartheta' = \vartheta'(p,q),$$

and let ψ be a given continuous function on the boundary of G. Our problem is to construct a generalized solution $z(x,y)$ of this equation in G which assumes the values ψ on the boundary.

This problem will also be solved with the aid of Schauder's fixed-point principle. The exposition will therefore be somewhat less detailed than in the preceding section.

Let γ denote the curve (whose projection is the boundary of the domain G), defined by the function ψ (the boundary values). Consider all convex surfaces $z = z(x,y)$, convex toward $z < 0$, whose projections coincide with G and have boundary γ. Denote the set of all such surfaces by Ω.

As in the preceding section, we shall define generalized mean and total curvatures and generalized area using the coefficients of the equation.

Let F be an arbitrary surface in Ω. We define a convex surface $A(F)$ in Ω with generalized total curvature ϑ equal to $h_F + \sigma_F$, by a construction similar to that of §2. There the surface was the limit of a sequence of convex polyhedra P whose boundaries γ_P converge to γ.

To carry out this construction of $A(F)$, we must first describe the construction of the polyhedra P. A necessary condition for this to be possible is

$$\int\int_{(pq)} \vartheta' \, dp \, dq > h_F(G) + \sigma_F(G).$$

Set

$$\bar{a}(\bar{p}, \bar{y}) = \max_{(qxz)} a(x, y, z, p, q),$$

$$\bar{b}'(\bar{p}, \bar{y}) = \max_{(\bar{q}\bar{x}z)} |b(\bar{x}, \bar{y}, z, \bar{p}, \bar{q})|,$$

$$\cdots\cdots\cdots\cdots\cdots$$

$$\bar{\varphi}(x, y) = \max_{(pqz)} \varphi(x, y, z, p, q),$$

where the maximum extends over x, y, $\bar{x}$, $\bar{y}$ in G, p, q, $\bar{p}$, $\bar{q}$ from $-\infty$ to $+\infty$, and $z \leqq \max \psi$. The condition for existence of the polyhedra P will then be satisfied if

$$\iint_G \bar{\varphi}\, dx\, dy + \iint_{(py)} \bar{a}\, dp\, dy + \ldots + \iint_{(\bar{q}\bar{x})} \bar{b}''\, d\bar{q}\, d\bar{x} < \infty,$$

$$\iint_{(pq)} \vartheta'\, dp\, dq = \infty.$$

However, the mere existence of the polyhedra P does not imply that their limit surface $A(F)$ is in the class Ω, since the latter surface may have a boundary distinct from γ. To determine conditions for the limit surface to have boundary γ, we appeal to the construction of §3.

Assume that, as $p^2 + q^2 \to \infty$,

$$\vartheta'(p,q) > \epsilon_0 / (p^2 + q^2), \qquad \epsilon_0 = \text{const} > 0.$$

Then the details of the construction ensure that the boundary of the limit surface of the polyhedra P coincides with γ, provided the curvature of the curve bounding G is strictly positive and, for any segment κ of the domain G,

$$\frac{1}{\sqrt{\delta}} (h_F(\varkappa) + \sigma_F(\varkappa)) < c < \infty,$$

where δ is the chord of the segment.

This condition will hold if we require that each of the functions

$$(*) \qquad \int_{-\infty}^{\infty} \bar{a}\, dp, \quad \int_{-\infty}^{\infty} \bar{b}'\, d\bar{p}, \ \ldots, \ \int_{-\infty}^{\infty} \bar{c}\, dq \ \text{and} \ \bar{\varphi}$$

be bounded by a constant.

Under these conditions, all the surfaces $A(F)$ are bounded and lie above a certain plane $z = z_0$. Let $\bar{\Omega}$ denote the subset consisting of all surfaces in Ω lying above the plane $z = z_0$. This subset is convex, since, together with any two surfaces $z = z_1(x, y)$ and $z = z_2(x, y)$, it contains the surface $z = \lambda z_1 + \mu z_2$ for $\lambda + \mu = 1$, $\lambda > 0$, $\mu > 0$.

To apply Schauder's principle we must prove that the operator A is continuous and the set $A(\overline{\Omega})$ compact.

The continuity of A follows from the uniqueness theorem for a surface with given boundary and given generalized total curvature (the case $\vartheta' = \vartheta'(p,q)$), and the weak convergence of generalized curvatures and areas of a sequence of convex surfaces. The compactness of $A(\overline{\Omega})$ is proved using condition (*).

Schauder's principle now implies that there exists a surface F with boundary γ, convex toward $z<0$, such that $\vartheta_F = h_F + \sigma_F$. The convex function $z(x,y)$ defining this surface is the required generalized solution of the boundary-value problem. Thus we have

THEOREM 1. *The Dirichlet problem for the strongly elliptic Monge-Ampère equation*

$$\vartheta'(rt - s^2) = ar + 2bs + ct + \varphi, \quad \vartheta' = \vartheta'(p,q),$$

in a domain G with boundary of positive curvature is solvable for any continuous boundary values ψ, if the following conditions hold:

1. $\vartheta' > \epsilon/(p^2+q^2)$, $\epsilon = \text{const} > 0$ *as* $p^2+q^2 \to \infty$.
2. *The functions*

$$\int_{-\infty}^{\infty} \overline{a}\, dp, \quad \int_{-\infty}^{\infty} \overline{b}'\, d\overline{p}, \ \ldots, \ \overline{\varphi}$$

are bounded.

THEOREM 1a. *Assume that the boundary of the convex domain G has positive curvature. Then the Dirichlet problem for the strongly elliptic Monge-Ampère equation*

$$\vartheta'(rt - s^2) = ar + 2bs + ct + \varphi$$

in G is solvable for arbitrary continuous boundary values, if the following conditions hold as $p^2+q^2 \to \infty$:

$$\vartheta' > \frac{\varepsilon}{p^2+q^2}; \quad a, \ |b|, \ c < \frac{N}{(p^2+q^2)^{\frac{1}{2}+\alpha}}; \quad \varphi < N,$$

where $\epsilon = \text{const} > 0$, $N = \text{const} < \infty$, $\alpha = \text{const} > 0$.

Theorem 1a follows from Theorem 1.

THEOREM 1b. *The statement of Theorem 1a is valid for the equation*

$$rt - s^2 = ar + 2bs + ct + \varphi$$

if the following conditions hold as $p^2+q^2 \to \infty$:

$$a,\ |b|,\ c < N(p^2+q^2)^{\frac{1}{2}-\alpha}, \quad \varphi < N(p^2+q^2),\ N<\infty,\ \alpha>0.$$

This theorem follows from Theorem 1a; it is sufficient to divide the equation by $1+p^2+q^2$.

Theorem 1c. *Theorems* 1a *and* 1b *remain valid for* $\alpha=0$ *if the curvature of the boundary of* G *is sufficiently large (and hence* G *itself is sufficiently small)*.

Theorem 1c is a direct corollary of Theorem 1.

It will be proved in §10 that, under very broad assumptions, the generalized solutions of a strongly elliptic Monge-Ampère equation with regular coefficients are regular solutions. The following uniqueness theorem is concerned with regular solutions.[4)]

Theorem 2. *In a domain* G *on the xy-plane, consider the strongly elliptic Monge-Ampère equation*

$$rt-s^2=ar+2bs+ct+\varphi$$

with regular coefficients such that

1) $\varphi_z \geqq 0$,
2) *the quadratic form*

$$a_z\xi^2+2b_x\xi\eta+c_zd\eta^2$$

is nonnegative.

Then, if two solutions $z_1(x,y)$ *and* $z_2(x,y)$, *both convex toward* $z<0$, *coincide on the boundary of* G, *they coincide throughout* G.

Proof. We rewrite the equation as follows:

$$(r-c)(t-a)-(s+b)^2=ac-b^2+\varphi.$$

We claim that $r-c$ and $t-a$ are positive for any solution $z(x,y)$ convex toward $z<0$.

In fact, $r-c$ and $t-a$ have the same sign, since $ac-b^2\geqq 0$ and $\varphi>0$. Moreover, if the coordinate system x, y is rotated, the signs of $r-c$ and $t-a$ do not change, since otherwise there would be a coordinate system in which $(r-c)(t-a)-(s+b)^2\leqq 0$, which is impossible.

Rotate the coordinate system in such a way that $s=0$ at the point in question. Now, if $r-c<0$ and $t-a<0$, it follows from $r,t>0$ and $a,c>0$ that

---

4) Essentially, the theorem to be proved in §10 yields uniqueness for generalized solutions.

$$|r - c| < c, \qquad |t - a| < a.$$

Hence

$$(r - c)(t - a) - (s + b)^2 < ac - b^2,$$

which is impossible.

Thus for any solution convex toward $z < 0$ we have $r - c > 0$ and $t - a > 0$.

Now substitute the solutions z_1 and z_2 into the equation, and subtract term by term; we get

$$Au_{xx} - 2Bu_{xy} + Cu_{yy}$$
$$= \frac{1}{2}(r_1 + r_2)(a_1 - a_2) + (s_1 + s_2)(b_1 - b_2) + \frac{1}{2}(t_1 + t_2)(c_1 - c_2) + \varphi_1 - \varphi_2,$$

where

$$A = \tfrac{1}{2}(t_1 + t_2 - a_1 - a_2), \quad B = \tfrac{1}{2}(s_1 + s_2 + b_1 + b_2),$$
$$C = \tfrac{1}{2}(r_1 + r_2 - c_1 - c_2),$$

or

$$Au_{xx} - 2Bu_{xy} + Cu_{yy} = D_p u_x + D_q u_y + D_z u,$$

where

$$D = \tfrac{1}{2}(a(r_1 + r_2) + 2b(s_1 + s_2) + c(t_1 + t_2)) + \varphi,$$

with the derivatives computed at suitable mean values of p, q, z.

Since the form $a_z\xi^2 + 2b_z\xi\eta + c_z\eta^2$ is nonnegative and the solutions z_1, z_2 convex functions, it follows that

$$a_z(r_1 + r_2) + 2b_z(s_1 + s_2) + c_z(t_1 + t_2) \geqq 0.$$

Consequently $D_z \geqq 0$.

The equation for $u(x, y)$ is elliptic. And since $D_z \geqq 0$, it follows that $u(x, y)$, which vanishes on the boundary of G, vanishes throughout G. This completes the proof.

§9. Regular solution of a special boundary-value problem for a strongly elliptic Monge-Ampère equation

Our proof that generalized solutions of strongly elliptic Monge-Ampère equations are regular involves the construction of a sequence of regular solutions converging to a given generalized solution. This construction will be accomplished with the aid of a special boundary-value problem, which we study in this section.

Consider a strongly elliptic Monge-Ampère equation

$$rt - s^2 = ar + 2bs + ct + \varphi,$$

with sufficiently regular coefficients, in a neighborhood of the origin. Let $\lambda(\rho)$ and $\mu(\rho)$ be two sufficiently regular functions defined by

$$\begin{array}{rrll} \mu=1, & \lambda=0 & \text{for} & \varepsilon \geqslant \rho > \varepsilon', \\ \mu=1, & 0 \leqslant \lambda \leqslant 1 & \text{for} & \varepsilon' \geqslant \rho > \varepsilon'', \\ 0 \leqslant \mu \leqslant 1, & \lambda = 1 & \text{for} & \varepsilon'' \geqslant \rho > \varepsilon''', \\ \mu = 0, & \lambda = 1 & \text{for} & \varepsilon''' \geqslant \rho > 0, \end{array}$$

and consider the equation

$$rt - s^2 = \bar{a}r + 2\bar{b}\bar{s} + \bar{c}t + \bar{\varphi}, \tag{*}$$

where

$$\bar{a}r + 2\bar{b}s + \bar{c}t = \lambda(\rho)(ar + 2bs + ct),$$
$$\bar{\varphi} = \lambda(\rho)\varphi + \mu(\rho), \quad \rho^2 = x^2 + y^2.$$

This equation is also strongly elliptic; we shall study it in the disk $G_\epsilon : x^2 + y^2 \leqq \epsilon^2$.

A regular function h defined on the circumference of the disk G_ϵ is said to be in class Ω_R if $|h| \leqq R$, and, if γ denotes the curve defined by the function h, there exists a surface through γ, convex toward $z < 0$, with generalized spherical image in the disk $\omega_R : p^2 + q^2 \leqq R^2$.

LEMMA. *Let the function*

$$\Phi = a\xi^2 + 2b\xi\eta + c\eta^2 + \varphi$$

of the variables ξ, η, x, y, z, p, q be nondecreasing in z and convex in p and q.

Then for any R and sufficiently small ϵ there exists in the disk G_ϵ a regular solution of equation (*), convex toward $z < 0$, which coincides on the circumference of G_ϵ with an arbitrary function h of class Ω_R.

The number ϵ is independent of the choice of the numbers ϵ', ϵ'', ϵ''', and of the functions λ, μ used to define the coefficients of equation (*).

As $\epsilon''' \to \epsilon$, the solution in question converges to a generalized solution of the original equation

$$rt - s^2 = ar + 2bs + ct + \varphi,$$

which assumes the same values h on the circumference of G_ϵ.

To prove the existence of the required solution it will suffice to establish a priori estimates for the maximum moduli of the solution and its first and second derivatives.

The independence of the number ϵ of the specific choice of the numbers ϵ', ϵ'', ϵ''' and the functions λ, μ will be proved if the estimates are established for small ϵ and arbitrary ϵ', ϵ'', ϵ''', λ, μ.

Even more: we assert that, in order to show that the solution of equation (*) converges as $\epsilon''' \to \epsilon$ to a generalized solution of the original equation with the same boundary values (h), it will suffice to prove that the a priori estimates for the first derivatives of the solution of equation (*) can be established independently of ϵ', ϵ'', ϵ''', λ, μ. Let us prove this assertion.

Let ϵ_k''' be a sequence of numbers ϵ''' converging to ϵ, z_k the corresponding solutions of equation (*). Since the solutions z_k coincide on the boundary of the disk G_ϵ and their derivatives are uniformly bounded, any two solutions differ by an arbitrarily small amount in a sufficiently small neighborhood of the circumference. Consequently the limit of the sequence z_k, if it exists, is equal to h on the circumference of G_ϵ.

We now show that the solutions z_k converge inside the disk G_ϵ. Take any $\bar{\epsilon} < \epsilon$. Then, if $\epsilon_k''', \epsilon_l''' > \bar{\epsilon}$, the functions z_k and z_l are solutions of the original equation in the disk $G_{\bar{\epsilon}}$, and hence $|z_k - z_l|$ assumes its maximum in the disk G_ϵ on the boundary of the disk $G_{\bar{\epsilon}}$ (see the proof of Theorem 2 in §8). But, as we have proved, for sufficiently small $\epsilon - \bar{\epsilon}$ the difference $|z_k - z_l|$ is arbitrarily small on the circumference of $G_{\bar{\epsilon}}$. Thus the solutions z_k form a convergent sequence. It is obvious that the limit function is a generalized solution of the original equation in the disk G_ϵ. Our assertion is thus proved.

We first estimate the modulus of the solution z of equation (*). Since the solution is convex toward $z < 0$, it assumes its maximum on the circumference of the disk G_ϵ, and hence $z \leqq R$. We now estimate z from below.

Define the function

$$\psi(p, q) = \max_{(x,y,z)} \{\bar{a}, |\bar{b}|, \bar{c}, \bar{\varphi}\},[5]$$

where the maximum extends over $(x, y) \in G_{\epsilon_0}$, $\epsilon_0 < \epsilon$ and $z \leqq R$. Equation (*) then implies the following inequality:

$$rt - s^2 \leqq \psi(p, q)(r + t + 1).$$

We divide this inequality by $\psi(1 + p^2 + q^2)$ and consider the following stronger version:

$$\frac{rt - s^2}{\psi(1 + p^2 + q^2)} \leqslant \frac{r}{1 + p^2} + \frac{t}{1 + q^2} + \frac{1}{1 + p^2 + q^2}.$$

---

[5] Since the function φ and the form $a\xi^2 + 2b\xi\eta + c\eta^2$ are nondecreasing in z, the function $\psi(p, q)$ is finite for any finite p, q.

Integrating this inequality over the surface F defined by the solution, and again strengthening the result, we get

$$\iint_{F^*} \frac{dp\,dq}{\psi(1+p^2+q^2)} \leqslant 4\pi\varepsilon + \varepsilon^2,$$

where the integration is performed over the generalized spherical image of F.

We now claim that $|z| < 2R$ for sufficiently small ϵ. Indeed, otherwise the domain F^* extends over the entire pq-plane as $\epsilon \to 0$, and hence the left-hand side of the inequality is bounded away from zero, while its right-hand side tends to zero. We thus have an a priori estimate for $|z|$.

It is not difficult to verify, furthermore, that this estimate may be assumed independent of the numbers ϵ', ϵ'', ϵ''' and functions λ and μ used to go from our equation to the original equation (*).

We now estimate the first derivatives of $z(x,y)$. Rewrite equation (*) as follows:

$$(r-\bar{c})(t-\bar{a})-(s+\bar{b})^2=\bar{K},$$

where $\bar{K}=\bar{\varphi}+\bar{a}\bar{c}-\bar{b}^2$.

By assumption the function h is in class Ω_R. This means that there exists a surface $z=h(x,y)$ through the contour defined by h, convex toward $z<0$, with generalized spherical image in the disk $\omega_R: p^2+q^2 \leqq R^2$. Consider the function

$$\tilde{z}=h(x,y)+\frac{1}{2\sqrt{\varepsilon}}(x^2+y^2-\varepsilon^2).$$

In the interior of the disk G_ϵ this function is bounded together with its first derivatives, and this is true uniformly with respect to the choice of the function h from Ω_R and of $\epsilon<1$. We have $\tilde{z}=h$ on the circumference of G_ϵ. We claim that for sufficiently small ϵ we have $\tilde{z}\leqq z$ in the disk G_ϵ for any function h.

Suppose the assertion false. Then $\tilde{z}-z$ assumes a positive maximum at some point P of the interior of G_ϵ.

Let $\tilde{K}$ denote the expression obtained by substituting $\tilde{z}(x,y)$ into the left-hand side of the equation, i.e.

$$\tilde{K}=(\tilde{r}-\tilde{c})(\tilde{t}-\tilde{a})-(\tilde{s}+\tilde{b})^2.$$

Subtracting the equation for $z(x,y)$ and setting $u=\tilde{z}-z$, we get

$$Au_{xx}-2Bu_{xy}+Cu_{yy}=A(\tilde{c}-\bar{c})-2B(\tilde{b}-\bar{b})+C(\tilde{a}-\bar{a})+\tilde{K}-K,$$

where

$$A = \frac{1}{2}(t + \tilde{t} - \tilde{a} - \bar{a}), \quad B = \frac{1}{2}(s + \tilde{s} + \tilde{b} + \bar{b}), \quad C = \frac{1}{2}(r + \tilde{r} - \tilde{c} - \bar{c}).$$

The quadratic form

$$(r - \bar{c})\,\xi^2 - 2\,(s + \bar{b})\,\xi\eta + (t - \bar{a})\,\eta^2$$

is positive definite. For sufficiently small ϵ (independently of h) the form

$$(\tilde{r} - \tilde{c})\,\xi^2 - 2\,(\tilde{s} + \tilde{b})\,\xi\eta + (\tilde{t} - \tilde{a})\,\eta^2$$

is also positive definite. Consequently the form

$$A\xi^2 - 2B\xi\eta + C\eta^2$$

is positive definite, and at the point P,

$$Au_{xx} - 2Bu_{xy} = Cu_{yy} \leqq 0.$$

Since the quadratic form $\bar{a}\xi^2 + 2\bar{b}\xi\eta + \bar{c}\eta^2$ is nondecreasing in z, the form

$$(\tilde{a} - \bar{a})\,\xi^2 + 2\,(\tilde{b} - \bar{b})\,\xi\eta + (\tilde{c} - \bar{c})\,\eta^2$$

is nonnegative at P. (At P we have $\tilde{p} = p$, $\tilde{q} = q$, $\tilde{z} > z$.) Hence

$$A\,(\tilde{c} - \bar{c}) - 2B\,(\tilde{b} - \bar{b}) + C\,(\tilde{a} - \bar{a}) \geqslant 0.$$

Since K is a nondecreasing function of z, its value at P cannot exceed $K(x, y, \tilde{z}, \tilde{p}, \tilde{q})$, and is therefore smaller than some constant independent of the specific choice of h in Ω_R and of ϵ. On the other hand, $\widetilde{K}$ is arbitrarily large for small ϵ, independently of h. Hence we see that for sufficiently small ϵ the difference $\widetilde{K} - K$ is positive, and this is a contradiction, since the left-hand side of the equation for u is nonpositive and its right-hand side positive.

Thus for sufficiently small ϵ we have $\tilde{z} - z \leqq 0$ in the disk G_ϵ. As was shown in §5, this makes it possible to estimate z on the circumference of G_ϵ, and hence throughout the disk, since $z(x, y)$ is a convex function.

Estimates for the second derivatives of the solution $z(x, y)$ will first be established on the circumference of G_ϵ. We shall use polar coordinates ρ, ϑ.

Since $z = h$ on the circumference of G_ϵ, it follows that $z_{\vartheta\vartheta} = h_{\vartheta\vartheta}$, so that we have an estimate for $|z_{\vartheta\vartheta}|$.

To estimate $z_{\rho\vartheta}$, note that the surface $z = \zeta(x, y)$, $\zeta = z_\vartheta$, is a surface of nonpositive curvature above the annulus $\epsilon'^2 \leqq x^2 + y^2 \leqq \epsilon^2$, since there $z(x, y)$ satisfies the equation $rt - s^2 = 1$. Hence, as in §5, we obtain an estimate for $|z_{\rho\vartheta}|$.

Derivation of an estimate for z_{pp} follows the reasoning of §5 for the equation $rt - s^2 = \varphi$ word for word. The presence of the term $\bar{a}r + 2\bar{b}s + \bar{c}t$ in equation (*) does not affect the reasoning, since it is nonnegative.

We thus have a priori estimates for the second derivatives of the solution on the circumference of the disk G_ϵ.

We now proceed to estimate the second derivatives in the entire disk G_ϵ.

Consider the auxiliary function $w = \lambda z''$, $\lambda = \exp(\alpha(x^2 + y^2))$, where z is differentiated in an arbitrary direction. In the disk G_ϵ this function assumes an absolute maximum w_0 at some point P, for some direction. If we can estimate w_0, this will furnish estimates for r and t:

$$r \leqq w_0/\min\lambda, \qquad t \leqq w_0/\min\lambda,$$

and then an estimate for s will follow from the equation.

If w assumes its maximum on the circumference of G_ϵ, then obviously $w_0 < 4\exp(\alpha\epsilon^2)M$, where M is the maximum of the moduli of r, s, t on the circumference.

It remains to estimate w_0 if P is an interior point of G_ϵ. Assume that the coordinate system is such that w assumes its maximum at P in the direction of the x-axis. Then the function $w = \lambda r$, $\lambda = \exp(\alpha(x^2 + y^2))$, also assumes its maximum at P, and moreover $s = 0$ at P, and $t \leqq r$.

Differentiating equation (*) twice with respect to x, we get

$$r_x(t-a) - 2s_x(s+b) + t_x(r-c) - a'_x r - 2b'_x s - c'_x t - \varphi'_x = 0,$$

$$r_{xx}(t-a) - 2s_{xx}(s+b) + t_{xx}(r-c) - (a''_{xx} r + 2b''_{xx} s + c''_{xx} t) -$$

$$(**) \qquad - 2(a'_x r_x + 2b'_x s_x + c'_x t_x) + 2(r_x t_x - s_x^2) - \varphi''_{xx} = 0.$$

To simplify the notation, we have omitted the bars over the coefficients of equation (*). To avoid misunderstandings, total differentiation is indicated by primes.

At the point P, where w assumes its maximum,

$$r_x = w\left(\frac{1}{\lambda}\right)_x = -\frac{r\lambda_x}{\lambda}, \qquad s_x = r_y = -\frac{r\lambda_y}{\lambda},$$

$$r_{xx} = \frac{w_{xx}}{\lambda} + r\lambda\left(\frac{1}{\lambda}\right)_{xx},$$

$$s_{xx} = r_{xy} = \frac{w_{xy}}{\lambda} + r\lambda\left(\frac{1}{\lambda}\right)_{xy},$$

$$t_{xx} = r_{yy} = \frac{w_{yy}}{\lambda} + r\lambda\left(\frac{1}{\lambda}\right)_{yy}.$$

Using the first equality of (**), we can bring $r_x t_x - s_x^2$ to the form

$$r_x t_x - s_x^2 = -\frac{1}{r}\left(r_x^2 (t-a) - 2s_x r_x (s+b) + s_x^2 (r-c)\right)$$
$$-\frac{c}{r(r-c)}\left(r_x^2 (t-a) - 2r_x s_x (s+b) + s_x^2 (r-c)\right) + \frac{Ar^2 r_x}{r-c},$$

where A is bounded.

Now, using the expressions derived above for r_x, r_y, r_{xx}, s_{xx} and t_{xx} at the point P, and observing that at this point $s=0$ and $t<r$, we can bring the second equality of (**) to the form

$$\frac{1}{\lambda}\left(w_{xx}(t-a) - 2w_{xy}(s+b) + w_{yy}(r-c)\right) - r^2(a_{pp}r + 2b_{pp}s + c_{pp}t)$$
$$(***) \qquad -\frac{c}{r(r-c)}\left(r_x^2 (t-a) - 2r_x s_x (s+b) + s_x^2 (r-c)\right)$$
$$-\frac{r}{\lambda}\left(\lambda_{xx}(t-a) - 2\lambda_{xy}(s+b) + \lambda_{yy}(r-c)\right) + B(1+\alpha\varepsilon)r^2 = 0,$$

where B is bounded by a constant independent of the second derivatives of z provided $r-c>1$.

Set $\alpha = 1/\epsilon$. Then the fourth term in (***) has order of magnitude r^2/ϵ for large r, while the fifth has order r^2. Hence, for sufficiently small fixed ϵ and r greater than some r_0, the sum of the last two terms in (***) is negative. It follows that $r \leqq r_0$ at the point P, for otherwise inequality (***) would be impossible.

Indeed,

$$w_{xx}(t-a) - 2w_{xy}(s+b) + w_{yy}(r-c) \leqq 0,$$

since w has a maximum at P. Further,

$$-r^2(a_{pp}r + 2b_{pp}s + c_{pp}t) \leqq 0$$

since the function $a\xi^2 + 2b\xi\eta + c\eta^2$ is convex in p, q. And, finally,

$$-\frac{c}{r(r-c)}\left(r_x^2 (t-a) - 2r_x s_x (s+b) + s_x^2 (r-c)\right) < 0,$$

since the form $(r-c)\xi^2 - 2(s+b)\xi\eta + (t-a)\eta^2$ is positive. Thus the left-hand side of (***) is negative for $r > r_0$. Hence r cannot exceed r_0 at the point P, and we get the estimate $w_0 \leqq r_0 e^{\epsilon}$. This completes the derivation of a priori estimates for the second derivatives.

§10. Regularity of generalized solutions of strongly elliptic Monge-Ampère equations with regular coefficients

In this section we prove, under very broad assumptions, that every generalized solution of a strongly elliptic Monge-Ampère equation

$$rt - s^2 = ar + 2bs + ct + \varphi$$

with regular coefficients is regular.

We shall study this equation in a domain G of the xy-plane. It is assumed that the regular coefficients depend on a parameter α in such a way that when $\alpha \to 0$ the coefficients and their derivatives of order up to three remain bounded for $(x,y) \in G$, $p^2+q^2+z^2 < R^2$, and $\varphi > c_0 > 0$ in the same domain.

Assume that z_α are solutions of the equation which converge as $\alpha \to 0$ to a convex function z_0. Obviously this function is a generalized solution of the limit equation.

Lemma. *Let H be any closed subset of the domain G. Then the second derivatives of the solutions z_α converging to z_0 satisfy estimates at points of H uniformly as $\alpha \to 0$.*

Proof. Let P be an arbitrary point of H. A straight line through P parallel to the z-axis cuts the surface $F_0 : z = z_0(x,y)$ at a point P_0, and the surface $F_\alpha : z = z_\alpha(x,y)$ at a point P_α. Consider any supporting plane at the point P_0, and displace it in the direction $z > 0$ till it touches the boundary of the surface F_α (we assume, as usual, that the surface F_0 is convex toward $z < 0$). Denote the plane in this position by σ_α.

Since the surface F_0 is strictly convex ($\varphi > c_0 > 0$), it follows that if α is sufficiently close to zero the distance of the point P_α from the plane σ_α in the direction of the z-axis exceeds some $\delta_0 > 0$ which is independent of the point P.

Let the equation of the plane σ_α be $z = \zeta_\alpha(x,y)$. Consider the auxiliary function $w = \delta\mu z_\alpha''$, where $\delta = \zeta_\alpha - z_\alpha$, $\mu = \exp(\beta z_\alpha')$, and the differentiation (indicated by primes) is in an arbitrary direction. We shall study this function in the domain G_P defined by the condition $\zeta_\alpha - z_\alpha > 0$. The function w assumes an absolute maximum at some interior point O of G_P, for some direction at this point. Take the point O as origin and the extremal direction as the direction of the x-axis. Then the function

$$w = \delta\mu r_\alpha, \quad \text{where} \quad \mu = \exp(\beta p_\alpha), \; p_\alpha = \partial z_\alpha / \partial x, \; r_\alpha = \partial^2 z_\alpha / \partial x^2,$$

assumes a maximum at the origin, where moreover $s = \beta q r_\alpha / 2$ and $t \leqq \exp(\beta(p_\alpha - q_\alpha))$.

Setting $\delta\mu = \lambda$, we have the following equalities at the point O:

$$r_x = -\frac{r\lambda_x}{\lambda}, \quad s_x = r_y = -\frac{r\lambda_y}{\lambda},$$

$$r_{xx} = \frac{w_{xx}}{\lambda} + w\left(\frac{1}{\lambda}\right)_{xx}, \quad r_{xy} = \frac{w_{xy}}{\lambda} + w\left(\frac{1}{\lambda}\right)_{xy}, \quad r_{yy} = \frac{w_{yy}}{\lambda} + w\left(\frac{1}{\lambda}\right)_{yy}$$

To simplify the notation, we omit the index α from now on.

Differentiating the original equation twice with respect to x, we get

$$(*)\quad \begin{aligned} &(r_x - c'_x)(t-a) + (r-c)(t_x - a'_x) - 2(s+b)(s_x + b_x) = K'_x, \\ &(r_{xx} - c'_{xx})(t-a) - 2(s_{xx} + b''_{xx})(s+b) + \\ &\quad + (t_{xx} - a''_{xx})(r-c) + 2(r_x - c'_x)(t_x - a'_x) - (s_x + b'_x)^2 = K''_{xx}, \\ &K = \varphi + ac - b^2. \end{aligned}$$

Using the expressions for the first and second derivatives of r at O, the expression for t_x derived from the first equality of (*), and the above estimates for s and t in terms of r at the point O, we can bring the second equality of (*) to the following form:

$$\begin{aligned} &\frac{1}{\lambda}\{w_{xx}(t-a) - 2w_{xy}(s+b) + w_{yy}(r-c)\} \\ &- \{(c_{pp}r^2 + 2c_{pq}rs + c_{qq}s^2)(t-a) - 2(b_{pp}r^2 + 2b_{pq}rs + b_{qq}s^2)(s+b) \\ &\qquad + (a_{pp}r^2 + 2a_{pq}rs + a_{qq}s^2)(r-c)\} \\ &\qquad - \frac{c}{r(r-c)}\{(t-a)r_x^2 + 2(s+b)r_x s_x + (r-c)s_x^2\} \\ &\qquad\qquad - \frac{\beta^2 w^2 \varphi}{\lambda^2} + \frac{1}{\lambda^2}(A\beta w^2 + B\beta w + C) = 0, \end{aligned}$$

where A, B, C are bounded by some constant M provided w is sufficiently large.

Suppose that $|A|$, $|B|$ and $|C|$ are smaller than M for $w \geqq w_0$. Take β so large that for $w \geqq w_0$

$$-\beta^2 w^2 \varphi + M(\beta w^2 + \beta w + 1) < 0.$$

We claim that $w < w_0$ in the domain G_P. Indeed, with the above choice of w_0 and β the second equality of (*) cannot hold at O. For after the above transformation of the equality the first term in curly brackets is nonpositive, since w assumes a maximum at O. The next term is nonpositive, since the form $a\xi^2 + 2b\xi\eta + c\eta^2$ is convex in the variables p, q. The third term is nonpositive because the form $(t-a)\xi^2 - 2(s+b)\xi\eta + (r-c)\eta^2$ is positive. The remaining terms are negative by virtue of our choice of w_0 and β.

Now that we have an estimate for w in G_P, estimates for r, s and t at the point P are readily found. It can be shown that the estimates derived in this way for the second derivatives r, s, t may be assumed uniform with respect to the choice of P in H and with respect to the parameter α (provided it is sufficiently close to 0).

THEOREM 1. *In a domain G on the xy-plane, consider the strongly elliptic Monge-Ampère equation*

$$rt - s^2 = ar + 2bs + ct + \varphi,$$

where the coefficients are differentiable with respect to p, q, z, and the function φ and form $a\xi^2 + 2b\xi\eta + c\eta^2$ are nondecreasing in z.

Then, if z and z' are generalized solutions of the equation, both convex toward $z < 0$, which coincide on the boundary of G, they coincide throughout G.

PROOF. In order to simplify the proof and lend clarity to the constructions, we shall assume that one of the solutions, say z', is regular, and the coefficients of the equation are independent of z. Both assumptions are not essential and do not play a critical role in the proof.

Consider the equation

$$rt - s^2 = (a + \epsilon)r + 2bs + (c + \epsilon)t + \varphi - \epsilon a - \epsilon c - \epsilon^2.$$

It has a regular solution $z'(x, y) + \epsilon(x^2 + y^2)/2$ and a generalized solution $z(x, y) = \epsilon(x^2 + y^2)/2$. The first of these assertions is verified by direct substitution. For the second, we proceed as follows.

By assumption, $z(x, y)$ is a generalized solution of the original equation. In essence this means that, for any function $\psi(x, y)$ continuous in the domain G and any sequence of regular convex functions z_α converging to z,

$$\iint_G (r_\alpha t_\alpha - s_\alpha^2 - ar_\alpha - 2bs_\alpha - ct_\alpha - \varphi)\,\psi\,dx\,dy \to 0.$$

Now replace z_α in the integrand by the function $\bar{z}_\alpha = z_\alpha + \epsilon(x^2 + y^2)/2$; we get

$$\iint_G (\bar{r}_\alpha \bar{t}_\alpha - \bar{s}_\alpha^2 - (a + \varepsilon)\bar{r}_\alpha - 2b\bar{s}_\alpha - (c + \varepsilon)\bar{t}_\alpha - \varphi$$
$$- \varepsilon a - \varepsilon c - \varepsilon^2)\,\psi\,dx\,dy \to 0.$$

But this means precisely that $\bar{z}_\alpha$ is a generalized solution of the transformed equation.

Without loss of generality we shall assume throughout the sequel that

$$a\xi^2 + 2b\xi\eta + c\eta^2 > \epsilon(\xi^2 + \eta^2)/2$$

and that the solutions z and z' have representations

$$z = z^* + \epsilon(x^2 + y^2)/2, \qquad z' = z'^* + \epsilon(x^2 + y^2)/2,$$

where z^* and z'^* are convex functions.

Suppose that the statement of the theorem is false, and the solutions do not coincide in G. To fix ideas, let $z' - z > 0$ at some point. Then $z' - z$ assumes a positive maximum m in G.

Let F and F' be the convex surfaces defined by the functions z and z', respectively. Displace the surface F' for a distance $m' < m$, close to m, in the direction $z < 0$. The surfaces F and F' will then intersect. The number m' may always be so chosen that the intersection consists of closed curves. Let γ be one of these curves.

Construct the convex hull of F and F' (with F' in its new position), and let Φ be the part of the convex hull which is convex toward $z < 0$. The surface Φ consists of pieces of the surfaces F and F' and developable surfaces separating these pieces. If m' is sufficiently close to m, the surface Φ touches F' along the boundary.

Let $\overline{F}$ denote the connected region of the surface F lying within the domain bounded by the curve γ and on the convex hull Φ. Let $\overline{G}^*$ be the spherical image of this region. Let $\overline{F}'$ denote the region on F' with the same spherical image $\overline{G}^*$. Obviously the surfaces $\overline{F}$ and $\overline{F}'$ have common supporting planes along their boundaries.

We now go over from the pointwise definition of the surfaces in cartesian coordinates x, y, z to the tangential representation in coordinates $\bar{x}$, $\bar{y}$, $\bar{z}$: the direction ratios of the supporting planes ($\bar{x}$ and $\bar{y}$) and the z-coordinate (with inverted sign) of the intercept of the supporting plane on the z-axis. The analytical relation between these coordinates is given by $z + \bar{z} = x\bar{x} + y\bar{y}$.

If the surface is regular and strictly convex, we have the following formulas:

$$\bar{x} = p, \quad \bar{y} = q, \quad x = \bar{p}, \quad y = \bar{q},$$

$$\bar{r} = \frac{1}{\Delta} r, \quad \bar{s} = \frac{1}{\Delta} s, \quad \bar{t} = \frac{1}{\Delta} t,$$

$$r = \frac{1}{\bar{\Delta}} \bar{r}, \quad s = \frac{1}{\bar{\Delta}} \bar{s}, \quad t = \frac{1}{\bar{\Delta}} \bar{t},$$

$$\Delta = rt - s^2, \quad \bar{\Delta} = \bar{r}\bar{t} - \bar{s}^2.$$

Transforming the given equation to tangential coordinates, we get

$$\varphi(\bar{r}\bar{t} - \bar{s}^2) + a\bar{r} + 2b\bar{s} + c\bar{t} - 1 = 0.$$

The function $\bar{z}'(\bar{x}, \bar{y})$ defining the surface F' in tangential coordinates obviously satisfies this equation. The same equation is satisfied, in the generalized sense, by the function $\bar{z}(\bar{x}, \bar{y})$ defining the surface F in tangential coordinates. That is to say, for any continuous function $\psi(\bar{x}, \bar{y})$ and sequence of regular convex surfaces $\widetilde{F}$ converging to F,

$$\iint_{\overline{G}^*} (\varphi(\tilde{r}\tilde{t} - \tilde{s}^2) + a\tilde{r} + 2b\tilde{s} + c\tilde{t} - 1)\, \psi \, d\bar{x}\, d\bar{y} \to 0,$$

where $\tilde{z}(\tilde{x}, \tilde{y})$ is the function defining $\tilde{F}$ in tangential coordinates.

By assumption the solutions z and z' have representations

$$z = z^* + \epsilon(x^2 + y^2)/2, \qquad z' = z'^* + \epsilon(x^2 + y^2)/2,$$

and hence the second derivatives of $\bar{z}'$ cannot exceed $1/\epsilon$:

$$\bar{r}' \leqq 1/\epsilon, \quad |\bar{s}'| \leqq 1/\epsilon, \quad \bar{t}' \leqq 1/\epsilon.$$

Such estimates also hold for the second derivatives of the generalized solution $\bar{z}$, wherever they exist, while the first derivatives satisfy Lipschitz conditions with coefficient $1/\epsilon$. This follows from the fact that the function z is the limit of a sequence of functions

$$\tilde{z} = \tilde{z}^* + \epsilon(x^2 + y^2)/2,$$

where $\tilde{z}^*$ are regular convex functions.

The Lipschitz condition for the first derivatives of $\bar{z}$ implies that $\bar{z}(\bar{x}, \bar{y})$ satisfies the equation

$$\varphi(\bar{r}\bar{t} - \bar{s}^2) + a\bar{r} + 2b\bar{s} + c\bar{t} - 1 = 0$$

almost everywhere.

Using the Hadamard lemma, we construct an equation for the difference $u = \bar{z}' - z$:

$$(\varphi\bar{t}_\vartheta + a)u_{xx} + 2(-\varphi\bar{s}_\vartheta + b)u_{xy} + (\varphi\bar{r}_\vartheta + a)u_{yy} + Ku_x + Lu_y = 0,$$

where the derivatives are computed at mean values

$$\bar{p}_\vartheta = \vartheta\bar{p} + (1 - \vartheta)\bar{p}', \cdots, \bar{t}_\vartheta = \vartheta\bar{t} + (1 - \vartheta)\bar{t}'.$$

This is a linear elliptic equation.

Since the displacement m' of the surface F' toward $z < 0$ was smaller than m, the difference $u = \bar{z}' - \bar{z}$ assumes a negative minimum within the domain G. But this is impossible, as can be seen by reasoning as in §6. This proves the theorem.

Theorem 2. *Let the coefficients of the strongly elliptic Monge-Ampère equation*

$$rt - s^2 = ar + 2bs + ct + \varphi$$

be regular (k *times differentiable*, $k \geqq 3$) *and satisfy the following conditions*:

1. *The function* φ *is nondecreasing in* z *and convex in* p, q.
2. *The quadratic form* $a\xi^2 + 2b\xi\eta + c\eta^2$ *is nondecreasing in* z *and convex in* p, q.

Then any generalized solution of the equation is regular ($k + 1$ *times differentiable*). *If the coefficients are analytic, the solution is analytic.*

PROOF. Let $\bar{z}(x,y)$ be a generalized solution of our equation in a domain G of the xy-plane, and O an arbitrary point in G. We shall prove that $\bar{z}(x,y)$ is regular in a neighborhood of O. Without loss of generality, we may assume that O is the origin.

Let $\bar{F}$ be the convex surface defined by the function $\bar{z}(x,y)$. Construct a sequence of convex analytic surfaces $\widetilde{F}$ converging to $\bar{F}$. Let $\tilde{h}$ be the convex function defining the surface $\widetilde{F}$. By the lemma of §9, if $\widetilde{F}$ is sufficiently close to $\bar{F}$ and ϵ sufficiently small, there exists a generalized solution $\tilde{z}$ of the equation in the disk $G_\epsilon : x^2+y^2 \leqq \epsilon^2$, assuming the values $\tilde{h}$ on the circumference of G_ϵ, which is the limit of regular solutions in any disk $G_{\epsilon'}$ of smaller radius.

It follows from the uniqueness theorem for generalized solutions (Theorem 1) that $\tilde{z} \to \bar{z}$ as $\widetilde{F} \to \bar{F}$. Thus the solution $\bar{z}$ is the limit of regular solutions in the disk $G_{\epsilon'}$.

By the lemma of §9, the second derivatives of the regular solutions converging to $\bar{z}$ satisfy a priori estimates, uniformly in $\widetilde{F} \to \bar{F}$, in some neighborhood of O.

With estimates available for the second derivatives, routine arguments yield estimates for the third and fourth derivatives of the regular solutions (Chapter II, §11). But this means that the solution $\bar{z}$ is thrice differentiable. Further regularity of $\bar{z}$ follows from the general theorem on the regularity of solutions of an elliptic equation with regular coefficients. This completes the proof.

CHAPTER IX

SURFACES OF BOUNDED EXTRINSIC CURVATURE

Aleksandrov's theory of manifolds of bounded curvature (Chapter I, §13), unlike the theory of manifolds of positive curvature, makes no reference to realization of these manifolds in space. Indeed, the problem of a geometric realization of these manifolds as surfaces in euclidean space, maintaining adequate connections between the metric of the manifold and the form of the surface, remains open to this day. Only a few classes of surfaces with metric of bounded curvature are known. An example is the class of all smooth surfaces for which the area of the spherical image, allowing for multiple covering, is bounded. The point is that bounded area of the spherical image is a natural and, in a sense, minimal precondition for a meaningful theory of surfaces satisfying such a weak regularity condition as smoothness.

Nash and Kuiper [**47, 38**] have shown that, under very broad assumptions, a Riemannian metric defined on a two-dimensional manifold can be realized on a smooth surface of euclidean three-space. Moreover, this realization is accomplished with the same degree of freedom as the topological embedding in space of the manifold on which the metric is defined. The results of Nash and Kuiper imply, among other things, that there exists a closed surface without self-intersections in euclidean space, which is homeomorphic to a torus and locally isometric to a plane.

It is self-evident that the fundamental theorem on the connection between intrinsic and extrinsic curvature cannot be retained in its usual form for smooth surfaces, even if their metrics are arbitrarily "well-behaved." If the surface is assumed to be merely smooth, even regularity of its metric does not ensure such a natural property as local convexity of a surface of positive Gauss curvature.

It thus becomes necessary to subject the surface to "extrinsic" restrictions over and above smoothness. The most geometrically natural condition is bounded area of the spherical image, which provides a basis for a simple and direct definition of the important concept of curvature. Smooth surfaces satisfying this additional condition will henceforth be called surfaces of bounded extrinsic curvature; our chapter is devoted to a systematic study of these surfaces, following the author's monograph [**63**].

§1. Continuous mappings of bounded variation

A surface Φ in euclidean three-space is said to be *smooth* if for any point and a suitable choice of rectangular cartesian coordinates x, y, z the surface is defined in a neighborhood of the point by an equation $z=\varphi(x,y)$, where $\varphi(x,y)$ is a continuous function with continuous first derivatives. Unless otherwise stated, the surface will always be assumed open, in the sense that the points of its boundary do not belong to it.

A set H on the surface is said to be closed if it is closed as a point set in space. Open sets are defined as complements of closed sets on the surface.

Let Φ and $\bar{\Phi}$ be smooth surfaces and f a continuous mapping of Φ onto $\bar{\Phi}$. It is clear that for any closed set F on the surface Φ the image $f(F)$ is a closed set on $\bar{\Phi}$, and hence it has a definite area (Lebesgue measure).

A mapping f of Φ onto $\bar{\Phi}$ is said to be a *mapping of bounded variation* if for any finite system of pairwise disjoint closed sets $F_1, \cdots, F_r$ on the surface Φ the sum of areas of their images on $\bar{\Phi}$

$$\mu f(F_1) + \mu f(F_2) + \cdots + \mu f(F_r)$$

is bounded by a constant independent of the choice of the sets or their number.

We now define the *absolute variation* of a continuous mapping of bounded variation. Let G be an open set on the surface Φ. The absolute variation of the mapping f on G is defined as

$$\sup(\mu f(F_1) + \mu f(F_2) + \cdots + \mu f(F_r))$$

over all systems of pairwise disjoint [closed] subsets $F_1, \cdots, F_r$ of G. The absolute variation of f on any set H on the surface Φ is defined as the infimum of the absolute variations of f on open sets containing H, and denoted by $v^0(H)$.

LEMMA 1. *Let f be a continuous mapping of a surface Φ onto a surface $\bar{\Phi}$, and H a set on Φ. Then the absolute variation of f on H cannot exceed the area of the image of H on $\bar{\Phi}$, i.e. $\mu f(H) \leq v^0(H)$. The set $f(H)$ is assumed to have a definite area (i.e. to be measurable).*

PROOF. We first consider a closed set H. Let G be any open set on Φ containing H. An obvious consequence of the definition of the absolute variation of f on an open set is $\mu f(H) \leqq v^0(G)$. But the absolute variation of f on H is the infimum of the absolute variations of f on open sets containing H. Therefore $\mu f(H) \leqq v^0(H)$.

Now consider an arbitrary set H. Let $F_1, F_2, \cdots$ be a monotonically increasing sequence of closed sets on Φ converging to Φ (as a set). Let F be a closed subset of $f(H)$ such that $\mu f(H) - \mu \bar{F} < \epsilon$ $(\epsilon > 0)$.

Let $\widetilde{F}_k$ denote the complete preimage of the set $\bar{F}$ in F_k. The set $\widetilde{F}_k$ is clearly closed.

Since the assertion has been proved for closed sets, we have $\mu f(\widetilde{F}_k) \leqq v^0(\widetilde{F}_k)$. Since $\widetilde{F}_k \subset H$, it follows that $v^0(\widetilde{F}_k) \leqq v^0(H)$. Therefore $\mu f(\widetilde{F}_k) \leqq v^0(H)$.

Since the monotonically increasing sequence of closed sets $f(\widetilde{F}_k)$ converges to $\bar{F}$, we have $\mu f(\widetilde{F}_k) \to \mu \bar{F}$. Hence $\mu \bar{F} \leqq v^0(H)$. Since ϵ is arbitrary it now follows that $\mu f(H) \leqq v^0(H)$.

This proves the lemma.

Theorem 1. *The absolute variation of a continuous mapping of a surface Φ onto a surface $\bar{\Phi}$ is a completely additive function on the ring of Borel sets on Φ, i.e. for any pairwise disjoint Borel sets H_k*

$$v^0\left(\sum_k H_k\right) = \sum_k v^0(H_k).$$

Proof. It will suffice to prove that absolute variation is an exterior Carathéodory measure, i.e.

1) $v^0(H) \geqq 0$ for any H;

2) $v^0(H_1) + v^0(H_2) = v^0(H_1 + H_2)$, if the distance between H_1 and H_2 is positive;

3) for any $H_1, H_2, \cdots$

$$v^0\left(\sum_k H_k\right) \leqslant \sum_k v^0(H_k).$$

The first assertion is obvious. To prove the second, we first consider open sets. Let G_1 and G_2 be two disjoint open sets on Φ. We must show that

$$v^0(G_1) + v^0(G_2) = v^0(G_1 + G_2).$$

Let $F_1, \cdots, F_n$ be arbitrary pairwise disjoint closed subsets of $G_1 + G_2$. Each F_k is the union of two closed subsets F_k' and F_k'', the first a subset of G_1, the second a subset of G_2. We have

$$\sum_k \mu f(F_k') \leqslant v^0(G_1), \quad \sum_k \mu f(F_k'') \leqslant v^0(G_2),$$

$$\sum_k \mu f(F_k) \leqslant \sum_k \mu f(F_k') + \sum_k \mu f(F_k'').$$

Hence
$$\sum_k \mu f(F_k) \leqslant v^0(G_1) + v^0(G_2).$$

And since the sets $F_1, \cdots, F_n$ were chosen arbitrarily, it follows that
$$v^0(G_1 + G_2) \leqq v^0(G_1) + v^0(G_2).$$

We now prove the inverse inequality. Let $F_1', \cdots, F_n'$ be arbitrary closed pairwise disjoint subsets of G_1, and $F_1'', \cdots, F_n''$ arbitrary closed pairwise disjoint subsets of G_2. Since the closed sets F_k', F_k'' are disjoint subsets of $G_1 + G_2$, it follows that
$$\sum_k \mu f(F_k') + \sum_k \mu f(F_k'') \leqslant v^0(G_1 + G_2).$$

Hence, since F_k' and F_k'' are arbitrary,
$$v^0(G_1) + v^0(G_2) \leqq v^0(G_1 + G_2).$$

Thus, if G_1 and G_2 are disjoint open sets on the surface Φ, then
$$v^0(G_1) + v^0(G_2) = v^0(G_1 + G_2).$$

Now let H_1 and H_2 be two arbitrary sets on Φ, the distance between which is positive. We must show that
$$v^0(H_1) + v^0(H_2) = v^0(H_1 + H_2).$$

Since the distance between H_1 and H_2 is positive, there exist disjoint open supersets G_1' and G_2' of H_1 and H_2, respectively. Moreover, it follows from the definition of absolute variation that for any $\epsilon > 0$ there exist open supersets G_1'', G_2'' and G'' of H_1, H_2 and $H_1 + H_2$, respectively, such that
$$v^0(G_1'') - v^0(H_1) < \varepsilon,$$
$$v^0(G_2'') - v^0(H_2) < \varepsilon,$$
$$v^0(G'') - v^0(H_1 + H_2) < \varepsilon.$$

Now define open sets G_1 and G_2 by
$$G_1 = G_1' \cap G_1'' \cap G'',$$
$$G_2 = G_2' \cap G_2'' \cap G''.$$

G_1 and G_2 are clearly disjoint supersets of H_1 and H_2, respectively, and
$$v_0(G_1) - v^0(H_1) < \varepsilon,$$
$$v^0(G_2) - v^0(H_2) < \varepsilon,$$
$$v^0(G_1 + G_2) - v^0(H_1 + H_2) < \varepsilon.$$

As we have already proved,

$$v^0(G_1) + v^0(G_2) = v^0(G_1 + G_2),$$

and ϵ is arbitrary; hence

$$v^0(H_1) + v^0(H_2) = v^0(H_1 + H_2).$$

Thus the second assertion is proved.

To verify the third condition, we first deal with the case of open sets. Let $G_1, G_2, \cdots$ be arbitrary open sets on the surface Φ. We must show that

$$v^0\left(\sum_k G_k\right) \leqslant \sum_k v^0(G_k).$$

Let $F_1, F_2, \cdots, F_r$ be an arbitrary finite system of disjoint closed subsets of $G = \sum G_k$. Define subsets F_k^i of each G_i by $F_k^i = G_i \cap F_k$.

Since the F_k^i are subsets of G_i and the distance between them is positive, it follows from the second condition that

$$\sum_k v^0(F_k^i) = v^0\left(\sum_k F_k^i\right) \leqslant v^0(G_i).$$

Furthermore, obviously

$$\mu f(F_k) \leqslant \sum_i \mu f(F_k^i).$$

By Lemma 1,

$$\mu f(F_k^i) \leqq v^0(F_k^i).$$

Summing these inequalities over i and comparing them with the preceding inequalities, we get

$$\mu f(F_k) \leqslant \sum_i v^0(F_k^i).$$

Now summing over k and recalling that

$$\sum_k v^0(F_k^i) \leqslant v^0(G_i),$$

we get

$$\sum_k \mu f(F_k) \leqslant \sum_i v^0(G_i).$$

Hence, since the disjoint sets F_k are arbitrary,

$$v^0\left(\sum_k G_k\right) \leqslant \sum_k v^0(G_k).$$

We now show that the third condition holds for arbitrary sets $H_1, H_2, \cdots$. Define open sets $G_1, G_2, \cdots$ such that

$$H_k \subset G_k, \quad v^0(G_k) - v^0(H_k) < \epsilon/2^k, \quad k = 1, 2, \cdots,$$

$$G = G_1 + G_2 + \cdots.$$

Since $\sum_k H_k \subset G$, and so

$$v^0\left(\sum_k H_k\right) \leqslant v^0(G),$$

while we have already proved that

$$v^0(G) \leqslant \sum_k v^0(G_k),$$

it now follows that

$$v^0\left(\sum_k H_k\right) \leqslant \sum_k v^0(H_k) + 2\varepsilon.$$

Since ϵ is arbitrary, this gives

$$v^0\left(\sum_k H_k\right) \leqslant \sum_k v^0(H_k).$$

Thus the absolute variation satisfies the three conditions for an exterior Carathéodory measure. Hence it is a completely additive function on the ring of Borel subsets on the surface Φ.

Let f be a continuous mapping of a surface Φ onto a surface $\overline{\Phi}$, and H any set on Φ. We define a function $n_H(Y)$, where Y is a point on $\overline{\Phi}$, as follows. If Y has finitely many preimages in H, then $n_H(Y)$ is the number of preimages. If the number of preimages of Y is infinite, $n_H(Y)$ is defined to be zero. We call $n_H(Y)$ the *multiplicity function* of the continuous mapping f on the set H. We are going to express the absolute variation of the mapping f in terms of the integral of its multiplicity function over the surface $\overline{\Phi}$.

LEMMA 2. *The absolute variation of a continuous mapping on any Borel set H is equal to the supremum of its absolute variations on closed subsets of H.*

PROOF. It follows from the definition of the absolute variation for any set H that there exists an open superset G of H such that $v^0(G) - v^0(H) < \epsilon$.

By the definition of variation on open sets, G contains a finite number of pairwise disjoint closed sets F_k such that

$$v^0(G) - \sum_k \mu f(F_k) < \varepsilon.$$

Let $\widetilde{F}_k$ denote the intersection of F_k with the complement of H. The absolute variation on the set $\sum_k \widetilde{F}_k$ is less than ϵ, since it is a subset of $G - H$. Hence there exists an open superset $\widetilde{G}$ of $\sum \widetilde{F}_k$ on which the absolute variation is less than ϵ. Let F_k' denote the intersection of F_k and $\widetilde{G}$, and let $F_k'' = F_k - F_k'$. Then the sets F_k'' are obviously closed subsets of H. We claim that

$$v^0(G) - v^0\left(\sum_k F_k''\right) < 2\varepsilon,$$

so that

$$v^0(H) - v^0\left(\sum_k F_k''\right) < 2\varepsilon.$$

Indeed,

$$v^0\left(\sum_k F_k\right) = v^0\left(\sum_k F_k'\right) + v^0\left(\sum_k F_k''\right).$$

But the left-hand side of this equality differs from $v^0(G)$ by at most ϵ, since

$$\sum_k \mu f(F_k) \leqslant v^0\left(\sum_k F_k\right) \leqslant v^0(G),$$

and

$$v^0(G) - \sum_k \mu f(F_k) < \varepsilon,$$

while the first term on the right-hand side cannot exceed ϵ, since $\sum_k F_k' \subset \widetilde{G}$ and $v^0(\widetilde{G}) < \epsilon$. This proves our assertion. Thus for any $\epsilon > 0$ there exists a closed subset F of H such that $v^0(H) - v^0(F) < \epsilon$. This completes the proof of Lemma 2.

LEMMA 3. *The multiplicity function $n_H(Y)$ is measurable for any Borel set H.*

PROOF. We first consider a closed set H. Let M^k denote the set of all points on the surface $\overline{\Phi}$ each having at least k preimages in H. To prove that $n_H(Y)$ is measurable, we need only prove that the sets M^k are all measurable.

Let M_n^k be the set of all points on $\overline{\Phi}$, each having k preimages in H the distance between which is at least $1/n$. The sets M_n^k are obviously closed, and $\sum_n M_n^k = M^k$. Hence each set M^k is an F_σ set, and is thus measurable.

Now consider an open set H. Let $F_1, F_2, \cdots$ be a monotonically increasing sequence of closed sets such that $H = F_1 + F_2 + \cdots$. We define a set $M_{s,n}^k$ on the surface $\overline{\Phi}$ as the set of all points on $\overline{\Phi}$ having at least k preimages in F_s the distance between which is at least $1/n$. The sets $M_{s,n}^k$ are closed, and $\sum_{s,n} M_{s,n}^k = M^k$. Hence M^k is again an F_σ set, and so is measurable.

Now consider an arbitrary Borel set H. By the definition of absolute variation on H and by Lemma 2, there exist an open superset G of H and a closed subset F of H such that

$$v^0(G) - \epsilon < v^0(H) < v^0(F) + \epsilon,$$

for any $\epsilon > 0$. We have $M^k(F) \subset M^k(H) \subset M^k(G)$. The sets $M^k(F)$ and $M^k(G)$ are measurable. Therefore in order to prove that $M^k(H)$ is measurable we need only show that the measure of the set $M^k(G) - M^k(F)$ is arbitrarily small if ϵ is sufficiently small.

Now, if Y is a point of the set $M^k(G) - M^k(F)$, it has a preimage in $G - F$. Therefore

$$M^k(G) - M^k(F) \subset f(G-F).$$

Hence

$$\mu(M^k(G) - M^k(F)) \leqq \mu f(G-F) \leqq v^0(G-F) < 2\epsilon.$$

This proves Lemma 3.

Lemma 4. *Let G be an open set on the surface Φ and M any measurable set on the surface $\overline{\Phi}$. Suppose that each point of M has at least p preimages in G. Let $\widetilde{M}$ denote the complete preimage of M in G. If $\mu(M) > m$, then $mp \leqq v^0(\widetilde{M})$.*

Proof. Let M_n denote the subset of all points of M which have p preimages in G the distance between which is at least $1/n$. It is clear that it will suffice to prove the assertion of the lemma for M_n, since $\mu(M_n) \to \mu(M)$ as $n \to \infty$, and $v^0(\widetilde{M}_n) \leqq v^0(\widetilde{M})$. Thus we may assume that each point in M has p preimages in G the distance between which is at least $\delta > 0$.

Let $\{A_k\}$ be a countable covering of G by pairwise disjoint F_σ sets of diameter less than δ. Let $\widetilde{m}_k$ denote the intersection of $\widetilde{M}$ and A_k, and m_k its image on the surface $\overline{\Phi}$. The sets m_k are clearly measurable and their union is M.

Define sets $m_{i_1,\cdots,i_p} = m_{i_1} \cap \cdots \cap m_{i_p}$, where the indices $i_1, \cdots, i_p$ are distinct. Obviously each point X in M is contained in at least one of these sets, since it has p preimages the distance between which is at least δ, and this distance is greater than the diameter of any set $\widetilde{m}_k$.

Now let $n_{i_1,\cdots,i_p}$ be pairwise disjoint measurable sets such that

$$n_{i_1, i_2, \ldots, i_p} \subset m_{i_1, i_2, \ldots, i_p},$$

$$\sum_{i_1, i_2, \ldots, i_p} n_{i_1, i_2, \ldots, i_p} = \sum_{i_1, i_2, \ldots, i_p} m_{i_1, i_2, \ldots, i_p}.$$

Obviously the sets $n_{i_1,\cdots,i_p}$ form a covering of M. Therefore

$$m < \sum_{i_1, \ldots, i_p} \mu n_{i_1, \ldots, i_p}.$$

Since each set $n_{i_1,\cdots,i_p}$ is contained in p sets $m_{i_1}, \cdots, m_{i_p}$, it follows that $\sum_k \mu(m_k) > m_p$. But

$$\sum_k \mu(m_k) \leqslant v^0(\widetilde{M}).$$

Consequently $mp \leqq v^0(\widetilde{M})$. This proves the lemma.

COROLLARY. *The set of points on the surface* $\overline{\Phi}$ *which have infinitely many preimages on* Φ *has measure zero.*

THEOREM 2. *For any Borel set* H *on* Φ,

$$v^0(H) = \int_{\overline{\Phi}} n_H(Y)\,dY,$$

where dY *is the area element on* $\overline{\Phi}$.

PROOF. By Lemma 3 the function $n_H(Y)$ is measurable. We must prove that it is summable. To this end we first note that $n_H(Y) \leqq n_\Phi(Y)$ for any H. Hence it will suffice to show that the function $n_\Phi(Y)$ is summable.

Let N^k be the set of points on $\overline{\Phi}$ each of which has exactly k preimages on Φ. The set N^k is the difference of two F_σ sets: $N^k = M^k - M^{k+1}$, where M^k is the set of points on $\overline{\Phi}$ each having at least k preimages on Φ. It is clear that the complete preimage $\widetilde{N}^k$ of the set N^k on Φ has a similar representation, and hence it is a Borel set. We have

$$\int_{\overline{\Phi}} n_\Phi(Y)\,dY = \sum_k k\mu(N^k).$$

But, by Lemma 4 we have $k\mu(N^k) \leqq v^0(\widetilde{N}^k)$. Hence, since absolute variation is completely additive,

$$\int_{\overline{\Phi}} n_\Phi(Y)\,dY \leqslant v^0(\Phi),$$

i.e. the function $n_\Phi(Y)$ is summable, and thus so is $n_H(Y)$ for any Borel set H.

Now consider the following set function $w(H)$ on the surface Φ:

$$w(H) = \int_{\overline{\Phi}} n_H(Y)\,dY.$$

It is completely additive on the ring of Borel sets. Indeed, let $H_1, H_2, \cdots$ be pairwise disjoint Borel sets. We must show that

$$w\left(\sum_k H_k\right) = \sum_k w(H_k).$$

First, note that

$$w(H_1+H_2+\cdots+H_n)=w(H_1)+w(H_2)+\cdots+w(H_n).$$

This follows in an obvious manner from properties of the multiplicity functions of pairwise disjoint sets. Namely, if the point Y has finitely many preimages on Φ, then

$$n_{H_1+\cdots+H_n}(Y)=n_{H_1}(Y)+\cdots+n_{H_n}(Y),$$

while the area of the set of points with infinitely many preimages is zero.

Denote $H^k=\sum_1^k H_i$ and $H=\sum_1^\infty H_i$. Then w is completely additive if

$$\int_{\bar{\Phi}} n_H(Y)\,dY=\lim_{k\to\infty}\int_{\bar{\Phi}} n_{H^k}(Y)\,dY.$$

Since $n_{H^k}(Y)$ is monotone increasing outside a set of measure zero, it follows that

$$\lim_{k\to\infty}\int_{\bar{\Phi}} n_{H^k}(Y)\,dY=\int_{\bar{\Phi}}\Big(\lim_{k\to\infty} n_{H^k}(Y)\Big)dY.$$

Hence, to prove that w is completely additive we need only show that the functions $n_H(Y)$ and $\lim n_{H^k}(Y)$ can differ only on a set of zero area.

Assume that on some set M

$$n_H(Y)>\lim_{k\to\infty} n_{H^k}(Y).$$

Then $n_H(Y)>n_{H^k}(Y)$ for any k. The number of preimages of a point $Y\in M$ in H cannot be finite, since then for sufficiently large k we would have $n_H(Y)=n_{H^k}(Y)$, while the area of the set of points with infinitely many preimages on Φ is zero. Thus, almost everywhere on $\bar{\Phi}$,

$$n_H(Y)=\lim_{k\to\infty} n_{H^k}(Y)$$

and so the set function $w(H)$ is completely additive.

It is well known that two completely additive set functions that coincide on arbitrary open sets must coincide on arbitrary Borel sets. To prove the theorem, therefore, it will suffice to show that $v^0(G)=w(G)$ for any open set G.

First, note that $w(G)\leqq v^0(G)$. This inequality was proved above for $G=\Phi$; the proof for arbitrary open G is exactly the same.

We now show that $w(G)\geqq v^0(G)$. Let $\epsilon>0$ and let $F_1,\cdots,F_n$ be a system of pairwise disjoint closed subsets of G such that

$$v^0(G)-\sum_k \mu_f(F_k)<\varepsilon.$$

Since the function w is additive, $w(G) \geqq w(F_1) + \cdots + w(F_n)$.

It follows from the definition of w that $w(F_k) \geqq \mu f(F_k)$. Therefore

$$w(G) \geqslant \sum_k \mu f(F_k) \geqslant v^0(G) - \varepsilon$$

and since ϵ is arbitrary, this gives $w(G) \geqq v^0(G)$.

Comparing this inequality with its inverse, proved above, we see that $w(G) = v^0(G)$. Q.E.D.

In conclusion, note the following lemma.

LEMMA 5. *If H is an arbitrary set on the surface Φ whose image on $\overline{\Phi}$ under the mapping f has zero area, then the absolute variation of f on H is zero.*

PROOF. Let $G_1, G_2, \cdots$ be a monotonically decreasing sequence of open sets on $\overline{\Phi}$ containing $f(H)$, such that $\mu(G_n) \to 0$ as $n \to \infty$.

Let $\widetilde{G}_n$ denote the complete preimage of G_n on Φ. $\widetilde{G}_n$ is an open superset of H. By Theorem 2,

$$v^0(\widetilde{G}_n) = \int_{\overline{\Phi}} n_{\widetilde{G}_n}(Y)\, dY.$$

As $n \to \infty$ the quantities $v^0(\widetilde{G}_n)$ decrease monotonically, and since all the $\widetilde{G}_n$ contain H it follows that $v^0(H) \leqq \lim_{n\to\infty} v^0(\widetilde{G}_n)$.

Since the functions $n_{\widetilde{G}_n}(Y)$ decrease monotonically as $n \to \infty$ outside a set of measure zero, it follows that

$$\lim_{n\to\infty} v^0(\widetilde{G}_n) = \int_{\overline{\Phi}} \left(\lim_{n\to\infty} n_{\widetilde{G}_n}(Y)\right) dY.$$

But the function $\lim_{n\to\infty} n_{\widetilde{G}_n}(Y)$ is nonzero only on a set of zero area. Hence $v^0(\widetilde{G}_n) \to 0$ as $n \to \infty$. Thus $v^0(H) = 0$. This proves Lemma 5.

§2. Positive, negative and total variation of a continuous mapping

A smooth surface Φ is said to be *orientable* if there exists a vector valued function $n(X)$, $X \in \Phi$, with the following properties:

1) $|n(X)| = 1$;

2) $n(X)$ is continuous;

3) the vector $n(X)$ is perpendicular to the tangent plane to the surface at the point X.

If a function $n(X)$ is defined on the surface Φ, the surface is said to be oriented. Obviously any orientable surface admits two orientations. If one of them is defined by a vector valued function $n(X)$, the other is defined by $-n(X)$.

Let Φ be an oriented surface, γ a simple curve on Φ bounding a domain homeomorphic to a disk. Then a direction of travel can be determined on the curve such that the following conditions hold:

1) If G_1 and G_2 are two disjoint domains bounded by curves γ_1 and γ_2, and the curves have a common arc γ_{12}, the orientation defines opposite directions of travel on γ_{12} as a subarc of γ_1 and γ_2, respectively.

2) If γ is a small simple contour around a point X, the direction of travel on γ and the direction of the normal $n(X)$ define a right-handed screw.

Under these conditions, the direction of travel is said to be induced by the orientation of the surface Φ.

Another equivalent definition of orientability is that a direction of travel can be defined for every simple curve bounding a domain homeomorphic to a disk in such a way that condition 1) holds. It is clear from this definition that orientability of surfaces is a topological invariant, i.e. surfaces homeomorphic to an orientable surface are orientable. A plane is obviously orientable.

Let γ be an oriented curve (i.e. a curve with given direction of travel) on an oriented plane, and A a point not on the curve. Let X_0 be an arbitrary point on γ. We define a function $\vartheta(X)$ on the curve γ as follows:

1) $\vartheta(X)$ is continuous at all points of the curve except X_0.

2) $\vartheta(X)$ is equal (up to a multiple of 2π) to the angle between the ray AX issuing from the point A to the point X on the curve and the initial ray AX_0.

It can be shown that the function $\vartheta(X)$ is uniquely determined by conditions 1) and 2), up to a constant term (a multiple of 2π). The variation of the function $\vartheta(X)$ as the point X_0 describes the curve, divided by 2π, is called the *degree* of the point A relative to the curve γ, denoted by $q_\gamma(A)$. The degree of a point is an integer which is independent of the choice of the point X_0 on the curve.

Intuitively speaking, the degree of a point A relative to a curve γ is the number of revolutions of a point on the curve about A when the curve is described in the indicated direction; this number is taken as positive if the orientation of γ is the same as that induced by the orientation of the plane, negative otherwise.

The degree of A relative to γ is invariant under continuous deformations of the curve under which the point A remains not on the curve. Similarly, it is invariant under a continuous variation of the position of the point A not on the curve.

If a point A has equal degrees relative to two curves γ_1 and γ_2, then one of the curves can be continuously deformed into the other, and at no stage of the deformation does the deformed curve pass through A.

Let P and Q be two arbitrary points on a closed oriented curve γ. They divide the curve γ into two parts, say γ_1 and γ_2. Let $\bar{\gamma}$ be an arbitrary curve connecting P and Q. The curves γ_1 and $\bar{\gamma}$, γ_2 and $\bar{\gamma}$ form two closed contours $\tilde{\gamma}_1$ and $\tilde{\gamma}_2$. Orient them in accordance with the directions of travel on γ_1 and γ_2 as subarcs of γ. Then, if A is any point not on the curves γ and $\bar{\gamma}$, its degree relative to γ is equal to the sum of its degrees relative to $\tilde{\gamma}_1$ and $\tilde{\gamma}_2$:

$$q_\gamma(A) = q_{\tilde{\gamma}_1}(A) + q_{\tilde{\gamma}_2}(A).$$

The concept of degree carries over in a natural way to an arbitrary surface homeomorphic to a plane. Let Φ be an oriented surface homeomorphic to a plane, γ an oriented contour on Φ and A a point on the surface, not on γ. Let f be a homeomorphism of Φ onto the plane. Orient the plane and the curve $f(\gamma)$ in accordance with the orientations of Φ and γ. Define the degree of A relative to the contour γ on the surface Φ as the degree of the point $f(A)$ relative to the contour $f(\gamma)$ on the plane. It turns out that this definition of the degree is independent of the specific homeomorphism f.

Let Φ and $\bar{\Phi}$ be smooth oriented surfaces homeomorphic to a plane, and f a continuous mapping of Φ onto $\bar{\Phi}$. Let us call a point on Φ *regular* relative to the mapping f if it has a neighborhood in which $f(Y) \neq f(X)$ for every point Y distinct from X. Any neighborhood homeomorphic to a disk and satisfying this condition will be called *normal*. By definition, every regular point has a normal neighborhood.

Let X be a point on a surface Φ, regular relative to a mapping f, and ω a normal neighborhood of X. In the neighborhood ω, consider an arbitrary oriented curve γ not passing through X, such that X has degree $+1$ relative to γ. The degree of the point $f(X)$ relative to the curve $f(\gamma)$ will be called the *index* of the point X relative to the mapping f, denoted by $j(X)$. The index is independent of the neighborhood ω and the curve γ, as we shall now show.

Let ω_1 and ω_2 be two normal neighborhoods of X, γ_1 and γ_2 closed oriented curves in ω_1 and ω_2 such that $q_{\gamma_1}(X) = q_{\gamma_2}(X) = +1$. We must show that

$$q_{f(\gamma_1)}(f(X)) = q_{f(\gamma_2)}(f(X)).$$

By a continuous deformation within $\omega_2 - X$, the curve γ_2 can be brought into any neighborhood of X; in particular, into ω_1. The degrees $q_{\gamma_2}(X)$ and $q_{f(\gamma_2)}(f(X)$ are invariant under this deformation. A suitable continuous deformation in $\omega_1 - X$ will now deform the curve γ_2 into γ_1, since the point X has the same degree relative to the curves γ_1 and

γ_2. But the degree $q_{f(\gamma_2)}(f(X))$ is also invariant under this deformation. Hence for the original curves γ_1 and γ_2 we have

$$q_{f(\gamma_1)}(f(X)) = q_{f(\gamma_2)}(f(X)).$$

This proves the assertion.

Lemma 1. *Let G be a domain on a surface Φ, homeomorphic to a disk, bounded by a simple curve γ, and X a point in G. Suppose that $f(Y) \neq f(X)$ for every point Y in G or on γ other than X. Then, if $|j(X)| > 1$, every point of the surface $\overline{\Phi}$ which can be connected to $f(X)$ by a continuous curve not cutting $f(\gamma)$ has at least two preimages in G.*

Proof. Let $\overline{Y}$ be a point on $\overline{\Phi}$ which can be connected to $f(X)$ without cutting the curve $f(\gamma)$. We first show that $\overline{Y}$ has at least one preimage in G. Suppose this is false. Contract the curve γ to the point X within $G-X$. Since by assumption $\overline{Y}$ has no preimages in G, it follows that at no stage of the deformation of γ does the curve $f(\gamma)$ pass through the point $\overline{Y}$, and hence it contracts to a point $f(X) \neq \overline{Y}$. Hence the degree of $\overline{Y}$ relative to the curve $f(\gamma)$ is zero. But, on the other hand, this degree is equal to $j(X) \neq 0$, since in the original position of the curve γ the point $f(X)$ can be connected by a continuous curve to $\overline{Y}$ without cutting $f(\gamma)$. This contradiction shows that $\overline{Y}$ must have a preimage in G.

Now suppose that $\overline{Y}$ has only one preimage in G, say $\widetilde{Y}$. The point $\widetilde{Y}$ is obviously regular, G is a regular neighborhood of $\widetilde{Y}$ and the index of $\widetilde{Y}$ is equal to that of X; hence $|j(\widetilde{Y})| > 1$. Since the proof that $\overline{Y}$ has preimages other than $\widetilde{Y}$ is rather involved, we first consider some auxiliary constructions.

Let a be a polygonal line issuing from a point O on a plane α; a has finitely many sides, and goes to infinity. Let γ be a closed polygonal line not containing the point O and cutting a in finitely many points. The lines a and γ are assumed to be identically oriented. The pattern of the intersections of γ and a can be represented by a string $aa\bar{a} \cdots a\bar{a}$ defined as follows. Moving along the line γ, whenever it cuts the line a enter a or $\bar{a}$ in the string, according to whether the line a is crossed from right to left or from left to right.

Two strings of this type will be called equivalent if they can be transformed into each other by elementary transformations. By an elementary transformation we mean any of the following operations: deletion of two consecutive entries a and $\bar{a}$; insertion of additional pairs of this type (the first and last elements of the string are regarded as consecutive);

transposition of the first element to the last place or the last element to the first place.

Obviously, after deletion of all pairs $a\bar{a}$ and $\bar{a}a$ the string of intersections of the two polygonal lines γ and a has the form $aa \cdots a$ or $\bar{a}\bar{a} \cdots \bar{a}$, where the number of elements (a or $\bar{a}$) is $|q_\gamma(O)|$.

Any continuous deformation of γ in the domain $\alpha - O$ either leaves its string of intersections with a unchanged or subjects it to an elementary transformation.

Now consider an arbitrary point A on the line a, dividing it into two lines—a finite line b and an infinite line c. With every closed polygonal line not containing the points O and A and cutting a in a finite number of points, associate a string of intersections $bc\bar{b} \cdots \bar{c}$ defined as before. By analogy with the preceding case, elementary transformations are deletion of pairs $(b\bar{b})$, $(c\bar{c})$, addition of such pairs, transposition of the first element to the last place and conversely.

Obviously, if a polygonal line γ can be deformed continuously in the domain $\alpha - O - A$ into a polygonal line γ' in such a way that the number of intersections with a is finite at each stage of the deformation, then the intersection strings of γ and γ' are equivalent, i.e. they can be obtained from each other by elementary transformations.

We can now proceed with the proof of the lemma. Without loss of generality we may assume that the surface $\overline{\Phi}$ is a plane. Let a be a simple polygonal line issuing from the point $f(X)$, passing through $\overline{Y}$, and going to infinity. Let b be the segment of the line between the points $f(X)$ and $\overline{Y}$, and c the remaining (infinite) segment.

Let γ' be a small simple curve bounding a domain G' homeomorphic to a disk, such that the point $\widetilde{Y}$ lies inside γ', X outside. Join an arbitrary point P on γ to an arbitrary point Q of γ' by a simple curve γ'' which does not pass through X, within the doubly connected domain bounded by the curves γ and γ'. Let $\bar{\gamma}$ denote the closed curve consisting of γ, γ' and the curve γ'' described twice. Obviously by a suitable continuous deformation within the domain $G - X - \widetilde{Y}$ the curve $\bar{\gamma}$ can be moved into any arbitrarily small neighborhood of the point X. Therefore the curve $f(\bar{\gamma})$ can be brought by a continuous deformation into any small neighborhood of the point $f(X)$, without passing at any time through the points $\overline{Y}$ or $f(X)$.

Now inscribe a polygonal line in the curve $f(\bar{\gamma})$, with sufficiently small sides. This line can also be brought by a continuous deformation in the domain $\overline{\Phi} - f(X) - \overline{Y}$ into an arbitrarily small neighborhood of

$f(X)$. To avoid introducing new notation, we shall assume that $f(\gamma)$ itself is a polygonal line.

It is easy to find a deformation of the polygonal line $f(\overline{\gamma})$ under which its segments $f(\gamma)$ and $f(\gamma')$ remain unchanged, while $f(\gamma'')$ goes into a polygonal line not cutting the line a. We shall therefore assume that the original line $f(\overline{\gamma})$ has this property. Then its string of intersections can be brought by elementary transformations to the form

$$ccc \cdots cb\bar{c}b\bar{c} \cdots b\bar{c}$$

for a suitable orientation of a. The part of the string up to the first element b is a reduced string of intersections of the line $f(\gamma)$, and the remainder is a reduced string of intersections of $f(\gamma')$. The number of c's in the first part and the number of pairs $b\bar{c}$ in the second part are equal to $|j(X)| = |j(\widetilde{Y})|$.

By a continuous deformation in the domain $\overline{\Phi} - f(X) - \overline{Y}$, the polygonal line $f(\overline{\gamma})$ can be moved into an arbitrarily small neighborhood of the point $f(X)$. Then the string of intersections of $f(\overline{\gamma})$, after simplification, is $bb \cdots b$, where the number of b's is $|j(X)|$. Obviously, the strings $bb \cdots b$ and $cc \cdots cb\bar{c} \cdots b\bar{c}$ are not equivalent. On the other hand, they must be equivalent, for they correspond to polygonal lines which can be transformed into each other by a continuous deformation in the domain $\overline{\Phi} - f(X) - \overline{Y}$. This contradiction completes the proof of Lemma 1.

Let Φ and $\overline{\Phi}$ be smooth oriented surfaces and f a continuous mapping of bounded variation mapping Φ onto $\overline{\Phi}$. We define the *positive* (*negative*) *variation* of the mapping f on a set H of Φ as the absolute variation of f on the subset of points of H which are regular relative to the mapping f and of positive (negative) index. The *total variation* of f on H is defined to be the difference between the positive and negative variations. We denote the positive, negative and total variations of f on H by $v^+(H)$, $v^-(H)$ and $v(H)$, respectively.

LEMMA 2. *The absolute variation of a mapping f on the set of points of F which are not regular relative to f is zero.*

PROOF. Let H be the set of points on Φ which are not regular relative to f, and $f(H)$ its image on the surface $\overline{\Phi}$. Since each point of $f(H)$ has infinitely many preimages on Φ, it follows from the Corollary to Lemma 4 in §1 that the area of $f(H)$ is zero. But then, by Lemma 1 of §1, the absolute variation of f vanishes on H. This proves Lemma 2.

LEMMA 3. *The set E of regular points on the surface Φ for which $|j(X)| > 1$ is at most countable.*

PROOF. Suppose the assertion false. Since the points of E are regular, there exists a number δ for each point X in E such that the δ-neighborhood of X contains no points with the same image $f(X)$ as X. Thus each point $X \in E$ is associated with a suitable number $\delta(X)$. It is obvious that for some $\delta_0 > 0$ there exists an uncountable set E_{δ_0} of points $X \in E$ such that $\delta(X) > \delta_0$.

Describe a "circle" γ_X of radius $\delta_0/2$ about each point $X \in E$. Let $\lambda(X)$ be the distance of the point $f(X)$ from the contour $f(\gamma_X)$. As before, we can extract from the set E_{δ_0} an uncountable subset $E_{\delta_0\lambda_0}$ for all points of which $\lambda(X) > \lambda_0 > 0$.

Since the set $E_{\delta_0\lambda_0}$ is uncountable, it has a point of accumulation A on the surface. Take two points X' and X'' of $E_{\delta_0\lambda_0}$, so close to A that the distance between them is smaller than $\delta_0/2$ and the distance between their images on $\overline{\Phi}$ smaller than λ_0. The points $f(X')$ and $f(X'')$ can then be connected by a continuous curve in $\overline{\Phi}$ not cutting the curve $f(\gamma_{X'})$. But then, by Lemma 1, there exist at least two points within the domain bounded by $\gamma_{X'}$ whose image is $f(X'')$. This is impossible, since the δ_0-neighborhood of X'' contains no points whose image is $f(X'')$ except X'' itself. This contradiction proves Lemma 3.

LEMMA 4. *Let H_j be the set of all points on Φ which are regular relative to the mapping f and have index j. Then H_j can be expressed as $H_j = B - \omega$, where B is a Borel set and ω a set of nonregular points.*

PROOF. Define a subset H_j^n of H_j by the condition that a point $X \in H_j$ is in H_j^n if the $1/n$-neighborhood of X contains no points with image $f(X)$ except X itself. We first show that any regular limit point Y of the set H_j^n is an element of H_j^n.

Let G be a normal neighborhood of Y of diameter less than $1/n$, γ a curve in this neighborhood relative to which the degree of Y is $+1$. If a point $X \in H_j^n$ is sufficiently close to Y, then G is a normal neighborhood of X, the degree of X relative to γ is $+1$, and the degree of its image $f(X)$ relative to the curve $f(\gamma)$ is equal to that of $f(Y)$. But this means that Y and X have equal indices, and so Y has index j and belongs to H_j^n.

Close the set H_j^n by adding the limit points not already in H_j^n, and denote the resulting set by $\overline{H}_j^n$. As we have proved, the only points of

H_j^n not in H_j^n are nonregular. Since $H_j = \sum_n H_j^n$ and the union of the H_j^n is a Borel set, it follows that H_j has the required representation. This proves Lemma 4.

THEOREM 1. *The positive, negative and total variations of a continuous mapping f of a surface Φ onto a surface $\overline{\Phi}$ are completely additive functions on the ring of Borel sets of Φ.*

PROOF. Let H be a Borel set and M an arbitrary set on which the absolute variation vanishes. Then $v^0(H-M) = v^0(H)$.

Indeed, $v^0(H-M) \leqq v^0(H)$, since $(H-M) \subset H$. We now prove the reverse inequality.

Since the absolute variation vanishes on M, it follows that for any $\epsilon > 0$ there exists an open superset G of M such that $v^0(G) < \epsilon$. We have

$$v^0(H-O) + v^0(G) \geqq v^0(H).$$

Hence,

$$v^0(H-M) \geqq v^0(H) - \epsilon.$$

Since ϵ is arbitrary, this gives us the desired inequality, and completes the proof of the assertion.

Let $H_1, H_2, \cdots$ be arbitrary pairwise disjoint Borel sets, and set $H = H_1 + H_2 + \cdots$. By definition $v^+(H) = v^0(H \cap \Omega^+)$, where Ω^+ is the set of points on Φ which are regular relative to f and have positive index. By Lemma 4, $\Omega^+ = B - \omega$, where B is a Borel set and ω a set containing only nonregular points.

By Lemma 2, the absolute variation vanishes on ω. Therefore $v^+(H) = v^0(H \cap B)$.

By the complete additivity of the absolute variation,

$$v^0(H \cap B) = \sum_k v^0(H_k \cap B),$$

and by what we have just proved

$$v^0(H_k \cap B) = v^0(H_k \cap \Omega^+).$$

Consequently

$$v^+(H) = \sum_k v^+(H_k).$$

The proof of complete additivity for the negative variation v^- is analogous. The complete additivity of v follows from that of v^+ and v^-. Q.E.D.

Theorem 2. *The positive, negative and total variations of a continuous mapping f of a surface Φ onto a surface $\bar{\Phi}$ have the following integral representations for any Borel set H:*

$$v^+(H) = \int_{\bar{\Phi}} n_H^+(Y)\,dY, \quad v^-(H) = \int_{\bar{\Phi}} n_H^-(Y)\,dY,$$

$$v(H) = \int_{\bar{\Phi}} (n_H^+(Y) - n_H^-(Y))\,dY,$$

where $n_H^+(Y)$ is the number of regular preimages of the point Y in H with index $+1$, and $n_H^-(Y)$ the number of regular preimages with index -1.

This theorem follows from Lemma 3 and Theorem 2 of §1.

Theorem 3. *Let G be a domain homeomorphic to a disk on a surface Φ, γ the curve bounding G; assume that the absolute variation of the mapping f vanishes on γ. Let $\bar{\gamma}$ be the image of γ on the surface $\bar{\Phi}$, and $q_{\bar{\gamma}}(Y)$ the degree of an arbitrary point Y on $\bar{\Phi}$ relative to the curve $\bar{\gamma}$. Then the total variation of f in the domain G is*

$$v(G) = \int_{\bar{\Phi}-\bar{\gamma}} q_{\bar{\gamma}}(Y)\,dY.$$

(The orientation of the curve $\bar{\gamma}$ is induced by that of γ, which in turn is induced by the orientation of Φ.)

Proof. We first observe that the function $q_{\bar{\gamma}}(Y)$ is measurable, since it is a constant integer on each component of the set $\bar{\Phi} - \bar{\gamma}$. Moreover, certain points may be deleted from the domain of integration $\bar{\Phi} - \bar{\gamma}$: all points Y among whose preimages on γ are nonregular points or regular points with index greater than unity in absolute value, and all points Y with infinitely many preimages. The set of deleted points has area zero.

Let Y be a point having $n^+(Y)$ preimages in G with index $+1$, $n^-(Y)$ preimages with index -1, and $n^0(Y)$ preimages with index 0. Divide the domain G into $n^+(Y)+n^-(Y)+n^0(Y)$ elementary domains G_k, each containing only one preimage X_k of the point Y. Let γ_k be a contour bounding the domain G_k and $\bar{\gamma}_k$ its image on $\bar{\Phi}$.

We have

$$q_{\bar{\gamma}}(Y) = \sum_k q_{\bar{\gamma}_k}(Y).$$

Since the domain G_k contains no preimages of Y other than X_k, it follows that

$$q_{\bar{\gamma}_k}(Y) = j(X_k).$$

Hence $q_{\bar{\gamma}}(Y) = n^+(Y) - n^-(Y)$. The proof is now completed by using the integral representation of the total variation established in the preceding theorem.

REMARK. The integral representation of Theorem 3 for the total variation is easily extended to multiply connected domains G. The integrand in the integral is then

$$q_{\bar{\gamma}}(Y) = q_{\bar{\gamma}_0}(Y) - q_{\bar{\gamma}_1}(Y) - \cdots - q_{\bar{\gamma}_r}(Y),$$

where γ_0 is the outer boundary of the domain and $\gamma_1, \cdots, \gamma_r$ the inner contours.

§3. Smooth surfaces of bounded extrinsic curvature

Let Φ be an oriented smooth surface and $n(X)$ the exterior unit normal vector defining the orientation. The *spherical image* of the surface is defined as its image under the mapping which takes each point X to the point of the unit sphere with radius vector $n(X)$.

A smooth surface Φ is called a *surface of bounded extrinsic curvature* if the area of its spherical image. allowing for multiple covering, is finite. The rigorous meaning of this descriptive condition is that for any finite system of pairwise disjoint closed sets on the surface the sum of areas of their images under the spherical mapping is uniformly bounded. Thus a surface has bounded extrinsic curvature if and only if its spherical mapping has bounded variation.

Let G be any open set on Φ, $F_1, \cdots, F_r$ arbitrary pairwise disjoint closed subsets of G and $\overline{F}_1, \cdots, \overline{F}_r$ their spherical images. We define the *absolute curvature* of the surface Φ on the set G as the supremum of the sums of areas of the sets $\overline{F}_k$, over all finite systems of pairwise disjoint closed subsets F_k of G. The absolute curvature of the surface on an arbitrary set H is defined as the infimum of its absolute curvatures on open supersets of H. Thus the absolute curvature of a surface on a set H is precisely the absolute variation of the spherical mapping of the surface on the set.

We list the principal properties of absolute curvature.

THEOREM 1. *The absolute curvature of a surface on an arbitrary set is not less than the area of the spherical image of the set. The absolute curvature vanishes if and only if the area of the spherical image vanishes.*

THEOREM 2. *The absolute curvature of a surface is a completely additive set function on the ring of Borel sets.*

THEOREM 3. *The absolute curvature of a surface on any Borel set H is*

$$\sigma^0(H) = \int_\Omega n(Y)\,dY,$$

where $n(Y)$ is the number of points of the surface whose image under the spherical mapping is the point Y of the unit sphere Ω, dY is the area element on the unit sphere, and the integration domain is the entire sphere.

All these properties follow directly from the corresponding properties of the absolute variation of a continuous mapping, since the spherical mapping of a smooth surface is continuous.

We now divide the points on a smooth surface into two classes. We call a point X of a smooth surface a *one-sided point* if a sufficiently small neighborhood of X on the surface lies on one side of the tangent plane at X. All other points on the surface are called *two-sided points*. Thus an arbitrarily small neighborhood of a two-sided point X contains points lying on either side of the tangent plane at X.

A point X on a smooth surface is said to be *regular* (relative to the spherical mapping) if a sufficiently small neighborhood of X contains no points at which the tangent plane is parallel to the tangent plane at X.

THEOREM 4. *The absolute curvature of a surface of bounded extrinsic curvature vanishes on the set of nonregular points.*

As before, this theorem follows from the properties of the absolute variation of the spherical mapping.

A further classification of the points of a smooth surface is made possible by the following theorem of Efimov [**32**].

THEOREM 5. *The intersection of a neighborhood of a regular point X on a smooth surface with the tangent plane at X is either a point (X) or an even number of simple curves issuing from X.*

PROOF. Let X be a regular point on the surface. First suppose that X is a one-sided point. We claim that the intersection of the tangent plane α at X with a sufficiently small neighborhood of X is the point X. Indeed, otherwise there exists a sequence of points X_k on the surface converging to X, such that each point of the sequence is in the plane α. Since X is one-sided, α must be a tangent plane at points X_k sufficiently close to X, and this contradicts the assumption that X is a regular point. Thus, if a regular point X is one-sided, the tangent plane to the surface at the point has only one point in common with a sufficiently small neighborhood of X, namely X itself.

Now let X be a two-sided point. We claim that the intersection of a neighborhood of X with the tangent plane at X consists of an even number of simple curves issuing from X.

Let U be a neighborhood of X containing no points with normals parallel to the normal at X. Let H denote the intersection of U with the tangent plane α at X, and $\overline{H} = H - X$. Since the tangent plane at each point of $\overline{H}$ is distinct from α, it follows that each component of the set $\overline{H}$ is a simple arc.

Let γ be a component of $\overline{H}$, and A an arbitrary fixed point on γ. Let $\lambda(Y)$ be the distance of a point Y on γ from X, and $\mu(Y)$ its distance from the boundary of the neighborhood U. Then, as the point Y describes the curve γ in either of the two directions from A, either $\lambda(Y) \to 0$ or $\mu(Y) \to 0$. Let us prove this statement.

Parametrize the curve γ with respect to arc length s, with origin at the point A (the curve γ is smooth, therefore rectifiable. over any subarc AY). The function $s(Y)$, considered on one side of A, may be either bounded or unbounded. We consider each case separately.

Assume that $s(Y)$ is bounded, and let a be its supremum. As $s(Y) \to a$, the point Y tends to a well-defined limit position Y_0. Y_0 cannot be an interior point of the domain $U - X$, since in a neighborhood of this point $\overline{H}$ is a simple curve which would contain Y_0 as an interior point. Thus Y_0 either coincides with X, in which case $\lambda(X) \to 0$, or lies on the boundary of U, in which case $\mu(Y) \to 0$.

Now consider the second case: $s(Y) \to \infty$ as Y moves away from A in the direction in question. If the statement concerning the functions λ and μ is false, there exists a sequence of points Y_k having the following properties:

1. $s(Y_k) \to \infty$ as $k \to \infty$.
2. $\lambda(Y_k) > \epsilon$ and $\mu(Y_k) > \epsilon$, where ϵ is a positive number.

Without loss of generality we may also assume that $s(Y_k) - s(Y_{k-1}) > 1$.

The set of points Y_k has an accumulation point Y_0, which is an interior point of the domain $U - X$. This point is obviously in $\overline{H}$, which, in a small neighborhood of the point Y_0, is a simple curve $\bar{\gamma}$ of finite length. And since each point Y_k is also in $\overline{H}$, all these points must lie on $\bar{\gamma}$ for sufficiently large k. Hence the curve $\bar{\gamma}$ is a subarc of γ. Since the curve $\bar{\gamma}$ has finite length, it cannot contain infinitely many points Y_k. This contradiction shows that a point moving along any component γ of $\overline{H}$ in either direction must approach either X or the boundary of H.

It is clear that each component of $\overline{H}$ along which a point approaches the boundary of U is at a positive distance from the point X.

If ϵ is sufficiently small, the set of all components of H which cut the ϵ-neighborhood of X is finite. To prove this, note first that there are no components γ of $\overline{H}$ along which points moving in both directions approach the point X. Indeed, any such component, completed by the point X, would bound a domain on the surface in U. And since the boundary of this domain lies in the plane α, it would obviously contain a point at which the tangent plane is parallel to α—the tangent plane at X. But this is impossible by the choice of the neighborhood U. Hence any component γ of the set $\overline{H}$ which cuts an ϵ-neighborhood of the point X must cut the boundary of the neighborhood.

Suppose that the number of components cutting an ϵ-neighborhood of X is infinite. On each such component γ consider a point Y on the boundary of the ϵ-neighborhood. Let Y_0 be an accumulation point of the points Y. Since the set $\overline{H}$ is a simple arc $\overline{\gamma}$ in a small neighborhood of Y_0, it follows that if Y is sufficiently close to Y_0, then Y must lie on $\overline{\gamma}$, and so γ contains Y_0. But this can be true only for one component. This contradiction shows that the number of components of $\overline{H}$ is finite, and so the immediate neighborhood of the point X contains only components which approach the point X in one direction and the boundary of the neighborhood U in the other. The point X is an endpoint of these components.

Thus the intersection of a neighborhood U of X with the tangent plane at X is the union of finitely many simple curves issuing from X.

The fact that the number of curves is even follows from the fact that they divide the neighborhood of X into "sectors," in such a way that adjacent sectors lie on different sides of the tangent plane at X. This completes the proof of the theorem.

A regular point X on a smooth surface is said to be *elliptic* if the intersection of a sufficiently small neighborhood of X with the tangent plane at X contains a single point, namely X.

X is said to be *parabolic* if the intersection of the tangent plane at X with the surface near X consists of two simple curves issuing from X.

X is said to be *hyperbolic* if the intersection of the tangent plane at X with the surface near X consists of four simple curves issuing from X.

All other regular points are called *flat points*. Near a flat point X, the intersection of the surface with the tangent plane consists of an even number (greater than four) of simple curves issuing from X.

Let X be any set of points on a smooth surface of bounded extrinsic curvature. We define the *positive* (*negative*) *curvature* of the surface on the set H as the absolute curvature of the surface on the subset of elliptic (hyperbolic) points in H. The *total curvature* of the surface on H is defined as the difference between the positive and negative curvatures.

THEOREM 6. *The positive, negative and total curvatures of a smooth surface of bounded extrinsic curvature are bounded, completely additive functions on the ring of Borel sets.*

THEOREM 7. *For any Borel set H the positive, negative and total curvatures of the surface have the following integral representations:*

$$\sigma^+(H) = \int_\Omega n_H^+(Y)\,dY, \quad \sigma^-(H) = \int_\Omega n_H^-(Y)\,dY,$$

$$\sigma(H) = \int_\Omega (n_H^+(Y) - n_H^-(Y))\,dY,$$

where $n_H^+(Y)$ is the number of elliptic points and $n_H^-(Y)$ the number of hyperbolic points in H whose spherical image is the point Y, dY is the area element on the unit sphere Ω, and the domain of integration is the entire sphere.

THEOREM 8. *Let G be a domain on the surface, homeomorphic to a disk, bounded by a simple curve γ the area of whose spherical image is zero. Let S be a point on the unit sphere, not contained in the spherical image of G. The total curvature of the surface in the domain G is*

$$\sigma(G) = \int_{\Omega - \bar{\gamma} - S} q_{\bar{\gamma}}(Y)\,dY,$$

where $q_{\bar{\gamma}}(Y)$ is the degree of the point Y on the sphere Ω punctured at S relative to the curve $\bar{\gamma}$ (the spherical image of γ), and the domain of integration is the set $\Omega - \bar{\gamma} - S$ on the unit sphere.

Theorems 6, 7 and 8 follow directly from the corresponding properties of the positive, negative and total variations (§2) applied to the spherical mapping, and from the following lemma 32

LEMMA. *Every regular point on a smooth surface has a well-defined index relative to the spherical mapping. If the unit sphere is suitably oriented, the indices of an elliptic, parabolic and hyperbolic point are $+1$, 0 and -1, respectively, and the index of a flat point is less than -1.*

PROOF. The fact that every regular point has a well-defined index relative to the spherical mapping follows from the fact that every such point is regular relative to the spherical mapping. Let us determine the index in each case.

By definition, the orientation of the surface is determined by a unit normal vector. We shall assume that the unit sphere is oriented by the unit vector $n(X)$ defining the orientation of the surface.

Let X be an elliptic point on the surface, α the tangent plane at X. Consider a plane parallel and close to α, in the halfspace (defined by α) which contains the surface. This plane cuts the surface near X in a smooth simple closed curve γ encircling the point X.

Let $\overline{X}$ be the spherical image of X and $\overline{\gamma}$ the spherical image of the contour γ. Let $t(Y)$ denote the semitangent to the curve γ at Y, and $n(Y)$ the exterior normal at Y. Project the contour $\overline{\gamma}$ onto the tangent plane to the unit sphere at $\overline{X}$. The ray $g(Y)$ issuing from $\overline{X}$ and passing through the point on the projection of $\overline{\gamma}$ corresponding to Y is parallel to $n(Y)$. The index of the point X is the total variation of the angle between the ray $g(Y)$ and some fixed direction, as Y describes the curve γ, divided by 2π. And since the rays $g(Y)$ and $t(Y)$ are perpendicular, this variation is equal to the total variation of the angle between the semitangent $t(Y)$ and a fixed direction in the plane. Since the latter variation is obviously 2π, the index of X is $+1$.

For regular two-sided points we proceed as follows. Let α_1 and α_2 be two parallel planes close to α, on different sides of α. The plane α cuts the surface in the neighborhood of X in $2m$ curves, which divide the neighborhood of X into $2m$ sectors. Each of the planes α_1 and α_2 cuts the surface in m curves. For one plane these curves lie in the even-numbered sectors, for the other, in the odd-numbered sectors. As α_1, $\alpha_2 \to \alpha$, these curves converge to the sides of the sectors.

Choose a point X_k on each curve γ_k in which the plane α cuts the surface, and join it by curves γ_k' and γ_k'' near X_k to the curves in which the planes α_1 and α_2 cut the surface, which lie in the sectors incident on γ_k. Thus the curves in which the planes α_1 and α_2 cut the surface, together with the curves γ_k' and γ_k'', form a contour γ encircling the point X. This contour will play the same role as the contour γ in the elliptic case.

Let $\overline{\gamma}$ be the spherical image of the contour γ and $\tilde{\gamma}$ its projection on the tangent plane to the sphere at $\overline{X}$. $g(Y)$ will denote the ray issuing from $\overline{X}$ and passing through the point of the curve $\tilde{\gamma}$ corresponding to a point Y on γ. Then the index of the point X is the total variation of the angle $\vartheta(Y)$ between $g(Y)$ and some fixed direction, when Y describes the contour γ, divided by 2π.

This variation can be expressed as the sum of the variations of the angle on the portions of the curve defined by the sectors. The variation of the angle $\vartheta(Y)$ on the part of the curve γ within the sector bounded by γ_k and γ_{k+1} is close to $\varphi_k - \pi$, where φ_k is the angle between the tangents to the curves γ_k and γ_{k+1} at the points X_k and X_{k+1}, respectively. This follows from the fact that $g(Y)$ is parallel to the normal to the curve in which the plane α_1 (or α_2) cuts the surface at the point Y.

The difference between the variation of $\vartheta(Y)$ over the portion of γ in the sector (γ_k, γ_{k+1}) and the angle $\varphi_k - \pi$ can be made arbitrarily small, provided the planes α_1 and α_2 are sufficiently close to α. Thus the index of the point X is arbitrarily close to the number

$$\frac{1}{2\pi}\sum_k (\varphi_k - \pi) = 1 - m.$$

Since it must be an integer, the index is exactly $1 - m$.

Hence the index of a parabolic point is zero, the index of a hyperbolic point is -1, and the index of a flat point is less than -1. This completes the proof of the lemma.

We shall now prove several theorems which will be applied in the following sections to investigation of surfaces of nonnegative and zero extrinsic curvature. Some of these theorems are also of independent interest.

We shall say that no "hummocks" can be cut from a surface Φ if no plane α cuts out a domain on the surface whose boundary lies wholly within the plane α. Obviously, if no hummocks can be cut from a surface it cannot contain elliptic points, and the positive curvature of the surface must vanish.

Theorem 9. *Let Φ be a smooth surface from which no hummocks can be cut, X a point on Φ and α the tangent plane at X.*

Then there are exactly two possible relative positions of X and the domains into which the plane α divides the surface Φ: a) *there are four domains whose boundary contains X, or* b) *the plane α cuts the surface in a continuum which contains the point X.*

Proof. Without loss of generality we may assume that the surface Φ is homeomorphic to a disk and can be projected in one-to-one fashion onto the plane α, which is the xy-plane. Let β be a plane through X perpendicular to α. It cuts the surface in some curve γ. We may assume that γ is not a straight-line segment in a neighborhood of X (for if γ is a straight-line segment near X for every plane β, the existence of the continuum in the statement of the theorem is obvious).

Let g be a straight line in the plane β which cuts the curve γ in two points A and B near X, is nowhere tangent to the curve, and cuts it in finitely many points. Without loss of generality, we may assume that there are no other points of intersection between A and B and the segment AB of g lies above the surface.

We claim that there is a plane δ through the straight line g such that, among the domains into which δ divides the surface, there are three whose boundaries contain at least one of the points A and B and are "contiguous" to the boundary of the surface Φ.

Let L, M, N be three points on g outside the surface: L is near A, outside the segment AB, and there are no points of the surface on the straight-line segment LA; M is on the segment AB; N is near B, outside the segment AB, and the straight-line segment BN does not cut the surface. Thus, under our assumptions concerning the relative position of the segment AB and the surface, M lies above the surface, L and N below it. Let $\bar{L}$, $\bar{M}$ and $\bar{N}$ denote the projections of the points L, M and N on the surface by straight lines parallel to the z-axis.

Let G_L, G_M and G_N be the domains into which the surface is divided by the plane δ, containing the points $\bar{L}$, $\bar{M}$ and $\bar{N}$, respectively. Obviously, if the domains G_L and G_N do not coincide, then G_L, G_M and G_N are precisely the domains whose existence we are trying to prove.

Let $\bar{\gamma}$ denote the subarc AB of the curve γ. Since no hummocks can be cut from the surface Φ, it follows that for any plane δ the point $\bar{M}$ can be connected to the boundary of the surface by a curve γ_δ lying entirely below the plane δ, such that $\bar{M}$ is the sole common point of γ_δ and $\bar{\gamma}$. We shall say that the curve γ_δ passes on the right (left) of $\bar{\gamma}$ if it lies on the right (left) of $\bar{\gamma}$ near the point $\bar{M}$.

Assume that for some plane δ there exist two curves γ_δ' and γ_δ'', passing respectively to the right and left of $\bar{\gamma}$. Obviously each of these curves may be assumed simple. Otherwise we need only take suitable subarcs of γ_δ' and γ_δ''.

The curves γ_δ' and γ_δ'' have no common points other than $\bar{M}$. Indeed, let P be the first point of intersection, proceeding along γ_δ from $\bar{M}$. The subarcs $\bar{M}P$ of the curves γ_δ' and γ_δ'' form a closed simple curve. It bounds a domain Φ' homeomorphic to a disk on the surface. One of the points A or B obviously belongs to Φ'. And since the boundary of the domain Φ' and the points A, B lie on opposite sides of the plane δ, it follows that the plane δ cuts off a hummock, which is impossible by assumption.

Thus the curve $\gamma_\delta' + \gamma_\delta''$ is simple. It divides the surface Φ into two parts. One of them contains the point A, and together with it $\bar{L}$; the other contains B and $\overline{N}$. Together with the points $\overline{L}$ and $\overline{N}$ these domains also contain G_L and G_N, which are therefore distinct. It remains to verify that there always exists a position of the plane δ in which there are two curves γ'_δ and γ''_δ with the desired property.

Rotate the plane δ through a small angle from its vertical position. In one direction there exists only a "right" curve γ_δ, in the other, only a "left" curve. Hence there exists a position δ_0 of the plane δ in which there exist planes arbitrarily close to δ_0 with right and left curves γ_δ.

Suppose that the curve γ_{δ_0} for the plane δ_0 itself lies on the right. Let δ_1 be a position of the plane δ close to δ_0, such that the curve γ_{δ_1} lies on the left. Since the curve γ_{δ_0} is strictly below the plane δ_0, the curve γ_{δ_0} must also be below δ_1, since δ_1 is close to δ_0. Thus for the plane δ_1 there exist curves γ_δ both on the right and on the left. But then the domains G_L, G_M and G_N are distinct.

Now consider a sequence of straight lines g converging to the tangent to γ at the point X. For each straight line g we have three disjoint domains G_L, G_M and G_N. When g converges to the tangent at X, the points L, M, N converge to X, and the plane δ_g defining these domains converges to the tangent plane at X.

The domain G_M is divided by the subarc $\overline{\gamma}$ of the curve γ into two domains, G'_M and G''_M. For each of the four domains $G = G_L$, G_N, G'_M or G''_M we can make two assumptions, which we call Ω and $\overline{\Omega}$.

Assumption Ω: For each $\epsilon > 0$ the ϵ-neighborhood of X in G contains a point Y whose distance from the tangent plane to the surface at X is at least $h_\epsilon > 0$, where h_ϵ depends only on ϵ and not on the plane defining G. *Assumption* $\overline{\Omega}$ is simply the negation of Assumption Ω.

We claim that if Assumption $\overline{\Omega}$ holds, there exists a continuum on the surface, containing the point X, along which the tangent plane is stationary, i.e. it coincides with the tangent plane at X.

Describe a small disk of radius $\epsilon/2$ about X on the surface, and let G_ϵ be the component of G inside this disk, sufficiently close to the point X. The angle between the tangent plane at an arbitrary point of G_ϵ and the plane α (the tangent plane at X) converges uniformly to zero for a suitable sequence of planes δ converging to α. Indeed, suppose that V is a point in G_ϵ such that the above angle is greater than $\vartheta_0 > 0$. Then there are points in G whose distance from the plane α is of the order of $\epsilon \tan \vartheta_0$, and this contradicts Assumption $\overline{\Omega}$.

Let μ_k be a sequence of positive numbers converging to zero. We define a sequence of sets M_k on the surface Φ: a point X is in M_k if its distance from the plane α is also at most μ_k. Any set M_k contains some

G_ϵ. Therefore M_k contains a component which cuts the boundary of the $\epsilon/2$-neighborhood U_ϵ of X and is arbitrarily close to X. Hence the intersection of all the sets M_k contains a continuum, issuing from X, along which the surface is tangent to the plane α.

Now let Assumption Ω hold for each of the four domains G. We shall show that there are then four domains, among those cut out on the surface by α, whose boundary contains the point X.

Displace the plane α upward and downward, parallel to itself, for a small distance h. Denote the two new planes by α^+ and α^- respectively. For sufficiently small h the domains G_L and G_N contain points above the plane α^+ and arbitrarily close to X, while the domains G'_M and G''_M contain points below α^- and arbitrarily close to X. Let $\widetilde{G}_L$, $\widetilde{G}_N$, $\widetilde{G}'_M$ and $\widetilde{G}''_M$ denote the domains on the surface Φ containing these points and lying above the planes α^+ and α^-, respectively. Now consider our plane δ, sufficiently close to α. Then the domains $\widetilde{G}_L$, $\widetilde{G}_N$, $\widetilde{G}'_M$ and $\widetilde{G}''_M$ are contained in the domains G_L, G_N, G'_M and G''_M and do not intersect. Each of the four domains $\widetilde{G}_L, \cdots, \widetilde{G}''_M$ is contiguous to the boundary of the surface Φ (otherwise it would constitute a hummock). As $h \to 0$, the domains $\widetilde{G}_L, \cdots, \widetilde{G}''_M$ converge to four disjoint domains, and the point X is on the boundary of each. This completes the proof.

Theorem 10. *Let Φ be a smooth surface from which no hummocks can be cut, and K a continuum on Φ whose spherical image is a point. Then the tangent plane to the surface along the continuum K is stationary, i.e. the surface has the same tangent plane at all points of K.*

Proof. Without loss of generality we may assume that the surface Φ is given by an equation $z = z(x,y)$ and at points of the continuum

$$\partial z/\partial x = \partial z/\partial y = 0.$$

Suppose that the assertion is false, and the continuum contains two points A and B with different z-coordinates, say $z_A < z_B$. Cut the surface Φ by a plane $\alpha_t : z = t,\ z_A \leqq t \leqq z_B$.

This plane contains at least one point C_t of the continuum K. Let M_t denote that component of the intersection of the plane α_t and the surface which contains C_t. By Theorem 9, M_t either contains a continuum along which $\partial z/\partial x = \partial z/\partial y = 0$, or it divides the surface into at least four domains. Kronrod [**37**] has shown that the set of t-values for which at least one of these possibilities holds has measure zero. It follows that the entire continuum K lies in a plane $z = \text{const}$, which is tangent to the surface at all points of K. Q.E.D.

THEOREM 11. *The positive curvature of a surface of bounded extrinsic curvature from which hummocks can be cut is not zero; in particular, the surface contains elliptic points.*

PROOF. Let α be a plane cutting a hummock S from the surface; S is a domain on the surface whose boundary lies wholly in the plane α. The surface S has tangent planes in all directions sufficiently close to that of α. Hence the spherical image of the set of one-sided points has nonzero area. Since the spherical image of the set of nonregular points of the surface has zero area and a regular one-sided point can only be elliptic, the spherical image of the set of elliptic points of the surface has nonzero area. Hence the positive curvature of the surface is not zero. Q.E.D.

THEOREM 12. *If a surface of bounded extrinsic curvature contains an elliptic (hyperbolic) point, then the positive (negative) curvature of the surface is not zero.*

If the surface contains a parabolic point, both positive and negative curvatures are not zero.

PROOF. The first statement of the theorem follows directly from Theorem 11, since hummocks can clearly be cut from a surface with elliptic points.

Now let X be a hyperbolic point. We must show that the negative curvature of the surface is not zero. Let ω be a small neighborhood of X, homeomorphic to a disk, bounded by a simple curve γ, such that at no point of ω or γ (other than X) is the tangent plane parallel to the tangent plane at X. Let $\overline{X}$ and $\overline{\gamma}$ denote the spherical images of X and γ, and U the domain on the sphere bounded by $\overline{\gamma}$ which contains X. The preimages of almost all points of U in ω can be only elliptic, hyperbolic and parabolic, and each point has only finitely many preimages. Now, since the degree of a point $Y \in U$ is equal to the sum of indices of the preimages on the one hand, and to -1 (the degree of $\overline{X}$) on the other, it follows that some of the preimages must be hyperbolic points. Thus the area of the spherical images of the set of hyperbolic points of the surface is at least the area of U. Hence the negative curvature of the surface is not zero.

Now let X be a parabolic point. Assume that there are elliptic points arbitrarily close to X. Then the positive curvature of the surface in an arbitrarily small neighborhood of the point X is not zero. It follows that any arbitrarily small neighborhood U^* of $\overline{X}$ contains a set M of positive area, each point of which has at least one elliptic preimage, while the other preimages can be only hyperbolic and parabolic points. Since the sum of indices of the preimages of a point $Y \in U$ in ω is zero (the degree of $\overline{X}$ relative to $\overline{\gamma}$) provided the neighborhood is sufficiently small, while its set of preimages contains an elliptic point, it follows that at least one preimage is a hyperbolic point. Thus, as was proved above, the negative curvature of the surface is also not zero.

Now assume that X has a small neighborhood containing no elliptic points. Without loss of generality we may assume that ω is such a neighborhood. By Theorem 11, no hummocks can be cut from the surface ω. But then, by Theorem 9, the point X cannot be parabolic, since the tangent plane to the surface at X divides the neighborhood into two domains, and there can be no flat continuum as in the statement of Theorem 9, since X is a regular point. This completes the proof.

THEOREM 13. *The total curvature of a closed surface Φ of bounded extrinsic curvature is $\sigma = 2\pi\chi(\Phi)$, where $\chi(\Phi)$ is the Euler characteristic of the surface.*

PROOF. For smooth piecewise analytic surfaces, this theorem follows, as is well known, from the Gauss-Bonnet Theorem.

Divide the surface Φ into small elementary domains G_k bounded by curves γ_k. Let X_k be a point in G_k, $\overline{X}_k$ its spherical image, and $\overline{X}_k^*$ the point on the sphere ω antipodal to $\overline{X}_k$. Let $\overline{\gamma}_k$ denote the spherical image of γ_k and $q_{\overline{\gamma}_k,\overline{X}^*}(Y)$ the degree of a point $Y \in \omega$ on the sphere ω punctured at the point $\overline{X}^*$. Then

$$\sigma(\Phi) = \sum_k \int_\omega q_{\overline{\gamma}_k,\, \overline{X}_k^*}(Y)\, dY.$$

It is obvious that the points $\overline{X}_k^*$ in this formula can be replaced by any sufficiently close points.

Let $\tilde{\Phi}$ be a smooth piecewise analytic surface close to Φ, such that the angles between the tangent planes to Φ and $\tilde{\Phi}$ at close points are small. Divide the surface $\tilde{\Phi}$ into elementary domains G_k by piecewise regular curves $\tilde{\gamma}_k$, close to the curves γ_k, such that the partition is topologically equivalent to the partition of the surface Φ into domains G_k. Suppose that the normals at corresponding points on the surfaces F and $\tilde{\Phi}$, lying on the boundary of at least three of the domains G, are parallel. The total curvature of the surface $\tilde{\Phi}$ is

$$\sigma(\tilde{\Phi}) = \sum_k \int_\omega q_{\bar{\tilde{\gamma}}_k, \bar{X}_k^*}(Y)\, dY.$$

We shall show that

$$\sum_k \int_\omega q_{\bar{\gamma}_k, \bar{X}_k^*}(Y)\, dY = \sum_k \int_\omega q_{\bar{\tilde{\gamma}}_k, \bar{X}_k^*}(Y)\, dY.$$

The topological equivalence of the partitions of the surfaces Φ and $\tilde{\Phi}$ induces a natural correspondence between subarcs on the curves $\bar{\gamma}_k$ and $\bar{\tilde{\gamma}}_k$. Let γ be a common subarc of the curves $\bar{\gamma}_i$ and $\bar{\gamma}_j$, $\tilde{\gamma}$ the corresponding common subarc of the curves $\bar{\tilde{\gamma}}_i$ and $\bar{\tilde{\gamma}}_j$. We have

$$\int_\omega q_{\bar{\gamma}_i}(Y)\, dY + \int_\omega q_{\bar{\gamma}_j}(Y)\, dY = \int_\omega \left(q_{\bar{\gamma}_i}(Y) + q_{\bar{\gamma}_j}(Y) \right) dY$$
$$= \int_\omega q_{\bar{\gamma}_i + \bar{\gamma}_j}(Y)\, dY = \int_\omega q_{\bar{\gamma}_i'}(Y)\, dY + \int_\omega q_{\bar{\gamma}_j'}(Y)\, dY,$$

where $\bar{\gamma}_i'$ and $\bar{\gamma}_j'$ are derived from $\bar{\gamma}_i$ and $\bar{\gamma}_j$ by replacing the arc γ by $\tilde{\gamma}$. In this way we can successively replace the subarcs of the curves $\bar{\gamma}_k$ by the corresponding subarcs of $\bar{\tilde{\gamma}}_k$, so that instead of the degree of a point relative to the curves $\bar{\gamma}_k$ we can consider its degree relative to $\bar{\tilde{\gamma}}_k$. This shows that the above two integrals are equal, i.e. the total curvatures of the surfaces Φ and $\tilde{\Phi}$ are equal.

Thus, in order to prove our theorem it will suffice to construct a piecewise analytic surface with the desired properties. Let $\epsilon > 0$ be small, and consider a sufficiently dense net of points A_k on the surface. Describe a ball ω_k of radius ϵ about each point A_k. Let T be the body formed by all the balls ω_k. If the net A_k is sufficiently dense, each component S of the boundary of T is a piecewise analytic surface close to Φ, and the angles between its tangent planes to the corresponding tangent planes of Φ are small. It remains to make this surface smooth.

Without loss of generality we may assume that no four of the spheres S_k bounding the balls ω_k intersect at one point. Consider a sufficiently small ball ω rolling over the surface S outside T. The resulting envelope is a certain surface $\widetilde{S}$. The required surface $\tilde{\Phi}$ will be a part of $\widetilde{S}$. It consists of parts of the spheres S_k, parts of tori for intersections of two spheres S_k, and parts of the spheres ω for intersections of three spheres S_k. It is clear that the construction can be carried out in such a way that the last condition to be met by $\tilde{\Phi}$ is satisfied: if two corresponding points of the surfaces Φ and $\tilde{\Phi}$ lie on the boundary of at least three elementary domains G, the normals at these points are parallel. This completes the proof of Theorem 13.

§4. Surfaces of zero extrinsic curvature

A smooth surface of bounded extrinsic curvature will be called a *surface of zero curvature* if its positive and negative curvature both vanish. The total curvature of such a surface obviously vanishes on any set. The converse also holds: If the total curvature of a surface vanishes on any set, then the surface is a surface of zero curvature. Even more can be proved: If the total curvature vanishes on any open set, then the surface is a surface of zero curvature. For, if the total curvature vanishes on any open set, then, by complete additivity, it vanishes on any Borel set. Hence the positive and negative curvatures also vanish on any Borel set. In particular, the positive and negative curvatures of the entire surface are zero.

In this section we shall determine the structure of surfaces of zero curvature.

THEOREM 1. *A surface of zero curvature is developable* (*i.e. locally isometric to a plane*) *and has the usual structure of developables: straight-line generators with stationary tangent plane along each generator.*

PROOF. We first show that *the spherical image of a surface of zero curvature has zero area*. Indeed, the surface can contain neither elliptic, hyperbolic, nor parabolic points, since, as we proved in the preceding section, a surface containing such points has either positive or negative curvature different from zero, which is impossible. Thus the absolute curvature of the surface is zero, and so the area of its spherical image vanishes.

In the sequel we shall confine ourselves to a sufficiently small region on the surface, assuming that it is defined by an equation $z = z(x, y)$ over some disk K in the xy-plane.

Let E_p denote the set of points on the surface at which $\partial z/\partial x = p$. Then *for almost all p the tangent plane along each component of E_p is stationary*. Indeed, the spherical image of E_p lies on an arc of a great circle κ_p. Since the area of the spherical image of the surface is zero, the linear measure of the spherical image of E_p on κ_p is zero for almost all p. Hence for each such p the spherical image of any component of E_p is a point. By Theorem 10 of §3 this implies that the tangent plane along each component of E_p is stationary. This proves our assertion.

Suppose that the tangent plane along each component of E_p is stationary. Then *each such component has nonempty intersection with the boundary of the surface*. Indeed, suppose that the assertion is false, and let κ be a component of E_p which does not cut the boundary of the surface. Let ϵ be a positive number, so small that the ϵ-neighborhood of κ is also disjoint from the boundary of the surface. There exists a partition of the set E_p into two closed subsets E_p' and E_p'', at a positive distance from each other, the first of which contains κ and is contained in its ϵ-neighborhood.

Since the set E_p' is at a positive distance from the boundary of the surface and from the set E_p'', there exists a simple closed curve γ separating E_p' from the boundary and not cutting E_p''. The domain G bounded by this curve is homeomorphic to a disk.

The difference $\partial z/\partial x - p$ has fixed sign along the curve γ. To fix ideas, suppose that $\partial z/\partial x > p + \delta$ $(\delta > 0)$. Let p' be an arbitrary number between p and $p+\delta$. The set $E_{p'}$ separates κ from γ. Consequently $E_{p'}$ contains a component $\kappa_{p'}$ separating κ from γ. This component is contained in the domain G. Let $\tilde{\kappa}$ denote the component of the partition of G by the set $\kappa_{p'}$ which lies in κ. $\tilde{\kappa}$ is obviously a domain with boundary $\kappa_{p'}$. Since p' is arbitrary, we may assume (using what was proved above) that the tangent plane along $\kappa_{p'}$ is stationary; denote it by α. The component of the partition of the surface by the plane α which contains a point of $\tilde{\kappa}$ not in α is a hummock. But no hummocks can be cut from the surface, since its positive curvature is zero. This proves our assertion.

To simplify the notation, we let Ω denote the set of all p satisfying the following two conditions:

1) The tangent plane along each component of E_p is stationary.

2) No component of E_p contains branch points (in particular, E_p contains no interior points).

As was shown above, the first condition holds for almost all p. The second can also be shown to hold for almost all p [**37**].

Let $p \in \Omega$. The set E_p divides the surface into domains. Let G be one of these, in which, say, $\partial z/\partial x > p$. Let κ be any component of the boundary of G. Obviously κ is contained in one component of E_p (or is itself a component). We claim that κ *cuts the boundary of the surface*. To prove this we shall employ the same reasoning as in the proof that each component of E_p cuts the boundary.

Let F be the entire boundary of the domain G. If our assertion is false, there exists a simple closed curve γ separating κ from the boundary of the surface and not cutting F. The curve γ lies in G. Indeed, let A be a point in G close to κ, for which $p' = \partial z/\partial x \in \Omega$. The component κ' defined by A in the set $E_{p'}$ lies in G and cuts the boundary of the surface, as proved above. Since the point A is close to κ, and therefore lies in the interior of the domain bounded by the curve γ, it follows that

κ' cuts γ. Thus the curve contains a point of G. But then the entire curve must pass within G, since it does not cut the boundary. We now proceed as before and prove that a hummock can be cut from the surface. Since this is impossible, our assertion is proved.

Let us continue our investigation of the component κ of the boundary of G. κ lies in some plane α which is tangent to the surface at each point of κ. We now apply to the surface a "flattening" transformation, as follows.

Let A be a point on the surface which is in neither G nor E_p. A must be separated by some component $\tilde{\kappa} \subset E_p$ from G. Let $\widetilde{G}$ be the component containing A in the partition of the surface by the set $\tilde{\kappa}$. Replace the domain $\widetilde{G}$ by a piece of the plane tangent to the surface along $\tilde{\kappa}$.

The new surface thus obtained is smooth, and its spherical image is contained in that of the original surface. By a finite or countable number of flattening transformations, we obtain a smooth surface of zero curvature such that $\partial z/\partial x \geqq p$, with strict inequality only in the domain G, which, together with its boundary, remains as it was on the original surface. The structure of the component κ may thus be studied on the transformed surface.

By contrast with the previous arguments, we shall assume here that κ is a component of the boundary of G on the open surface (without its boundary curve). Let $\tilde{\Phi}$ denote the domain on the plane α (the tangent plane along the component κ) into which the surface is projected by straight lines parallel to the z-axis, and $\widetilde{G}$ the domain corresponding to G. The boundary of $\widetilde{G}$ clearly coincides with that of G, and so κ is a component of the boundary of $\widetilde{G}$.

Let κ^* denote the component of the set E_p, on the flattened surface, containing the component κ of the boundary of G. *The component κ^* lies in the plane α.* Indeed, since the spherical image of E_p has linear measure zero, the spherical image of κ^* is a point. Hence κ^* lies in the plane tangent to the surface at all points of the set κ^*. But κ is contained in κ^*, and so this plane is α.

The set κ lies on the boundary of κ^*. For if A is a point of κ, there are points of G in any neighborhood of A, so that it is indeed on the boundary of κ^*.

Let g be an arbitrary straight line in the plane α, parallel to the xz-plane and cutting κ^*. The intersection of g and κ^* is either a point or a segment. This follows from the fact that on the flattened surface we have $\partial z/\partial x \geqq p$, and p is a direction ratio of the plane α. Thus *the component κ^* is a union of straight-line segments and points lying on a continuous family of straight lines in α, parallel to the xz-plane.*

The intersection of the closure of the set κ^* with the boundary of the surface is connected. Indeed, otherwise κ^* divides the surface into at least two domains, each containing points of G. But this is impossible since G is connected.

Since the intersection of the closure of κ^* and the boundary of the surface is connected, it follows that the boundary of κ^* is connected; hence *the entire boundary of κ^* reduces to κ.*

Let W_{xz} denote the family of straight lines in the plane α parallel to the xz-plane, and W_{xz}^* the subset of W_{xz} consisting of the straight lines cutting κ^*. Let g be any line in W_{xz}^*, A_g^- its first and A_g^+ its last point of intersection with κ^*, counting from $x<0$ to $x>0$ along g. It may happen that one or even both of these points do not exist. This is the case when one or both endpoints of the segment in which g cuts the set κ^* belong to the boundary of the surface. The points A_g^- and A_g^+ define rays g^- and g^+ on the straight line g, directed toward $x<0$ and $x>0$, respectively. Denote the sets of rays g^- and g^+ by W_{xz}^- and W_{xz}^+, respectively.

Each component w of the set W_{xz}^- (or W_{xz}^+) is an open subset of W_{xz}^- (or W_{xz}^+), and hence is homeomorphic to a segment if W_{xz}^* does not reduce to a single straight line. Let g_1 and g_2 be the straight lines bounding w, and g any line in w lying between g_1 and g_2. We claim that *the set of points A_g^- (or A_g^+) is a convex curve, convex toward $x>0$ (or $x<0$).* We shall prove this assertion for a component w of W_{xz}^-.

Suppose that the assertion is false. Then there exist three straight lines g', g'', g''' satisfying the following conditions:

1) g'' lies between g' and g'''.

2) There exists a straight line t in the plane α which separates the point $A_{g''}^-$ from the points $A_{g'}^-$ and $A_{g'''}^-$, and moreover $A_{g''}^-$ lies on the side of t toward $x<0$.

The lines g_1 and g_2 bound a strip S in the plane α. Let S^- denote the set of points in the strip lying on rays g^-. Let Φ_S be the part of the surface projected by straight lines parallel to the z-axis onto the strip S. Transform the surface Φ_S, leaving unchanged the region corresponding (in the above projection) to S^-, and replacing the remainder by the set $S-S^-$. Denote the transformed surface by $\overline{\Phi}_S$. The surface $\overline{\Phi}_S$ is smooth and has zero curvature; hence no hummocks can be cut from it.

We now make an essential observation concerning the surface $\overline{\Phi}_S$. For sufficiently large x, we have $\partial z/\partial x = p$ (for the surface $\overline{\Phi}_S$ "merges" with the plane α for large positive x), and moreover $\partial z/\partial x \geqq p$ throughout the surface; hence the surface $\overline{\Phi}_S$ lies below the plane α. Furthermore, let D be the domain on $\overline{\Phi}_S$ whose projection on the strip S is bounded by the straight line t and the rays g'^{-} and g'''^{-}. Except for a certain portion on the straight line t, the boundary of this domain lies strictly below the plane α.

Let β be a plane through the straight line t, close to α, such that the point $A_{g''}^{-}$ is above β. Let R_β be the component containing $A_{g''}^{-}$ in the partition of the surface $\overline{\Phi}_S$ by the plane β. The point $A_{g''}^{-}$ lies in the domain D. And since no hummocks can be cut from the surface, R_β must cut the boundary of D. But this is impossible if β is sufficiently close to α. Indeed, R_β cannot cut the part of the boundary of D corresponding to t, since R_β lies above one of the halfplanes of β defined by t. R_β cannot cut the remainder of the boundary, for it lies strictly below the plane α and therefore also below β, provided the latter is close to α. This contradiction proves our assertion.

We now show that the number of components of each set W_{xz}^{-} and W_{xz}^{+} cannot be "too large." In fact, there are at most three components. Indeed, if the number of components is greater than three, there exist two components w and w' bounded by straight lines which are not boundary lines of W_{xz}. The components w and w' define domains T and T', respectively, on the surface, which are projected by straight lines parallel to the z-axis onto w and w'. The boundary of each of the domains T and T' is in κ. But each of these domains contains points of G, and hence κ partitions G, which is impossible.

Since the number of components of the sets W_{xz}^{-} and W_{xz}^{+} is at most three, it follows that κ is the union of a finite number of convex curves c which can be projected in one-to-one fashion onto the y-axis, and straight-line segments parallel to the xz-plane. Since κ contains no branch points and divides the surface into two domains, we conclude that *κ is a simple curve with endpoints on the boundary of the surface.*

We now prove that *any convex curve c defined as above by a component w of the set W_{xz}^{-} (or W_{xz}^{+}) is a straight-line segment.* If this is not true, the curve c contains a point M and two points K, L close to and on either side of M, such that the ray g_M issuing from M in the plane α, parallel to the xz-plane and pointing toward $x<0$, is cut by the segment KL. Let $\tilde{\gamma}$ be the curve in which the plane through M parallel to the xz-plane cuts the original surface.

Take the point $\widetilde{M}$ on the curve $\tilde{\gamma}$, on the side $x<0$, such that $\partial z/\partial x \neq p$. Such a point exists in an arbitrarily small neighborhood of M, since otherwise κ would contain a straight-line segment parallel to the xz-plane with endpoint at M, and the latter would be a branch point for κ, which is impossible. To fix ideas, suppose that $\partial z/\partial x = p_M > p$ at $\widetilde{M}$.

Let p_1^* and p_2^* be two numbers in Ω, close to p, such that $p_M > p_1^* > p > p_2^*$. Let G_1 and G_2 be the open sets on the surface in which $\partial z/\partial x < p_1^*$ and $\partial z/\partial x > p_2^*$, respectively. κ is obviously in the intersection of these sets. Let G_1^* and G_2^* denote the components of the sets G_1 and G_2 containing the curve κ. Since $G_1^* \cap G_2^*$ separates the point $\widetilde{M}$ from κ, there exists a component κ^* of the boundary of $G_1^* \cap G_2^*$ which separates $\widetilde{M}$ from κ. κ^* is obviously a component of the boundary of one of the sets G_1^* or G_2^*.

If p_1^* and p_2^* are sufficiently close to p, the component κ^* lies in an arbitrarily small neighborhood of κ. As we have proved, κ^* is a simple curve in some plane α^* tangent to the surface at all points of κ^*.

Let G^* denote the domain G_1^* or G_2^* whose boundary contains κ^*. Flatten the surface outside the domain G^*, as we did previously outside G. Let $\bar{c}$ be the projection of the segment KL of the curve c on the plane α^*, in the direction of the z-axis. Let g be a ray issuing from any point on $\bar{c}$ in α^*, parallel to the xz-plane and pointing toward $x < 0$. If κ^* is sufficiently close to κ, this ray must cut κ^* at some point A_g. Otherwise we could obviously connect the point of c to $\widetilde{M}$ without cutting κ^*.

As we have proved, the set of points A_g is a convex curve c^*, convex toward $x < 0$. When p_1^* and p_2^* tend to p, this curve tends to c. But the curve c is strictly convex toward $x > 0$ and hence cannot be the limit of c^*. This contradiction proves our assertion. Thus every arc c of the curve κ is a straight-line segment, and the curve itself is a simple polygonal line with endpoints on the boundary of the surface.

Finally, we prove that κ is simply a straight-line segment. Suppose that κ is a polygon and let Q be one of its vertices. Let δ be a vertical plane through Q, which separates the sides of the polygon incident at Q, such that the curve γ in which δ cuts the surface does not contain straight-line segments with endpoints at Q. Such a plane clearly exists. Otherwise one easily shows that there is a straight-line segment on the surface containing Q, along which the tangent plane is stationary, and which does not coincide with the sides of the polygon κ incident at Q. This segment must be in κ, but κ contains no branch points.

Introduce a new coordinate system, retaining the original xy-plane but taking δ as the new xz-plane. Now, in order to prove that κ has no "corner" at Q, it suffices to repeat word for word the previous proof that the convex arcs c of the curve κ are straight-line segments. We thus conclude that κ is a straight-line segment with endpoints on the boundary of the surface.

Let Q be an arbitrary point of the surface. Then *there are exactly two possibilities*: a) *there is a straight-line segment through* Q, *with endpoints on the boundary of the surface, along which the tangent plane is stationary*, or b) *the point* Q *has a plane neighborhood.*

Suppose that Q has no plane neighborhood. Then for any $\epsilon > 0$ there always exists a vertical plane δ through Q whose intersection with the surface in the ϵ-neighborhood of Q is not a straight-line segment. Otherwise the ϵ-neighborhood of Q on the surface would be part of a plane. Take the plane δ as the xz-plane. Let $\partial z/\partial x = p_Q$ at the point Q. Then there is a point A, whose distance from Q is at most ϵ, at which $\partial z/\partial x = p_A \neq p_Q$.

Consider any p of Ω in the interval (p_A, p_Q). The set E_p separates A from Q. Let G_A be the domain defined on the surface by E_p which contains the point A. The point Q is separated from A by some component κ of the boundary of G_A. We have proved that κ is a straight-line segment with endpoints on the boundary of the surface and stationary tangent plane. Since it separates A from Q, the distance from κ to Q cannot exceed ϵ. But ϵ is arbitrary, and hence it follows that there is a straight-line segment through Q with endpoints on the boundary of the surface and stationary tangent plane.

We now claim that *if the point* Q *has no plane neighborhood, the straight-line segment through* Q *is unique.*

The fact that the above straight-line segment is unique will follow from a more general assertion, which will also be needed: Let P and Q be two points on the surface, at least one of which, say Q, has no plane neighborhood; let g_P and g_Q be straight-line segments through P and Q on which the tangent plane is stationary. Then, if the segments g_P and g_Q have a common point, they must coincide.

Suppose that g_P and g_Q are distinct. Since Q has no plane neighborhood, there exists a point A arbitrarily close to Q which also has no plane neighborhood and is not in the tangent plane at Q. By what we have proved, there is a straight-line segment g_A through A. If A is sufficiently close to Q, the segment g_A cuts either g_P or g_Q, and so lies in the tangent plane of the straight lines g_P and g_Q; this is impossible by the choice of A.

We can now prove that *a surface of zero curvature is locally isometric to a plane; i.e. each point has a neighborhood isometric to a plane domain.*

Let O be an arbitrary point on the surface. If O has a plane neighborhood, we are done. Assume that O has no plane neighborhood. Then there is a unique straight-line generator g_O through O. Without loss of generality we may assume that O is the origin, the tangent plane to the surface at O is the xy-plane, and the straight-line generator lies on the x-axis.

Let γ be the curve in which the plane $x=0$ cuts the surface. F will denote the set of points on the curve γ close to O ($|y| \leq \epsilon$) each of which has no plane neighborhood. There is a unique straight-line generator g_Q through each point Q of F.

Since the generator g_Q is unique, it must depend continuously on Q in the set F. Since different straight lines g_Q are disjoint, the direction of g_Q at points Q close to O is close to that of g_O.

The complement G of the set F on γ is an open set. It is a union of *straight-line* segments. Let δ be one of them. Its endpoints Q_1 and Q_2 are in F. We claim that *the region of the surface between the straight lines* g_{Q_1} *and* g_{Q_2} *is a plane domain,* provided that all arguments refer to a sufficiently small neighborhood of the curve γ.

Indeed, the tangent plane is stationary along the segment δ and the generators g_{Q_1} and g_{Q_2}. Hence δ, g_{Q_1} and g_{Q_2} lie in one plane, α say. If our assertion is false, there exists a point A between g_{Q_1} and g_{Q_2} which has no plane neighborhood and is not on the plane α. Let g_A be the straight-line generator through A. It cuts either δ, g_{Q_1} and g_{Q_2}, and so must lie in the plane α, which is impossible by our choice of the point A. Thus the surface between the generators g_{Q_1} and g_{Q_2} is indeed a plane domain.

The straight-line generators are as yet defined only for points Q of the set F. We can now define straight-line generators g_Q through points Q of G, such that

1) g_Q depends continuously on Q, and
2) straight-line generators through different points Q are disjoint.

Let γ_1 and γ_2 be the curves in which the planes $x=\epsilon_1$ and $x=-\epsilon_1$ cut the surface (where ϵ_1 is small). The straight-line generators through the endpoints of the curve γ, together with the curves γ_1 and γ_2, define a neighborhood D of the point O on the surface. We claim that this neighborhood is isometric to a plane domain.

Let $A_1, \cdots, A_n$ be a sufficiently dense sequence of points on the curve γ. Let B_i and C_i be the points at which the generators g_{A_i} cut the curves γ_1 and γ_2. Consider the polyhedron P formed by the triangles $B_1C_1C_2$, $B_1B_2C_2$, $B_2C_2C_3, \cdots$. As the density of the points A_i on γ increases, the polyhedron P converges to the domain D of the surface, while its faces converge to the tangent planes of the surface. It follows that the intrinsic metric of the polyhedron converges to the metric of the surface.

But the polyhedron P can obviously be developed on a plane. Thus the domain D is isometric to a plane domain. Our assertion is proved, and this completes the proof of Theorem 1.

THEOREM 2. *A complete surface with zero curvature is a cylinder.*

PROOF. It will clearly suffice to prove that any two straight-line generators of the surface, through points which have no plane neighborhoods, are parallel.

Let g_1 and g_2 be two such generators. Since the surface is complete, g_1 and g_2 can be extended indefinitely, i.e. they are complete straight lines. They cannot intersect, since they pass through points having no plane neighborhoods. Thus it will suffice to show that g_1 and g_2 cannot be skew lines.

Join any two points on g_1 and g_2 by a simple curve γ on the surface. Now gradually develop the surface onto a plane. In the process the plane will be covered exactly once. The straight lines g_1 and g_2, as geodesics on the surface, go into certain straight lines $\bar{g}_1$ and $\bar{g}_2$ on the plane. Since the straight lines g_1 and g_2 do not intersect on the surface, the same holds for $\bar{g}_1$ and $\bar{g}_2$, which are therefore parallel.

Consider two points $\bar{P}_1$ and $\bar{P}_2$ on $\bar{g}_1$ and $\bar{g}_2$, and suppose that they move away to infinity from some initial position, in such a way that the distance between them remains bounded. This is possible because the lines $\bar{g}_1$ and $\bar{g}_2$ are parallel. The corresponding points P_1 and P_2 on the generators g_1 and g_2 will also go to infinity. The distance between P_1 and P_2 in space, a fortiori on the surface, tends to infinity, since by assumption the straight lines g_1 and g_2 are skew. This contradiction completes the proof of the theorem.

§5. Surfaces of nonnegative extrinsic curvature

A surface Φ of bounded extrinsic curvature is called a surface of *nonnegative curvature* if its negative curvature is zero.

Every regular point on a surface of nonnegative curvature is elliptic, since, as we established in §1, a surface on which there is at least one regular point with index less than unity must have negative curvature zero.

A necessary condition for a surface to be a surface of nonnegative curvature is that the total curvature be nonnegative on any set of points on the surface; a sufficient condition is that the total curvature be nonnegative on all open sets.

In this section we intend to determine the extrinsic structure of surfaces of nonnegative curvature. We shall assume in so doing that the positive curvature is not zero—the case in which the positive curvature vanishes was considered in the preceding section.

LEMMA 1. *Let X be an elliptic point on a surface of nonnegative curvature, and $\overline{X}$ its spherical image. Then the points X and $\overline{X}$ have neighborhoods G and $\overline{G}$, respectively, such that almost all points of $\overline{G}$ have exactly one preimage in G.*

PROOF. Let G be a neighborhood of X, homeomorphic to a disk, bounded by a simple curve γ, and containing no points with normals parallel to the normal at X (other than X itself). Let $\overline{G}$ be the component containing $\overline{X}$ in the partition of the unit sphere by the spherical image $\overline{\gamma}$ of γ. We claim that G and $\overline{G}$ are the required neighborhoods.

Indeed, almost all points of $\overline{G}$ have regular preimages in G, i.e. elliptic points; moreover, almost all points have only finitely many preimages in G. Let $\overline{Y}$ be one of these points. The degree of $\overline{Y}$ relative to $\overline{\gamma}$ is equal to that of $\overline{X}$, which is $+1$. And since the degree of $\overline{Y}$ is equal to the sum of indices of its preimages, each of which is $+1$, there must be only one preimage. Q.E.D.

LEMMA 2. *A surface of nonnegative curvature is convex in the neighborhood of each elliptic point, i.e. every elliptic point has a neighborhood which is a convex surface.*

PROOF. Let X be an elliptic point of the surface and α the tangent plane at X. X has a neighborhood G on the surface which lies on one side of the plane α, and X is the only point of the neighborhood in α.

Let β be a plane close to α which cuts G. Since there are no points close to X with normals parallel to the normal at X, it follows that β cuts G in a simple closed curve γ^* bounding a domain G^* homeomorphic to a disk, which contains X and lies between the planes α and β. Construct the convex hull of the domain G^*, and let $\overline{G}^*$ be the part of the convex hull lying outside the plane β. We shall show that all elliptic points of G sufficiently close to X are in $\overline{G}^*$.

Indeed, let Y be an elliptic point close to X, not in $\overline{G}^*$. It has a small neighborhood U which is also not contained in $\overline{G}^*$. The spherical image of U has positive area. If Y is sufficiently close to X, the spherical image of U is contained in the spherical image of the intersection of G^* and $\overline{G}^*$. Thus, near the spherical image $\overline{X}$ of X there exists a set of positive area each point of which has at least two preimages. This contradicts Lemma 1.

Thus all elliptic points near X are in $\overline{G}^*$. It follows that the plane β can be assumed to be so close to α that all elliptic points of G^* are in $\overline{G}^*$. We claim that if this is so, then $G^* = \overline{G}^*$.

The part of G^* not in $\overline{G}^*$ consists of developable surfaces (it contains no regular points). Let g be any straight-line generator of this surface. One of the endpoints of g is in $\overline{G}^*$. But g is tangent to $\overline{G}^*$ at this point, and so g is in $\overline{G}^*$ by virtue of the convexity of $\overline{G}^*$. This contradiction proves the lemma.

THEOREM 1. *Let Φ be a surface of nonnegative curvature and α a plane cutting the surface. Then each of the domains into which the plane divides Φ, with boundary lying in α, is a convex surface, i.e. a domain on the boundary of a convex body.*

PROOF. Let Φ' be one of the domains defined by the plane α on the surface. We must show that Φ' is a convex surface. Without loss of generality we may assume that the number of points of Φ at which the tangent plane is parallel to α is finite, and moreover all these points are elliptic. Indeed, there exist planes α' arbitrarily close to α which have this property. If we succeed in proving the theorem for α', a suitable limit procedure then establishes the assertion for α.

We claim that *the intersection of the surface Φ' with any plane β parallel to α is a simple closed curve*. Let β_0 be a plane parallel to α, tangent to the surface Φ', such that the entire surface Φ' lies between α and β_0. In the general case this plane may touch the surface Φ' at several points (elliptic points, since nonregular points are excluded by assumption). Denote these points by $A_1, \cdots, A_r$.

Displace the plane β parallel to itself, from the position β_0 to the position α. Since the surface Φ' is convex in a neighborhood of each point A_i, it follows that the intersection of any plane β close to β_0 with Φ' (denote this intersection by χ_β) consists of r simple curves $\kappa_1, \cdots, \kappa_r$. This structure of the intersection of β and Φ' obtains until β becomes tangent to Φ' at a new point (or points).

How does χ_β change when this happens? None of its components can contract to a point, since otherwise the contracting component would trace out a closed surface isolated from the rest of the surface Φ'. Neither can two components κ_i, κ_j come into contact without both contracting to a point, since then the point of contact of κ_i and κ_j would be a point of tangency for the surface Φ' and the plane β, and this point, being elliptic, is an isolated point of χ_β. The only remaining possibility is that at the time in question χ_β is enlarged by a finite number of new points. Thus, as the plane β passes through a point of tangency with the surface Φ', the number of components increases.

When the plane β finally coincides with α, each of its components κ_s traces out a surface Φ_s. No two of these surfaces can intersect in points of the surface Φ'. But the surface Φ' is connected, and so there can be only one surface Φ_s. Hence for every plane β between α and β_0 the set χ_β consists of a single component: a simple closed curve. We now show that *the curve* χ_β *is which the plane* β *cuts the surface* Φ' *is convex*. Suppose that χ_β is not convex. Then there is a straight line g in the plane β which divides χ_β into at least three parts. Let χ_β' and χ_β'' be the two parts of the curve on the same side of g.

Let Φ_β denote the region cut from the surface Φ' by the plane β and lying on the same side as the plane β_0. Let δ be a halfplane through g, in the halfspace defined by β which contains the surface Φ_β. If the halfspace δ is strongly inclined toward the curves χ_β' and χ_β'', it necessarily divides the surface Φ_β into two components Φ_δ' and Φ_δ'', each containing one of the curves χ_β' and χ_β''.

Now rotate δ continuously about the axis g, enlarging Φ_δ' and Φ_δ''. Obviously at some stage of this transformation Φ_δ' and Φ_δ'' will merge to form a single component. Let δ^* be the corresponding position of the halfplane δ. Construct a halfplane $\bar{\delta}$ satisfying the following conditions:

1) The halfplane $\bar{\delta}$ is close to δ^*, and its boundary line lies in the plane β.

2) The components of the partition of the surface Φ_β containing the curves χ_β' and χ_β'' are distinct.

3) There exist only finitely many points on the surface Φ_β at which the tangent planes are parallel to $\bar{\delta}$, and all these points are elliptic.

Now displace $\bar{\delta}$ parallel to itself, enlarging the components $\Phi_{\bar{\delta}}'$ and $\Phi_{\bar{\delta}}''$; since $\bar{\delta}$ is close to δ^*, these components rapidly merge. The beginning of this merging process cannot occur in the plane β, by virtue of the choice of the straight line g. At the same time, it cannot occur at an interior point of the surface, for the reason set forth in previous arguments. Hence we have a contradiction, and our assertion is proved: the curve χ_β is convex. Consequently the boundary of Φ' in the plane α is a convex curve.

Let $\bar{\Phi}'$ be the union of the surface Φ' with the convex domain on the plane α bounded by the boundary of Φ'. We claim that *the closed surface* $\bar{\Phi}'$ *is convex*. To prove this it will suffice to show that almost all planes cutting $\bar{\Phi}'$ intersect it in closed convex curves.

Let π be a plane with the following property: The surface Φ' has only finitely many tangent planes parallel to π and the corresponding points of tangency are elliptic. Obviously almost all planes have this property. We claim that the plane π cuts the surface $\overline{\Phi}'$ in a closed convex curve.

The proof of this assertion proceeds in two steps. In the first, one proves that the intersection of the plane π and the surface $\overline{\Phi}'$ consists of a single closed curve c_π. This part of the proof proceeds exactly as in the proof that the intersection χ_β of the surface Φ' with a plane β parallel to α consists of a single closed curve. In the second step one shows that the curve c_π is convex. This is a word-for-word repetition of the proof that the curve χ_β is convex. This completes the proof.

THEOREM 2. *A complete surface Φ of nonnegative and nonzero curvature is either a closed convex surface or an unbounded convex surface.*

PROOF. Since the positive curvature of the surface Φ is not zero, its spherical image must contain a point which has finitely many preimages, all elliptic points. Let A be one of these preimages. Let α be the tangent plane at A. We displace the plane α parallel to itself and study the behavior of the component Φ_α of the resulting partition of the surface Φ which contains A. By Theorem 1, Φ_α is always a convex surface. Its boundary γ_α cannot contain points with tangent planes parallel to α, unless γ_α contracts to a point.

If γ_α contracts to a point, then the closed convex surface that it describes during the displacement of α must be Φ itself. Otherwise we would have a contradiction to the connectedness of Φ.

If the curve γ_α does not contract to a point, it describes an unbounded convex surface which, as before, must constitute the entire surface Φ. This completes the proof.

§6. The threaded surface

In §7 we shall prove that an arbitrary surface of bounded extrinsic curvature can be approximated by regular surfaces whose extrinsic curvatures converge to that of the approximated surface. In this section we consider a special ("threaded") surface which, after suitable smoothing, will yield the required approximation.

In a strip $a \leqq x \leqq b$ on the xy-plane, consider two rectifiable curves γ and $\bar{\gamma}$, which can be projected in one-to-one fashion onto the x-axis and are therefore defined by equations $y = \varphi(x)$ and $y = \bar{\varphi}(x)$, $a \leqq x \leqq b$.

Suppose that the curve γ lies above $\bar{\gamma}$, i.e. that $\bar{\varphi}(x) \leqq \varphi(x)$ for all x in the above interval. Finally, let A and B be two points on the straight lines $x = a$, $x = b$ between the curves γ and $\bar{\gamma}$. Let G denote the closed domain in the xy-plane bounded by the curves γ, $\bar{\gamma}$ and the straight lines $x = a$, $x = b$.

LEMMA 1. *There exists a shortest curve joining the points A and B in the closed domain G. This curve is unique; it can be projected in one-to-one fashion onto the x-axis, and so can be defined by an equation $y = \tilde{\varphi}(x)$, $a \leqq x \leqq b$.*

PROOF. The existence of a shortest join follows from the compactness of the set of curves of bounded length joining two given points in a closed domain, and from the semicontinuity of the lengths of curves (if γ_n are curves converging to a curve γ_0 and their lengths converge to a number l, then the length of γ_0 is at most l).

Let $\tilde{\gamma}$ denote a shortest join. We now show that $\tilde{\gamma}$ can be projected in one-to-one fashion onto the x-axis. First, the curve $\tilde{\gamma}$ does not intersect itself, since otherwise its length could be reduced.

Let g be an arbitrary straight line parallel to the y-axis between the lines $x = a$ and $x = b$. Let P_1 denote the first, P_2 the last point of the curve $\tilde{\gamma}$ on the line g, counting along the curve from A to B. Our goal will be achieved if we can show that $P_1 \equiv P_2$ for any straight line g.

Suppose that P_1 and P_2 are distinct for some g. It is then obvious that the part of $\tilde{\gamma}$ lying between these points must be a straight-line segment joining them.

Let Q_1 be the midpoint of the segment P_1P_2. Let P be a point on the curve $\tilde{\gamma}$, close to P_1 but not on P_1P_2. Let Q be a point completing the three points P, P_1, Q_1 to a parallelogram. Then, if P is sufficiently close to P_1, the straight-line segments PQ and Q_1Q are in G. Hence the arc PP_1 of the curve $\tilde{\gamma}$ must be a straight-line segment, for otherwise the arc PQ_1 of the curve could be shortened, replacing it by the polygonal line PQQ_1. But then the straight-line segment PQ_1 lies in G, and we can thus shorten the curve $\tilde{\gamma}$ by replacing the arc PQ_1 by the corresponding straight-line segment. This contradiction proves our assertion. Thus the curve can be projected in one-to-one fashion onto the x-axis.

We now show that the curve $\tilde{\gamma}$ is unique. Suppose that there are two shortest joins, $\tilde{\gamma}_1$ and $\tilde{\gamma}_2$. Both curves of course have the same length l. Parametrize both curves $\tilde{\gamma}_1$ and $\tilde{\gamma}_2$ with respect to the sum of their arc lengths, measured from the point A. The curves are then defined by equations

$$\tilde{\gamma}_1: \quad x = x_1(s), \quad y = y_1(s) \qquad (0 \leqslant s \leqslant 2l),$$

$$\tilde{\gamma}_2: \quad x = x_2(s), \quad y = y_2(s) \qquad (0 \leqslant s \leqslant 2l),$$

where the functions $x_1(s)$, $x_2(s)$, $y_1(s)$, $y_2(s)$ satisfy a Lipschitz condition, and by our choice of parameter $x_1(s) = x_2(s)$.

Consider the curve $\tilde{\gamma}$ defined by the equations

$$x = \tfrac{1}{2}(x_1(s) + x_2(s)), \qquad y = \tfrac{1}{2}(y_1(s) + y_2(s)).$$

This curve is obviously a rectifiable curve in the domain G, since the functions $x_1(s) + x_2(s)$ and $y_1(s) + y_2(s)$ satisfy a Lipschitz condition and are therefore of bounded variation. We claim that its length $\bar{l}$ is at most l, the length of $\tilde{\gamma}_1$ and $\tilde{\gamma}_2$. We have

$$\int_0^{2l} \sqrt{x_1'^2 + y_1'^2}\, ds = l, \quad \int_0^{2l} \sqrt{x_2'^2 + y_2'^2}\, ds = l,$$

$$\int_0^{2l} \left\{ \left(\frac{x_1' + x_2'}{2}\right)^2 + \left(\frac{y_1' + y_2'}{2}\right)^2 \right\}^{1/2} ds = \bar{l},$$

$$\int_a^b \left\{ \frac{\sqrt{x_1'^2 + y_1'^2} + \sqrt{x_2'^2 + y_2'^2}}{2} - \left[\left(\frac{x_1' + x_2'}{2}\right)^2 + \left(\frac{y_1' + y_2'}{2}\right)^2 \right]^{1/2} \right\} ds = l - \bar{l}.$$

Since the function $\sqrt{u^2 + v^2}$ is convex, the integrand in the last formula is nonnegative. Therefore $l \geqq \bar{l}$. But since $\tilde{\gamma}_1$ and $\tilde{\gamma}_2$ are shortest joins, we must have $\bar{l} \geqq l$. Hence $l = \bar{l}$. Thus we have the integral of a nonnegative function equal to zero, and this implies that the integrand must vanish almost everywhere. Hence $x_1' = x_2'$ and $y_1' = y_2'$.

Since the functions $x_1(s)$, $x_2(s)$, $y_1(s)$, $y_2(s)$ satisfy a Lipschitz condition and the corresponding functions are equal at $s = 0$, the fact that their derivatives are equal implies that the functions themselves are equal. Thus the curves $\tilde{\gamma}_1$ and $\tilde{\gamma}_2$ coincide, and we have a contradiction. This completes the proof of Lemma 1.

Lemma 2. *The shortest join $\tilde{\gamma}$ of Lemma* 1 *depends continuously on the curves γ and $\bar{\gamma}$ bounding the domain G and on the positions of the points A and B on the straight lines $x = a$ and $x = b$.*

Proof. Essentially we have to prove the following statement. Given sequences of curves γ_n, $\bar{\gamma}_n$ converging to curves γ, $\bar{\gamma}$, respectively, and a sequence of pairs of points A_n, B_n converging to A, B, the corresponding shortest curves converge to $\tilde{\gamma}$.

If the assertion is false, there exists a subsequence of curves $\tilde{\gamma}_n$ not converging to $\tilde{\gamma}$. Without loss of generality we may assume that it converges to some curve $\tilde{\gamma}_0$. The curve $\tilde{\gamma}_0$ is clearly a shortest join for the points A and B in G, and this contradicts the uniqueness of $\tilde{\gamma}$ (Lemma 1). This proves the lemma.

LEMMA 3. *Assume that each of the curves γ and $\bar{\gamma}$ consists of a finite number of arcs of algebraic curves, and the angle facing G between the semitangents at each point of contact of adjacent arcs is at most π. Then the curve $\tilde{\gamma}$ is smooth. That part of $\tilde{\gamma}$ lying strictly between γ and $\bar{\gamma}$ consists of a finite number of straight-line segments, and the remainder consists of convex arcs of algebraic curves, convex toward G.*

PROOF. We first prove that if the angle facing G between the semitangents at a point P separating two adjacent arcs of γ (or $\bar{\gamma}$) is smaller than π, then P cannot be on $\tilde{\gamma}$. Note that by assumption P cannot be common to γ and $\bar{\gamma}$.

Let g be a ray issuing from P into the domain G. Let M and N be two points on γ, close to and on different sides of P. If M and N are sufficiently close to P, the segment MN will cut the ray g at some point Q, an interior point of G. Without loss of generality we may assume that M and N are the only points of γ on the segment MN. In order to reach P, the curve $\tilde{\gamma}$ must cut the segment MN. Let M_1 and N_1 be the first and last points of intersection of $\tilde{\gamma}$ and MN. If we replace the arc M_1N_1 of the curve $\tilde{\gamma}$ by the straight-line segment M_1N_1, we get a new curve lying wholly within G but of smaller length, which is absurd.

We now show that the portion of the curve $\tilde{\gamma}$ in the interior of G consists of finitely many straight-line segments.

Let h be a connected component of the set of points on $\tilde{\gamma}$ which are interior points of G. Let P be an arbitrary point of h. Let M and N be points on different sides of P. Since P is an interior point of G, the straight-line segment MN will lie in G, provided M and N are sufficiently close to P. Hence this segment must be in h. And since this is the case for any point P, it follows that h is a straight-line segment, and its endpoints must be on the boundary of G, i.e. on one of the curves γ, $\bar{\gamma}$ or the points A, B.

Suppose that an endpoint P of the segment h is a point of the curve γ (or $\bar{\gamma}$). As we have proved, P is a smooth point of γ. We claim that the segment h lies on the tangent to γ at P. If this is false, obviously P cannot be a point of $\bar{\gamma}$. Let g_P be a vertical ray issuing from P into G, and g a vertical ray through the other endpoint of h; let Q be the intersection of g and γ. Let M be a point on h, not one of the endpoints. Let N be the first point of intersection of the straight-line MQ with the curve γ, counting from M to Q.

Replace the subarc PN of the curve γ by the polygonal line PMN. Then the new domain G, bounded by the new curve γ and the previous curve $\bar{\gamma}$, is a subdomain of the old domain G. Thus the curve $\tilde{\gamma}$ is still a shortest join, and moreover in the new domain G the curve $\tilde{\gamma}$ cannot pass through P, as was shown above. Thus the segment h is tangent to γ at P.

Since a finite number of algebraic curves can have only finitely many common tangents at different points, the number of straight-line segments h of the curve $\tilde{\gamma}$ is finite. Thus the remainder of $\tilde{\gamma}$ is the union of a finite number of arcs of algebraic curves lying on γ and $\bar{\gamma}$. Let c be one of these arcs.

The arc c can be divided into a finite number of convex curves c_k, each satisfying one of the following conditions:

1) c_k is a subarc of γ, but not of $\bar{\gamma}$;
2) c_k is a subarc of $\bar{\gamma}$, but not of γ;
3) c_k is a subarc of both γ and $\bar{\gamma}$.

We now show that in each case the arc c_k is convex toward G (regarding a straight-line segment as convex in both senses). Indeed, in the first and second cases, if c_k is convex toward the complement of G, then the length of the curve $\tilde{\gamma}$ can be reduced in an obvious way. In the third case we need only regard c_k as a subarc of the corresponding curve γ or $\bar{\gamma}$ to see that it is convex toward G. This completes the proof of Lemma 3.

Consider two finite families of algebraic surfaces $\{F_k\}$ and $\{\bar{F}_k\}$, satisfying the following conditions:

1) Each of the surfaces F_k and $\bar{F}_k$ can be projected in one-to-one fashion onto the xy-plane.

2) The projection of the union of the surfaces in the family $\{F_k\}$ covers the rectangle $a \leqq x \leqq b$, $c \leqq y \leqq d$. The same holds for the family $\{\bar{F}_k\}$.

3) If P is a point on the boundary of some surface F_i $(\bar{F}_i)$ and the projection of P lies in the above rectangle, then there exists an interior point Q on some surface F_j $(\bar{F}_j)$ which has the same projection as P on the xy-plane and lies below (above) the point P.

4) The points of intersection of any straight line parallel to the z-axis with the surfaces F_k do not lie below its points of intersection with the surfaces $\bar{F}_k$.

Let $z=f_k(x,y)$ and $z=\bar{f}_k(x,y)$ be the equations of the surfaces F_k and $\bar{F}_k$, respectively. We define two functions $\varphi(x,y)$ and $\bar{\varphi}(x,y)$ in the rectangle $a \leqq x \leqq b$, $c \leqq y \leqq d$ by the conditions

$$\varphi(x, y)=\min_k f_k(x, y), \quad \bar{\varphi}(x, y)=\max_k \bar{f}_k(x, y).$$

These functions are clearly continuous and piecewise-algebraic.

Let Φ and $\overline{\Phi}$ denote the piecewise algebraic surfaces defined by

$$\begin{matrix} z=\varphi(x, y), \\ z=\overline{\varphi}(x, y) \end{matrix} \qquad (a\leqslant x\leqslant b, \quad c\leqslant y\leqslant d).$$

These surfaces possess some important properties. First, the surface Φ lies above the surface $\overline{\Phi}$. Second, any plane $y=\eta$ $(c \leqq \eta \leqq d)$ cuts the surfaces Φ and $\overline{\Phi}$ in curves γ_η and $\overline{\gamma}_\eta$ which are unions of finitely many algebraic curves; moreover, at common points of adjacent arcs of γ_η the apex of the angle between the semitangents points toward $z>0$; the apices of the analogous angles for the curve $\overline{\gamma}_\eta$ point toward $z<0$.

Let γ_1 and γ_2 be two algebraic curves in the planes $x=a$ and $x=b$, lying between the surfaces Φ and $\overline{\Phi}$, both of which can be projected in one-to-one fashion onto the y-axis. Let A_η and B_η denote the points of intersection of γ_1 and γ_2 with the plane $y=\eta$. Join the points A_η and B_η in the plane $y=\eta$ by a shortest arc $\tilde{\gamma}_\eta$ lying between the curves γ_η and $\overline{\gamma}_\eta$.

Lemma 4. *As η varies from c to d, the curve $\tilde{\gamma}_\eta$ describes a piecewise analytic surface $\tilde{\Phi}$ (the threaded surface) which can be projected in one-to-one fashion onto the rectangle $a \leqq x \leqq b$, $c \leqq x \leqq d$.*

That part of the surface $\tilde{\Phi}$ lying strictly between the surfaces Φ and $\overline{\Phi}$ consists of analytic developable surfaces, while the remainder consists of pieces of the analytic surfaces F_k and $\overline{F}_k$.

The surface $\tilde{\Phi}$ can fail to be smooth only along a finite number of curves $\tilde{\gamma}_\eta$; moreover, if a point P at which the surface is not smooth belongs to Φ $(\overline{\Phi})$, then the dihedral angle between the tangent halfplanes to Φ $(\overline{\Phi})$ at this point, facing the region bounded by the surfaces $\overline{\Phi}$ and Φ, is smaller than π.

Proof. The fact that the surface traced out by the curve $\tilde{\gamma}_\eta$ can be projected in one-to-one fashion onto the rectangle follows from the fact that the curve $\tilde{\gamma}_\eta$ can be thus projected onto the straight line $y=\eta$ on the xy-plane and depends continuously on the curves γ_η, $\overline{\gamma}_\eta$ and the points A_η, B_η.

That part of the surface $\tilde{\Phi}$ lying strictly within the domain bounded by the surfaces Φ and $\overline{\Phi}$ is traced out by straight-line segments of the curve $\tilde{\gamma}_\eta$, which is tangent at its endpoints to the algebraic surfaces or cuts the analytic curves γ_1, γ_2. Hence this part of the surface indeed consists of analytic developables. The number of these developables is finite, since the number of algebraic surfaces F_k and $\overline{F}_k$ is finite.

That the surface $\tilde{\Phi}$ fails to be smooth on a finite number of curves $\tilde{\gamma}_\eta$ follows from the fact that Φ and $\overline{\Phi}$ are finite unions of algebraic surfaces.

Let P be a point at which $\tilde{\Phi}$ is not smooth, which lies on the surface Φ. Let $\vartheta(P)$ be the angle between the tangent halfplanes to Φ at this point, facing the region bounded by Φ and $\overline{\Phi}$. We must show that $\vartheta(P) \leqq \pi$.

Indeed, P is a smooth point on one of the surfaces F_k. In a neighborhood of P, the surface Φ lies below F_k; the same therefore holds a fortiori for the surface $\tilde{\Phi}$. Hence the angle $\vartheta(P)$ can never be greater than π facing the region in question. The case in which P lies on $\overline{\Phi}$ is treated similarly. This completes the proof of Lemma 4.

Let Φ be a piecewise analytic surface which can be projected in one-to-one fashion onto the rectangle

$$a \leqq x \leqq b, \quad c \leqq y \leqq d \qquad (c < 0 < d).$$

Let $z = \varphi(x,y)$ be the equation of the surface, where $\varphi(x,y)$ is a piecewise analytic function in the rectangle, satisfying the following conditions:

1) The derivative $\partial\varphi/\partial x$ is continuous throughout the rectangle $a \leqq x \leqq b$, $c \leqq y \leqq d$.

2) The derivative $\partial\varphi/\partial y$ can be discontinuous only at the x-axis, and the size of the jump there cannot exceed ϵ.

3) At each point of the x-axis for which $\partial^2\varphi/\partial x^2$ exists,

$$\partial^2\varphi/\partial x^2 \Delta(\partial\varphi/\partial y) \leqq 0,$$

where $\Delta(\partial\varphi/\partial y)$ is the increment of the derivative $\partial\varphi/\partial y$ when the surface crosses the x-axis at this point from $y < 0$ to $y > 0$.

LEMMA 5. *The surface $\Phi: z = \varphi(x,y)$, $a \leqq x \leqq b$, $c \leqq y \leqq d$, satisfying the above conditions can be approximated by an analytic surface Φ^*: $z = \varphi^*(x,y)$, $a \leqq x \leqq b$, $c \leqq y \leqq d$, with the following properties*:

1) $|\varphi(x,y) - \varphi^*(x,y)| < \epsilon_1$ *at all points (x,y) of the rectangle.*

2) *At all points of the rectangle for which the derivatives $\partial\varphi/\partial x$ and $\partial\varphi/\partial y$ exist,*

$$|\partial\varphi/\partial x - \partial\varphi^*/\partial x| < \epsilon_2, \quad |\partial\varphi/\partial y - \partial\varphi^*/\partial y| < \epsilon_2.$$

3) *The positive curvature of the surface Φ in its domain of regularity differs from that of the surface Φ^* by less than ϵ_3.*

The numbers ϵ_1 and ϵ_3 may be taken arbitrarily small, while ϵ_2 tends to zero with ϵ (the maximum discontinuity of the surface Φ).

PROOF. First note that it will suffice to construct a piecewise analytic surface possessing properties 1), 2) and 3). Approximation of the latter surface by an analytic surface with the same properties is a routine matter.

Let λ be a small positive number. We define a function $\psi(x,y)$ in the rectangle $a \leqq x \leqq b$, $c \leqq y \leqq d$ by

$$\psi(x, y) = \varphi(x, y) \quad \text{for} \quad |y| \geqslant \lambda,$$

$$\psi(x, y) = \alpha(x) y^3 + \beta(x) y^2 + \gamma(x) y + \delta(x) \quad \text{for} \quad |y| \leqslant \lambda,$$

where $\alpha(x)$, $\beta(x)$, $\gamma(x)$ and $\delta(x)$ are so chosen that for $|y| = \lambda$

$$\varphi(x,y) = \psi(x,y) \quad \text{and} \quad \partial\varphi(x,y)/\partial y = \partial\psi(x,y)/\partial y.$$

We claim that for sufficiently small λ the function $\psi(x,y)$ is smooth and piecewise analytic, and possesses the properties 1), 2) and 3) listed in the statement of the lemma.

The above conditions determine the functions $\alpha(x)$, $\beta(x)$, $\gamma(x)$ and $\delta(x)$ uniquely. Omitting the details, we give the final result:

$$\alpha = \frac{1}{4\lambda^3}\{\lambda(\varphi_y^+ + \varphi_y^-) - (\varphi^+ - \varphi^-)\}; \quad \beta = \frac{1}{4\lambda}(\varphi_y^+ - \varphi_y^-);$$

$$\gamma = \frac{1}{4\lambda}\{3(\varphi^+ - \varphi^-) - \lambda(\varphi_y^+ + \varphi_y^-)\}; \quad \delta = \frac{1}{4}\{2(\varphi^+ + \varphi^-) - \lambda(\varphi_y^+ - \varphi_y^-)\},$$

where the superscripts plus and minus indicate that the second argument of the function y takes the values $+\lambda$ and $-\lambda$, respectively.

Substituting these functions α, β, γ and δ in $\psi(x,y)$, we get the following expression for the function in the domain $|y| \leqq \lambda$:

$$\psi(x, y) = \frac{1}{2}(\varphi^+ + \varphi^-) + (\varphi^+ - \varphi^-)\left(-\frac{y^3}{4\lambda^3} + \frac{3y}{4\lambda}\right)$$
$$+ (\varphi_y^+ + \varphi_y^-)\left(\frac{y^3}{4\lambda^2} - \frac{y}{4}\right) + (\varphi_y^+ - \varphi_y^-)\left(\frac{y^2}{4\lambda} - \frac{\lambda}{4}\right).$$

Since $\varphi^+ - \varphi^-$ is of first order in λ for small λ, it follows that when $|y| \leqq \lambda$,

$$\psi(x,y) = \tfrac{1}{2}(\varphi^+ + \varphi^-) + O(\lambda),$$

where $O(\lambda)$ is a function of first order in λ. Hence by the continuity of $\varphi(x,y)$ we conclude that $|\varphi(x,y) - \psi(x,y)|$ is arbitrarily small if λ is sufficiently small.

Differentiating the function $\psi(x,y)$ with respect to x, and noting that $|\varphi_{xy}^+ - \varphi_{xy}^-|$ is of first order in λ, we see by identical reasoning that $|\varphi_x(x,y) - \psi_x(x,y)|$ is arbitrarily small if λ is small.

Now consider $|\varphi_y(x,y) - \psi_y(x,y)|$. We have

$$\psi_y(x, y) = (\varphi^+ - \varphi^-)\left(-\frac{3y^2}{4\lambda^3} + \frac{3}{4\lambda}\right) + (\varphi_y^+ + \varphi_y^-)\left(\frac{3y^2}{4\lambda^2} - \frac{1}{4}\right) + (\varphi_y^+ - \varphi_y^-)\left(\frac{y}{2\lambda}\right).$$

For small λ,

$$\varphi^+ = \varphi^0 + \lambda\varphi_y^+ + O(\lambda^2), \quad \varphi^- = \varphi^0 - \lambda\varphi_y^- + O(\lambda^2),$$

where φ^0 denotes $\varphi(x,0)$. Therefore

$$\psi_y = \frac{1}{2}(\varphi_y^+ + \varphi_y^-) + \frac{y}{2\lambda}(\varphi_y^+ - \varphi_y^-) + O(\lambda).$$

Hence it is clear that $|\varphi_y(x,y) - \psi_y(x,y)|$ tends to zero with λ and ϵ.

Finally, we prove that the change in positive curvature from the surface $z = \varphi(x,y)$ to the surface $z = \psi(x,y)$ can be made arbitrarily small, provided λ is sufficiently small.

First, the surfaces $z = \varphi(x,y)$ and $z = \psi(x,y)$ are distinct only in the domain $|y| < \lambda$. In this domain the positive curvature of the surface $z = \varphi(x,y)$ is small for small λ. Thus we must prove that the positive curvature of the surface $z = \psi(x,y)$ is small for small λ in the domain $|y| < \lambda$.

The positive curvature of the surface $z = \psi(x,y)$ in the domain $|y| \leqq \lambda$ is

$$\omega^+ = \int\int \frac{\psi_{xx}\psi_{yy} - \psi_{xy}^2}{(1 + \psi_x^2 + \psi_y^2)^{3/2}}\, dx\, dy,$$

where the domain of the integration is that part of the domain $|y| \leqq \lambda$ in which the integrand is positive.

In order to estimate the above integral, we examine the second derivatives of $\psi(x,y)$ for $|y| \leqq \lambda$. We shall first prove that when $|y| \leqq \lambda$ the derivatives ψ_{xx}, ψ_{xy} and $\lambda\psi_{yy}$ are uniformly bounded. To this end we again use the explicit expression for $\psi(x,y)$ in the domain $|y| \leqq \lambda$.

Differentiating the expression for $\psi(x,y)$ twice with respect to x, we conclude that ψ_{xx} is bounded, as a simple consequence of the fact that the second and third derivatives of φ are uniformly bounded in its domain of analyticity.

Differentiating $\psi(x,y)$ with respect to x and y, we get

$$\psi_{xy}(x, y) = (\varphi_x^+ - \varphi_x^-)\left(-\frac{3y^2}{4\lambda^3} + \frac{3}{4\lambda}\right) + (\varphi_{xy}^+ + \varphi_{xy}^-)\left(\frac{3y^2}{4\lambda^2} - \frac{1}{4}\right) + (\varphi_{xy}^+ - \varphi_{xy}^-)\frac{y}{2\lambda}.$$

The second and third terms of this expression are bounded, since the second derivatives of φ are bounded. Since the function $\varphi_x(x,y)$ is continuous, it follows that the first term can be expressed as

$$\left(\frac{1}{\lambda}\int_{-\lambda}^{\lambda}\varphi_{xy}\,dy\right)\left(-\frac{3y^2}{4\lambda^2}+\frac{3}{4}\right).$$

The first factor in this product is bounded because φ_{xy} is bounded; the second is trivially bounded. Hence the derivative ψ_{xy} is bounded.

All these arguments apply, naturally, only to points of the domain at which the derivatives exist. However, the derivatives exist at almost all points of the domain of definition of φ, and so λ may be suitably chosen, e.g., so that the required derivatives of the functions φ exist almost everywhere on the straight lines $y = \pm\lambda$.

We now show that $\lambda\psi_{yy}$ is bounded. Differentiating the expression for ψ twice with respect to y, we get

$$\lambda\psi_{yy} = -\frac{3y}{2\lambda^2}(\varphi^+ - \varphi^-) + \frac{3y}{2\lambda}(\varphi_y^+ + \varphi_y^-) + \frac{1}{2}(\varphi_y^+ - \varphi_y^-).$$

The second and third terms are bounded, since the derivatives of φ are bounded. The first term can be expressed as

$$-\frac{3y}{2\lambda}\left(\frac{1}{\lambda}\int_{-\lambda}^{\lambda}\varphi_y\,dy\right)$$

and is therefore bounded for the same reason. Thus $\lambda\psi_{yy}$ is bounded.

The surface $z = \varphi(x,y)$ consists of finitely many analytic surfaces, separated by analytic curves. Let $A_1, A_2, \cdots, A_r$ be the points on the x-axis lying on the boundaries of the domains in which φ is analytic toward $y > 0$, and $\bar{A}_1, \cdots, \bar{A}_s$ the points lying on the boundaries of the domains of analyticity toward $y < 0$. Both these sets of points are finite. Let $B_1, B_2, \cdots, B_t$ be the points on the x-axis at which φ_{xx} vanishes, but does not vanish identically in a neighborhood. The number of these points is finite, since the surface φ is piecewise analytic.

Let each point A_i, $\bar{A}_i$ and B_i be enclosed in a small interval δ, so that the total length of all the intervals is at most ϵ'. Denote the union of the intervals δ by D.

The complement of the set D with respect to the interval (a,b) on the x-axis is a union of intervals $\bar{\delta}$. In a sufficiently small semineighborhood of each interval $\bar{\delta}$, in both regions $y > 0$ and $y < 0$, the function $\varphi(x,y)$ is analytic. Choose λ so small that the function $\varphi(x,y)$ is analytic within each rectangle with side $\bar{\delta}$ and parallel side on $y = \pm\lambda$.

The set of points of the intervals $\bar{\delta}$ can be divided into two subsets

D' and D''. D' contains all points at which $|\varphi_{xx}| < \epsilon''$, and D'' all those at which $|\varphi_{xx}| \geqq \epsilon''$.

Let D_λ, D'_λ and D''_λ denote the sets of points of the rectangle $a \leqq x \leqq b$, $|y| \leqq \lambda$ whose projections on the x-axis lie in the sets D, D' and D'', respectively. We set

$$\frac{\psi_{xx}\psi_{yy} - \psi_{xy}^2}{\left(1 + \psi_x^2 + \psi_y^2\right)^{3/2}} = \Omega.$$

Finally, let G_λ denote the set of points (x, y) at which $|y| \leqq \lambda$ and $\Omega > 0$. Then

$$\omega^+ = \int_{G_\lambda} \Omega\, dx\, dy = \int_{G_\lambda \cap D_\lambda} \Omega\, dx\, dy + \int_{G_\lambda \cap D'_\lambda} \Omega\, dx\, dy + \int_{G_\lambda \cap D''_\lambda} \Omega\, dx\, dy.$$

The third integral on the right-hand side vanishes for sufficiently small λ, since $G_\lambda \cap D''_\lambda$ is empty. Indeed, in each component of D''_λ the function φ is analytic for $y \leqq 0$ and $y \geqq 0$; hence for sufficiently small λ

$$\psi_{xx}(x, y) = \varphi_{xx}(x, 0) + O(\lambda),$$

$$\psi_{yy}(x, y) = \frac{1}{2\lambda}(\varphi_y(x, +0), -\varphi_y(x, -0)) + O(1).$$

By assumption

$$\varphi_{xx}(x, 0)\,(\varphi_y(x, +0) - \varphi_y(x, -0)) < 0.$$

Thus for sufficiently small λ we have $\Omega < 0$ in D''_λ, and so D''_λ contains no points of G_λ.

We now estimate

$$\int_{G_\lambda \cap D'_\lambda} \Omega\, dx\, dy.$$

For the same reason as in the case of D''_λ, the above expressions for ψ_{xx} and ψ_{yy} are valid in D'_λ. And since $|\varphi_{xx}| < \epsilon''$ in D', for small λ we have

$$|\psi_{xx}\psi_{yy}| < c\epsilon''/\lambda$$

in D'_λ, where c is a constant independent of λ and ϵ''. The mixed derivative ψ_{xy} was estimated above. Since the area of the domain of integration $G_\lambda \cap D'_\lambda$ is at most $(b - a)\lambda$, we conclude that the integral

$$\int_{G_\lambda \cap D'_\lambda} \Omega\, dx\, dy$$

is bounded by ϵ'', and hence tends to zero with ϵ''.

Finally, let us estimate the integral

$$\int_{G_\lambda \cap D_\lambda} \Omega \, dx \, dy.$$

Since the functions ψ_{xx}, $\lambda\psi_{yy}$ and ψ_{xy} are uniformly bounded for $|y| \leqq \lambda$, the integrand is of order at most $1/\lambda$. The area of the domain of integration is at most $\epsilon'\lambda$. Hence the integral tends to zero together with ϵ'.

Summarizing, we see that for sufficiently small λ the positive curvature of the surface $z = \psi(x,y)$ in the domain $|y| \leqq \lambda$ is arbitrarily small. This completes the proof of the lemma.

LEMMA 6. *Let $z = \varphi(x,y)$, $a \leqq x \leqq b$, $c \leqq y \leqq d$, be the threaded surface whose existence was proved in Lemma 4. Then there exists an analytic surface $z = \psi(x,y)$, $a \leqq x \leqq b$, $c \leqq y \leqq d$, satisfying the following conditions:*

1) $|\varphi(x,y) - \psi(x,y)| < \epsilon_1$.

2) $|\partial\varphi/\partial x - \partial\psi/\partial x| < \epsilon_2$, $|\partial\varphi/\partial y - \partial\psi/\partial y| < \epsilon_2$.

3) *For any set M_φ in a domain of regularity of the surface φ,*

$$|\sigma^+(M_\varphi) - \sigma^+(M_\psi)| < \epsilon_3,$$

where M_ψ is the projection of M_φ on the surface ψ by straight lines parallel to the z-axis.

The numbers ϵ_1 and ϵ_3 may be taken arbitrarily small, while ϵ_2 is small if the jumps of the derivatives of φ at points of discontinuity are small.

This lemma follows from Lemma 5.

§7. Approximation of surfaces of bounded extrinsic curvature by regular surfaces

Before proving our approximation theorem for surfaces of bounded extrinsic curvature, we need several lemmas on straight-line generators of a smooth surface.

Let us call a straight-line segment g on a smooth surface *isolated* if there is no sequence of straight-line segments on the surface converging to g.

LEMMA 1. *The set of isolated segments on a surface is at most countable.*

PROOF. Let us associate with each straight-line segment on the surface a sextuple of numbers $x_1, \cdots, x_6$, the first three being the cartesian coordinates of one of its endpoints, the last three those of its other endpoint. We thus have a mapping of the segments on the surface onto points $(x_1, \cdots, x_6)$ of euclidean six-space.

Let H be the set of points in six-space corresponding to straight-line segments on the surface. Obviously, if g is isolated, the corresponding point $X_g \in H$ is an isolated point. But the set of isolated points of H is at most countable. This proves the lemma.

Let g be a nonisolated straight-line segment on the surface. We claim that if any sequence g_n of straight-line segments converging to g contains segments arbitrarily close to and coplanar with g, then the tangent plane to the surface is stationary along g.

Without loss of generality we may assume that g lies in the xz-plane. Let A and B be two arbitrary interior points of g. The planes through these points perpendicular to the x-axis cut the surface in two curves, say γ_A and γ_B. Let A_n and B_n be the points in which γ_A and γ_B cut a segment g_n coplanar with g.

Set

$$\xi_A^n = \begin{cases} \dfrac{z(A_n) - z(A)}{y(A_n) - y(A)}, & \text{if} \quad A_n \neq A, \\ \partial z / \partial y \,|_A, & \text{if} \quad A_n = A, \end{cases}$$

$$\xi_B^n = \begin{cases} \dfrac{z(B_n) - z(B)}{y(B_n) - y(B)}, & \text{if} \quad B_n \neq B, \\ \partial z / \partial y \,|_B, & \text{if} \quad B_n = B. \end{cases}$$

Since $\xi_A^n = \xi_B^n$, while as $n \to \infty$

$$\xi_A^n \to \partial z / \partial y \,|_A \quad \text{and} \quad \xi_B^n \to \partial z / \partial y \,|_B,$$

it follows that $\partial z / \partial y \,|_A = \partial z / \partial y \,|_B$.

Since it is obvious that the derivatives $\partial z / \partial x$ at A and B are equal, our assertion is proved.

Lemma 2. *If the angle between the normals to the surface at the endpoints of a nonisolated straight-line segment g is less than ϑ, then the angle between the normals at any other two points of the segment is also less than ϑ.*

Proof. We shall assume that the segment g lies in the xz-plane. The statement of the lemma is clearly true if every sequence g_n converging to g contains segments arbitrarily close to and coplanar with g, since then the tangent plane is stationary along g.

Let A and B be points on g close to its endpoints, and let C be the point of the segment dividing the distance between A and B in the ratio $\lambda : (1 - \lambda)$. As before, we consider planes through the points A, B, C perpendicular to the x-axis, and denote by A_k, B_k, C_k the points of intersection of the planes with a segment g_k. The remark at the beginning of the proof indicates that we may assume that $A_k \neq A$, $B_k \neq B$ and $C_k \neq C$.

To prove the lemma it will clearly suffice to show that the derivative $\partial z/\partial y$ at the point C is bounded between its values at A and B.

Set

$$\xi_A^n = \frac{z(A_n) - z(A)}{y(A_n) - y(A)}, \quad \xi_B^n = \frac{z(B_n) - z(B)}{y(B_n) - y(B)}, \quad \xi_C^n = \frac{z(C_n) - z(C)}{y(C_n) - y(C)}.$$

Geometric arguments show that ξ_C^n lies between ξ_A^n and ξ_B^n. This may also be verified arithmetically, using the known relations between the coordinates of three collinear points.

Hence, letting $n \to \infty$, we see that the derivative $\partial z/\partial y|_C$ is indeed bounded between $\partial z/\partial y|_A$ and $\partial z/\partial y|_B$. This proves the lemma.

LEMMA 3. *If the tangent plane is not stationary along a nonisolated straight-line segment, then the quantity*

$$\mu_{A,B}^{n,m} = \frac{y(B_n) - y(B_m)}{y(A_n) - y(A_m)}$$

tends to a well-defined nonzero limit as $n, m \to \infty$.

PROOF. First, for sufficiently large m and n the segments g_m and g_n cannot intersect, for this would immediately imply that the tangent plane is stationary along g.

Consider the three quotients

$$\alpha = \frac{z(A_n) - z(A_m)}{y(A_n) - y(A_m)}, \quad \beta = \frac{z(B_n) - z(B_m)}{y(B_n) - y(B_m)}, \quad \gamma = \frac{z(C_n) - z(C_m)}{y(C_n) - y(C_m)}.$$

In view of the known relations between the coordinates of the points A_k, B_k, C_k:

$$x(C_k) = \lambda x(A_k) + (1 - \lambda)\, x(B_k),$$
$$y(C_k) = \lambda y(A_k) + (1 - \lambda)\, y(B_k),$$
$$z(C_k) = \lambda z(A_k) + (1 - \lambda)\, z(B_k),$$

we get

$$\mu_{A,B}^{n,m} = -\frac{\lambda(\alpha - \gamma)}{(1 - \lambda)(\beta - \gamma)}.$$

As $m, n \to \infty$,

$$\alpha \to \partial z/\partial y|_A, \quad \beta \to \partial z/\partial y|_B, \quad \gamma \to \partial z/\partial y|_C.$$

And since

$$\partial z/\partial y|_A \neq \partial z/\partial y|_C \neq \partial z/\partial y|_B,$$

it follows that, as $m, n \to \infty$,

$$\mu_{A,B}^{m,n} \to \mu_{A,B} = -\frac{\lambda\,(\partial z/\partial y\,|_A - \partial z/\partial y\,|_C)}{(1-\lambda)\,(\partial z/\partial y\,|_B - \partial z/\partial y\,|_C)}\,.$$

Q.E.D.

Let us associate with each nonisolated segment g parallel to the xz-plane a number ϑ_g, defined as the angle between the normals to the surface at its endpoints. If ϑ_g is not zero, we associate another number μ_g with g, defined as the limit of $\mu^g_{A,B}$ as A and B approach the endpoints of g.

LEMMA 4. *The set G_ϵ of segments g of length at least $l > 0$ such that $\vartheta_g \geqq \epsilon > 0$ and $|\mu_g - 1| \geqq \epsilon$, is finite.*

PROOF. Assume that G_ϵ is infinite. Then we can extract a convergent sequence g_n whose limit is some segment g_0 and for which $\vartheta_{g_n} \to \vartheta_{g_0}$. Since $\vartheta_{g_n} \geqq \epsilon$, we also have $\vartheta_{g_0} \geqq \epsilon$. Hence μ_{g_0} exists, and it is the limit of the sequence μ_{g_n}; this implies that $|\mu_{g_0} - 1| \geqq \epsilon$.

On the other hand, the segment g_0 is the limit of segments g_n parallel to the xz-plane, and so $\mu_{g_0} = 1$. This contradiction proves the lemma.

Let $g \in G_\epsilon$ be a segment on a surface F; suppose that g lies in a plane $y = t = \text{const}$. Let g_P be a straight line through each point P of the segment, in the plane $x = x_P$, tangent to the surface F at the point P. The segments g_P $(P \in g)$ generate an analytic developable surface F_g (a hyperbolic paraboloid).

Indeed, the slope k_P of the straight line g_P at an arbitrary point P dividing the distance between the endpoints in the ratio $\lambda : (1-\lambda)$ is given by

$$k_P = \frac{\lambda\mu_g k_1 + (1-\lambda)\,k_2}{\lambda\mu_g + (1-\lambda)}$$

where k_1 and k_2 are the slopes of the straight lines corresponding to the endpoints, and thus k_P is an analytic function of λ. Hence the surface F_g is analytic.

We now construct two curves γ^+ and γ^- in the plane $y = t$, as follows. Let $\bar{A}$ and $\bar{B}$ be two points on the segment g, close to its endpoints A and B. Displace A and B toward $z > 0$ for a distance $\bar{h}$. The curve γ^+ is the union of the straight-line segment $\bar{A}\bar{B}$ and two parabolas tangent to g at the points $\bar{A}$ and $\bar{B}$ and passing through the points $A_{\bar{h}}$ and $B_{\bar{h}}$ obtained by displacing A and B. The curve γ^- is similar, except that A and B are displaced toward $z < 0$.

We now construct surfaces F_g^+ and F_g^-. Construct a parabola through each point of the curve γ^+, in the plane $x = x_P$, whose slope at P is equal to the slope of the straight-line generator of the surface F_g lying in the plane $x = x_P$, and such that the coefficient of y^2 is $\bar{\epsilon}/\bar{h}$. These parabolas generate a smooth piecewise analytic surface F_g^+. The construction of F_g^- is similar, except that the curve γ^+ is replaced by γ^-, and the coefficient of y^2 is $\bar{\epsilon}/\bar{h}$.

Displace the surface F_g toward $z > 0$ for the distance $\bar{h}$. A certain part of the surface F_g^+ will then lie below the surface F_g. Denote this part by $\widetilde{F}_g^+$. This surface has the following properties:

1. If $\bar{h}$ is sufficiently small and $\bar{\epsilon}$ is fixed, the boundary of the surface lies above the surface F.

2. The projection of $\widetilde{F}_g^+$ on the xy-plane lies in an $\bar{h}/\sqrt{\bar{\epsilon}}$-neighborhood of the projection of the segment g.

3. The absolute curvature of the surface $\widetilde{F}_g^+$ is of first order in $\bar{\epsilon}$.

All these properties of the surface $\widetilde{F}_g^+$ can be verified by direct calculation. The surface $\widetilde{F}_g^-$ is defined similarly, using F_g^-, and it possesses analogous properties.

Displace the surface F for a small distance $h \ll \bar{h}$ toward $z > 0$ and $z < 0$, to get new surfaces F^+ and F^-, respectively, such that the boundary of the surface $\widetilde{F}_g^+$ lies above F^+, the boundary of $\widetilde{F}_g^-$ below F^-. Let $\widetilde{G}_g'$ be the component containing $g_{\bar{A},\bar{B}}$ in the partition of the surface $\widetilde{F}_g^+$ by the surface F^+. Now modify the surface F^+, replacing the domain in F^+ onto which $\widetilde{G}_g^+$ is projected by straight lines parallel to the z-axis by the domain $\widetilde{G}_g^+$ itself. Call the new surface $\widetilde{F}^+$. The surface $\widetilde{F}^-$ is constructed by an analogous modification of $\widetilde{F}_g^-$.

LEMMA 5. *Let M be the set of straight-line segments of length at least l, parallel to the xz-plane and lying between the surfaces $\widetilde{F}^+$ and $\widetilde{F}^-$. Then there exists $d > 0$ such that for any $\bar{\epsilon} > 0$ and sufficiently small $\bar{h}$ the d-neighborhood of the segment g contains no segments of M other than g.*

PROOF. Suppose the assertion false, and let g' be a segment in M in the d-neighborhood of g. Let P' and Q' be points on g' dividing it into three equal parts, and S' a point between them dividing the segment $P'Q'$ in the ratio $\lambda : (1 - \lambda)$. Construct planes through each of these points parallel to the yz-plane. For sufficiently small d these planes will cut the segment g in points P, Q and S, respectively.

Consider the three quotients

$$\xi_A = \frac{z(A) - z(\bar{A})}{y(A) - y(\bar{A})}, \quad \xi_B = \frac{z(B) - z(\bar{B})}{y(B) - y(\bar{B})}, \quad \xi_C = \frac{z(C) - z(\bar{C})}{y(C) - y(\bar{C})}.$$

If d and $\bar{h}/\sqrt{\bar{\epsilon}}$ are sufficiently small, these quotients are close to $\partial z/\partial y|_A$,

$\partial z/\partial y|_B$ and $\partial z/\partial y|_C$, respectively. Consequently

$$\xi_C = \frac{\lambda\mu_g\xi_A + (1-\lambda)\,\xi_B}{\gamma\mu_g + (1-\lambda)} + \tilde{\varepsilon},$$

where $\tilde{\epsilon}$ is arbitrarily small if d and $\bar{h}/\sqrt{\bar{\epsilon}}$ are small. On the other hand, it is obvious that

$$\xi_C = \lambda\xi_A + (1-\lambda)\,\xi_B.$$

Since $|\xi_A - \xi_B|$ is bounded away from zero $(\vartheta_g \neq 0)$ and $\mu_g \neq 0$, it follows that for $\lambda = \frac{1}{2}$ and sufficiently small $\bar{\epsilon}$ these two expressions for ξ_C cannot be arbitrarily close. This contradiction proves the lemma.

THEOREM. *Let F be a surface of bounded extrinsic curvature and X an arbitrary point on F. Then X has a neighborhood U_X such that there exists a sequence of regular, even analytic surfaces F_n converging to F in U_X together with their spherical images; the positive curvatures of these surfaces converge weakly to that of F.*

A rigorous statement of this theorem is as follows. Convergence of the surfaces F_n to F in the neighborhood U_X means that there exists a homeomorphism f_n of F onto F_n, defined in U_X. The theorem asserts that, first, for sufficiently large n the distance between corresponding points of F and F_n and the angles between the tangents to the surfaces at these points are arbitrarily small.

Second, the homeomorphism f_n maps every set $E \subset U_X$ onto some set $f_n(E)$ on F_n. Let $\sigma_n^+(E)$ be its positive curvature. Then, according to our theorem, for any continuous function $\psi(Y)$ defined in U_X,

$$\lim_{n\to\infty} \int_{U_X} \psi(Y)\, d\sigma_n^+(E(Y)) = \int_{U_X} \psi(Y)\, d\sigma^+(E(Y)).$$

PROOF. Let F be a smooth surface of bounded extrinsic curvature. If the tangent plane to the surface at X is taken as the xy-plane, a sufficiently small neighborhood of X on the surface is defined by an equation $z = \varphi(x,y)$, $a \leqq x \leqq b$, $c \leqq y \leqq d$, where $\varphi(x,y)$ is continuous and continuously differentiable in the neighborhood.

Since the set of isolated segments on the surface is at most countable, we may assume without loss of generality that none of them is parallel to the xy-plane.

For each segment g of the set G_ϵ (Lemma 4), construct the piecewise analytic surfaces $\widetilde{F}_g^+$ and $\widetilde{F}_g^-$ for sufficiently small $\bar{\epsilon}$.

Divide the rectangle $\Delta: a \leqq x \leqq b$, $c \leqq y \leqq d$ by straight lines parallel to the coordinate axes into small rectangles δ such that the absolute curvature of the surface on the set of points projected onto the sides and vertices of the rectangles is zero. Let F_δ be the domain on the surface F whose projection is the rectangle δ. Let F'_δ be the region of the convex hull of F_δ convex toward $z > 0$, and F''_δ the region convex toward $z < 0$.

Surfaces F'_{δ_1} and F'_{δ_2} over rectangles δ_1 and δ_2 with a common side always have a common boundary curve (a convex curve projected onto the common side of the rectangles), so that the union of the surfaces F'_δ is a certain convex surface F' which can be projected in one-to-one fashion onto the rectangle Δ. The surfaces F''_δ have an analogous property.

Let H be any set of points on the surface F. Then the positive curvature of the surface F on the set H is not less than the sum of positive curvatures of the surfaces F'_δ and F''_δ on the sets projected onto H by straight lines parallel to the z-axis. This follows directly from the fact that each one-sided point on F'_δ (or F''_δ), not contained in a straight-line segment, is projected onto a one-sided point on F.

Approximate each surface F'_δ by an algebraic convex surface $\widetilde{F}'_\delta$ satisfying the following conditions:

1) The total curvature of $\widetilde{F}'_\delta$ is close to the total curvature of F'_δ.

2) If P is a boundary point of one of the surfaces $\widetilde{F}'_\delta$, then at least one of the adjacent surfaces $\widetilde{F}'_\delta$ contains an interior point lying above P.

3) No two surfaces $\widetilde{F}'_\delta$ are tangent to each other, so that they either have no common points or intersect in algebraic curves.

It is clearly easy to carry out this construction.

The surfaces $\widetilde{F}'_\delta$ divide the region $a \leqq x \leqq b$, $c \leqq y \leqq d$ in space into two regions. Let E' be the region which contains points with arbitrarily large z-values. Let F' denote the part of the boundary of E' consisting of domains on the algebraic surfaces $\widetilde{F}'_\delta$. F' is a piecewise algebraic surface, which can be projected in one-to-one fashion onto the rectangle Δ.

Similarly, construct a piecewise algebraic surface F'' from the algebraic surfaces $\widetilde{F}''_\delta$ approximating the surface F''_δ.

Displace the surface F'' upward for a small distance h, and the surface F' downward for the same distance. Since h is small, the boundary of each surface $\widetilde{F}_g^+$ will lie above the surface F''. Replace the domains on the surface F'' cut out by the surfaces $\widetilde{F}_g^+$ by the corresponding domains of the latter surfaces. Call the resulting surface Φ. The analogous surface constructed from F' and the $\widetilde{F}_g^-$ will be called $\overline{\Phi}$.

It is clear that the size of the small rectangles may be so chosen that for a given displacement h the surface Φ will lie above $\overline{\Phi}$.

Approximate the curves in which the surface F cuts the planes $x=a$ and $x=b$ by algebraic curves $\tilde{\gamma}_1$, $\tilde{\gamma}_2$ and construct the threaded surface $\tilde{\Phi}$ for these curves and the surfaces Φ, $\overline{\Phi}$. We claim that for a suitable choice of the various parameters determining the structure of the surface $\tilde{\Phi}$ it will possess the following properties:

1) The surface $\tilde{\Phi}$ is close to F.

2) The angles between the tangent planes to $\tilde{\Phi}$ and the tangent planes to F at corresponding points are small.

The first assertion is obvious, for h is small and the distance between corresponding points of F and $\tilde{\Phi}$ (i.e. points lying on the same vertical) is at most h.

Suppose now that the second assertion is false. Then each surface $\tilde{\Phi}$ contains a point $\widetilde{Q}$ such that the angle between the tangent plane at $\widetilde{Q}$ and the tangent plane to F at the corresponding point Q is greater than $\tilde{\epsilon}>0$, where $\tilde{\epsilon}$ is the same for all $\tilde{\Phi}$.

We may assume that the point $\widetilde{Q}$ is on neither Φ nor $\overline{\Phi}$, since the tangent planes to these surfaces are near the tangent planes to F, by construction. Thus $\widetilde{Q}$ must lie on a straight-line segment g which is either tangent at one (or both) endpoint(s) to the surfaces Φ, $\overline{\Phi}$ and/or has one (or both) endpoints on $\tilde{\gamma}_1$, $\tilde{\gamma}_2$.

For the same reasons, the distances of the point $\widetilde{Q}$ from the endpoints of the segment $\tilde{g}$ (which contains it) are bounded from below.

As the surface $\tilde{\Phi}$ approaches arbitrarily close to F, the segment $\tilde{g}$ may be assumed to converge to some segment g_0 (or part of a segment) on the surface F, and the point $\widetilde{Q}$ to an interior point Q_0 of g_0. As regards the segment g_0, there are only three possibilities: 1) $\vartheta_{g_0}\geqq\epsilon$, $|\mu_{g_0}-1|\geqq\epsilon$; 2) $\vartheta_{g_0}<\epsilon$; 3) $|\mu_{g_0}-1|<\epsilon$. But we shall show that the determining parameters of the surfaces $\tilde{\Phi}$ converging to F may be so chosen that none of these possibilities can hold for the limit segment g_0.

By Lemma 4 it is easy to ensure that the first case is impossible.

Suppose that g_0 satisfies the second condition ($\vartheta_{g_0}<\epsilon$). Then the variation of the derivative $\partial\varphi/\partial y$ along the segment g_0 is at most ϵ, i.e. the derivative remains constant up to a quantity ϵ. The value of $\partial\tilde{\varphi}/\partial y$ at the point $\widetilde{Q}$ is bounded between its values at the endpoints of the segment $\tilde{g}$, which are close to the values of $\partial\varphi/\partial y$ at the corresponding points of g_0. It follows that if $\tilde{\Phi}$ is sufficiently close to F the derivatives $\partial\varphi/\partial y$ and $\partial\tilde{\varphi}/\partial y$ at the points Q and $\widetilde{Q}$, respectively, differ by a quantity of first order in ϵ, and since ϵ is arbitrary this quantity may be assumed to be smaller than $\tilde{\epsilon}$. Thus we need only take ϵ sufficiently small in order to exclude the second possibility.

Now suppose that the segment g_0 satisfies the third condition, $|\mu_{g_0} - 1| < \epsilon$. We shall show that this is impossible for sufficiently small ϵ. When $|\mu_{g_0} - 1| < \epsilon$, the derivatives $\partial\varphi/\partial y$ and $\partial\tilde{\varphi}/\partial y$ vary by the same amount along the segments g_0 and $\tilde{g}$, respectively, up to a quantity of first order in ϵ. Now, since these derivatives are close to each other at the endpoints of the segments, they are also close to each other at points dividing the segments in equal ratios. Thus, if ϵ is small in comparison with $\tilde{\epsilon}$, the second possibility cannot hold for the segment g_0, and we have a contradiction. Hence there exists a surface $\tilde{\Phi}$ which is close to F together with its tangent planes.

For each surface $\tilde{\Phi}$, construct an analytic surface $\tilde{\Phi}^*$ according to Lemma 6 of §6. We claim that the family of analytic surfaces $\tilde{\Phi}^*$ contains a sequence of surfaces satisfying the conditions of the theorem.

The fact that the family of surfaces $\tilde{\Phi}^*$ contains a sequence $\tilde{\Phi}_n^*$ converging to F together with their spherical images is obvious. We need only prove that the sequence can be so chosen that the positive curvatures converge weakly as required.

For any $\epsilon^+ > 0$, if the surface $\tilde{\Phi}^*$ is sufficiently close to F, then $\sigma^+(\tilde{\Phi}^*) > \sigma^+(F) - \epsilon^+$. Let us prove this statement.

Let H_k be a finite family of closed pairwise disjoint sets on F whose sum of areas of spherical images is close to the absolute curvature of F.

We define a set W_n of one-sided points on F as follows. A point X belongs to W_n if it has a neighborhood $U(X)$ such that

1) the entire neighborhood $U(X)$ lies on one side of the tangent plane at X;

2) the boundary of $U(X)$ is at a distance at least $1/n$ from the tangent plane;

3) the diameter of $U(X)$ is at most the distance between the set $\sum H_k$ and the boundary of the surface F (which corresponds to the boundary of the rectangle Δ).

$\{W_n\}$ is an [increasing] sequence of closed sets. For sufficiently large n the sum of areas of the spherical images of the sets $W_n \cap H_k$ is close to the positive curvature of F, since $\sum_1^\infty W_n$ contains all elliptic points of the set $\sum_k H_k$.

Let the surface $\tilde{\Phi}^*$ be so close to F that it lies in an ϵ^*-neighborhood of F. If ϵ^* is small in comparison with $1/n$, then for each point $X \in W_n \cap H_k$ there is a one-sided point $\widetilde{X}^*$ on $\tilde{\Phi}^*$ such that

1) the distance between X and $\widetilde{X}^*$ tends to zero with ϵ^*;

2) the tangent planes to F and $\tilde{\Phi}^*$ at the points X and $\widetilde{X}^*$ are parallel.

Let M_k denote the set of all such points $\widetilde{X}^*$ corresponding to the set $W_n \cap H_k$. For sufficiently small ϵ^* the sets M_k are pairwise disjoint. The sum of areas of their spherical images is at least the sum of areas of the spherical images of the sets $W_n \cap H_k$, which is close to the positive curvature of the surface, by construction. Hence we see that $\sigma^+(\tilde{\Phi}^*) > \sigma^+(F) - \epsilon^+$ for any $\epsilon^+ > 0$ and $\tilde{\Phi}^*$ sufficiently close to F.

The distribution of the positive curvature on the surface $\tilde{\Phi}^*$ is almost the same as its distribution on the threaded surface $\tilde{\Phi}$, from which $\tilde{\Phi}^*$ is derived by a smoothing procedure. The positive curvature of the threaded surface $\tilde{\Phi}$ is concentrated in the domains in which it is in contact with the surfaces Φ and $\overline{\Phi}$.

Set $\sigma^+(\tilde{\Phi}_\delta^*) - \sigma^+(F_\delta) = \eta_\delta$. Then, for the same reason, the sum of all possible nonzero values of η_δ is smaller than ϵ^+. Thus, in view of the preceding inequality, the sum of absolute values of the η_δ is at most $2\epsilon^+$:

$$\sum_\delta |\sigma^+(\tilde{\Phi}_\delta^*) - \sigma^+(F^\delta)| < 2\varepsilon^+.$$

Let us denote the positive curvatures of the surfaces F and $\tilde{\Phi}^*$ on sets projected onto a subset E of the rectangle Δ by $\sigma_F^+(E)$ and $\sigma_{\tilde{\Phi}^*}^+(E)$, respectively.

Let f be any continuous function defined in Δ. Consider the difference

$$R = \int_\Delta f\, d\sigma_F^+ - \int_\Delta f\, d\sigma_{\tilde{\Phi}^*}^+.$$

Obviously R may be expressed as

$$R = \sum_\delta f_\delta' \sigma_F^+(\delta) - \sum_\delta f_\delta'' \sigma_{\tilde{\Phi}^*}^+(\delta),$$

where f_δ' and f_δ'' are certain values of f in the rectangle δ. It is now easy to estimate R, for

$$R = \sum_\delta f_\delta'' \left(\sigma_F^+(\delta) - \sigma_{\tilde{\Phi}^*}^+(\delta)\right) + \sum_\delta \sigma_F^+(\delta)\left(f_\delta' - f_\delta''\right)$$

The first sum is small since ϵ^+ is arbitrarily small (in the inequality for the positive curvatures established above). The second sum is small since the function f is uniformly continuous.

Hence we may assume that the positive curvatures of the surfaces $\tilde{\Phi}_n^*$ converge weakly to the positive curvature of F. This completes the proof of the theorem.

§8. Surfaces of bounded extrinsic curvature as manifolds of bounded intrinsic curvature

In this section we shall prove that every surface of bounded extrinsic curvature is a manifold of bounded intrinsic curvature in the sense of Aleksandrov. We first recall the basic definitions and theorems for manifolds of bounded intrinsic curvature.

A manifold R of bounded intrinsic curvature is defined by two axioms:

1. R is a manifold with intrinsic metric.

2. Each point of R has a neighborhood such that, for any finite family of pairwise nonoverlapping triangles in the neighborhood, the sum of absolute values of their excesses is uniformly bounded.

These axioms need some explanation. Let R be a metric manifold with metric ρ, i.e. a metric space which is a manifold. Then the length of any curve $X(t)$, $a \leqq t \leqq b$, in R can be defined in the usual manner:

$$l = \sup \sum \rho(X(t_k), X(t_{k-1})), \quad a \leqq t_1 \leqq t_2 \leqq \cdots \leqq t_n \leqq b.$$

The metric of a metric manifold R is said to be intrinsic if the distance $\rho(X, Y)$ between any two points X and Y of R is equal to the infimum of the lengths of all curves connecting the points.

A curve connecting points X and Y in R is said to be a segment (shortest join) if its length is at most that of any other curve connecting the points. A geodesic is defined as a curve which is locally a segment.

Segments in manifolds of bounded curvature possess the following properties:

1. For every point A in R and every neighborhood U of A, there exists a neighborhood V of A such that any segment in R connecting any two points X and Y of V lies in U.

2. A segment is homeomorphic to an interval on the real line; hence two sides can be defined near any interior point of a segment.

3. The length of a segment is equal to the distance between its endpoints in the metric ρ.

We define a triangle ABC in a manifold R as the figure formed by three segments γ_{AB}, γ_{BC}, γ_{AC} connecting the three points A, B, C in pairs. The points A, B, C are called the vertices of the triangle; the segments are its sides.

Two triangles T and T' are said to be nonoverlapping if there exist disjoint domains G' and G'' homeomorphic to a disk whose closures contain T' and T'', respectively.

Let γ_1 and γ_2 be two curves in R issuing from a point O. Let X and

Y be variable points on γ_1 and γ_2, respectively. Construct a plane triangle with sides $\rho(O,X)$, $\rho(O,Y)$, $\rho(X,Y)$. Let $\vartheta(X,Y)$ be the angle of this triangle opposite the side $\rho(X,Y)$. We define the upper angle between the curves γ_1 and γ_2 at O as the upper limit of the angles $\vartheta(X,Y)$ as $X,Y \to O$. The angle between γ_1 and γ_2 at O is the limit of $\vartheta(X,Y)$ as $X,Y \to O$. The angle between curves at their common point may not exist, but the upper angle always exists.

The upper angle of the triangle ABC at the vertex A will be called the upper angle between the sides γ_{AB} and γ_{AC} at the point A. The excess of the triangle is the difference between the sum of its upper angles and π.

In a manifold of bounded intrinsic curvature, any two segments issuing from a single point form a definite angle in the sense of the above definition. This angle obviously coincides with the upper angle.

Let G be an open set in a manifold R of bounded intrinsic curvature. The positive (negative) curvature of R on the set G is defined to be the supremum (infimum with inverted sign) of the positive (negative) sums of excesses of pairwise nonoverlapping triangles in G, and is denoted by $\omega^+(G)$ $(\omega^-(G))$.

The positive and negative curvatures of a manifold R on any set M are defined by

$$\omega^+(M) = \inf_{G \supset M} \omega^+(G), \qquad \omega^-(M) = \inf_{G \supset M} \omega^-(G).$$

The total curvature of R on M is the difference between the positive and negative curvatures of R on M. The absolute curvature of R on M is the sum of the positive and negative curvatures on M.

The positive, negative, total and absolute curvatures are finite on compact sets and completely additive on the ring of Borel sets.

A metric manifold R is said to be Riemannian if a neighborhood of each of its points admits a parametrization u, v (i.e. a homeomorphism into the plane with cartesian coordinates u, v) in which the metric in the neighborhood is defined by a positive definite quadratic form

$$ds^2 = E du^2 + 2F du\, dv + G dv^2$$

with distances between points calculated in the usual manner (as the infimum of the lengths of the curves connecting the points).

If a Riemannian manifold admits a parametrization (u,v) such that the coefficients E, F, G of the form ds^2 are twice continuously differentiable, it is called a manifold of bounded curvature.

The positive and negative curvatures of a Riemannian manifold on any set M are

$$\omega^+ = \int_{K>0} K\, d\sigma, \qquad \omega^- = -\int_{K<0} K\, d\sigma,$$

where K is the Gauss curvature, $d\sigma$ the area element of the manifold, and the domains of integration are the subsets of M defined by $K>0$ and $K<0$.

Let ρ_n be a sequence of Riemannian metrics on a manifold R, which converge to a metric ρ in the following sense: For any $\epsilon>0$, sufficiently large n, and any two points X and Y in R, we have $|\rho(X,Y)-\rho_n(X,Y)| <\epsilon$. Then if the absolute curvatures of the Riemannian manifolds $R_{\rho n}$ (the manifold R with the metric ρ_n) are uniformly bounded, the manifold R_ρ is a manifold of bounded curvature. Moreover, if the positive (negative) curvatures of the manifolds R_{ρ_n} are bounded above by ω_0^+ (ω_0^-), then the positive (negative) curvature of R_ρ is also at most ω_0^+ (ω_0^-). The sequence of manifolds R_{ρ_n} contains a subsequence of manifolds whose total curvatures converge weakly to the total curvature of R_ρ.

The total curvature of a closed manifold with Euler characteristic x is $2\pi\chi$.

Let Φ be a smooth surface which can be projected in one-to-one fashion onto the xy-plane and is given by an equation $z=\varphi(x,y)$, $a\leqq x\leqq b$, $c\leqq y\leqq d$, where φ is continuous and continuously differentiable in the rectangle. Let Φ_n be a sequence of smooth surfaces, which can also be projected one-to-one onto the rectangle $a\leqq x\leqq b$, $c\leqq y\leqq d$, and converge together with their tangent planes to Φ. We claim that the intrinsic metrics of the surfaces Φ_n converge to the intrinsic metric of Φ, where correspondence between the points of the surfaces is defined by projection in the direction of the z-axis.

Suppose that the surface Φ_n is defined by an equation $z=\varphi_n(x,y)$, $a\leqq x\leqq b$, $c\leqq y\leqq d$. Let A and B be two arbitrary points on Φ, and connect them by a curve γ on the surface. The length of this curve is

$$l_\gamma = \int_\gamma \left\{\left(1+\varphi_x^2\right) dx^2 + 2\,\varphi_x\varphi_y\, dx\, dy + \left(1+\varphi_y^2\right) dy^2\right\}^{\frac{1}{2}}.$$

Since the derivatives φ_x and φ_y are continuous, their absolute values are less than some constant c. Hence

$$l_\gamma < \sqrt{1+2c^2}\cdot l_{\bar\gamma},$$

where $l_{\bar{\gamma}}$ is the length of the curve $\bar{\gamma}$, the projection of γ on the xy-plane. If γ is a curve lying in a plane perpendicular to the xy-plane, then

$$l_\gamma < \sqrt{1+2c^2}\, d,$$

where d is the diagonal of the rectangle onto which the surface is projected. Hence it follows that the distance between any two points A and B on the surface is at most $\sqrt{1+2c^2}d$.

Since the surfaces Φ_n converge together with their tangent planes to Φ, the functions $\varphi_n(x,y)$ converge uniformly together with their first derivatives to the function $\varphi(x,y)$. Hence for sufficiently large n

$$\left|\frac{\partial \varphi_n}{\partial x}\right| < c \text{ and } \left|\frac{\partial \varphi_n}{\partial y}\right| < c.$$

Hence, as in the case of the surface Φ, we conclude that the distance between any two points on Φ_n is less than $\sqrt{1+2c^2}d$.

Let $\epsilon > 0$. We must show that for sufficiently large n the distance between any two points A and B on Φ differs from the distance between the corresponding points A_n and B_n on Φ_n by less than ϵ.

By the definition of the metric ρ, there exists a curve γ joining the points A and B on Φ such that $|l_\gamma - \rho(A,B)| < \epsilon'$. For sufficiently large n,

$$\left|\frac{\partial \varphi}{\partial x} - \frac{\partial \varphi_n}{\partial x}\right| < \varepsilon'', \qquad \left|\frac{\partial \varphi}{\partial y} - \frac{\partial \varphi_n}{\partial y}\right| < \varepsilon''.$$

Using the easily checked inequality

$$|\sqrt{\alpha} - \sqrt{\beta}| \leqslant \sqrt{|\alpha - \beta|},$$

we see that

$$|l_\gamma - l_{\gamma_n}| < 2\sqrt{c\varepsilon''}\, l_{\bar{\gamma}},$$

where l_{γ_n} is the length of the curve γ_n on Φ_n which corresponds to the curve γ on Φ, and $l_{\bar{\gamma}}$ is the length of the projection of γ on the xy-plane. Since $l_{\bar{\gamma}} \leqq l_\gamma$,

$$|l_\gamma - l_{\gamma_n}| < 2\sqrt{c\varepsilon''}\, l_\gamma.$$

Hence, since l_γ differs from $\rho(A,B)$ by at most ϵ', while $l_{\gamma_n} \geqq \rho_n(A_n, B_n)$, it follows from the fact that ϵ' and ϵ'' are arbitrary that $\rho_n(A_n,B_n) - \rho(A,B) < \epsilon$.

Similarly, one proves the inequality $\rho(A,B) - \rho_n(A_n,B_n) < \epsilon$. To do this, take a curve γ_n on the surface Φ_n joining the points A_n and B_n, such that

$$|l_{\gamma_n} - \rho(A_n, B_n)| < \epsilon',$$

and then use similar arguments. This proves our assertion.

THEOREM 1. *A surface of bounded extrinsic curvature is a manifold of bounded intrinsic curvature.*

PROOF. Let Φ be a surface of bounded extrinsic curvature and O an arbitrary point on the surface. Define rectangular cartesian coordinates in space, with O as origin and the tangent plane to the surface at O as the xy-plane. Then some neighborhood G of the point O on the surface can be defined by the equation $z = \varphi(x,y)$, $a \leqq x \leqq b$, $c \leqq y \leqq d$.

By the approximation theorem of §7 there exists a sequence of analytic surfaces $\Phi_n : z = \varphi_n(x,y)$, $a \leqq x \leqq b$, $c \leqq y \leqq d$, with uniformly bounded positive curvatures, converging to Φ together with their tangent planes in G. As we have proved, the metrics of Φ_n converge to the metric of Φ in G.

Let κ be a square on the xy-plane, with center O and side δ. Let A_n, B_n, C_n and D_n be the points on the surface Φ_n whose projections are the vertices of the square κ. We claim that for small δ and all sufficiently large n the four segments connecting the points A_n, B_n, C_n, D_n in pairs on Φ_n, in the same way as the sides of κ connect its vertices, bound a domain G_n^* homeomorphic to a disk, whose projection on the xy-plane covers a disk $x^2 + y^2 \leqq \delta^*$, where δ^* is a small positive number.

Let γ_n be one of the above four segments, say that connecting the points A_n and B_n. We shall show that for small δ and large n its projection on the xy-plane is in a λ-neighborhood of the corresponding side of the square, where λ is small in comparison with δ. If this is not so, there exist $\epsilon > 0$, a sequence of numbers $\delta_k \to 0$ and a sequence of surfaces Φ_{n_k} such that the projections of the segments γ_{n_k} on the xy-plane do not remain within the $\delta\epsilon$-neighborhood of the corresponding side of the square. But then the length of the curve $\bar{\gamma}_{n_k}$ (the projection of γ_{n_k} on the xy-plane) is at least $(1 + \epsilon^2)\delta$. Hence, a fortiori, $l(\gamma_{n_k}) > (1 + \epsilon^2)\delta$.

On the other hand, the length of the curve connecting A_{n_k} and B_{n_k} on Φ_{n_k} and projected onto the side of the square, as was shown above, is at most $\sqrt{1 + 2c^2}\delta$, where c is the maximum modulus of the derivatives $\partial\varphi_{n_k}/\partial x$ and $\partial\varphi_{n_k}/\partial y$ in the square κ. And since by assumption $\partial\varphi/\partial x = \partial\varphi/\partial y = 0$, at the point O, it follows that for small δ and sufficiently large n_k

$$\sqrt{1+2c^2\delta} < (1+\epsilon^2)\delta.$$

This contradiction shows that for small δ and sufficiently large n the projections of the segments γ_n on the xy-plane remain within a λ-neighborhood of the corresponding side of κ, where λ is small in comparison with δ.

Let δ be so small that λ may be taken equal to $\delta/4$. We shall show that the segments connecting A_n, B_n, C_n, D_n form a simple curve, which therefore bounds a domain homeomorphic to a disk. First note that the two segments $\gamma_{A_nB_n}$ and $\gamma_{C_nD_n}$ have no common point, and the same is true of the segments $\gamma_{A_nD_n}$ and $\gamma_{B_nC_n}$. This is obvious, for otherwise their projections on the xy-plane would also intersect, which is impossible $(\lambda = \delta/4)$.

Moreover, it is easy to see that the segments $\gamma_{A_nD_n}$ and $\gamma_{A_nB_n}$ have only one common point. Indeed, otherwise, by the inclusion property of segments on a regular surface, one of the segments would be included in the other. But this is impossible, since the projection of each segment lies in a $\delta/4$-neighborhood of the corresponding side of the square.

Analogous conclusions hold for the other pairs of segments: $\gamma_{B_nA_n}$ and $\gamma_{B_nC_n}$, $\gamma_{C_nB_n}$ and $\gamma_{D_nC_n}$, $\gamma_{D_nB_n}$ and $\gamma_{D_nA_n}$.

Thus the four segments $\gamma_{A_nB_n}, \cdots, \gamma_{D_nA_n}$ form a closed curve bounding a domain G_n^* homeomorphic to a disk on the surface Φ_n. The projection of each of the domains G_n^* on the xy-plane covers the disk $x^2+y^2 \leqq (\delta/4)^2$, since the projection of each of the segments bounding G_n^* remains at a distance at least $\delta/4$ from the point O.

By the approximation theorem, the positive curvatures of the surfaces Φ_n are bounded by a constant C. This constant is a fortiori an upper bound for the positive curvatures of the surfaces Φ_n in the domains G_n^*.

By the Gauss-Bonnet Theorem the absolute value of the total curvature of each quadrangle G_n^* is at most 4π. Hence the negative curvature of each domain G_n^* is at most $4\pi + C^+$. Thus the absolute curvature of G_n on the surface Φ_n is at most $4\pi + 2C^+$. This bound is valid a fortiori for the domains $\tilde{\Phi}_n$ on the surfaces Φ_n whose projections coincide with the disk $x^2+y^2 < \delta^2/16$.

Since the absolute curvatures of the surfaces $\tilde{\Phi}_n$ are uniformly bounded, it follows that the limit surface $\tilde{\Phi}$ (the domain on Φ whose projection is the disk $x^2+y^2 < \delta^2/16$) is a manifold of bounded curvature. The point O on the surface Φ was chosen arbitrarily, and therefore Φ is a manifold of bounded curvature. This completes the proof.

§9. Relation between intrinsic and extrinsic curvature. Gauss's Theorem

In the regular case, Gauss's Theorem establishes the relation between the integral curvature of any domain on a surface and the area of its spherical image. In this section we shall prove a corresponding theorem for surfaces of bounded extrinsic curvature. The proof of this theorem will be preceded by several lemmas.

LEMMA 1. *Let F be a surface of bounded extrinsic curvature, G any open set on F with boundary γ in the surface (the surface itself is assumed open), and H any closed subset of G. The spherical images of G and its boundary are assumed disjoint.*

Then there exists an infinite sequence of open subsets $G_1, G_2, \cdots$ of G, all containing H as a subset, with the following properties:

1) *Each G_k is the union of finitely many domains (in general, multiply connected) bounded by simple curves.*
2) *The spherical image $\bar{\gamma}_k$ of the boundary of any set G_k has zero area.*
3) *The complete preimages of $\bar{\gamma}_i$ and $\bar{\gamma}_j$ in G are disjoint.*

PROOF. Without loss of generality we may assume that the surface F can be projected in one-to-one fashion onto the xy-plane. Divide the xy-plane into small squares δ and let F^δ denote the closed set on the surface whose projection is the square δ. Let M^δ be the closed set consisting of all the sets F^δ which cut the boundary γ of G.

If the squares δ are sufficiently small, the spherical images of H and M^δ are disjoint.

The open set $G - M^\delta$ is the union of finitely many domains (generally multiply connected) bounded by simple curves. Approximate the boundary of each of the domains by simple curves whose spherical images have zero area. Denote the union of the resulting domains by G_1. Obviously the approximation of the boundaries can be effected in such a way that G_1 contains H and the spherical images of H and $G - G_1$ are disjoint.

We now construct G_2. Let M_1 be the set of all points of G whose spherical images lie in the spherical image of $G - G_1$. By construction, $M_1 \cap H = 0$. The set G_2 is now constructed in the same way as G_1, with G replaced by $G_1 - M_1$. Now let M_2 be the set of all points of G whose spherical images lie in the spherical image of $G - G_2$; $M_2 \cap H = 0$, and construct G_3 in the same way, etc.

It is clear that the resulting sequence of sets G_k possesses the required properties 1), 2) and 3), and the lemma is proved.

$$\Big|\int\limits_{\delta_{\bar{\gamma}_k}} q_{\bar{\gamma}_k}(Y)\,dY\Big| \to 0 \quad \text{as} \quad \delta \to 0.$$

there exist arbitrarily large n such that

$$\Big|\int\limits_{\delta_{\bar{\gamma}_k^n}} q_{\bar{\gamma}_k^n}(Y)\,dY\Big| < \frac{\varepsilon}{2},$$

s precisely the set whose existence we wish to prove.
se that the assertion of the lemma is false. Then for any k and
r all sufficiently large n,

$$\Big|\int\limits_{\delta_{\bar{\gamma}_k^n}} q_{\bar{\gamma}_k^n}(Y)\,dY\Big| \geqslant \frac{\varepsilon}{2}.$$

e an arbitrary natural number. Since the sets $\tilde{\gamma}_k$ are pairwise
it follows that for sufficiently small δ the sets $\delta_{\bar{\gamma}_k^n}$, $k = 1, \cdots, m$,
pairwise disjoint. For sufficiently large n the absolute curvature
et $\tilde{\delta}_{\bar{\gamma}_k^n}$ $(k \leq m)$ is greater than $\epsilon/2$. Therefore the absolute curvature
greater than $m\epsilon/2$ for sufficiently large n. But this is impossible,
is arbitrary, while the absolute curvatures of the surfaces F_n
ormly bounded by assumption. This contradiction completes
f of Lemma 2.

3. *Let F be a surface of bounded extrinsic curvature and G an*
n F satisfying the assumptions of Lemma 1. Let F_n be a sequence
surfaces with uniformly bounded positive and negative curvatures,
g to F together with their spherical images.
or any $\epsilon > 0$ at least one of the domains G_k constructed in Lemma
property that there exist arbitrarily large values of n such that
$\omega(G_k^n)| < \epsilon$.

Via the homeomorphism defining the convergence of F_n to
rvatures of the surfaces F_n may be regarded as set functions
ecisely, if E is an arbitrary set on F and E_n the corresponding
$_n$, we set

$$\omega_n(E) = \omega(E_n), \ \omega_n^+(E) = \omega^+(E_n), \ \omega_n^-(E) = \omega^-(E_n).$$

e that the complete preimage $\tilde{\gamma}_k$ of the curve $\bar{\gamma}_k$ has a δ-
hood $\delta_{\tilde{\gamma}_k}$ such that there exist arbitrarily large values of n for

LEMMA 2. *Let F be a surface of bounde*
open set on F satisfying the assumptions of
of regular surfaces with uniformly bounded
to F together with their spherical images.

Then, for any $\epsilon > 0$, at least one of the sets
in Lemma 1 is such that for infinitely many

$$|\sigma(G_k) - \sigma(G_k^n)|$$

where G_k^n is the set on F_n corresponding to

PROOF. To simplify the notation, we
G_k consists of a single domain homeomorp
curve γ_k. Then the total curvatures of the
surfaces F and F_n are

$$\sigma(G_k) = \int_\omega q_{\bar{\gamma}_k}(Y)\,dY, \quad \sigma(G_k^n) =$$

where q denotes the degree of an arbitrary
relative to the indicated curve, and dY is the

Let $\delta_{\bar{\gamma}_k}$ and $\delta_{\bar{\gamma}_k^n}$ denote the δ-neighborho
$\bar{\gamma}_k^n$, respectively. Then, since the spherical i
which implies that the $\bar{\gamma}_k^n$ converge uniforml
that

$$\int_{\omega - \delta_{\bar{\gamma}_k^n}} q_{\bar{\gamma}_k^n}(Y)\,dY \to \int_{\omega - \delta_{\bar{\gamma}_k}} q_{\bar{\gamma}_k}$$

Let $\tilde{\gamma}_k$ be the complete preimage of the curve
absolute curvature of $\tilde{\gamma}_k$ is zero, since the sp
curve $\bar{\gamma}_k$, whose area is zero. Let U be an ope
absolute curvature is arbitrarily small. For
complete preimage $\tilde{\delta}_{\bar{\gamma}_k}$ of the set $\delta_{\bar{\gamma}_k}$ lies in the
its absolute curvature is arbitrarily small.

The quantity

$$\left|\int_{\delta_{\bar{\gamma}_k}} q_{\bar{\gamma}_k}(Y)\,dY\right|$$

cannot exceed the absolute curvature of $\tilde{\delta}_{\bar{\gamma}_k}$
$-n^-(Y)$ for almost all Y, where $n^+(Y)$ is the n
and $n^-(Y)$ the number of hyperbolic points in $\tilde{\delta}$
is Y. Consequently

Now, if

then G_k

Supp

$\delta > 0$,

Let m
disjoin
are als
of each
of F_n is
since
are un
the pr

LEM
open s
of regu
conver

The
1 has
$|\omega(G_k)$

PRO
F, th
on F.
set o

Ass
neigh
whic

$$\omega^+\left(\delta_{\tilde{\gamma}_k}\right) < \frac{\varepsilon}{2}, \quad \omega^-\left(\delta_{\tilde{\gamma}_k}\right) < \frac{\varepsilon}{2},$$

$$(*) \qquad \omega_n^+\left(\delta_{\tilde{\gamma}_k}\right) < \frac{\varepsilon}{2}, \quad \omega_n^-\left(\delta_{\tilde{\gamma}_k}\right) < \frac{\varepsilon}{2}.$$

We claim that there are then arbitrarily large values of n such that

$$|\omega(G^k) - \omega_n(G^k)| < \epsilon.$$

Indeed, without loss of generality we may assume that the total intrinsic curvature of F_n converges weakly to the total curvature of F. This means that for any continuous function f defined on F

$$\int_F f\, d\omega_n \to \int_F f\, d\omega.$$

Define a function f by

$$f(X) = 1 \quad \text{if} \quad X \in G_k - \delta_{\tilde{\gamma}_k},$$

$$f(X) = 0, \quad \text{if} \quad X \in F - G_k - \delta_{\tilde{\gamma}_k},$$

$$f(X) = \frac{\mu(X)}{\lambda(X) + \mu(X)}, \quad \text{if} \quad X \in \delta_{\tilde{\gamma}_k}.$$

The function f is clearly continuous.

For sufficiently large n satisfying inequalities (*),

$$\left|\int_F f\, d\omega_n - \omega_n(G_k)\right| < \frac{\varepsilon}{2}, \quad \left|\int_F f\, d\omega - \omega(G_k)\right| < \frac{\varepsilon}{2}.$$

Hence $|\omega(G_k) - \omega_n(G_k)| < \epsilon$ for these values of n. Thus, if our assumption that there exists a δ-neighborhood of $\tilde{\gamma}_k$ as stipulated above is valid, then the domain G_k has the desired property.

Suppose the assumption false for every curve $\tilde{\gamma}_k$. Then for any $\delta > 0$, any k, and sufficiently large n, at least one of the numbers $\omega^+(\delta_{\tilde{\gamma}_k})$, $\omega^-(\delta_{\tilde{\gamma}_k})$, $\omega_n^+(\delta_{\tilde{\gamma}_k})$ or $\omega^-(\delta_{\tilde{\gamma}_k})$ is not smaller than $\epsilon/2$.

Let m be any natural number. Take δ so small that the sets $\delta_{\tilde{\gamma}_k}$ $(k \leqq m)$ are disjoint. Set $M = \delta_{\tilde{\gamma}_1} + \cdots + \delta_{\tilde{\gamma}_m}$. Then there exist n such that at least one of the numbers $\omega^+(M)$, $\omega^-(M)$, $\omega_n^+(M)$ or $\omega_n^-(M)$ is not smaller than $m\epsilon/8$. But this is impossible, since m is arbitrary and the set functions in question are uniformly bounded. This proves the lemma.

LEMMA 4. *Let F be a surface of bounded extrinsic curvature and G an open set with boundary γ on the surface. Assume that the spherical images of G and γ are disjoint. Let F_n be a sequence of regular surfaces with* *uniformly bounded absolute curvatures, converging to F together with their spherical images. Then for any closed set $H \subset G$ and number $\epsilon > 0$ there is an open set G^* containing H and contained in G such that there exist arbitrarily large values of n for which*

$$|\sigma(G^*) - \sigma(G^{*n})| < \epsilon, \quad |\omega(G^*) - \omega(G^{*n})| < \epsilon.$$

PROOF. Construct the sequence of open sets G_k of Lemma 1. By Lemma 2 there exists a set G_{k_1} such that for infinitely many n

$$|\sigma(G_{k_1}) - \sigma(G_{k_1}^n)| < \epsilon.$$

Omitting G_{k_1} from the sequence G_k, we get a new sequence, from which we extract G_{k_2} and so on. Thus there exist an infinite number of sets G_{k_s} with this property. By Lemma 3 at least one of the sets G_{k_s} is such that for infinitely many n

$$|\sigma(G_{k_s}) - \sigma(G_{k_s}^n)| < \varepsilon, \quad |\omega(G_{k_s}) - \omega(G_{k_s}^n)| < \varepsilon.$$

This set G_{k_s} is precisely the open set G^* whose existence we were to prove.

LEMMA 5. *Let G be a small domain on a surface F of bounded extrinsic curvature, γ its boundary, $\bar{\gamma}$ the spherical image of γ and $\tilde{\gamma}$ its complete preimage in G.*

Then the total intrinsic and total extrinsic curvatures of the open set $G - \tilde{\gamma}$ are equal.

PROOF. Let H be a closed subset of $G - \tilde{\gamma}$ such that

$$|\sigma^0(G - \tilde{\gamma}) - \sigma^0(H)| < \epsilon, \quad |\omega^0(G - \tilde{\gamma}) - \omega^0(H)| < \epsilon,$$

where ϵ is any positive number.

We construct a sequence of analytic surfaces F_n with uniformly bounded absolute curvatures, converging to F together with their spherical images. By Lemma 4 there exists an open subset G^* of $G - \tilde{\gamma}$, containing H, such that

$$|\sigma(G^*) - \sigma(G^{n*})| < \epsilon, \quad |\omega(G^*) - \omega(G^{n*})| < \epsilon.$$

By the choice of H,

$$|\sigma(G^*) - \sigma(G - \tilde{\gamma})| < \epsilon, \quad |\omega(G^*) - \omega(G - \tilde{\gamma})| < \epsilon.$$

Since Gauss's Theorem is valid for the surfaces F_n, we have $\sigma(G^{n*}) = \omega(G^{n*})$, and thus the above four inequalities directly imply

$$|\sigma(G - \tilde{\gamma}) - \omega(G - \tilde{\gamma})| < 4\epsilon.$$

And since ϵ is arbitrary, it follows that $\sigma(G - \tilde{\gamma}) = \omega(G - \tilde{\gamma})$. Q.E.D.

Lemma 6. *Let the area of the spherical image of the boundary γ of an open set G be zero. Then $\omega(G) \leqq \sigma(G)$.*

Proof. Let $\tilde{\gamma}$ be the complete preimage of the spherical image of γ. By Lemma 5, $\sigma(G-\tilde{\gamma}) = \omega(G-\tilde{\gamma})$; but $\sigma(G-\tilde{\gamma}) = \sigma(G)$. Hence it will suffice to show that $\omega(G) \leqq \omega(G-\tilde{\gamma})$.

Let $\delta_{\tilde{\gamma}}$ denote a δ-neighborhood of the set $\tilde{\gamma}$. It is the limit of the corresponding sets on the surfaces F_n (§7), which have arbitrarily small positive curvatures when δ is sufficiently small. Therefore the positive intrinsic curvature of the limit surface on $\delta_{\tilde{\gamma}}$ is small for small δ. Consequently the total intrinsic curvature of the set $G \cap \tilde{\gamma}$ is nonpositive, and, since ω is completely additive, $\omega(G) \leqq \omega(G-\tilde{\gamma})$. This proves the lemma.

Lemma 7. *For any Borel set H on a surface of bounded extrinsic curvature, $\omega(H) \leqq \sigma(H)$.*

Proof. We first prove the assertion for an arbitrary open set G. Lemma 6 gives the result only for sufficiently small open sets, since it employs regular approximation of the surface, which is a local property.

Let F be a closed subset of G such that the absolute intrinsic and extrinsic curvatures of the set $G-F$ are smaller than ϵ. Divide the space into small cubes by planes parallel to the coordinate planes, such that the absolute intrinsic and extrinsic curvatures of the set of points on the surface lying on the dividing planes are zero. This is easily done, since the absolute curvatures are bounded and completely additive. If the cubes are sufficiently small, Lemma 6 will be applicable to the domain cut from the surface by each cube.

Let G^* be the union of all those intersections of cubes with the surface which contain points of F. Obviously, if the cubes are small, we have $F \subset G^* \subset G$. Combined with the complete additivity of the curvatures, Lemma 6 gives $\omega(G^*) \leqq \sigma(G^*)$. And since

$$|\omega(G) - \omega(G^*)| < \epsilon, \quad |\sigma(G) - \sigma(G^*)| < \epsilon,$$

and ϵ is arbitrary, it follows that $\omega(G) \leqq \sigma(G)$.

It now follows from standard theorems on additive set functions that this inequality is valid not only for open sets but for arbitrary Borel sets.

Lemma 8. *Let H be a Borel set on a surface of bounded extrinsic curvature such that $\omega(H) = \sigma(H)$. Then $\omega(H') = \sigma(H')$ for any Borel subset H' of H.*

In fact, by Lemma 7,

$$\omega(H') \leqq \sigma(H'), \quad \omega(H - H') \leqq \sigma(H - H').$$

Hence in view of the additivity of the functions ω, σ and the equality $\omega(H) = \sigma(H)$ we get $\omega(H') = \sigma(H')$.

Theorem 1. *Every regular point X on a surface of bounded extrinsic curvature has a neighborhood such that $\omega(H) = \sigma(H)$ for any Borel subset H of this neighborhood.*

Proof. Since X is a regular point, it has a neighborhood G such that neither G nor its boundary γ contains points at which the tangent planes are parallel to the tangent plane at X. Let $\bar{\gamma}$ be the spherical image of the boundary of G and $\tilde{\gamma}$ the complete preimage of $\bar{\gamma}$ on the surface. The open set $G - \tilde{\gamma}$ contains X and is therefore a neighborhood of X.

By Lemma 5 we have $\omega(G - \tilde{\gamma}) = \sigma(G - \tilde{\gamma})$. And now, by Lemma 8, for any Borel set $H \subset G - \tilde{\gamma}$ we have $\omega(H) = \sigma(H)$. This proves the theorem.

Theorem 2. *There exists an open set G on a surface of bounded extrinsic curvature, containing all regular points of the surface, such that $\omega(H) = \sigma(H)$ for any Borel subset H of G.*

In particular, if the surface is quasiregular, i.e. every point on the surface is regular, this equality holds for any Borel set H.

Proof. Let $G(X)$ be the neighborhood of X whose existence is proved in Theorem 1. Set $G = \sum_{(X)} G(X)$, where the summation extends over all regular points of the surface. The family of neighborhoods $G(X)$ contains a countable set of neighborhoods covering G. Let these be $G_1, G_2, \cdots$ and let H be an arbitrary Borel subset of G. Set

$$H_k = H \cap (G_k - G_{k-1} - \cdots - G_1).$$

Obviously $H_k \subset G_k$ and $\sum_1^\infty H_k = H$. By Theorem 1 and the complete additivity of ω and σ we have $\omega(H) = \sigma(H)$. Q.E.D.

Theorem 3. *For any Borel set H on a closed surface of bounded extrinsic curvature, $\omega(H) = \sigma(H)$.*

Proof. The total intrinsic and extrinsic curvatures of a closed surface of bounded extrinsic curvature are both equal to $2\pi\chi$, where χ is the Euler characteristic of the surface. By Lemma 8 this implies that the total intrinsic and extrinsic curvatures are equal for any Borel set on the surface. Q.E.D.

THEOREM 4. *For any Borel set H on a surface F of bounded extrinsic curvature,*

$$\omega^+(H) = \sigma^+(H), \quad \omega^-(H) = \sigma^-(H).$$

PROOF. Let E be the set of elliptic points on the surface. Then $\omega^+(F-E) = 0$. In fact, otherwise there would be a subset $M \subset F-E$ such that $\omega(M) > 0$, and this is impossible since $\sigma(M) = 0$ and $\omega(M) \leqq \sigma(M)$.

On the other hand, $\omega^-(E) = 0$, since otherwise there would be a subset N of E such that $\omega^-(N) < 0$. But since $N \subset E$, we must have $\omega(N) = \sigma(N) > 0$.

Now

$$\sigma^+(F) = \sigma(E) = \omega(E) = \omega^+(\Phi).$$

By Lemma 8, this implies that $\sigma^+(H) = \omega^+(H)$ for any H. And since $\sigma(H) \geqq \omega(H)$, it follows that $\omega^-(H) \leqq \sigma^-(H)$. This completes the proof.

THEOREM 5. *Let F be a surface of bounded extrinsic curvature and X an arbitrary point on F. Then there exists a neighborhood U of X and an infinite sequence of analytic surfaces F_n with the following properties:*

1. *The surfaces F_n converge to F in U together with their spherical images.*

2. *The positive extrinsic curvatures of F_n converge weakly to the positive extrinsic curvature of F in U.*

3. *The positive, negative and total intrinsic curvatures of F_n converge weakly to the corresponding curvatures of F in U.*

If the point X is regular, then the negative and total extrinsic curvatures of F_n also converge weakly to the corresponding curvatures of F in U.

PROOF. On the basis of our previous results, we may assume that the sequence F_n already possesses properties 1 and 2, and moreover the total intrinsic curvatures of F_n converge weakly to the total intrinsic curvature of F.

By Theorem 4 the positive intrinsic curvatures of F_n also converge weakly, and hence the negative intrinsic curvatures also converge weakly (since the total and positive curvatures converge).

By similar reasoning, if X is a regular point the negative and total extrinsic curvatures of F_n must converge to the corresponding curvatures of F. This completes the proof of Theorem 5.

We now consider surfaces of nonnegative intrinsic and bounded extrinsic curvature. We begin with surfaces of zero intrinsic curvature, i.e. surfaces whose positive and negative intrinsic curvatures vanish. Such a surface can contain neither elliptic nor hyperbolic points.

Indeed, were the surface to contain a hyperbolic point, its negative extrinsic curvature would be different from zero. Consequently the surface would contain an uncountable set of hyperbolic points, with nonzero area of spherical image.

On the set of hyperbolic points of the surface, the total intrinsic and total extrinsic curvatures are equal. And since the extrinsic curvature is known to be negative, the same holds for the intrinsic curvature. But this is impossible, since the total curvature on any set of points on the surface must vanish. Thus there can be no hyperbolic points on a surface with zero intrinsic curvature.

Similar reasoning shows that a surface of zero intrinsic curvature can contain no elliptic points. Hence the positive and negative extrinsic curvatures of such a surface are zero. Hence, using the theorems of §4, we get

THEOREM 6. *Any surface of bounded extrinsic curvature and zero intrinsic curvature is developable* (*i.e. locally isometric to a plane*). *It has the usual structure of a regular developable*: *straight-line generators with stationary tangent plane along each generator.*

THEOREM 7. *A complete surface with zero intrinsic curvature and bounded extrinsic curvature is a cylinder.*

We now consider surfaces of nonzero positive intrinsic curvature and zero negative intrinsic curvature. There can be no hyperbolic points on such a surface; the proof is identical with that for the case of surface of zero curvature.

Thus surfaces of nonnegative intrinsic curvature are also surfaces of nonnegative extrinsic curvature. Hence the following theorems are valid (§5).

THEOREM 8. *A surface with positive intrinsic curvature and bounded extrinsic curvature, whose boundary lies in a plane, is a convex surface* (*i.e. a domain on the boundary of a convex body*).

THEOREM 9. *A complete surface with bounded extrinsic curvature and positive intrinsic curvature is either a closed convex surface or an unbounded convex surface.*

As we mentioned at the beginning of this chapter, Nash and Kuiper have proved the existence of a great variety of different C^1-isometric embeddings of a given two-dimensional manifold with regular metric

[**47**, **38**]. Their results imply, among other things, that the sphere admits isometric mappings onto a smooth surface F of arbitrarily small diameter. This indicates that smoothness of a surface is too weak a condition to yield truly significant relations between the intrinsic metric of a surface and its extrinsic form. The situation changes drastically if smoothness is replaced by two-fold differentiability. It is natural to ask: for what values of $\alpha > 0$ are the relations between the intrinsic metric and the extrinsic form of surfaces of class $C^{1,\alpha}$ as rich as for regular twice-differentiable surfaces, and for what values of α are these relations weakened to the same degree as for C^1 surfaces? This question has been studied in a series of papers by Borisov [**24**]. His main result is the following theorem.

Any surface of class $C^{1,\alpha}$ with intrinsic metric of fixed sign (the curvatures of all open sets are not zero and have the same sign) is a surface of bounded extrinsic curvature for any $\alpha > 2/3$. Conversely, for sufficiently small α (for example, $\alpha < 1/13$) even analyticity of the metric does not ensure bounded extrinsic curvature.

One result of this theorem, in combination with the results of §5, is that a surface of class $C^{1,\alpha}$, $\alpha > 2/3$, with positive intrinsic curvature is locally convex. Moreover, by the theorem on regularity of a convex surface with regular metric, we have the following result: If a surface of class $C^{1,\alpha}$, $\alpha > 2/3$, has regular metric and positive Gauss curvature, then the surface is regular. As we mentioned above, for sufficiently small α the metric can even be analytic without the surface being regular.

APPENDIX

UNSOLVED PROBLEMS

Young geometers often experience difficulties in choosing problems for research. We therefore present a few unsolved problems in this Appendix, most of them with hints for solution.

1. Prove the following assertion: The spherical image of a geodesic on a convex surface is a rectifiable curve, i.e. every interior point of the geodesic has a neighborhood whose spherical image has finite length.

We suggest the following approach. First note that it is sufficient to prove rectifiability for the spherical image of a small arc of the geodesic. We may therefore assume that the convex surface is closed. It can always be completed to a complete surface, in such a way that a small arc of a segment (shortest join) remains a segment. Moreover, we may assume that the segment under consideration can be continued as a segment beyond at least one of its endpoints. This is true for any arc of a segment. Thus we may confine attention to a segment γ on a closed convex surface F, which can be continued as a segment beyond one of its endpoints.

Construct a sequence of polyhedra P_n converging to the surface F. Without loss of generality we may assume that the endpoints A and B of the segment lie on all polyhedra P_n. Let γ_n be the segment on P_n connecting A and B. By the inclusion property for segments (Chapter I, §3) the sequence γ_n converges to γ. The spherical images γ_n^* of the segments γ_n converge to the spherical image γ^* of γ. To prove that γ^* is rectifiable, it now suffices to show that the lengths of the curves γ_n^* are uniformly bounded.

Let $\alpha_1, \alpha_2, \cdots$ be the faces of the polyhedron P_n through which the segment γ_n passes, $E_1, E_2, \cdots$ the halfspaces defined by the planes of the faces $\alpha_1, \alpha_2, \cdots$. The intersection of the halfspaces E_n is a solid polyhedron P_n'. Let T be a cube containing all the polyhedra P_n. The intersection of the cube T and the polyhedron P_n' is a certain polyhedron Q_n contained in T. The segment γ_n of P_n lies on the polyhedron Q_n. Since P_n is contained in Q_n, it follows that γ_n is also a segment on Q_n (by Busemann's theorem).

Let $k_1, k_2, \cdots$ be the edges of the polyhedron Q_n which cut the segment γ_n, $\delta_1, \delta_2, \cdots$ their lengths and $\vartheta_1, \vartheta_2, \cdots$ the exterior angles at the edges $k_1, k_2, \cdots$. Then the length of the spherical image of γ_n, i.e. the length of γ_n^*, is equal to $s_n = \vartheta_1 + \vartheta_2 + \cdots$. The quantity $\vartheta_1\delta_1 + \vartheta_2\delta_2 + \cdots$ can be estimated in terms of the integral mean curvature of the polyhedron Q_n, hence in terms of the edge of the cube T containing Q_n. It follows that if $\delta_k > c_0 > 0$ for all k, then $s_n \leqq C(T)/c_0$.

Now assume that the polyhedron Q_n has edges k_s of length less than ϵ_n, and the sum of exterior angles ϑ_s over these edges is greater than σ_n, where $\epsilon_n \rightarrow 0$ and $\sigma_n \rightarrow \infty$ as $n \rightarrow \infty$. One proves that for sufficiently large n the curve γ_n cannot be a segment on Q_n. The reason is that for small ϵ_n and large σ_n a "large amount of curvature" is concentrated near γ_n. Therefore the length of γ_n cannot be the absolute minimum of the lengths of curves connecting A and B on the polyhedron Q_n.

2. Prove the following theorem.

A convex surface, homeomorphic to a disk, with nonnegative (nonpositive) integral geodesic curvature (i.g.c.) along the boundary, can be applied to any isometric surface (i.e. continuously bent into it).

We indicate an approach to the proof. First consider a polyhedron. Let P_1 be a convex polyhedron homeomorphic to a disk, whose angles at the boundary vertices are $\geqq \pi$ (the i.g.c. along the boundary is nonpositive). Let P_2 be a convex polyhedron isometric to P_1. We must show that P_1 can be applied to P_2. Let Q_1 and Q_2 be the convex hulls of P_1 and P_2. They are closed polyhedra. The polyhedron Q_i is the union of P_i and some polyhedron P_i' isometric to a convex plane polygon. Let A and B be two vertices of the polyhedron Q_i on the boundary of the polyhedron P_i, α and β the curvature at these vertices. Connect A and B by a segment γ within the domain P_i'. (This is possible, since P_i' is a convex domain.) Now take two plane triangles with base l equal to the length of γ, and angles $\alpha' \leqq \alpha$, $\beta' \leqq \beta$ at the base. Glue these triangles together along their lateral sides, and glue the bases to the polyhedron Q_i cut along the segment γ. By the Gluing Theorem there exists a closed polyhedron Q_i' which realizes the polyhedral metric obtained by gluing the triangles to the cut. Subjecting the angles α' and β' to a continuous variation from zero to α and β, respectively, we get a continuous deformation of Q_i' (because of monotypy). The domain on Q_i' corresponding under isometry to P_i then undergoes a continuous bending.

Now take two other vertices on the boundary of P_i, or one vertex on the boundary and a new vertex generated by the above gluing procedure.

Connect these vertices by a segment, cut the polyhedron along the segment, and glue two new triangles to the cut. A finite number of repetitions of this procedure transforms the original polyhedron Q_i into a polyhedron $\overline{Q}_i$ which is the union of a polyhedron isometric to P_i and a polyhedron $\widetilde{P}_i$ isometric to a cone. The boundaries of $\widetilde{P}_1$ and $\widetilde{P}_2$ have equal i.g.c. It follows easily that they are isometric. The monotypy theorem for polyhedra then implies that $\overline{Q}_1$ and $\overline{Q}_2$ are congruent, and hence so are the domains on them isometric to P_1 and P_2. We have thus transformed the original polyhedra P_1 and P_2 by a continuous bending into congruent polyhedra. Hence each can be applied to the other.

In order to proceed now from polyhedra to general convex surfaces, one uses simultaneous approximation of isometric convex surfaces by isometric polyhedra and the monotopy theorem for general convex surfaces.

Now assume that the angles at the boundary vertices of the polyhedra P_i do not exceed π. We again consider the convex hulls Q_i. Let $\overline{P}_i$ be the polyhedron completing P_i to the closed polyhedron Q_i. To prove that P_1 can be applied to P_2, it suffices to show that $\overline{P}_1$ can be applied to $\overline{P}_2$, with the conditions governing its contact with P_1 observed at each stage of the deformation. This is the purpose of the following constructions.

By cutting and gluing triangles, one transforms the polyhedron Q_i into a polyhedron $\widetilde{Q}_i$ containing a domain $\widetilde{P}_i$ isometric to $\overline{P}_i$; the remainder of $\widetilde{Q}_i$ is a surface V_i isometric to a cone (Figure 35). Now flatten out the "leaves" of the polyhedron V_i, preserving their convexity. This transforms the domain $\widetilde{P}_i$ completing the polyhedron V_i to the convex hull into some domain $\widetilde{P}_i'$, while the polyhedron V_i becomes a domain on some polyhedral angle V_i'. Now apply the angle V_1' to the angle V_2'. When this is done, the domain $\widetilde{P}_1'$ is applied to P_2'. The result is a continuous deformation of $\overline{P}_1$ into $\overline{P}_2$, with the conditions governing its contact with P_1 observed at each stage of the deformation. One now proceeds as before to general convex surfaces.

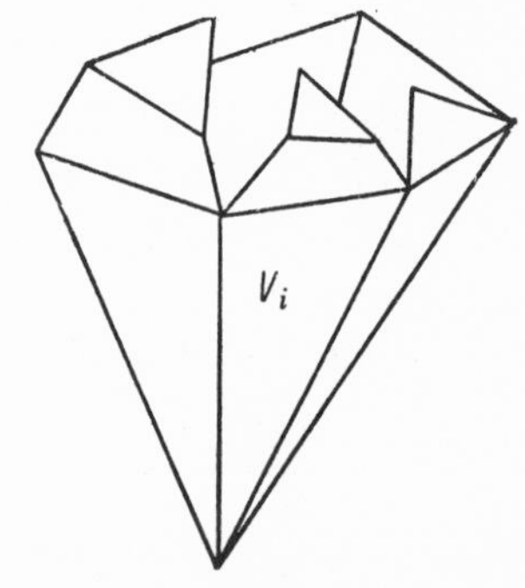

FIGURE 35

3. In §4 of Chapter IV we derived formulas associating with any

pair of isometric surfaces in elliptic space a pair of isometric surfaces in euclidean space. Conversely, each pair of isometric surfaces in euclidean space goes into a pair of isometric surfaces in elliptic space. These formulas involve a vector parameter e_0. Let F_1 and F_2 be two isometric surfaces in euclidean space. Determine the corresponding surfaces Φ_1 and Φ_2 in elliptic space with parameter $e_0 = e_0'$. Now proceed from the surfaces Φ_1 and Φ_2 to a pair of isometric surfaces F_1' and F_2' in euclidean space, using the formulas with parameter $e_0 = e_0''$. Derive formulas setting up the correspondence between the isometric surfaces F_1 and F_2 and the surfaces F_1' and F_2'. Find conditions under which the convexity of the surfaces F_1 and F_2 implies that of F_1' and F_2'. Varying the parameters e_0' and e_0'', and also the relative position of the surfaces F_1 and F_2, consider the problem of transforming a pair of unbounded isometric convex surfaces into a pair of bounded isometric surfaces. In particular, determine whether the monotypy problem for unbounded convex surfaces can be reduced in this way to the monotypy problem for closed convex surfaces or for convex surfaces with fixed boundary.

4. As in the elliptic case, studied in §4 of Chapter IV, formulas can be determined which associate with each pair of isometric surfaces in hyperbolic space a pair of isometric surfaces in euclidean space. Study this correspondence. In particular, find conditions under which convexity of the surfaces in hyperbolic space implies convexity of the corresponding surfaces in euclidean space. Can the monotypy problem for surfaces in hyperbolic space be reduced to the monotypy problem for euclidean space? Some partial results in this direction have been obtained by Gajubov [**34**], but they are far from complete.

5. An incomplete convex metric defined in a domain G is in general not realizable as a convex surface, for the simple reason that the total (integral) curvature of the manifold G with this metric may exceed 4π, while the curvature of a convex surface is always $\leqq 4\pi$. However, there are grounds for the assertion that, under very broad assumptions, this metric is realizable as a *locally* convex surface, i.e. a surface each point of which has a neighborhood which is a convex surface.

Here are some considerations on this problem. Let G be a doubly connected domain (homeomorphic to a circular annulus). Assume that the contours γ_1 and γ_2 bounding the domain are geodesic polygons in the given metric. Divide each side δ_1 of the polygon γ_1 into two by its midpoint P_1, and identify points on γ_1 equidistant from P_1. Do the same for the sides δ_2 of γ_2. The result is a closed manifold R whose curvature is nonnegative everywhere except for two points A_1 and A_2, at which the vertices of the polygons γ_1 and γ_2 are identified.

The manifold R is isometrically embeddable into the locally euclidean space considered in §11 of Chapter VI, with the points A_1 and A_2 on the z-axis. The required realization of R as a locally convex surface is now obtained by mapping the locally euclidean space into euclidean space, identifying geometrically identical points of these spaces.

6. A convex metric M defined in a domain G, homeomorphic to a disk, with boundary γ of nonnegative i.g.c. in the metric M, is realizable as a certain convex cap F. Consider the problem of realizability of a convex metric M defined in a domain G homeomorphic to a disk, whose boundary γ lies on a given surface Φ.

One attack on this problem is as follows. To simplify matters, assume that Φ is an unbounded surface which can be projected in one-to-one fashion onto the entire xy-plane. Let E^+ denote the region of space lying above the surface Φ. Construct a Riemannian space R from two mirror-symmetric copies of the euclidean region E^+ and a regular intermediate layer δ, such that when the thickness of the layer δ tends to zero the space R becomes a metric space R_0 consisting of the two copies of E^+ glued together along the boundary surface Φ. The constructed space R must be symmetric with respect to some totally geodesic surface σ within the layer δ, and the regions E^+ must be symmetric to each other with respect to σ.

Now form a closed manifold M' homeomorphic to a sphere, from two oppositely oriented copies of the manifold M and a regular layer h separating them in such a way that M' admits an inner symmetry with respect to a closed geodesic γ within the layer h, and moreover the two copies of M correspond by symmetry.

The manifold M' is now realized in the space R as a closed surface F' (it is assumed that this can be done). In view of the symmetry of the manifold M' and the space R, this realization can be so constructed that the geodesic γ lies in the surface σ and the surface F' is symmetric with respect to σ. Now letting the thickness of the layers δ and h tend to zero, we get a closed surface F_0 in the space R_0. The domain on this surface lying in the region E^+ furnishes the required realization of the manifold M as a convex surface with boundary on the surface Φ.

7. Complete the proof of the rigidity of multiply-connected locally convex surfaces in a Riemannian space, as indicated in §12 of Chapter VI.

8. Let F_1 and F_2 be two isometric, identically oriented convex surfaces. Assume that corresponding unit vectors on the surfaces satisfy the relation $\tau_1 + \tau_2 \neq 0$. Then the vector valued function $r = \frac{1}{2}(r_1 + r_2)$, where r_1 and r_2 are the radius vectors of corresponding points on F_1 and F_2, defines a convex surface F. Under certain additional assumptions this was proved in Chapters I and II.

The vector field $z = r_1 - r_2$ is a bending field for the surface F. Using this relation between a pair of isometric surfaces and the infinitesimal bendings of the mean surface F, many monotypy problems for convex surfaces can be reduced to the problem of the rigidity of the mean surface.

The following question is natural in this context. Can the surfaces F_1 and F_2 always be so placed that $\tau_1 + \tau_2 \neq 0$ for directions corresponding under the isometry? This is apparently the case for almost all (in measure) relative positions. Prove this assertion.

9. Consider the problem of the existence of a closed convex surface satisfying the equation $f(R_1, R_2, n) = \varphi(n)$, where R_1, R_2 are the principal radii of curvature and n the unit normal to the surface.

This problem can be attacked by the methods of §5 of Chapter VII.

10. Consider the existence problem for a convex surface F whose spherical image coincides with a given convex domain ω on the unit sphere, whose supporting function $H(n)$ coincides with a given continuous function on the boundary of the spherical image, and whose principal radii of curvature at each interior point of the surface satisfy the equation $f(R_1, R_2) = \varphi(n)$, where $f(R_1, R_2) \equiv g(R_1 R_2, R_1 + R_2)$ is strictly monotone in R_1 and R_2, i.e. $\partial f / \partial R_1 > 0$ and $\partial f / \partial R_2 > 0$.

Suppose that the domain ω is in the upper hemisphere $x^2 + y^2 + z^2 = 1$, $z > 0$. Set $h(x, y) = H(x, y, 1)$, where H is the supporting function of the required surface. The function h satisfies an elliptic equation

$$\Phi(h_{11}, h_{12}, h_{22}, x, y) = 0$$

(see §4, Chapter VII). The existence problem for F reduces to the solvability of the equation $\Phi = 0$. This can be treated on the basis of Bernšteĭn's theorem, first deriving a priori estimates for the posited solution and its first and second derivatives. One first considers the case of analytic functions f, φ and a domain ω bounded by an analytic contour.

For considerations relating to the derivation of a priori estimates, see §§3 and 5 of Chapter VII. We remark that if h_1 and h_2 are solutions of the equations $f = \varphi_1$ and $f = \varphi_2$, then $h_1 - h_2$ cannot assume a maximum in the interior of ω if $\varphi_1 \leqq \varphi_2$. To proceed from analytic to regular data, it suffices to establish a priori estimates for the second derivatives at interior points (see §3 of Chapter VII).

11. It is well known that metric duality can be defined in elliptic space. Let Φ be a convex surface in elliptic space and Φ' the surface polar to Φ (the surface Φ' is the envelope of the polars of the points of Φ). There is a natural correspondence between the points of Φ and Φ': any point P on Φ is associated with the point at which the surface Φ' is tangent to the polar of P. If the curvature K of the space is unity, the extrinsic curvature of Φ on an arbitrary set M is equal to the area of the corresponding set M' on the surface Φ'.

Delete some plane from the elliptic space, and interpret the remaining region on the three-dimensional hemisphere

$$x_0^2 + x_1^2 + x_2^2 + x_3^2 = 1, \quad x_0 > 0.$$

Let Φ be a closed convex surface in the spherical zone $0 < x_0 < \epsilon$. The polar surface Φ' lies in the ϵ-neighborhood of the pole $P(0,0,0,1)$.

The line element of the surface Φ can be expressed as

$$ds^2 = ds_0^2 + \lambda d\sigma^2,$$

where ds_0^2 is the line element of the unit sphere and $\lambda \to 0$ as $\epsilon \to 0$. The extrinsic curvature of the surface Φ is $K_e = \lambda\varphi_\sigma + O(\lambda^2)$, where φ_σ depends on the quadratic form σ.

Project a neighborhood of the pole P of the hemisphere onto the tangent hyperplane E of the hemisphere at P. Let $\overline{\Phi}'$ be the projection of the surface Φ' onto the euclidean space E. Subject the surface $\overline{\Phi}'$ to a similarity mapping with ratio of similitude $1/\lambda$ and let $\epsilon \to 0$. The curvature of the limit surface is $1/\varphi_\sigma$.

Using this construction, derive a new solution of Minkowski's problem, based on the theorem which states that a given metric ds^2 can be realized on a convex surface in elliptic space.

12. In §7 of Chapter VIII we considered the existence of a closed convex surface with given generalized curvature. Analytic interpretation of the result leads to a theorem on the solvability of a certain equation of a very general type defined on the sphere.

Consider the one-dimensional analog of this problem, relaxing the requirement that the curvature be positive. This yields a certain theorem on the existence of a closed curve with given generalized curvature. Analytically speaking, this implies the existence of a periodic solution of an equation $y'' = \varphi(x, y, y')$, where the function φ is periodic in x. For what classes of equations, i.e. for what functions φ, does the geometric theorem guarantee the existence of periodic solutions? Generalize the result to systems of equations

$$y_i'' = \varphi(x, y_1, \cdots, y_n, y_1', \cdots, y_n'), \quad i = 1, 2, \cdots, n.$$

The one-dimensional analog of Minkowski's problem is this: Prove that there exists a closed curve with given radius of curvature $R(\vartheta)$, as a function of the angle of rotation ϑ of the tangent. The problem has a solution if

$$\oint_0^{2\pi} e^{i\vartheta} R(\vartheta)\, d\vartheta = 0.$$

This condition always holds if $R(\vartheta + \pi) = R(\vartheta)$. The supporting function $p(\vartheta)$ of the required curve has a simple expression:

$$p(\vartheta) = \int_0^{\vartheta} e^{i(\vartheta + \tau)} R(\tau)\, d\tau.$$

Using Schauder's fixed-point principle, as in the existence proof for solutions of strongly elliptic Monge-Ampère equations (§8 of Chapter VIII), prove the most general possible theorem on the existence of a closed curve with given generalized length. Interpret the result in terms of the existence of periodic solutions of the equation $y'' = \varphi(x, y, y')$, where φ is periodic in x. Employing geometric ideas, study the problem analytically, under the broadest possible assumptions on the equation. Consider the case of systems of equations.

BIBLIOGRAPHY

G. M. ADELSON-WELSKY

1. *Généralisation d'un théorème géométrique de M. Serge Bernstein,* C. R. (Dokl.) Acad. Sci. URSS **49** (1945), 391-392. MR **8**, 91.

A. D. ALEKSANDROV

2. *Intrinsic geometry of convex surfaces,* OGIZ, Moscow, 1948; German transl., Akademie Verlag, Berlin, 1955. MR **10**, 619; **17**, 74.
3. *Convex polyhedra,* GITTL, Moscow, 1950; German transl., Math. Lehrbücher und Monographien, Forschungsinstitut für Math. II. Abteilung: Math. Monographien, Band VIII, Akademie-Verlag, Berlin, 1958. MR **12**, 732; **19**, 1192.
4. *On a class of closed surfaces,* Mat. Sb. **4** **(46)** (1938), 69-77. (Russian; German summary)
5. *Smoothness of a convex surface of bounded Gaussian curvature,* C. R. (Dokl.) Acad. Sci. URSS **36** (1942), 195-199. MR **4**, 169.
6. *Existence and uniqueness of a convex surface with a given integral curvature,* C. R. (Dokl.) Acad. Sci. URSS **35** (1942), 131-134. MR **4**, 169.
7. *Almost everywhere existence of the second differential of a convex function and some properties of convex surfaces connected with it,* Učen. Zap. Leningrad. Gos. Univ. **37** Mat. **3** (1939), 3-35. (Russian) MR **2**, 155.
8. *Dirichlet's problem for the equation* $\operatorname{Det}\|z_{ij}\| = \Phi(z_1, \cdots, z_n, z, x_1, \cdots, x_n)$. I, Vestnik Leningrad. Univ. Ser. Mat. Meh. Astr. **13** (1958), no. 1, 5-24. (Russian) MR **20** # 3385.
9. *Some theorems on partial differential equations of the second order,* Vestnik Leningrad. Univ. **9** (1954), no. 8, 3-17. (Russian) MR **17**, 493.
10. *Curvature of convex surfaces,* C. R. (Dokl.) Acad. Sci. URSS **50** (1945), 23-26. MR **14**, 577.
11. *The inner metric of a convex surface in a space of constant curvature,* C. R. (Dokl.) Acad. Sci. URSS **45** (1944), 3-6. MR **7**, 167.
12. *On infinitesimal deformations of irregular surfaces,* Mat. Sb. **1** **(43)** (1936), 307-322. (Russian; German summary)
13. *Sur les théorèmes d'unicité pour les surfaces fermees,* C. R. (Dokl.) Acad. Sci. URSS **22** (1939), 99-102.
14. *On the work of S. E. Cohn-Vossen,* Uspehi Mat. Nauk **2** (1947), no. 3 (19), 107-141. (Russian) MR **9**, 485.
15. *Uniqueness theorems for surfaces in the large.* I, Vestnik Leningrad. Univ. **11** (1956), no. 19, 5-17; English transl., Amer. Math. Soc. Transl. (2) **21** (1962), 341-353.
16. *On the theory of mixed volumes of convex bodies.* III. *Extension of two theorems of Minkowski on convex polyhedra to arbitrary convex bodies,* Mat. Sb. **3** **(45)** (1938), 27-46. (Russian; German summary)

A. D. ALEKSANDROV AND E. P. SEN'KIN

17. *On the rigidity of convex surfaces,* Vestnik Leningrad. Univ. **10** (1955), no. 8, 3-13; supplement, ibid. **11** (1956), no. 1, 104-106. (Russian) MR **17**, 295; 998.

A. D. ALEKSANDROV AND V. A. ZALGALLER

18. *Two-dimensional manifolds of bounded curvature,* Trudy Mat. Inst. Steklov. **63** (1962); English transl., *Intrinsic geometry of surfaces,* Transl. Math. Monographs, vol. 15, Amer. Math. Soc., Providence, R. I., 1967. MR **27** # 1911.

I. JA. BAKEL'MAN

19. *Geometric methods of solution of elliptic equations*, "Nauka", Moscow, 1965. (Russian) MR **33** #2933.

S. N. BERNŠTEĬN

20. *Renforcement de mon théorème sur les surfaces à courbure négative*, Izv. Akad. Nauk SSSR Ser. Mat. **6** (1942), 285-290. MR **5**, 14.

21. *Investigation and integration of second order partial differential equations of elliptic type*, Soobšč. Har'kov. Mat. Obšč. (2) **11** (1910), 1-164. (Russian)

W. BLASCHKE

22. *Vorlesungen über Differentialgeometrie und geometrische Grundlagen von Einsteins Relativitätstheorie*. Band I: *Elementare Differentialgeometrie*, Springer, Berlin, 1924; reprint, Dover, New York, 1945; Russian transl., ONTI, Moscow, 1935. MR **7**, 391.

T. BONNESEN AND W. FENCHEL

23. *Theorie der konvexen Körper*, Springer, Berlin, 1934; reprint, Chelsea, New York, 1948.

JU. F. BORISOV

24. *The parallel translation on a smooth surface*. I, II, III, Vestnik Leningrad. Univ. **13** (1958), no. 7, 160-171; ibid. **13** (1958), no. 19, 45-54; ibid. **14** (1959), no. 1, 34-50; errata, ibid. **15** (1960), no. 19, 127-129. (Russian) MR **21** #3032; #3033; #3034; **24** #A1078.

H. BUSEMANN AND W. FELLER

25. *Krümmungseigenschaften konvexer Flächen*, Acta Math. **66** (1936), 1-47.

É. CARTAN

26. *Leçons sur la géométrie des espaces de Riemann*, Gauthier-Villars, Paris, 1928; Russian transl., ONTI, Moscow, 1936.

S. E. COHN-VOSSEN

27. *Bending of surfaces in the large*, Uspehi Mat. Nauk **1** (1936), 33-76; reprinted in [**28**], pp. 19-86. (Russian) MR **22** #4039.

28. *Some problems of differential geometry in the large*, Fizmatgiz, Moscow, 1959. (Russian) MR **22** #4039.

I. A. DANELIČ

29. *The uniqueness of certain convex surfaces in Lobačevskiĭ space*, Dokl. Akad. Nauk SSSR **115** (1957), 217-219. (Russian) MR **19**, 979.

A. A. DUBROVIN

30. *Regularity of a convex surface with a regular metric in spaces of constant curvature*, Ukrain. Geometr. Sb. Vyp. **1** (1965), 16-27. (Russian) MR **35** #6102.

N. V. EFIMOV

31. *Qualitative problems of the theory of deformations of surfaces*, Uspehi Mat. Nauk **3** (1948), no. 2 (24), 47-158; English transl., Amer. Math. Soc. Transl. (1) **6** (1962), 274-423. MR **10**, 324.

32. *Qualitative problems of the theory of deformations of surfaces in the small*, Trudy Mat. Inst. Steklov. **30** (1949). (Russian) MR **12**, 531.

33. *On rigidity in the small*, Dokl. Akad. Nauk SSSR **60** (1948), 761-764. (Russian) MR **10**, 324.

G. N. GAJUBOV

34. *Some questions on the theory of surfaces in hyperbolic space*, Taškent. Gos. Univ. Naučn. Trudy Vyp. 245 (1964), 17-22. (Russian) MR **37** #2141.

E. HEINZ

35. *Neue a-priori Abschätzungen für den Ortsvektor einer Fläche positiver Gaussscher Krümmung durch ihr Linienelement*, Math. Z. **74** (1960), 129-157. MR **22** #7089.

36. *Über die Differentialungleichung* $0 < \alpha \leqq rt - s^2 \leqq \beta < \infty$, Math. Z. **72** (1959/60), 107-126. MR **22** #7088.

A. S. KRONROD

37. *On functions of two variables,* Uspehi Mat. Nauk **5** (1950), no. 1 (35), 24-134. (Russian) MR **11**, 648.

N. KUIPER

38. *On C^1-isometric imbeddings.* I, Nederl. Akad. Wetensch. Proc. Ser. A **58** = Indag. Math. **17** (1955), 545-556; Russian transl., Matematika **1** (1957), no. 2, 17-28. MR **17**, 782.

A. S. LEĬBIN

39. *On the deformability of convex surfaces with a boundary,* Uspehi Mat. Nauk **5** (1950), no. 5 (39), 149-159. (Russian) MR **12**, 733.

H. LEWY

40. *A priori limitations for solutions of Monge-Ampère equations,* Trans. Amer. Math. Soc. **37** (1935), 417-434; Russian transl., Uspehi Mat. Nauk **3** (1948), no. 2 (24), 191-215. MR **10**, 41.

I. M. LIBERMAN

41. *Geodesic lines on convex surfaces,* C. R. (Dokl.) Acad. Sci. URSS **32** (1941), 310-313. MR **6**, 100.

A. I. MEDJANIK

42. *On the central symmetry of a closed strictly convex surface,* Ukrain. Geometr. Sb. Vyp. 2 (1966), 52-58. (Russian) MR **35** #3612.

T. MINAGAWA

43. *Remarks on the infinitesimal rigidity of closed convex surfaces,* Kōdai Math. Sem. Rep. **8** (1956), 41-48. MR **18**, 503.

H. MINKOWSKI

44. *Volumen und Oberfläche,* Math. Ann. **57** (1903), 447-495.

C. MIRANDA

45. *Equazioni alle derivate parziali di tipo ellittico,* Ergebnisse der Mathematik und ihrer Grenzgebiete, Heft 2, Springer-Verlag, Berlin, 1955; English transl., Springer-Verlag, Berlin, 1970; Russian transl., IL, Moscow, 1957. MR **19**, 421.

46. *Su un problema di Minkowski,* Rend. Sem. Mat. Roma **3** (1939), 96-108. MR **1**, 86.

J. F. NASH, JR.

47. *C^1-isometric imbeddings,* Ann. of Math. (2) **60** (1954), 383-396; Russian transl., Matematika **1** (1957), no. 2, 3-16. MR **16**, 515.

L. NIRENBERG

48. *On nonlinear elliptic partial differential equations and Hölder continuity,* Comm. Pure Appl. Math. **6** (1953), 103-156; Addendum, 395; Russian transl., Matematika **3** (1959), no. 3, 9-55. MR **16**, 367.

49. *The Weyl and Minkowski problems in differential geometry in the large,* Comm. Pure Appl. Math. **6** (1953), 337-394. MR **15**, 347.

S. P. OLOVJANIŠNIKOV

50. *Généralisation du théorème de Cauchy sur les polyèdres convexes,* Mat. Sb. **18** **(60)** (1946), 441-446. (Russian; French summary) MR **8**, 169.

51. *On the bending of infinite convex surfaces,* Mat. Sb. **18** **(60)** (1946), 429-440. (Russian; English summary) MR **8**, 169.

A. V. POGORELOV

52. *Quasi-geodesic lines on a convex surface,* Mat. Sb. **25** **(62)** (1949), 275-306; English transl., Amer. Math. Soc. Transl. (1) **6** (1962), 430-473. MR **11**, 201.

53. *Deformation of convex surfaces,* GITTL, Moscow, 1951. (Russian) MR **14**, 400; 1278.

54. *Some results relating to geometry in the large,* Izdat. Har'kov. Gos. Univ., Kharkov, 1961. (Russian)

55. *The rigidity of convex surfaces,* Trudy Mat. Inst. Steklov. **29** (1949). (Russian) MR **12**, 437.

56. *Uniqueness determination of general convex surfaces*, Monografii Instituta Matematiki, vyp. II, Akad. Nauk Ukrain. SSR, Kiev, 1952. (Russian) MR **16**, 162.

57. *Infinitesimal bending of general convex surfaces*, Izdat. Har'kov. Gos. Univ., Kharkov, 1959. (Russian) MR **22** #2960.

58. *On the regularity of convex surfaces with a regular metric in spaces of constant curvature*, Dokl. Akad. Nauk SSSR **122** (1958), 186-187. (Russian) MR **20** #5508.

59. *Some questions in geometry in the large in a Riemannian space*, Izdat. Har'kov. Gos. Univ., Kharkov, 1957. (Russian) MR **20** #4303.

60. *Extension of a general uniqueness theorem of A. D. Aleksandrov to the case of non-analytic surfaces*, Dokl. Akad. Nauk SSSR **62** (1948), 297-299. (Russian) MR **10**, 325.

61. *Regularity of a convex surface with given Gaussian curvature*, Mat. Sb. **31** (**73**) (1952), 88-103. (Russian) MR **14**, 679.

62. *Monge-Ampère equations of elliptic type*, Izdat. Har'kov. Gos. Univ., Kharkov, 1960; English transl., Noordhoff, Groningen, 1964. MR **23** #A1137; **31** #4993.

63. *Surfaces of bounded outer curvature*, Izdat. Har'kov. Gos. Univ., Kharkov, 1956. (Russian)

64. *Topics in the theory of surfaces in an elliptic space*, Izdat. Har'kov. Gos. Univ., Kharkov, 1960. (Russian) MR **22** #8460.

Ju. G. Rešetnjak

65. *Isothermal coordinates on manifolds of bounded curvature*. I, II, Sibirsk. Mat. Ž. **1** (1960), 88-116. 248-276. (Russian) MR **23** #A4072.

R. Sauer

66. *Infinitesimale Verbiegungen zueinander projektiver Flächen*, Math. Ann. **111** (1935), 71-82.

J. Schauder

67. *Über lineare elliptische Differentialgleichungen zweiter Ordnung*, Math. Z. **38** (1934), 257-282.

I. N. Vekua

68. *Systems of differential equations of the first order of elliptic type and boundary value problems, with an application to the theory of shells*, Mat. Sb. **31** (**73**) (1952), 217-314. (Russian) MR **15**, 230.

69. *Generalized analytic functions*, Fizmatgiz, Moscow, 1959; English transl., Pergamon Press, London; Addison-Wesley, Reading, Mass., 1962. MR **21** #7288; **27** #321.

Ju. A. Volkov

70. *Stability of the solution of Minkowski's problem*, Vestnik Leningrad. Univ. Ser. Mat. Meh. Astronom. **18** (1963), no. 1, 33-43. (Russian) MR **26** #4258.

H. Weyl

71. *Über die Bestimmung einer geschlossenen konvexen Fläche durch ihr Linienelement*, Vierteljschr. Naturforsch. Ges. Zürich **61** (1916), 40-72; Russian transl., Uspehi Mat. Nauk **3** (1948), no. 2 (24), 159-190. MR **10**, 60.

V. A. Zalgaller

72. *On curves with curvature of bounded variation on a convex surface*, Mat. Sb. **26** (**68**) (1950), 205-214. (Russian) MR **11**, 681.

AUTHOR INDEX

SUBJECT INDEX